T0187768

THE ROUTLEDGE HANDBOOK OF LATIN AMERICAN DEVELOPMENT

The Routledge Handbook of Latin American Development seeks to engage with comprehensive, contemporary, and critical theoretical debates on Latin American development. The volume draws on contributions from across the humanities and social sciences and, unlike earlier volumes of this kind, explicitly highlights the disruptions to the field being brought by a range of anti-capitalist, decolonial, feminist, and ontological intellectual contributions.

The chapters consider in depth the harms and suffering caused by various oppressive forces, as well as the creative and often revolutionary ways in which ordinary Latin Americans resist, fight back, and work to construct development defined broadly as the struggle for a better and more dignified life. The book covers many key themes including development policy and practice; neoliberalism and its aftermath; the role played by social movements in cities and rural areas; the politics of water, oil, and other environmental resources; indigenous and Afro-descendant rights; and the struggles for gender equality.

With contributions from authors working in Latin America, the US and Canada, Europe, and New Zealand at a range of universities and other organizations, the handbook is an invaluable resource for students and teachers in development studies, Latin American studies, cultural studies, human geography, anthropology, sociology, political science, and economics, as well as for activists and development practitioners.

Julie Cupples is Professor of Human Geography and Cultural Studies at the University of Edinburgh in the UK.

Marcela Palomino-Schalscha is Lecturer in Geography and Development Studies at Victoria University of Wellington in New Zealand.

Manuel Prieto is Researcher at the Institute of Archaeology and Anthropology (IAA) at Universidad Católica del Norte in San Pedro de Atacama, Chile.

THE ROUTLEDGE HANDBOOK OF LATIN AMERICAN DEVELOPMENT

Edited by Julie Cupples, Marcela Palomino-Schalscha, and Manuel Prieto

LONDON AND NEW YORK

First published 2019
by Routledge
2 Park Square, Milton Park, Abingdon, Oxon OX14 4RN

and by Routledge
605 Third Avenue, New York, NY 10017

First issued in paperback 2020

Routledge is an imprint of the Taylor & Francis Group, an informa business

British Library Cataloguing-in-Publication Data
A catalogue record for this book is available from the British Library

Library of Congress Cataloging-in-Publication Data
Names: Cupples, Julie, editor. | Prieto, M. (Manuel), editor. |
Palomino–Schalscha, Marcela, editor.
Title: The Routledge handbook of Latin American development /
edited by Julie Cupples, Manuel Prieto and Marcela Palomino–Schalscha.
Description: London ; New York : Routledge, 2019. | Includes
bibliographical references and index.
Identifiers: LCCN 2018035221 | ISBN 9781138060739 (hbk : alk. paper) |
ISBN 9781315162935 (ebk) | ISBN 9781351669672 (mobi/kindle)
Subjects: LCSH: Economic development—Latin America. | Latin
America—Economic conditions. | Latin America—Foreign economic
relations. | Latin America—Economic policy.
Classification: LCC HC125 .R678 2019 | DDC 338.98—dc23
LC record available at https://lccn.loc.gov/2018035221

ISBN 13: 978–0–367–73238–7 (pbk)
ISBN 13: 978–1–138–06073–9 (hbk)

Typeset in Bembo
by Apex CoVantage, LLC

A lxs estudiantes autoconvocadxs de Nicaragua

CONTENTS

FIGURES

TABLES

BOXES

EDITORS

Julie Cupples is Professor of Human Geography and Cultural Studies at the University of Edinburgh in the UK. She is also a member of the Latin American Executive and the Centre for Contemporary Latin American Studies and Chair of the Human Geography Research Group. She works in Nicaragua, Costa Rica, Colombia, and Mexico and has published on a range of themes, including gender and sexuality, disasters, elections, energy politics, and indigenous and Afro-descendant media. She has authored and edited five books, *Latin American Development* (Routledge, 2013), *Mediated Geographies and Geographies of Media* (Springer, 2015, with Susan Mains and Chris Lukinbeal), *Communications/Media/Geographies* (Routledge, 2017, with Paul Adams, Kevin Glynn, André Jansson and Shaun Moores), *Shifting Nicaraguan Mediascapes: Authoritarianism and the Struggle for Social Justice* (Springer, 2018, with Kevin Glynn), and *Unsettling Eurocentrism in the Westernized University* (Routledge, 2018, with Ramón Grosfoguel).

Marcela Palomino-Schalscha is Lecturer in Geography and Development Studies at Victoria University of Wellington in New Zealand. Her research interests lie at the intersection of social geography, development studies, and political ecology, with a special emphasis on Indigenous rights. Most of her work is located in Latin America, where she theorises the politics of scale and place, diverse and solidarity economies, decolonisation, identity politics, Indigenous tourism, development, neoliberalism, and relational ontologies. More recently, she has also embarked on the use of arpilleras, textiles with political content, as more-than-textual research methods to explore the experience of refugee-background and migrant Latin American women in New Zealand. She is the co-editor of the forthcoming *Indigenous Places and Colonial Spaces: The Politics of Intertwined Relations* (Routledge, 2019). She is also Co-editor of *ACME: An International E-Journal for Critical Geographies*.

Manuel Prieto is Researcher at the Institute of Archaeology and Anthropology (IAA) at Universidad Católica del Norte in San Pedro de Atacama, Chile and Associate Researcher at the Center for Indigenous and Intercultural Research (CIIR). His research centres on the intersections of political ecology, cultural ecology, political geography, and environmental science. His research examines the socio-natural transformation associated with water marketization, the process of state formations, local environmental knowledge, and indigenous identities. His most recent research focuses on high-altitude Andean peatlands. His work has been funded by Conicyt, Fulbright, and the Inter-American Foundation.

CONTRIBUTORS

Gabriela Alvarez Minte is Development Practitioner from Chile and has many years of experience in the international cooperation system. She has done consultancies for several UN agencies, and was the Gender Advisor for Plan International UK. Prior to Plan, she worked as Programme Specialist for the Latin American and Caribbean Section at UN Women at HQ in New York, USA, and previously in the same position in UNIFEM. She worked for several years in an NGO in Chile. She holds a Social Anthropology degree from Universidad de Chile, a PhD from Birkbeck, University of London, and a MSc in Sociology from the University of Oxford.

Marcos Andrade-Flores was born in Bolivia. He studied Physics at Universidad Mayor de San Andres (UMSA) and later Atmospheric Sciences at University of Maryland, College Park (UMD). He did his postdoctoral research at the Joint Center for Earth Systems Technology, a centre formed between NASA Goddard Space Flight Center and the University of Maryland, Baltimore County. He is currently Director of the Laboratory for Atmospheric Physics at the Institute for Physics Research (UMSA) and Adjunct Associate Professor at the Department of Atmospheric and Oceanic Science (UMD). His research interests include atmospheric aerosols, greenhouse gases, and precipitation in the Central Andes.

Javier Arellano-Yanguas is Research Fellow and Lecturer at the Centre for Applied Ethics at the University of Deusto (Bilbao, Spain). He is currently Director of the Centre. Javier holds a PhD in Development Studies from the Institute of Development Studies (University of Sussex) and a degree in Religious Studies from the University of Deusto. His work focuses on the political economy of natural resources led development, social conflicts, social accountability, and the interactions between religion and development. Most of this research is done in Andean countries and encompasses both quantitative and qualitative approaches.

Maurizio Atzeni is Researcher at CEIL/CONICET (Centro de Estudios e Investigaciones Laborales – Labor Studies Research Centre of the Argentinian National Research Council) based in Buenos Aires, having previously held positions at Loughborough and De Montfort Universities in the UK. He has published extensively on labour-related issues. He is the author of *Workplace Conflict: Mobilization and Solidarity in Argentina* (Palgrave, 2010) and of *Workers and Labour in a Globalised Capitalism* (Macmillan, 2014), a book that analyzes labour from an interdisciplinary perspective and is currently in translation to Chinese and Spanish. Maurizio serves on the editorial board of *Work, Employment and Society* and of the *Journal of Labor and Society*.

Anna Ayuso has a PhD in International Law and a Master's in European Studies from the Universidad Autónoma de Barcelona (UAB). Since 2002, she has been Senior Research Fellow on Latin America issues in CIDOB and is former coordinator of the International Cooperation Area (1995–2001). She is also Associate Professor in International Law at the UAB, Visiting Teacher at the Institut Barcelona d'Estudis Internacionals (IBEI), and a member of the Area of Freedom, Security and Justice research group (AFSJ) in International Law Department at UAB. She sits on the editorial boards of *Revista CIDOB d'Afers Internacionals*, the *International Journal Mural of the Universidade do Estado do Rio de Janeiro*, and the *Comillas Journal of International Relations*. She has held visiting positions at the Fundacao Getulio Vargas, Colegio de Mexico, School of Oriental and African Studies, University of Sussex, and Deutsches Institut für Entwicklungspolitik.

Florence E. Babb is the Anthony Harrington Distinguished Professor of Anthropology at the University of North Carolina at Chapel Hill, where she is affiliated with the Institute for the Study of the Americas and the Sexuality Studies Program. She specializes in gender and sexuality as well as race and class in Latin America. Her most recent book *Women's Place in the Andes: Engaging Decolonial Feminist Anthropology* (University of California Press, 2018) examines feminist debates of the last few decades concerning Andean women, race, and indigeneity – debates in which she participated and now considers in the critical context of decolonizing anthropologies.

Anthony Bebbington is Australia Laureate Fellow in the School of Geography at the University of Melbourne and Milton P. and Alice C. Higgins Professor of Environment and Society in the Graduate School of Geography, Clark University. His research has addressed: agriculture, livelihoods, and rural development; social movements, NGOs, and policy processes; and environmental governance and extractive industries. He is a director of Oxfam America, a research associate at RIMISP-Latin American Centre for Rural Development, based in Chile, an elected member of the National Academy of Sciences and the American Academy of Arts and Sciences, and has been a Guggenheim Fellow.

Beth Bee is Assistant Professor in the Department of Geography, Planning and Environment at East Carolina University. Her research explores the theoretical and empirical intersections between feminist theory, climate change, and rural livelihoods in Mexico. For example, she has investigated the ways that knowledge production and gendered relations of power shape adaptive capacity and food security in the face of climatic uncertainty. More recently, she has also investigated the multiple forms of power and inequities embedded in forestry conservation projects that comprise Mexico's Reducing Emissions, Deforestation, and Degradation (REDD+) early-action activities.

Jennifer Bickham Mendez is Professor of Sociology and Director of Global Studies at the College of William and Mary. She is the author of *From the Revolution to the Maquiladoras: Gender, Labor and Globalization in Nicaragua* (Duke University Press, 2005), and she and Nancy Naples are co-editors of *Border Politics: Social Movements, Collective Identity, and Globalization* (NYU Press, 2015). Her publications have appeared in a variety of academic journals, including *Social*

Problems, Ethnic and Racial Studies, Gender and Society, and *Mobilization* as well as in numerous edited volumes. Her current work focuses on the experiences of Latino/a immigrants in Williamsburg, Virginia, and their struggles for security, inclusion, and belonging.

Rutgerd Boelens is Professor of Water Governance and Social Justice at Wageningen University; Professor of Political Ecology of Water in Latin America with CEDLA, University of Amsterdam; and Visiting Professor at the Catholic University of Peru and the Central University of Ecuador. He directs the international Justicia Hídrica/Water Justice alliance (www.jus ticiahidrica.org). His research focuses on political ecology, water rights, legal pluralism, cultural politics, governmentality, and social mobilization. Among his latest books are *Water Justice* (with Perreault and Vos, Cambridge University Press, 2018), *Water, Power and Identity. The Cultural Politics of Water in the Andes* (Routledge, 2015), and *Out of the Mainstream: Water Rights, Politics and Identity* (with Getches and Guevara-Gil, Earthscan, 2010).

Sarah Bradshaw is a feminist scholar-practitioner. She is Professor of Gender and Sustainable Development in the School of Law at Middlesex University. Her research focuses on Latin America and seeks to better understand gendered experiences of poverty and promote the realization of gendered rights. She is also interested in gendered experiences of disasters and published the first book that considers the nexus between *Gender, Development and Disasters* (Elgar, 2013). She combines research with practice, having lobbied around World Bank policies, advocated for the inclusion of gendered rights in UN processes, and engaged in intergovernmental negotiations around international policy frameworks.

Joe Bryan is Associate Professor of Geography at the University of Colorado, Boulder. He has worked with indigenous peoples on mapping projects in Nicaragua, Honduras, Chile, Mexico, and the United States. He is the co-author, with Denis Wood, of *Weaponizing Maps: Indigenous Peoples and Counterinsurgency in the Americas* (Guilford, 2015).

Deborah Bush is a black Creole woman from the Caribbean Coast of Nicaragua. She has an undergraduate degree in Sociology and a Master's in Social Anthropology. She has completed postgraduate courses in globalization, identity, migration and autonomy, and community forestry. She is currently the Delegate of the Instituto Nicaragüense de Cultura (Nicaraguan Institute of Culture) in the North Caribbean Autonomous Region (RACCN). She is the founding member and co-president of Afro's Voices Center of Nicaragua (AVOCENIC) in Puerto Cabezas, an organization that develops community activism and builds alliances with regional and international organizations and universities. Deborah is also a member of the International Commission of the *Coloquio Internacional Afrodescendiente (International Afrodescendant Colloquium)*.

Shaun Bush is a black woman born and raised in the city of Puerto Cabezas, Nicaragua. Professionally she is a registered nurse with a speciality in Obstetrical-Gynaecological Health and Labour-Delivery. She also holds a Master's in International Social Welfare and Health Policy. She is Founding Member and Secretary of the Afro's Voices Center of Nicaragua (AVOCENIC), Founding Member and co-host of the first black TV programme in Puerto Cabezas "Black/Creoles: Building together our wellbeing," and former coordinator of projects in public and community health at the University of the Autonomous Regions of the Caribbean Coast of Nicaragua (URACCAN).

Kendall Cayasso-Dixon is an Afro-Costa Rican activist, teacher, journalist, reporter, and musician, based in Limón in Costa Rica where he works to defend and promote Afro-descendant culture and struggles. He has a degree in Ecological Tourism from the Universidad de Costa Rica and teaches in the Centro Educativo San Marcos in Limón. Inspired by his mother, who is a leading Afro-Costa Rican activist in Limón, in 2007 Kendall became an active member of the Universal Negro Improvement Association (UNIA) where he has supported the communication and outreach activities of this organization. He is also a reporter for Prensamérica Internacional and the CEO of Townbook Limón, an organization that seeks to visibilize Afro-descendant culture through the use of media and new technologies. Kendall is also an accomplished musician and he manages a group called Di Gud Frendz Mixup that blends the ancestral rhythms of the calypso of Limón with Reggae, Dancehall, Soca, and Funk.

Sylvia Chant is Professor of Development Geography at the London School of Economics and Political Science, where she directs the MSc in Urbanisation and Development. A specialist in gender and development, with particular interests in female-headed households and the "feminisation of poverty," Sylvia has conducted field research in Mexico, Costa Rica, the Philippines, and Gambia. Her latest books include *Gender, Generation and Poverty: Exploring the 'Feminisation of Poverty' in Africa, Asia and Latin America* (Elgar, 2007) and *Cities, Slums and Gender in the Global South* (Routledge, 2016, with Cathy McIlwaine). Sylvia is currently serving as a member of the Expert Advisory Group for UN Women's *Progress of the World's Women 2018*.

Guadalupe Correa-Cabrera is Associate Professor in the Schar School of Policy and Government at George Mason University. Her areas of expertise are Mexico-US relations, organized crime, immigration, border security, and human trafficking. Her newest book is titled *Los Zetas Inc.: Criminal Corporations, Energy, and Civil War in Mexico* (University of Texas Press, 2017). She is Past President of the Association for Borderlands Studies (ABS). She is also Global Fellow at the Woodrow Wilson International Center for Scholars and Non-resident Scholar at the Baker Institute's Mexico Center (Rice University).

Piergiorgio Di Giminiani is Associate Professor of Anthropology at the Pontificia Universidad Católica de Chile. He's the author of *Sentient Lands: Indigeneity, Property and Political Imagination in Neoliberal Chile* (2018), an analysis of indigenous land politics in Mapuche areas of Southern Chile. His new book project focuses on forest conservation in Chile, in which through a focus on networks of collaboration linking settlers, indigenous farmers, state agencies, NGOs, and scientists, he explores the ways in which different forms of world-making coexist, entangle, and enter in conflict.

Fábio Duarte is Research Scientist at the Massachusetts Institute of Technology (Senseable City Lab), and Professor at the Pontificia Universidade Católica do Paraná (PUCPR), Curitiba, Brazil. Duarte's books include *Unplugging the City: The Urban Phenomenon and Its Sociotechnical Controversies* (Routledge, 2018).

Rodolfo Elbert is Researcher at Argentina's Consejo Nacional de Investigaciones Científicas y Técnicas and Director of the "Programa de Investigación sobre Análisis de Clases Sociales" at the Instituto de Investigaciones Gino Germani (Universidad de Buenos Aires). His work focuses on the linkages between informality, labour, and class in Latin America. His research has been published in *Current Sociology*, *Critical Sociology*, and *Latin American Perspectives*, among other journals. He teaches on research methods and contemporary sociological theory in the Department of Sociology of the Universidad de Buenos Aires and is currently a council member at the Labor Studies and Class Relations Section of the Latin American Studies Association.

Laura J. Enríquez is Professor of Sociology at the University of California at Berkeley. She has published extensively on the topic of social transformation in Latin America, including several articles focused on struggles around agrarian change in contemporary Venezuela. Her most recent book is *Reactions to the Market: Small Farmers in the Economic Reshaping of Nicaragua, Cuba, Russia, and China*. Her current work has branched out to address Latin American emigration to Europe.

Camila Esguerra Muelle is a Postdoctoral Researcher at CIDER (Centro Interdisciplinario de Estudios sobre Desarrollo – Interdisciplinary Centre of Studies on Development) at the Universidad de Los Andes. They have a PhD in Humanities (sobresaliente cum laude) from the Universidad Carlos III in Madrid, an MA in Gender and Diversity from the University of Oviedo, an MA in Gender and Ethnicity from Utrecht University, and an undergraduate degree in Anthropology from the Universidad Nacional de Colombia. They are affiliated with the CIDER research group at Universidad de los Andes, with GIEG (Grupo Interdisciplinario de Estudios de Género – Interdisciplinary Gender Studies Group) at the Universidad Nacional de Colombia and the Visual Studies Group at Pontificia Universidad Javeriana.

Mary Finley-Brook has taught Geography, Environmental Studies, and Global Studies at the University of Richmond since 2006. Researching nature–society interactions and territorial rights, she has published in journals including *Energy Research and Social Science*, *Annals of the American Association of Geographers*, *Geopolitics*, *Water Alternatives*, *International Forestry Review*, *AlterNative*, *Mesoamérica*, *Bulletin of Latin American Research*, and *Journal of Latin American Geography*. Her current work focuses on climate and energy justice in marginalized communities in the Western hemisphere.

Robert Fletcher is Associate Professor in the Sociology of Development and Change group at Wageningen University in the Netherlands. His research interests include conservation, development, tourism, climate change, globalization, and resistance and social movements. He is the author of *Romancing the Wild: Cultural Dimensions of Ecotourism* (Duke University, 2014) and co-editor of *NatureTM Inc.: Environmental Conservation in the Neoliberal Age* (University of Arizona, 2014).

Jasmine Gideon is Senior Lecturer in Development Studies at Birkbeck, University of London. Her research interests are centred around the gendered political economy of health in Latin America with a specific focus on three central elements: gender and health, globalization and development, and transnational migration and health. She is currently looking at questions of health and well-being among Chilean exiles in the UK as well as the gendered dimensions of privatization in the Chilean health sector. She is the author of *Gender, Globalization and Health in a Latin American Context*, published in 2014 by Palgrave Macmillan and the editor of the *Handbook on Gender and Health* published in 2016 by Edward Elgar.

Jere Gilles is Associate Professor of Rural Sociology at the University of Missouri whose work focuses on natural resource management, agricultural development, and the development of appropriate technologies. His research focuses on developing ways of combining local and scientific knowledge through stakeholder participation in order to improve the processes of technology development and extension. Previous research has focused on overgrazing and desertification, genetically modified maize, and the management of irrigation systems. More recently, his attention has been focused on evaluating methods for improving forecasts in the Bolivian Altiplano and understanding climate adaptation strategies of small producers in the region.

Charlotte Gleghorn holds a Lectureship in Latin American Film Studies at the University of Edinburgh. She obtained a PhD from the University of Liverpool (2009) with a thesis on women's filmmaking from Argentina and Brazil and has published in journals and several edited volumes on Latin American cinema, including the *Blackwell-Wiley Companion to Latin American Cinema* (2017). She is co-investigator on the AHRC International Networking Grant 'Afro-Latin (In)Visibility and the UN Decade,' investigating Afro-descendant filmmaking, and is currently embarking on an AHRC-funded Fellowship on Indigenous Filmmaking in Latin America.

Kevin Glynn is Associate Professor in the Department of Geography and Environmental Sciences at Northumbria University in the UK. He has also taught at universities in the US and Aotearoa/New Zealand, where he co-founded and directed the country's only degree program in Cultural Studies at the University of Canterbury. He has published widely in media studies, cultural studies, and critical and cultural geography. He is author of *Tabloid Culture: Trash Taste, Popular Power, and the Transformation of American Television* (Duke University Press), and co-author of *Communications/Media/Geographies* (Routledge). His most recent book, co-authored with Julie Cupples, is *Shifting Nicaraguan Mediascapes: Authoritarianism and the Struggle for Social Justice* (Springer). His work has also appeared in many leading international journals and anthologies.

Eduardo Gudynas is Senior Researcher at the Centro Latinoamerican de Ecología Social (Latin American Center of Social Ecology CLAES), based in Uruguay. He is an expert on Latin American environmental issues and social movements. He has conducted research on extractivisms and their impact on development and the environment as well as on the concept of *buen vivir* and alternatives to development. His books include *Extractivismos y corrupción. Anatomía de una íntima relación* (2017), *Extractivismos. Ecología, economía y política de un modo de entender el desarrollo y la Naturaleza* (2015), *Derechos de la Naturaleza. Etica biocéntrica y políticas ambientales* (2014), *El mandato ecológico* (2009), and *Ecología, economía y ética del desarrollo sostenible* (2004). In 2015 he was selected by esglobal as being among the 50 most influential intellectuals in Latin America and Spain. He blogs at www.accionyreaccion.com

Grant Gutierrez is a graduate student in the Ecology, Evolution, Ecosystems and Society PhD program at Dartmouth College. His research examines the role of social movements in shaping energy politics in Chile, particularly focused on debates concerning green energy and watershed conservation. His fieldwork combines ethnographic methods with approaches from activist anthropology.

Matthew Gutmann is Professor of Anthropology at Brown University. His books include *The Meanings of Macho: Being a Man in Mexico City*; *The Romance of Democracy: Compliant Defiance in Mexico City*; *Fixing Men: Sex, Birth Control and AIDS in Mexico*; *Breaking Ranks: Iraq Veterans Speak Out against the War* (with Catherine Lutz); and *Global Latin America: Into the 21st Century* (edited with Jeffrey Lesser). He is completing *Men Are Animals: An Anthropology of Sex, Violence,* *and Biobabble.* Gutmann has a Master's in Public Health, and in 2008 he won the Eileen Basker Memorial Award for the best scholarly study on gender and health.

Charles R. Hale is the SAGE Sara Miller McCune Dean of Social Sciences at UC Santa Barbara. He is the author of *Resistance and Contradiction: Miskitu Indians and the Nicaraguan State, 1894–1987* (1994); and *"Más que un indio . . .": Racial Ambivalence and Neoliberal Multiculturalism in Guatemala* (2006); the editor of *Engaging Contradictions: Theory, Politics and Methods of Activist Scholarship* (2008); co-editor (with Lynn Stephen) of *Otros Saberes: Collaborative Research with Black and Indigenous Peoples in Latin America* (2014); and the author of articles on activist scholarship, identity politics, racism, resistance to neoliberalism among indigenous and Afro-descendant peoples. He was the director of LLILAS Benson Latin American Studies and Collections at University of Texas-Austin from 2009 to 2016 and president of the Latin American Studies Association from 2006 to 2007.

George Henríquez Cayasso has an undergraduate degree in Business Administration and Hotel Hospitality Management and a Master's degree in Gender, Ethnicity and Cultural Citizenship. He is a black Kriol activist, freelancer, and entrepreneur, with a specific interest in gender, interculturality, conflict resolution, autonomy, advocacy, and inter-ethnic alliances. For the past few years, he has been engaged in community work with Afro-descendant and indigenous peoples, trying to raise awareness of the Autonomy Law, of intercultural-bilingual education, and of territoriality as a strategy for the consolidation of autonomy beyond institutionality. His work aims to promote the visibility of the people that live on the Nicaragua Caribbean Coast, also known as la Moskitia.

Barbara Hogenboom is Professor of Latin American Studies at the University of Amsterdam (UvA), and Director of the Centre for Latin American Research and Documentation (CEDLA-UvA). Barbara Hogenboom's field of study is the politics and governance of development and environment. Her research focuses on the clashing values and interests at play across scales in relation with the use of natural resources in Latin America. Among her recent co-authored publications are *Environmental Governance in Latin America* (Palgrave Macmillan, 2016); The Extractive Imperative in Latin America (special issue of *The Extractive Industries and Society*, 2016); *Latin America Facing China: South-South Relations beyond the Washington Consensus* (Berghahn Books, 2012); The New Politics of Mineral Extraction in Latin America (special issue of *Journal of Developing Societies*, 2012).

Osvaldo Jordan Ramos holds a Master of Arts in Latin American Studies and a PhD in Political Science from the University of Florida. He has worked as a consultant for academic institutions, non-governmental associations, and intergovernmental organizations on environmental conservation, indigenous rights, and public participation. His research has also focused on indigenous politics, environmental conflicts, and climate change vulnerability, being active in a number of environmental and human rights organizations. In 2001, he was a founder of Alianza para la Conservacion y el Desarrollo (ACD), a Panama-based nonprofit that supports the protection of indigenous territories and environmental justice.

Cristóbal Kay is Emeritus Professor at the International Institute of Social Studies, Erasmus University Rotterdam and at FLACSO, Quito, Ecuador. He is also Professorial Research Associate, Department of Development Studies at SOAS, University of London. His research is in the fields of development theory and rural development studies. He has been the editor of the

European Journal of Development Research and a co-editor of the *European Review of Latin and Caribbean Studies*. He is currently an editor of the *Journal of Agrarian Change*. He is the author of *Latin American Theories of Development and Underdevelopment* and has written articles on Raúl Prebisch, Celso Furtado, Solon Barraclough, Willem Assies, and André Gunder Frank.

Dixie Lee Smith is from Bilwi/Puerto Cabezas, Nicaragua. He is an Afro-descendant academic at the University of the Autonomous Regions of the Caribbean Coast of Nicaragua (URACCAN), where he directs IEPA (Instituto para el Estudio y Promoción de la Autonomía/ Institute for the Study and Promotion of Autonomy). Dixie is a civil leader who has served as a member of the advisory council of indigenous and Afro-descendant peoples, created by the United Nations Development Program (UNDP) in Nicaragua. In 2010, he co-founded AVO-CENIC (Afro's Voices Center of Nicaragua) whose purpose is to promote the collective rights of the Afro-descendant people of Nicaragua. Dixie has also done research on Afro-descendant cultural traditions and practices in Nicaragua.

Brian Linneker is Independent Scholar and Freelance Senior Researcher in Economic Geography. He has worked for over 25 years on poverty, vulnerability, and social exclusion for the UK government departments, the UK international and Latin American national NGOs and civil society organizations, and within various academic institutions including the London School of Economics and Political Science, King's College London, Birkbeck College, Queen Mary University of London, and Middlesex University.

Melanie Lombard is Lecturer in the Department of Urban Studies and Planning at the University of Sheffield. Her research agenda involves connecting the built environment to social processes through exploring the everyday activities that construct cities, with a focus on urban informality, and land and conflict. She has explored these themes in cities in Latin America (Mexico and Colombia) and Europe (UK). She is currently undertaking research on the effects of the Colombian peace process on low-income neighbourhoods. She has published articles in journals including *Urban Studies*, *Progress in Planning*, *Environment and Planning D: Society and Space*, and the *International Journal of Urban and Regional Research*.

Marcelo Lopes de Souza is Professor at the Department of Geography of the Federal University of Rio de Janeiro (UFRJ), Brazil. He acted as Academic Visitor or Visiting Professor at several universities in Europe (Germany, United Kingdom, and Spain) and Latin America (Mexico). He has published 11 books and more than one hundred papers and book chapters in several languages covering subjects such as urban theory, the spatial dimension of social movements, and political ecology (focusing especially on environmental justice). He is one of the editors of the Brazilian urban studies journal *Cidades*, being also Associate Editor of *City* (published by Routledge).

Ernesto López-Morales holds a PhD degree in Urban Planning from the Development Planning Unit, University College London, and currently works as Associate Professor at the University of Chile. Over the last decade, his academic work has focused on gentrification and urban dispossession in the Global North and South, as he has not only successfully and plausibly applied the gentrification concept beyond the North-Atlantic domain, but also reinterpreted and bolstered existing theory with critical evidence carefully analyzed in Latin America. Since 2005, he has been engaged in empirical research on gentrification in inner-city areas of Santiago, Chile, with further comparative empirical research into several Latin American cases conducted since 2011.

Angus Lyall is a PhD candidate in Geography at the University of North Carolina at Chapel Hill and a member of the *Colectivo de Geografía Crítica de Ecuador*. He is an economic and cultural geographer who examines the institutions and cultural politics of territorial governance in Ecuador. His current research centres on indigenous engagements with oil-driven development and urbanization in the Amazon. He also has ongoing projects on the politics of work in enclaves of cut-flower production and transformations in rural politics related to tourism development in the Andes.

Andrés Malamud (PhD European University Institute, 2003) is Senior Research Fellow at the Institute of Social Sciences of the University of Lisbon. He is Recurring Visiting Professor at universities in Argentina, Brazil, Italy, and Spain, and was Visiting Researcher at the Max Planck Institute for Comparative Public Law and International Law (Heidelberg) and the University of Maryland, College Park. He conducts research, and has published extensively, on comparative regional integration, foreign policy, democracy and political institutions, EU Studies, and Latin American politics.

Clara Marticorena is Sociologist with a Master's degree in Labour Sciences and a PhD in Social Sciences from the Universidad de Buenos Aires. She currently works as Researcher at CEIL/CONICET (Centro de Estudios e Investigaciones Laborales – Labour Studies Research Centre of the Argentinian National Research Council) and also teaches at the University of Buenos Aires. She is a specialist in labour conditions, labour relations, and union actions. She is the author of *Trabajo y negociación colectiva. Los trabajadores en la industria argentina, de los '90 a la posconvertibilidad* (Imago Mundi, 2014). She has written many book chapters and published several papers in different national and international academic journals, such as *Perfiles Latinoamericanos, Estudios de Sociología*, and *Estudios del Trabajo*. She coordinates a research project about the dynamics of collective bargaining and its relation with labour conflicts in contemporary Argentina at the Faculty of Social Sciences at the University of Buenos Aires.

Javier Martínez-Contreras is Lecturer and Research Fellow at the Centre for Applied Ethics at the University of Deusto (Bilbao, Spain). He is currently the Coordinator of the Humanities Degree at the Faculty of Social and Human Sciences. He holds a PhD in Philosophy from the University of Deusto, a degree in Philosophy from the same university, and another one in Religious Studies from the Pontifical University of Salamanca. His work focuses on topics of contemporary Philosophy, especially in the areas of Political Philosophy and Hermeneutics, and issues related to fundamental and applied Ethics.

Daniel Mato is Principal Researcher at CONICET (National Council for Scientific and Technical Research) and the Universidad Nacional Tres de Febrero (UNTREF) in Argentina. He is also Director of the UNESCO Chair on Higher Education and Indigenous and Afro-descendant Peoples in Latin America. Between 1979 and 2010, he was Professor at the Universidad Central de Venezuela. Working in collaboration with indigenous and Afro-descendant intellectuals and organizations throughout Latin America, he has published widely on cultural diversity and higher education. His main achievements include his academic leadership in the UNESCO International Institute for Higher Education in Latin America and the Caribbean (2007–2010), the creation of the Programme on Higher Education and Indigenous and Afro-descendant Peoples in Latin America at UNTREF (2012), and the establishment of the Latin American Inter-University Network on Higher Education and Indigenous and Afro-descendant Peoples (2014).

Cathy McIlwaine is Professor of Development Geography in the Department of Geography, King's College London. In addition to researching gender and development issues in the Global South for many years, Cathy has also worked on international and transnational migration in London with a specific focus on the Latin American community in relation to transnational livelihoods, citizenship, and political participation among migrants from a gendered perspective. She has published ten books including *Cities, Slums and Gender in the Global South* (with Sylvia Chant [2016], Routledge), *Cross-Border Migration among Latin Americans* ([edited] [2011] Palgrave Macmillan), and *Global Cities at Work: New Migrant Divisions of Labour* (with Jane Wills, Kavita Datta, Jo Herbert, Jon May, and Yara Evans) ([2010] Pluto) as well as over 40 journal papers. She is a trustee at the charity Latin Elephant and an advisor for the Latin American Women's Rights Service.

Jerónimo Montero Bressán is a full-time Researcher at Argentina's Consejo Nacional de Investigaciones Científicas y Técnicas (CONICET). He is based at the Universidad Nacional de San Martín (IDAES/UNSAM). His research is focused on the changes in the geography of fashion production and consumption over the last 50 years, and their consequences over labour. He also teaches economic geography at the Faculty of Economics, Universidad de Buenos Aires.

Cecilia Moreno Rojas has an undergraduate degree in Sociology and an MA in Corporate Communication from the Universidad Santa María la Antigua in Panamá. She has also completed postgraduate courses at Universidad de Río de Janeiro, Florida Internacional University, CSUCA, and Louisville University. She is Founder and Executive Director of the Centro de la Mujer Panameña (Centre of the Panamanian Woman) and the Founder and current national Coordinator of the Red de Mujeres Afrodescendientes de Panamá (Network of Afrodescendant Women of Panama). She has served on many committees working to defend Afrodescendant rights, including the Consejo Nacional de la Etnia Negra Panameña (National Council of the Black Panamanian Ethnicity), Comisión Nacional Contra la Discriminación en Panamá (National Commission against Discrimination in Panama), and the Organización Negra Centroamericana/Central American Black Organization (ONECA/CABO). She has also represented Panama in many international meetings and summits, including the historic World Conference against Racism (Durban, 2001).

Peter Motavalli is Professor in Soil Nutrient Management in the School of Natural Resources at the University of Missouri in Columbia, Missouri (USA). He received his PhD in soil fertility and plant nutrition from Cornell University in 1989 and was a faculty member for four years at the University of Guam prior to going to the University of Missouri in 1999. Dr. Motavalli has conducted research in multiple countries including India, South Africa, Bolivia, Brazil, and the Sudan. His research focuses on the development of appropriate and sustainable agricultural fertilization practices that increase crop production but limit environmental contamination.

Laura A. Ogden is Associate Professor of Anthropology at Dartmouth College and president of the Anthropology and Environment Society. She has conducted ethnographic research in the Florida Everglades, with urban communities in the United States, and is currently working on a long-term project in Tierra del Fuego, Chile. She is the author of *Swamplife: People, Gators and Mangroves Entangled in the Everglades* and is writing a new book entitled "Traces of Being: An Alternative Archive of the Present."

Diana Ojeda is Associate Professor at the Instituto de Estudios Sociales y Culturales Pensar, Pontificia Universidad Javeriana in Bogota, Colombia. She holds a PhD in Geography from Clark University (Worcester, MA, USA). Her work addresses processes of dispossession, environmental destruction and recovery, and state formation from a feminist political ecology perspective. Her work has been published by journals such as *The Journal of Peasant Studies, Geoforum* and *Gender, Place and Culture*. She also co-authored the graphic novel *Caminos Condenados* (2016 Cohete Cómics).

Tiffany L. Page is Lecturer in International and Area Studies at the University of California Berkeley. She received a BA in Economics, as well as a MA and PhD in Sociology from U.C. Berkeley. Based off of her doctoral research on the politics of the agrarian reform in Venezuela under the Chávez government, she published "Can the State Create *Campesinos*? A comparative analysis of the Venezuelan and Cuban repeasantization programmes" in the *Journal of Agrarian Change*. She also authored a chapter on race and ethnicity in Venezuela for the *International Handbook of the Demography of Race and Ethnicity*.

Ramón Emilio Perea Lemos has a degree in Psychology from the Universidad San Buenaventura in Medellín and has worked with Afro-descendant communities for over 15 years. As co-founder of Carabantú association and active member of the Proceso de Comunidades Negras (PCN), he has developed community work with Afro-descendant communities in Colombia which has gained both national and international recognition. Among the projects he has been involved in, his contribution to founding the Sindicato de Mujeres Afrocolombianas Empleadas del Servicio Doméstico (Union of Afro-Colombian Domestic Service Employees) is of particular note, along with his role in establishing the International Festival of Black Communitarian Cinema – Kunta Kinte (Festival Internacional de Cine Comunitario Afro "Kunta Kinte," FICCA KUNTA KINTE).

Tom Perreault is DellPlain Professor of Latin American Geography at Syracuse University. His research and teaching interests are in political ecology, agrarian political economy, and rural development. In particular, his work focuses on the intersections of extractive industries, water governance, environmental justice, and indigenous/campesino political movements in the central Andes and western Amazon regions. He has published over 60 journal articles and book chapters, and has authored or edited four books, including *Water Justice* (Cambridge, 2018), *The Handbook of Political Ecology* (Routledge, 2015), and *Minería, Agua y Justicia Social en los Andes: Experiencias Comparativas de Perú y Bolivia* (PIEB/CBC, 2014).

Nancy Postero is Professor of Anthropology at UC San Diego. Her research focuses on the intersection of race, politics, and economics in Latin America, and specifically in Bolivia. She is the author of *Now We Are Citizens* (Stanford, 2007), *The Struggle for Indigenous Rights*, with Leon Zamosc (Sussex, 2004), *Neoliberalism Interrupted*, with Mark Goodale (Stanford, 2013), and *The Indigenous State: Race, Politics, and Performance in Plurinational Bolivia* (California, 2018).

Natalia Quiroga Díaz is Academic Coordinator for the Master's in Social Economy and Lecturer and Researcher at the Instituto del Conurbano at the Universidad Nacional de General Sarmiento in Argentina. She is also the co-ordinator of the Clacso research group Emancipatory Feminist Economy. She has an economics degree from the Universidad Nacional de Colombia, a degree in Regional Development from the Universidad de los Andes and a Master's in Social Economy from the Universidad Nacional de General Sarmiento. She is currently completing a PhD in Social Anthropology at the Universidad de San Martín. Her research is positioned at the intersection of social and popular economies and perspectives from decolonial feminism.

Laura T. Raynolds is Professor of Sociology and Director of the Center for Fair and Alternative Trade at Colorado State University, USA. As one of the world's foremost scholars of fair trade and alternative agro-food networks, her work advances our understanding of globalization, development, social movements, and shifting production relations, drawing on field-research in Latin America and the Caribbean. She has authored numerous highly cited chapters and articles in *World Development, Sociologia Ruralis, Agriculture and Human Values*, and *Journal of Rural Studies* and edited *Fair Trade: The Challenges of Transforming Globalization* and *The Handbook of Research on Fair Trade*.

Raquel Ribeiro holds a BA in Journalism and Communication Sciences (Universidade Nova de Lisboa, Portugal), followed by a PhD in Hispanic Studies (University of Liverpool). She was the first recipient in the Humanities of the Nottingham Advanced Research Fellowship (2010–2012), at the University of Nottingham, where she developed the postdoctoral project on the cultural representations of the Cubans in the Angolan war. She is a member of the Cuba Research Forum at the University of Nottingham, and she taught Brazilian Literature at the University of Oxford before joining the University of Edinburgh in 2014. Raquel is a permanent arts freelance correspondent and literary critic for the Portuguese newspaper *Público* since 2001 and has been the awarded the Beca Gabriel García Márquez in Cultural Journalism by the Fundación Nuevo Periodismo Iberoamericano, in Colombia (2013). As a creative writer, she has published two novels and several short-stories.

Dennis Rodgers is Research Professor in the Department of Anthropology and Sociology at the Graduate Institute of International and Development Studies, Geneva, Switzerland, and Visiting Professor in International Development Studies at the University of Amsterdam, the Netherlands. A social anthropologist by training, his research focuses broadly on issues relating to the political economy of development, including in particular the dynamics of conflict and violence in cities in Latin America (Nicaragua, Argentina) and South Asia (India). He was recently awarded a European Research Council Advanced Grant for a 5-year project on "Gangs, Gangsters, and Ganglands: Towards a Comparative Global Ethnography" (GANGS), which aims to systematically compare gang dynamics in Nicaragua, South Africa and France.

Maurício Rombaldi is Brazilian and Professor of the Postgraduate Programme in Sociology at the Universidade Federal da Paraíba (UFPB). He holds a Master's and a PhD in Sociology from the Universidade de São Paulo (USP). He completed his postdoctoral studies at the Universidade Federal do Rio Grande do Sul (UFRGS). Focused on the intersection between culture and politics, his current line of research lies in analyses of globalization, labour relations, and trade unionism. Particularly noteworthy are his studies on the telecommunications, metallurgy, and construction sectors in the context of Labour Reform in Brazil and on the relationships between trade union internationalization and labour regulation.

Tara Ruttenberg is a PhD Candidate in Development Studies at Wageningen University. She holds a Master's in International Peace Studies from the UN-mandated University for Peace, as well as a Bachelor of Science in Foreign Service and a Certificate in Latin American Studies from Georgetown University. Tara is a writer by trade and teaches on post-development, critical sustainability studies, and wellbeing economics for graduate and undergraduate students, with an emphasis on decolonizing sustainable surfing tourism. You can read more of Tara's work at www.tarantulasurf.com.

Megan Ryburn is an LSE Fellow in Human Geography in the Department of Geography and Environment at the London School of Economics. She obtained her PhD from Queen Mary University of London, where she held a Principal's Studentship jointly funded by the School of Geography and the School of Politics and International Relations. Prior to that, she completed an MPhil in Latin American Studies at the University of Cambridge. Megan's work focuses on migration and citizenship in Latin America. Her first book, *Uncertain Citizenship: Everyday Practices of Bolivian Migrants in Chile*, is forthcoming [2018] with the University of California Press.

Julia Soul is Anthropologist and Researcher at CEIL (Centro de Estudios e Investigaciones Laborales – Labor Studies Research Center) at CONICET (National Scientific Research Council) in Argentina. Her research themes are: working class recent history – changes and continuities in everyday lives, local and international unionism and organizations, collective identities, and organizational and political traditions. She currently conducts a research project about the changes and continuities in labour relationships in Brazil and Argentina. She is the author of "SOMISEROS: la conformación y el devenir de un grupo obrero desde una perspectiva socio-antropológica" (2014) and other papers published in *Dialectical Anthropology*, *Sociologia del Lavoro*, *Nueva Antropología*. She is also a member of TEL (Labor Studies Workshop), a nonprofit organization for educational and collaborative research with union activists, delegates, and organizers.

Kate Swanson is Associate Professor in the Department of Geography at San Diego State University, California. She earned her PhD from the University of Toronto, Canada, and has published widely on the topic of the urban informal sector in Latin America. While her research interests are broad, she currently focuses on migration in Latin America and the US/Mexico border region.

Sergio Tischler is Professor and Researcher at the "Alfonso Vélez Pliego" Graduate School of Sociology of the Institute of Social Sciences and Humanities at the Benemérita Autonomous University of Puebla, where together with John Holloway he is coordinator of the Permanent Subjectivity and Critical Theory Seminar. He is the author of *Memoria, tiempo y sujeto; Tiempo y emancipación: Mijaíl Bajtín y Walter Benjamin en la Selva Landona* and *Revolución y destotalización* and the co-editor of *What is to be Done?: Leninism, Anti-Leninist Marxism and the Question of Revolution Today* (with Werner Bonefeld) and *Negativity and Revolution. Adorno and Political Activism* (with John Holloway and Fernando Matamoros).

Corinne Valdivia is Professor of Agricultural and Applied Economics in the Division of Applied Social Sciences (DASS), College of Agriculture Food and Natural Resource (CAFNR) at the University of Missouri (MU). She teaches graduate and undergraduate courses in International Agricultural Development and Policy and is Director of the Graduate Interdisciplinary International Development Minor at MU. Her research focuses on transformational changes, such as climate change, migration, globalization, and innovations, and translational approach to working with people and communities to negotiate and adapt to these changes in the Andes, East Africa, and rural communities of USA's Midwest.

Gabriela Valdivia is Associate Professor of Geography at the University of North Carolina, Chapel Hill. Her work focuses on the political ecology of natural resource governance in Latin

America: how states, firms, and civil society appropriate and transform resources to meet their interests, and how capturing and putting resources to work transforms cultural and ecological communities. Her most recent project, *Crude Entanglements*, draws on feminist political ecology to examine the affective dimensions of oil production. She is co-author of the book *Oil, Revolution, and Indigenous Citizenship in Ecuadorian Amazonia*, which examines the political ecology of the Ecuadorian petro-state since the turn of the century.

Zulma Valencia Casildo is President of the Organización de Desarrollo Étnico Comunitario (ODECO) based in La Ceiba, Honduras. ODECO is an NGO that works to promote Afro-Honduran rights and cultural diversity. She has a degree in Business Administration from the Instituto San Isidro in La Ceiba. She has participated in many Afro-descendant meetings and workshops in Honduras and internationally, including the III World Conference against Racism held in Durban, South Africa in 2001, along with the Latin American preconference in Santiago de Chile. She has also coordinated a number of Afro-descendant development projects. These include projects to promote land and environmental protection in Afro-Honduran communities (Defensa de la Tierra en Comunidades Afrohondureñas), to develop leadership training in human rights (Escuela de Formación de Lideres Afrodescendientes en Derechos Humanos), and to facilitate Afro-descendant political participation (Monitoreo y Evaluación del Proyecto Democracia, Gobernabilidad y Participación Política de las Comunidades Afrohondureñas).

Gregory Weeks is Associate Dean for Academic Affairs, College of Liberal Arts and Sciences, University of North Carolina at Charlotte. He has published several books and dozens of articles on Latin American politics, US-Latin American relations, and Latino immigration. His textbook *Understanding Latin American Politics* will soon be available as an Open Access book online and then will be updated as a second edition. He is editor of the academic journal *The Latin Americanist* and writes regularly on his blog Two Weeks Notice: A Latin American Politics Blog (http://weeksnotice.blogspot.com).

Nefratiri Weeks is a Sociology PhD student at Colorado State University, concentrating on political economy and the institutionalization of unequal global trade relations, and Research Assistant and Graduate Student Associate at the Center for Fair & Alternative Trade. She studies the mechanisms through which inequality is maintained in the globalized capitalist economy, focusing on how it fails in fair distribution, perpetuates inequalities, and maintains unsustainable production and consumption. Her Master's thesis examines the intersection of Fair Trade certification and ethical finance in the Fairtrade Access Fund, revealing the potential for enhancing credit outcomes for smallholder farmers in peripheral nations.

Katie Willis is Professor of Human Geography at Royal Holloway, University of London. Her main research areas are gender, development and migration, with a particular focus on Mexico, China and Singapore. She is Vice-President (Expeditions and Fieldwork) at the Royal Geographical Society (with the Institute of British Geographers) and the author of *Theories and Practices of Development* (3rd edition, Routledge, 2019).

Karina Yager is Assistant Professor in Sustainability Studies in the School of Marine and Atmospheric Sciences at Stony Brook University, New York. Yager specializes in interdisciplinary research aimed at monitoring the impacts of climate change in mountain regions, while also understanding the human dimensions of unprecedented socio-ecological change. Her current NASA ROSES and CONICYT (*Comisión Nacional de Investigación Científica y Tecnológica,*

Gobierno de Chile) research is focused on deciphering climate and societal drivers of land-cover land-use change in the Andes of South America. Yager's research combines remote sensing analysis with alpine vegetation studies and ethnographic fieldwork with Andean pastoralists.

George Yúdice is Professor of Modern Languages and Literatures and Latin American Studies at the University of Miami. He is the author of *Cultural Policy* (with Toby Miller, Sage, 2002); *The Expediency of Culture: Uses of Culture in the Global Era* (Duke University Press, 2003); *Nuevas tecnologías, música y experiencia* (Barcelona: Gedisa, 2007); *Culturas emergentes en el mundo hispano de Estados Unidos* (Madrid: Fundación Alternativas, 2009). He edited, translated, and introduced Néstor García Canclini's *Consumers and Citizens: Globalization and Multicultural Conflicts* (University of Minnesota Press, 2001) and *Imagined globalization* (Duke UP, 2014; originally Paidós, 1999). He is the editor (as well author of the introduction and an essay on the impact of digital technologies on policies) of *Políticas Culturais para a Diversidade: lacunas inquietantes*, *Revista do Observatório do Itau Cultural*, N° 20 (May 2016). He has published over 150 articles on cultural policy, music and audiovisual industries, new media, literary criticism, and rethinking aesthetics in the age of social media. He is on the editorial board of *International Journal of Cultural Policy*.

Aram Ziai is Heisenberg-Professor of Development and Postcolonial Studies at the University of Kassel. After studying sociology, political science, history, and English literature, he taught international relations and development studies at the universities of Aachen, Magdeburg, Hamburg, Amsterdam, Bonn, Vienna, and Accra. His areas of research are post-development and postcolonial approaches, development theory and policy, and relations of power in global political economy.

ACKNOWLEDGEMENTS

Julie, Manuel, and Marcela would like to thank all of the people that have made it possible for us to complete a volume of this kind. Thanks go to Routledge, especially to Andrew Mould for inviting Julie to embark on this project and to Egle Zigaite for editorial assistance. We'd also like to thank all of the people that have supported our research and fieldwork in Latin America over the years, especially the Miskito and black Creole people of Nicaragua and the Mapuche-Pewenche people and the Atacameño people of Chile. In particular, Marcela would like to thank her family for their love and support and Julie and Manuel for their patience, understanding, and solidarity in the tough times.

Thanks to all the entities and institutions that have funded our research in Latin America too, including the Antipode Foundation, CONICYT, Fulbright, the Inter-American Foundation, the Tinker Foundation, AHRC, ESRC, the British Council, the Marsden Fund of the Royal Society of New Zealand, Department of Geography at the University of Canterbury, and Victoria University of Wellington.

Finally, special thanks go to all of our wonderful contributors for the depth and breadth you've brought to this project. We've learned a lot from your work and it has been a joy to work with such a large group of people committed to social and political justice and sophisticated intellectual enquiry in the part of the world we now refer to as Latin America.

LATIN AMERICAN DEVELOPMENT

Editors' introduction

Julie Cupples, Manuel Prieto, and Marcela Palomino-Schalscha

Embarking on a handbook of Latin American development at this particular moment is an ethically fraught, impossible and yet necessary task. Ethically fraught because we write in a conjuncture in which the concept of "development" is being claimed, contested, rejected, and remade in complex and contradictory ways by scholars, activists, practitioners, beneficiaries, as well as those who experience it in their everyday lives. As a result, we cannot say with any certainty whether development is good or bad and whether it is something to move beyond or something to struggle for. We also cannot definitively say what development includes and excludes. Its parameters might well include any struggle for human (and indeed nonhuman) dignity, integrity, and rights. Development is made in the boardrooms of the World Bank and in the microspaces of everyday life (Cupples, Glynn, and Larios, 2007). For some, development still means the pursuit of economic growth and this pursuit frequently results in outcomes that are not socially or environmentally sustainable. This task is also virtually impossible because we are dealing with a vast, historically complex, and heterogeneous continent that has come to be known as "Latin America" as a result of European conquest. While recognizing that this naming is a problematic geographical and colonial construction and indeed welcoming other decolonial naming strategies, such as Abya Yala, Afro-Latino-América, América and Nuestra América (see Escobar, 2018; Mignolo, 2005; Kusch, 2010[1970]); Martí, 2002[1891]), Latin America is used by convention by the region's inhabitants and by outsiders and it is also productive of anti-colonial, decolonial, anti-capitalist, and feminist solidarities, that include the recognition of a shared history of European colonialism and conquest that has left profound political, cultural, social, economic, spiritual, and linguistic legacies that must still be confronted and negotiated in everyday and institutional spaces. Latin America mobilizes ways of knowing and produces forms of belonging that undo and transcend its violent colonial history. This book is necessary because "development" continues to be evoked by all kinds of political actors, both dominant and subordinated ones, and therefore remains central to the struggles over capital, land, resources, gender, sexuality, race, and nonhuman ontologies. As Joel Wainwright (2008: 11), drawing on Spivak's Derridean insights, writes, development is "a site of great epistemic violence" but rejecting development is not morally possible as it remains "the hegemonic domination for our responsibility." Wainwright (2008) describes development as an "aporia," that is, an inevitable paradox, which exhorts us to fight for development (or in his words makes it impossible for us not to "desire" it) while keeping in view development's violences and inadequacies. So this book is written in a spirit of responsibility towards the forms of inequality and oppressions that development has mostly failed to address, but is still called upon to do so.

While it is important to recognize that Latin Americans do operate on the margins of the colonial power matrix from a positionality that Enrique Dussel (2003) has referred to as "relative exteriority," the Conquest is however a force to be reckoned with, as if it had not happened, we would not have the concept of "Latin American development." The region now known as Latin America was the first part of the world to be colonized by Europeans, a colonization that brought genocide, disease, resource destruction, European gender orders, evangelization, capitalism, and racism to the continent. All of these colonial phenomena are still in evidence today, although the ways in which they manifest themselves and are contested are highly dynamic. While most of Latin America became formally independent from European powers before other parts of the Global South were colonized, Latin America's post-independence struggles have been long-term ones, proving that independence does not lead to decolonization but rather changes the conditions in which ongoing and long-term struggles for decolonization are fought. We note at the same time that there are *still* colonies in Latin America – Puerto Rico and French Guiana are still colonial possessions of the United States and France respectively, and there is also a significant amount of what we might call internal colonialism and ongoing struggles for independence and autonomy. For many who live on Easter Island, on the Caribbean Coast of Nicaragua (in the former Mosquitia), and in the islands of San Andrés and Providencia, the nation-states of Chile, Nicaragua, and Colombia are frequently seen as colonizers. Latin America is also the first region of the Global South to be designated as the Third World and as a region requiring development, understood in the geopolitical context of the post-war era as development that is administered from North to South, from experts to beneficiaries and as involving (superior) Eurocentric and scientific knowledges that would replace (inferior) indigenous and local knowledges. And it was the first Global South region to both embrace and be forced to adopt neoliberal structural policies, both prior to (Chile) and especially after the Mexican default of 1982.

In the geopolitical context of the post-Second World War world, development studies as a field came into being and Northern and Eurocentric scholars of all kinds – economists, demographers, agronomists, political scientists, anthropologists, and geographers and a range of government, non-governmental, national, and supranational entities, including the United Nations, the International Monetary Fund (IMF), and the World Bank, began to concern themselves with the question of Latin American development. As Arturo Escobar's (1995) work so amply captured, Latin America was constructed as a problem space – unstable, corrupt, racist, patriarchal, diseased, traditional, malnourished, illiterate, violent – that required Eurocentric expertise in order to become modern, prosperous, and civilized. It did however continue to be viewed simultaneously as a land ripe for extraction that could fuel the wealth of an affluent minority both inside and outside the region. Development was embraced and promoted by dominant elites because it would enable colonial activities to continue while sounding more appealing. Indigenous, local, and popular knowledges were all seen as either irrelevant, inferior, primitive, or as a hindrance to the pursuit of development, except for perhaps when they were romanticized and commodified for promoting tourism. Latin America, its peoples and its lands, was then profoundly harmed by development in a multitude of ways but could never be contained by "development." In part, this is many other things were done or thought that had nothing to do with development, and also because the age of development was also the age of the contestation of development. Development as uncontested modernization had a very short intellectual shelf life, in part because Latin American thinkers such as José Carlos Mariátegui (1997[1928]) had already developed very influential theoretical accounts of Latin American realities that emphasized the need to bring Marxist thought into dialogue with local knowledges. Furthermore, it rapidly became apparent to those who cared to look that "development" was being mobilized, subverted, rearticulated, and appropriated by the subaltern and put to alternative uses. And in Latin America perhaps more so than anywhere, it also was clear that the development project was an imperialist one in so many ways, not least because the US government continued to intervene militarily in Latin America in the second half of the 20th century, just as it had in the first half, driven largely by anticommunist obsession and a desire to protect economic interests. There was extensive CIA and US government support for brutal dictators and coups d'état as well as involvement in the overthrow of democratically elected presidents, such as Jacobo Arbenz in Guatemala in 1954, José María Velasco Ibarra in Ecuador in 1961, and Salvador Allende in Chile in 1973. The US set up military bases in Cuba, Panama, Honduras, and Ecuador, assassinated Che Guevara in Bolivia, trained Latin American soldiers in how to deploy torture techniques against insurgents at the School of the Americas, mined Nicaragua's harbours, and assisted in a coup to oust President Aristide of Haiti. US military aid to El Salvador and Guatemala in response to armed revolutionary struggles resulted in some of the most tragic human rights abuses the world has ever seen. The US was also the main driver (with important allies throughout the continent) behind the socially devastating and economically debatable structural adjustment policies, aptly referred to as the Washington consensus, imposed on much the continent in the wake of the debt crisis. So development became as much a means and a lens through which to critique US imperialism and intervention as much as it was invented in order to facilitate it.

Latin American development also became a question of ethics not just because of harmful military interventions but also because it became increasingly apparent that "development" was also far more beneficial to scholars and practitioners than it was for those supposedly in need of development. By the early 1990s, development studies had been polemically described as irrelevant (Edwards, 1989), a critique that would take hold and coalesce into the idea of postdevelopment. Furthermore, the mobilizations against the IMF and the World Bank and the devastating premises of the Washington consensus had intensified and the neoliberal commonsense was becoming unstuck. And indigenous peoples, *campesinos*, and women were marching, blockading, occupying, and protesting in a highly visible way.

Indeed, the development age has coincided with some of the most courageous and extraordinary social movements and political struggles the world has ever witnessed. Anyone wishing to understand how struggles for a better world are enacted and fought and how they play out needs to look to Latin America which has given us armed revolutionary movements in Cuba, Nicaragua, Guatemala, El Salvador, and Mexico; movements led by women to protest at human rights abuses such as the Madres de La Plaza de Mayo in Argentina and the Grupo de Apoyo Mutuo in Guatemala; movements led by indigenous peoples such as CONAIE in Ecuador and the Zapatistas in Mexico; movements to fight for land rights such as the MST in Brazil; the movements to fight against water privatization such as the recovered factories movement in Argentina; the organization of communal kitchens in the countries of the southern cone; and the movements to defend communities from mining in Peru, Colombia, and Guatemala. These movements have produced a vast literature since the 1980s, enhancing global understandings of how we protect lives and livelihoods, claim space, grow food, feed children, protect turtles, dispose of garbage, own and occupy land, gender and queer development politics, construct identities, collaborate and distribute wealth, and also why we fail in endeavours to do these things in ways that are equitable and serve to promote well-being.

These struggles are also hugely impactful and have shaped the global development industry in important ways. Thanks to the bottom-up struggles and contestations of activists, many of them now work the corridors of power, in the United Nations and other development agencies and in National Assemblies, and they have given rise to a range of international development instruments (UN Decades, Conventions and Declarations, World Conferences, Permanent Forum of Indigenous Peoples, MDGs, SDGs, and so on). The presence of the formerly subordinated in such spaces of power is, as much of this book recognizes, shot through with contradictions, disappointments, and political setbacks.

Of course, social movements and other more progressive forces exist because of entrenched capitalist and racist interests that have fuelled and continue to fuel economic inequality, environmental destruction, and the criminalization of activists. Indeed, while Latin America has some of the world's most inspirational social movements, there are many depressing continuities. Indeed, it is not hard to document the failure of existing development models in a range of processes: caudillismo and corruption; forced migrations; persistent gang violence, homicide and feminicide; drug trafficking and addiction; misogyny and machismo; homophobia and transphobia; displacement and dispossession; and deforestation, extractivism, and pollution. Much violence continues to be state-led or at the very least state-endorsed. Three Latin American countries, El Salvador, Honduras, and Venezuela, have the highest murder rates in the world, *by far*, even exceeding figures for the United States (see Kuang Keng, 2016). In Mexico, it is very dangerous to work as a journalist or run for office, in Honduras and Colombia it is very dangerous to be an environmental activist, in Nicaragua it is very dangerous to oppose the current government in power, and in Guatemala you might die because you live close to an active volcano but your life is not deemed worthy of protection by the state. Latin American history, including very recent history, is full of those whose lives were taken because they opposed power and embraced an understanding of development that threatened those in power and those who pursue socially and environmentally destructive development projects. Berta Cáceres, the Ayotzinapa 48, Angel Gahona, and Temistocles Machado were all murdered for their beliefs, their activism, and their work. They were denied the right to freedom of speech and expression and to an extent their deaths drive home the catastrophic failure of development in Honduras, Mexico, Nicaragua, and Colombia respectively. Re-imagining development remains politically risky in many parts of the continent. The 20th century was particularly brutal in terms of human rights abuses. So far, the 21st century is proving no less brutal, even though the global conjuncture has shifted, and experiences in Latin America (as well as those elsewhere) worryingly point to the possibility of the emergence of a post-human rights era. Indeed, we write this introduction as young unarmed Nicaraguan students are gunned down by state paramilitary forces with weapons of war and Honduran and Salvadoran children are taken from their parents by US border guards and kept in cages and impunity is everywhere.

One thing that is clear however is that experiences and struggles in Latin America have subverted the idea of development as a mode from expertise delivered from North to South. While Latin America has always been a source of political inspiration to outsiders, this inspiration is taking on a new intensity as Latin American activists, especially those of indigenous and African descent, those with non-dominant gender and sexual identities, those fighting ecologically destructive practices in the city and the countryside, and low-income urban dwellers, are asserting innovative and non-Eurocentric ways of dealing with the profound economic, ecological, and epistemological crisis in which that the world is immersed. Ideas surrounding decoloniality, interculturality, *buen vivir*, plurinationality, queer and trans critiques, and nonhuman ontologies, all amply discussed on the pages of this book, are key sources of scholarly, activist, indigenous, and popular inspiration. As John Holloway writes of Bolivia in the foreword to Raúl Zibechi's (2010: xv) book on contemporary political struggles in Bolivia:

Time has done a somersault. Bolivia used to be seen as a backward, underdeveloped country which could hope, if it was lucky, to attain the development of a country like Germany one day in the future. Perhaps even now there are some people who still think like that. But, as the disintegration of the capitalist world becomes more and more obvious, more and more frightening, the flow of time-hope-space is reversed. For more and more Europeans, Latin America has become a land of hope. And now as we read of the movements in Bolivia, we say not "poor people, have they any hope of catching up with us?" but rather "how wonderful! Can we Germany (or wherever) possibly hope to do something like that? Can we ever aspire to act like the people of Cochabamba or El Alto?"

It is apparent that scholars, activists, and policymakers outside Latin America, including those in the Global North or the so-called developed countries, now have more to learn from Latin America than Latin America has to learn from them. Indeed, we can see all kinds of development alternatives and alternatives to development in a range of creative practices, in education, transport, health, and work, often forged and shaped from the bottom-up. Such practices call into question the reified sets of social relations that allow for oppressions and reveal there is nothing inevitable (but much that is highly resilient) about colonial/capitalist modernity. Studying and analysing Latin American social movements as modes of development intervention is politically and intellectually stimulating. It forces us to engage with questions of fairness and citizenship in ways that tend to illuminate struggles outside of Latin America too. We are forced to think carefully about who gets to own, to speak, to occupy, and to belong and what can and can't be bought and sold on the market. We are thus confronted with the structural modes of injustice in which current arrangements are embedded and most importantly how things might be different. Students, researchers, activists, politicians, journalists along with those who work more formally in the development industry have much to gain if they learn from and with those that work to promote Latin American development in a bottom-up way. And of course, many key moments including the Zapatista uprising and the Cochabamba water war have had global repercussions. It is however just as important to understand the struggles and movements for development that start out as progressive or revolutionary and go horribly wrong. Here we might include the Shining Path of Peru, the FARC of Colombia, and the growing authoritarianism and emphasis on harmful extractivisms within the Pink Tide governments.

Development's theoretical trajectories

While development as a top-down form of external intervention is now widely discredited but persists in a range of institutions especially in the Global North as well as in dominant discourses, development remains a social and cultural force to be reckoned with because of the ways in which it has been harnessed and rearticulated by ordinary people and put to diverse uses. Its internal inconsistency and the ongoing contestations that surround its meanings mean that development has been made sense of intellectually in a range of ways. Development is also quite a promiscuous arena of study in terms of its disciplinary affiliations. While its origins are clearly rooted in area studies and development studies as invented initially in the Global North, development now moves fluidly across a range of academic fields Europe and the Americas and is studied by scholars and researchers across the humanities and social sciences, including anthropology, economics, geography, history, political science, and sociology and it influences and is influenced by interdisciplinary studies including Latin American studies and cultural studies. Therefore, dominant liberal and capitalist approaches coexist with more critical approaches that emerge from Marxism, political economy, and political ecology. Many theoretical approaches have also been subject to high degrees of hybridization where Eurocentric and external theories have been reworked or indigenized by Latin American approaches (both Eurocentric and non-Eurocentric varieties). For example, engagements with poststructuralism led to a focus on the power of development discourses and a turn to the concept of postdevelopment, a move which proved to be highly polemical in the 1990s but is probably less so today. Furthermore, these approaches have been further complexified by the so-called decolonial option (otherwise known as the MCD [Modernity/Coloniality/Decoloniality] paradigm), in which scholars are taking seriously indigenous and Afro-descendant epistemologies and worldviews, recognizing that colonial practices and attitudes (coloniality) continue to shape development policies and projects, and engaging more fully with the geopolitics and body-politics of knowledge in important ways (see Mignolo and Tlostanova, 2006). In particular, the decolonial option has reworked our understandings of modernity revealing its dark side in terms of its entanglements with capitalist exploitation, colonialism, and racism (see Mignolo, 2011). The decolonial approach is inseparable from other forces of liberation and so some scholars are reading decolonization through more radical lenses that brings in gender, queer identities, and intersectionality in important ways, approaches that expand what we think of as development and the necessary routes out of poverty and oppression. The decolonial option is also inspired by and also inspires what we might call an ontological turn in development studies which is encouraging us to think of development beyond the human and beyond Eurocentric nature–culture binaries and acknowledge and account for non-human agency and liveliness.

Structure of the volume

We are three editors that have all worked with themes related to Latin American development broadly defined. We are at different stages in our careers and work in different parts of the world, namely in universities in the UK, Chile, and New Zealand. Two of us are Latin American and one of us is European. But all of us have had the privilege to be exposed to decolonial thought and practice in our work, particularly as a result of working closely with indigenous, peasant, and Afro-descendant communities and low-income urban neighbourhoods in Chile, Nicaragua, Colombia, and Costa Rica and engaging in depth with indigenous, decolonial, and ontological thought produced elsewhere in the continent and beyond. So, we share a particular commitment to using our collaborations and our scholarship to create a different kind of world. So this handbook does not seek to be objective, rather it is written in a spirit of commitment to liberatory transformation (for both humans and non-humans) in the recognition that dominant development has done grave harm but that other worlds are indeed possible and are already being enacted in Bangkukuk Taik, Chiapas, Cochabamba, El Chocó, El Wallmapu and in many other places. We also write in a belief that you can't be neutral in the face of extractivism, indigenous land dispossession, state-led violence, or the oppression of women or LGTBIQ people. We have tried as much as possible to produce a book that includes a diverse set of authors that are located in different disciplines, in different parts of the world, and are at different career stages. Our authors are based in universities and other organizations in Argentina, Brazil, Chile, Colombia, Costa Rica, Ecuador, Germany, Honduras, Mexico, Netherlands, Nicaragua, Panama, Portugal, Spain, the United Kingdom, and United States. We have also tried to produce a book in the spirit of Enrique Dussel's (2003) philosophy of liberation. In other words, we have selected authors that in their diversity write in some way on the side of the subordinated and the dispossessed and whose research to varying degrees has included "an enquiry into the lived experience of those involved" (Dussel, 2003: 182).

As with any project of this nature, our volume cannot cover all areas of Latin American development and there are inevitable gaps. These gaps result from the scope and size of the project, our own theoretical concerns and indeed biases, and from the networks in which we move. A different set of editors would have produced a different volume. Our volume is unashamedly critical – by critical we mean broadly anti-capitalist, feminist, and decolonial – in its approach. We do not seek to rescue development from its critics, nor do we give a platform to those promoting development as usual, based on modernist forms of extractivism and Eurocentric viewpoints. We do however seek to capture some of the contradictory outcomes that result from the development project. In attempt to provide a sense of organizational coherence, we have assembled our chapters in a series of sections that are both somewhat arbitrary and overlapping, in the sense that many of them could appear in other sections. Some core themes such as neoliberalism, gender, and indigenous rights and knowledges appear in and move across all the sections to varying degrees and do so because we don't think these issues can be confined to a single section. The book contains 47 chapters organized into seven different sections.

Part I seeks to provide a set of debates that open the volume, providing important theoretical introductions to those less familiar with Latin American development and a set of provocations and new insights for those that are. Chapter 1 by Cristóbal Kay is a detailed overview of modernization and dependency theory with a particular focus on the ways in which these approaches have been debated and contested by Latin American intellectuals. In Chapter 2, George Yúdice traces the relationships between culture and development from the early 20th century to the present day, with a focus on questions of transculturation, cultural imperialism, hybridity, and the cultural and creative industries. In Chapter 3, Nancy Postero puts indigeneity at the heart of debates on development in Latin America, focusing in particular on how indigenous peoples have both been folded into development at different historical moments, as well as discussing the challenges they have posed to the development project. In Chapter 4, Camila Esguerra Muelle provides us with an introduction to decolonial theory through a feminist lens, with a focus on gender, sexuality, displacement, and migration and on the forms of decolonial resistance and rebellion migratory processes engender. In Chapter 5, Aram Ziai discusses theories of post-development, detailing the intellectual arguments that support and critique it, ending with a set of reflections on how such approaches can help to think about inequalities in a more inclusive way. Chapter 6, by Charles Hale, contains a set of reflections on the influential debates that surround the idea of neoliberal multiculturalism and the complexities and contradictions that accompany the embrace of multiculturalism by dominant elites. He brings these ideas up to date, suggesting we are now witnessing the demise of the era of neoliberal multiculturalism which creates opportunities and challenges for racial-cultural politics. In Chapter 7, Laura Enríquez and Tiffany L. Page consider the rise and fall of the Pink Tide, focusing on the centre-left and left-wing governments that came to power in much of the continent in the past two decades following substantial social movement mobilization. The chapter evaluates the performance of the Pink Tide governments and considers the factors that led to their demise. The final chapter of the first section by Javier Arellano-Vargas and Javier Martínez-Contreras considers the complex role of the church and religion in development processes. It focuses primarily on the role of the Catholic Church as both a progressive and reactionary force in the continent.

Part II shifts our focus to questions of globalization and international relations, exploring global development policies, migratory flows, forms of regional integration and Latin America's diplomatic and economic relations with other parts of the world. In Chapter 9, Tara Ruttenberg considers the legacies of Latin America's deeply transformative engagements with the

International Monetary Fund, the World Bank, and the World Trade Organization and how social movements have responded to their negative social and environmental effects. While important continuities with neoliberal development remain, Ruttenberg also explores how it is challenged by critical alternatives including the politics of *buen vivir*. Chapter 10, written by Katie Willis, contains an assessment of a key recent international development instrument, the Sustainable Development Goals (SDGs), looking at how it has been adopted within the continent. Focusing in particular on targets to reduce inequality and protect the environment, she discusses the degree to which the SDGs represent a departure from top-down externally driven development. In Chapter 11, Guadalupe Correa-Cabrera discusses the War on Drugs from a development perspective, including the role played by anti-narcotic and security policies and discusses who wins and who loses from the global drug trade. Migration is the central theme of Chapter 12 written by Cathy McIlwaine and Megan Ryburn who trace the nature of migration flows within and beyond Latin America and situate these in the context of socio-economic and political dynamics in the region. The final four chapters in this section focus on international relations. In Chapter 13, Andrés Malamud focuses on regional integration within Latin America and provides an overview of the most influential regional organizations in the continent. Chapter 14 by Gregory Weeks focuses on US-Latin America relations, Chapter 15 by Barbara Hogenboom on China-Latin America relations, and Chapter 16 by Anna Ayuso on relations between the EU and Latin America. Together these chapters outline changing geopolitical, diplomatic, and trading relations, all of which have implications for development dynamics.

Part III of the book explores a number of decolonial political and cultural struggles in the continent. It examines the multiple ways in which development has been contested in Latin America. The chapters cover a wide range of empirical contexts, from more than human politics to counter-mapping. The chapters in this section reflect on the room of manoeuvre that can be opened against the colonial project reproduced by development in Latin America, but also the contradictions and paradoxes faced by different emancipatory projects. The first chapter in this section by Laura Ogden and Grant Gutierrez reviews the multiple ways non-human life became connected through colonialism. They argue that the nonhuman's political agency poses contradictions to the colonial project in Latin American, highlighting how Indigenous and other social justice movements have been key actors in this reorganization of the political. The following chapter, by Daniel Mato studies the diversity of institutional arrangements through which intercultural universities has been organized in Latin America. Mato focuses on the modes of learning developed in these institutions, the way they relate to indigenous and Afro-descendant peoples, and the conflicts between these institutions and the states. In Chapter 19, Piergiorgio Di Giminiani examines the principal outcomes of indigenous activism in Latin America. The chapter focuses on how citizenship rights, indigeneity, and world-making are inevitably interwoven within political processes. The next chapter discusses black and Afro-descendant rights and struggles. The chapter is a collaborative piece by Deborah Bush, Shaun Bush, Kendall Cayasso-Dixon, Julie Cupples, Charlotte Gleghorn, Kevin Glynn, George Henríquez Cayasso, Dixie Lee Smith, Cecilia Moreno Rojas, Ramón Perea Lemos, Raquel Ribeiro, and Zulma Valencia Casildo. The authors critically examine the structural racism and invisibilization that Afro-Latin Americans have suffered, and discuss how activists have responded to disadvantages and discrimination. The chapter explores the challenges and contradictions that remain within the context of neoliberal multiculturalism and new forms of coloniality. The chapter that follows, by Sergio Tischler, provides a summary of Zapatista thought and how it has contributed to the anti-capitalist movement. Tischler explains the Zapatista idea of revolution by exploring the concepts of the anti-capitalist subject and the notion of autonomy, and also by problematizing the concept of revolution within the state form. The section's closing chapter, by Joe Bryan, explores a contradiction faced by counter-mapping: how despite its challenges to dominant forms of power it is a project rooted in hegemonic forms of conceiving and understanding space, extending their reach and dominance. In order to negotiate this paradox, Bryan argues that maps must also be continually contested, challenged, and resisted.

While gender and sexuality appear across a number of chapters, Part IV has four chapters that deal in depth with these themes focusing on their implications in terms of both cultural politics and social policy. Chapter 23 is authored by Sarah Bradshaw, Sylvia Chant, and Brian Linneker and explores the gendered and contradictory outcomes of anti-poverty policies such as Conditional Cash Transfers. The chapter discusses these outcomes in the context of gender-based violence and an escalation of feminized poverty. Chapter 24 by Jasmine Gideon and Gabriela Alvarez Minte focuses on the intersections between gender, health, and neoliberalism and discusses the gendered consequences of health privatization with a focus on the case of Chile. The next chapter by Matthew Gutmann outlines both successes and obstacles to involving men in development processes, highlighting in particular the question of health, sexuality, gender-based violence, migration, governance, and citizenship. In Chapter 26, Florence Babb turns our attention to LGBTQ human and legal rights and the history of nonheteronormative sexual expression in Latin America. She provides specific details on LGBTQ social movements from Nicaragua.

The fifth section of our volume provides a number of insights into Latin American labour and campesino movements in both cities and rural areas. Tony Bebbington opens the section with a chapter that theorizes rural social movements, emphasizing the dangers faced by environmental defenders. Bebbington argues that rural social movements are never merely rural as they deal with issues and flows that articulate the rural and the urban and therefore are fundamental in reflections on Latin American development. The next chapter by Maurizio Atzeni, Rodolfo Elbert, Clara Marticorena, Jerónimo Montero Bressán, and Julia Soul provides us with a detailed introduction to labour movements and trade union organizing. The authors identity three distinct political courses taken by labour movements – stable neoliberalism, stable popular reformism, and conservative shift – and outline the situation in Mexico, Chile, Ecuador, Bolivia, Venezuela, Argentina, and Brazil. In Chapter 29, Maurício Rombaldi examines the labour dynamics of sporting mega-events including the 2014 World Cup and the 2016 Olympic Games. The chapter demonstrates the forms of labour precarity that result from such events but also discusses the strategies adopted by unions globally to defend workers' rights. Chapter 30, by Kate Swanson, focuses on a ubiquitous category of Latin American worker, namely the urban street vendor. She looks at how these workers are threatened by urban revitalization strategies that often result in their displacement as well as the ways in which these workers and their allies resist through a "right to the city" approach. The next chapter written by Jennifer Bickham Mendez tackles the question of maquiladora labour, with a focus on working conditions and the difficulties of unionization in these factories. She also looks at how women have organized in order to address the challenges they face as low-paid factory workers. The final chapter in this section by Laura Raynolds and Nefratiri Weeks looks at campesino movements and their involvement in the global fair trade market. Despite a number of contradictions, the authors argue that fair trade certification is able to foster development in both the short and long term.

Part VI of the book includes nine chapters that critically examine the problem of development in relation to environmental issues. Studying a wide range of empirical cases, from land-grabbing to food security, the chapters in this section reveal how the different approaches to forms of development in Latin America are unquestionably a problem related to land, resources, and environmental struggles. In the opening chapter, Eduardo Gudynas argues that the concept of modes of appropriation is necessary for analysing the strategies of development that depend on the extraction of natural resources. He describes these modes as articulated with the modes of production in the sense that they organize the process through which the modes of production appropriate the raw materials. In his opinion, the concept of modes of appropriation allows not only a more precise explanation of the problems of development and extraction in Latin America, but can also contribute to political transformation. The second chapter in this section is written by Diana Ojeda who explores how the history of development in Latin America is also one of landgrabbing. In her analysis, she insists on the importance of the historical processes of differentiation, exclusion, and dispossession and how these get sedimented in landscapes. Chapter 35 by Rob Fletcher deals with the theme of conservation and focuses on the rise of and challenges faced by protected areas throughout Latin America. Fletcher explains how the initial "fortress"-style management has been complemented by community-based conservation strategies and market-based instruments. He argues that this model is now challenged by expanded raw material extraction motivated by increasing trade relations with East Asia and elsewhere. Chapter 36 is by Tom Perreault who critically examines mining and its importance for Latin American development. After a detailed historical analysis, he demonstrates that in Latin America mining has been crucial to economic and political planning and has been located at the centre of social movement mobilization. The next chapter discusses the politics of water. Rutgerd Boelens explores how in Latin America water management is constrained by abstract, universalist, and de-humanized models, but also imaginaries that reproduce romantic ideas about how water is locally managed. In his chapter, Boelens calls for a multi-actor, multiscalar contextual, grounded, and relational approach to water justice. In Chapter 38, Mary Finley-Brook and Osvaldo Jordan utilize eight case studies for illustrating the physical, territorial, structural, and other forms of violence generated by the transitions away from petroleum to cleaner energy sources in Latin America. The next chapter is by Gabriela Valdivia and Angus Lyall who trace how certain tendencies and practices of oil rule have become habits of the Latin American oil complex. In so doing, they develop a historical examination of the petro-politics in oil-producing Latin American countries and in the spaces of oil logistics (what they called oil frontiers). They close the chapter by exploring the role of social movements in opening up the possibility of changing existing habits of oil rule. Chapter 40 by Beth Bee delivers an overview of the diverse factors that have moulded food security and sovereignty in Latin America. Bee studies the present-day debates and emerging issues, with particular emphasis on gender and climate change. The last chapter in this section by Corinne Valdivia and Karina Yager discusses the issue of climate change and discusses how rural communities in the Andes are experiencing and negotiating its effects and the uncertainties it brings. It provides empirical details from three different regions in Bolivia and Peru.

The final section of the book considers the tensions, contradictions, and opportunities associated with urban development in Latin America. The section contains six chapters, organized around the problems of gentrification, gang violence, informal settlements, mobility, environmental injustice, and gender. The section starts with a chapter by Ernesto López Morales in which the author explores the multiple phenomena that have influenced gentrification processes in Latin American cities. He argues that gentrification is the violent result of the strategies developed by technocrats, the state, and the private sector that aim at reproducing and expanding real state capital and service industry capital. In order to do so, he develops a comparative study based on four cities, Mexico City, Buenos Aires, Rio de Janeiro, and Santiago. In the next chapter, Dennis Rodgers critically explores the nature of gangs and gang-related violence in Latin America. In his analysis, he argues that gangs, rather than being anti-developmental, reveal the persistent unequal political economy of Latin America. This chapter is followed by one by Melanie Lombard who presents a detailed summary of key debates emerging from

Latin America's informal settlements. The chapter reviews the main characteristics of informal settlements in the region, explores the question of how this phenomenon has informed policy and practice, and explores alternative narratives to pessimistic viewpoints for informal settlements. In Chapter 45, Fábio Duarte focuses on urban mobility in Latin America through the study of innovative transportation solutions first proposed in Latin America that have been influential worldwide. He argues that, notwithstanding all the major urban mobility problems faced by Latin America, there are spaces for implementing and adapting some creative solutions. In the next chapter, Marcelo Lopes de Souza argues that the concept of environmental injustice is crucial for analysing the contradictions of the Latin American urbanization process. This concept also invites a politicized examination of the paradoxes and limitations of the idea of development. Lopes de Souza studies three representative environmental injustice problems: waste dumping, exclusionary forms of 'environmental protection,' and unequal access to water. In the final chapter, Natalia Quiroga reflects on the mobilizations of women that took place in Latin America in 2017 and 2018. She argues that this phenomenon is a part of a broader social articulation for resisting and rejecting extractivism, the financialization of the economy, and a system that disrespects the female body. Quiroga encourages us to focus on the feminization of politics, a process that aims to reposition women in power and to prevent the woman's body from becoming a sacrifice zone for neoliberalism.

References

Cupples, J., Glynn, K. and Larios, I. (2007) Hybrid cultures of postdevelopment: The struggle for popular hegemony in rural Nicaragua. *Annals of the Association of American Geographers* 97(4): 786–801.

Dussel, E. (2003) *Philosophy of Liberation*. Eugene: Wipf and Stock.

Edwards, M. (1989) The irrelevance of development studies. *Third World Quarterly* 11(1): 116–135.

Escobar, A. (1995) *Encountering Development: The Making and Unmaking of the Third World*. Princeton, NJ: Princeton University Press.

Escobar, A. (2018) *Otro posible es posible: Caminando hacia las transiciones desde Abya Yala/Afro/Latino-América*. Bogotá: Ediciones desde abajo.

Kuang Keng, K. S. (2016) Map: Here are countries with the world's highest murder rates. *PRI*, 27 June. www.pri.org/stories/2016-06-27/map-here-are-countries-worlds-highest-murder-rates (Accessed 18 June 2018).

Kusch, R. (2010[1970]) *Indigenous and Popular Thinking in América*. Durham, NC: Duke University Press.

Mariátegui, J. C. (1997[1928]) *Seven Interpretive Essays on Peruvian Reality*. Austin: University of Texas Press.

Martí, J. (2002[1891]) *Nuestra América*. Jalisco: Universidad de Guadalajara.

Mignolo, W. (2005) *The Idea of Latin America*. Malden, MA: Wiley-Blackwell.

———. (2011) *The Darker Side of Western Modernity: Global Futures, Decolonial Options*. Durham, NC: Duke University Press.

Mignolo, W. and Tlostanova, M. V. (2006) Theorizing from the borders: Shifting to geo- and body-politics of knowledge. *European Journal of Social Theory* 9(2): 205–221.

Wainwright, J. (2008) *Decolonizing Development: Colonial Power and the Maya*. Malden, MA: Wiley-Blackwell.

Zibechi, R. (2010) *Dispersing Power: Social Movements as Anti-State Forces*. Oakland, CA: AK Press.

PART I

Debates and provocations

1

MODERNIZATION AND DEPENDENCY THEORY

Cristóbal Kay

In this chapter I discuss the main ideas of the theories of modernization and dependency from the perspective of the Latin American region. Both theories have to be set against the background of the Cold War and the competing ideologies of capitalism and communism. Modernization theory emerged in the North during the 1950s and 1960s and was largely absorbed uncritically at the time by social scientists and policy makers in Latin America. From the late 1960s it was fiercely challenged by the *dependentistas*, as the dependency theorists and followers were often called. While modernization theorists aimed to develop and strengthen the capitalist system in the Third World, distinct strands within dependency theory aimed either to overthrow it and start a process of transition to socialism (a Marxist strand), or to reform the underdeveloped structures of the region and reforming the international economic system towards a more equitable relation between North and South (a structuralist position). I will first examine modernization theory (MT), proceed to the analysis of dependency theory (DT), and finish with some conclusions.

Modernization theory

After the Second World War the process of decolonization gathered pace. Social scientists were encouraged by governments, international institutions, aid agencies, and policy makers to undertake research into the so-called 'backward' countries or 'non-Western societies' with a view to designing and implementing development strategies and policies which overcame the various economic and social obstacles to development. For economists these barriers boiled down to finding ways to increase the rate of savings and investment while sociologists focused on changing social norms and cultural values. Development was seen as a process of transition from a traditional to a modern economy and society.

The development goal for traditional societies was seen as involving a process of modernization that followed the footsteps of the West by adopting the experience of the developed countries as a model. In this dualistic typology of traditional and modern, traditional societies were defined as simple and undifferentiated societies with a large rural subsistence economy based on family labour, employing primitive technology which explained their low productivity and poverty. They were also characterized by low levels of literacy, health, and political participation. By contrast modern societies were depicted as complex and differentiated, with an industrial economy geared to the market, based on wage labour and the adoption of scientific technology which explains their high productivity and standard of living. They enjoyed high levels of literacy, health, and political participation. While modern societies are characterized by universalism (actors judge each other according to general principles in their personal relationships), achievement orientation (individuals are judged according to their actions), and self-oriented (when private interest dominated actors' behaviour), the opposite is the case for traditional societies who display particularism (actors judge each other according to their personal relationships), adscription (individuals are judged according to their given status), and collective orientation (when collective interests are given priority). Countries were accordingly located by modernization theory in a unidirectional continuum between these two extremes. Modernization theorists dangled the promise before the underdeveloped countries that they could catch up with the developed countries by replicating their experience.

One characteristic which modernization theory regarded as blocking development in traditional societies was traditionalism itself, whose fatalistic approach to life ('things have always been as they are') discouraged people from taking action. Meanwhile in modern societies there was held to be a willingness to overcome obstacles and embrace change while maintaining certain traditions. They were characterized as having upward social mobility, equal opportunities, rule of law, individual rights, and freedoms. Thus for traditional societies to progress, prevailing attitudes, beliefs, and cultural and social values had to be displaced by modern viewpoints through the diffusion of scientific knowledge, technical capabilities, organizational and institutional capacities, and so on. Development aid from rich to poor countries would facilitate the process of transition.

Different authors within modernization theory emphasized different factors for modernization to take place. Wilbert Moore (1963) underscored value changes towards individual merit, social mobility, science, and political participation. In turn, David Lerner (1958) stressed transformations in personality, David McClelland (1961) highlighted such psychological factors as the desire to achieve, and Everett Hagen (1962) expressed this in terms of developing an entrepreneurial and innovative spirit. Bert Hoselitz (1960) uses the pattern variables of Talcott Parson for his analysis of social change and economic development.

The prominent political scientist, Samuel Huntington (1968), argued that the main objective of modernization should be the promotion of political order. He was concerned that the political institutions in developing countries would be unable to cope with the increasing political participation and social demands unleashed by a process of modernization and that this could result in the eventual breakdown of the political system. Hence his overriding concern was for securing political order and stability above all other objectives. His analysis was interpreted (by some) as justifying strong and even authoritarian governments. Huntington also differed with some of the tenets of MT as, in his view, all societies combined to different degrees both traditional and modern elements. Thus all societies are in transition. He thus critiqued mainstream MT for being too static and unable to explain the actual processes of change.

In sum, modernization theorists emphasized that social, cultural, psychological, and political factors are also important for economic development. In this way, the non-economic social sciences were put ultimately at the service of economic objectives, although economists were seen as having the upper hand in the design of development strategies. This secondary role of sociological and cultural factors has since been challenged, especially with the rise of postmodern, postcolonial, and cultural studies which largely dismissed economics and the whole idea of development.

One of the most popular and probably most influential analyses of economic growth within a modernization perspective is that of the economic historian Walt W. Rostow (1960). He argued that there are five stages in the modernization process which all countries must sooner or later follow; (1) the traditional society, (2) the preconditions for take-off, (3) take-off, (4) the drive to maturity, and (5) the age of mass-consumption. Societies are situated at different points on this continuum. Traditional societies are located within the first two stages and modern societies in the last two stages. The take-off stage was the key turning point towards modernization. The preconditions for take-off is a protracted stage of over a century, involving the formation of a vigorous commercial and entrepreneurial class whose savings finance the industrialization process. By contrast the take-off stage is a much shorter period of two or three decades during which the rate of investment increases substantially enabling the rate of economic growth to exceed the rate of population growth and for income per capita to rise for the first time in history.

One of the main attempts to apply Rostow's stages of growth theory to Latin America was made by Guido di Tella and Manuel Zymelman (1973). Their depiction of six stages of growth in the case of Argentina were similar to Rostow's but with some modifications as in their analysis reaching a certain stage did not automatically secure the transition to the next. In the case of Argentina, they inserted an additional stage between Rostow's (3) and (4) called 'delay' as Argentina failed to adjust fast enough to the change from an agricultural export economy to an industrial economy. Although the country had reached in their view the self-sustained stage (1933–1952) they characterized the subsequent period as one of readjustment rather than leading to the age of mass consumption. Meanwhile Gino Germani (1962) distinguished four stages of economic growth in Latin America which are traditional society, weakening of traditional society, dual society, and, finally, mass society. For each stage he analyzed the economic, societal, and political characteristics.

In the late 1960s André Gunder Frank (1972a, orig. 1967) launched a devastating critique of MT. He forcefully rejected its traditional-modern dualism, unilinearity, its adoption of neo-Parsonian social pattern variables and neo-Weberian cultural and psychological categories. Furthermore, he argued that it was empirically faulty, theoretically derisory, and useless as a policy tool for underdeveloped countries (Frank, 1969). But his main point of difference with MT concerned the common assumption that underdevelopment was an original state and that the rich countries had developed in isolation, ignoring the ways in which development and underdevelopment are part of a single process in the formation of the world capitalist system since the 15th century. Indeed, I would contend that MT regarded change as being largely determined internally, ignoring the impact of colonialism and imperialism on underdeveloped countries or assuming these to be benign. Empirical evidence has shown discontinuities and reversals in the development process and that rather than a single evolutionist path of development there is a variety of transitions to modernity.

While some Latin American thinkers merely replicated the ideas of the modernization theorists there was one notable exception. Gino Germani (1981) introduced some ideas of his own by creatively adapting MT to the particular conditions of Latin America. For example, he argued that social forms belonging to different historical periods coexist in any process of transition thereby creating conflicts as some spheres change faster than others. There are lags, backlogs, and various asynchronies, as he calls them, such as between regions, institutions, social groups, and motivations. He is also concerned with the "demonstration effect" of advanced societies which can create aspirations for a level and type of consumption in the transition societies which cannot be met and can therefore lead to mass mobilizations and crisis. But some of the abovementioned shortcomings of MT also apply to him. From a Marxist perspective he fails to discuss class, exploitation, and class conflicts, as well as ethnic conflicts, which can be very acute in underdeveloped countries.

Varieties of dependency theory

Background

DT arose at a particular time and context within Latin America. The theories of imperialism and the ideas of the Latin American structuralist school of development were two key influences. The structuralist school arose out of the Economic Commission for Latin America and the Caribbean (ECLAC), a United Nations body established in Santiago, Chile, in the late 1940s, whose driving force was the Argentinian Raúl Prebisch (Kay, 2006). Although not generally considered to be a dependency theorist, Prebisch was certainly one of its precursors. He wrote the influential *Economic Survey of Latin America 1949* (ECLA, 1951) which the famous thinker Albert Hirschman (1961: 13) appropriately dubbed the 'ECLA Manifesto.' Prebisch's centre-periphery paradigm exposed how the international trade system worked to the benefit of centre countries (the developed rich countries of the North) and to the detriment of the periphery (underdeveloped, developing, or poor countries of the South) exacerbating inequality between them. The centre countries retained most of the benefits of their industrial technological progress (through increased profits and/or wages) and profited from the (relatively) lower prices of the primary commodities they imported from the periphery. The periphery, by contrast, transferred part or all of its increased productivity in the form of the primary commodities they exported to the centre while at the same time as being unable to benefit from the increases in productivity of the industrial commodities exported by the centre. This phenomenon, which flatly contradicted Ricardian orthodox international trade theory on comparative advantages, is referred to in the economic literature as the 'Prebisch-Singer thesis' on the periphery's deterioration of the terms of trade, i.e. the price of primary commodities exported by the periphery falls more quickly (or rises more slowly) than the price of the industrial commodities exported by the centre.

Prebisch's conceptualization of the world economy and his revelation of the systemic unequal relationship between centre and periphery profoundly changed how the problem of development should be addressed. It constituted a very different way of looking at the problem of development from MT and from orthodox economic theory (Kay, 1989). In his last book, published five years before his death, he attempts to develop a theory of 'peripheral capitalism' which is written in the best structuralist tradition that he pioneered and where he introduces aspects of DT while also attacking the emergent neoliberal paradigm (Prebisch, 1981).

Differences and commonalities within DT

The early influence of the structuralist school on DT aside, it is my contention that there is not one, but several distinct versions of DT so that strictly speaking one should talk of dependency theories in the plural rather than the singular. This broad array of ideas which emerged in the mid-1960s reached its widest influence in Latin America and worldwide during the 1970s. Its demise started with the economic crisis of the 1980s and the rise of neoliberalism, first through the pursuit of World Bank- and IMF-designed structural adjustment programmes and later with the Washington Consensus in the 1990s. In this transformed context, some dependency thinkers embraced world system theory, others, known as neo-structuralists returned to their structuralist roots with modifications, and yet others argued that aspects of DT acquired new relevance with the spread of neoliberal capitalism globally (see Gwynne and Kay, 2000).

I distinguish mainly two strands within DT, one with origins in structuralism and the other in Marxist theories of imperialism. Both arose as a critique of the process of import-substitution industrialization (ISI) that most Latin American countries (especially the larger ones) had been following since the Second World War or earlier. In the early decades growth rates under ISI rates were high and industry became the most dynamic sector of the economy. Later growth rates declined as ISI became 'exhausted.' Expectations of ISI relating to employment, income distribution, technological diffusion, and export orientations never fully materialized. What particularly provoked and irked progressive critics at the time was that ISI led to increasing control by foreign capital over industry rather than greater national control over the economy. Foreign capital not only retained a high degree of control of the traditional primary product export sector, particular mining and plantation agriculture, but began to dominate the new industrial sector as well.

The key difference between the structuralist and Marxist strands in DT is political. While the former sought to surmount dependency and underdevelopment by reforming the capitalist system, the latter argued that the shackles of dependency could only be surmounted by overthrowing capitalism, if necessary by revolutionary means, and replacing it with socialism. Another key difference pertains to the theoretical apparatus deployed. While structuralists used the analytical and conceptual tools of development theory, albeit of a heterodox kind, Marxists relied on historical materialism and the labour theory of value.

While both strands of DT share a comparable definition of dependency, the problematic of dependence cannot be reduced to and encapsulated in a single definition as will become clear later. Sunkel (1973: 136) writing from a structuralist perspective argues:

> Development and underdevelopment should therefore be understood as partial but interdependent structures, which form part of a single whole. The main difference between the two structures is that the developed one, due basically to its endogenous growth capacity, is the dominant structure, while the underdeveloped structure, due largely to the induced character of its dynamism, is a dependent one. This applies both to whole countries and to regions, social groups and activities within a single country.

Writing from a Marxist perspective, Dos Santos (1973: 76) notes:

> Dependence is a conditioning situation in which the economies of one group of countries are conditioned by the development and expansion of others. A relationship of interdependence between two or more economies or between such economies and the world trading system becomes a dependent relationship when some countries can expand through self-impulsion while others, being in a dependent position, can only expand as a reflection of the expansion of the dominant countries, which may have positive or negative effects on their immediate development. In either case, the basic situation of dependence causes these countries to be both backward and exploited.

As can be seen both views emphasize the interdependence between the developed and underdeveloped countries and the absence of an autonomous capacity for growth in dependent countries. The dependency situation arose with colonialism when Latin America became incorporated into the emerging world capitalist system. Due to this colonial dependent relationship, which restructured the economy, society, and polity of the region, the resulting process of transformation created a distinctive dynamic from that of the dominant colonial power, a dynamic which the independence of the republics did not fundamentally alter. Hence, contrary to Rostow's stages of growth thesis and to some modernization theorists, Latin America's process of change has a distinctive origin and dynamic from that of the developed countries. As formulated by Sunkel (1972: 520): "Development and underdevelopment . . . are simultaneous processes: the two faces of the historical evolution of capitalism."

Meanwhile Frank (1972b: 19–20, his emphasis) from his marxisant position writes:

> The point of departure for any credible analysis of Latin American reality must be its fundamental determinant, which Latin Americans have come to recognize and now call *dependence*. This dependence is the result of the historical development and contemporary structure of world capitalism, to which Latin America is subordinated, and the economic, political, social, and cultural policies generated by the resulting class structure, especially by the class interests of the dominant bourgeoisie. It is important to understand, therefore, that throughout the historical process, dependence is not simply an "external" relation between Latin America and its world capitalist metropolis but equally an "internal," indeed integral, condition of Latin American society itself.

What is notable is that Frank does not claim to be one of the founders of DT as regarded by some analysts, especially in the English-speaking world. I also detect in Frank a preference for the term underdevelopment rather than dependence, as will be discussed further on.

While I position Cardoso overall as a reformist, he straddles both DT approaches. For Cardoso and Faletto (1979: 15):

> The concept of dependence tries to give meaning to a series of events and situations that occur together, and to make empirical situations understandable in terms of the way internal and external structural components are linked. In this approach, the external is also expressed as a particular type of relation between social groups and classes within the underdeveloped nations. For this reason it is worth focusing the analysis of dependence on its internal manifestations.

As can be observed Frank, as well as Cardoso and Faletto, emphasizes the importance of analysing the relationships between the external and internal factors in the dependency situation, although Frank in his work tends to stress the external and Cardoso and Faletto the internal.

While distinguishing two main strands in DT, authors within each strand have focused or prioritized different aspects. Thus I will next refer to the particular contribution to DT by some of its key thinkers.

Structuralist views on dependency

I focus on the work of Osvaldo Sunkel, Celso Furtado, and Fernando Henrique Cardoso as representative of the structuralist DT. Sunkel's (1969) concern focused on the increasing penetration of foreign capital. He argued that this hindered rather than helped to overcome the difficulties arising from ISI. Foreign industries, often branches of transnational corporations, did not shift production to the export market but merely took advantage of the protective tariff barriers introduced under ISI by producing for the domestic market alone. By transferring profits and royalties to their country of origin and above all by neglecting the export market, they exacerbated the foreign exchange constraint that was 'strangling' the economy. By producing mainly consumer goods for the domestic market, their actions exacerbated balance of payment problems as the required intermediate and capital goods such as machinery, tools, and spare parts, had all to be imported and paid for in foreign exchange. Without those essential imports the industrialization process would grind to a halt.

Sunkel (1973) was particularly disturbed by the increasing domination of foreign trans-national corporations that he saw as deepening Latin America's dependence and leading to national disintegration. By taking over or bankrupting locally owned industries, the presence of transnational corporations weakened the ability and willingness of the national bourgeoisie to pursue a national development strategy, encouraging them to ally themselves instead to foreign capital. Transnationalization fragmented society as only a minority of the middle and work-ing class was incorporated into this transnational network, the majority being excluded and marginalized in the informal economy. Over time those who benefited from the intrusion of the transnational corporations absorbed their social, political, and cultural values and became disengaged from the less privileged majority. In Sunkel's view these transnationalized segments of society increasingly shaped public policy in a way which did not accord with the national interest.

The focus of Furtado (1973), another key thinker in this strand, fell on what he referred to as 'dependent consumption patterns.' He argued that it was the consumption patterns of those social classes with significant purchasing power that determined the structure of industrial pro-duction. Their consumption habits were much influenced by those existing in the developed countries so that they demanded almost as many brands and models of cars, televisions, refrig-erators and washing machines, and so on, as consumers in the rich countries. The production of these goods was relatively capital intensive, necessitating the importation of machinery and equipment, and sometimes even inputs like steel as only rarely were intermediate and capital goods, with their higher level of investment, skilled labour, and more complex technology, pro-duced in the country. This created a wasteful and inefficient industrial structure as the domestic market was too small for the industry to work at full capacity and achieve economies of scale thereby requiring increasing protection.

In short, the dependent consumption patterns led to the premature diversification of indus-try, which required far more capital and foreign exchange than would have been the case if there had been a more equal distribution of income and hence a mass market for industry. An industry geared towards the needs of the majority of the population would have been more labour inten-sive and less import demanding. Instead all the opposite was the case as a capital-intensive indus-try favoured those whose earnings came from owning capital. The level of employment was also lower than it would have been otherwise, resulting in less wage income for workers. This meant that a larger proportion of the national income went to the capitalist and a smaller proportion to the workers hereby perpetuating the concentration of income and the premature consumption pattern. In this way a vicious circle of dependence and underdevelopment persisted (Kay, 2005a).

Fernando Henrique Cardoso and Enzo Faletto (1969) wrote one of the most influential books on DT while both authors were working at the United Nations *Instituto Latinoameri-cano de Planificación Económica y Social* (ILPES) which was established as a sister organization of ECLAC in 1962 under the direction of Prebisch. Given the significance of this text it is surpris-ing that it took a decade to publish the English translation though with a new preface and a post scriptum. In the preface to the English translation, which appeared a decade later, they clarify that, although the dependent countries differ from the central capitalist countries, they do not pretend to propose a 'theory of dependent capitalism' and instead prefer to speak of 'situations of dependency' (Cardoso and Faletto, 1979: xxiii). In emphasizing the diversity of dependency relations, they differentiate themselves from some Marxist analyses of dependency which sought to find the common features of dependency or even its specific laws of capital accumulation (Palma, 1978). While Frank sought 'unity in diversity' Cardoso and Faletto sought 'diversity in unity.' Contrary to the structuralists, Cardoso and Faletto (1979: xxiv) do not propose autono-mous capitalism as an alternative to dependency but rather a process of transition to socialism similar to the Marxists. Cardoso often used Marxist categories in his analyses as well as being sympathetic if not close, to the ideas emanating from ECLAC.

The main aim of Cardoso and Faletto was to provide a sociological and political analysis of the situations of dependence, thereby enriching the centre-periphery analysis of the writings of ECLAC which focused at the time on the economic aspects of the region. They characterized the region as undergoing a process of 'dependent development,' distancing themselves from some dependency interpretations, such as that of Furtado and Frank, which they characterized as being 'stagnationist.' While hampered by dependency, they emphasized that most Latin American economies did grow and in some cases quite substantially, hence the title of their book 'dependency and development.' They also argued that following the crisis of ISI the developmentalist alliance between the industrial bourgeoisie and sectors of the middle class and working class under the aegis of the developmental state broke down and a 'new dependency' arose under a corporatist-authoritarian State controlled by a militarized technocratic bureaucracy.

Cardoso, reflecting on the military regime after the coup d'état of 1964 in Brazil, characterized it as opening an 'associated-dependent development' process. He argued that the intrusion of foreign transnational corporations created a new international division of labour whose interests to some extent "become compatible with the internal prosperity of the dependent countries" thereby stimulating development (Cardoso, 1973: 149). The authoritarian regimes thus established a new alliance with foreign corporations but under conditions of dependence.

The coup d'état in Brazil was followed by military take-overs in several Latin American countries leading Guillermo O'Donnell (1973) to characterize these regimes as 'bureaucratic authoritarian.' He explained their rise as a way of tackling the 'exhaustion' of the first stage of ISI as the subsequent production of intermediate- and capital-goods industries required a far higher rate of capital investment. The foreign multinational corporations could provide part of the required investment and complex technology; the authoritarian aspect of the State was required for controlling the industrial working class, who had grown in number, become better organized, and had achieved higher wages and access to social benefits under the previous developmentalist State. It had become a political force which the military and capitalist class wished to curb and possibly destroy so as to reduce the costs of labour thereby raising profits and enabling the next stage in the process of industrialization.

Marxist views on dependency

Turning now to the Marxist strand of DT, whose analysis of dependency was influenced by Marxist theories of imperialism such as those of Lenin, Luxemburg, Bukharin, and Hilferding. However, the Marxist *dependentistas* found the theories of imperialism wanting inasmuch as their focus was restricted to the imperial countries and did not take into account the developments from the perspective of the dominated countries. It is this shortcoming which they sought to overcome by discussing the internal dynamics of the dependent countries within the context of the world system which was shaped by the dominant countries. As representative of this dependency view I will refer to the ideas of Theotonio Dos Santos, Ruy Mauro Marini, and André Gunder Frank.

Dos Santos (1968) refers to the 'new character of dependency' in Latin America. Initially in the early stages of ISI, dependency had centred on foreign capital's control of the country's natural resources and exports. With the industrialization of the region the dependency relation changed although this was not understood at the time. The developmentalist State fully supported the industrialization process by undertaking infrastructural investment, creating a development corporation for establishing state enterprises or joint ventures in those sectors where private capital hesitated to invest such as electricity, steel, and petrochemical plants; setting up the protectionist tariff barriers, state development banks, technical colleges, development plans, and so on. It was anticipated that by industrializing the dependence on raw material exports would be overcome and that the State would acquire a greater degree of autonomy to shape the country's development path. But, this was not to be. On the contrary, owing to foreign exchange constraints, dependence on foreign capital and above all on the foreign corporations which owned the country's key industries increased. Hence, Dos Santos's use of the term 'the new character of dependency.'

The lack of a capital goods industry meant that the dependent countries did not have full control over their economy as they had to import the equipment, machinery, and spare parts for their consumer- and intermediate-goods industries. They were thus 'disarticulated' economies as they could not complete internally the full cycle of capital and were unable to generate their own technological progress. However, in his analysis of Brazil, Dos Santos observed that industrialization did lead to some limited production of capital goods, but that further progress was stymied by the reluctance of foreign corporations to transfer technology and by the inability to develop an indigenous technological capacity in the country. Dos Santos identified this technological dependence as the key impediment for dependent countries to achieve an 'articulated' economy and a relative degree of autonomy over their development process. He was aware that absolute autonomy was not attainable given the interdependencies of the world system. In any case he was striving for the creation of a world socialist system in which neo-imperialism and dependency would be overcome.

Ruy Mauro Marini (1973) focused his analysis of dependence on unequal exchange and overexploitation of labour. In their relations with dependent countries, the dominant countries modify and recreate the production relations of the subordinated countries in order to ensure the continuity of the dependency relationship. The dependence favours the dominant countries because through unequal exchange they capture part of the surplus value produced by workers in dependent countries. While there are certain similarities between the theory of unequal exchange and the Prebisch-Singer thesis on the deterioration of the terms of trade, there are some key differences. One difference being that the theory of unequal exchange uses the labour theory of value, which is not the case with the Prebisch-Singer thesis. Another difference is that the theory of the unequal exchange is more general since it does not depend on the type of products exchanged between the dominant countries and the dependent one, meanwhile the deterioration of the terms of trade arises from the peripheral countries exporting primary products and importing industrial products from the countries of the centre. Therefore, even when some dependent countries manage to export industrial products, there is still a transfer of value to developed countries. Due to this unequal exchange the capitalist class of dependent countries, so as to maintain their rate of profit, over-exploit workers by increasing the hours of the working day, reducing wages, and/or increasing the intensity of work thereby capturing a greater amount of the value produced by the workers.

Marini introduces the concept of sub-imperialism which is linked to his analysis of the Brazilian political economy after the military coup of 1964 which O'Donnell had characterized as bureaucratic authoritarianism. As Brazil's capitalist development was hampered by insufficient internal demand, thereby limiting the growth of profits and capital accumulation, one solution to this restriction was to expand military spending and engage in sub-imperialist practices. Brazil's large economy, geography, and population as well as it geopolitical importance meant it could use its power to engage in sub-imperialist practices with smaller and weaker countries as well as gaining some concessions from the imperialist countries and foreign corporations regarding access to modern technology and export markets. Authoritarian policies also kept wages in check and allowed income inequalities which fed this distorted industrial structure to continue. Furthermore, the military could develop an arms industry to stimulate the development of an as yet small capital goods sector thereby enhancing the technological capacities of the country as well as providing an outlet for the realization of capital.

Serra and Cardoso (1978) unleashed a fierce debate with their critique of Marini. They disagreed with his analysis of unequal exchange and his position that this inevitably leads to a fall in the rate of profit and therefore necessitates the super-exploitation of labour in the dependent countries. They also questioned his thesis on sub-imperialism as they held that Brazil did not face a problem of insufficient demand for its products. Finally, they disagreed with Marini's political conclusion that the dilemma facing Latin America was that of fascism or socialism, a position also held by Dos Santos. In brief, Serra and Cardoso did not share Marini's pessimistic view on under-consumption, stagnation, unequal exchange, super-exploitation, and sub-imperialism. In their view economic development could be achieved albeit of a dependent kind. They characterized his analysis as being economistic, voluntarist, and politically reductionist. Marini (1978) rejected these criticisms on theoretical and empirical grounds and in turn indicted Serra and Cardoso of engaging in sociologism, political reductionism, and neo-developmentalism. Furthermore, he charged them with class collaborationism by wanting to build an alliance between the bourgeoisie and the working class under the hegemony of the bourgeoisie and transnational capital with a statist sub-imperialist programme. This debate raised many important issues about Latin America's development and its future. What is remarkable is that Marini was perhaps the first person to foresee the neo-developmentalist character of Serra's and Cardoso's critique. Later both played a prominent role in Brazilian politics. Cardoso was elected twice to the presidency of Brazil and during his government from 1995–2003 he implemented a neo-developmentalist strategy within a global neoliberal context.

Turning now to Frank who is probably the best known dependency writer, particularly through his expression 'the development of underdevelopment,' the title of his seminal article published in the independent socialist magazine *Monthly Review*. The article was translated into several languages, published in about 20 journals, and reproduced in many edited books. In this article Frank (1966: 18) argued that the metropolitan countries had underdeveloped the satellite countries and that the linkages between metropolitan and satellite countries worked constantly to recreate the underdevelopment process of the underdeveloped countries. Hence his use of the phrase 'the development of underdevelopment.' It was only by breaking those linkages, through a socialist revolution, that a path of development could be created. Frank's source of inspiration at the time was the Cuban Revolution, like that of a whole generation of activists in the 1960s. Thus his article and subsequent writings found an avid and devoted audience who further diffused his work.

As Frank published in English, and as only few of the writings of the other *dependentistas* were translated, or only after several years, Frank's work was taken as representing DT in the English-speaking world. This has had the unfortunate consequence that the variety and richness of the dependency writings by Latin American thinkers remained largely unknown, producing a partial and even distorted view of DT. Furthermore, Frank was reluctant to use the term dependency, preferring instead to employ his concept of the development of underdevelopment (Kay, 2005b). When the term dependency does appear in his writing, it tends to figure only in the title or to be placed between inverted commas or in italics as in Frank (1970: 20). In his view the word dependence had become a euphemism. Indeed, he later declared dependence as dead (Frank, 1972c, 1974). In this 1972 article, he shifted his attention to the process of world accumulation and, to my mind, signalled his final transition to world system theory.

I thus consider him as one of its pioneers together with Immanuel Wallerstein, Samir Amin, and Giovanni Arrighi. His analysis of the 'development of underdevelopment' is best considered as a forerunner of world system theory, as might also Prebisch's centre-periphery paradigm and the dependency writings (Kay, 2011).

Frank's work also revived a debate within Marxism on the modes of production. Scholars and left-wing politicians had already engaged in a debate in the first half of the 20th century on how to characterize the mode of production in Latin America. Most authors characterized it as feudal or semi-feudal on the grounds that the dominant hacienda or latifundia system relied on servile tenant labour or coerced labour relations. In their view free wage labour only began to emerge with industrialization, particularly since the 1930s. Frank (1967) challenged that interpretation by arguing that the region had begun to be incorporated into the emerging world capitalist system since the Iberian conquest and hence the mode of production was already capitalist in the colonial period. With his radical intervention Frank directed his critique at MT as well as at the predominant Marxist interpretation of the time.

Frank's thesis has been forcefully challenged, fuelling the wider Marxist controversies on modes of production and their articulation which raged worldwide during the 1960s and 1970s. The communist parties were particularly keen to dismiss Frank's view for political reasons as their strategy was to build alliances with the 'progressive' bourgeoisie against the 'feudal' landed oligarchy which had dominated since colonial times. Here we find echoes of the 1920s debate between Mariátegui and Haya de la Torre. By contrast more radical left-wing political and social organizations argued that, given the dependent capitalist character of the countries and the close association of the bourgeoisie with transnational capital, only a socialist revolution could break the dependency relationship which would liberate them from exploitation and lead to an autonomous, articulated, and inclusive development process. Ernesto Laclau's (1971) critique, from a non-communist perspective, is the more substantial one. He faults Frank for over emphasizing the circulation of commodities and the region's participation in the world market in his characterization of a mode of production and neglecting the importance of relations of production. As, at the time, pre-capitalist relations of production were still prevalent in the rural areas of many Latin American countries they could not be characterized as being fully capitalist. Furthermore, these pre-capitalist relations were not only compatible with production for the world market but were even intensified by it. In some of these, at times dogmatic, debates Frank was accused of not being a Marxist. He replied straight forwardly that he never claimed to be a Marxist. Whatever the verdict, he undoubtedly was considerably influenced by US and Latin American Marxist thinkers and activists and by the Cuban Revolution, while being highly critical of the Soviet system, and in turn he influenced many Latin American Marxists.

Conclusions

In this chapter I have analyzed modernization and dependency theories from a Latin American perspective. MT as applied to the developing regions in disciplines such as the sociology and politics of development was mainly a product of Northern social scientists. With a few notable exceptions, such as Gino Germani who modified it for the Latin American case, it was largely uncritically absorbed by Southern colleagues. But after Frank's devastating critique and that of the dependency thinkers, MT did not prosper although at times authors, either knowingly or unknowingly, adopt aspects of MT. Meanwhile DT had its own spectacular rise and decline, although it has been resurrected by several authors, see Osorio (2004), Sotelo (2005), Beigel

(2006), Martins and Sotelo (2009), Munck (2013), Olave (2016), Seabra (2016), Castillo (2017), and Delgado Wise and Veltmeyer (2018), among others.

The significance of DT lies in the fact that for the first time in the history of the social sciences ideas emanating from Latin America, diffused in a flurry of publications, achieved global influence. DT emerged at a time when the Cuban Revolution and Che Guevara inspired revolutionaries throughout the world, when people were protesting against the Vietnam War and the 1968 student revolt swept through several countries in the North. DT exposed the inequalities of the global system and held out the prospect of a new world. It also influenced political parties and even some governments, particularly that of Salvador Allende in Chile whose socialist government was overthrown in 1973 by general Augusto Pinochet with the support of the US administration.

Whatever the strengths and weaknesses of the propositions emanating from this diverse school of thought it will be remembered for being the first major challenge to the Northern-centric or Eurocentric character of the social sciences. It gave fresh impetus to the social sciences and boosted a new generation of social scientists to think creatively about development problems from the perspective of the South and beyond. In the words of the *maestro*:"Development policy must be based on an authentic interpretation of the Latin American reality. In the theories we have received and continue to receive from the great centres there is often a false pretence of universality. It is especially up to us, people from the periphery, to contribute to correct those theories and introduce in them the dynamic elements necessary to approach our reality" (Prebisch, 1982: 150, orig. 1963, my own translation).

References

Beigel, F. (2006) Vida, muerte y resurrección de las "teorías de la dependencia". In CLACSO (ed) *Crítica y Teoría en el Pensamiento Social Latinoamericano*. Buenos Aires: CLACSO, pp. 287–326.

Cardoso, F. H. (1973) Associated-dependent development: Theoretical and practical implications. In A. Stepan (ed) *Authoritarian Brazil: Origins, Policies, and Future*. New Haven: Yale University Press, pp. 142–176.

Cardoso, F. H. and Faletto, E. (1969) *Dependencia y Desarrollo en América Latina: Ensayo de Interpretación Sociológica*. Mexico City: Siglo Veintiuno Editores.

Cardoso, F. H. and Faletto, E. (1979) *Dependency and Development in Latin America*. Berkeley: University of California Press.

Castillo, D. (2017) (ed) Actualidad de la Teoría de la Dependencia en América Latina? Barcelona: Editorial Anthropos.

Delgado Wise, R. and Veltmeyer, H. (2018) Development and social change in Latin America. In R. Munck and H. Fagan (eds) *Handbook on Development and Social Change*. London: Edward Elgar, pp. 228–247.

Dos Santos, T. (1968) *El Nuevo Carácter de la Dependencia*. Santiago: Cuadernos de Estudios Socio Económicos, No.10, CESO, Universidad de Chile.

Dos Santos, T. (1973) The crisis of development theory and the problem of dependence in Latin America. In H. Bernstein (ed) *Underdevelopment and Development: The Third World Today*. Harmondsworth: Penguin Books, pp. 57–80.

ECLA (1951) *Economic Survey of Latin America 1949*. New York: United Nations Department of Economic Affairs.

Frank, A. G. (1966) The development of underdevelopment. *Monthly Review* 18(4): 17–31.

———. (1967) *Capitalism and Underdevelopment in Latin America*. New York: Monthly Review Press.

———. (1969) *Latin America: Underdevelopment or Revolution*. New York: Monthly Review Press.

———. (1970) *Lumpenburguesía: Lumpendesarrollo, Dependencia, Clase y Política en Latinoamérica*. Santiago: Editorial Prensa Latinoamericana.

———. (1972a, orig. 1967) Sociology of development and underdevelopment of sociology. In J. D. Cockcroft, A. G. Frank and D. L. Johnson (eds) *Dependence and Underdevelopment: Latin America's Political Economy*. Garden City, NY: Anchor Books, pp. 321–397.

———. (1972b) Economic dependence, class structure, and underdevelopment policy. In J. D. Cockcroft, A. G. Frank and D. L. Johnson (eds) *Dependence and Underdevelopment: Latin America's Political Economy*. Garden City, NY: Anchor Books, pp. 19–45.

———. (1972c) La dependencia ha muerto, viva la dependencia y la lucha de clases. *Sociedad y Desarrollo* 3: 217–234.

———. (1974) Dependence is dead, long live dependence and the class struggle: An answer to critics. *Latin American Perspectives* 1(1): 87–106.

Furtado, C. (1973) The concept of external dependence in the study of underdevelopment. In C. K. Wilber (ed) *The Political Economy of Development and Underdevelopment*. New York: Random House, pp. 118–123.

Germani, G. (1962) *Política y Sociedad en una Época de Transición: De la Sociedad Tradicional a la Sociedad de Masas*, Buenos Aires: Editorial Paidos.

———. (1981) *The Sociology of Modernization: Studies on Its Historical Aspects with Special Regard to the Latin American Case*. New Brunswick, NJ: Transaction Books.

Gwynne, R. N. and Kay, C. (2000) Relevance of structuralist and dependency theories in the neoliberal period: A Latin American perspective. *Journal of Developing Societies* 16(1): 49–70.

Hagen, E. (1962) *On the Theory of Social Change*. Homewood: Dorsey Press.

Hirschman, A. O. (1961) Ideologies of economic development in Latin America. In A. O. Hirschman (ed) *Latin American Issues*. New York: Twentieth Century Fund, pp. 3–42.

Hoselitz, B. F. (1960) *Sociological Aspects of Economic Growth*. Chicago: Free Press.

Huntington, S. (1968) *Political Order in Changing Societies*. New Haven: Yale University Press.

Kay, C. (1989) *Latin American Theories of Development and Underdevelopment*. London: Routledge.

Kay, C. (2005a) Celso Furtado: Pioneer of structuralist development theory. *Development and Change* 36(6): 1201–1207.

———. (2005b) André Gunder Frank: From the development of underdevelopment to the world system. *Development and Change* 36(6): 1173–1179.

———. (2006) Raúl Prebisch (1901–1986). In D. Simon (ed) *Fifty Key Thinkers on Development*. London: Routledge, pp. 199–205.

———. (2011) André Gunder Frank: 'Unity in Diversity' from the development of underdevelopment to the world system. *New Political Economy* 16(4): 523–538.

Laclau, E. (1971) Feudalism and capitalism in Latin America. *New Left Review* 67: 19–38.

Lerner, D. (1958) *The Passing of Traditional Society*. New York: Free Press.

Marini, R. M. (1973) *Dialéctica de la Dependencia*. Mexico City: Ediciones Era.

———. (1978) Las razones del neodesarrollismo (respuesta a F. H. Cardoso y J. Serra). *Revista Mexicana de Sociología* 40(E): 57–106.

Martins, C. E. and Sotelo, A. (2009) (eds) *A América Latina e os Desafios da Globalização*. São Paulo: Boitempo Editorial.

McClelland, D. (1961) *The Achieving Society*. Princeton, NJ: D. van Nostrand.

Moore, W. E. (1963) *Social Change*. Englewood Cliffs: Prentice-Hall.

Munck, R. (2013) *Rethinking Latin America*. New York: Palgrave Macmillan.

O'Donnell, G. (1973) *Modernization and Bureaucratic-Authoritarianism. Studies in South American Politics*. Berkeley: Institute of International Studies and University of California Press.

Olave, P. (2016) (ed) *A 40 Años de Dialéctica de la Dependencia*. Mexico City: UNAM.

Osorio, J. (2004) *Crítica de la Economía Vulgar: Reproducción del Capital y Dependencia*. Mexico City: Miguel Ángel Porrúa.

Palma, G. (1978) Dependency: A formal theory of underdevelopment or a methodology for the analysis of concrete situations of dependency? *World Development* 6(7-8): 881–924.

Prebisch, R. (1981) *Capitalismo Periférico: Crisis y Transformación*. Mexico City: Fondo de Cultura Económica.

———. (1982, orig. 1963) *Hacia una dinámica del desarrollo latinoamericano*. In A. Gurrieri (ed) *La Obra de Prebisch en la CEPAL*, Segunda Parte. Mexico City: Fondo de Cultura Económica, pp. 137–227.

Rostow, W. W. (1960) *The Stages of Economic Growth: A Non-Communist Manifesto*. Cambridge: Cambridge University Press.

Seabra, R. L. (2016) (ed) *Dependência e Marxismo: Contribuições ao Debate Latino-Americano*. Florianópolis: Editora Insular.

Serra, J. and Cardoso, F. H. (1978) Las desventuras de la dialéctica de la dependencia. *Revista Mexicana de Sociología* 40(E): 9–55.

Sotelo, A. (2005) *América Latina: De Crisis de Paradigmas. La Teoría de la Dependencia en el Siglo XXI.* Mexico City: Plaza y Valdés.

Sunkel, O. (1969) National development policy and external dependence in Latin America. *Journal of Development Studies* 6(1): 23–48.

———. (1972) Big business and 'dependencia': A Latin American view. *Foreign Affairs* 50(3): 517–531.

———. (1973) Transnational capitalism and national disintegration in Latin America. *Social and Economic Studies* 22(1): 132–176.

Tella, G. di and Zymelman, M. (1973) *Las Etapas de Desarrollo Económico Argentino.* Buenos Aires: Editorial Paidos.

2

CULTURE AND DEVELOPMENT IN LATIN AMERICA

George Yúdice

Introduction

Since the independence of the Latin American nations, the idea of development among intellectuals has fluctuated between two tendencies. On the one hand, the creation of institutions following the lead of England, France, or the United States under positivist premises regarding education, science, material progress, industrialization, with the understanding that most fellow citizens, particularly indigenous peoples, Afro-descendants, mixed-race peoples, and rural folk were not apt learners. On the other hand, a few intellectuals, like José Martí, placed great emphasis on deepening knowledge of home-grown ways of doing things. This fluctuation continues to this day, with neoliberals and globalizers, on the one hand, promoting a liberal cosmopolitanism, and on the other hand, decolonial activists seeking to work from indigenous ways of knowing. A third position is that of those who seek to network across differences, advancing local, national, and regional projects. With regard to the definition of culture, this chapter takes as its point of departure the so-called anthropological view adopted by UNESCO: the representations, symbols, values, and practices by which a community reproduces itself.[1] In the 1980s and 1990s there emerged an economic understanding of culture, in large part oriented toward the cultural and creative industries (CCI), especially those that exploit copyright. A major challenge has been to mediate between these two tendencies. As thin as these thumbnail sketches of culture and development are, it should already be evident that they relate to each other, as in the developmentalist view that traditional cultures hold back development, or as in the contrary view that culture, like the environment, is a necessary factor in achieving a sustainable development that does not exhaust resources or endanger cultural and natural ecologies. There are positions as well that see sustainable development as an alibi for a kinder and gentler despoliation of culture and nature. And finally, the Internet, social media, and OTT streaming platforms are transforming what we understand by culture and development. We shall comment on all of these positions in the course of this essay.

Another point to keep in mind is that there are many cultural matrices among the 20 or more countries that constitute the region, and significant variety within each nation. What all these countries have in common is colonialism, a legacy that to this day has not been eradicated, despite the independent status of most of them.[2] Moreover, as decolonial scholars argue, it is not possible to speak of modernity or modernization, the lynchpin of development, without the other side of the coin: coloniality.[3] Its continued legacy is what accounts for the subordination of native peoples and Afro-descendants, as well as many mixed-race people, and the relegation of their epistemologies to inferiority or superstition. Until very recently, the cultures of these subordinated peoples were considered to account for Latin America's underdevelopment. Indeed, many Europeans and North Americans had condescending views of Latin Americans even into the late 20th century.

From a developmentalist perspective, there certainly were advances to be lauded. In the wake of the Great Depression, populist governments – Lázaro Cárdenas (Mexico, 1934–1940), Getulio Vargas (Brazil, 1930–1945 and 1951–1954), and Juan Domingo Perón (Argentina, 1946–1955) – turned to import-substitution industrialization as economic crisis and then World War II created favourable circumstances for this new economic model. As we will see, this model goes hand in glove with a series of cultural changes. On the one hand, as increasing numbers of the popular classes entered the workforce, the populist governments fomented nationalist expressions of culture, evident in the promotion of samba in Brazil (Raphael, 1980) or Mexican Golden Age cinema (1933–1964) with its focus on the Mexican Revolution, rural themes featuring mariachis and *ranchera* songs, and urban problems with the onset of modernization.

Having ushered in rapid economic growth in the 1940s and 1950s, the import-substitution industrialization model began to show its weaknesses by the 1960s. Commodity prices fell after the Korean War of 1953, exports could not keep up with imports thus leading to a balance of payments crisis. Inflation made exports even less competitive, and an increased workforce sought salary increases and labour protections, which were met by state repression (Ward, 2004). Indeed, in Brazil the military staged a coup d'état in 1964 to brake the rise of worker activism. State-led industrialization actually deepened inequality inherited from the unequal distribution of land during colonial times by concentrating industrial and financial capital (Bulmer-Thomas, 2003: 10). The 1973 coup d'état against Salvador Allende was another expression of economic elites seeking to suppress labour activism and socialist policies.

The failures of industrialization and economic development, which looked so promising in the 1940s and 1950s, were attributed in large part by modernization theorists such as W.W. Rostow (1960) to the inertia of traditional values and institutions: prevalence of primary economic activities, undifferentiated social roles and political structure, little social mobility, traditional and hierarchical sources of authority. According to development theorists, these features slowed the process of moving toward a modern society characterized by differentiated social roles and political structures, high social mobility, high productivity and individual capacities for achievement, rational sources of authority (Valenzuela and Valenzuela, 1978). While it is not the purpose of this chapter to delve into the economic analyses of development, it is nonetheless necessary to understand the relations that social scientists of development established between the cultural matrix of a society (traditions, values, etc.) and its capacity to generate economic growth. This point is clearly brought home in Seymour Martin Lipset's 1963 analysis of Latin American elites, whom he faults for their lack of entrepreneurship, due in large part to their Catholicism, as opposed to the Weberian causal connection between the Protestant ethic of northern Europeans and economic development and modernization. For Lipset, culture clearly plays an important role in underdevelopment: "the comparative evidence from the various nations of the Americas sustains the generalization that cultural values are among the major factors which affect the potentiality for economic development" (Lipset, 1963, 30; cit. in Valenzuela and Valenzuela, 1978: 541).

In response to modernization theory, a generation of quite heterogenous dependency theorists (see Kay, this volume) countered that development in core countries and underdevelopment in peripheral ones are determined by the latter's historical insertion into what Wallerstein

(1974) called the world system, such that the former set the terms of labour, extraction of raw materials, and trade in the latter. Synthesizing this copious literature, Valenzuela and Valenzuela (1978) state that it isn't values or attitudes that contribute to underdevelopment but, rather, dependency "produces an opportunity structure such that personal gain for dominant groups and entrepreneurial elements is not conducive to the collective gain of balanced development" (Ibid.: 545). This does not mean that external factors exclusively shape the economy but that dependency entails a particular fit between the external factors of the world system and "internal variables [that] may very well reinforce the pattern of external linkages" (Ibid., 546). That fit varies so that there are different instantiations of dependency. The most recent opportunity structure at the time that dependency theory emerged was transnational capitalism, made possible by innovations in communications, transportation, and technology, which reset the conditions for insertion into the world economy.

Modernization and dependency were also categories in the cultural sphere, particularly in literature, the visual arts, and media. In this regard, as Jean Franco (1975: 66) argues regarding Latin American literature, its difference cannot be seen simply as a "continuous and unresolved opposition between the universal and the regional." What dependency theory established for the economy, i.e., the specific conditions that account for the fit into the world system, also makes sense, *mutatis mutandis*, in the cultural sphere. Rather than see the cultural differences of Latin American literature as aberrant or as versions of the Western canon (e.g., as has been done with writers like Borges, Vargas Llosa, or García Márquez), Franco suggests that these differences make visible the "hidden ideological assumptions which are seen as natural and normative in the metropolis" (Ibid., 67). Literature, especially fiction, played a major role in discussions of modernization among Latin American intellectuals, already evident in the debate regarding *modernismo*, which will be reprised in the 1960s and 1970s regarding the Latin American New Novel and the concept of transculturation, both of which are discussed below.

More importantly, the relation to the world system affects the conditions of cultural production, orienting critical attention and sales in the cultural market to those genres and forms of representation that can have greater uptake internationally. Writing in 1975, Franco focuses on US technological influence, particularly in communications, which established, with few exceptions, the norms according to which the media sphere – including genres, styles, and forms of circulation – was shaped. Dependency in this area was addressed in the 1960s by the New World Information and Communication Order (NWICO), which followed from a set of critiques of Western mainstream media made by intellectuals and activists from developing nations. But Franco's major focus is on literature, hence we shall deal with this first.

In the 1970s, there emerged a debate over the subservience or autonomy of *modernista* literature in Spanish America.[4] In the above quoted essay, Franco dismisses the poetry of Rubén Darío and other *modernistas* (with the exception of José Martí) – who crowded their poetry and stories with princesses, luxury commodities, chinoiseries, classical references, and other refinements – as the expression of a will to belong to the brotherhood of metropolitan writers (Ibid.: 73). A decade and a half earlier, Juan Marinello (1959: 26), an important Cuban critic, had cast Darío as an alienated writer: "The movement captained by Rubén Darío was an American phenomenon but not in the service of our peoples. . . . It was the dazzling vehicle of a repudiable evasion, the brilliant mining of a malnourishing vein." Françoise Perus (1976: 117), in turn, characterizes Darío and the modernistas as ideologically connected to the oligarchies who, "in the absence of a bourgeois-democratic revolution that would eliminate them in a radical way, stubbornly persist." Angel Rama (1970, 1985) and Noé Jitrik (1978), on the contrary, argue that far from reproducing a backward-looking perspective, Darío's style established a literary autonomy that achieved on the symbolic level what was not possible on the economic or social levels.[5]

It is not surprising that the notion of a Latin American cultural autonomy would emerge at the same time as dependency theory. In the very same period, one can also find claims for cultural autonomy in reflections on the "New Latin American Novel," also known as the "Boom," with its radical experimentalism. In his 1969 book on the new novel, Carlos Fuentes writes that literature expresses "our own model of progress" (98). The Boom novelists and their boosters felt that Latin America had finally come into its own, overcoming dependency by naming it in their newly invented (literary) language. There was also the sense that Castro's revolutionary triumph in 1959 had brought liberatory possibilities that spread out to other spheres. Roberto Fernández Retamar, director of Cuba's premier cultural institution, Casa de las Américas, saw in the new novel a reprise of the achievements of Darío and the modernistas; the novelists "seem to do for the novel what the Hispanoamerican modernistas did for the poetry of their region" (1972: 322). Indeed, in his best known work, "Caliban: Notes Toward a Discussion of Culture in Our America," written in 1971, Fernández Retamar begins by countering the colonialist assumption that Latin American culture is derivative and argues for ending Latin America's "irremediable colonial condition" (1989: 3).

In the 1970s, the Non-Aligned Movement, comprised of what World Systems theory labelled peripheral nations, sought autonomy from the two superpowers of the Cold War, and to that end led a global campaign under the auspices of UNESCO for a New World Information and Communication Order. NWICO was the cultural arm of the New International Economic Order that Third World countries espoused at the United Nations Conference on Trade and Development (UNCTAD). NWICO sought to counter biases in reporting on developing nations, filtered by Western news agencies, and to establish a balance in the flow of TV programs. It demanded technology transfer so that these countries could develop robust communications systems. It also sought to counter the manufacture of desire for consumer goods through advertising (MacBride and Roach, 1994).[6] Latin American anti-imperialist intellectuals were quite active in this movement: Ariel Dorfman and Armand Mattelart (1991[1971]) in Chile; Héctor Schmucler (1975) in Argentina; José Marques de Melo (1971) in Brazil; Antonio Pasquali (1963) in Venezuela.

Most emblematic is Dorfman and Mattelart's 1971 *How to Read Donald Duck: Imperialist Ideology in the Disney Comic*, in which they deconstruct the apparently innocent entertainment value of Disney comics and other US entertainment products, demonstrating that there are political objectives and consequences in their dissemination. In their preface to the English language edition, Dorfman and Mattelart point out that two items were not subject to the US embargo on the socialist government of Salvador Allende: "planes, tanks, ships and technical assistance for the Chilean armed forces; and magazines, TV serials, advertising, and public opinion polls for the Chilean mass media, which continued, for the most part, to be in the hands of the small group which was losing its privileges" (1991: 9). David Kunzle's introduction to the 1991 English language edition makes the important point that "culture," which in this case means primarily the media and content industries protected by copyright, is the United States' second net export, and cites a 1990 *Time* magazine article that claims: "Today culture may be the country's most important product, the real source of economic power and its political influence in the world" (Ibid: 11). Indeed, this aspect of culture, as export, will constitute an important part of what I argue below, but before elaborating on this issue, it is important to consider other arguments regarding what Dorfman and Mattelart call cultural imperialism, for they also make a claim to, if not autonomy, certainly innovations in Latin American culture.

Dorfman and Mattelart deploy a "hypodermic needle" theory of media influence, whereby audiences are brainwashed by what they see and hear. The assumption is that if audiences read comics or view TV series in which characters are individualistic and only interested in profit, they will acquire those values and ways of being. Audience reception is, of course, much more complex, which is not an argument for privileging US or European media over local production. As we will see below, what is at stake is not only a negative media trade balance or that audiences are brainwashed but that the media sphere can be monopolized by stories and other expressions that refer to realities from elsewhere, thus providing the grist of what people discuss in daily conversations: "Did you see the latest episode of xxx, did you hear the latest song by yyy?" This jury is still out on whether OTT services like Netflix, Apple TV, and Amazon prime or music streaming services like Spotify, Pandora, and Apple Music or video sharing sites like YouTube effectively circulate the diversity of local media and stories along with international fare. Today, cultural policies aim for diversity rather than an exclusive focus on national culture; increasingly, this also seems to be the goal of the new media: to foster a greater heterogeneity in conversations that take place in social media as well as face-to-face encounters. In the conclusion, I shall discuss how the development of media and cultural delivery platforms has changed and how that change factors into understandings of cultural development in Latin America. But before doing so, it is important to examine a line of argument about cultural development that does not see the penetration by foreign media and cultures as producing a backwardness in Latin American cultures, but rather a reality that since the conquest of America has resulted in the emergence of creative hybridities.

This line of argument can be illustrated by briefly considering the work of José María Arguedas, Peruvian writer, linguist, and ethnologist. In the 1950s, he was Chief of the Folklore Section of the Ministry of Education of Peru, and in that capacity he operated as a mediator between local Andean musics and the new technologies that came with US music. García Liendo (2017: 146) writes that Arguedas did "not seek to freeze traditional culture within technology, but encouraged its metamorphosis and recognized the possibilities offered by mass culture." Moreover, this influence was not limited to cosmopolitan coastal cities like Lima but travelled in both directions with the migrants from the predominantly indigenous highlands of Peru, encouraging greater exchange between the capital and the provinces, between migrants and their places of origin (Ibid.: 145; Arguedas, 1996[1971]). According to García Liendo, Arguedas did not reject mass culture in favour of an idealized popular culture; instead, he saw in it the possibility of using the communicative potential of the then new technologies for the development of Peruvian culture, while minimizing the contents that come packaged in North American technology (Idem). In other words, traditional culture could survive and indeed increase its scope as it circulated nationally, transformed via the appropriation of new technologies. Arguedas did not seek to disalienate the popular classes because he did not see them as alienated. Instead, he sought to help them create and legitimize a cultural counter-hegemonic space.

It is possible to bring together several approaches to development discussed thus far: Rama's discussion of Darío's compensatory symbolic modernization and dependency theory's premise that capitalism underdevelops the periphery, generating a dual social structure whereby elites are joined at the hip with the metropolis and the popular classes hold to traditional values. Both approaches reflect a situation of combined and uneven development, that is, of the "amalgam of archaic with more contemporary forms" (Trotsky, 1967: 432; cit. in WReC, 2015: 6). During the heyday of dependency theory in the 1960s and 1970s, the possible options for breaking free from dependency seemed to be the full integration of popular sectors via industrial labour, although that option foundered with the failure of import industrialization, or revolution, which became more difficult after the Chilean coup in 1973, the US embargo on Cuba, and the US-backed counter-insurgency strategies that brought the Central American national liberation civil wars in the 1980s to the negotiating table without any gains for popular groups. While the international success of the Boom novelists – Julio Cortázar, Carlos Fuentes, Gabriel García

Márquez, and Mario Vargas Llosa – made it possible to argue along with Fernández Retamar that Latin America did indeed not only have culture but a highly developed one, that success did not translate to recognition of the subaltern popular classes, who, it was argued, held to traditional values.

In 1974, Angel Rama published an essay on the transcultural processes in Latin American narrative, subsequently augmented in a 1982 book, in which he argued that a generation of regional writers – José María Arguedas in the Andean Highlands and coastal shantytowns of Peru, Juan Rulfo in Jalisco, Mexico, João Guimaraes Rosa in Minas Gerais, Brazil, Gabriel García Márquez in the Caribbean region of Colombia – elaborated creative formal solutions for giving renewed vigour to local traditions that had been reduced by critics to conservative folklore in reaction to the cosmopolitan technical urban prowess of urban vanguards and the Latin American New Novel. Rama characterized these solutions as "formulations that enable the external influence to be absorbed and dissolved as a simple ferment within broader artistic structures in which the problematic [i.e., the centre/periphery dialectic] and the peculiar flavours that they have been guarding can be translated" (1974: 14). Rama adopted the term transculturation from Fernando Ortiz's 1940 classic study *Cuban Counterpoint: Sugar and Tobacco*, for whom the term captures better than the Anglo-American "acculturation" (acquisition of a new culture), the various processes (loss and acquisition) at play, but most importantly "the creation of new cultural phenomena, which might be called 'neoculturation'" (Ortiz, 1995: 103). It is clear that Rama discerns in the regional novelists an original solution to the conundrums of combined and uneven development that social scientists had not seen. He argues that at the same time as this solution registers the idiosyncratic values of a nation's past, it also harnesses "a creative energy that self-confidently acts on its particular inheritance as well as on external influences and in that capacity confirms an original elaboration, . . . it finds proof of the existence of a specific, living, creative, distinct society, which dwells . . . in the most secluded layers of the interior" (Rama, 1974: 17).

Rama found the basis of that "specific, living, creative, distinct society" in the ways in which the regional transcultural novelists were able to "recover [] the structures of oral and popular narrative" in contrast to the modernizing tendencies that drove the contemporary novel and in particular the Latin American New Novel. García Márquez, the most celebrated of the Boom writers, is included by Rama in this study because his great innovation, magical realism, enables him to "stylistically resolve the articulation of the verisimilitude and historicity of the events and the marvellous, from whose perspective the characters see those real events" (2008: 53).

Rama sees these transculturators as mediators between different cultures found within the boundaries of the nation-state, who are sensitive to the ways in which their societies are modernized (Ibid., 118). Indeed, he seems to attribute to them something akin to the Romantic visionary genius who could see the spiritual meaning in nature, with the difference that culture replaces nature and the nation replaces spirit in transculturation. Rama writes: "beyond their personal skills, there acts upon them quite strongly the specific situation in which the culture to which they belong is located as well as the patterns according to which it is modernized" (Idem.). But as we see in the case of Arguedas, the mediator, protagonist of Rama's book, is not necessarily a transcendentalist Romantic, but someone who understands cultural diversity and seeks to promote, without folklorizing, the practices of those who have a different vision of the world, whether they are indigenous or migrants whose culture is undergoing change.

This mediating role is even more evident in Arguedas' work with music precisely when migrants from the Andean highlands settled in the coastal cities where mass culture was transforming the cultural landscape, exercising an influence that literature would never have. Far from seeking to protect a pristine Andean culture, Arguedas sought to facilitate the massification of

Andean musics. In a society already shaped by mass culture, the counter-hegemonic sonorities of massified Andean fusion musics could provide alternatives to the invasion of "TV, radio, film and frivolous literature," particularly for the middle classes "who do not have other cultural anchors" (cit. in García Liendo, 2017: 394–395). Already in the 1960s, Arguedas was working to provide a space in which Peruvians could develop a cultural resilience with inputs from within any of their regional cultures as well as from abroad.

While Arguedas sought to give Andean culture, albeit a hybridized one, protagonism within the framework of the Peruvian nation-state, subsequently hybridity discourse moved to a more transnational scale. Néstor García Canclini (1993 [1982]) demonstrates how Mexican indigenous cultures reconverted under the impact of capitalism, not closing off possibilities but opening them, despite the subordinate status of indigenous peoples. Like Arguedas, García Canclini is attuned to the technologies and media at the disposal of indigenous peoples and their cultural production, especially crafts and *fiestas*, but he takes the discussion an extra step by focusing on the range of complex mediations (religious beliefs, bureaucratic agencies, markets, media, and tourism) that provide a range of heterogeneous channellings that make that production hybrid. Also like Arguedas, García Canclini argues that the recovery of cultural identity is a misplaced idea; the encounter of traditional practices with the various forces of capitalist modernity leads not to accommodation but to emergent and contestatory expressions (1993: 84). His later work examines a wealth of examples from art, literature, crafts, and urban cultures to demonstrate that traditions have not been rendered obsolete but that they are in transition (2005 [1990]). "Reconversion prolongs their existence" (155).

It is precisely this focus on the various forces in relation to which culture reconverts, as well as the assimilation of culture to the market rationality of free trade agreements in the 1990s, which leads García Canclini and other cultural analysts to intervene in cultural policy debates. The book that he edited with Guillermo Guevara Niebla (1992) diagnoses the likely impact of the North American Free Trade Agreement on education, cultural industries, technological innovation, intellectual property and copyright, and tourism. This focus on the force field in which culture takes shape, with local as well as international impacts, shows that the relationship between culture and power is much broader and more complex than the representation of traditions, identity groups, or even social movements. These intellectuals understood that the United States intended to get control over culture and nature through the terms of trade, which override national laws. Questions of national cultural identity were no longer enough in order to understand the workings of culture in an era of neoliberal globalization.

Until the 1980s and 1990s, Latin American cultural policy-making institutions (ministries, secretariats, and councils) had largely focused on supporting Eurocentric arts (national theatres, symphony orchestras, museums) that would make them conversant in a league of modern nations, and on folklore and heritage, which would provide ballast for national identity. We have already seen how Dorfman and Mattelart sought to protect national culture from the juggernaut of Western media. García Canclini is one of a handful of Latin American cultural analysts who have sought to guide cultural policy away from national protectionism and towards a recognition of diversity, or more accurately, speaking from many places at once.[7] García Canclini (2001 [1995]) went on to extend the discussion of hybridity in the globalized market setting of the late 20th century to what an effective citizenship might look like. To this end, he makes a series of recommendations for a regional, Latin American media and cultural space, as well as establishes the parameters of a democratic interculturality. This means acknowledging heterogeneity in the design of policies to promote the relation between local traditions and the cultural and creative industries, which largely operate according to international standards. We can see this process at work in a 2002 meeting that García Canclini coordinated of 50 Latin American and

Spanish policy designers, cultural administrators, analysts, and politicians to formulate policies for cultural development. The resulting set of recommendations presages the 2005 UNESCO Convention on the Protection and Promotion of the Diversity of Cultural Expressions, extending the notion of diversity beyond ethnicity and gender to a range of other issues, such as scale: small regions and countries or small and medium enterprises should be granted affirmative action vis-à-vis large countries or corporations (García Canclini, 2002). These recommendations, as well as the 2005 UNESCO Convention, aimed to put a brake on neoliberal free trade.[8]

This focus on hybridity, heterogeneity, diversity, and interculturality was also interpreted ecologically, in analogy with the need to safeguard biodiversity for the survival of the planet. Indeed, as Arturo Escobar (2012) notes, indigenous and inter-religious movements in Latin America have sought to establish culture as the fourth pillar of development. This proposal is in consonance with United Cities and Local Governments' Agenda 21, which includes culture as the fourth pillar of sustainable development, along with economic growth, social equality, and environmental balance (UCLG, 2010). Agenda 21 defines sustainable development as economic growth with social inclusion, environmental balance, and "the development of the cultural sector itself (i.e. heritage, creativity, cultural industries, crafts, cultural tourism)" and the transversal collaboration of culture with other sectors (UCLG, 2010: 4). Indeed, several Latin American cities have been quite active in UCLG and have made great strides in formulating cultural policy within this ecological framework.

The best known Latin American cases for the inclusion of culture in integral development policies are the Colombian cities of Bogotá and Medellín. The latter's Secretariat of Culture was created in 2002 and became a key piece of the government plan of Sergio Fajardo, mayor of Medellín between 1 January 2004 and 31 December 2007, to implement the principles of his *Compromiso ciudadano* (citizen commitment) platform: make the city more livable by strengthening security, focusing on programs to increase equality and social cohesion through the reform of urban space and transport, especially in the poorer and less accessible areas; improve education; facilitate citizen participation in public management; achieve maximum administrative transparency; and orient culture towards peaceful coexistence (Escobar Arango, 2007). The government of Fajardo was characterized by the transversality of projects for citizen empowerment. The most iconic example are the Library Parks, also part of a broader strategy to integrate the city. They link with schools, day-care centres, sports centres, cultural centres, and gardens that facilitate participation in activities for the poorer sectors of the city because they are located at major transport intersections and provide access to media (Melguizo, 2011). The intersectorial and transversal character of urban revitalization is articulated with aesthetic, cultural, and educational processes. As architect Alejandro Echeverri, who directed the revitalization project as Secretary of Urban Development, explains, a true transformation cannot be achieved by investing only in physical or material infrastructure. Local residents are encouraged to participate actively in the program, orienting its policies (Gerbase, 2013). In Medellín, the fragmented parts of the city are bridged by establishing a shared governance between the state and citizens.

To be sure, there also are urban cultural policies that seek to position certain cities like Buenos Aires or Rio de Janeiro squarely in the international cultural marketplace. From the mid-1990s on, creative cities policies circulated around the globe, encouraging urban policy-makers to invest in the cultural and creative industries, and in particular those industries protected by copyright. There was a double impetus: on the one hand, investment in cultural infrastructure, as in the urban revitalization of Barcelona for the 1992 Olympics (García, 2012) and the so-called "Bilbao effect" of contracting renowned starchitects to attract tourism and business to previously deteriorated deindustrialized cities (Yúdice, 2009). On the other hand, Tony Blair's

New Labour "Cool Britannia" cultural policy, focused on the creative industries in the context of trade-in-culture (or copyright) in the era of free trade, had a ripple effect throughout the world. Its 1998 Creative Industries Mapping Document (DCMS, 1998) was taken as a reference in other cities (e.g., Bogotá) and in a series of other analyses of this motley cluster of sectors, such as those of UNCTAD (2008) and the Inter-American Development Bank (Buitrago Restrepo and Duque Márquez, 2013). The 2002 Bogotá creative industries mapping document is a virtual clone of the British document in terms of methodology and sectors included: architecture, art, performing arts, crafts, film and video, software design, fashion design, graphic design, industrial design, textile design, photography, books, brochures, newspapers and magazines, music, heritage, advertising, television, and radio (Departamento de Diseño, 2002). The inclusion of advertising and software should raise eyebrows regarding how culture (art, performing arts, film, books, music, heritage) is reconstituted through the notion of creativity, but it is clear that the international trade context encouraged repackaging these heterogeneous sectors as one policy priority. In Latin America we see the creation of creative city clusters in Buenos Aires (Palermo Hollywood, La Boca, Colegiales, and Barracas); Rio de Janeiro (Porto Maravilha); Mexico City (the gathering of small trendsetters enterprises in the Historic Center, revitalized by Carlos Slim Heliu, as well as in Roma and La Condesa) (Olivera Martínez, 2015; García Canclini et al., 2012).

Yet, as one of the strongest advocates of a sui generis Brazilian creative economy – Ana Carla Fonseca Reis – put it, "it is not worth very much to stimulate the growth of sectors that generate astronomical revenues from intellectual property rights, if the creation of that wealth is not accompanied by a better distribution of income, driven by a socioeconomic inclusion that takes advantage of fundamental symbolic benefits, such as those of democratic access, valuing diversity, and the strengthening of national identity" (Reis, 2007: 293). As in the case of Medellín, we see this focus in the Brazilian Ministry of Culture, under the leadership of music megastar Gilberto Gil (2003–2008) and his successor Juca Ferreira (2008–2010). Arguably Latin America's most innovative cultural policy, the Points of Culture program sought to do exactly as Reis argues: provide democratic access, value diversity, and strengthen a myriad of identities and not one national identity. A community cultural organization is selected by a committee of peers and receives funding, computer equipment, Internet connectivity, and inclusion in a network of other points, enabling communication on common issues that they confront.

It is important to acknowledge, as well, that Gilberto Gil, already a seasoned city councilman in his hometown of Salvador da Bahia, with abundant expertise in cultural policy, drew on human rights activism since the dictatorship years as well as the vibrant cultural activism throughout Brazil when he created the Cultura Viva department, which houses the Points of Culture. The Points of Culture, a priority action of Cultura Viva, are the nodal points of a horizontal network of articulation, reception, and dissemination of cultural initiatives.[9] They do not have a single model, nor do they necessarily have physical facilities, programming, or activities. A common feature among them is the transversality of culture since they necessarily involve actions that cross sectorial boundaries.

Célio Turino, who was selected by Gil to direct the Points of Culture program, explains that the selection and renewal of the points of culture at the local level leads to the strengthening of the commitment to the community (Turino, 2010: 36). Turino sees the Points of Culture as nodes that are linked, in contrast to the social divisions that characterize society in terms of class and race. The numerous points of culture make that diversity visible, not only from a symbolic point of view (which in itself is already an important achievement) but also as a process that can generate a new economy (Ibid.: 57). The points of culture make visible the living heritage of the communities, that's why the large platform that hosts them was named Cultura Viva (living culture), emphasizing the idea that not only professionals produce culture but also people in their day to day activities. This program revolutionized cultural policy in Brazil, and as we shall see, throughout Latin America.

Brazil's Points of Culture program, together with innovations in municipal cultural policies in Medellín, was a crucial inspiration for establishing Cultura Viva Comunitaria (CVC), a network of community cultural initiatives throughout Latin America. The immediate prehistory of CVC begins in December 2009 in Mar del Plata, Argentina, where a number of cultural networks and leaders of cultural organizations and movements from half a dozen countries met at the First International Congress of Culture for Social Transformation, organized by the Cultural Institute of the Province of Buenos Aires with the collaboration of the Federal Investment Council. It is not by chance that these actors came together; many got to know each other in the context of the left-turn in Latin America in the new millennium, which prioritized the protagonism of the popular classes and marginalized groups such as Afrodescendants and indigenous peoples. For example, the World Cultural Forum – whose first meeting took place in São Paulo in 2004 under the auspices of then new Minister of Culture Gilberto Gil – was inspired by the World Social Forum, which began in Brazil in 2001 but had its roots in progressive movements that sought alternatives to global hegemony under neoliberal policies whose effects were particularly damaging to the more disadvantaged sectors of the population. The participants in these forums sought to empower the disadvantaged through art and cultural practice, not as spectators but as active participants. Among the best known were Jorge Melguizo, then Secretary of Social Development of Medellín, and that city's former Secretary of Culture, and Célio Turino.

While CVC is not a government program – it is a network of civil society organizations – the one hundred organizations from most countries of Latin America and the Caribbean that came together in Medellín in October of 2010 formed *Plataforma Puente*, an umbrella organization that lobbies national and municipal governments to legislate policies on art, culture, education, social transformation, sustainable development, and other goals and to designate at least 0.1% of national budgets to support the processes of living community cultures. These goals are consistent with those of a transnational alter-globalization movement:

- To strengthen and multiply popular cultural organizations in Latin America
- To gain the institutional and legal recognition on the basis of their legitimacy as protagonists in the construction of peoples' identity
- To obtain economic and institutional support for them from the state
- To promote the "Points of Culture" policy in Latin America
- To build networks of popular cultural organizations in Latin America for sovereignty over natural resources, fair distribution of wealth and democracy

(Cultura Viva Comunitaria, 2013)

Points of culture programs have been created in several Latin American cities and countries: Antofagasta-Chile, Argentina, Uruguay, Paraguay, Peru, Costa Rica, El Salvador, Guatemala, and Spain. They are members of Iberculturaviva, which is the platform within the Iberoamerican General Secretariat (a UN-like organization of countries from Latin America, Spain, and Portugal), from which the Points of Culture movement continues to lobby governments.[10] Thus far, CVC is a success story of bottom-up efforts of local community cultural organizations that have networked to defend cultural rights, not only through lobbying but also capacity-building initiatives in cultural management and policy and legislation design.

Conclusion[11]

While Latin American countries have made significant headway in democratizing diverse forms of culture, in the past two decades new technologies of information and communication are radically transforming Latin America's media landscape, promising diverse content, including user-generated content, that presumably will satisfy any preference, and increased digital access to the world's peoples. These promises are made not by governments as was the case with telephony and the television spectrum or with regard to cultural subventions like those of the Points of Culture. No, the promises are made by rather immodest corporations such as Google and Facebook. Latin American governments are ill-prepared, both legally and financially, to support domestic equivalents, which would contribute to tax revenues. At the same time, the question arises as to who should provide a seemingly public service, such as the Internet, which is nowadays the equivalent of classic utilities of the 20th century, such as water and power, and which increasingly carries all kinds of culture (film, TV shows, theatre, opera, music, books, videogames, etc.). Most states have very little to say in this regard, except for China, which has put up a Great Firewall to Facebook, Google, and other Western platforms.

In Latin America, there is very little resistance to global new media penetration. In fact, governments have made deals to have huge corporations like Facebook provide free service in poor areas, handing over what is most valuable in the new media: data, which enable corporations to formulate predictors of behaviour, not only concerning consumption, but even with regard to voting and security. We might say that Latin American nations are at the mercy of these global conglomerates.

Culture is transversal, necessarily intertwined with technology, media, enterprise, politics, etc. It is also globalized. The United States and its allies in the new turn in neoliberalism – cognitive-experience-and-affect capitalism – seek to establish and strengthen world trade laws through the intergovernmental institutions they dominate (WTO, WIPO, etc.). Latin America is a very unequal region in cultural commerce: according to PriceWaterhouseCoopers (2012), the media and entertainment market for 2015 was estimated at 6% of the world market. But that does not mean that Latin American ventures exported this percentage; that's the size that dominant companies can take advantage of. The export market would be more similar to the percentage of Latin American companies included in the Forbes list of the world's largest companies, which is more or less 3% (Wright and Pasquali, 2015). Of course, the most important criteria to be taken into consideration in assessing Latin America's market share, from a cultural viewpoint, are the cultural offerings, in the broadest sense of what might be meant by culture, and this has not been measured in these reports on cultural industries or media and entertainment. But even when the measure shows that the market is larger, the arguments regarding the power of these new media platforms on which the cultural sector is developed are very important and they have the ability to monitor and guide development. In addition to thinking about safeguarding small local markets in the region, which is fundamental, one must also think about the regional macro-market, since the dominant companies aim to capture it precisely in these terms, as a region. There have been attempts to create regional markets, such as Mercosur or the Central American Common Market, and even cultural markets such as MICSUR, the Cultural Industries Market of South America, but they have not yet devised policies for dealing with the new media. That is the current challenge.

At a more local, yet regionally networked level, 17 media labs sited in cities in Argentina, Brazil, Chile, Colombia, Ecuador, Mexico, Spain, and Uruguay are working intensively to "solve[] social problems with (digital, social, ancestral) technologies and innovative methodologies, through the involvement of the community affected."[12] Networked in Innovación Ciudadana, like Iberculturaviva, a platform created within the Iberoamerican General Secretariat, the network has mapped another 5,000 citizen innovation initiatives in 32 cities in 16 Iberoamerican countries that are working to create alternatives for more inclusive, ethical, sustainable, participatory, and habitable urban experiences.[13] Aside from working with local communities, the labs also seek to relay new forms of self-organization to municipal governments.[14] A next step will be for this network to mobilize its efforts at the national and regional levels, making common cause with efforts like MICSUR.

Notes

1 This essay does not pretend to provide exhaustive definitions of the terms development and culture. Regarding the former, I refer the reader to Escobar (2011); regarding the latter, see Yúdice (2003, 2007).

2 While I include Puerto Rico as a Latin American country, it is a hybrid "Free Associated State," with Puerto Ricans recognized as citizens of the United States but disenfranchised at the national level. Moreover, Puerto Rico is a Spanish-speaking nation.

3 See the essays in Mignolo and Escobar (2010).

4 *Modernismo* is the label given to stylistic innovations in Spanish American literature between 1880 and 1920. The Nicaraguan poet and journalist, Rubén Darío (1867–1916), considered its greatest exponent, embodied most of the characteristics of this style: aesthetic refinement, sensuality, classical mythological references, cosmopolitanism evident in Hellenisms, Gallicisms and settings in far away places, emphasizing the flight from the mundanity of life in Latin America. Cuban poet, journalist, and political activist José Martí (1853–1895) is considered Darío's contrary, whose writing focuses on Nuestra America's (i.e., Our [Latin] America's) coming to maturity vis-à-vis Europe and the United States as well as on narrating the novelties and problems of the modern world of social and technological developments (for an excellent treatment of these issues see Ramos [1989]). Both writers lived in metropolitan centres (Paris and New York, respectively) for many years.

5 For a detailed critique of this debate, and in particular the arguments made by Jitrik, see Yúdice (1984).

6 NWICO declined in effectiveness by the early 1980s, in great part due to the opposition of the United States and Britain, both of which withdrew from UNESCO in 1985 citing anti-Western bias stoked by Soviet influence and radical Third Worldism. Additionally, NWICO's nationalist focus on cultural protectionism did not jibe with the diversity of media demand by its citizens, something that is finally addressed in the 2005 UNESCO Convention for the Protection and Promotion of the Diversity of Cultural Expressions, which is addressed below.

7 It is important to distinguish "speaking from many places at once," or heterogeneity, from the notion of *mestizaje* or biological and cultural fusion that some critics thought to be at work in García Canclini's notion of hybridity. In response to a critique by the Peruvian intellectual Antonio Cornejo Polar, García Canclini embraced the former's identification of hybridization with non-dialectical heterogeneity (2005).

8 This Convention itself could not provide effective restraint on the World Trade Organization's laws, practices and policies, or place sufficiently strong obligations on signatories to comply with its principles (Neil 2006), but it nevertheless provided an impetus for many countries (and cities) to give access to groups historically excluded from funding and exposure.

9 See http://culturaviva.org.br/programa-cultura-viva/.

10 See http://iberculturaviva.org/?lang=es. I have also been told that Ecuador is negotiating entry into Iberculturaviva.

11 These brief comments draw on Yúdice (2017).

12 www.innovacionciudadana.org/en/#about.

13 www.innovacionciudadana.org/mapeo-de-la-innovacion-ciudadana/.

14 I have described the workings of the MediaLab-Prado, a founding member of this network, in Yúdice (2018).

References

Arguedas, J. M. (1996[1971]) *El zorro de arriba y el zorro de abajo.* 2nd Ed. Madrid: ALLCA XX, Colección Archivos.

Buitrago Restrepo, F. and Duque Márquez, I. (2013) *The Orange Economy: An Infinite Opportunity.* Washington, DC: Inter-American Development Bank. https://publications.iadb.org/bitstream/handle/

11319/3659/BID_The_Orange_Economy%20Final.pdf?sequence=7&isAllowed=y (Accessed 4 January 2018).

Bulmer-Thomas, V. (2003) *The Economic History of Latin America since Independence.* 2nd Ed. Cambridge: Cambridge University Press.

Cultura Viva Comunitaria. (2013) *1er Congreso Cultura Viva Comunitaria – Documento de convocatoria,* 2 March. https://tomatecolectivo.wordpress.com/2013/03/03/convocatoria-1er-congreso-cultura-viva-comunitaria/ (Accessed 4 January 2018).

DCMS. (1998) *Creative Industries Mapping Document 1998.* London: DCMS. http://webarchive.nationalarchives.gov.uk/+/www.culture.gov.uk/reference_library/publications/4740.aspx (Accessed 4 January 2018).

Departamento de Diseño de la Facultad de Arquitectura y Diseño, CEDE (Centro de Estudios de Desarrollo Económico) of the Universidad de los Andes, and the British Council. (2002) *Mapeo de las Industrias Creativas en Bogotá.* Bogotá: British Council. www.britishcouncil.org.co/mapeo.pdf (Accessed 4 January 2018).

Dorfman, A. and Mattelart, A. (1991[1971]) *How to Read Donald Duck: Imperialist Ideology in the Disney Comic.* New York: I.G. Editions.

Escobar, A. (2007) (ed) *Del miedo a la esperanza. Alcaldía de Medellín 2004 | 2007.* http://acimedellin.org/wp-content/uploads/publicaciones/del-miedo-a-la-esperanza-2014.pdf (Accessed 4 January 2018).

———. (2011) *Encountering Development: The Making and Unmaking of the Third World.* 2nd Ed. Princeton, NJ: Princeton University Press.

———. (2012) Cultura y diferencia: La ontología política del campo de Cultura y Desarrollo. *Wale'keru. Revista de investigación en cultura y desarrollo,* 2. https://dugi-doc.udg.edu/handle/10256/7724

Escobar Arango, D. (2007) (ed) *Del miedo a la esperanza. Alcaldía de Medellín 2004 | 2007.* http://acimedellin.org/wp-content/uploads/publicaciones/del-miedo-a-la-esperanza-2014.pdf (Accessed 4 January 2018).

Fernández Retamar, R. (1972) Intercomunicación y nueva literatura en Nuestra América. In C. Fernández Moreno (ed) *América Latina en su literatura.* Paris: UNESCO, pp. 317–331.

———. (1989) Caliban: Notes toward a discussion of culture in our America. In *Caliban and Other Essays.* Trans. Edward Baker. Minneapolis: University of Minnesota Press, pp. 3–45.

Franco, J. (1975) Dependency theory and literary history: The case of Latin America. *The Minnesota Review* 5 (Fall): 65–80.

Fuentes, C. (1969) *La nueva novela hispanoamericana.* Mexico: Joaquín Mortiz.

García, B. (2012) *The Olympic Games and Cultural Policy.* New York: Routledge.

García Canclini, N. (1993) *Transforming Modernity: Popular Culture in Mexico.* Trans. Lidia Lozano. Austin: University of Texas Press.

———. (2001) *Consumers and Citizens: Globalization and Multicultural Conflicts.* Trans. George Yúdice. Minneapolis: University of Minnesota Press.

———. (2002) (ed) *Iberoamérica 2002: Diagnóstico y propuestas para el desarrollo cultural.* México: Santillana.

———. (2005) *Hybrid Cultures: Strategies for Entering and Leaving Modernity.* New Ed. Trans. Cristopher L. Chiappari and Sylvia L. López. Minneapolis: University of Minnesota Press.

———, Cruces, F. and Urteaga, M. (2012) *Jóvenes, culturas urbanas y redes digitales: prácticas emergentes en las artes, las editoriales y la música.* Barcelona: Ariel; Madrid: Fundación Telefónica.

García Liendo, J. (2017) *El intelectual y la cultura de masas argumentos latinoamericanos en torno a Ángel Rama y Jose María Arguedas.* West Lafayette, IN: Purdue University Press.

Gerbase, F. (2013) O urbanismo social do arquiteto Alejandro Echeverri na transformação de Medellín, Interview with Alejandro Echeverri. *Jornal O Globo,* 23 de septiembre. http://oglobo.globo.com/rio/o-urbanismo-social-do-arquiteto-alejandro-echeverri-na-transformacao-de-medellin-10113541 (Accessed 4 January 2018).

Guevara Niebla, G. and García Canclini, N. (1992) (eds) *La Educación y la cultura ante el tratado de libre comercio.* Mexico: Nueva Imagen.

Jitrik, Noé. (1978) *Las contradicciones del modernismo: productividad poética y situación sociológica.* Mexico: Colegio de México.

Lipset, S. M. and Solari, A. (1967) (eds) *Elites in Latin America.* New York: Oxford University Press.

MacBride, S. and Roach, C. (1994) The New International Information Order. In G. Gerbner, H. Mowlana and K. Nordenstreng (eds) *The Global Media Debate: Its Rise, Fall, and Renewal.* Norwood: Ablex, pp. 3–11.

Marinello, J. (1959) *Sobre el modernismo: Polémica y definición.* Mexico: Universidad Nacional Autónoma de México.

Melguizo, J. (2011) Parques-Biblioteca de Medellín: Da Engenharia à Jardinagem Cultural. In A. C. Fonseca Reis (ed) *Cultura e Transformação Urbana: Anais do seminário internacional*. São Paulo: SESC Belenzinho, pp. 26–32. http://garimpodesolucoes.com.br/wp-content/uploads/2012/10/Cultura-e-Transforma%C3%A7%C3%A3o-Urbana.pdf (Accessed 4 January 2018).

Melo, J. M. de. (1971) *Comunicação, opinião, desenvolvimento*. Petrópolis: Editora Vozes.

Mignolo, W. and Escobar, A. (2010) (eds) *Globalization and the Decolonial Option*. New York: Routledge.

Neil, G. (2006) Assessing the effectiveness of UNESCO's new Convention on Cultural Diversity. *Global Media and Communication* 2(2): 257–262.

Olivera Martínez, P. E. (2015) Gentrificación en la Ciudad de México. In V. Delgadillo, I. Díaz and L. Salinas (eds) *Perspectivas del estudio de la gentrificación en México y América Latina*. México: UNAM, Instituto de Geografía, pp. 91–110.

Ortiz, F. (1995) *Cuban Counterpoint: Tobacco and Sugar*. Trans. Harriet de Onís. Durham, NC: Duke University Press.

Pasquali, A. (1963) *Comunicación y cultura de masas*. Monte Ávila: Caracas.

Perus, F. (1976) *Literatura y sociedad en América Latina: el modernismo*. Havana: Casa de las Americas.

PriceWaterhouseCoopers PWC. (2012) *Global Entertainment and Media Outlook 2012–2016*. www.career catalysts.com/pdf/PwCOutlook2012-Industry%20overview%20%283%29.pdf (Accessed 4 August 2018).

Rama, A. (1970) *Rubén Darío y el modernismo (circunstancia socioeconómica de un arte americano)*. Caracas: Ediciones de la Biblioteca de la Universidad Central de Venezuela.

———. (1974) Los procesos de transculturación en la narrativa latinoamericana. *Revista de Literatura Hispanoamericana* 5(April) (Universidad del Zulia, Venezuela).

———. (1985) *Las máscaras democráticas del modernismo*. Montevideo: Fundación Angel Rama; Arca Editorial.

———. (2008) *Transculturación narrativa en América Latina* [1982]. 2nd Ed. Buenos Aires: Ediciones El Andariego.

Ramos, J. (1989) *Desencuentros de la modernidad en América Latina*. México: Fondo de Cultura Económica.

Raphael, A. (1980) *Samba and Social Control: Popular Culture and Racial Democracy in Río de Janeiro*. Tesis de doctorado. Columbia University.

Rostow, W. W. (1960) *The Stages of Economic Growth: A Non-Communist Manifesto*. Cambridge: Cambridge University Press.

Schmucler, H. (1975) La investigación sobre comunicación masiva. *Comunicación y Cultura* 4.

Trotsky, L. 1967. *History of the Russian Revolution* [1932–1933]. Vol. 1. London: Sphere Books.

Turino, C. (2010) *Ponto de Cultura "Culture Point": Brazil from bottom up*. 2nd Ed. São Paulo: Anita Garibaldi.

UCLG. (2010) *Culture: Fourth Pillar of Sustainable Government*. Barcelona. www.agenda21culture.net/sites/default/files/files/documents/en/zz_culture4pillarsd_eng.pdf (Accessed 4 January 2018).

UNCTAD. (2008) *Creative Economy Report 2008: The Challenge of Assessing the Creative Economy: Toward informed policy-making*. Geneva: UNCTAD. http://unctad.org/en/docs/ditc20082cer_en.pdf (Accessed 4 January 2018).

Valenzuela, A. and Valenzuela, S. (1978) Modernization and dependency: Alternative perspectives in the study of Latin American underdevelopment. *Comparative Politics* 10(4): 535–557.

Wallerstein, I. (1974) *The Modern World System: Capitalist Agriculture and the Origins of the European World Economy in the Sixteenth Century*. New York: Academic Press.

Ward, J. (2004) *Latin America: Development and Conflict since 1945*. 2nd Ed. London; New York: Routledge.

WReC Warwick Research Collective. (2015) *Combined and Uneven Development: Towards a New Theory of World-Literature*. Liverpool: Liverpool University Press.

Wright, G. and Pasquali, V. (2015) *World's Largest Companies*. 19 November. www.gfmag.com/global-data/economic-data/largest-companies (Accessed 4 August 2018).

Yúdice, G. (1984) Las contradicciones del análisis telquelista aplicado a la literatura hispanoamericana: *Las contradicciones del modernismo* de Noé Jitrik. *Letras* (Lima) 90: 57–80.

———. (2003) *The Expediency of Culture: Uses of Culture in the Global Era*. Durham, NC: Duke University Press.

———. (2007) Culture. In B. Burgett and G. Hendler (eds) *Keywords in American Cultural Studies*. New York: New York University Press, pp. 68–72.

———. (2009) Culture-based urban development in Rio de Janeiro. In R. Biron (ed) *City/Art: The Urban Scene in Latin America*. Durham, NC: Duke University Press, pp. 211–231.

———. (2017) The challenges of the new media scene for public policies. In V. Durrer, T. Miller and D. O'Brien (eds) *The Routledge Companion to Global Cultural Policy*. Oxford: Routledge, pp. 282–396.

———. (2018) For a new institutional paradigm. *Atlántica: Journal of Art and Thought*. www.revistaatlantica.com/en/contribution/towards-new-institutional-paradigm/. (Accessed 4 August 2018).

3

INDIGENOUS DEVELOPMENT IN LATIN AMERICA

Nancy Postero

Introduction

I begin by noting that all three of the terms in the title of this chapter are deeply contested. Who counts as indigenous? How is the category defined, enforced, and mobilized? Scholars understand that encountering indigeneity is not to "describe it as it really is," but to "explore how difference is produced culturally and politically" (García, 2008: 217). It is both a historically contingent formulation that changes over time, and a relational concept that emerges from a contested field of difference and sameness (de la Cadena and Starn, 2007: 4; Postero, 2013: 108). Development, as the chapters in this volume demonstrate, is equally complex. Is it a discourse producing "underdevelopment" (Escobar, 2011) or a set of economic practices aimed at alleviating poverty and empowering marginalized populations? Is it a familiar and understandable wish to live better, or the name given to a capitalist system of commodity production deployed by powerful actors who are destroying the planet in the process (Dinzey-Flores, 2018: 166)? Who determines the goals and the beneficiaries (McMichael, 2010)? Even the notion of Latin America can and has been contested. Is it an idea (Mignolo, 2009)? Is it a geographical region south of the Rio Grande united by history? Do its borders extend into the US along with the diaspora of migrants inhabiting transnational labour circuits (Zilberg, 2004)? Do its populations share enough to make it a meaningful term of analysis (Goodale and Postero, 2013)? It can be argued that Latin America only came into being in opposition to the West, as a site where the West engaged in a struggle to tame the savage Other (Hall, 1996; Trouillot, 1991).

While I do not pretend to answer all these questions, in this chapter, I hope to add to these important and complex debates by thinking about how development and indigeneity play out in entangled ways in Latin America. I argue that their meanings are co-constitutive, having been formulated in tandem over centuries of contestation, exploitation, and violence. What it has meant traditionally to be "developed" in Latin America is to be different from, opposed to, and superior to native peoples and their visions of life and society. This formulation ignores the fact that this development was built with and on indigenous labour and resources. The complement to this is that, as Hall (1996) has famously argued, the category of indigeneity also only has meaning in opposition to the West, as its underdeveloped Other, but also its alternative. Thus, indigenous people, who remain among the continent's poorest residents, are the object of development, but they are also sometimes held up as the only remaining solution to it. As Nancy Fraser (1997: 15) pointed out long ago, economic and cultural injustice are fundamentally related: discursive categories and practices are underpinned by material supports, and economic institutions operate through culturally meaningful frameworks. Yet, the ways they are articulated constantly shift, depending upon the conjunctures in which they are lived. In this chapter, I briefly trace the history of colonial and capitalist development in the region to highlight the ways America's native peoples were folded into each era's notion of development and how they posed challenges at every stage. I hope to show the deeply political implications of struggles over economies, labour, land, and the distribution of the benefits of what we now might call development.

Precolonial and colonial era

Precolonial indigenous societies had their own economies and values, which, like all societies, dynamically transformed as they faced change (Postero, 2007a). The breadth of their diversity is impossible to catalogue here, but we can acknowledge that the largest civilizations, the Incas, Aztecs, and Mayans, were complex societies with extensive populations, systems of production and distribution, imperial military, and all the religious and cultural apparatuses of European societies. First-hand accounts of Spanish warriors remind us that these Europeans were stunned by their encounters with the beauty and organization of Tenochtitlán, a city of a quarter million well-fed residents (along with slaves from their imperial conquests) (see Diaz, 1963). Thus, indigenous Latin Americans were not "underdeveloped" when the conquistadors arrived, but were forced into poverty through war, massacres, slavery, dispossession, exploitative labour practices, and disease (Wright, 1992). As scholars have widely documented, native peoples and African slaves provided the labour for the plantations that produced sugar, rum, cotton, and indigo (Mintz, 1986; Wolf, 2010); the extraction of forest products like rubber and nuts (Taussig, 1987); and the deadly mines that filled European coffers with gold and silver (Galeano, 1973). Native peoples grew food, made cloth, carried out all the reproductive labour, and transported these products, making the colonial system functional. In his widely read treatise, *The Open Veins of Latin America* (1973), Galeano showed how these riches funded the Industrial Revolution in Europe, setting into place an enduring system of inequality in which the West benefitted from the resources and suffering of Latin America. The blood and sweat of native peoples were transformed into the capital that kick-started modern capitalism. Dependency theorists argued this imbalance explained the ongoing disparities between metropole and periphery (Cardoso and Faletto, 1979), offering a characterization of "underdevelopment" that still rings true today.

Yet, historians caution us not to see indigenous peoples as mere victims of colonial exploitation. Steve Stern's work (1993) in what is now Peru demonstrated that native peoples were also active economic and political agents, owners of mines, merchants, and crafts people, many of them competing with Spanish colonizers. Local people used the Spanish courts to fight back against colonial practices, often forcing them to alter their policies. Colonial economic systems relied on indigenous traditional forms of organization to exist and expand (Barragán, 2015). Douglas Cope (1994) showed how racial schemes intended to keep native peoples and Afro-descendants in particular jobs and spaces to enable their exploitation, often broke down, as people intermarried and escaped spatial enclosure. The pictures contemporary historians give us show that despite the horrors of colonization, some indigenous elites gained wealth and status, inserting themselves into the economy.

Republican period

In the 19th century, Latin Americans waged wars of independence from Spain and Portugal, establishing new nations ruled by creoles, the descendants of the European colonizers. These new leaders justified their new Latin American nations in part by opposing the brutal colonial treatment of indigenous peoples, promising to establish more rational liberal societies based on Enlightenment notions. Yet, the existential question of race complicated this intention. Deep racial fears still motivated the rulers. The 1790s saw massive organized resistance to European rule – from the insurrections in the Andes to the revolution in Haiti. In Bolivia at the end of the 1899 war between Conservatives and Liberals, Andean leader Zárate Willka's troops massacred a group of white soldiers. As Nancy Egan (2007) showed, the resulting public Mohoza trial aired the deep distrust whites had; even the most fervent Liberals failed to see the universal humanity in the indigenous defendants. These attitudes of racism were linked to and justified economic regimes. Across the region, while slavery had disappeared, other forms of servitude, like *ponjeage* or debt bondage, took their place (Bernstein, 2000). As Larson (2004: 13) notes in her comprehensive treatment of the Liberal era across the region, the big paradox was how to impose universal definitions of free labour and citizenship, while at the same time creating categories of difference that would set limits on these ideals, and allow continued domination over the Indian Other and their labour. Indigenous practices were seen as obstacles to modernity. Instead, modernizers imagined indigenous peoples as leaving behind their communal societies to join in the nation's development project. In the Bolivian case, as elsewhere, the solution was to make communal property illegal, throwing land up for grabs for the criollo elite while also opening a path to modernity, as rural natives were encouraged to become small-scale farmers. The result, however, was the *latifundio* system of large landholding, in which indigenous people served not as slaves but as landless peons. While many mounted vigorous resistance, many others were integrated into this system in a position of deep disadvantage.

The modern 20th century: indigenismo and mestizaje

In the 20th century, Latin America's nation-states sought other means to integrate their indigenous populations to their development projects. In part, this was the result of widespread populist revolutions marking the deep dissatisfaction the region's people held about the decades of failed development projects. Unequal land tenure and deepening class divisions showed the need for new ways to address the structural inequalities that marked the region. In some places, popularly elected governments undertook reforms to address these concerns, often with encouragement and financial backing from the US. The Alliance for Progress, President Kennedy's aid project, targeted rural peasants likely to rise up in revolutions, hoping to seduce them with rural development projects to address their poverty. In others, Leftist sectors resisted, sometimes engaging in guerrilla wars. This produced a violent push-back from conservative forces, and a wave of military dictatorships across the region. While most countries have made a transition to formal democracy, it is clear that "the twinned legacies of revolutionary struggle and the violence of state repression continue to shape arguments for and against alternative models of social change and forms of governance in contemporary Latin America" (Goodale and Postero, 2013: 7). That is, many of the battles that produced this cycle of conflict – about what forms of development nations should pursue, and more importantly, who should benefit – continue in the present moment.

These debates over development were often articulated in a complex register of race and class. In Mexico, for example, after the 1910 Revolution, a new discourse about Mexican national identity emerged. The discourse of *indigenismo* replaced the brutal suppression of native peoples with a narrative of the *raza cósmica*, the cosmic race resulting from the mixture of indigenous and European peoples. Indigeneity was recognized, yet placed in the past, as a glorious foundation of the progressive and modern *mestizo*, who would lead the country to a new form of economic and social development (Knight, 1990). Friedlander's (1975) work in the 1970s demonstrated the effects this model had on rural indigenous people: they were devalued and left out of the modern world, shamed by their links to traditional practices. Elsewhere, indigeneity also gave way to other power-laden categories. In Bolivia, after a coalition of miners, peasants, and petty bourgeoisie led a successful revolution against the oligarchy in 1952, the new revolutionary state formulated a development model to incorporate all these sectors. There, as in other countries in the region, land reform, universal suffrage, and general education sought to benefit the poor and rural citizens, who were identified in terms of class relations rather than race/ethnicity. Indians became peasants, and they organized in *sindicatos* (unions) to demand their rights. The modern state's development model brought these formerly excluded segments into its embrace through a vertical model of patronage and assistance. Across the region, peasants, miners, and factory workers produced for the national market, and rural people streamed into cities to become the labour force for new forms of industrialized national production. They built homes in the barren outskirts of what became the megacities like Mexico City and São Paulo, forming new working classes (Holston, 2008).

But race never left the scene. The "poor Indian" still needed help to advance. George Foster (1965) studied the peasant (read indigenous) people of central Mexico, arguing their cultural models, especially one he called " the image of limited good," limited their ability to enter the rational modern market. State policies especially targeted indigenous women and their homes as sites of intervention, teaching them hygiene and enforcing Western notions of family structures (Larson, 2005; Stephenson, 1999). As Laura Gotkowitz (2007) argues, race and gender became central sites of nation-making, as indigenous people, especially indigenous women, were disciplined to be proper members of a modern nation-state. However, indigeneity did not melt away under the pressures of the modernization development model. June Nash's (1979) early work with tin miners in Bolivia showed how these proletarians also engaged their indigenous cultural values to understand their work as a meaningful contribution to the cosmos as well as the nation. Her images of Andean miners worshipping the subterranean deities through rituals of tobacco and coca leaves showed the endurance of indigenous values. In Mexico, Guillermo Bonfil Batalla (1996) pushed back against the inevitability of *mestizaje*, arguing that indigenous Mesoamerican society, what he termed *México profundo*, or deep Mexico, was still present, underlying all contemporary life. He thought modern industrialized *mestizo* Mexico was an illusion, doomed to failure if it did not take into account indigenous understandings of the land.

Neoliberal multiculturalism

The late 20th century brought a new capitalist development model: neoliberalism. David Harvey's oft-cited definition captures its gist:

> Neoliberalism is in the first instance a theory of political economic practices that proposes that human well-being can best be advanced by liberating individual entrepreneurial freedom and skills within an institutional framework characterized by strong private property rights, free markets, and free trade. The role of the state is to create and preserve an institutional framework appropriate to such practices.
>
> *(Harvey, 2005: 2)*

Throughout the 1980s and 90s, Latin American countries adopted neoliberal policies, restructuring their economies to adhere to the orthodox norms of what was called the "Washington Consensus." In Chile, the shift to neoliberalism was imposed by a military dictatorship, responding in part to the Leftist development models introduced by socialist President Allende. Naomi Klein (2005) described the violent imposition of neoliberalism there as a form of "disaster capitalism" because Pinochet was able to force the unpopular model on a population already stunned by the military coup. In other places, countries were forced to accept the new policies as part of the conditions of World Bank and the International Monetary Fund loans to repay foreign debt. Called "Structural Adjustment Programs," these policies forced the privatization of public enterprises, radical cutbacks on social spending, fiscal and monetary reforms, and the end to tariffs and subsidies for national industries. As many scholars have documented, this had drastic impacts on the rural populations of the region, as their agricultural products were replaced by cheap imports (see Ugarteche, 1999). For Mexican rural indigenous people, for instance, corn, the staff of Mayan life, was now imported from large agribusinesses in the US. These policies were accompanied by an increased emphasis on natural resource extraction, as neoliberal governments opened their economies to transnational companies. Mines, roads, dams, and hydrocarbon exploration spread across indigenous lands from the Andes to the Amazon, leading to substantial resistance (Sawyer, 2004; Hindery, 2013; Perreault, 2005). Large-scale agricultural businesses expanded in the lowlands. Global commodities, like soy, safflower, cattle, and sugar cane, destroy vulnerable tropical forests, spread pesticides, and use enormous quantities of water, affecting the lands and livelihoods of lowland indigenous groups (Fabricant and Postero, 2015).

Neoliberalism was not just economic policy and practices, however. Much as modernization was paired with the cultural discourse of *mestizaje* to enlist popular participation in that development model, in the neoliberal era, the market was paired with a cultural formulation called multiculturalism. In the late 1980s and early 1990s, indigenous peoples began to organize in many parts of the region, renewing their cultural identity as indigenous (rather than peasants), and making demands for lands, political participation, and substantive citizenship (Postero and Zamosc, 2004). This was part of a larger international push by the indigenous movement, which received substantial support from international institutions. The International Labor Organization's Convention 169 (1989) declared that states should recognize indigenous people, their cultures, and their territories, and this was followed up by the UN Declaration of the Rights of Indigenous Peoples (UNDRIP) in 2007. "Neoliberal multiculturalism" refers to the policies of neoliberal states to respond to this movement, to recognize the cultural differences of their diverse populations, and to articulate those segments to its national development project. If in past regimes indigeneity was seen as an obstacle to progress, in this new era, indigeneity was recognized, institutionalized, and made functional to global capitalist economic systems (Hale, 2002; Postero, 2007b: 14–18). In his analysis of reforms in Guatemala, Charles Hale concluded that neoliberalism includes a seductive cultural project. He identified a form of governmentality through which citizens – be they individuals or indigenous collectivities – take on the responsibility of resolving their problems, governing themselves in accordance with the logics of global capitalism. The result, what Hale calls the "menace" of neoliberal multiculturalism, is that those indigenous people who conduct (2002) themselves with this logic are rewarded and empowered. He terms them "*indios permitidos*," or authorized Indians, (a term drawn from Rivera Cusicanqui). Unruly conflict-prone Indians are condemned to the racialized spaces of poverty and social exclusion (Hale, 2004). Patricia Richards (2013) showed the extreme version of this in neoliberal Chile, where Mapuche indigenous people have pushed back against land dispossession and the dams and large-scale forest plantations on their former lands. These "*indios prohibidos*," or unauthorized Indians, are treated as terrorists, their lands militarized, and those seen as violators given harsh jail sentences.

Neoliberal reforms took many forms. Bilingual or intercultural education taught indigenous languages. States surveyed and titled collectively held indigenous territories, giving indigenous people some renewed control over their lands (Anthias and Radcliffe, 2015). In others, indigenous organizations were recognized as appropriate representatives in local development projects. For instance, in the Bolivian case I studied in the early 1990s, the new Law of Popular Participation allowed indigenous leaders, selected through traditional norms called *usos y costumbres*, to determine how municipal funds should be allocated. I found that the reforms were insufficient to overcome the overarching racism and the domination of white-*mestizo* political parties. Yet, in the long run, I argued, indigenous people were able to utilize some aspects of the reforms to gain political power (Postero, 2007b). Winning positions in local towns, indigenous people and other popular sectors were able to join forces, build a new political party, and elect the first indigenous president, Evo Morales, in 2005. Morales campaigned on an anti-neoliberalism platform, promising to push back the economic policies that gave away the country's resources to transnational corporations. I discuss below the efforts to "decolonize" Bolivia's society and economy.

There were also other significant forms of resistance to neoliberal development visions. As local rural communities felt the impacts of neoliberal trade policies, farmers and indigenous producers joined together to advocate for sustainable and just food production systems. Producers pushed for "fair trade" systems that took the middlemen out of the commodity chains, leaving more for the producers themselves (Moberg, 2010). Forest certification projects guaranteed more environmentally friendly production processes and fair labour conditions. The Via Campesina organization brought together many of these concerns, pushing for food sovereignty, a notion that combined food security for vulnerable farmers and peasants, with the right of farmers and states to make decisions about agricultural policy. It prioritizes local food production and protecting local peoples and their lands and water over the rights of transnational corporations (Via Campesina, 2003; McMichael, 2006). A more radical form of resistance emerged in Mexico, as the Zapatista army rose up to protest the NAFTA agreements and the effects of neoliberalism on Mexico's indigenous peasant populations. Over the next decades, the Zapatista autonomous communities in resistance created a new political and economic model, which respects indigenous culture and practices an alternative to market-driven development (Earle and Simonelli, 2005; Dinerstein and Deneulin, 2012). As Alicia Swords (2010) shows, their development model engages political education and community organizing to enable the development of cooperatives, push for gender equality, and promote resistance to neoliberal policies.

Challenges today

In the current era, as in previous ones, indigenous responses to globalized capitalist models of development vary widely. I focus in this last section on two opposing poles, but remind the reader that there is a wide spectrum in between.

The first pole I characterize as engaging with capitalism. Here I refer to the large numbers of indigenous people who, willingly or not, work in the global system – as producers, merchants, or servants. One example is the diaspora of Mexican indigenous workers who have migrated to the US to work in the agricultural fields. Lynn Stephen (2007) has documented their migration, showing the harrowing journeys North, the difficult conditions in which they work, and the constant fear of surveillance and discipline from the US government – the latter, I note, increasingly daily under the Trump administration (Stephen, 2007). For many, this is a development model: their children receive education and chances not possible in rural Mexico ravaged by neoliberalism. Their remittances to their home communities are also a form of development, enabling cross-border organizations that sustain indigenous cultural practices and intervene in local political struggles at home (Blackwell, 2015). Indigenous peoples also serve across the region in the burgeoning tourist industry. Castellanos (2010) documents the migration of Mayan peoples into the service industry for Cancún's hotels, showing how their labour is fundamental to the whole industry. But the growing eco-ethno-tourism industry also relies on indigenous participants to demonstrate their authenticity, to share their "ancient wisdom" with visitors (see Córdoba Azcárate, 2011). Meisch (2002) describes how Otavalo musicians from Ecuador travel the world selling their music and recordings, packaged to appeal to Western notions of authentic indigenous culture, including feathers and mountains. Yet, despite their globe-trotting entrepreneurship, she argues the living they make allows them to sustain a strong and meaningful cultural identity at home, one even the youth continue to support. The final example is the rapidly growing middle class of Aymara merchants in Bolivia, who are engaged in transnational circuits of commerce connecting China, Bolivia, and Brazil. Nico Tassi and his colleagues (2013) have shown how these merchants utilize kinship relations and Aymara understandings of growth and abundance to build their businesses, asserting a proud sense of Aymara identity in their extravagant colourful homes.

The other pole is the important push from indigenous organizations towards autonomy and decolonization. Since the arrival of the European colonizers, native peoples have been resisting the destruction of their own political and cultural institutions. Since the insurrections of the colonial period, there have been calls for autonomy and self-determination. As mentioned above, the international indigenous movement codified many of these long-held demands in ILO 169 and the UNDRIP, but while nation-states signed onto these declarations, in practice indigenous communities remained inserted into states. Demands for real self-government were seen as challenges to national sovereignty, and so indigenous peoples used other less threatening frameworks, like human rights or cultural rights (Engle, 2010). In the last decade or so, however, indigenous organizations and their allies have articulated a strong discourse of decolonization, calling for an end to the colonialism that discriminates against indigenous epistemologies, practices, and forms of government (see Walsh, 2007). They have urged the decolonization of nation-states, to allow for indigenous structures of self-government (see Postero, 2017). This has been linked to a decolonized notion of development, arguing that instead of continuing the mad dash for consumption that is destroying the planet and its climate, development should be driven by "*buen vivir*" or living well, an indigenous form of sustainable life that protects collective society as well as nature (Radcliffe, 2012). In both Bolivia and Ecuador, this discourse is now codified in the constitution, which calls the state to enact this alternative post-neoliberal form of development.

Constitutional reforms have also created spaces for experiments in political autonomy. This is a critical turn for development, as indigenous peoples are beginning to make decisions about what forms of development can and cannot take place in their communities. Scholars are now studying the diverse forms this trend is taking – from *usos y costumbres* in Oaxaca, Mexico to the new *autonomía indígena originaria campesina* (AIOC) in Bolivia (Tockman, Cameron, and Plata, 2015). While these new structures are the site of great hope, it is important to recognize their limitations, as national governments continue to assert sovereignty over subsoil rights, granting oil concessions to transnational corporations for exploitation of natural resources. Understandably, a central site of struggle has become "free, prior, and informed consent" (FPIC), which both international instruments and national constitutions guarantee whenever local communities might be impacted by development. These rights are regularly disregarded, but even when there are formal consultation processes, the radical power imbalances between international companies and the state, on one hand, and small local communities, on the other, make it very difficult for communities to resist the projects (Schilling Vacaflor, 2011). Thus, a central concern for development studies now is how indigenous communities can engage in "indigenous resource governance." This term recognizes that indigenous communities are agentive actors, negotiating between powerful entities. In some cases, they may resist. The Sarayacu community in Ecuador is perhaps the most emblematic case, having taken their objections to petroleum exploration on their lands to the Inter American Court of Human Rights and won (Cultural Survival, 2012). But other communities make strategic decisions to participate in extractivism, negotiating directly with TNCs for funding and resources. Penelope Anthias (2018) documents the case of Guaranís in Bolivia, who defined autonomy not in terms of their relation to the state, but in terms of their ability to negotiate with petroleum companies. What notions of development will these autonomous indigenous communities enact? How will they engage and challenge global capitalism as they work to benefit their own communities?

Conclusion

In every era, indigenous peoples have been deeply enmeshed in the development models of the day, pulled into exploitative relations with capitalism, engaging it, resisting it, and posing alternatives to it. Their relation to development in each era is the result of particular articulations of economic, political, and cultural forces, which shift over time. Indigenous "difference" has been mobilized by states and the market to formulate visions of progress to enlist the productive labours of the nation. In the contemporary era, indigenous communities have gained new power to enact their own visions of development. But their visions must be carried out within the constraints of larger forces – global capital and the sovereign nation-state. The case of Bolivia is a cautionary tale. There, despite an indigenous president who espouses decolonization and *vivir bien*, and a constitution sworn to protect indigenous peoples and their autonomy, the state continues and expands a national developmental model based on extractivism, and continues to sacrifice indigenous lands to this model (Postero, 2017). Yet, as Sarah Radcliffe (2015: 278) concludes in her analysis of decolonization and development in Ecuador, indigenous grounded ontologies – their experiences and understandings of the world – offer a profound challenge to mainstream development. The indigenous women she worked with argue that to make their visions of living well into reality requires a thoroughgoing political revolution, decolonizing the state and the institutions of development. Will that be the next era of indigenous development? It appears that the particular conjunctures in which we live today – the terrifying implications of climate change; the continuing lack of social equality due to the excesses and failures of global capitalism; and the increasing political conflicts that result – may be leading to a renewed valuation of indigeneity as offering solutions to the world's problems. But, as we have seen throughout this essay, states and markets have continuously found ways to articulate these challenges to their interests.

References

Anthias, P. (2018) *Limits to Decolonization: Indigeneity, Territory, and Hydrocarbon Politics in the Bolivian Chaco.* Ithaca, NY: Cornell University Press.

———. and Radcliffe, S. (2015) The ethno-environmental fix and its limits: Indigenous land titling and the production of not-quite-neoliberal natures in Bolivia. *Geoforum* 64: 257–269.

Barragán, R. R. (2015) Ladrones, pequeños empresarios o trabajadores independientes? K'ajchas, trapiches y plata en el cerro de Potosí en el siglo XVIII. *Nuevo Mundo*, March 20. http://journals.openedition.org/nuevomundo/67938 (Accessed 15 February 2018).

Bernstein, H. (2000) Colonialism, capitalism, development. In T. Allen and A. Thomas (eds) *Poverty and Development into the 21st Century*. Oxford: Oxford University Press, pp. 241–270.

Blackwell, M. (2015) Geographies of difference: Transborder organizing and indigenous women's activism. *Social Justice* 42(3/4): 137–154.

Bonfil Batalla, G. (1996) *México Profundo: Reclaiming a Civilization*. Trans. Philip Dennis. Austin: University of Texas Press.

Cardoso, F. H. and Faletto, E. (1979) *Dependency and Development in Latin America*. Berkeley, CA: University of California Press.

Castellanos, M. B. (2010) *A Return to Servitude: Maya Migration and the Tourist Trade in Cancún*. Minneapolis: University of Minnesota Press.

Cope, R. D. (1994) *The Limits of Racial Domination: Plebeian Society in Colonial Mexico City, 1660–1720*. Madison: University of Wisconsin Press.

Córdoba Azcárate, M. (2011) 'Thanks God, this is not Cancun!' Alternative tourism imaginaries in Yucatán (Mexico). *Journal of Tourism and Cultural Change* 9(3): 183–200.

Cultural Survival. (2012) *Confirming rights: Inter-American Court Ruling Marks Key Victory for Sarayaku People in Ecuador*. www.culturalsurvival.org/publications/cultural-survival-quarterly/confirming-rights-inter-american-court-ruling-marks-key (Accessed 20 June 2018).

de la Cadena, M. and Starn, O. (2007) *Indigenous Experience Today*. Oxford: Berg.

Diaz, B. (1963) *The Conquest of New Spain*. London: Penguin Books.

Dinerstein, A. C. and Deneulin, S. (2012) Hope movements: Naming mobilization in a post-development world. *Development and Change* 43(2): 585–602.

Dinzey-Flores, Z. (2018) The development paradox. *NACLA* 50(2): 163–169.

Earle, D. and Simonelli, J. M. (2005) *Uprising of Hope: Sharing the Zapatista Journey to Alternative Development*. Walnut Creek, CA: Rowman Altamira.

Egan, N. (2007) *Citizenship, Race, and Criminalization: The Proceso Mohoza, 1899–1905*. Working Papers, Center for Iberian and Latin American Studies, UC San Diego. http://escholarship.org/uc/item/6680j2k8

Engle, K. (2010) *The Elusive Promise of Indigenous Development: Rights, Culture, Strategy*. Durham, NC: Duke University Press.

Escobar, A. (2011) *Encountering Development: The Making and Unmaking of the Third World*. Princeton, NJ: Princeton University Press.

Fabricant, N. and Postero, N. (2015) Sacrificing indigenous bodies and lands: The political-economic history of Lowland Bolivia in light of the recent TIPNIS debate. *Journal of Latin American and Caribbean Anthropology* 20(3): 452–474.

Foster, G. (1965) Peasant society and the image of limited good. *American Anthropologist* 67(2): 300–323.

Fraser, N. (1997) *Justus Interruptus: Critical Reflections on the "Postsocialist" Condition*. London: Routledge.

Friedlander, J. (1975) *Being Indian in Hueyapan: A Study of Forced Identity in Contemporary Mexico*. New York: St. Martin's Press.

Galeano, E. (1973) *The Open Veins of Latin America*. New York: Monthly Review Press.

García, M. E. (2008) Indigenous encounters in contemporary Peru. *Latin American and Caribbean Studies* 3(3): 217–226.

Goodale, M. and Postero, N. (2013) Revolution and retrenchment, illuminating the present in Latin America. In M. Goodale and N. Postero (eds) *Neoliberalism, Interrupted: Social Change and Contested Governance in Contemporary Latin America*. Stanford, CA: Stanford University Press, pp. 1–22.

Gotkowitz, L. (2007) *A Revolution for Our Rights, Indigenous Struggles for Land and Justice in Bolivia, 1880–1952*. Durham, NC: Duke University Press.

Hale, C. R. (2002) Does multiculturalism menace? Governance, cultural rights, and the politics of identity in Guatemala. *Journal of Latin American Studies* 34: 485–524.

———. (2004) Rethinking indigenous politics in the era of the 'Indio Permitido'. *NACLA Report on the Americas* 38(2): 16–20.

Hall, S. (1996) The West and the rest. In S. Hall, D. Held, D. Hubert and K. Thompson (eds) *Modernity: An Introduction to Modern Societies*. Malden, MA: Wiley-Blackwell, pp. 185–227.

Harvey, D. (2005) *A Brief History of Neoliberalism*. Oxford: Oxford University Press.

Hindery, D. (2013) *From Enron to Evo: Pipeline Politics, Global Environmentalism, and Indigenous Rights in Bolivia*. Tucson: University of Arizona Press.

Holston, J. (2008) *Insurgent Citizenship: Disjunctions of Democracy and Modernity in Brazil*. Princeton, NJ: Princeton University Press.

Klein, N. (2005) The rise of disaster capitalism. *The Nation*, May 2, pp. 9–11. www.thenation.com/article/rise-disaster-capitalism/ (Accessed 7 September 2015).

Knight, A. (1990) Racism, revolution, and indigenismo: Mexico, 1910–1940. In R. Graham (ed) *The Idea of Race in Latin America, 1870–1940*. Austin: University of Texas Press, pp. 71–113.

Larson, B. (2004) *Trials of Nation Making: Liberalism, Race, and Ethnicity in the Andes, 1810–1910*. Cambridge: Cambridge University Press.

———. (2005) Capturing Indian bodies, hearths, and minds: The gendered politics of rural school reform in Bolivia, 1920s – 1940s. In A. Canessa (ed) *Natives Making Nation: Gender, Indigeneity, and the State in the Andes*. Tucson: University of Arizona Press, pp. 32–59.

McMichael, P. (2006) Reframing development, global peasant movements and the new agrarian question. *Critical Journal of Development Studies* 27(4): 471–483.

———. (2010) (ed) *Contesting Development: Critical Struggles for Social Change*. New York: Routledge.

Meisch, L. (2002) *Andean Entrepreneurs*. Austin: University of Texas Press.

Mignolo, W. D. (2009) *The Idea of Latin America*. Malden, MA: John Wiley & Sons.

Mintz, S. (1986) *Sweetness and Power: The Place of Sugar in Modern History*. London: Penguin Books.

Moberg, M. (2010) A new world? Neoliberalism and fair trade farming in the Eastern Caribbean. In S. Lyon and M. Moberg (eds) *Fair Trade and Social Justice*. New York: New York University Press.

Nash, J. (1979) *We Eat the Mines and the Mines Eat Us: Dependency and Exploitation in Bolivian Tin Mines*. New York: Columbia University Press.

Perreault, T. (2005) Geographies of neoliberalism in Latin America. *Environment and Planning A* 37(2): 191–201.

Postero, N. (2007a) Andean utopias in Evo Morales's Bolivia. *Latin American and Caribbean Ethnic Studies* 2(1): 1–28.

———. (2007b) *Now We are Citizens, Indigenous Politics in Postmulticultural Bolivia*. Stanford, CA: Stanford University Press.

———. (2013) Negotiating identity. *Latin American and Caribbean Ethnic Studies* 8(2): 107–121.

———. (2017) *The Indigenous State: Race, Politics, and Performance in Plurinational Bolivia*. Los Angeles: University of California Press.

Postero, N. and Zamosc, L. (2004) Indigenous movements and the Indian question in Latin America. In N. Postero and L. Zamosc (eds) *The Struggle for Indigenous Rights in Latin America*. Brighton: Sussex Academic Press, pp. 1–31.

Radcliffe, S. (2012) Development for a postneoliberal era? *Sumak Kawsay*, Living well, and the limits to decolonization in Ecuador. *Geoforum* 43: 240–249.

———. (2015) *Dilemmas of Difference: Indigenous Women and the Limits of Postcolonial Development Policy*. Durham, NC: Duke University Press.

Richards, P. (2013) *Race and The Chilean Miracle*. Minneapolis: University of Minnesota Press.

Sawyer, S. (2004) *Crude Chronicles: Indigenous Politics, Multinational Oil, and Neoliberalism in Ecuador*. Durham, NC: Duke University Press.

Schilling-Vacaflor, A. (2011) Bolivia's new constitution: Towards participatory democracy and political pluralism? *European Review of Latin American and Caribbean Studies* 90: 3–22.

Stephen, L. (2007) *Transborder Lives: Indigenous Oaxacans in Mexico, California, and Oregon*. Durham, NC: Duke University Press.

Stephenson, M. (1999) *Gender and Modernity in Andean Bolivia*. Austin: University of Texas Press.

Stern, S. J. (1993) The challenge of conquest in wider perspective, and Prologue: Paradigms of conquest: History, historiography, and politics. In S. J. Stern (ed) *Peru's Indian Peoples and the Challenges of Spanish Conquest*. Madison: University of Wisconsin Press, pp. xv–liii.

Sword, A. (2010) Teaching against neo-liberalism in Chiapas, Mexico: Gendered resistance via neo-Zapatista network politics. In P. McMichael (ed) *Contesting Development: Critical Struggles for Social Change*. New York: Routledge, pp. 132–146.

Tassi, N., Medeiros, C., Rodríguez-Carmona, A. and Ferrufino, G. (2013) *"Hacer plata sin plata." El desborde de los comerciantes populares en Bolivia*. La Paz: Fundación PIEB.

Taussig, M. (1987) *Shamanism, Colonialism and the Wild Man: A Study in Terror and Healing*. Chicago: University of Chicago Press.

Tockman, J., Cameron, J. and Plata, W. (2015) New institutions of indigenous self-governance in Bolivia: Between autonomy and self-discipline. *Journal of Latin American and Caribbean Ethnic Studies* 10(1): 37–59.

Trouillot, M. R. (1991) Anthropology and the savage slot: The poetics and politics of Otherness. In R. G. Fox (ed) *Recapturing Anthropology*. Santa Fe: School of American Research.

Ugarteche, O. (1999) The structural adjustment stranglehold. *NACLA* July/August: 21–23.

Via Campesina. (2003) Food sovereignty. *La Via Campesina*, 15 January. https://viacampesina.org/en/food-sovereignty/ (Accessed 20 June 2018).

Walsh, C. (2007) Shifting the geopolitics of critical knowledge: Decolonial thought and cultural studies 'Others' in the Andes. *Cultural Studies* 21(2–3): 224–239.

Wolf, E. (2010) *Europe and the People Without History*. Los Angeles: University of California Press.

Wright, R. (1992) *Stolen Continents, The Americas Through Indian Eyes Since 1492*. Boston, MA: Houghton Miflin.

Zilberg, E. (2004) Fools banished from the kingdom: Remapping geographies of gang violence between the Americas (Los Angeles and San Salvador). *American Quarterly* 56(3): 759–779.

4

COLONIALITY, COLONIALISM, AND DECOLONIALITY

Gender, sexuality, and migration

Camila Esguerra Muelle
Translated by Anna Holloway

Introduction

The multiple histories of the region known today as Latin America are permeated by a series of colonial and neocolonial operations taking place since the 16th century and even before that. The processes of formation of the nation-states in the (former) colonies of Latin America are the result of successive and complex processes of colonial expansion that began many centuries ago and have had long-lasting effects on the region's historical and cultural processes. They have been characterized by the establishment of dictatorships and military – or even paramilitary or parastate – regimes during the 20th and the beginning of the 21st century.

We are talking about a "long-term colonial horizon" (Rivera Cusicanqui, 1993: 58) consisting of a multidimensional violence that is based on the operation of the matrix of domination and causes expropriation, exclusion, and disciplining. In these territories, this "horizon" has produced an endemic state of war with occasional ceasefires or detentes of armed conflict or military, armed, or paramilitary repression. In this context, there has been an instrumentalization of a discursive construct and a repertoire of violence that are specifically racist, misogynistic, or lesbo-, homo-, and transphobic (Esguerra Muelle, 2015) and have, to a great extent, led to the exodus, uprootings, migrations, displacements, and exiles that characterize globalizing modernity.

In turn, this modern colonialist discourse walks hand in hand with a discourse of development that is, to paraphrase Lugones and Mignolo, "the clear face" of capitalist modernity. The colonial system in general, and the modern sex-gender colonial system in particular, are operated through international agendas of development and gender, with actors that seem to enact an epic showdown between supranational organizations and theocratic states or international religious organizations (Esguerra Muelle, 2017). In this chapter I argue that the uprooting, displacement, and exile that are generated in this context are the way for colonial relations to reproduce their meanings in the present with unexpected force. They also represent the "dark and concealed face" of globalization.

There have been significant theoretical contributions to this issue since the beginning of the 20th century that allow us to understand and confront these colonial operations, that have been taken up in part by the so-called Modernity/Coloniality/Decoloniality group. In this chapter, I take a closer look at what – based on the thought developed in the course of many centuries in the Americas and other parts of the world – Lugones (2008) has rounded up in her theory on the "coloniality of gender" and Rivera Cusicanqui (2010a, 2010b) in her *teoría descolonial*[1] (in Spanish "descolonial" very different to "decolonial") theory.

Decoloniality/Decolonization

> There is a syntagmatic surface of the present where one can see syntagmas of the deep pre-Hispanic past that fuel resistance.
>
> *(Rivera Cusicanqui in Santos, 2015: 83, my translation)*

According to Grosfoguel (2006: 17), the main notions developed by theorists from Latin America and the Caribbean who have carried out a re-reading of postcolonial theories, insufficient to explain the historical and social becoming of Latin America and the Caribbean (Rivera Cusicanqui, 1993: 58), can be summarized as follows: (1) "Geopolitics of knowledge," coined by Dussel; (2) "Body-politics of knowledge," inspired by the works of Fanon and Anzaldúa; (3) "Coloniality of power," developed by Quijano; and (4) "The colonial difference," coined by Walter Mignolo. To this classification we must add the "coloniality of being," formulated by Maldonado-Torres (2007), and the "coloniality of gender" developed by Lugones (2008), to whom I will make further reference in the following section.

According to Quijano (2000), coloniality is constituted within a series of mechanisms of symbolic regulation, instituted by colonialism through various forms of cultural, epistemic, political, and military violence and also – might we add – of technologies of terror, that in response face the opposition of a variety of forms of resistance and decolonial strategies that operate at the micro- and mesopolitical level. That is, the seed of decoloniality is in modernity itself. Furthermore, in the words of Hall (1996: 249), who does not belong to the MCD group, coloniality is the process of "expansion, exploration, conquest, colonization, and imperial hegemonisation which constituted 'the outer face,' the constitutive outside European and then Western capitalist modernity after 1492." This idea of 1492 being the starting point of European modernity and of the constitution of the "European ego" through the "concealment" (*encubrimiento*)[2] of what is today know as the Americas, had already been formulated by Dussel (1995) in his 1992 Frankfurt conferences.

Colonial relations are constitutive of Western modernity – founded on a rhetoric of salvation, progress, "civilisation," and development – just as decoloniality emerges at the same time as the modern westernizing project but works against it in indigenous and Afro-Caribbean thought (Mignolo, 2007: 27). The "colonial turn," a concept developed by Puerto Rican philosopher Maldonado-Torres (2007), complements the category of "decolonisation" used by social sciences at the end of the 20th century. For Césaire, who speaks from the Afro-Caribbean experience, decolonization does not consist in asserting "one's own" with a fundamentalist and provincialist intention or from a false nationalism or autochthonous pride, but rather in affirming a concrete universalism that congregates singularity, as opposed to the abstract universalism of Western modernity (Grosfoguel, 2007: 72). This concrete universalism is a constant exercise of negotiation of interculturality, while abstract universalism accommodates itself to the facade of multiculturality. According to Grosfoguel (2007: 76) "the decolonisation of the notion of Western, euro-centered universality is crucial if one wants to achieve the Zapatista slogan of building 'a world where many worlds fit'."

Therefore, decoloniality opposes the need to conclude the unfinished project of modernity; it is a project of an impure epistemic approach, *mulato, mestizo*: transmodern. Transmodernity is the concept proposed by Dussel to transcend the Eurocentric Western version of modernity and complete the unfinished project of decolozisation (Grosfoguel, 2007: 72) and not of modernity, as Habermas had posited. In this sense, the "decolonial turn" proposed by Nelson Maldonado-Torres (2007) is an epistemic turn not only for social sciences but also for modern institutions such as law, the academy, or art, amongst others (Mignolo, 2007: 21).

However, Rivera Cusicanqui prefers the term "internal colonialism," originally coined by Gonzales Casanova. Rivera Cusicanqui establishes that, beyond the original and more economicist meaning of this concept, internal colonialism is a form of domination: "I do not use the idea of the coloniality of power and such, because coloniality is a state, an abstract entity. Colonialism, on the contrary, is a type of asset that is embedded in subjectivity (. . .) We are all colonised" (Rivera Cusicanqui in Santos, 2015: 83).

Therefore, Rivera Cusicanqui speaks of "decolonisation" and not "decoloniality" and, in doing so, inaugurates decolonial thought. On the one hand, according to my interpretation of the author's work, decolonization is a *putting into practice* forms of confronting epistemic violence and colonialism in general – I assume in its dimensions of knowledge, power, being, gender, and colonial difference – through practical and personal, i.e. impure, political wagers, leaving aside, for example, the Left-Right dichotomy as the only possible political situation. On the other hand, she calls to forms of knowledge and thought that are still to be rediscovered, forms that are being constructed in the spaces of everyday life in Latin America and the Caribbean.

Decolonization is linked to micropractices as well as to proposals for structural changes and the incorporation of knowledges, given the toll not only of colonization, but also of the "archaising" recolonization performed by the colonial elites. It is to bet on a modernity pertaining to the Americas in order to confront a Westernized "archaicising" modernity, as Rivera Cusicanqui would say. This would be a modernity based on recovering practices of resistance employed during the colony through recalling, for example, the networks used for the circulation of goods across long distances, or the networks of production and the markets of the densely populated urban centres of Latin America before and during the colony. Networks that were mutilated by the rentier and tax-collecting voracity of the Spanish crown against, for example, the project of political, religious, and historical self-determination of the Katari-Amaru, as well as their forms of imagination and representation, which constitute an example of indigenous modernity (Rivera Cusicanqui, 2010a: 53–54).

On the other hand, "decoloniality," according to Rivera Cusicanqui (2010a), is an entelechy forged by the intellectuals of the so-called group of decolonial or post-Western studies, or the Latino project Coloniality/Modernity, with the consent of the industries of knowledge of the North and the West, where "neologisms such as '*decolonial*,' '*transmodernity*,' '*eco-si-mía*' proliferate" (*ibid.*, 64) (. . .) "without altering anything of the relations of force in the 'palaces' of empire" (*ibid.* 57, 58) and without modifying the co-production between capitalism, liberalism, civilizing discourse, and the conservation of the privileges of the republican colonial elites.

Before continuing with the discussion, I would like to point out that, from a "heterarchic" perspective (Kontopoulos, 1993), colonial relations do not only consist in the cultural, economic, and administrative control and domination of powers or metropolises over countries that are considered peripheral. Within this logic, coloniality and colonialism have a dimension that is geopolitical – and, of course, necropolitical (Mbembe, 2011) – but also meso-, micro-, and even anatomo-political. This means that colonialism, as a means or as an end, has made use of institutions as well as of bodies, and that is where racism and gender coloniality come into play. The Native American, Chicano, and Afro-Caribbean feminisms have reflected not only on racist and capitalist relations, but also on the oppression enforced through other systems that are inherently violent, such as gender.

Coloniality of gender [of sexuality] and decolonial/descolonial feminism

Part of the Western feminist discourse has represented non-Western women as subjects outside historical, political, and cultural structures, as a homogeneous collective with no history or political position, instead of analysing the ways in which women are constructed through precisely these structures (Mohanty, 1991). In an exercise of hetero-determination, this discourse adopts a generalizing approach to women of the so-called "Third World," imposing upon them a stamp of homogenizing similarity and one-dimensional poverty (Femenías, 2007: 24). This leads to the production of forms of neo-colonization of some women over others (Mohanty, 1991: 350–352) through a cultural and geopolitical localization within alterity, poverty, and exoticism.

At the same time, as Femenías (2007: 11) points out, "we know that a considerable part of (. . .) postcolonial thought sees us as the 'others'." According to Mohanty (1991: 334), some feminist writings have colonized the materiality and historical differences of the lives of women from the so-called Third World – and the same has occurred with Afro-descendant (hooks, 1984) and indigenous women around the world. This produces a monolithic and homogeneous "Third World woman," an image that has been arbitrarily constructed and sustains the Western humanist discourse, for Woman (singular and capitalized) is white and Western, as well as heterosexual and cisgender.

For their part, theorists that define themselves as feminist and decolonial deploy a theoretical and political proposal originating from many sources: first the interpellating tensions and contributions of Afro-descendant, indigenous, and Chicano women to the feminist and women's movements in Latin America and the Caribbean (Abya Yala) and, second, the critique of the so-called gender perspective in the developmentist agenda of international cooperation as a new process of Westernization and loss of autonomy (Rivera Cusicanqui, 2014) through guidelines for the funding of NGOs and states. The latter constitute new devices of the global geopolitical order (Mendoza, 2014), promoted by the UN and criticized by autonomous feminism, established by the VI Feminist Conference of Latin America and the Caribbean (El Salvador, 1993) and the VII Conference (Cartagena de Chile, 1996) that led to the Declaration of Autonomous Feminism (Espinosa, Gómez Correal, and Ochoa, 2014: 23–24).

However, according to Espinosa et al. (*ibid*: 25), thanks to the simultaneous participation of indigenous feminist Maya and Xinca women, of the movement of Community Feminists in Bolivia and Guatemala, and of black and Afro-descendant women in struggles for ethnic-racial identity and against racism during and even before the 1990s, those who made the initial demands for the autonomy-institutionality rupture in autonomous feminism went on to reflect on how the sexuality, race, and class order had to be opposed with a feminist agenda in Abya Yala[3]; black women, particularly, produced a critical discourse against racism, the history of slavery and the wiping out of the history of Afro-descendant women, and, with it, against the long process of colonization. To this body of knowledge, I believe we should also add the contributions of women or "women" that also challenge this identity category through practices of cross-dressing, cultural translation and bilingualism, including the things that they have done and said within indigenous and anarchist movements (Masiello, 1997).

For her part, Curiel (2007) has asserted that postcolonial, or better, decolonial, theory – established by the Modernity/Coloniality/Decoloniality group and the Coloniality Working Group[4] – is elitist and androcentric, and ignores the theorizing of feminist movements in the Americas – Chicano, migrant, of colour, Afro-Latin, indigenous, and Black Feminism – on colonial domination. For her part, Rivera Cusicanqui (2010b) shows how the colonial imposition of a Western patriarchal regime entails the establishment of borders and of private property in detriment of indigenous peoples, and particularly women, in what today is Bolivia.

On the other hand, various academics who work with feminist decolonial theory have been criticized for their positions with lesbian feminism in which they are also active. The critique centres around the non-inclusion of trans women and men and non-binary people in certain lesbian feminist spaces. Ramírez and Castellanos (2013) show how Latin American feminist lesbianism – at least in its "official" discursive production – has positioned trans lives and bodies outside of its sphere of action. This exclusion explains the absence of a more sophisticated analysis that would allow the category "lesbian" not only to challenge heteronormative sexuality, but also the cisgender gender order. As Witting (1992) noted, "lesbians are not women." Furthermore, I argue that these positions do not recognize that the cisgender order is at the hardest core of the coloniality of gender neither do they recognize that the category "lesbian" needs to be called into question, as it is a white, urban, and Western label that maintains the homosexuality-heterosexuality binary which means it needs to be subject to a decolonial feminist interpretation. This framework already accounts for transfeminisms, which according to my own understanding, pose a challenge to the borders established by categorizing order of the hegemonic modern gender and sexuality system.

It is however María Lugones who makes the most significant theoretical leap, with her trenchant critique of Quijano for ignoring the co-constitutive relationship of race and gender within colonialism/coloniality/modernity,[5] and on the basis of which she develops her concept of "colonial modern gender system" which I present and analyze from a critical perspective. As Lugones explains, Quijano, for whom "struggles for the control of 'sexual access, its resources, and products' define the sex/gender sphere and revolve around the axes of coloniality and modernity" (Lugones, 2014: 58, 61), is not aware of the fact that he has naturalized the hegemonic gender system. As a result, when making his analysis from the standpoint of the coloniality of power, he takes for granted that the precolonial sex-gender systems were identical to the systems of the colonizers – binary, androcentric, dimorphic, and heterosexual – and had a central, universal character as the organizers of modernity. Lugones draws here on the work of Oyewúmi who notes that the existence and significance of gender are not always the same as in the West.

For Lugones (2014), within the colonial modern gender system the ontological locus of "being a woman" can only be occupied by white women as the reproducers of the "race," which, at the same time, guarantees their attribute of "being human." Thus, Lugones clarifies how the coloniality of gender operates as a matrix of domination, where race and gender mutually produce one another:

> The intersection of "woman" and "black" lacks a place for the black woman, precisely because neither "woman" nor "black" include her (. . .) Only in perceiving gender and race as interwoven or tightly fused together can we truly see women of colour.
>
> *(Lugones, 2014: 61)*

Therefore, to think of the colonial modern gender system does not only involve the consideration of gender relations and sexuality within colonialism and coloniality; it also means reflecting on how gender and sexuality organize modernity – not only the system of reproduction but the material and symbolic modern social organization in general.

However, there is something that Lugones does not fully develop in her thesis on the colonial modern gender system. Perhaps unintentionally, she once again falls into the trap of analysing an inherently dimorphic, binary, and heterosexual gender, while she seems to perform an artificial conceptual separation between sex and gender. It is striking that she begins her analysis by saying "I investigate the intersection of race, class, gender, and sexuality in order to understand the alarming indifference of men towards the violence that is systematically inflicted upon women of colour" (Lugones, 2014: 57).

In her sharp intersectional race/gender approach, she seems to omit that the binary, dimorphic, and heterosexual order – to which I would also add cisgender – is also inseparable from the race/gender/class axis. While she acknowledges that "the term 'woman'," without any specification of fusion, is itself devoid of meaning or racist – as Sojourner Truth pointed out in the 19th century – she sometimes forgets to analyze that the subjects produced and subalternized by the sexual and gender order are not only cisgender women of colour, but also persons, not necessarily women, nor cisgender, nor heterosexual, nor white; or, better, not only what the Western ethos calls woman.

The author does refer generally, even if using the same discursive terms of the hegemonic sex-gender system, to the precolonial existence of "homosexual" – nothing more modern and eurocentric than the term homosexual – and intersexual persons – where the imprecise and insufficient translation of the colonizing language should say "transgender" or maybe "intersex." However, she fails to mention that the term "woman" with no other specification is also cisgenderist and hetero-centered, and even transphobic and lesbophobic, and that "woman" is not only an ontic colonial locus but also a disputed terrain.

The decolonization of gender and sexuality has also included the re-appropriation or reproduction of sex-gender and sexuality systems that precede the "concealment" (1492 onwards) and the constant questioning of the norms of the heterosexual and cisgenderist colonial regime, even when to challenge these norms is to betray and trade-off the structures of meaning of the modern hegemonic sex-gender system. It also conceals the fact that concealing and archaicizing colonialism/coloniality has, from its beginning and until today, physically, materially, and symbolically eliminated people of flesh and blood – not only women – who have not obeyed the rules, not only of heterosexuality but also of cisgenderism.

To understand the inseparability of cisgenderism/heterosexuality/race, one must recall the interest shown by the sexological discursive construct, when it was still in the making, for so-called inverted as well as Afro-descendant, African, and black (anafemales) (Somerville, 2000), who were even less likely than white women to be conceded the attributes of "reason," "civilisation," "adulthood," pertaining to an *ego conquiro* and an *ego cogito* that are fundamentally masculine, white, and adult. In the *Sexual Inversion in Women*, written around 1890 and published in the United States in 1895 and in England in 1997, Havellock Ellis tries to show that "inversion" (for he preferred not to use the hybrid term of homosexuality), and particularly "feminine inversion," is a congenital anomaly that is not only psychological but also anatomical. He engaged in the task of showing that "inverted" women (or, in contemporary terms, *lesbians*) had a common anatomical feature with African and African-American women, namely a prominent vulva and clitoris, and focused his studies on "inverted" female bodies and on African and African-American women, establishing an explanatory racist model that intersected with the "pathologisation" and "bestialisation" of female homoeroticism (Somerville, 2000: 27–30). Within a racist, cisgenderist, and hetero-centered framework, those who defy these rules or do not abide by them have been constructed by racism and scientific sexism as "non-human."

At this point, it is important to return to Wittig (1992: 32) who points out that the category "'woman' [and man, I would add] has meaning only in heterosexual systems of thought and heterosexual economic systems. Lesbians are not women." This approach is crucial when thinking of the coloniality of hegemonic modern gender, for the oppressive and colonial gender system is ultimately based on "hetero thought," and the *lesbian* subject [to which I would add trans and non-binary persons] constitutes a rupture in this system. *Lesbian* understood not as the woman who has affective or erotic relations with other women, but as a being who is liminal or external to the sex-gender system that is organized on the basis of "the copula that dialectically unites" "man" and "woman" (Wittig, 1992: 29). This is a dialectic that constitutes a binary semiosis which closes itself off and leads to all the mechanisms of social and "biological" reproduction (*ibid*). Wittig's view is that reproduction is the basis for the heterosexual economy and the role of social reproduction that is assigned to women, which is perhaps the greatest burden of their lives.

That is why a long-lasting connection must be made between gender systems or, generally speaking, the multiple forms of social organization in the Americas (Esguerra Muelle, 2006) before the European invasion, and the long history of struggles for the decolonization of gender by women, men, and transgender persons, as well as lesbians (to use the terms employed in colonial language).

Thus, in the organization of the world by hetero thought, pertaining to the Western *ethos*, women are in charge of all social reproduction – which must not be confused with reproductive work but rather considered as the insoluble fusion of productive and reproductive work – without any type of acknowledgement or remuneration for the material and emotional surplus value created through a work that has no guarantees and has come to be considered the "natural" obligation of women.

This system, which produces and automatically devalues feminine and feminized subjects, is founded on the organization of sex on the basis of gender discourses, the expropriation of female sexuality through its symbolic construction, and the imposition of a forced heterosexuality and a prescriptive cisgenderism. It is my opinion that forced heterosexuality and prescriptive cisgenderism are the two main institutions of the modern hegemonic colonial sex system, which organizes the existence of people on the basis of the ontologized establishment of a discursively defined coherence between sex, gender, and desire.

What I call "prescriptive cisgenderism" consists in the political institution established as a means – and as part of the ends – of the colonial modern project to ensure the continuation of sexual dimorphism (Laqueur, 1990) established during the 17th century in Europe and of the gender binary, in order to guarantee not only the regime of heterosexuality but also the sexual, racial, and international division of labour. The latter is pursued through "global care chains" (Hochschild, 2000) or transnational networks of care,[6] the vertical and horizontal segregation of labour, as well as the subordination of the feminine and the equation of the masculine with the universal (Arango, 2004).

The heterosexual and cisgenderist regime results in a lack of awareness of the importance of current struggles against these two systems as decolonial rebellions. It also works to erase political memory of gender and sexuality, and of the existence of precolonial, non-binary, non-heterosexual sex-gender systems.

Throughout histories and within different cultures, there have been many categories that have defined "feminine homoeroticism," "transgenderism," and "intersexuality." These Western notions are named differently in the indigenous languages of the Americas, in ways that cannot be literally translated. There is revealing literature – from the chronicles of European invaders to archives and even anthropological or historical research – that confirm the existence of sex-gender systems that diverge from the Western model in different times and places and persist until today as a form of resistance or decolonization. These include the cases of "bardajismo" (Williams,1992; Roscoe, 1988; Lucena Samoral, 1966) and "mati work" (Wekker, 2006), to name but a few examples of a genealogy that has not been sufficiently studied.

Therefore, to extend Lugones' (2008) thinking, it is my understanding that the colonial modern sex-gender introduces a political economy of the oppression of female and feminized – and therefore devalued – subjects, as well as subjects with unintelligible gender identities or non-reproductive and non-heterosexual sexualities, or subjects who have been hypersexualized through racist, ableist, or ageist discourses, amongst others. It is a system that operates through a semiotic and syntagmatic set of naturalized asymmetries of domination.

We could say that in globalization we witness an exodus and global traffic not only of women but also of "pornifiable" bodies (Preciado, 2008) and feminized subjects, expressed globally as "trafficking in women" (Rubin, 1996). Migrant women from the so-called developing countries are often reinserted into transnational branches of care, which define their situation in "global cities" and "circuits of survival that have emerged in response to the deepening misery of the Global south" (Sassen, 2004: 255). This misery is the product of colonial relations, from the unequal capacity to compete in the liberalized global market to corruption in both the government and private sector. Kofman (2000: 138) argues that there is an increase in professionally qualified women who become migrants; we know that many of these women and qualified feminized persons face underemployment in the course of their migration through the transnationalization of care and of precarious employment.

At the same time, the heterosexual and cisgenderist regime expels women, trans or non-binary persons, and lesbians from its territories through wars, regimes of terror or military rule, or simply through symbolic violence and a discursivity that always has material effects. It also expels them from the territory of the modern nation-state, a device that alternates between closing its borders – surveilled by geopolitical, biopolitical, necropolitical, and micropolitical regimes so as to prevent their circulation – with opening them to reinsert these feminized, i.e. devalued, subjects into the transnational networks of care and of precarious labour in general, thus creating a transnational colonial, sexual, and racial order of labour and a coloniality of power. These feminized subjects often remain suspended or adrift for long periods of time, if not forever, in a Frontera/Borderland (Anzaldúa, 1987; Esguerra Muelle, 2014) that does not acknowledge their "modern existence" as "citizens," that dehumanizes, bestializes, criminalizes them, and governs them through a colonial regime which, at the same time, expels persons and closes borders.

Conclusions

Just as the Modernity/Coloniality/Decoloniality group revealed the insufficiency of postcolonial theory in understanding how the continent that is today known as the Americas articulates the colonial modern project, María Lugones and Silvia Rivera Cusicanqui formulate decolonial theories and theories on the coloniality of gender that turn the paradigms established by the MCD group upside down.

The possibilities posited by the theory of gender coloniality and decoloniality must be developed, precisely to insist on our narrative that the heterosexual and cisgenderist regime is at the very centre of coloniality; if there is something that has been concealed in the Americas it is the sexuality and gender order, classified as loathsome sin and *pecatum mutum* (Esguerra Muelle, 2002).

At the same time, it is important to see that phenomena such as exile, migration, displacement, or uprooting, which constitute a moment that is almost painfully spectacular in the positioning of the colonial modern Western project, entail a permanent reproduction of the relations of the colonial system in its dimensions of power, knowledge, being, gender, and sexuality, and that those migrating movements are also responses of resistance and rebellion in the face of the Western colonialist project.

Notes

1 In Latin America, a distinction is drawn between the *teoría descolonial* of Silvia Rivera Cusicanqui and other feminist writers and the decolonial theory that emerges from the Modernity/Coloniality/ Decoloniality (MCD) research paradigm. Translation into English does not allow for a distinction between decolonial and descolonial.
2 *Tr. Note* The Spanish word *encubrimiento*, which literally means "concealment," is a play of words with *descubrimiento*, the Spanish word for "discovery", and a direct reference to the continent not being "discovered" but rather covered up.
3 The kuna, cuna, or tule name for the continent that was colonized and named America by the colonists.
4 Based at Binghampton University
5 Breny Mendoza (2014: 98–100) has also produced an important criticism of the positions of Quijano and Dussel, that I have not included here.
6 In the context of my postdoctoral research project, consisting in an ethnography of different locations such as Cartagena, Bogotá, Cali, Medellín, Barcelona, and Madrid on the transnationalised care regime, I have begun to provisionally speak of the "[trans]national networks of care" in order to provide a more multidimensional and less linear idea than the one projected by the notion of "global care chains."

References

Anzaldúa, G. (1987) *Borderlands/La frontera*. San Francisco: Aunt Lute Book Company.
Arango, L. G. (2004) *Mujeres, trabajo y tecnología en tiempos globalizados*. Bogotá: Series Cuadernos CES, N° 5.
Curiel, O. (2007) Crítica poscolonial desde las prácticas del feminismo antirracista. *Nómadas* 26: 92–101.
Dussel, E. (1995) *The Invention of the Americas. Eclipse of "the Other" and the Myth of Modernity*. New York: Continuum Publishing.
Esguerra Muelle, C. (2002) *Del pecatum mutum al orgullo de ser lesbiana: Grupo Triángulo Negro de Bogotá, 1996–1999*. Unpublished undergraduate dissertation Universidad Nacional de Colombia. http://bdig ital.unal.edu.co/58137/
———. (2006) Lo innominado lo innominable y el nombramiento. In M. Viveros, C. Rivera and M. Rodríguez (eds) *De mujeres hombres y otras ficciones: Género y sexualidad en América Latina*. Bogotá: CES- Grupo TM, pp. 247–281.
———. (2014) Dislocación y borderland: Una mirada oblicua desde el feminismo descolonial al entramado migración, régimen heterosexual, (pos)colonialidad y globalización. *Universitas Humanística* 78: 137–161.
———. (2015) Mujeres imaginadas: Mujeres migrantes, mujeres exiliadas y sexualidades *no normativas*. Unpublished PhD dissertation in Humanities. Universidad Carlos III de Madrid https://dialnet.uniri oja.es/servlet/tesis?codigo=73091
———. (2017) Cómo hacer necropolíticas en casa: Ideología de género y acuerdos de paz en Colombia. *Sexualidad y Salud Social* 27: 172–198.
Espinosa, Y., Gómez Correal, D. and Ochoa, K. (2014) Introducción. In Y. Espinosa, D. Gómez Correal, K. Ochoa (eds) *Tejiendo de Otro Modo: Feminismo, epistemología y apuestas descoloniales en Abya Yala*. Popayán: Universidad del Cauca, pp. 13–40.
Femenías, M. L. (2007) Esbozo de un feminismo latinoamericano. *Revista Estudios Feministas* 15(1): 11–25.
Grosfoguel, R. (2006) La descolonización de la economía-política y los estudios postcoloniales: Transmodernidad, pensamiento fronterizo y colonialidad global. *Tabula Rasa* 4: 17–48.
———. (2007) Descolonizando los universalismos occidentales: el pluri-versalismo transmoderno decolonial desde Aimé Césaire hasta los Zapatistas. In S. Castro-Gómez and Ramón Grosfoguel (eds) *El giro decolonial: Reflexiones para una diversidad epistémica más allá del capitalismo global*. Bogotá: Iesco-Pensar-Siglo del Hombre Editores, pp. 63–78.
Hall, S. (1996) When was "the post-colonial"? Thinking at the limit. In I. Chambers and L. Curti (eds) *The Post-Colonial Question*. London: Routledge, pp. 242–260.
Hochschild, A. (2000) Global care chains and emotional surplus value. In W. Hutton and A. Giddens (eds) *On the Edge: Living with Global Capitalism*. London: Jonathan Cape, pp. 130–146.
hooks, b. (1984) *Feminist Theory from Margin to Centre*. Boston, MA: South End Press.

Kofman, E. (2000) Beyond a reductionist analysis of female migrants in global European cities: The unskilled, deskilled and professional. In M. H. Marchand and A. Sisson Runyan (eds) *Gender and Global Restructuring: Sightings, Sites and Resistances*. London; New York: Routledge, pp. 129–139.

Kontopoulos, K. (1993) *The Logics of Social Structure*. Cambridge: Cambridge University Press.

Laqueur, T. (1990) *Making Sex: Body and Gender from the Greeks to Freud*. Cambridge, MA: Harvard University Press.

Lucena Samoral, M. (1966) Bardaje entre una tribu guahibo del Tomo. *Revista Colombiana de Antropología* 14: 261–266.

Lugones, M. (2008) Colonialidad y género. *Tabula Rasa* 9: 73–101.

———. (2014) Colonialidad y género. In Y. Espinosa, D. Gómez Correal and K. Ochoa (eds) *Tejiendo de Otro Modo: Feminismo, epistemología y apuestas descoloniales en Abya Yala*. Popayán: Universidad del Cauca, pp. 57–74.

Maldonado-Torres, N. (2007) Sobre la colonialidad del ser: contribuciones al desarrollo de un concepto. In S. Castro-Gómez and R. Grosfoguel (eds) *El giro decolonial. Reflexiones para una diversidad epistémica más allá del capitalismo global*. Bogotá: Iesco-Pensar-Siglo del Hombre Editores, pp. 127–167.

Masiello, F. (1997) Las mujeres como agentes dobles en la historia. *Debate Feminista* 16: 251–271.

Mbembe, A. (2011) *Necropolítica*. Trans. Elisabeth Falomir Archambault. Madrid: Melusina.

Mendoza, B. (2014) La epistemología del sur, la colonialidad del género y el feminismo latinoamericano. In Espinosa Yuderkys, Gómez Correal Diana and Ochoa Karina (eds) *Tejiendo de Otro Modo: Feminismo, epistemología y apuestas descoloniales en Abya Yala*. Popayán: Universidad del Cauca, pp. 91–104.

Mignolo, W. (2007) El pensamiento decolonial: desprendimiento y apertura. Un manifiesto. In S. Castro-Gómez and Ramón Grosfoguel (eds) *El giro decolonial: Reflexiones para una diversidad epistémica más allá del capitalismo global*. Bogotá: Iesco-Pensar-Siglo del Hombre Editores, pp. 25–46.

Mohanty, C. T. (1991) Under Western eyes: Feminist scholarship and colonial discourses. In C. Mohanty, A. Russo and L. Torres (eds) *Third World Women and the Politics of Feminism*. Bloomington: Indiana University Press, pp. 51–80.

Preciado, B. (2008) *Testo Yonky*. Madrid: Espasa Calpe.

Quijano, A. (2000) Coloniality of power, Eurocentrism, and Latin America. *Neplanta* 1(3): 533–580.

Ramírez A. L. and Castellanos, E. (2013) Autorizar una voz para desautorizar un cuerpo: producción discursiva del lesbianismo feminista official. *Íconos. Revista de Ciencias Sociales* 45: 41–57.

Rivera Cusicanqui, S. (1993) La raíz: colonizadores y colonizados. In S. Rivera Cusicanqui and R. Barrios (eds) *Violencias encubiertas en Bolivia*. La Paz: CIPCA; Aruwiyiri, pp. 25–139.

———. (2010a) Ch'ixinakax utxiwa: una reflexión sobre prácticas y discursos descolonizadores – 1ª Ed-Buenos Aires: Tinta Limón.

———. (2010b) *Violencias (re)encubiertas en Bolivia*. La Paz: Ediciones La Mirada Salvaje.

———. (2014) La noción de 'derecho' o las paradojas de la modernidad postcolonial: indígenas y mujeres en Bolivia. In Y. Espinosa, D. Gómez Correal and K. Ochoa (eds) *Tejiendo de Otro Modo: Feminismo, epistemología y apuestas descoloniales en Abya Yala*. Popayán: Universidad del Cauca, pp. 91–104.

Roscoe, W. (1988) *Strange Country This: Images of Berdaches and Warrior Women in Living the Spirit: A Gay American Indian Anthology*. New York: St. Martins Press.

Rubin, G. (1996) El tráfico de mujeres: notas sobre la economía política del sexo. In M. Lamas (ed) *El género la construcción cultural de la diferencia sexual*. México: PUEG, pp. 35–96.

Santos, B. de S. (2015) *Revueltas de indignación y otras conversas*. La Paz: Alice-Intermon Oxfam.

Sassen, S. (2004) Global cities and survival circuits. In B. Ehrenreich and A. R. Hochschild (eds) *Global Woman: Nannies, Maids, and Sex Workers in the New Economy*. New York: Metropolitan Books, pp. 254–274.

Somerville, S (2000) *Queering the Color Line: Race and the Invention of Homosexuality in American Culture*. Durham, NC: Duke University Press.

Wekker, G. (2006) *The Politics of Passion. Women's Sexual Culture in the Afro-Surinamese Diaspora*. New York: Columbia University Press.

Williams, W. (1992) *The Spirit and the Flesh, Sexual Diversity in the American Indian Culture*. Boston, MA: Beacon Press.

Wittig, M. (1992) *Straight Mind and Other Essays*. Boston, MA: Beacon Press.

5

POST-DEVELOPMENT

Aram Ziai

Introduction

During the last two decades, a novel and controversial approach in development theory has been discussed, but also been increasingly accepted in academic debate: the post-development (PD) school. Whereas earlier criticisms, such as the dependency school, had criticized development theory and policy usually with a view to devising better theories and policies of development (e.g. Hayter, 1971; Kay, this volume), the post-development school explicitly refused to do so, engaging in destructive instead of constructive criticism. In its first landmark publication, Gustavo Esteva (1992: 6) called development an "unburied corpse . . . from which every kind of pest has started to spread." And Wolfgang Sachs (1992b: 1), in the introduction to the volume, proclaimed that "[t]he idea of development stands like a ruin in the intellectual landscape" and found that "the time is ripe to write its obituary." Post-development can thus be seen as a fundamental critique, one which intended to lay to rest 'development' and called for **'alternatives to development'** instead of **'alternative development'** (Escobar, 1995: 215). In this chapter, I explore what the authors meant by that and why they opposed the concept and practice of development so vehemently. First, I deal with the historical origins of the approach before engaging a number of its central arguments and the alternatives it proposes. At the end of this chapter, I also discuss some criticisms which have been raised in opposition to post-development and reflect on the importance of this school of thought.

The origins of post-development

While the first post-development publications emerged in the 1980s (Esteva, 1985; Escobar, 1985; Rahnema, 1985; Rist/Sabelli, 1986; Latouche, 1986), the approach has been influenced by three earlier bodies of work: the writings of Ivan Illich, Michel Foucault, and anti-colonial writers like Mohandas Gandhi and Frantz Fanon. In order to better understand and contextualize the approach, a look at their arguments is useful. All of them unsettle the notion that 'development' is a good thing, and that the 'less developed societies' should become 'developed.'

During the 1960s and 70s, Ivan Illich (1971, 1997), an Austrian theologian, criticized the institutions of Western industrial modernity because they would teach people to be dependent: dependent on doctors for healing, dependent on the school for education, dependent on the church for faith and spirituality. Concerning development aid in Latin America, he argued that an understanding of progress as the spread of these institutions was counterproductive, because it produced needs that could not be fulfilled for the majority of the population (e.g. for Coca-Cola, advanced surgery, and high school education) while neglecting goods and services more suited to their situation (e.g. vehicles which can handle rough terrain and can easily be repaired, clean water, healing assistants, communal storage, and public transport). In terms of technology, not only goods produced as commodities were problematic (i.e. those which are produced to be sold on the market for profit – a view common among socialists), but all goods which would either come with the price of dependency on experts or which could be provided only for a privileged minority. Development policy would thus damage institutions over which people themselves have control and increasingly subject them to institutions which threatened their autonomy and produced poverty in the sense of unfulfilled needs and dependence. Illich outlined a politics of 'conviviality,' based on tools and institutions which enhance people's autonomy and self-help capacity without having these drawbacks (e.g. bicycles, phones, mail), as well as on restraint in energy consumption.

The writings of French philosopher and historian Michel Foucault were hardly concerned with issues of the Global South. In his most popular writings, he deals with questions of knowledge and power, and with the construction of what counts as normal and true. He shows that what is accepted as true – even in the sciences – is dependent on the historical, social, and political contexts, on a 'regime of truth' of this society and epoch. In this way he cautions us that what is currently seen as true and acceptable is not so irrespective of history and place: it may easily be seen and mad and monstrous in the next century – or in a different discourse. The concept of discourse, understood here as a system of representation which is linked to relations of power and has consequences for behaviour, is used by many PD scholars.

While the writings of Mohandas Ghandi (1909) and Frantz Fanon (1961) were concerned with different struggles of anti-colonial resistance (against British rule in India and French rule in Algeria) and they held opposing views concerning the question of violence, both were agreed on one point: the assumed cultural or civilizational superiority of Europe was based on Eurocentric conceit and it would be a grave mistake if the aspirations of anti-colonial movements were confined to reproducing the models of state and society of the colonizers.

A fundamental critique of 'development'

What distinguished PD from previous critiques in development studies was that it did not intend to improve the attempts to bring about 'development,' but questioned this very objective, advocated the "rejection of the entire paradigm," and instead called for "alternatives to development" (Escobar, 1995: 215). But what exactly is meant by 'development' in this context? A close look reveals that the signifier is linked to different (yet connected) signifieds in PD. (1) 'Development' is criticized as an ideology of the West, promising material affluence to decolonizing countries in Africa, Asia, and Latin America in order to prevent them from joining the communist camp and maintaining a colonial division of labour (Rahnema, 1997: 379). (2) It is criticized as a failed project that seeks to universalize the way of life of the 'developed' countries on a global scale which has for the overwhelming majority of affected people led to the "progressive modernization of poverty" (Esteva, 1985: 79). (3) It is criticized as a Eurocentric and hierarchical construct defining non-Western, non-modern, non-industrialized ways of life as inferior and in need of 'development' (in the sense of social change as has occurred in the 'developed' countries) (Esteva, 1992: 7). (4) It is criticized as an economic rationality centred around accumulation, a capitalist logic of privileging activities earning money through the market (and devaluing all other forms of social existence), and the idea of the homo economicus (whose needs for consumption are infinite) (Esteva, 1992: 7–19). (5) It is criticized as a concept legitimizing interventions into the lives of people defined as 'less developed' as justified in the name of a higher, evolutionary goal or simply the common good defined by people claiming expert knowledge (Sachs, 1999: 7).

But why does PD argue that the epoch of 'development' has ended? Sachs gives four reasons: (1) The evidence that the ecological consequences of the 'developed' way of life in terms of resource consumption and environmental destruction were such that this way cannot be treated as a model to be replicated. (2) After the end of the Cold War there was no geopolitical motivation for the West to uphold the 'promise of development' to countries in the South. (3) The increasing gap between rich and poor countries during the 'development decades' had rendered the promise implausible. (4) More and more people had realized that successful 'development' entailed the Westernization of the world, leading to a global monoculture (Sachs, 1992b: 2–4).

According to the post-development school, 'alternative development' is not enough, because it reproduces the idea that the majority of the world's population is 'underdeveloped' and needs to live like the West. So what about the 'alternatives to development' that PD envisions? Esteva (1992: 21) claims that because of the exclusion produced by the project of 'development,' ordinary men and women would recover "their own definition of needs" as well as "autonomous ways of living." They would be creating alternatives to 'development,' i.e. to the universal Western models of the economy, politics, and knowledge, and reclaiming the commons, often falling back on indigenous or traditional concepts and practices. For him, the Zapatistas in Chiapas/Mexico provide a good example of such alternatives. Escobar (2008) sees them in the Process of Black Communities (PCN) in the Pacific region of Colombia. But when looking closely enough, one realizes that alternative concepts of what a good society looks like and alternative practices of organizing it exist in almost every society: *buen vivir* in Ecuador and Bolivia, Ubuntu in South Africa, Swaraj in India, Gharbzadegi in Iran, or Décroissance in France are some of the most well-known examples. (For a cautious attempt at a limited comparison, see Ziai, 2015).

Core arguments of post-development

The first core argument of PD is that 'underdevelopment' was 'invented' by US president Truman in his inaugural address in 1949, as part of a political campaign to maintain or increase Western influence in Africa, Asia, and Latin America (now dubbed 'underdeveloped areas') in the context of the Cold War and processes of decolonization. This was to be done through a 'program of development' consisting of investments, technical progress, and aid and promising improvements in the standard of living (*The invention of underdevelopment*) (Esteva, 1992; Escobar, 1995; Rist, 1997: 69–72). The thrust of this argument is that we must acknowledge the historicity of the concept and locate it in the geopolitical context in which it emerged, leading some to talk about an 'ideology of development' which legitimizes neocolonialism (Rahnema, 1997: ix).

Another central argument is that the content of the concept of 'development' is ambiguous, contested, and far from clear. Because in the second half of the 20th century the concept underwent frequent redefinitions and was used to refer to just about any kind of measure officially intended to improve people's lives, from building roads to economic reforms, environmental protection or empowering women, PD authors argue, its contours are blurred and it has become a shapeless, amoeba-like concept (*Development as amoeba*) (Sachs, 1992b: 1–5; Esteva, 1985: 79).

The subtitle of the development dictionary (Sachs, 1992a) indicates another central argument: that knowledge about and representations of the social world are not neutral, but have a certain perspective and imply relations of power. Knowledge about 'development' therefore always implies a claim on how other ('underdeveloped') people should live, how their lives can be improved, and thus a justification of intervention (*Knowledge as power*).[1] Some PD writers go further and argue that these interventions in the context of development cooperation suggest that problems of poverty can be solved by technical knowledge, ignoring relations of power and oppression and thus depoliticizing poverty (Ferguson, 1994: 256; Escobar, 1995: 143; Sachs, 1990: 9). And the – very few – feminist PD authors add that this universal and technical developmentalist knowledge was not neutral but closely related to patriarchy and its dichotomous thinking (Shiva, 1992, see also Shiva, 1989; Mies and Shiva, 1993; Saunders, 2002).

By characterizing the majority of the world population as primarily 'underdeveloped,' the West (comprised mainly of Europe and its former settler colonies in North America, Australia, and New Zealand) defined its own society as standing at the top of a universal scale of progress and other societies as deficient, perceiving difference merely as backwardness and disregarding non-Western cultures, contexts, and conceptions of a good society. This is possible only by employing Eurocentric standards as universal standards (*Eurocentrism*).[2]

Many PD writers claim that the promises of 'development' turned out to be empty, and that the processes, projects, and programmes undertaken in its name did not lead to an improvement in standards of living, but to increased inequality, experiences of exclusion, and impoverishment at least for significant parts of the population (*Development as impoverishment*). The main reason is that subsistence economies were undermined or destroyed, while the modern cash economy did not provide secure livelihoods for all (Sachs, 1992b: 3; Esteva, 1992: 13; Escobar, 1995: 4; Rahnema, 1997: x). Again, PD feminists have pointed out that women suffered most from the enclosure of the commons and the 'war against subsistence' (Bennholdt-Thomsen and Mies, 1999; Sittirak, 1998).

According to some PD authors, the worldview of economics and its notions of markets, production, and labour should be seen as a contingent expression of cultural values, not as a universal science (*Economics as culture*). They argue that this worldview is focused on growth, productivity, and the satisfaction of infinite needs for material goods and based on the model of economic man (homo economicus), but that it is very short-sighted or ignorant of different kinds of economic activity. For example, it overlooks the relation to mother earth or Pachamama, one's ancestors, or the community which often play an important role in agricultural labour and determine why and how people work. It also ignores the idea of sufficiency, the idea of having enough and requiring no more goods, or ideas of hospitality, solidarity, and sharing, which are at odds with the maximization of personal gain. They criticize the reductionist view of work and on what is perceived as valuable and productive and what is not (Esteva, 1992: 17–20; Escobar, 1995; Sachs, 1990: 18–22).

The last central argument of PD concerns the claim that an increasing number of people in the Global South, disappointed by the promise and excluded from the project of 'development,' resist Westernization, reject this economic worldview, engage in alternatives, and turn to models of the economy, politics, and knowledge based on local culture and difference or at least on hybrid models, that are not striving to catch up with the 'developed' countries (*Alternatives to Development*)(Esteva, 1992: 20–22; Escobar, 1995: ch. 6). In their portrayal of local alliances between municipalities and civil society in Northern Léon in Nicaragua, Cupples, Glynn, and Larios (2007) give an example of such hybrid models.

Noble savages: the debate on PD

In the ensuing debate[3] PD has been confronted with a number of criticisms, the most important of which are reiterated here. The first concerns the unconditional rejection of modernity and 'development.' This criticism urges us not to overlooks the successes and emancipatory dimensions that what is often referred to as 'development' has brought, such as the rise in life expectancy and reduction in child mortality throughout the so-called Third World. Instead of throwing out the baby with the bath water, these critics argue that it is necessary to differentiate between the positive and negative consequences of development projects and modern science – as some of them did improve the lives of people in the South (Corbridge, 1998: 144).

The second criticism is focused on the idea of noble savages and the ways that the 'alternatives to development' are often romanticized. The picture of traditional subsistence communities drawn by PD was seen as romanticizing because it ignores or downplays relations of domination and exploitation in these communities and because it assumes that their inhabitants are not interested in Westernization and material goods. In this respect, PD indeed provided the 'last refuge of the noble savage,' while evidence indicated that the marginalized in the South were frequently more interested in 'access to development,' and less in its rejection (Kiely, 1999: 44; Storey, 2000: 42).

The third criticism concerns the question of cultural relativism. The relativist rejection of allegedly Eurocentric universal standards of a good society in PD leads, in the view of some critics, to indifference towards oppression and misery and prevents critique 'from outside.' The promotion of traditional culture – conceived in essentialist terms as static – allows local elites to frame their dismissal of modern practices that challenge their privileges as anti-imperialist (Knippenberg and Schuurman, 1994: 95; Nanda, 1999: 11).

The fourth criticism revolves around the idea of paternalism. PD has also been criticized for preaching an ethics of sufficiency from a paternalist affluent perspective, claiming to know better about the needs of the poor (and their legitimacy) than they themselves. So in the end, again a certain model of society (traditional subsistence communities) is prescribed as the right one (Knippenberg and Schuurman, 1994: 95; Cowen and Shenton, 1996: 69).

An opposing criticism has also been voiced in terms of Pontius-Pilate politics. PD has been accused of failing to provide any concrete alternatives, there was 'critique but no construction' (Nederveen Pieterse, 1998: 366). Instead of a political programme, there was merely political agnosticism confined to supporting social movements in the South, thus washing its hands of political decisions and engaging in Pontius-Pilate-politics (Nederveen Pieterse, 1998: 366; Nederveen Pieterse, 2000: 182; Kiely, 1999: 45f).

The final criticism is concerned with methodological deficits in the employment and selective application of discourse analysis, related to an attitude exhibited towards grassroots movements. The emphasis on differences and discontinuities so central in Foucault's work was replaced by essentialisms and the construction of a monolithic discourse of 'development' in PD, turning discourse analysis from a method into an ideological platform (Nederveen Pieterse, 1998: 363; Nederveen Pieterse, 2000: 180).

While here is not the place to thoroughly discuss these arguments, a more in-depth-analysis concluded that all of these criticisms were justified in regard to some PD texts, but not all (Ziai, 2004). It is therefore useful to make clear that not all variants of PD reject all elements of modernity and many endorse cultural hybridization. Sceptical PD scholarship is critical of cultural traditions and much of it abstains from articulating desirable models of society, employing instead a dynamic, constructivist concept of culture. Neo-populist PD, however, does reject modern industrial society altogether and promotes the return to (often idealized) subsistence communities, employing an essentialist concept of culture. Whereas sceptical PD thus leads to a radical democratic position, neo-populist PD potentially has reactionary consequences, as it is able to dismiss people's desire for 'development' as the result of ideology and manipulation, based on privileged knowledge of their 'real' needs – and aligning themselves to a position that is dangerously close to that of the 'development experts' they criticize so sharply. It is therefore more valuable to talk about a sceptical and a neo-populist PD discourse (see Ziai, 2004).

One last point of criticism has to be mentioned, as it is both significant and (in this form) rarely mentioned. It concerns the agency of the South and its ability to provoke and appropriate discourses, as Frederick Cooper (1997: 84) argues: while "development ideology was originally supposed to sustain empire, not facilitate the transfer of power," the discourse still provided "something trade union and political leaders in Africa could engage with, appropriate, and turn back." Therefore "[m]uch as one can read the universalism of development discourse as a form of European particularism imposed abroad, it could also be read . . . as a rejection of the fundamental premises of colonial rule, a firm assertion of people of all races to participate in global politics and lay claim to a globally defined standard of living" (Cooper, 1997: 84). It is noteworthy that this element, the vision of global social equality, has been quietly buried with the disillusionment over the 'lost decade of development,' the ascent of neoliberalism, and the abandonment of the 'development' promise in the neoliberal discourse of 'globalisation.' However, as Cupples, Glynn, and Larios (2007) in their study on Nicaragua demonstrate, there is an ongoing contestation over the meaning of 'development' and poverty reduction and an appropriation of these terms through place-based social movements even in a period of ongoing neoliberal hegemony.

Slaying the 'development monster': PD's response to the criticisms

Although post-development authors concede that their proclamation about the end of the era of 'development' was premature, they reaffirm the necessity of performing a "decolonization of the imagination" (Rist, 2012: 273; Sachs, 2010: vi, ix; Escobar, 2012: vii). Responding to the criticism that post-development had presented 'development' as monolithic while in fact it was heterogeneous and contested, Escobar concedes that the critics were right. But, he continues, they "fail to acknowledge . . . that their own project of analysing the contestation of development on the ground was in great part made possible by the deconstruction of development discourse." Post-development's project had been to 'slay the development monster,' i.e. to break the consensus about 'development' being necessary, self-evident, positive, and unquestionable and thus pave the way for more nuanced analyses. Romanticization, however, was an accusation used against any and all visions of societies which transcend the current hegemonic model (Escobar, 2000: 12f).

Sachs (2010: viii) also engages with the idea of contestation and admits:

> we had not really appreciated the extent to which the development idea has been charged with hopes for redress and self-affirmation. It certainly was an invention of the West, as we showed at length, but not just an imposition on the rest. On the contrary, as the desire for recognition and equity is framed in terms of the civilizational model of the powerful nations, the South has emerged as the staunchest defender of development.

This resonates with Cooper's criticism mentioned above concerning the discourse of 'development' that could also (!) be read as a discourse of rights.

Another interesting response comes from Rist (2012) who engages with the criticism of cultural relativism and support for groups who disrespect human rights. He argues:

> It may well be true that certain movements which oppose 'development' have scant regard for certain articles in the Declaration of Human Rights; or that they force boys to look after goats instead of going to school; or that, as in the case of our grandmothers

in Europe, they do not allow women to go out of the house 'bareheaded'. Neverthe-less, if respect for the values linked to modernity is the only criterion for judging the social order, what should be said of our own society, which amid general indifference is increasing the numbers of those excluded in the name of economic growth? And what of the wars that cause countless victims, especially civilians, in the name of democracy and human rights?

(Rist, 2012: 276)

The argument can of course be interpreted as countering the accusation of abusive prac-tices in one group with a reference to abusive practices in some other group – not an entirely convincing argument. However, it could also be interpreted not as indifference towards abusive practices, but as a refusal to attribute these practices to the non-Western Other while exempting the Western Self, allowing us to affirm our 'civilisational superiority'.

The contribution of post-development to development studies[4]

The contribution of PD is twofold: first in a critique of different kinds of power relations and second in its potential contribution to a non-Eurocentric and more power-sensitive theory of positive social change. PD points to the naturalization of the universal scale: the first and maybe most fundamental achievement of PD is the insight that the categories and strategies of 'development' imply a certain perspective which is contingent – in contrast to being the natural and normal way of seeing things. That societies can be compared according to their 'level of development,' that there are 'developed' and 'less developed' countries, and that the latter can be found in Africa, Asia, and Latin America and are in need of 'development,' development experts, development projects, and development aid provided by the former, are assumptions that are by no means self-evident (Escobar, 1995: 12, 39). They constitute a historically specific way of looking at different societies and global inequality, naturalizing the norms and historical processes of the European Self.

Second, it reveals the process of othering and the problematization of deviance. The natu-ralization of the Self enables the problematization of the Other as deviant. The universal scale allows us to measure and compare according to a Eurocentric norm and to thus define the majority of humanity as 'underdeveloped.' The Other is not seen as different, but as a deficient version of the Self, which is why development discourse operates by identifying deviance from the norm as inferiority ('underdevelopment,' 'illiterate,' 'unemployed,' etc.) (Escobar, 1995: 41).

It also shows how the promise of betterment functions as a mechanism of legitimation. Even if the prescriptions of development discourse do not work, the development industry claims to have learnt from its earlier mistakes, reaffirming its knowledge on how to improve other people's lives, but now rendering visible a new aspect hitherto neglected (poverty, basic needs, women, the environment, the market, good governance, ownership, and so on). Failures thus lead only to a reformulation of the promise (Sachs, 1990: 6; Escobar, 1995: 58).

Furthermore, it points to the hierarchization of different types of knowledge (and some-times also cultures and values), with one type (universally applicable expert knowledge, typically claimed by trustees) being privileged and the other (local, 'unscientific' knowledge) denigrated (DuBois, 1991: 7).

Additionally, PD highlights the depoliticization which often takes place in the discourse of 'development.' The discourse of 'development,' at least the one employed by most develop-ment agencies, assumes that 'development' is something that benefits everyone and therefore no one can object to, something removed from conflicts over political and economic questions, portraying positive social change as a technical matter related exclusively to the presence or absence of knowledge, technology, and capital. Simply put, this discourse wants to help the poor without hurting the rich (on a national and international level). It has to do so in order to gain support and legitimacy, but in doing so neglects an analysis of the structural causes of poverty and depoliticizes the conflicts and divisions in society (Ferguson, 1994: 256; Escobar, 1995: 45).

Finally, PD's rather small feminist current has revealed that 'development' is in fact not gender-neutral: its privileging of rationality, productivity, technology, and mastery over nature has clearly masculine connotations. According to Shiva, the "treatment of nature as a resource which acquires value only in exploitation for economic growth has been central to the project of development" and is certainly not unrelated to the "fundamental dichotomizing between male and female, mind and matter, objective and subjective, the rational and the emotional" which has been a central feature of European modernity at least since Bacon (Shiva, 1992: 211).

This critical potential of PD can be used also by persons not sharing PD's commitment to 'alternatives to development,' which is why it gained so much currency in the discipline (see Ziai, 2017). However, PD's full potential unfolds only if these alternatives are taken seriously, allowing it to contribute to a reformulation of a theory of emancipatory social change beyond 'development.' The problem of development theory's impasse lies in the fact that it has not sufficiently dealt with the implications of pluralism. Nederveen Pieterse (2010: 214) correctly remarks: "The idea of development as a single forward path . . . or generalizing across developing societies lies well behind us." And he is also right in stating that since the crisis, "[m]ainstream approaches have coopted elements of alternative development like participation" (Nederveen Pieterse, 2010: 184). But the implication of this rejection of 'development as a single path' has not been fully recognized. Kothari and Minogue (2002: 9) agree with Nederveen Pieterse that "forms of alternative development have become institutionalized as part of mainstream development," but argue that this type of alternative development "does not redefine development, but instead questions its modalities, agency and procedures. . . . It is still ultimately about the achievement of Western modernity by developing and transitional countries." This means that alternative development merely looks for different roads to arrive at the same goal. However, if this goal is unambiguous and defined by modern, industrial capitalist societies, then even alternative development remains firmly grounded in the Western, or more precisely: hegemonic models of politics (nation-state and liberal democracy), the economy (neoliberal, globalized capitalism), and knowledge (Western science). Thus we are still assuming a single conception of a good society and the potential of non-Western alternatives to these models to improve human well-being remains untapped. If we take the imperative of 'development pluralism' (Nederveen Pieterse, 2010: 214) seriously, we need to consider these non-Western alternatives as well in order to redefine development, taking into account not only different paths to modernity but different ideas of a good life altogether. This is why the Post-Development approach which questions these hegemonic models and promotes non-Western alternatives could play a central role in a reinvention of development theory beyond the impasse.

Of course, such a theory would have to embrace the plurality of knowledges (being aware of its limits and of other ways of knowing) and operate in the words of Santos (2014: 44) as a 'rearguard theory' – theorizing the practice of social movements instead of prescribing it like vanguard theories use to do. However, PD can be used for different ends: for criticizing discourses and practices imbued with Eurocentrism and relations of power and thinking about global inequality beyond the discourse of 'development,' or for theorizing contemporary struggles and envisioning different futures based on non-capitalist values, communal ownership, and a more humble relation of human beings to nature. While the strong demand for more pragmatic solutions also needs to be accommodated, in the light of climate change, resource depletion, and a world economy still structured by neoliberalism, the turn towards convivial, low-energy local alternatives to global capitalism (we might call it a radical PD position) is far from absurd.

Notes

1 "Though development has no content, it does possess one function: it allows any intervention to be sanctified in the name of a higher goal" (Sachs, 1992b: 4). Li (2007) talks about the 'will to improve' in this context, Cowen and Shenton (1996) about 'trusteeship.'
2 "development cannot be separated from the idea that all peoples of the planet are moving along one single track towards maturity, exemplified by the nations 'running in front.' In this view, Tuaregs, Zapotecos or Rajashtanis are not seen as living diverse and non-comparable ways of human existence, but as somehow lacking in terms of what has been achieved by the advanced countries" (Sachs, 1992b: 3). "The metaphor of development gave global hegemony to a purely Western genealogy of history, robbing peoples of different cultures of the opportunity to define the forms of their social life" (Esteva, 1992: 9).
3 For an overview over the debate see Simon (2006), Ziai (2007), McGregor (2009), Matthews (2010), Sidaway (2013), Gudynas (2017). For a general introduction to PD see Ziai (2016a).
4 For a longer version of the arguments in the first part of this section see Ziai (2016b), Ch. 15.

References

Bennholdt-Thomsen, V. and Mies, M. (1999) *The Subsistence Perspective: Beyond the Globalised Economy*. London: Zed Books.

Cooper, F. (1997) Modernizing bureaucrats, backward Africans, and the development concept. In F. Cooper and R. Packard (eds) *International Development and the Social Sciences: Essays on the History and Politics of Knowledge*. Berkeley, CA and Los Angeles: University of California Press, pp. 64–92.

Corbridge, S. (1998) 'Beneath the pavement only soil': The poverty of post-development. *Journal of Development Studies* 34(6): 138–148.

Cowen, M. and Shenton, R. (1996) *Doctrines of Development*. London: Routledge.

Cupples, J., Glynn, K. and Larios, I. (2007) Hybrid cultures of postdevelopment: The struggle for popular hegemony in rural Nicaragua. *Annals of the Association of American Geographers* 97(4): 786–801.

DuBois, M. (1991) The governance of the Third World: A Foucauldian perspective on power relations in development. *Alternatives* 16(1): 1–30.

Escobar, A. (1985) Discourse and power in development: Michel Foucault and the relevance of his work to the Third World. *Alternatives* 10: 377–400.

———. (1995) *Encountering Development: The Making and Unmaking of the Third World*. Princeton, NJ: Princeton University Press.

———. (2000) Beyond the search for a paradigm? Post-development and beyond. *Development* 43: 11–14.

———. (2008) *Territories of Difference: Place, Movements, Life, Redes*. Durham, NC: Duke University Press.

———. (2012) *Encountering Development: The Making and Unmaking of the Third World*. 2nd Ed. Princeton, NJ: Princeton University Press.

Esteva, G. (1985) Development: Metaphor, myth, threat. *Development: Seeds of Change* 3: 78–79.

———. (1992) Development. In W. Sachs (ed) *The Development Dictionary: A Guide to Knowledge as Power*. London: Zed Books, pp. 6–25.

Fanon, F. (1961) *The Wretched of the Earth*. London: MacGibbon and Kee.

Ferguson, J. (1994) *The Anti-Politics Machine: 'Development', Depoliticization and Bureaucratic Power in Lesotho*. Minneapolis: University of Minnesota Press.

Gandhi, M. (1909) *Indian Home Rule or Hind Swaraj*. Ahmedabad: Navajivan Publishing House.

Gudynas, E. (2017) Post-development and other critiques of the roots of development. In H. Veltmeyer and P. Bowles (eds) *The Essential Guide to Critical Development Studies*. London: Routledge, pp. 84–93.

Hayter, T. (1971) *Aid as Imperialism*. Harmondsworth: Penguin Books.

Illich, I. (1971) *Celebration of Awareness*. London: Marion Boyars.

———. (1997) Development as planned poverty. In M. Rahnema and V. Bawtree (eds) *The Post-Development Reader*. London: Zed Books, 94–101.

Kiely, R. (1999) The last refuge of the noble savage? A critical assessment of post-development theory. *The European Journal of Development Research* 11(1): 30–55.

Knippenberg, L. and Schuurmann, F. (1994) Blinded by rainbows: Anti-modernist and modernist deconstructions of development. In F. Schuurman (ed) *Current Issues in Development Studies: Global Aspects of Agency and Structure, Nijmegen Studies in Development and Social Change Vol.21*. Saarbruecken: Verlag für Entwicklungspolitik Breitenbach, pp. 90–106.

Kothari, U. and Minogue, M. (2002) (eds) *Development Theory and Practice: Critical Perspectives*. Houndmills: Palgrave Macmillan.

Latouche, S. (1986) *Faut-il réfuser le développement? Essai sur l'anti-économique du tiers monde*. Paris: Presses Universitaires de France.

Li, T. (2007) *The Will to Improve: Governmentality, Development, and the Practice of Politics*. Durham, NC: Duke University Press.

Matthews, S. (2010) Postdevelopment theory. In *Oxford Research Encyclopaedia of International Studies*, online. http://internationalstudies.oxfordre.com/view/10.1093/acrefore/9780190846626.001.0001/acrefore-9780190846626-e-39

McGregor, A. (2009) New possibilities? Shifts in post-development theory and practice. *Geography Compass* 3(5): 1688–1702.

Mies, M. and Shiva, V. (1993) *Ecofeminism*. London: Zed Books.

Nanda, M. (1999) Who needs post-development? Discourses of difference, green revolution and agrarian populism in India. *Journal of Developing Societies* 15(1): 5–31.

Nederveen Pieterse, J. (1998) My paradigm or yours? Alternative development, post-development, reflexive development. *Development and Change* 29: 343–373.

———. (2000) After post-development. *Third World Quarterly* 20(1): 175–191.

———. (2010) *Development Theory: Deconstructions/Reconstructions*. 2nd eE. London: Sage.

Rahnema, M. (1985) NGOs – Sifting the wheat from the caff. *Development: Seeds of Change* 3: 68–71.

———. (1997) Introduction. In M. Rahnema and V. Bawtree (eds) *The Post-Development Reader*. London: Zed Books, pp. ix–xix.

Rahnema, M. and Bawtree, V. (1997) (eds) *The Post-Development Reader*. London: Zed Books.

Rist, G. (1997) *The History of Development: From Western Origins to Global Faith*. London: Zed Books.

———. (2012) *The History of Development: From Western Origins to Global Faith*. 4th Ed. London: Zed Books.

———. and Sabelli, F. (1986) (eds) *Il y était une fois le développement. . . .* Lausanne: editions d'en bas.

Sachs, W. (1990) The archaeology of the development idea. *Interculture* 23(4), Nr. 109: 1–37.

———. (1992a) (ed) *The Development Dictionary: A Guide to Knowledge as Power*. London: Zed Books.

———. (1992b) Introduction. In W. Sachs (ed) *The Development Dictionary: A Guide to Knowledge as Power*. London: Zed Books, pp. 1–5.

———. (1999) *Planet Dialectics: Explorations in Environment and Development*. London: Zed Books.

———. (2010) (ed) *The Development Dictionary: A Guide to Knowledge as Power*. 2nd Ed. London: Zed Books.

Santos, B. D. S. (2014) *Epistemologies of the South: Justice against Epistemicide*. Boulder, CO: Paradigm Publishers.

Saunders, K. (2002) (ed) *Feminist Post-Development Thought: Rethinking Modernity, Post-Colonialism and Representation*. London: Zed Books.

Shiva, V. (1989) *Staying Alive: Women, Ecology, and Development*. London: Zed Books.

———. (1992) Resources. In W. Sachs (ed) *The Development Dictionary: A Guide to Knowledge as Power*. London: Zed Books, pp. 206–218.

Sidaway, J. (2013) Post-Development. In V. Desai and R. Potter (eds) *The Companion to Development Studies*. 2nd Ed. London: Routledge, pp. 16–20.

Simon, D. (2006) Separated by common ground? Bringing (post)development and (post)colonialism together. *The Geographical Journal* 172(1): 10–21.

Sittirak, S. (1998) *The Daughters of Development: Women in a Changing Environment*. London: Zed Books.

Storey, A. (2000) Post-development theory: Romanticism and Pontius Pilate politics. *Development (SID)* 43(4): 40–46.

Ziai, A. (2004) The ambivalence of post-development: Between reactionary populism and radical democracy. *Third World Quarterly* 25(6): 1045–1061.

———. (2007) (ed) *Exploring Post-Development: Theory and Practice, Problems and Perspectives*. London: Routledge.

———. (2015) Post-development concepts? Buen Vivir, Ubuntu and Degrowth. In B. D. S. Santos and T. Cunha (eds) *Epistemologies of the South: South-South, South-North and North-South Global Learnings. Other Economies*. Coimbra: CES, pp. 143–154.

———. (2016a) Post-development and alternatives to development. In P. Haslam and J. Schafer and P. Beaudet (eds) *Introduction to International Development: Approaches, Actors, and Issues*. 3rd Ed. Oxford: Oxford University Press, pp. 65–83.

———. (2016b) *Development Discourse and Global History: From Colonialism to the Sustainable Development Goals*. London: Routledge.

———. (2017) "I am not a postdevelopmentalist, but . . ." – The influence of post-development on development studies. *Third World Quarterly* 38(12): 2719–2734.

6

NEOLIBERAL MULTICULTURALISM

Charles R. Hale

Introduction

The concept of "neoliberal multiculturalism" has been in circulation for two decades, and the phenomenon it describes for perhaps a decade more. In my reckoning, Slavok Zizek can be credited with the first iteration in print (1997), followed by Speed and Collier (2000), though neither used that precise wording. By the mid-2000s, a number of scholars, myself included, were using the term, in many cases influenced by one another, although possibly also through independent invention (Hale, 2002; Hale, 2005; Hale and Millaman, 2006); Melamed, 2006; Overmyer-Velázquez, 2010; Postero, 2007; Rivera Cusicanqui, 2004; Speed, 2005). It would be a mistake, however, to assume that the first circulation in academic circles marks the time when people developed the concept – a point that Silvia Rivera Cusicanqui (2010) has made forcefully with regard to related ideas of "decolonial" theory. In my experience, for example, the concept took shape in dialogue with Maya intellectuals in Guatemala in the late-1990s, as we puzzled over why the state, which had inflicted massive genocidal violence on the Maya a few years earlier, would be so readily responsive to Maya demands for cultural rights. Over the next two decades, the central message – that rights grounded in cultural difference and neoliberal economic policies are mutually enabling rather than opposed – was at first received with scepticism and critique (e.g. Van Cott, 2006), especially among those who had been theorizing liberalism's communitarian turn with most enthusiasm (e.g. Kymlicka, 1995; Kymlicka and Banting, 2006). As of this writing, critique has given way to widespread use of the term, with debate focusing instead on the extent to which the bearers of multicultural rights have generally suffered from neoliberal polices, or made them work toward broader goals of empowerment and self-determination (for an example of the latter position, see: Kymlicka, 2013). Ironically, terminological consensus has emerged just as the phenomenon itself has begun to recede, displaced by a new, even more menacing, mode of governance.[1]

Making sense of the "neo" in neoliberalism

Many scholars have bemoaned the lack of precision in the term neoliberalism, which leads them to miss or downplay the remarkable evolution of its meaning, from a principled defence, to a "vernacularized" trenchant critique of capitalism. In response to the dominance of Keynesian macro-economic theory in the inter-and post-war years of the last century, a group of conservative social scientists from Europe and the US, led by Frederick Von Hayek, founded the Mont Pelerin Society in 1947, to develop and propagate a rigorously liberal alternative. According to one account, in formative activities over the preceding decade participants debated various epitomizing phrases, and in one meeting in 1938, someone floated the idea of adding the "neo" as a means to encapsulate their twin goals of recuperating and renovating classic liberal principles (Romero Sotelo, 2016). David Harvey's (2005) critical history of neoliberalism also emphasizes its dual character: a utopian ideology of individual realization through freedom from government constraints, and a practical strategy to halt and reverse the redistributive tendencies of Keynesian economics. The 1973 military coup against the socialist government in Chile put the latter on brutal display, as acolytes of Milton Friedman – who served as President of the Mount Pelerin Society (1970–1972) – worked closely with the Pinochet dictatorship, sending the infamous "Chicago Boys" there to give in situ advice on how to remake the Chilean economy under the "ideal" conditions of near-total repression of dissent. These origins as state policy should not overshadow, however, the other side of Harvey's formulation. The ascent of neoliberal doctrine through electoral victory in Great Britain (1979) and the United States (1980) came deeply articulated with a utopian message that acquired "considerable ideological and intellectual authority outside the realm of the state proper" by associating unfettered markets and downsized government with economic empowerment and individual betterment (Hall, 1988: 47).

The key to understanding neoliberalism's venacularization as an epithet, the encapsulating culprit of economic inequality and social suffering, at least in Latin America, is its repackaging as the "Washington Consensus." When economist John Williamson (1990) first coined that phrase in 1989, global institutions such as the World Bank and the IMF had lined up behind these principles, now familiar to all: privatization of state-owned industry, regularization of property rights, opening of national borders to capital flows, reduced clout of organized labour, strict macro-economic fiscal discipline, etc. Whether convinced on their own accord or nudged along by coercive powers of the global institutions, Latin American governments fell into line: from Salinas de Gortari in Mexico (1988–1994) to Raúl Alfonsín in Argentina (1983–1989). Although in Central America the revolutionary government (in Nicaragua) and movements (in El Salvador and Guatemala) delayed the shift somewhat, by the early 1990s the entire region had followed suit. But whatever utopian (or even minimally consensual) allure these policies once held did not last: although specific histories vary in timing and intensity, widespread protest quickly led to modifications in the standard package nearly everywhere. In this crucial sense, the Pinochet dictatorship was both a precursor and an outlier. Rather than brute repression, neoliberalism in Latin America generally came with a commitment to the procedural trappings of electoral democracy, with material incentives for adherents and selective repression for those who resisted, and justifications that combined remnants of the utopian allure with bare-bones pragmatism ("the global economy is making us do it"). As time went on, the vernacular meaning of the term won out nearly everywhere. While Sebastian Edwards (1995), a leading proponent of neoliberal reforms, could subtitle his 1995 essay on the first wave, "from despair to hope," by the beginning of the new century, even proponents had to use a different epitomizing term. This is the first great paradox of the "neoliberal turn" in Latin America: by the time the principles gained widespread endorsement as government policy, no politician could win an election as a self-described "neoliberal."

The second great paradox is that neoliberalism, as a form of governance, became most effective when it incorporated policies inimical to its own founding principles. In Williamson's original rendition of the Washington Consensus, the ten points include no reference to rights (except in the negative, with regard to organized labour), not to mention collective rights grounded in cultural difference. Political theorist Will Kymlicka (2013) argues forcefully that in the OECD countries "multiculturalism" – understood as a variant of liberalism that recognizes collective rights – preceded the neoliberal turn by more than a decade. He explains the rise of this "social liberalism" largely as a response to political mobilization from below by peoples marginalized by classical liberal notions of citizenship, and notes a series of examples, such as the Black Panthers and the American Indian Movement in the United States and the Front de Liberation du Quebec in Canada (2013: 102). While it is surely a stretch to associate these particular movements with something recognizable as the demand for multiculturalism, his note on historical sequence, in the north, may still be correct. But if so, this stands as a warning not to let the Global North become an implicit universal. In Latin America, the sequence was the inverse: premises of monocultural citizenship held sway through into the early 1990s in most places, well after the neoliberal turn was underway. This makes the explanatory puzzle more challenging: with a few exceptions, it would be difficult to argue for Latin America, as Kymlicka does for the OECD countries, that when elites adopted the principles of neoliberal governance, multiculturalism was already an embedded fixture in the political landscape.

Using constitutional recognition of indigenous peoples as a significant, if imperfect, gauge of the onset of state-endorsed multiculturalism, we can locate its beginning in Latin America in the late 1980s, with the preponderance of major reforms in the 90s, and the first decade of the new century (Van Cott, 2006).[2] One exception, where multicultural reforms did precede the neoliberal turn was Nicaragua, in that they formed part of broader negotiations to end a three year internal armed conflict in the mid-1980s. While in all cases significant mobilizations preceded recognition of the rights in question – and here Nicaragua is paradigmatic – it is unlikely that mobilizations alone, in the absence of powerful allies and other inducements, would have been enough to effect the changes in question. A full explanation of the extraordinary achievements of indigenous and black rights in this period falls beyond the scope of this entry; the factors that enhanced the power and scope of these mobilizations include: (1) general failure of the traditional left to achieve its objectives, combined with deepening frustration with left-wing "tutelage"; (2) weakened corporatist ties as a result of market-oriented reforms, which both posed new threats to black and indigenous peoples, while freeing them from governmental constraints (Yashar, 2005); (3) support from global civil society (environmental, human rights, cultural rights, etc.), which emerged as an influential force in this period (Paschel, 2016); and (4) a general ethos of "the right to have rights" (Dagnino, 1998), which benefited civil society-based organizing across the board. Especially given the fierce opposition of most traditional elites to the demands that animated these mobilizations, and the general understanding of neoliberalism as limited to the ten points of the Washington Consensus, it is no surprise that early analyses portrayed a pitched battle, with multicultural reforms advancing only as neoliberal gatekeepers reluctantly ceded ground.

By the end of the century, however, this portrayal had become seriously incomplete and misleading. Five distinct categories of individual and institutional actors, generally aligned with and empowered by the neoliberal turn, came to embrace some version of multiculturalism, following a logic that went deeper than pragmatic responses to external political pressure. In none of these realms was the convergence seamless or uncontested: in each there were others who refused the fusion, which in turn produced an experience of controversy and contingency, and a distinction between "progressive" dominant actors, who were willing to endorse some facets of collective rights claims, and "recalcitrant" ones, who were not. Yet amid this controversy a common ethos seeped into both: that cultural difference is a constituent feature of global capitalist modernity; and to be fully attuned with these conditions – i.e. fully "modern" – cultural difference must be recognized and respected. Fundamentalists of the Mont Pelerin Society, having contributed mightily to the advance of neoliberalism, would again become minority voices, as neoliberals began to endorse the multicultural turn.

Although chronologies are tricky since the drama unfolds in different ways across diverse national spaces, it is safe to say that the multilateral development banks were among the earliest "unlikely" proponents of neoliberal multiculturalism. The shift did not happen quickly or easily. Throughout the 1990s, within the World Bank, for example, one heard accounts of pitched battles between World Bank economists, and an upstart category of functionaries glossed as "sociologists," who had begun to question the premises of market fundamentalism, and promote collective rights as a means to advance the newly minted notion of "ethno-development" (Davis and Partridge, 1994). The most well-known protagonist among these Bank sociologists was Shelton H. Davis, who had strong street creds with collective rights advocates, as an outspoken critique of World Bank-funded mega-projects in the Brazilian Amazon (Davis, 1977), and later, of state-perpetrated genocide in Guatemala (Davis, Hodson, and Swartzchild, 1983). The Bank economists resisted, but gradually submitted, and by the new century multiculturalism had gone mainstream in Bank doctrine and practice. A parallel story unfolded in the Inter-American Development Bank (IDB), which became a major supporter of "ethno-development" among both indigenous and Afro-descendants of Latin America, as well ancillary initiatives, such as the program "*Todos Contamos*" ("We all Count"), to correct chronic under- and mis-counting of these peoples in national censuses. The number of institutional actors in this realm could be much larger, depending on how the criterion "aligned with and empowered by the neoliberal turn" is interpreted. But these two were thought leaders who, without ceasing to be bastions of neoliberal doctrine, turned substantively multicultural.

The second realm is authorities and institutions of the state, both elected officials and professionals. Latin America is the emblematic region, because the shift is so recent and pervasive. Most remarkably, in a region characterized by deep, continual, and often violent political polarization between left and right, differences on this issue revolved around rhetorical styles and strategies of incorporation, rather than basic tenets. While it did make a difference for indigenous and Afro-descendant peoples' quality of life when a class-based program of redistribution – the hallmark of leftist ideologies – achieved state backing, before multiculturalism, both left and right affirmed some variant of *mestizaje's* "all-inclusive ideology of exclusion" (Stutzman, 1981), which left problems of race-based marginalization and hierarchy largely unaddressed. For example, Cuban Revolution spokespeople claimed for decades that socialism had conquered racism, but had to backtrack in the new century, when Afro-Cubans finally gained the right to identify and organize as such. It is perplexing that the Mexican revolution earns pride of place in Van Cott's (2006) chronology of "constitutional recognition of indigenous peoples" (in 1917 – some 70 years before the others!), given that 20th-century Mexico became a showcase of *mestizo* nationalism's racist and exclusionary consequences. While it is possible that a deeply inculcated *mestizaje* ideology made state functionaries more receptive to the multicultural turn than their US counterparts, when indigenous and Black movements agitated for rights, that ideology became a principal target of their critique.

As Van Cott's data demonstrates (2006: 274), once the shift began it took the region by fire: not only constitutional recognition, but a series of other collective rights grounded in cultural difference, and related international legal instruments, such as ratification of Convention 169. Equally important was the widespread opening of what Mayans in Guatemala would later derisively call *ventanillas* (little windows) in government institutions, and the diversification of political parties, in competition for who had the strongest claim to a *rostro maya* (Mayan face). There are no doubt inspiring stories to tell about people who worked on each category of collective rights (bilingual education, territory, political representation, free informed prior consent, etc.), principled quasi-insiders, like Shelton Davis in the World Bank, with ties to civil society movements, who deployed effective strategies to prevail against the odds. Yet these individual achievements come embedded in a broader context, in which respect for cultural rights is framed not only as the proper path for "modern" states to follow, but as a strategy to burnish state legitimacy, with little risk of encouraging serious challenge to state power. This is the only way to explain, for example, the Colombian constitutional reforms of 1991, which extended indigenous rights much further than a calculated response to political pressure would have required; the same goes for Guatemala's indigenous rights agreement, reached in 1996, in which Maya representatives barely had a voice. Something was in the air.

The third realm is the judicial arena, which took significant steps in this same period to embrace what has come to be known as "judicial pluralism." The inclusion of this realm may be questioned by some, on the grounds that the judicial establishment in Latin America is not explicitly aligned with the principles of neoliberalism. This view, I contend, does not give sufficient weight to neoliberalism's transformation from a narrowly conceived mandate for economic reform (Williamson's ten points), to a full-fledged project of governance. As Marilyn Strathern's influential book attests (2000), "audit culture" penetrated and reshaped many realms – from the courts, to organized religions, to institutions of higher education – such that neoliberal premises became normalized commonsense. With regard to the judicial arena in particular, this argument points to a strong affinity between the Western positivist law tradition and classic liberalism, and a generally receptive stance (or at least an absence of outright antagonism) toward the neoliberal turn. Subsequent transformations fit the overall storyline quite well: strong opposition at first, on the grounds that fundamental judicial principles would be abrogated, followed by support from select "progressive" judges and scholars, with gradual inroads, and eventually, more widespread endorsement. While there are still many holdouts, by the 2010s the idea of cultural difference as a legal principle – both as a rationale for certain actions (e.g. "cultural defense"), and the basis for collective rights – had won the day, with little threat to the neoliberal economic status quo.

The fourth realm – corporate rhetoric and practice – is both more straightforward and more complex than the first three. In some respects, corporations were onto the recognition of cultural difference long before the others, and made the shift more seamlessly. This goes a long ways in explaining why the early fears that global capitalism would generate a process of cultural homogenization – the "McWorld" effect – turned out to be so wrong (Barber, 1995). A fascinating interview with Coca-Cola's Director of International Advertising, for example, points to a shift, sometime in the late 1970s, when the soft-drink giant made a conscious decision to go "multicultural" in its advertising strategy (O'Barr, 1989). Shortly thereafter, the idea of culturally differentiated marketing niches, both within countries and between them, became commonplace. While market-driven cultural recognition has no intrinsic connection with cultural rights, once the "multicultural door" is opened the sharp distinction between the two tends to blur. Overtly racist or exclusionary corporate images are bad for business; cultural niche advertising can act to reinforce the identity in question (Comaroff and Comaroff, 2001; Davila, 2001); consumer-driven corporate responsibility programs generate internal policy changes that mandate greater respect for marginalized cultures.[3] While this realm is clearly "aligned with and empowered by" neoliberal principles, a more difficult question is whether the corporate embrace of multiculturalism is anything more than an instrumental manoeuvre to bolster the bottom line. Either way, it seems accurate to place corporate endorsement of the "multicultural turn" on one end of a continuum, rather than in a completely distinct category.

The fifth and final realm – individual and institutional actors who wage the struggle from within the neoliberal establishment, without endorsing or submitting to its premises – brings us to the analytical crux. The challenge here is the claim that the fusion is nothing but a strategic ploy to advance substantive principles of cultural rights, appropriating the discourse and tools of the neoliberal establishment, knowing all the while that at some point the two paths will diverge. Indigenous rights leaders, solidary scholars, race-conscious state functionaries, "insiders" the likes of Shelton Davis could all fit this description, such that a principal claim of neoliberal multiculturalism – that the fusion deeply shapes subjectivities and political outcomes – would be called into question. Yet for every stalwart actor who enters with his or her eyes wide open and succeeds in using the tools of neoliberal governance against the system, there are others who enter having submitted to the dominant logic from the start, and still others, who enter fully intending to resist, but gradually submit to the gentle persuasion of neoliberal governance's perks and pleasures. A proper reading of the closely related idea of the space of the *indio permitido* (Hale, 2004) reinforces this message of radical contingency. Such spaces come with a series of built-in constraints, consistent with the limited package of rights that neoliberal multicultural-ism allows. The outcome – what happens when people choose to occupy the space – is an open empirical question, not a preordained conclusion.

My own experience, along with colleagues of the Caribbean Central American Research Council (CCARC), in two participatory mapping projects in support of indigenous and Afro-descendant rights to territory, illustrates this last point. CCARC bid for both projects, in response to RfP's emitted by the Nicaraguan government (in 1996), and the Honduran government (in 1999), both funded by the World Bank. In keeping with its multicultural turn, the Bank had determined that national-level development goals in both countries would be best served by a process to regularize land tenure in the vast areas on the north (Honduras) and east (Nicaragua) Caribbean coast, historically occupied by indigenous and Afro-descendant peoples. Fully aware that the intention of these programs was to demarcate small islands of collective territory, in a sea of property to be bought and sold on the market, we accepted the challenge, and worked closely with organizations representing those peoples to produce "counter-maps," which might advance their more expansive territorial claims. In the short-term, this work did contribute to large collective territories in both sites, though the process had its share of contradictions and unintended consequences (Gordon, Gurdian, and Hale, 2003). In the medium term, we prob-ably overestimated the benefits that secure territorial rights would bring, and underestimated the entanglements (Hale, 2011; Bryan and Wood, 2015). A general conclusion follows: enter-ing the spaces opened by neoliberal multiculturalism with a radical critique, and a strategy to "use the system against the system," is an essential first step, but has no guarantees, except perhaps the guarantee that the consequences, which reveal themselves over time, will never map easily onto the original intentions.

With the exception of this final realm, these descriptions beg a crucial analytical question: why did these diverse actors embrace neoliberal multiculturalism in the first place? In each realm they were deeply divided, such that the eventual ascent of neoliberal multiculturalism was any-thing but a foregone conclusion. In engaging this question, ethnographies of dominant actors' thinking and practice are especially crucial. One critique that has emerged from such analyses is that the theory of neoliberal multiculturalism paints an overly Machiavellian picture: a cabal of neoliberals, scheming to break the momentum of rights movements by coopting select leaders while turning the rest away; granting a limited package of rights, as the most effective means to refuse and proscribe the rights that really matter. Arturo Escobar summarizes this critique, in his analysis of Afro-Colombians' achievement of extensive territorial rights, with the wry observa-tion that this confers "upon the state a coherence and prescience it rarely has" (Escobar, 2008: 159). However, while we do need to check this image of the "coherent and prescient" cabal, it is equally crucial not to underestimate dominant actors' capacity to evolve and shift in their thinking, in response to a combination of lessons learned, pragmatic responses to pressure and resistance, and broader ideological currents that become part of their commonsense.

All these elements are present, in varying degrees, in the four principal lines of explanation for the shift. The first builds on the key distinction developed by Nancy Fraser (1997) and others, between "recognition" and "redistribution." Although collective rights pose a direct challenge to the individualist precepts of classic liberalism, they are easily reconciled with the neoliberal capitalist order as we know it. As experience soon demonstrated that the recognition of cultural rights had few serious consequences for the redistribution of societal income (not to mention wealth), these rights became less painful for elites to endorse. A second area of learning involved the governance effects of recognition. As Speed and Collier (2000) argued early on, when the State grants rights, the beneficiaries come under the purview of State management in new ways. This standard pact of liberal citizenship takes on an especially menacing dimension when the subjects in question face State-endorsed racial subordination, which persists after cultural rights have been granted. A third element, more proactive and ideological, involves a recasting of identity politics, to highlight a convergence between cultural rights and neoliberal principles. The World Bank and other development institutions have promoted these ideas vigorously, with key phrases such as "ethnic entrepreneurship," or "ethnicity as social capital," drawing an implicit contrast with ethnicity's counter-productive (even pathological) features, the first yielding "*propuesta*" and the second "*protesta*." The message here is an adaptation of the familiar ideological pillar of capitalism: if bearers of cultural rights play by the rules, and use their identity as social capital, they stand to benefit from full participation in the neoliberal order. Like all such tenets, it is woefully misleading as a general proposition, but carries a grain of truth just large enough to hold a profound allure. The final explanation is ineffable, but especially compelling in combination with the other three. By associating neoliberal economic principles with modern economic management, the modernity template extended to encompass the polity as well. Modern polities keep the military in the barracks, follow the procedural rules of electoral democracy, and recognize their societies' cultural pluralism. By the turn of the century, States that clung to the 19th-century precepts of a culturally homogenous citizenship could be persuasively represented as woefully backward. This logic was dramatically on display in the iconic Inter-American Human Rights Court trial, Awas Tingni v. the State of Nicaragua (2002): lawyers for the community brilliantly cinched their case with the argument that, in contrast to the lamentable racism of Nicaraguan state officials, modern states recognize cultural difference. The judges (from six different Latin American countries), nodded their heads solemnly, ruling unanimously in favour of the indigenous community (Anaya and Grossman, 2002; Hale, 2006).

To move from the enumeration of these four explanatory factors, to a global explanation for elite endorsement of the multicultural turn, would gloss over much particularity and variation. Considering only the Latin American region, for example, there is ample variation in the timing, scope, and content of elite endorsement, corresponding to such factors as the character of national political cultures, the relative size of indigenous and Afro-descendant populations, the patterns of politicization of their identities, and their strategies of mobilization – to name just a few. In addition, as Juliet Hooker (2005) and Tianna Paschel (2016), among others, have brought to our attention, an important distinction between demands for cultural rights and demands for racial equality tends to infuse the distinct histories of racialization, identify formation, and political mobilization of indigenous and Afro-descendant peoples (see also Giminiani, this volume). Appropriately deployed, the notion of neoliberal multiculturalism should be an invitation to explore this heterogeneity, driven by two central propositions: (1) dominant actors and institutions aligned with and empowered by the principles of neoliberalism also tend to endorse a limited cluster of multicultural rights; (2) the resulting fusion – neoliberal multiculturalism – leaves unchallenged, and often more deeply entrenched than ever, pervasive relations of structured racial inequality.

These two propositions combine to yield a third: neoliberal multiculturalism is best understood not simply as an unlikely fusion of cultural rights and economic policy mandates, but rather, as a new project of governance, displacing the previous one based on 19th-century ("classic") liberal principles of individual citizenship. By highlighting a governance project, we direct attention not only to the effects of the policies, but crucially, to the accompanying processes of subject formation. Here again, rather than preordained outcomes, the theory should motivate empirical investigation, framed by a central question: have protagonists of black and indigenous struggles normalized the limited package of cultural rights that dominant institutions allow, viewing this as all that is possible, or even, all that they justly deserve? Have they come to endorse the key hegemonic premise that their people's empowerment is best achieved from within the neoliberal establishment? Originally some of us framed this "subject formation" question through the lens of Fraser's "recognition" vs "redistribution" dichotomy. However, as Duggan (2003) pointed out early on, the dichotomy rests on the dubious assumption that representations have no material consequences; moreover, as cultural rights extended to encompass collective rights to territory – what Offen (2003) called the "territorial turn" – the dichotomy lost much of its explanatory value. A more robust framing focuses on persisting relations of structured racial inequality, after the shift to multiculturalism, and a normalization of these relations as inevitable, if not just. A key series of empirical questions follow: if these spaces are designed to produce this normalization, what happens to those who opt to occupy them? Do they submit to the dominant logic, taking it on as their own, distancing themselves from the rest? Or do they manage to subvert it, seizing the opportunity for articulation with the broadly defined needs and interests of the majority who they in some sense represent? Whatever the specific rights, these questions focus crucially on the variable outcomes of rights recognition, positing that, in the absence of vigorous counter-efforts, the dominant logic of these spaces is likely to prevail.

If indeed the era of neoliberal multiculturalism is coming to a close, this will provide prime conditions for historical analysis of these propositions. After three decades of dramatic expansion of cultural rights (1988–2017), in the context of ever more deeply embedded neoliberal economic principles and persistent racialized structural inequality, how have the protagonists of these rights – their organizations, collective voice, and political horizons – fared? Posing the question in this way also raises the counterfactual: however badly they have fared, would it have been worse without the multicultural turn? My sense is that conditions have improved over these three decades in certain respects: indigenous and afro-descendent peoples have been empowered by the affirmation that they have rights, they have benefited from a vast increase in leaders with higher education, from widespread political experience at all levels of government, and much expanded dialogues around their people's political horizons. Yet these positive outcomes have come paired with a series of crippling negatives: deepened internal divisions; subordination to governance projects made more effective because they are "humanized" by the inclusion of cultural rights; persisting racialized inequality that no longer can be explained by the absence of rights, and therefore invites "blame the victim" reasoning; organizations deeply dependent on foreign funding, with attendant de-politicization. As historical analysis considers both sets of patterns, one key variation will be the so-called "post-neoliberal regimes" such as Morales' Bolivia and Correa's Ecuador. On the one hand, especially in Bolivia, policies of neoliberal multiculturalism in the previous regime did allow opponents to gather strength and bide their time for right moment to seek (and achieve!) its overthrow. On the other hand, these "post-neoliberal" regimes clearly have not shaken free from the larger dilemmas of neoliberal governance, as evidence by the widespread conflicts in recent years with large sectors of indigenous and black citizens.

Is the era of neoliberal multiculturalism actually coming to a close? After nearly three decades of steady (if uneven) expansion in the codification of cultural rights, the current panorama in the Americas offers few cases where this expansion continues, and a growing trend of retrenchment. Rights movements are on the defensive, increasingly criminalized for mobilizations to defend and exercise rights already gained. In many cases these mobilizations involve direct protest – against the rising threats of extractivist economies, for example – because claims to rights generate so little traction. The rapidly increasing concentration of wealth – a direct outcome of three decades of neoliberal policies – has predictably racialized contours; moreover, extractivsim exerts increasing pressures on the resource-rich territories that black and indigenous peoples recently gained. The cumulative effects of neoliberal multiculturalism also left its mark on prospects for political mobilization: deepened divisions between the small group of black and indigenous peoples who have benefitted from the neoliberal turn, and the vast majority who have not; backlash by dominant culture actors who argue that concessions have been excessive, such that rights are re-signified as accusations of "reverse racism." This "they-have-gone-too-far" argument, if it were not so spurious, could be taken as evidence that multicultural rights have indeed generated significant material advances. But especially when exercised by lower- and middle-class members of the dominant culture, the backlash is motived more by perceived threats to the symbolic props of white/*mestizo* privilege than by actual material deprivation (Hooker, 2017).

Firm conclusions would be premature, because the emerging era of racial recalcitrance is just beginning to take shape (Calla, Hale, and Mullings, 2017). First impressions point to dire consequences for racialized peoples, with fewer openings, and a threatening atmosphere of rights denial, repression, and criminalization of protest. In lieu of conclusions, I close this entry with a series of questions, generated by the juxtaposition of the two eras:

- State-recognized rights to territory and autonomy, the crowning achievement for racialized peoples under neoliberal multiculturalism, are apt to be the first to come under attack in the coming era. What are the prospects for achieving or defending autonomy and combatting racism, *without* recourse to state-endorsed rights?
- If one of the major shortcomings of indigenous and black mobilizations for rights under neoliberal multiculturalism was the inability to confront racial capitalism head on, what are the prospects for doing so in the coming era, when its precepts are defended more baldly, and prospects for gaining countervailing rights are increasingly remote?
- A prominent reflection on the coming era suggests that "cultural rights" were always part of the problem, and that a keyword of anti-systemic struggle in the future will be "ontological" rather than "cultural" politics (e.g. Blaser, 2010; de la Cadena, 2015; Escobar, 2013; Ogden and Gutierrez, this volume). What possibilities does the "ontological turn" open, and what doors does it close?
- Under neoliberal multiculturalism, the prevailing mode of anti-systemic struggle was counter-hegemonic: use the dominant "language of contention" (Roseberry, 1994) to open and occupy spaces within the system, and then push for more. What are the modes of and prospects for counter-hegemonic struggle in the emerging era of racial recalcitrance?

Although at times misinterpreted as a gloomy depiction of an iron cage, in which the bearers of cultural rights were unwittingly trapped, the notion of neoliberal multiculturalism – at least in my understanding – was conceived with precisely the opposite valence. It called for political strategies to claim cultural rights, to occupy spaces and seize opportunities that rights offered, while taking the "menace" of this gambit fully into account. We now need historically informed analysis, to assess when, why, and to what extent these strategies worked; and with even greater urgency, we need "histories of the present," to map the coming era, mark its most noxious features, and address, once again, the perennial question: what is to be done?

Notes

1 Further evidence of the term's entry into the mainstream comes from Benjamin Dixon's video blog of political commentary. See, for example, "Neoliberal Multiculturalism Is NOT New" (5/31/16) www.youtube.com/watch?v=YcW9F2FYj8o
2 According to Van Cott's usefully compiled Latin America-wide data, there are six "outlier" polities of the 20 total, where constitutional recognition occurred earlier. Three of these (Belize, Guyana, and Panama) follow Caribbean patterns, and one (Costa Rica) followed patterns more closely linked to the US. The other two (Honduras and El Salvador), are enigmas, somewhat less important due to their small size. If blackness rather than indigeneity were the point of reference, the timetable would be considerably later, and the rights achieved more limited.
3 I recently served as co-PI in a study of the forest industry in Chile, their relations with Mapuche indigenous communities, and role of indigenous rights certification by the Forest Stewardship Council (FSC). While the overall conclusions of the study are not especially sanguine, the research does document precisely these effects: a "multicultural turn" in corporate culture, induced by the need to achieve FSC certification. See, Hale and Millaman (2017).

References

Anaya, S. J. and Grossman, C. (2002) The case of Awas Tingni v. Nicaragua: A new step in the international law of indigenous peoples. *Arizona Journal of International and Comparative Law* 19(1): 1–15.
Barber, B. R. (1995) *Jihad vs. McWorld: How the Planet Is Both Falling Apart and Coming Together and What This Means for Democracy*. New York: Random House.
Blaser, M. (2010) *Storytelling Globalization from the Chaco and Beyond*. Durham, NC: Duke University Press.
Bryan, J. and Wood, D. (2015) *Weaponizing Maps: Indigenous Peoples and Counterinsurgency in the Americas*. New York: Guilford Press.
Calla, P., Hale, C. R. and Mullings, L. (2017) Race matters in dangerous times. *NACLA* 49(1): 81–89.
Comaroff, J. and Comaroff, J. L. (2001) *Millennial Capitalism and the Culture of Neoliberalism*. Durham, NC: Duke University Press.
Dagnino, E. (1998) The cultural politics of citizenship, democracy, and the state. In S. E. Alvarez, E. Dagnino and A. Escobar (eds) *Cultures of Politics, Politics of Cultures: Re-Visioning Latin American Social Movements*. Boulder, CO: Westview Press, pp. 33–63.
Davila, A. (2001) *Latinos, Inc.: The Marketing and Making of a People*. Berkeley, CA: University of California Press.
Davis, S. H. (1977) *Victims of the Miracle: Development and the Indians of Brazil*. Cambridge: Cambridge University Press.
———, Hodson, J. and Swartzchild, N. (1983) *Witnesses to Political Violence in Guatemala: The Suppression of a Rural Development Movement*. Boston, MA: Oxfam America.
———. and Partridge, W. (1994) Promoting the development of indigenous people in Latin America. *Finance and Development* March: 38–40.
de la Cadena, M. (2015) *Earth Beings: Ecologies of Practice Across Andean Worlds*. Durham, NC: Duke University Press.
Duggan, L. (2003) *The Twilight of Equality? Neoliberalism, Cultural Politics, and the Attack on Democracy*. Boston, MA: Beacon Press.
Edwards, S. (1995) *Crisis and Reform in Latin America: From Despair to Hope*. Oxford: Oxford University Press.
Escobar, A. (2008) *Territories of Difference: Place, Movements, Life, Redes*. Durham, NC: Duke University Press.
———. (2013) *Territorios de diferencia: La ontología política de los 'derechos al territorio'*. Documento preparado Para el Segundo Taller Internacional SOGIP. Los Pueblos Indígenas y sus Derechos a la Tierra: Política Agraria y Usos, Conservación, e Industrias Extractivas. www.sogip.ehess.fr: Naciones Unidas
Fraser, N. (1997) *Justice Interruptus: Critical Reflections on the "Postsocialist" Condition*. New York: Routledge.

Gordon, E. T., Gurdian, G. C. and Hale, C. R. (2003) Rights, resources and the social memory of struggle: Reflections on a study of indigenous and black community land rights on Nicaragua's Atlantic Coast. *Human Organization* 62(4): 369–381.

Hale, C. R. (2002) Does multiculturalism menace? Governance, cultural rights and the politics of identity in Guatemala. *Journal of Latin American Studies* 34: 485–524.

———. (2004) Rethinking indigenous politics in the era of the "indio permitido". *NACLA* 38(1): 16–20.

———. (2005) Neoliberal multiculturalism: The remaking of cultural rights and racial dominance in Central America. *Polar* 28(1): 10–28.

———. (2006) Activist research v. cultural critique: Indigenous land rights and the contradictions of politically engaged anthropology. *Cultural Anthropology* 21(1): 96–120.

———. (2011) Resistencia para que? Territory, autonomy and neoliberal entanglements in the 'empty spaces' of Central America. *Economy and Society* 40(2): 184–210.

——— and Millaman, R. (2006) Cultural agency and political struggle in the era of the 'indio permitido'. In D. Sommer (ed) *Cultural Agency in the Americas*. Durham, NC: Duke University Press, pp. 281–304.

———. (2017) *La Industria Forestal de Chile, la Certificación FSC y las Comunidades Mapuche*. Temuco: Forest Stewardship Council.

Hall, S. (1988) The toad in the garden: Thatcherism among the theorists. In C. Nelson and L. Grossberg (eds) *Marxism and the Interpretation of Culture*. Urbana: University of Illinois Press, pp. 35–57.

Harvey, D. (2005) *A Brief History of Neoliberalism*. Oxford: Oxford University Press.

Hooker, J. (2005) Indigenous inclusion/black exclusion: Race, ethnicity and multicultural citizenship in Latin America. *Journal of Latin American Studies* 37: 1–26.

———. (2017) Black protest/white grievance: On the problem of white political imaginations not shaped by loss. *South Atlantic Quarterly* 116(3): 483–504.

Kymlicka, W. (1995) *Multicultural Citizenship*. Oxford: Oxford University Press.

———. (2013) Neoliberal multiculturalism? In P. Hall and M. Lamont (eds) *Social Resilience in the Neoliberal Era*. Cambridge: Cambridge University Press, pp. 99–126.

——— and Banting, K. (2006) *Multiculturalism and the Welfare State: Recognition and Redistribution in Contemporary Democracies*. Oxford: Oxford Unviersity Press.

Melamed, J. (2006) The spirit of neoliberalism: From racial liberalism to neoliberal multiculturalism. *Social Text* 24(4): 1–24.

O'Barr, W. (1989) The airbrushing of culture: An insider looks At global advertising. Interview with M. Moreira. *Public Culture* 2(1): 1–19.

Offen, K. H. (2003) The territorial turn: Making black territories in Pacific Colombia. *Journal of Latin American Geography* 2(1): 43–73.

Overmyer-Velázquez, R. (2010) *Folkloric Poverty: Neoliberal Multiculturalism in Mexico*. University Park, PA: Pennsylvania State University Press.

Paschel, T. S. (2016) *Becoming Black Political Subjects: Movements and Ethno-Racial Rights in Colombia and Brazil*. Princeton, NJ: Princeton University Press.

Postero, N. (2007) *Now We Are Citizens, Indigenous Politics in Post-Multicultural Bolivia*. Stanford, CA: Stanford University Press.

Rivera Cusicanqui, S. (2004) Reclaiming the nation. *NACLA* 39(3): 19–23.

———. (2010) *Chhixinakax utxiwa: Una reflexion sobre practicas y discursos descolonizadores*. Buenos Aires: Tinto Limón.

Romero Sotelo, M. E. (2016) *Los Origenes del Neoliberalismo en Mexico: La Escuela Austriaca*. Mexico: FCE-UNAM.

Roseberry, W. (1994) Hegemony and the language of contention. In G. M. Joseph and D. Nugent (eds) *Everyday Forms of State Formation*. Durham, NC: Duke University Press, pp. 355–366.

Speed, S. (2005) Dangerous discourses: Human rights and multiculturalism in Mexico. *Polar* 28(1): 29–51.

——— and Collier, J. (2000) Limiting indigenous autonomy in Chiapas, Mexico: The state government's use of human rights. *Human Rights Quarterly* 22(4): 877–905.

Strathern, M. (2000) *Audit Cultures: Anthropological Studies in Accountability, Ethics, and the Academy*. New York: Routledge.

Stutzman, R. (1981) El mestizaje: An all-inclusive ideology of exclusion. In N. Whitten (ed) *Cultural Transformations and Ethnicity in Modern Ecuador*. Urbana: University of Illinois Press, pp. 45–93.

Van Cott, D. L. (2006) Multiculturalism versus neoliberalism in Latin America. In W. Kymlicka and K. Banting (eds) *Multiculturalism and the Welfare State: Recognition and Redistribution in Contemporary Democracies*. Oxford: Oxford University Press, pp. 272–296.

Williamson, J. (1990) What Washington means by policy reform. In J. Williamson (ed) *Latin American Adjustment: How Much Has Happened*. Washington, DC: Institute for International Economics, pp. 7–20.

Yashar, D. (2005) *Contesting Citizenship in Latin America: The Rise of Indigenous Movements and the Postliberal Challenge*. Cambridge: Cambridge University Press.

Zizek, S. (1997) Multiculturalism, or, the cultural logic of multinational capitalism. *New Left Review* (225): 28–51.

7

THE RISE AND FALL OF THE PINK TIDE

Laura J. Enríquez and Tiffany L. Page

Introduction

Starting in 1998, a number of presidential elections in Latin America gave rise to left and center-left governments. This was a striking occurrence for several reasons: throughout the region's history, revolutionary governments had mostly come to power through armed struggle; and this wave of electoral victories by progressive political figures came after at least a decade during which the right had claimed an "end of ideology" (i.e. of the legitimacy of a leftist alternative). These electoral results were largely a reaction against the social impact of the policies of the conservative governments of the 1980s and 1990s. There have been quite substantial differences in the nature of the distinct "Pink Tide"[1] governments. They have ranged from governments with more radical agendas and political styles – including Venezuela, Bolivia, and Ecuador, to those that are more moderate in their programs and discourse – including Brazil (under Lula), Argentina (under the Kirchners), Nicaragua, El Salvador, and Uruguay. The depth of these differences will be discussed below.

Despite their discrepancies, their emergence suggested that there was "an alternative," in contrast to what Margaret Thatcher – one of the main proponents of neoliberalism – had stated. It also suggested that if they stood united they would have strength to push back against the principal global hegemon (the US) and the international lending community. Hence, in addition to a variety of new policies put in place by these regimes, they also established a number of region-wide institutions that sought to represent another path from that preached by the Global North (e.g. ALBA – the Bolivarian Alternative Trade Alliance; the Banco del Sur; and the Community of Latin American and Caribbean States – a political body that does not include the US and Canada).

Although several of these governments have since been replaced by right-wing governments, others maintain an important presence in the region. In the pages that follow, we describe the rise of the Pink Tide states, their social, political, and economic projects, and the principal weaknesses that emerged within those projects. In so doing, we draw on the specific experience of some of these states so as to illustrate these dynamics, as well as the similarities between them and variation across them. Our selection of the cases described stems from their particularly clear representation of the dynamic in question.

The rise of the Pink Tide

Neoliberal policies spread throughout Latin America in the wake of the debt crises of the 1980s through condition-tied loans from the IMF and the World Bank (WB). These policies included currency stabilization, trade and capital liberalization, an emphasis on increasing exports, attracting foreign capital, deregulation, privatization, and fiscal austerity (Veltmeyer and Petras, 2010). They resulted in economic displacement, reduced access to social programs, an expansion of the informal sector, falling wages, and an increase in poverty and inequality. Riots erupted during the 1980s and 1990s to protest the policies being implemented, and social movements emerged to sustain the anti-neoliberal challenge and develop alternatives. The social backlash to these policies and the unresponsiveness and repression of mobilization by the governments implementing them undercut the legitimacy of the major political actors and parties of the time, leading to the introduction of new political actors who sparked the rise of the Pink Tide. For example, in Uruguay, the election of Tabaré Vázquez in 2004 marked a clear departure from the reigning political elite as he was not a member of either of the political parties that had dominated Uruguayan politics since independence. And, similar patterns prevailed in the other countries in this group of nations.

Looking more closely at the Bolivian case, between 1985 and 2005, its government implemented neoliberal reforms under the direction of the WB. Policies to attract foreign capital and privatize the mining and hydrocarbons sectors increased the control of transnational corporations over the country's natural resources and the revenues they generated. Meanwhile, royalties and taxes collected by the state from the extraction of these natural resources declined. These policies also stimulated an intensification of extractive activities, which caused environmental destruction, thereby harming the livelihoods of indigenous groups living in the area. State regulatory agencies favoured the interests of the transnational corporations (TNCs). And, the WB refused to allow local communities to monitor the hydrocarbon projects. In addition, privatization of the state-owned mining company resulted in the loss of jobs for 25,000 miners, and many turned to the cultivation of coca as a livelihood. According to Hindery and Hecht (2013: 31), "[b]y the World Bank's own accounts, extreme poverty increased from 36.5% in 1997 – when capitalization occurred – to 41.3% in 2002." Consequently, popular opposition to these policies – which were implemented without consultation with the public, or transparency or accountability – grew. There was a feeling that the country's resources were coming increasingly under the control, and being used for the benefit, of foreign actors to the detriment of average Bolivians (Hindery and Hecht, 2013).

Anti-neoliberal mobilization occurred in the 1990s and increased between 2000–2005. Strong indigenous movements emerged. The struggles focused on control over natural resources that indigenous communities directly relied on for their livelihood (e.g., land, water), as well as those that could potentially generate state revenues (e.g., hydrocarbons). In 1997, protests ensued when the WB insisted that the water systems in La Paz and Cochabamba be privatized. The latter came to be known as La Guerra del Agua, or the War over Water. This was followed by protests against the US-supported eradication of coca. In 2003, riots broke out after the government implemented IMF-supported measures, including a tax increase and austerity measures. Due to the government's repressive response and a plan to ship natural gas to California (which came to be known as La Guerra del Gas, or the War over Gas), President Sánchez de Lozada lost legitimacy and was forced out of office. Popular demands focused on dismantling neoliberal policies, the formation of a constituent assembly to write a new constitution, and the nationalization of natural gas. Evo Morales, who had emerged as a leader in the coca grower movement, was elected President in 2005 promising the re-nationalization of hydrocarbons, as well as a new constitution that would give recognition and voice to the indigenous people. The new constitution ended up going as far as granting nature intrinsic rights and enshrining the idea of Vivir Bien, or Living Well, as a reconceptualization of development that challenges the modernization paradigm of Vivir Mejor, or Living Better (Hindery and Hecht, 2013).

Similarly, when economic crisis hit Ecuador in the early 1980s, stabilization, austerity, and deregulation measures were implemented. Rural indigenous communities were negatively impacted by the elimination of price controls on agricultural inputs and subsidies. As in Bolivia, indigenous movements played an important role in the mobilization against neoliberalism. The first national mobilization of indigenous people in Ecuador occurred in 1990. Another mass mobilization occurred in 1994 when the legislature tried to pass an agrarian law that would have benefitted large-scale agricultural producers to the detriment of indigenous farmers (Yashar, 2005). Inequality, poverty, and debt servicing costs were increasing. The economic situation led to a significant out migration as Ecuadorians sought economic opportunity elsewhere. During a financial crisis in 2000, the government bailed out the banks, which were widely viewed as corrupt institutions. This move failed to stop the crisis and the IMF stepped in and required the implementation of further austerity measures, which included the elimination of subsidies for household energy consumption. Under IMF direction, the government froze bank accounts, launched plans to privatize state industries, and replaced the country's currency with the US dollar (Mijeski and Beck, 2011). In response, the indigenous population helped overthrow the government in 2000, but then they were pushed aside by their military allies and the new civilian president continued to implement neoliberal policies (Yashar, 2005).

In sum, the Ecuadorian population lost confidence in the country's political institutions and actors due to corruption scandals and what were viewed as poor economic decisions. A crisis of legitimacy in the political system ensued, opening the way for the election of Rafael Correa to the presidency in 2006. Correa was a political outsider who critiqued policies that he viewed as a "continuation of Ecuador's history of economic dependence and subordination to outsiders" (Conaghan, 2011: 265). In both Ecuador and Bolivia, an anti-imperialist discourse resonated with the population after years of WB- and IMF-imposed policies.

Projects of the Pink Tide

There has been great variation between the Pink Tide governments in the degree to which they have sought to change their economies and societies. They range from Venezuela – where profound transformation was legislated and partially enacted, to Argentina – where access to social services grew somewhat while the overall neoliberal economic orientation remained. In this section we sketch the social, economic, and political projects of these two governments, as well as their outcomes, in order to illustrate this diversity.

As early as the ratification of Venezuela's new constitution in late 1999, a vision of a distinct economic structure was outlined (República Bolivariana de Venezuela, 1999). It would consist of a mixed economy composed of cooperative, private, and state sectors. The objectives of this structure were several-fold, including to expand employment – given the large pool of unemployed, and to increase production in the hands of economic sectors sympathetic to the overall project of the Chávez government. This latter objective became especially important in the wake of the opposition-organized economic strike of 2002/2003 (Ellner, 2008).

But, the Chávez government also aimed to diversify the economy, so as to decrease the country's dependence on oil exports (República Bolivariana de Venezuela, 2001). The implementation of an agrarian reform was central to this effort, so as to increase local food production and, potentially, exports of agricultural goods. The further development of petrochemical products, local production of pipes for the oil industry, and the creation of a ship-building industry were also on the agenda.

Moreover, the Constitution envisioned major changes in the political sphere. Multi-party representative democracy was to be complemented by participatory democracy. The idea of combining the two was to ensure that the voices of those who had previously been margin-alized, would now find avenues of expression. Politically engaged organizations emerged in multiple spheres as part of this expanded idea of democracy, from farmers organizations to com-munity radio stations (Fernandes, 2010).

Then, as of 2010, an initiative to develop communes, which would be both economic and political units, took shape. They would come to incorporate cooperatives and communal coun-cils – which were set up to allow for grassroots participation in municipal-level planning – among other bodies within their reach. The objective was for communes to take on increasingly larger development projects and assume growing political weight in decision-making (República Bolivariana de Venezuela, 2010).

And, alongside the traditional government bureaucracies a set of new institutions were cre-ated starting in 2003 – the *Misiones* – which were geared toward expanding the provision of social services, such as health care and education, as well as providing inputs to farmers to facilitate the growth of agricultural production (Ellner, 2008). The *Misiones* were also a means of getting around bottlenecks in pre-existing state institutions, which had been designed with other objectives in mind. All of these initiatives, as well as the reshaping of the economy, were supposed to reduce income inequality and poverty.

The program and projects of the Chávez government produced positive social outcomes. Inequality was reduced significantly: the Gini Index dropped from .49 in 1998 to .40 in 2012 (one of the lowest in Latin America)[2]; the poverty rate was reduced dramatically: from the 86% it had reached by 1996 (with extreme poverty affecting 65% of the population) to 23.9% by 2012 (with extreme poverty affecting 9.7% of the population)[3]; and official unemployment dropped from 11.3% in 1998 to 7.8% in 2013.[4] At least through the early 2010s, Venezuela's government was successful in terms of its social objectives.

The picture in terms of economic transformation was less positive. The early 2000s wit-nessed tremendous growth in the cooperative sector, increasing from 877 cooperatives in 1998, to between 30,000 and 60,000 by 2009 (Piñeiro Harnecker, 2009: 309). But, by 2010 the weaknesses of that sector led the government to back away from its promotion and communes became the newly prioritized model. Alongside the latter, the state enterprise sector benefited from the growing expropriation of private enterprises during the second half of the 2000s (Azzellini, 2011).

With regard to economic diversification, the drive to expand agricultural production was only partially successful (Enríquez and Newman, 2016). Significant resources were put into this effort, and the investment created a livelihood in farming for some who had none previously. Diversification in other areas was more limited (Hammond, 2011). Hence, the economy con-tinued to be heavily dependent on the production and export of oil.

On the political front, the outcomes of the project were also mixed. The participatory model created new avenues for incorporation of the popular sectors into decision-making processes. One example was participatory budgeting, which – because of its legitimacy, was even prac-tised to some degree in municipalities governed by opposition political parties (Hetland, 2014). Chávez maintained a high-level of support within the electorate – he was reelected by notable margins in 2000 (after the new constitution was ratified); in 2004 (in a recall referendum); and in 2006 and 2012 in regular presidential elections. However, his hand-picked successor, Nicolás Maduro, was less popular, and as the opposition grew in strength (as evidenced in the legislative elections of 2015, in which it took the majority of the seats), the spaces for dialogue diminished and the political sphere became starkly polarized.

In contrast, the social, economic, and political project under the Kirchners in Argentina (Néstor Kirchner 2003–2007 and Cristina Fernández de Kirchner 2007–2015) was more moderate than the projects of the governments in Venezuela and Bolivia. The Kirchners' economic model was described as neo-developmental, as it contained elements of both populism and neoliberalism (Wylde, 2011). They selectively applied protectionist measures, intervened in the economy, and focused on agroindustry to drive growth.

In general, the Kirchners were more cautious in their dealings with the private sector than the more radical Pink Tide governments. They did little to reverse the privatizations carried out under Carlos Menem's government of the 1990s. They did, however, increase regulations on the private sector, which at times put them in conflict with domestic capital. And, as part of an effort to raise government revenue, they increased export taxes on soybeans, a major export crop. During Kirchner's government, export taxes were raised from 23.5% to 35% (Lapegna, 2017: 320). When Fernández tried to further increase export taxes on soybeans, though, media conglomerates backed soy producers and helped defeat the initiative (Etchemendy and Garay, 2011).

The Kirchners' relationships with TNCs were also less confrontational than those of the more radical left governments and, in some cases, they were on friendly terms. While Argentine capital controlled most of the soy production, global agribusiness controlled the agro-export chains, and Fernández maintained close relationships with Monsanto and Cargill (Lapegna, 2017). In contrast to Venezuela's agricultural policy, which prioritized the production of food for domestic consumption, channelled resources to smaller-scale producers, and experimented with agroecology, in Argentina the Kirchner governments supported agribusiness, the expansion of soy production for export, and agricultural biotechnology. These policies were initiated during the 1990s and the Kirchners continued them, seeking only to gain greater rents from this economic activity in order to increase state revenue.

TNCs were also involved in the expansion of the mining sector in Argentina – a country where mining had not been a major activity prior to the neoliberal economic reforms. Kirchner did not reverse the mining policies put in place in the 1990s, and in fact intensified extraction. The Kirchners viewed mining activity as "an engine of regional development and a source of national sovereignty" (López and Vértiz, 2015: 163). The mining sector was seen as a way to create jobs and provide cheap minerals needed in other parts of Argentina's economy. Yet, its activity had negative social and environmental impacts, and social movements emerged to challenge these TNC-led, government-supported mega-mining projects (López and Vértiz, 2015).

The Kirchners' political project was also more moderate than the project in Venezuela. They came to power through the Peronist Party, the same party as former President Menem who had implemented the neoliberal reforms in the 1990s. And, Fernández' running mate in 2007 was from the Radical Civic Union, a political party that historically had been the opposition party to the Peronist Party (Lapegna, 2017: 322). The election of Kirchner through one of the traditional political parties differs from what transpired in Venezuela, where there was a complete collapse of the traditional mainstream political parties. Also in contrast to the "radical left" governments in Venezuela, Bolivia, and Ecuador, the Pink Tide government in Argentina did not write a new constitution. And, it did not attempt to create institutions to facilitate a more participatory form of democracy.

Nonetheless, Argentina – like Venezuela – sought to reduce poverty and increase employment opportunities. Kirchner put measures in place that raised wages, appointed union-friendly officials to positions in the state, reversed some of the labour flexibilization measures enacted during the 1990s, and strengthened the role of national labour federations in collective bargaining

(Etchemendy and Garay, 2011). The government also provided subsidies to stabilize the price of certain goods to ensure popular access. In addition, the Kirchners increased social transfer programs, and expanded access to social services and pensions. According to Wylde (2011), Kirchner's social programs were largely meant to help only those who were at the very bottom. The nationalization of the pension system (a reversal of the partial privatization in 1994), however, expanded coverage to nearly the entire population. The government also launched some infrastructure projects, including the construction of public housing, and involved members of unemployed groups in the construction.

While Kirchner was confronted with high levels of poverty and unemployment when he was elected, he benefited from the fact that Argentina had begun to recover from its 2001 economic crisis at the beginning of his administration (Etchemendy and Garay, 2011). Under Kirchner, Argentina's GDP growth fluctuated between 8% and 9% from 2004 to 2007. The World Bank named Argentina "the top performer in the region in reducing poverty and boosting shared prosperity between 2004 and 2008. Incomes of the bottom 40% grew at an annualized rate of 11.8% compared to average income growth of 7.6%."[5] Unemployment fell from 17.6% in 2003 to 8.4% in 2009 (Etchemendy and Garay, 2011: 296). With the 2008 global recession, the more rapid growth of incomes at the bottom slowed down. Yet, as of 2014, the poverty rate was down to 12.7% (Ibid).[6] Inequality also dropped: the Gini Index decreased from .50 in 2004 to .43 in 2014. But, Argentina's exports remained heavily reliant on primary sector goods, specifically soy and soy products, which were subject to price fluctuations in international commodity markets.

The waning power of the Pink Tide

By the mid-2010s several of the Pink Tide states had been replaced by right-wing governments, through elections and other means. In Argentina, Cristina Fernández was followed by Mauricio Macri, whose electoral platform for the presidency in 2015 positioned him as a proponent of neoliberalism. And, in Brazil, the Worker Party's President – Dilma Rousseff – was removed from office in a politically orchestrated manoeuvre in 2015 and replaced by Michel Temer. Temer is a far-right politician. A number of factors contributed to this situation across the Pink Tide states, four of which we discuss here.

The 2000s were characterized by a "boom" in the prices of the key commodities of a number of these countries, following the commodity "bust" of the 1980s and 1990s. According to the World Bank (2009: 51), "the real prices of energy and metals more than doubled [between 2003 and 2008], while the real price of internationally traded food commodities increased 75 percent." Given that the economies of some of the Pink Tide States – particularly Venezuela, Bolivia, Ecuador, and Argentina – were heavily dependent on commodity exports, the vacillations of this market had a dramatic impact on the prospects of their governments and societies. The boom meant that they had vastly more resources with which to engage in development and redistribution. For example, Venezuela's *Misiones* were largely funded with revenues from oil exports, which made up about 95% of its exports in 2015.[7] Revenues from oil have also been important for Ecuador's economy, making up about 58% of total exports in 2012 (Mateo and García, 2014). Correa's government took advantage of the commodity boom to invest in infrastructure and social programs. In fact, in 2014 social spending by the Ecuadoran government was almost 3.5 times its 2007 level (*World Politics Review*, 2017). Social investments made there also led to a notable drop in poverty and extreme poverty, by 13% and 9%, respectively; and inequality was reduced by 8%.

However, the "bust" that followed this boom wreaked economic havoc, especially where the dependence on commodity exports was the strongest. Venezuela represents this dynamic in the extreme, with its GDP vacillating wildly between 2003 and 2015: its GDP grew 18.3% in 2004, fell by 4.7% in 2009, rose 4.2% in 2012, and fell again by 6.9% in 2015.[8] This pattern followed the international price of oil. It made planning extremely difficult and had a catastrophic effect on the economy. When the price of oil (and the GDP) fell, it made the imports of goods not produced locally more difficult because of the lack for foreign exchange earnings. Since shortages were the inevitable result, inflation took off: while it had averaged around 30% in the 2000s, it shot up to 48.5% in 2014 and 180.9% in 2015.[9] This hurt all sectors of society, but most especially those with fixed and low incomes. Not surprisingly, the economic downturn undercut support for the Maduro government, as was evident in the elections of December of 2015.

Given its dependence on oil exports, Ecuador's economy also experienced tremendous fluctuations between 2003 and 2015. However, reflecting the country's more varied export repertoire, the fluctuations were less extreme: GDP growth was 8.2% in 2004, it dropped to negative 1.1% in 2009, rose to 7.9% in 2011, and fell to negative 1.3% in 2015.[10] Its ability to import was also less affected, and inflation rates did not rise. Hence, the impact on political support for the government was less, but still notable: while Correa had won by wide margins in 2006 and 2009, his successor – Lenín Moreno – won with only 51% of the vote in the second round of elections in 2017. In the wake of those elections, Moreno adopted a more conciliatory position toward business and allowed an anti-corruption drive to lead to the removal of his vice-president. His position generated a split in the ruling party – the Country Alliance – that Correa had founded, but also appeared to have stimulated a dramatic increase in Moreno's popularity.

A second factor that came into play where there was a turn away from a Pink Tide government was a distancing in the relationship between the political party leadership and the social movement base that had brought them to power. One case where this pattern was evident was Brazil. The government of Luiz Inácio Lula Da Silva (Lula) was elected in 2002 on the Workers' Party (PT) platform. The PT, which Lula had been an early leader of, was formed in 1980 (during Brazil's military dictatorship) as a joint effort of a broad swath of civil society and social movement organizations. The party, working hand in hand with these mostly leftist organizations, had been a significant organizing force of the opposition during the two decades leading up to his election. Nonetheless, once in office, Lula and his successor – Dilma Rousseff – oversaw a government that prioritized economic stabilization and was quite moderate in its reform initiatives. The outcome was that, while poverty was reduced, inequality remained as extreme as ever (Zibechi, 2016). Once Lula's unwillingness to steer a more radical course was clear, civil society and social movement organizations began to separate themselves from him and from the PT (Hochstetler, 2008).

Hence, when massive mobilizations emerged in 2013, which were triggered by a rise in bus fares, the social movement organizations at the helm were not affiliated with – nor necessarily sympathetic to – the PT (Friendly, 2017). The mobilizations quickly expanded to a denunciation of a wide variety of issues, including the restructuring of public education, the state of health care, and government corruption. And, the right-wing opposition seized the initiative to frame the PT government as the problem, positioning itself to benefit from the impeachment/coup that followed.

Loss of support from social movements also occurred in the case of Ecuador. Ecuador's social movements – and principally its indigenous movements – played an important role in paving the way for Correa to come to power and initially were strong advocates of him. They, however, distanced themselves from Correa as they came to believe that he was not interested in implementing structural reforms that would create a more participatory and equal society. His agricultural policy, for example, privileged large-scale production, not smaller farmers like those at the core of Ecuador's powerful indigenous movements. They also criticized him for his focus on economic growth driven by environmentally contaminating extractive activities. Moreover, activists who challenged these extractive activities were charged with offences such as terrorism and sabotage. As he lost the backing of various social movements on the left, he became more reliant on the unorganized urban lower-class – who benefited from social programs – for his electoral base. Not only was this a more precarious base, but also the social movements that abandoned him were the very movements that had toppled multiple presidents in the decade preceding Correa's election.

A third factor was the regaining of strength by the right in some of the Pink Tide countries. The neoliberal era had a contradictory impact on the right. On the one hand, it lost political power in the countries considered part of the Pink Tide. On the other hand, the decades of neoliberal policies had, in many cases, fortified the economic base of the right. Initially, due to the vast popular mobilization that brought the Pink Tide governments to power, the increasing government revenues from the commodities boom, and the use of this revenue for social programs, the right was not able to successfully undermine these governments. Over time, however, as some countries experienced economic problems due to over reliance on particular export commodities (and subsequent falls in their prices) and the social bases that had brought these governments to power became less mobilized in support of them, the right was able to exert more power. Furthermore, in the neoliberal era the media had become ever more concentrated. According to Díaz Echenique, Ozollo, and Vivares (2011: 203), "[d]uring the final quarter of the twentieth century, the region as a whole witnessed the consolidation of diverse oligopolies of media communication, controlled by large transnational companies." The highly concentrated media conglomerates, in alliance with other actors on the right, allowed the latter to project their narrative, and limit competing framings.

Fernández' loss of the 2008 "farm war" in Argentina is a prime example of the right using its economic strength and control of the media to increase its political influence and expand its political power within the legislative branch. In Argentina, neoliberal policies of liberalization, privatization, and deregulation were implemented in the agricultural sector under Menem, and this ended up strengthening the country's large economic groups. As the price of international grains increased, Argentina's countryside was increasingly converted into massive tracts of genetically modified soy, which further concentrated land. Given the country's dependence on soy exports to fuel its economic growth, the owners of these large farms, while numerically small, had significant economic power. When Fernández tried to increase export taxes on soy, these producers waged a successful four-month long battle to prevent it. They used a variety of tactics including a national strike and road blockades, which created food shortages (Díaz Echenique, Ozollo, and Vivares, 2011). According to Cannon (2016: 125), "[c]osts to the economy were high … with an estimated US$3.4 billion being lost in total earnings." A key ally that played a pivotal role in their success was the media, which was able to spin a particular story about the conflict that provided legitimacy to the struggle waged by the agro-export producers, projected a sense of political crisis, and, over time, won over a diverse group of actors within the country. The right successfully defeated the agro-export tax with Fernández' very own Vice-President – who was from the opposition party – turning on her, to place the decisive vote in opposition to it. The "farm war" was a major setback for the governing coalition and led to a loss of electoral support in the 2009 legislative elections. Not only did the governing coalition's electoral support fall from 60% to 30%, it also lost control of Congress (Díaz Echenique, Ozollo, and Vivares, 2011: 204). And, in 2015 the right won the presidency with the election of Mauricio Macri.

Finally, and briefly, it must be mentioned that the US has not been a neutral actor vis-à-vis these states. It has supported the opposition press, strategized with and funded opposition candidates, endorsed illegal to highly questionable political manoeuvres by the right, and used its diplomatic might to attempt to undercut – if not simply oust – each of these regimes (Cannon, 2016; Díaz Echenique, Ozollo, and Vivares, 2011). Had it not engaged in these activities, the path forward for the Pink Tide states would certainly have been less strewn with obstacles.

Conclusion

Despite the loss of the presidency to right-wing parties in several of the Pink Tide states, and the political (and economic) crisis in another (Venezuela), the left – whether of a radical or moderate nature – is still the dominant political actor in most of the remainder of these Latin American states. Yet, its weakening in these senses suggests a number of things regarding its alternative approach to development. The first is that it is critical to move beyond the production of a few, key primary products – especially those derived from extractive industries – given the economic volatility and environmental problems they bring with them. Until now, government policy makers have often considered undertaking such a move to be impossible while trying to increase public spending to bring about greater social justice. However, not doing so has cast these regimes as prioritizing the growth of foreign exchange earnings over protecting the health, well-being, and prevailing lifeways of the population. Further transformation of the economy *will* pose new dilemmas, as it typically requires redistribution of productive resources that entails a greater degree of confrontation with the dominant economic class. But, redistribution will also undercut the economic strength of that class, which should provide breathing room for the change that is sought.

And, second, maintaining the engagement of those who brought these states to power is essential if they are not to be won over by the opposition in the inevitable moments of difficulty that result from modifications in the structure and orientation of the economy. Ideally, their ongoing engagement can come, among other ways, through their involvement in the new economic activities and novel forms of organizing production that emerge from this effort. In addition, in cases like Bolivia, the contradiction between the conceptualization of development enshrined in the new constitution – Vivir Bien – and the government's continued (and expanded) reliance on extractive industries produced numerous conflicts between the government and part of its base. Its failure to realize a fundamental shift in this regard resulted in the government's loss of popularity. Only when these governments' base is effectively included in the construction of a genuinely alternative approach to economic growth will uninterrupted pursuit of a sustainable, socially just model of development be possible.

Notes

1 According to the "Urban Dictionary," this term originated in the 1920s and referred to the spread of communism or socialism. In contrast, the *Oxford Reference* dates its usage to the electoral victory of Hugo Chávez Frías as Venezuela's president in 1998 and states that it describes the multiple left-of-center governments that were elected to power in Latin America starting then.
2 World Bank data, 2017 (www.worldbank.org/en/country/venezuela/overview). Accessed on 17 May 2017
3 The figures for 1996 are from Buxton (2003: 121), and for 2012 from CEPAL (2013: 52).
4 CEPAL (http://estadisticas.cepal.org/cepalstat/Perfil_Nacional_Economico.html?pais=VEN&idioma= english). Accessed on 17 May 2017.
5 World Bank, 2009, www.worldbank.org/en/country/argentina/overview. Accessed on 16 August 2017.
6 Poverty in this calculation was defined as living on less than $4/day.
7 Calculated from OPEC, "Venezuela Facts and Figures. See www.opec.org/opec_web/en/about_us/171. htm. Accessed on 19 May 2017.

8 CEPAL data, 2017. See http://estadisticas.cepal.org/cepalstat/Perfil_Nacional_Economico.html?pais= VEN&idioma=english. Accessed on 20 May 2017. As of this writing, neither CEPAL nor the World Bank have statistics available for Venezuela for 2016; however, according to Trading Economics, the country's GDP fell by 18.6% in 2016. See https://tradingeconomics.com/venezuela/gdp-growth-annual. Accessed on 17 January 2018.

9 Ibid. for figures up to and including 2015. See Footnote 8 with regard to these sources for later dates; however, Trading Economics posits that inflation had reached a monthly rate of 85% by December of 2016. See https://tradingeconomics.com/venezuela/gdp-growth-annual. Accessed 17 January 2018.

10 CEPAL data, 2017. See http://estadisticas.cepal.org/cepalstat/Perfil_Nacional_Economico.html?pais= ECU&idioma=english. Accessed on 20 May 2017.

References

Azzellini, D. (2011) De las cooperativas a las Empresas de Propiedad Social Directa en el proceso Venezolano. In C. Piñiero-Harnecker (ed) *Cooperativas y socialismo: Una mirada desde Cuba*. Havana, Cuba: Editorial Caminos, pp. 301–320.

Buxton, J. (2003) Economic policy and the rise of Hugo Chávez. In S. Ellner and D. Hellinger (eds) *Venezuelan Politics in the Chávez Era*. Boulder, CO: Lynne Rienner, pp. 113–130.

Cannon, B. (2016) *The Right in Latin America: Elite Power, Hegemony and the Struggle for the State*. New York; London: Routledge.

CEPAL. (2013) *Social Panorama of Latin America*. Santiago, Chile: United Nations.

Conaghan, C. (2011) Ecuador: Rafael Correa and the citizens' revolution. In S. Levitsky and K. Roberts (eds) *The Resurgence of the Latin American Left*. Baltimore: Johns Hopkins University Press, pp. 260–282.

Díaz Echenique, L., Ozollo, J. and Vivares, E. (2011) The new Argentine right and the Cristina Fernández administration. In F. Dominguez, G. Lievesley and S. Ludlam (eds) *Right-Wing Politics in the New Latin America: Reaction and Revolt*. London; New York: Zed Books, pp. 194–209.

Ellner, S. (2008) *Rethinking Venezuelan Politics*. Boulder, CO: Lynne Rienner.

Enríquez, L. and Newman, S. (2016) The conflicted state and Agrarian transformation in Pink Tide Venezuela. *Journal of Agrarian Change* 16(4): 594–626.

Etchemendy, S. and Garay, C. (2011) Argentina: Left populism in comparative perspective, 2003–2009. In S. Levitsky and K. Roberts (eds) *The Resurgence of the Latin American Left*. Baltimore: Johns Hopkins University Press, pp. 283–305.

Fernandes, S. (2010) *Who Can Stop the Drums? Urban Social Movements in Chávez's Venezuela*. Durham, NC: Duke University Press.

Friendly, A. (2017) Urban policy, social movements, and the right to the city in Brazil. *Latin American Perspectives* 44(2): 132–148.

Hammond, J. (2011) The resource curse and oil revenues in Angola and Venezuela. *Science & Society* 75(3): 348–378.

Hetland, G. (2014) The crooked line: From populist mobilization to participatory democracy in Chávez-Era Venezuela. *Qualitative Sociology* 37(4): 373–401.

Hindery, D. and Hecht, S. (2013) *From Enron to Evo: Pipeline Politics, Global Environmentalism, and Indigenous Rights in Bolivia*. Tucson: University of Arizona Press.

Hochstetler, K. (2008) Organized Civil Society in Lula's Brazil. In P. Kingstone and T. Power (eds) *Democratic Brazil Revisited*. Pittsburgh, PA: University of Pittsburgh Press, pp. 33–53.

Lapegna, P. (2017) The political economy of the agro-export boom under the Kirchners: Hegemony and passive revolution in Argentina. *Journal of Agrarian Change* 17: 313–329.

Lopez, E. and Vértiz, F. (2015) Extractivism, transnational capital, and subaltern struggles in Latin America. *Latin American Perspectives* 42(5): 152–168.

Mateo, J. and García, S. (2014) The oil sector in Ecuador: 2000–2010. *Problemas del Desarrollo: Revista Latinoamerican a de Economía* 45(177).

Mijeski, K. and Beck, S. (2011) *Pachakutik and the Rise and Decline of the Ecuadorian Indigenous Movement*. Athens: Ohio University Press.

Piñeiro Harnecker, C. (2009) Workplace democracy and social consciousness: A study of Venezuelan cooperatives. *Science & Society* 73(3): 309–339.

República Bolivariana de Venezuela. (1999) *Constitución de la República Bolivariana de Venezuela*. Gaceta Oficial, 30 December.

———. (2001) *Lineas Generales del Plan de Desarrollo Economico y Social de la Nacion, 2001–2008*, September.

———. (2010) *Ley Organica de las Comunas*. Gaceta Oficial, 21 December.

Veltmeyer, H. and Petras, J. (2010) Social structure and change in Latin America. In J. K. Black (ed) *Latin America: Its Problems and Its Promise: A Multidisciplinary Introduction*. Boulder, CO: Westview Press, pp. 119–137.

World Bank. (2009) *Global Economic Prospects: Commodities at the Crossroads*. Washington, DC: World Bank.

World Politics Review (2017) *Ecuador Finds Progress Made During Boom Years Is Unsustainable*. See www.worldpoliticsreview.com/trend-lines/19142/ecuador-finds-progress-made-during-boom-years-is-unsustainable. Accessed on 19 May 2017

Wylde, C. (2011) State, society and markets in Argentina: The political economy of *neodesarrollismo* under Néstor Kirchner, 2003–2007. *Bulletin of Latin American Research* 30(4): 436–452.

Yashar, D. (2005) *Contesting Citizenship in Latin America: The Rise of Indigenous Movements and the Post-Liberal Challenge*. Cambridge: Cambridge University Press.

Zibechi, R. (2016) Progressive fatigue? Coming to terms with the Latin American left's New 'Coyuntura'. *NACLA Report on the Americas* 48(1): 22–27.

8

RELIGION AND DEVELOPMENT

Javier Arellano-Yanguas and Javier Martínez-Contreras

Introduction

Religion influences public life and development in Latin America in multiple ways, from policies on reproductive health, to the daily life of the peasantry, to the activity of thousands of social organizations, to indigenous peoples' struggles for self-government. This influence was nurtured in the colonial period, when the Catholic Church provided a strong legitimating religious discourse for the ruling political authorities that, in turn, facilitated the permeation of the Church into social, economic, and political life. In numerous Latin American countries, the continuity of the political elites after independence and their links to the Church explains the perdurable historical influence of Catholicism.

Some decades ago, sociological theory predicted that the advance of economic development, education, and the adoption of modern lifestyles would lead to the decline of the importance of religion in the lives of individuals and in the public sphere (Bellah, 1976; Berger, 1969; Luckmann, 1967). Accordingly, the lens of religion was practically absent from the field of development studies. Religion was perceived as a relic lacking in capacity to structure society. At most, religion deserved attention due to its tendency to hinder modernization.[1]

Over the last decades, these predictions about the decline of religion have not been fulfilled. Europe's rapid path towards secularization is now seen as an exception rather than the norm. Religion has been remarkably pervasive in many world regions. In Latin America, the number of people who identify with a religion continues to be above 96% (Pew Research Center, 2014).[2] The religious switching from Catholicism to evangelical Protestant churches is the most significant change in the Latin American religious landscape. Nevertheless, despite a decrease in affiliation of 13 percentage points between 1995 and 2013, 67 to 69% of Latin Americans continue to identify as Catholic (Latinobarómetro, 2014; Pew Research Center, 2014).

The decline of the secularization theory has gone hand in hand with an increase in academic interest in the relationship between development and religion. In the past decade, academics have explicitly addressed the role of faith-based organizations in the promotion of development (Clarke, 2007; Olson, 2008a; Tomalin, 2012) and the influence of religion on development generally (Deneulin and Bano, 2009; Deneulin and Rakodi, 2011; Haynes, 2007; Lunn, 2009; Olson, 2006; Tomalin, 2015). In the Latin American case, the relationship between religion and development is addressed indirectly in the analysis of the influence of religion on politics

(Lehmann, 1992; Levine, 1981, 1988; Mainwaring, 1986; Olson, 2008b); democratization, protest, and social change (Levine, 1985, 1986; Levine and Mainwaring, 1989; Mainwaring and Wilde, 1989; Olson, 2006; Peterson, Vásquez, and Williams, 2001; B. Smith, 1975); the promotion of human rights (Levine, 2006; Wilde, 2015); the emergence of indigenous mobilizations (Cleary and Steigenga, 2004); and the role of faith-based organizations (Arellano-Yanguas, 2015; Olson, 2008a). Academic interest in the topic was especially high in the 1970–1980s. Changes in the traditional political and social positions of the Catholic Church due to the influence of liberation theology fueled that interest. After that period, the academic study of religion in relation to development lost momentum. Over the last two decades, the study of religion in Latin America has focused on the transformations of the religious landscape due to the growing importance of Pentecostalism across the continent, but with a diminished interest in investigating the social and political consequences of those transformations.

In this chapter, we assume a historical stance to forward an alternative analysis of the relationship between religion and development. Following Duneulin and Bano's approach (2009), we try to escape an instrumental logic in which religion is evaluated in terms of its contribution to models of development that have been defined independently of religious reference. The chapter focuses on how the involvement of religious people and institutions in social and political initiatives aimed at the promotion of justice, the eradication of poverty, the defence of human rights, and the protection of the environment has changed religion itself, catalyzing profound institutional, spiritual, and theological transformations. In turn, those transformations have reinforced the social commitment of Catholics across communities.

This chapter analyzes the historical transformation of the Catholic Church. The continuity, diffusion, and importance of the Church in Latin America allow us to undertake a longue durée approach. We argue that until recent decades, Catholicism enjoyed the monopoly of religion and was internally monolithic. Since the 1960s, and more clearly after Vatican Council II, a significant number of Catholics have engaged in a myriad of social projects and initiatives. The combination of new theological approaches and involvement at the grassroots has nurtured new ways of thinking and living the religious experience. Liberation theology and its associated developmental proposals epitomize those processes. The new situation has spurred divisions within the Church, generating dissent and space for a pluralistic understanding of its own nature and of the social models correlated with its doctrine. Pluralism has generated diverse and frequently conflicting visions and proposals. While some Church groups support a progressive agenda in favour of poor people, indigenous groups, and the defence of the environment, others seek the restoration of a more conservative and unified Church. The election of Pope Francis reinforces the position of the groups most committed to social transformation, but internal pluralism remains the key feature of Francis' approach.

The Catholic Church's social and political stance: a historical perspective

The Church's evolving positions on social and political issues in Latin America has been determined by its location in society, the groups it serves and to which it relates, and the religious doctrine that prevails at each historical moment.

The ecclesial presence during the conquest was marked by a general connivance with the Spanish and Portuguese empires. Religious doctrine was used to legitimize colonial domination over indigenous populations. However, despite this generalized subordination to political and military power, there was space for doctrinal discussion and plural practices, as some friars and priests dissented to the behaviour of the new rulers and took stances in defence of indigenous groups. This dissent disappeared in later centuries. The strong institutionalization, the close link with the colonial rulers and with the new national elites after independence, and the growing doctrinal control that gathered momentum in the Council of Trent (1545–1563) leading to the definition of the Papal infallibility in Vatican I (1869–1870) reinforced homogenization. There was little room for dissent (Dussel, 1983). In the 20th century, social and doctrinal changes created the conditions for the renewal of the Church and for the emergence of a Catholic commitment to social transformation to coexist with the traditional conservative approach.

The colonial Church: between domination and dissent

The expansion of Catholicism was one of the main engines of the conquest of Latin America. Spaniards "firmly believed that God had provided them with the means to subjugate and convert pagans to Christianity" (Penyak and Petry, 2006: 17). This position of spiritual domination was transferred to the social order. Indigenous people were reduced to slaves in the system of *encomiendas* (lands and indigenous people given to Spaniards as property). However, in those early moments of the conquest, some ecclesial figures raised their voices to denounce abuses and to defend indigenous peoples' rights. As early as 1511, Antonio de Montesinos, a Dominican friar, angered colonial authorities of *La Española*[3] by preaching the rights of indigenous people, the illegitimacy of slavery, and the need to restore their property. In the following decades, other ecclesiastics, among whom stands out Bartolomé de las Casas, also Dominican, defended those challenges to the colonial powers. His theological and political controversy with Ginés de Sepúlveda (1550–1551) in Valladolid in front of a Pontifical envoy marked opposing interpretations of the mission of the Spanish and Portuguese colonizers and the role of the Church: domination (Sepulveda) versus accompaniment and respect for indigenous people (Las Casas) (Buey, 1992). The former became the hegemonic practice in the following centuries, and the Church did not question the brutal methods to which the local populations were subjugated.

Between the 17th and 19th centuries, the Church became the wealthiest and largest landowner in Latin America (Schwaller, 1985). Moreover, the Church developed a dense network of convents, churches, universities, schools, hospitals, and orphanages that structured the social life of the colony. Religious orders and institutions were powerful actors and assimilated the worldview of rulers (Klaiber, 1992: 340). They put religion at the service of political power to maintain order and prevent revolts. Similar to the early years of the conquest, some Catholic leaders and religious orders challenged the communion between political and religious powers. For example, the Jesuit missions were established to protect indigenous populations from enslavement and the forced labour of *encomiendas* (Dolan, 1992).

The 18th century was marked by controversies between liberals and conservatives, preparing the path for the independence movements from the beginning of the 19th century to Cuba's independence in 1898. Those political events fractured the Church. Bishops and priests were bitterly divided over independence, seminaries were closed, the number of priests declined, and dioceses became vacant. During that period, Church property was confiscated to finance the operations of both sides, and Church influence was weakened. Despite these difficulties, the Church emerged from the independence processes maintaining an important social position and was frequently allied with conservative groups (Dussel, 1992).

Social Catholicism and popular experiments

During the 19th century, the Church gave way to two divergent lines that developed during the 20th century and continue to the present day. On one hand, the Catholic counter-reformation gained momentum, reinforcing centralization, doctrinal uniformity, conservatism, and authoritarianism. The definition of papal infallibility by the Vatican I Council epitomizes this tendency. The Church emerges from Vatican I turning its back on modernity, closing the door to internal doctrinal plurality, and allied with conservative political groups. On the other hand, at the end of the 20th century, the growing popularity of Marxism among industrial workers shook the Church. In 1891, Pope Leon XIII reacted by publishing *Rerum Novarum*, an encyclical responding to the struggles of workers. The document initiated a new stream, 'social Catholicism,' which was crucial in associating the growing commitment of Catholic groups in Latin America with the progress and modernization of their countries (Penyak and Petry, 2006). The Church was closed to any doctrinal innovation, but open to 'controlled' social experiments aimed at counteracting the popularity of revolutionary forces.

In the first decades of the 20th century, trade unions were founded in Mexico and Colombia, as were universities in such cities as Bogotá (1937), Medellin (1945), Sao Paulo (1947), and Buenos Aires and Cordoba (1950). Political parties, like El Salvador's Christian Democratic Party (1960) and Chile's National Phalanx Party (1938) were also established. These organizations promoted Catholicism and defended social justice. Although they were debtors of their upper- and middle-class backgrounds, they generated space for interaction between the traditional Church and popular movements, representing the transition from an older, elitist Catholicism to the more modern and pluralistic Catholicism of Vatican II (Klaiber, 1996: 28).

The Second Vatican Council (1962–1965) brought change to the whole Church. The Council called for an updating of the Church, asking leaders and adherents to acknowledge the contributions of modernity, to respect the autonomy of society, and to work to realize the hopes and aspirations of humanity. In Latin America, this new ecclesial atmosphere was felt in 1968 during the second meeting of the Conference of Latin American Bishops (CELAM) in Medellin, Colombia. The conference focused on two topics the bishops had never before considered: the grinding poverty of most of the Latin American population and the political instability affecting all countries moving between military dictatorships and oppressive regimes (Consejo Episcopal Latinoamericano, 1997). The content of the documents produced in that meeting represented a remarkable break with the historical trajectory of the Church, which had previously identified with the economic and political elites' interests. The documents criticized the ruling classes, capitalism, chronic poverty and social injustice, and called attention to the material and social conditions of the impoverished population of Latin America, establishing what was called "the preferential option for the poor" (Gutiérrez, 1990).

This new approach had two mutually reinforcing effects that transformed the traditional role of the Church in Latin America. On one hand, groups of Catholics multiplied their presence at the grassroots level with a strong commitment to the cause of popular movements. They went beyond the usual charity-based approach, adopting a reflection on how to change the structural conditions that provoke and nourish poverty. The popular education movement, inspired by Paulo Freire, was a good example of this new approach, but also experiences like the community of Solentiname (Nicaragua) or the "Comunidades Eclesiales de Base" extended across the continent (Goldfrank and Rowell, 2012).[4] On the other hand, the theological centres of Latin America were the cradles of liberation theology that influenced Catholic thought beyond the continent. The work of Gustavo Gutiérrez (1971) heralded this new way of thinking about God. He argued that the Church's mission is to accompany the process of liberating God's people and that the poor are the main actors in that process. The combination of these effects generated important changes in the Church and society.

The reinvigoration of the Church's engagement in social and political struggles did not happen without challenges and conflicts, external and internal. Catholic groups participated, with different weight, in the revolutionary experiments that abounded in Latin America in the 1960s and 1970s. In most cases, their initial enthusiasm about the possibilities of social transformation became frustrated by the realpolitik and the tendency in many of those experiments to lean towards authoritarianism. More frequently, Catholic groups and authorities were involved in the promotion of social justice and the defence of human rights. The traditional elites and the conservative factions of the Church saw in the new position of those groups a betrayal of the traditional alliance between the Church and secular power. Dramatic events in Central America during the 1970s and 1980s illustrate those conflicts. Numerous pastoral agents, priests, and bishops were killed in order to derail their commitment to the poor. Conservative elites, previously close to the Church, promoted their assassination, as was the case in El Salvador for Monseñor Romero in 1980, and Ignacio Ellacuría and other Jesuits in 1989.

In the 1980s and 1990s, the convergence of the exhaustion of revolutionary processes, the expansion of Pentecostalism, and the papacy of John Paul II catalyzed important changes in the Catholic Church that weakened its institutional commitment to social transformation. In the political arena, those decades were marked by neoliberalism, economic adjustment, and, in the best cases, the reconstruction of democratic institutions. The second Latin American Bishops Conference was held in Puebla, Mexico in 1979. There, the conservative section of the Church, led by Cardinal Alfonso López Trujillo, criticized and rejected liberation theology, especially Gustavo Gutierrez's work (Comblin, 1993). However, the debates did not support its general rejection. The Conference backed the ecclesial concern over the social and structural problems of Latin American society expressed in the previous Conference. Moreover, the bishops widened the previous agenda, showing their commitment to the defence of human rights and democracy. Observers said that by incorporating these new issues some bishops aimed to weaken the social importance of liberation theology.

In the 1990s, the Catholic hierarchy, under the leadership of John Paul II, blamed the ecclesial opening after Vatican II for the loss of faithful who passed to the ranks of Pentecostalism. The solution was to promote a conservative counter-reform that reinforced the leverage of conservative groups such as Opus Dei, Legion of Christ, Sodalitium of Christian Life, and Neo-catechumenal Way (Chesnut, 2010). In contrast, the 1990s saw the weakening of the "Comunidades Eclesiales de Base" and the erosion of the mobilization capacity of liberation theology (Comblin, 1993).

Church and social transformation in the new millennium: new agenda, new role, and papal commitment

The Latin American Church that entered the new millennium was politically and socially weaker than in the past, but also more plural and collaborative (Romero, 2009). The Church's customary coexistence of conservative and socially transformative forces continued, but internal conflicts were less virulent than in the post-council decades. From the perspective of ecclesial engagement in social transformation, the progressive ranks present three significant changes: (i) the incorporation of the defence of human rights and the protection of the environment into the social justice agenda; (ii) the new role of the Church as companion of grassroots organizations; and (iii) the impulse of the Pope to promote the Church's social commitments. These three changes imply a renewal of the Church's commitment to justice and social transformation, although doubts persist about its ability to sustain those commitments in the future. In parallel, the Church has also faced new crises.

The incorporation of the defence of human rights and the protection of the environment into the conventional agenda on social justice and explicit religious discourses has been a valuable contribution of the Church to the legitimacy of the struggles of peasant communities, indigenous peoples, environmentalists, and other subordinate groups against the growing power of corporations and their allied governments.

In the early 1990s, the environment became a key issue on the international developmental agenda. The Rio Summit in 1992 signalled the emergence of a post-Cold War order less concerned with the conventional right-left cleavage. Initially, the progressive Latin American Church was not especially keen to work on the green agenda, which it saw as less radical than the promotion of social justice. Thus, care for the Earth and the environment was completely absent from the main compilation of liberation theology published in 1990 (Ellacuría and Sobrino, 1990). However, the readiness of international donors and NGOs to support an environmental agenda prompted Church-related organizations to assume an environmental focus. Thus, initially, the commitment to human rights and the environment was seen as external to the Church – something the Church did, but not essential to its mission (Arellano-Yanguas, 2000). In the 1990s, growing Church involvement in these issues raised the question of the religious meaning of this type of engagement. Pastoral agents working on these topics discovered that there was strong link between their faith and their work. Moreover, they realized that environmental injustice is deeply interconnected with social injustice, ethnic and gender discrimination, and the violation of human rights (Carruthers, 2008). In that context, the emphasis on God as the God of life, one of the central concepts of liberation theology, catalyzed the formation of a religious discourse on rights and the environment (Levine, 2006: 131).

Church agents quickly realized that peasants and indigenous populations strongly embraced the idea of God as the creator and champion of the environment, because it reconnected religion with the defence of their livelihood and traditional spirituality. Additionally, the human rights framework allowed the Church to link its doctrine on inalienable human dignity with a secular discourse that was widely accepted. Thus, the Church could offer peasants, indigenous people, and other vulnerable groups a powerful discourse to confront abuses and social injustice. Beyond the personal involvement of nuns, priests, and other pastoral agents, institutionally this new perspective materialized in the creation of numerous offices and organizations for the defence of human rights and the environment that have been very active in supporting local populations in their resistance against mining and oil operations, dams, and other infrastructure projects in Guatemala, Honduras, Peru, Bolivia, and elsewhere.

In its recent role as companion of grassroots organizations, the Church has developed an extensive network of institutions that encompass the majority of Latin American geography, such as parishes, schools, social centres, health facilities, and radio stations. Historically, that institutional power gave the Church a social leadership role. That was also the case when sections of the Church got involved in the struggles for social justice in the 1970s. Bishops, priests, and nuns took on a clear leadership role. They felt a responsibility to sensitize the poor, to be the voice of the voiceless, to organize resistance and alternatives, and, in extreme cases, to offer their lives in martyrdom for the causes of the poor (Smith, 1991). However, in the 1980s and 1990s, the crisis of the left and the animosity of the Vatican towards liberation theology and its followers led to a revision of the strategy. The pastoral agents who had been mobilized through liberation theology took refuge in their work at the grassroots level. They animated thousands of social projects and humble initiatives that fostered the embeddedness of the Church in a new, less visible, way. That strategy, which initially appears as a retreat, had unexpected reinvigorating effects (Smith, 2002).

The principal effect was the change in the Church's role. Bishops, priests, and nuns shifted in their roles as leaders to companions. They understood that liberation theology is about the poor being the main actors in the historical process of liberation rather than its passive beneficiaries

(Gutiérrez, 1971). Thus, they developed a spirituality centred on accompanying the people in their own processes and assuming a subsidiary role regarding the leadership of lay people (Arellano-Yanguas, 2014). Rather than leading, they responded to the demands of the grassroots organizations they had worked with for a long time, sometimes even supporting the creation of new organizations.

The involvement of the Church had two different but equally important meanings. From an internal perspective, given the pervasiveness of religiosity in the rural Latin American mindset, Church backing legitimized grassroots organizations' demands in the eyes of the public. Moreover, the Church helped to link local mobilizations and struggles to national and transnational activist networks, amplifying their impact (Theije, 2006). The Church frequently provided the material infrastructure necessary to sustain mobilization over time.

The new Catholic position has, in turn, nurtured a new type of spirituality that puts the needs of subaltern groups first, thereby relativizing the need to control the processes to defend or reinforce the Church's social centrality. This new way of living the faith correlates with a new doctrine – a new way of understanding the Church and its mission. Of course, these processes and the groups involved in them are not the whole of the Church. They are not even the majority, because conservative Catholic groups continue to be very important and powerful. However, they are significant enough to make their influence felt. Today, many Catholics understand that their faith and work for justice are intrinsically connected. That is also the current papal doctrine.

The current papal standpoint is also a novelty. The election of Jorge Mario Bergoglio as Pope Francis in 2013 was surprising. His Latin American origin and what many perceived as his personal simplicity was the first surprise, but that was not all. Within a few months, his socially radical discourse and his vindication of liberation theology astonished many. Bergoglio had a personal record of austere living and closeness to the poor, but he had not demonstrated progressive ideological thinking. Rather, many feared the opposite – that the new Pope would represent the continuity of a doctrinally conservative social Catholicism. That impression quickly dissipated. His speeches consistently contain references to social justice and calls to take care of people and the Earth, our common home. Moreover, he has held meetings in the Vatican with popular social movements from around the world, and he has forwarded radical criticisms of the hegemonic economic and social models that articulate society. From the doctrinal perspective, in 2015 his first encyclical, *Laudato Si*, highlighted the main points of the new Pope's thought. The commitment to save the Earth, to liberate the poor, to change the patterns of consumption, and to transform the relationships with other people and, more generally, with all living beings, is not a mere practical translation of an immutable faith into actions. For the Pope, this commitment transforms the way Catholics understand their faith: there is no real faith without that transformative impulse. Pope Francis's approach closes a theological and historical circle: real faith calls for social engagement that, in turn and in parallel, transforms the world, the believers themselves, and, most radically, the image of God and the way faith is understood.

The unexpected appointment of this Pope has been a boost for groups of Catholics in Latin America that are seriously committed to social and ecological causes. After decades of rowing against the tide, the emergence of a richer agenda of issues, the tight interaction with grassroots organizations, and the new climate in the Vatican have opened spaces for the Church to collaborate more actively with other groups.

Nevertheless, not everything is positive within the Church. Catholicism currently faces three challenges that might undermine its capacity to contribute substantially to social transformation in Latin America. The first relates to its loss of institutional capacity due to the lack of new priests to replace the generations that prompted the change in the Church after the Second

Vatican Council, and the continuing loss of Catholic faithful who turn to evangelical groups. The second is its difficult coexistence with the growing pressure for gender equality. On some issues, internal pluralism has allowed the Church to satisfy different social sectors. However, regarding gender demands, its patriarchal structure and the belligerent positioning of the bishops in many countries against what they have pejoratively called 'gender ideology' have alienated the Church from a large number of women. Finally, the clumsy response to sexual abuse scandals by some members of the Church hierarchy has undermined the Church's credibility and legitimacy. These three phenomena weaken the Church and may limit its social impact and its ability to make significant contributions to social transformation.

Conclusion

This chapter has discussed some of the multiple connections between religion and development in Latin America. The traditional approach to the analysis of this relationship has been unilateral: the influence of religion on development. We have assumed a dialectical approach to show how the Catholic Church's involvement in the promotion of social justice issues, the defence of human rights, and the protection of the environment has also changed the way in which the Church understands its mission and faith. Initially, the chapter shows how the history of Latin American societies cannot be fully grasped without considering the social, political, and economic role of religion and its institutions. From the time of conquest, the Catholic Church has played a relevant role. Generally, the Church has been close to and has supported conservative ruling groups. Nevertheless, this conservative social position has not been unanimous. At different points in history, dissenting Catholic groups have advocated for the Church to side with subordinate social groups and their defenders. These different social approaches have generated internal ecclesial tensions.

The history of the Latin American Catholic Church during the last half century is probably the most relevant in terms of ecclesial transformation. After a long period of monolithic doctrinal uniformity, the period after the second war world, and more importantly after Vatican Council II, the Church entered a period of experimentation in which many Catholic groups started to collaborate in social or political projects to generate structural transformation. They were animated by new theological approaches that, combined with participation at the grassroots, nurtured new ways of thinking and living the religious experience. Liberation theology and its associated developmental proposals epitomize those processes. The new situation spurred divisions within the Church, generating dissent and space for a more pluralist understanding of its own nature and the nature of prevailing social models correlated with its doctrine. However, after a period of transformative enthusiasm, the ecclesial conservative turn in the 1980s and 1990s generated bewilderment but also a deeper commitment of some Catholic groups and new ways of engagement in which the grassroots groups are the protagonists of emancipatory processes. The election of Pope Francis has reinforced the position of Catholic groups committed to Earth's care, human rights, promotion of social justice, and the defence of indigenous peoples. However, the persistence of some important challenges raises doubts about the ability of the Church to sustain significant activity in favour of inclusive social transformation.

Notes

1 See, for example Rostow (1960).
2 Competing secularizing and evangelizing forces are present simultaneously in all Latin American countries. However, there are important differences between countries. While the percentage of people who

identified with a religion is above 95% in Paraguay, Ecuador, Bolivia, and Peru, it is only 63% in Uruguay (Pew Research Center, 2014).

3 Currently the Dominican Republic.

4 In 1965, the Nicaraguan priest, poet, and social activist Ernesto Cardenal established on the Solentiname archipelago (Nicaragua) an alternative religious community inspired by the Christian liberation theology.

References

Arellano-Yanguas, J. (2000) *Ecología en Perspectiva Salvífica*. Bilbao: Universidad de Deusto.

———. (2014) Religion and resistance to extraction in rural Peru: Is the Catholic Church following the people? *Latin American Research Review* 49(S): 61–80.

———. (2015) From preaching to listening: Extractive industries, communities, and the Church in rural Peru. In A. Wilde (ed) *Religious Responses to Violence*. Notre Dame, IN: University of Notre Dame Press, pp. 311–338.

Bellah, R. N. (1976) *Beyond Belief: Essays on Religion in a Post-Traditional World*. New York: Harper and Row.

Berger, P. (1969) *A Rumour of Angels: Modern society and the Rediscovery of the Supernatural*. London: Allen Lane.

Buey, F. F. (1992) La controversia entre Ginés de Sepúlveda y Bartolomé de las Casas: Una revisión. *Boletín Americanista* 42: 301–347.

Carruthers, D. V. (2008) *Environmental Justice in Latin America: Problems, Promise, and Practice*. Cambridge, MA: MIT Press.

Clarke, G. (2007) Agents of transformation? Donors, faith-based organisations and international development. *Third World Quarterly* 28(1): 77–96.

Cleary, E. L. and Steigenga, T. J. (2004) (eds) *Resurgent Voices in Latin America: Indigenous Peoples, Political Mobilization and Religious Change*. New Brunswick, NJ: Rutgers University Press.

Comblin, J. (1993) La iglesia Latinoamericana desde Puebla a Santo Domingo. In J. Comblin, J. I. Gonzalez Faus and J. Sobrino (eds) *Cambio Social y Pensamiento Social Cristiano en América Latina*. Madrid: Trotta, pp. 29–56.

Consejo Episcopal Latinoamericano. (1997) *Medellín, Reflexiones en el CELAM*. Madrid: Biblioteca de Autores Cristianos.

Chesnut, A. (2010) Conservative Christian competitors: Pentecostals and charismatic Catholics in Latin America's new religious economy. *SAIS Review of International Affairs* 30(1): 91–103.

Deneulin, S. and Bano, M. (2009) *Religion in Development: Rewriting the Secular Script*. London: Zed Books.

——— and Rakodi, C. (2011) Revisiting religion: Development studies thirty years on. *World Development* 39(1): 45–54.

Dolan, J. P. (1992) *The American Catholic Experience: A History from Colonial Times to the Present*. Notre Dame, IN: University of Notre Dame Press.

Dussel, E. (1983) Historia de la Iglesia en América Latina: Una interpretación. *Revista de História* 115: 61–87.

———. (1992) *Historia General de la Iglesia en America Latina: Medio Milenio de Coloniaje y Liberacion, 1492–1992*. Madrid: Mundo Negro.

Ellacuría, I. and Sobrino, J. (eds) (1990) *Mysterium Liberationis: Conceptos Fundamentales de la Teología de la Liberación*. Madrid: Trotta.

Goldfrank, B. and Rowell, N. (2012) Church, state, and human rights in Latin America. *Politics, Religion & Ideology* 13(1): 25–51.

Gutiérrez, G. (1971) *Teología de la Liberación: Perspectivas*. Lima: Centro de Estudios y Publicaciones.

———. (1990) Pobres y opción fundamental. In I. Ellacuría and J. Sobrino (eds) *Mysterium Liberationis: Conceptos Fundamentales de la Teología de la Liberación (Vol. I)*. Madrid: Trotta, pp. 303–321.

Haynes, J. (2007) *Religion and Development: Conflict or Cooperation?* Basingstoke: Palgrave Macmillan.

Klaiber, J. L. (1992) *The Catholic Church in Peru, 1821–1825: A Social History*. Washington, DC: Catholic University Press.

———. (1996) Catholic Action. In B. Tenenbaum (ed) *Encyclopedia of Latin American History and Culture*. New York: Charles Scribner's Sons.

Latinobarómetro, C. (2014) *Las Religiones en Tiempos del Papa Francisco*. Santiago de Chile: Latinobarómetro.

Lehmann, D. (1992) *Democracy and Development in Latin America: Economics, Politics and Religion in the Post-war Period*. Philadelphia, PA: Temple University Press.

Levine, D. (1981) *Religion and Politics in Latin America: The Catholic Church in Venezuela and Colombia*. Princeton, NJ: Princeton University Press.

———. (1985) Review: Religion and politics: Drawing lines, understanding change. *Latin American Research Review* 20(1): 185–201.

———. (1986) *Religion and Political Conflict in Latin America*. Chapel Hill: University of North Carolina Press Books.

———. (1988) Assessing the impacts of liberation theology in Latin America. *The Review of Politics* 50(2): 241–263.

———. (2006) Religious transformations and the language of rights in Latin America. *Taiwan Journal of Democracy* 2(2): 117–141.

——— and Mainwaring, S. (1989) Religion and popular protest in Latin America: Contrasting experiences. In S. Eckstein (ed) *Power and Popular Protest: Latin American Social Movements*. Berkley: University of California Press, pp. 203–240.

Luckmann, T. (1967) *The Invisible Religion: The Problem of Religion in Modern Society*. Basingstoke: Palgrave Macmillan.

Lunn, J. (2009) The role of religion, spirituality and faith in development: A critical theory approach. *Third World Quarterly* 30(5): 937–951.

Mainwaring, S. (1986) *The Catholic Church and Politics in Brazil, 1916–1985*. Stanford, CA: Stanford University Press.

——— and Wilde, A. (1989) *The Progressive Church in Latin America*. Notre Dame, IN: University of Notre Dame Press.

Olson, E. (2006) Development, transnational religion, and the power of ideas in the high provinces of Cusco, Peru. *Environment and Planning A* 38(5): 885.

———. (2008a) Common belief, contested meanings: Development and faith-based organisational culture. *Tijdschrift voor economische en sociale geografie* 99(4): 393–405.

———. (2008b) Confounding neoliberalism: Priests, privatization and social justice in the Peruvian Andes. In A. Smith, A. Stenning and K. Willis (eds) *Social Justice and Neoliberalism: Global Perspectives*. London: Zed Books, pp. 39–60.

Penyak, L. and Petry, W. (2006) *Religion in Latin America: A Documentary History*. New York: Orbis books.

Peterson, A. L., Vásquez, M. A. and Williams, P. J. (2001) *Christianity, Social Change, and Globalization in the Americas*. Piscataway: Rutgers University Press.

Pew Research Center. (2014) *Religion in Latin America: Widespread Change in a Historically Catholic Region*. Washington, DC: Pew Research Center.

Romero, C. (2009) Religion and public spaces: Catholicism and civil society in Peru. In F. Hagopian (ed) *Religious Pluralism, Democracy, and the Catholic Church in Latin America*. Notre Dame, IN: University of Notre Dame Press.

Rostow, W. (1960) *The Stages of Economic Growth*. Cambridge: Cambridge University Press.

Schwaller, J. (1985) *Origins of Church Wealth in Mexico: Ecclesiastical Revenues and Church Finances, 1523–1600*. Alburqueque: University of New Mexico Press.

Smith, B. (1975) Religion and social change: Classical theories and new formulations in the context of recent developments in atin America. *Latin American Research Review* 10(2): 3–34.

Smith, C. (1991) *The Emergence of Liberation Theology: Radical Religion and Social Movement Theory*. Chicago: University of Chicago Press.

———. (2002) "Las Casas" as theological counteroffensive: An interpretation of Gustavo Gutiérrez's "Las casas: In search of the poor of Jesus Christ". *Journal for the Scientific Study of Religion* 41(1): 69–73.

Theije, M. D. (2006) Local protest and transnational Catholicism in Brazil. *Focaal* 47: 77–89.

Tomalin, E. (2012) Thinking about faith-based organisations in development: Where have we got to and what next? *Development in practice* 22(5–6): 689–703.

———. (2015) *The Routledge Handbook on Religions and Global Development*. London: Routledge.

Wilde, A. (2015) *Religious Responses to Violence: Human Rights in Latin America Past and Present*. Notre Dame, IN: University of Notre Dame Press.

PART II

Globalization, international relations, and development

9

POST-NEOLIBERALISM AND LATIN AMERICA

Beyond the IMF, World Bank, and WTO?

Tara Ruttenberg

Economic dependence: yesterday and today

In many ways, Latin America's role in the global capitalist economy has changed little from the late 19th century, when the region's export-oriented growth model was established alongside the development of its raw materials sector – notably reliant on foreign capital and effectively solidifying the dependent relationships between Latin American export economies and the foreign interests and markets that both control these export industries and rely on them for their own economic flourishing (Skidmore and Smith, 2005: 44). While these dependent relationships became increasingly diversified with the complexities of modernity and industrialization, they can be traced as an evolutionary outgrowth of the colonial period when the economic grandeur of the Iberian Peninsula relied on the extraction of the Earth's riches in the form of silver and gold and agricultural goods, made possible by the exploitative slave labour of the New World's native peoples. In the decades following formal independence from Spain in the early 19th century, any remaining whispers of hope for the Bolivarian dream of a "liberated and unified" Latin America (Ali, 2006: 127) dissipated northward into silent resignation as the region saw its economic dependence on the Peninsula replaced by that of the "free markets" of the United States and Europe. As modern industrialization required more and more raw materials from the South, Latin America's position as supplier of natural resources, commodities, and cheap labour to the globalized capitalist system was all but set in stone. While formally free from colonial domination, market imperialism ensured a new round of neocolonialism for at least a century to come.

Flash forward to the early 1990s, the end of the "lost decade" in Latin American economic history and the height of the debt crisis; a time when the Washington Consensus reigned supreme and the region was hard-pressed to accept the policies of the Bretton Woods institutions as their only saving grace from financial crisis. These policies ushered in the full-fledged assault on public ownership, trade protectionism, and state economic intervention that would come to define the neoliberal project of the late 20th century. With Latin America's highly lucrative export and extraction industries in the hands of foreign corporations in collusion with national elite interests exploiting low-paid labour, and social services increasingly privatized without safety nets for the income-poor, the start of the 21st century was a grim period for the majority of people in the Americas, ushering in a new wave of anti-imperial sentiment as poverty, insecurity, and income inequality skyrocketed (Green, 2003; Roddick, 1988).

While the story told here is perhaps oversimplified in historical detail, the experience of Latin America's long-standing economic dependence and colonial-turned-neocolonial relations of production has persisted throughout the history of the region's insertion into the global economy dating back to the 16th century under the Spanish empire. As power politics transferred economic and geopolitical hegemony from Europe to the United States in the centuries that followed, Washington wielded coercive political, economic, military, and ideological measures to consolidate US influence as a neo-imperial power vis-à-vis the Americas (Pearce, 1982). In contemporary times, these historically situated power relationships came to define the experience of economic development in Latin America throughout the second half of the 20th century.

The IMF, World Bank, and WTO in Latin America

In the post-World War II period, the international financial and development agendas colluded among the Bretton Woods institutions created in 1944, including the International Monetary Fund (IMF), World Bank, and World Trade Organization (WTO) (previously the Global Agreement on Tariffs and Trade – GATT), responsible for designing and imposing a conditional policy framework intended to open new markets for trade and finance while guaranteeing the Global North's privileged access to Latin American resources and labour. The International Monetary Fund is a financial organization comprising 189 member countries, whose policies follow a liberal economic framework to foster and sustain economic growth, financial and trade liberalization, as well as global monetary cooperation. The World Bank is an international banking organization that acts as an arbiter in distributing economic aid and offering loans as part of an economic growth-based approach to global poverty alleviation. Finally, the World Trade Organization is the global entity designed to determine and monitor international trade in goods and services among member countries. Together, the IMF, World Bank, and WTO were the most powerful international organizations determining global economic policy and international development objectives in this period. From the outset, international development objectives sought new and expanded commercial and financial markets, industrial modernization of the Global South in the image of the West, capital accumulation for elite interests, and economic growth as a driver for the global capitalist economy. As a resource-rich region with vast market potential for consumption, investment, and savings, Latin America was fertile ground for the growth-oriented objectives of the international development agenda (Peet, 2003; Woods, 2006).

As economic growth stagnated throughout Latin America in the later stages of import-substitution industrialization (ISI), resulting primarily from domestic market saturation and high costs associated with importing expensive industrial machinery and other production inputs, the IMF responded by offering loan packages to Latin American governments in attempt to stimulate growth, setting a number of conditional policy requirements that became the standards for international financial flows into Latin America (Skidmore and Smith, 2005). As policies shifted from import substitution industrialization (ISI) in the post-WWII decades to trade liberalization, resource privatization, and export-led economic growth strategies in the 1980s and 90s, the Washington Consensus – a set of ten economic policy recommendations adopted by the Bretton Woods institutions – held significant influence over Latin America's governmental decisions on public spending, fiscal and monetary policy, tariffs and trade, as well as labour rights, property ownership, and corporate regulations. Specifically, the policies promoted by the Washington Consensus and imposed by the IMF, World Bank, and WTO include (Williamson, 2004):

1   Fiscal discipline in effort to remedy inflation and balance of payments deficits
2   Reordering public expenditure priorities away from non-merit subsidies and cash transfers to the income-poor
3   Tax reform
4   Liberalizing interest rates
5   Competitive exchange rate
6   Trade liberalization
7   Liberalization of inward foreign direct investment
8   Privatization of resources and services
9   Deregulation on entry and exit barriers for production, trade, and investment
10 Property rights (land privatization)

It is important to note that while the Washington Consensus was perhaps the most influential set of policy objectives informing neoliberalism in Latin America, the term itself was coined in 1989 after a long decade of neoliberal adjustments already implemented in the region throughout the 1980s, aligned with the international development policies espoused by the Organization for Economic Cooperation and Development, founded in 1948 (Panizza, 2009). As such, the Washington Consensus can be understood as evolving from the policy architecture defined by the broader international development agenda beginning in the post-WWII period and evolving slowly throughout the second half of the 20th century.

Latin America's debt crisis peaked in the early 1980s, when foreign commercial banks responded to global recession by raising interest rates and calling-in outstanding loans at a time when regional economic growth had stagnated or declined, budget deficits were rampant, and Latin America, as a region, held $327 billion in debt to US creditors (FDIC, 1997). In 1982, Mexico was unable to repay its 80-billion-dollar debt, and 15 other countries in Latin America followed-suit shortly thereafter, requiring significant debt rescheduling and conditional loan restructuring as a strategy to mitigate crisis. The majority of banks ceased lending to Latin America as if overnight, resulting in deep and sudden recession across the Americas. Washington then negotiated a financial "rescue" strategy, leveraging funds from international creditors, including the IMF as a primary lender, to offer emergency loans in exchange for Latin American governments agreeing to significant structural economic reforms as a means toward reversing budget deficits (Griffith-Jones and Sunkel, 1989).

Particularly detrimental in this period were the structural adjustment policies (SAPs) implemented as a last-ditch effort to boost economic growth, as Latin American governments struggled unsuccessfully to repay hefty multilateral loans, despite adhering to the policy conditions dictated by Washington. Structural adjustment entailed a further hollowing-out of the state via the privatization of public services like health care and education, plus devastating cuts in social welfare and the elimination of safety nets for the income-poor (Brown et al., 2000; Sparr, 1993). The effects on poverty and social inequality were extreme, resulting in a widening gap between the rich and income-poor, rising unemployment, an exacerbation of extreme poverty, and a diminishing middle class. By the end of the 20th century, Latin America had become the most income-unequal region in the world, violence surged as a result, and anti-neoliberal sentiment was ripening quickly toward political change.

Beginning in the 1990s, a number of regional and multi-lateral free trade arrangements – including the North American Free Trade Agreement (NAFTA), the Central American Free Trade Agreement (CAFTA), and the Free Trade Area of the Americas (FTAA or ALCA for its acronym in Spanish) – were negotiated under the same legal framework as the WTO, with contradictory protectionist clauses favouring US agricultural and corporate interests, while Latin

American nations were forced to liberalize to the detriment of small-scale farmers, small-to-medium sized businesses, and labour unions who suffered the consequences. Production and consumption patterns have transformed dramatically as a result, with many Latin American countries now importing much of what they once produced as staple crops for subsistence living. Under NAFTA, for example, Mexico – once a thriving producer of a diverse variety of corn, an integral part of the Mexican diet and culture – became the world's largest corn importer (primarily) from the United States, whose highly subsidized and almost entirely genetically modified corn industry benefits from the one-sided protectionist clauses favouring US agroindustry vis-à-vis small producers in Mexico and elsewhere in the Americas (Rhoda and Burton, 2012; Fox and Haight, 2011; Carlsen, 2011).

With international loans conditioned on the adoption of IMF requirements, Latin American governments had little room for negotiation or choice in the matter if they were to receive the loans they needed to finance development platforms and ambitious infrastructure and public works projects. Similarly, the economic ideology behind the Washington Consensus was enticing, with neoliberalism promising a "rising tide to lift all boats," founded on neo-classical economics' infamous "trickle-down" theory, implying that economic growth was a panacea for socioeconomic wellbeing, and that increased GDP would "trickle-down" through social classes for the betterment of all, particularly the income-poor (Dollar and Kraay, 2001, 2002). Unfortunately for the majority of Latin Americans, neoliberal theory left much to be desired where the realities of social welfare, income equality, and access to social services and dignified livelihoods were concerned. A rising tide did not, in fact, lift all boats. Instead, "redistributive effects and increasing social inequality have in fact been such a persistent feature of neoliberalization as to be regarded as structural to the whole project" (Harvey, 2005: 16).

From the macro to the micro levels, the Bretton Woods institutions defined Latin America's economic development experience throughout the second half of the 20th century. The conditional policy-loan framework of the IMF determined a narrow vision for continued economic dependence on and dominance by Washington, making policy independence for Latin American governments virtually impossible with so much debt accrued over the years. Foreign development funding by the World Bank focused on income-generating projects at the micro-level to encourage both savings and consumption among Latin America's income-poor as both a poverty alleviation strategy, and a means of economic growth via wider market access for consumer goods and financial services. Finally, the WTO defined the terms for trade arrangements that shifted Latin American economies away from subsistence farming toward urban manufacturing, textiles, resource extraction, and large-scale foreign-owned agriculture, transforming the employment landscape and livelihood opportunities across the Americas. Together, as the architects and administrators of the neoliberal economic development agenda, the IMF, World Bank, and WTO institutionalized the economic growth-for-development paradigm as the cornerstone of Latin America's development policy priorities, both past and present.

The neoliberal policy experience in Latin America, particularly as a result of exorbitant debt and structural adjustment cuts to social spending in the 1980s and 90s, was responsible for many of the region's social challenges at the turn of the current century, and persisting today – including rising poverty, income inequality, and violent crime (Huber and Solt, 2004; Kingstone, 2011; Jaramillo and Saavedra, 2010). Palma's (2011) review of income deciles highlights the growing income share of the wealthiest 10% as that of the poorest 10% diminished in Latin America during the years of neoliberal economic policy. Similarly, Petras, Veltmeyer, and Vieux (1997); Green (2003); and Grugel and Riggirozzi (2009, 2012) all "separately report the increasing levels of poverty and indigence over the [neoliberal] period from 1980 to 1999" (Kingstone, 2011: 76).

A significant body of research emphasizes the correlation between income inequality and violent crime and conflict (Casas-Zamora and Dammert, 2012; Centro Internacional de Informacion e Investigaciones para la Paz, 2000; Bourguignon, 2001; Fajnzylber, Lederman, and Loayza, 2002), helping us understand why Latin American countries have experienced rising violence in recent decades. At the start of the 2000s, the region of Latin America and the Caribbean (LAC) was the most income-unequal and one of the most violent regions in the world, with over a third of its population living in poverty. LAC income inequality peaked at a Gini coefficient of .53 (Lopez-Calva and Lustig, 2010), and of the top ten countries with the highest homicide rates in the world, eight of the ten are located in LAC (UNODC, 2014). The region has "long exhibited the highest income concentration in the world" (Casas-Zamora and Dammert, 2012), and in 2003, the richest income decile in LAC earned 48% of total income, while the poorest 10% earned 1.6% of total income (Ferreira et al., 2004). In 2001, 36% of the population of LAC lived below the poverty line (Leipziger, 2001).

While the quantitative expression of neoliberalism's effects on poverty, inequality, and violence in Latin America is staggering in statistics, qualitative perspectives offer a more nuanced view into the lived experiences of citizens suffering the often-indirect nature of the impact of neoliberalism, what Reygadas (2006: 139) refers to as "inequality by expropriation and inequality by disconnectedness." As a result of trade liberalization favouring large corporations over small producers and elite plundering of public wealth under privatization, Latin Americans experienced the centrifugal income effects of wealth accumulating among the already wealthy vis-à-vis those "disconnected" social sectors lacking the financial, technological, and educational resources necessary to improve their economic situation. Those most negatively affected by neoliberalism's prioritization of macroeconomic concerns over social wellbeing included: "wage workers, *campesinos* and small farmers, small and medium entrepreneurs oriented to the domestic market, women and young people" (Vilas, 2006: 237), with excessive unemployment resulting from the Washington Consensus' economic restructuring agenda implemented in the 1990s (Grugel and Riggirozzi, 2012; Kingstone, 2011). Lustig (1999, as cited in Lopez-Calva and Lustig, 2010:6) comments: "The unequalizing effect of the [debt] crisis was compounded because safety nets for the poor and vulnerable were conspicuously absent (or poorly designed and inadequate) in the Washington-led structural adjustment programs in the 1980s." As the research demonstrates, neoliberal policies – including privatization, trade liberalization, structural adjustment, and elimination of social safety nets – exacerbated poverty and inequality in Latin America, contributing to increased violence, sowing resentment among excluded social sectors, and paving the way for the emergence of leftist Latin America's post-neoliberal policy framework, explored in detail to follow.

Post-neoliberalism in Latin America: toward a new development paradigm?

At the start of the 21st century, the socioeconomic failures of the Washington Consensus and neoliberal development strategies imposed by the Bretton Woods institutions were clear, and Latin America embarked upon an era of socio-political change founded on widespread anti-neoliberal sentiment, unearthing leftist ideologies grounded in the anti-imperialist legacies of Jose Martí and Simon Bolívar. In what is referred to as Latin America's "Pink Tide" or "left turn," leftist leaders campaigning on anti-neoliberal platforms won democratic elections across the region, including in Venezuela, Brazil, Bolivia, Ecuador, Argentina, Uruguay, Paraguay, El Salvador, Honduras, and Nicaragua. Their socialist-leaning intentions were clear, and resonated with a wide majority of Latin Americans whose quality of life suffered under neoliberal development policies. Social movements demanded alternatives to structural adjustment, and social wellbeing became a political priority, with governments implementing policies geared toward income redistribution and increased social spending to alleviate extreme poverty and reduce income inequality (Cameron and Hershberg, 2010).

As Washington's ideological and socioeconomic influence waned in the Americas, the IMF, World Bank, and WTO lost credibility as social movements and leftist governments coalesced around new strategies for development and social wellbeing. Reversing natural resource privatization and structural adjustment policies implemented in the 1980s and 90s, governments sought to nationalize foreign- and privately owned industries like oil, water, and natural gas as a means of public revenue for social spending on education and health care, as well as welfare and food programs for the income-poor. In Bolivia and Ecuador in particular, the rights of indigenous peoples and the environment have played a central role in both countries' newly articulated constitutions and national development plans, manifest through the concept and cosmovision of *buen vivir*, promoting social wellbeing in harmony with nature. While the inclusive political manifestation of *buen vivir* represents socio-ideological innovation and holds revolutionary potential in incorporating the rights of nature and previously marginalized peoples into domestic development priorities, serious limitations persist as both Ecuador and Bolivia continue to pursue post-neoliberal policy objectives reliant upon extractive resource industries fuelling economic growth, dictated by the still quite neoliberal growth-for-development paradigm (Ruttenberg, 2013a).

In 2002, Argentina (in the midst of a devastating financial crisis) surprised the world by defaulting on $100 billion in outstanding debt with the IMF, prioritizing national socioeconomic needs over creditor contracts, while setting a new precedent for Latin American economic sovereignty (Hebert and Schreger, 2015). Ecuador quickly followed suit when President Rafael Correa deemed his country's 3.9-billion-dollar debt immoral, and therefore illegitimate, and defaulted on payment. In 2006, Argentina and Brazil repaid their remaining debt in full, refusing future loans and entering into regional financial agreements instead, effectually freeing themselves of neoliberal policy conditions attached to IMF financing. And as an affront to WTO authority in the region, Latin America's newly elected leftist leaders strengthened existing, and ventured into new, regional trade and economic integration agreements, including ALBA, MERCOSUR, BANCOSUR, and UNASUR, exclusive of US leadership or involvement (albeit with varying degrees of success). This, at a time when the WTO's international credibility hit a series of crippling weak points, particularly in 2000 and 2008 when negotiations stalled due to intense pressure and protests from civil society calling for fairer trade to benefit disenfranchised farmers, small producers, and landless peoples of the Global South, at odds with multinational corporate interests and the economic power politics of wealthy nations.

While governments have separately responded to national issues of poverty, inequality, employment, debt, investment, inflation, taxation, land and labour rights, social expenditure, and environmental conservation in diverse ways, there are a number of similarities worthy of reflection and analysis. Scholars have defined a post-neoliberal era in contemporary leftist Latin America, characterized by a revival of the state in development and a policy platform founded on export- and extraction-led growth complemented by public expenditure toward income redistribution. The common identifying tenets of post-neoliberalism include (Ruttenberg, 2012, 2013b):

- Nationalizing natural resources and/or negotiating highly favourable resource ownership/ tax agreements to increase government revenue
- Tax reform to close loopholes and ensure proper collection

- Repaying, defaulting, or de-prioritizing foreign debt repayment
- Social policies targeting income redistribution and poverty reduction: cash transfers, social protection, and services
- Counter-cyclical economic policies: investment in infrastructure, emergency employment plans, stimulus for business, and social programs
- Social public expenditure on health, education, housing, agriculture, employment, poverty reduction, etc. funded by export-led growth, particularly dependent on commodities
- Prioritizing domestic production and labour demand
- Regional integration based on cooperation and solidarity to solidify the post-neoliberal economic and political framework

As a result of post-neoliberal policies implemented in the 2000s, and particularly due to increased social spending coupled with the favourable commodity boom, the start of the 2010s offered a beacon of hope where poverty and inequality indicators were concerned, led in large part by the example of the region's most "contestatory" leftist governments (Argentina, Bolivia, Ecuador, Nicaragua, and Venezuela) (Lustig, 2009; Kingstone, 2011). According to ECLAC regional statistics for Latin America, the poverty rate dropped from 43.8% (with extreme poverty at 18.5%) in 1999 to 30.4% (12.3% extreme poverty) in 2011 (ECLAC, n.d.). Similarly, a 2012 World Bank report concluded that since 1997, more than fifty million people had risen into the middle class in Latin America, representing 30% of the total population – nearly equal to the region's percentage of people living in poverty at the time (Watts, 2012). Birdsall and Szekely (2003) emphasize that social policies that strengthened human capital, most notably in education and health care contributed most to reducing poverty and inequality. Lopez-Calva and Lustig (2010) add that a decrease in the earnings gap between skilled and low-skilled labour, along with increasing financial transfers to the poor, including both government transfers and foreign remittances, contributed most to declining poverty and inequality at the time.

While quantitative indicators demonstrated socioeconomic improvement in poverty reduction, employment, greater income equality, and economic growth attributed to post-neoliberal policies in the first decade of the 21st century, 174 million Latin Americans still lived in income poverty and 73 million in extreme poverty in 2011 (ECLAC, n.d.). Similarly, significant disparity persists in access to quality social services like health care and education, reflecting the legacy of neoliberalism's heavy social toll throughout the region, whereby "poorer families have no choice but to put their children in low-standard schools and their sick in poorly funded hospitals, while the middle class spends substantial sums on private education and health care" (Watts, 2012). In addition, improvements in income indicators do not present a comprehensive understanding of the human face of poverty or social wellbeing, and the quantitative income-based data is both oversimplified and incomplete in that regard.

Consistent with the neoliberal aims of the Bretton Woods institutions and the fundamental objectives of the international development agenda, Latin America continues to pursue income-oriented strategies for poverty alleviation and social wellbeing, based on the economic growth-for-development paradigm – a policy strategy explicit in the invention of "Development" for the "third-world" in the post-WWII period (Escobar, 1995), and persistent in today's post-neoliberal era. As the fields of post-development and wellbeing economics are apt in illuminating, the twin development objectives of human wellbeing and environmental sustainability are incompatible with neoliberalism's market-oriented, capitalist, and resource-based modes of production and the subsequent levels of global consumption required to sustain economic growth. Moreover, income levels are not synonymous with social wellbeing, and per capita GDP is not a useful or effective indicator for determining quality of life. The place-based, culturally embedded, environmental, and individual nature of understanding human wellbeing presents a fundamental challenge to the economic growth-for-development paradigm on which both neoliberal and post-neoliberal socioeconomic objectives are founded. As such, exploring alternatives to growth-based development strategies and evaluating the qualitative elements of social wellbeing vis-a-vis post-neoliberal policy are two complementary areas in need of deeper inquiry.

Of similar concern, and vital to the critique of Latin America's post-neoliberal development experience, is the region's continued reliance on an export-led and extraction-based growth model whose heavy environmental toll and quickening resource depletion represent an unsustainable source of revenue for social expenditure beyond the short-term. Despite the waning regional influence of the IMF, World Bank, and WTO, post-neoliberal Latin America's socioeconomic model is intrinsically tied to economic growth, with development policy priorities presently inseparable from the region's externally dependent, commodity-based economies. In legacy and practice, the neoliberal tenets of Bretton Woods' interventions have become institutionalized in the post-neoliberal project and disturbingly presented as a seemingly independent or somehow alternative economic model, even couched under the misleading guise of a would-be "socialism for the 21st century" (espoused by Venezuela's Hugo Chavez, in particular). In reality, rather than representing a socialist alternative, consistence with neoliberal objectives as opposed to fundamental change in the economic growth model is foundational to the contemporary post-neoliberal period in Latin American development.

As we near the end of the second decade of the 21st century, Latin America sits at an important juncture in determining the future of its socioeconomic history. While vociferous in anti-imperial rhetoric and progressive in leveraging natural resources for national economic interests, post-neoliberal Latin America has so far demonstrated a pragmatic continuity with the region's centuries-long history of economic dependence and contemporary neoliberal economic growth-for-development paradigm, consistent with the policy objectives of the Bretton Woods institutions, Washington Consensus, and the neoliberal international development agenda. Greater economic autonomy and significant change in the region's socioeconomic model will require a structural shift away from the unsustainable growth model and a fundamental transformation in the relations of production, moving away from extractive, capitalist enterprise toward a post-development, post-extractivist future (Escobar, 2012; Gudynas, 2011) aligned with social wellbeing priorities – perhaps as envisioned by the framework of *buen vivir*, as both socio-environmental concept and broader cosmovision (Gudynas and Acosta, 2011; Gudynas, 2010) – a paradigm for "good living" in harmony with nature.

References

Ali, T. (2006) *Pirates of the Caribbean: Axis of Hope*. New York: Verso.

Birdsall, N. and Szekely, M. (2003) Bootstraps, not Band-Aids: Poverty, equity and social policy. In J. Williamson and P. Kuczynski (eds) *After the Washington Consensus: Restarting Growth and Reform in Latin America*. Washington, DC: Peterson Institute for International Economics, pp. 49–73.

Bourguignon, F. (2001) Crime as a social cost of poverty and inequality: A review focusing on developing countries. In S. Yusuf et al. (eds) *Facets of Globalization: International and Local Dimensions of Development* Washington, DC: World Bank, pp. 171–192.

Brown, E., Milward, B., Mohan, G. and Zack-Williams, A. (2000) *Structural Adjustment: Theory, Practice and Impacts*. New York: Routledge.

Cameron, E. and Hershberg, E. (2010) (eds) *Latin America's Left Turns: Politics, Policies & Trajectories of Change*. Boulder, CO: Lynne Rienner.

Carlsen, L. (2011) NAFTA is starving Mexico. *Foreign Policy in Focus*. Washington, DC: The Institute for Policy Studies. www.fpif.org/nafta_is_starving_mexico/

Casas-Zamora, K. and Dammert, L. (2012) *Public Security Challenges in the Americas*. Washington, DC: Brookings Institution Press. www.brookings.edu/~/media/research/files/reports/2012/7/07%20 summit%20of%20the%20americas/07%20public%20security%20casas%20zamora%20dammert

Centro Internacional de Informacion e Investigaciones para la Paz (2000) *El Estado de la Paz y la Evolucion de las Violencias en America Latina*. Montevideo: Centro Internacional de Informacion e Investigaciones para la Paz, pp. 97–138.

Dollar, D. and Kraay, A. (2001) Trade, growth and poverty. *Finance and Development* 38(3). www.imf.org/external/pubs/ft/fandd/2001/09/dollar.htm (Accessed 10 May 2018).

———. (2002) Growth is good for the poor. *Journal of Economic Growth* 7(3): 195–225.

ECLAC. (n.d.) *Latin America (18 Countries): People Living in Poverty and Indigence – Around 2002, 2010 and 2011, Public Information and Web Services Section*. ECLAC, no date. www.cepal.org/prensa/noticias/comunicados/9/48459/tabla-pobreza-indigencia-en.pdf

Escobar, A. (1995) *Encountering Development: The Making and Unmaking of the Third World*. Princeton, NJ: Princeton University Press.

———. (2012) Post-extractivismo y pluriverso. *América Latina en Movimiento*, No 473. www.alainet.org/es/active/53567

Fajnzylber, P., Lederman, D. and Loayza, N. (2002) Inequality and violent crime. *Journal of Law and Economics* 45(1): 1–40.

FDIC. (1997) The LDC Debt Crisis. In *History of the Eighties – Lessons for the Future, Volume I: An Examination of the Banking Crises of the 1980s and Early 1990s*. Washington, DC: Federal Deposit Insurance Corporation.

Ferreira, F., de Feranti, D., Perry, G. and Walton, M. (2004) *Inequality in Latin America: Breaking with History?* Washington, DC: World Bank.

Fox, J. and Haight, L. (2011) *Subsidizing inequality: Mexican corn policy since NAFTA*. Washington, DC: Woodrow Wilson International Center for Scholars. www.wilsoncenter.org/publication/subsidizing-inequality-mexican-corn-policy-nafta-0

Green, D. (2003) *Silent Revolution: The Rise and Crisis of Market Economics In Latin America*. 2nd Ed. New York: Monthly Review Press.

Griffith-Jones, S. and Sunkel, O. (1989) *Debt and Development Crises in Latin America: The End of an Illusion*. New York: Oxford University Press.

Grugel, J. and Riggirozzi, P. (2009) (eds) *Governance After Neoliberalism in Latin America*. Basingstoke: Palgrave Macmillan.

———. (2012) Post-neoliberalism in Latin America: Rebuilding and Reclaiming the State after Crisis. *Development and Change* 43(1): 1–21.

Gudynas, E. (2010) *Buen Vivir: Un necesario relanzamiento. Politica y Economia*. Centro Latino Americano de Ecología Social (CLAES), December. www.politicayeconomia.com/2010/12/buen-vivir-un-necesario-relanzamiento/

———. (2011) Más allá del nuevo extractivismo: transiciones sostenibles y alternativas al desarrollo. In F. Wanderley (ed) *El desarrollo en cuestión: Reflexiones desde América Latina*. La Paz: Oxfam y CIDES UMSA, pp. 379–410.

——— and Acosta, A. (2011) El buen vivir mas allá del desarrollo. *Qué Hacer* 181. Lima: DESCO. www.desco.org.pe/apc-aa-files/6172746963756c6f735f5f5f5f5f5f5f/11_Gudynas_181.pdf

Harvey, D. (2005) *A Brief History of Neoliberalism*. New York: Oxford University Press.

Hebert, B. and Schreger, J. (2015) *The Costs of Sovereign Default: Evidence from Argentina*. Stanford University Press. https://people.stanford.edu/bhebert/sites/default/files/hebertschregerdec2015.pdf (Accessed 10 May 2018).

Huber, E. and Solt, F. (2004) Successes and failures of neoliberalism. *Latin American Research Review* 39(3): 150–164.

Jaramillo, M. and Saavedra, J. (2010) Inequality in post-structural reform Peru: The role of market forces and public policy. In L. Lopez-Calva and N. Lustig (eds) *Declining Inequality in Latin America: A Decade of Progress?* New York: UNDP and Washington, DC: Brookings Institution Press, pp. 218–244.

Kingstone, P. (2011) *The Political Economy of Latin America*. New York: Routledge.

Leipziger, D. (2001) The unfinished poverty agenda: Why Latin America and the Caribbean Lag Behind. *Finance & Development* 38(1). www.imf.org/external/pubs/ft/fandd/2001/03/leipzige.htm

Lopez-Calva, L. and Lustig, N. (2010) Explaining the decline in inequality in Latin America: Technological change, educational upgrading, and democracy. In L. Lopez-Calva and N. Lustig (eds) *Declining*

Inequality in Latin America: A Decade of Progress? New York: UNDP and Washington DC: Brookings Institution Press, pp. 1–24.

Lustig, N. (2009) *Poverty, Inequality, and the New Left in Latin America.* Washington, DC: Woodrow Wilson International Center for Scholars. 5.

Palma, J. (2011) Homogeneous middles vs. heterogeneous tails, and the end of the inverted-U: It's all about the share of the rich. *Development and Change* 42(1): 87–153.

Panizza, F. (2009) *Contemporary Latin America: Development and Democracy Beyond the Washington Consensus.* New York: Zed Books.

Pearce, J. (1982) *Under the Eagle: United States Intervention in Central America and the Caribbean.* London: Latin America Bureau.

Peet, R. (2003) *Unholy Trinity: The IMF, World Bank and WTO.* New York: Zed Books.

Petras, J., Veltmeyerj, H. and Vieux, S. (1997) *Neoliberalism and Class Conflict in Latin America.* London: Palgrave Macmillan.

Reygadas, L. (2006) Latin America: Persistent inequality and recent transformations. In E. Hershberg and F. Rosen (eds) *Latin America After Neoliberalism: Turning the Tide in the 21st Century?* New York: The New Press, pp. 120–143.

Rhoda, R. and Burton, T. (2012) Mexico, the home of corn, is now the world's largest corn importer. *Geo-Mexico: The Geography and Dynamics of Modern Mexico.* 16 April. http://geo-mexico.com/?p=6370

Roddick, J. (1988) *Dance of the Millions.* New York: Monthly Review Press.

Ruttenberg, T. (2012) Policies to match the rhetoric: Buen Vivir in Ecuador. *The Americas Blog.* Center for Economic and Policy Research, November 6. www.cepr.net/index.php/blogs/the-americas-blog/policies-to-match-the-rhetoric-buen-viver-in-ecuador

———. (2013a) Wellbeing economics and *Buen Vivir:* Development alternatives for inclusive human security. *PRAXIS: The Fletcher Journal of Human Security.* http://fletcher.tufts.edu/Praxis/~/media/Fletcher/Microsites/praxis/xxviii/article4_Ruttenberg_BuenVivir.pdf

———. (2013b) Economic and social policy in post-neoliberal Latin America: Analyzing impact on poverty, inequality and social wellbeing. *Ciencia & Tropico* 35(1). http://periodicos.fundaj.gov.br/index.php/CIT/article/view/1456

Skidmore, T. and Smith, P. (2005) *Modern Latin America.* New York: Oxford University Press.

Sparr, P. (1993) *Mortgaging Women's Lives: Feminist Critiques of Structural Adjustment.* London: Zed Books.

United Nations Office on Drugs and Crime (UNODC). (2014) Global Study on Homicide: Trends, Context, Data. Vienna: UNOD.

Vilas, C. (2006) The Left in South America and the resurgence of national-popular regimes. In E. Hershberg and F. Rosen (eds) *Latin America After Neoliberalism: Turning the Tide in the 21st Century?* New York: The New Press, pp. 232–251.

Watts, J. (2012) Latin America's income inequality falling, says World Bank. *The Guardian,* 13 November. www.guardian.co.uk/world/2012/nov/13/latin-america-income-iniquality-falling (Accessed 15 May 2015).

Williamson, J. (2004) *A Short History of the Washington Consensus.* Paper commissioned by Fundación CIDOB for 'From the Washington Consensus towards a new Global Governance'. Barcelona, 24–25 September. https://piie.com/publications/papers/williamson0904-2.pdf (Accessed 10 May 2018).

Woods, N. (2006) *The Globalizers: The IMF, the World Bank, and Their Borrowers* (Cornell Studies in Money). Ithaca, NY: Cornell University Press.

10

THE SUSTAINABLE DEVELOPMENT GOALS

Katie Willis

Introduction

The Sustainable Development Goals (SDGs), agreed by the United Nations General Assembly in September 2015, are the latest manifestation of what Gillian Hart has termed 'big D development' (Hart, 2001); that is, intentional interventions to achieve progress and modernity. Since the Second World War there have been different forms of international cooperation to bring 'development' to economically poorer parts of the world, to improve living standards and reduce marginalization. These attempts at development have been criticized for their top-down, Northern-centric perspectives, and how they classify peoples and places of the Global South according to what they lack rather than what they have (Escobar, 1995; Sachs, 1992).

The Millennium Development Goals (MDGs) ran from 2000–2015 with eight goals. These focused largely on poverty alleviation and raising the living standards of the world's poorest (Willis, 2016). Latin America's overall indicators of human development and income per capita meant that it was much less of a focus during the MDG period than other regions of the Global South, most notably Sub-Saharan Africa, South and East Asia. Over the MDG period, Latin America made significant progress against many of the MDGs and achieved many of the targets (Nicolai et al., 2016; UNDP, 2015). However, the pattern of success was not uniform in terms of both the goals and the countries of the region. For example, targets for reductions in extreme poverty and hunger were met at a regional level, but Paraguay did not meet the target for the reduction in malnutrition for children under five (MDG Track, 2018). Across the region maternal mortality rates remained higher than target by 2015, despite having fallen by 40% over the MDG period (UNDP, 2015). At sub-national levels, there were also significant social and spatial differences, for example in Mexico between urban and isolated rural areas in access to schooling and health services, and between indigenous and non-indigenous girls in access to education (INEGI, 2013).

Continuing with a global development goal format for the post-MDG period, plans for the design and implementation of Sustainable Development Goals were launched at the United Nations Rio+20 conference in 2012. In a major change from the MDG formulation process, the United Nations launched a global consultation process called MY World which invited participants to identify the six most important issues from a list of 16. They also had the option of adding other ideas. By 2015 over 9.7 million people had completed the survey, with approximately 80% collected offline through campaigns run by grassroots organizations (UN Millennium Campaign, 2015). Given the survey methodology it is important to stress that the survey was not representative of the global population in either demographic (it was dominated by participants aged under 30) or region. As an indication of intent and a signal of the importance of participation and inclusion, it can be seen as successful.

The SDGs encompass a much broader notion of development than the MDGs. Rather than focus on what populations lack, there is a much stronger engagement with what a sustainable future would look like. This includes a much stronger focus on the environment than the MDGs, but also significant recognition of diversity and inequalities, and how the SDG agenda should ensure that no one is left behind as development is achieved. Bringing in this broader definition of development, as well as acknowledging the findings from the SDG consultation process, has meant a final list of 17 goals and 169 targets (see Table 10.1). Because the focus is on more than poverty alleviation, the SDGs are aimed at all governments, not just as partners in international development cooperation, although this is covered in SDG 17, but as responsible for achieving the SDGs in their own countries.

This chapter is divided into three main sections. The first section discusses Latin American involvement in the framing of the SDGs, particularly through the global consultation process. The second section moves on to examine the institutions and mechanisms through which the SDGs are being implemented in the region. As Satterthwaite (22018: 408) observes, the SDGs provide little information as to how they are to be implemented and by whom. He also highlights the limited discussion of whether current systems and structures are fit to deliver on the

Table 10.1 Sustainable Development Goals

SDG		Number of Targets
1	End poverty in all its forms everywhere	7
2	End hunger, achieve food security and improved nutrition, and promote sustainable agriculture	8
3	Ensure healthy lives and promote well-being for all at all ages	13
4	Ensure inclusive and quality education for all and promote lifelong learning	10
5	Achieve gender equality and empower all women and girls	9
6	Ensure access to water and sanitation for all	8
7	Ensure access to affordable, reliable, sustainable, and modern energy for all	5
8	Promote inclusive and sustainable economic growth, employment, and decent work for all	12
9	Build resilient infrastructure, promote sustainable industrialization, and foster innovation	8
10	Reduce inequalities within and between countries	10
11	Make cities inclusive, safe, resilient, and sustainable	10
12	Ensure sustainable consumption and production patterns	11
13	Take urgent action to combat climate change and its impacts	5
14	Conserve and sustainably use the oceans, seas, and marine resources	10
15	Sustainably manage forests, combat desertification, halt and reverse land degradation, halt biodiversity loss	12
16	Promote just, peaceful, and inclusive societies	12
17	Revitalize the global partnership for sustainable development	19

Source: United Nations (2018).

SDGs, questioning whether "the national governments and international agencies that have failed to meet so many goals and targets in the past can now transform their approaches and effectiveness" (Satterthwaite, 2018: 408). Issues of structural change have also been raised by the UN Economic Commission for Latin America and the Caribbean (ECLAC), along with the role of community and grassroots participation. The roles of national and local governments, regional organizations such as ECLAC, the private sector, and civil society in delivering the SDGs will be discussed in the chapter's second section. Finally, the chapter focuses on two aspects of the SDGs which have particular resonance in the region; reducing inequality and environmental protection. It considers how the SDG process can be used to achieve both these objectives and what obstacles there are to success. If the SDGs are to be more than a list of unrealistic aspirations or the latest incarnation of a blueprint for development success, then a recognition of how the goals are prioritized, interpreted, and implemented in particular contexts is vital.

Framing the post-2015 development agenda

While the SDGs are now global goals, as part of the MY World consultation process which informed the final list, regional differences were apparent. This is unsurprising given the diversity of economic situations, political systems, cultural and social norms, and environmental conditions across the world, but it is vital to acknowledge such differences in development priorities.

Of the over 9.7 million people who responded to the MY World survey, approximately 2.2 million were based in Latin America. However, this number was significantly skewed by the over 1.9 million who responded in Mexico. Of the other Latin American countries, all had some respondents to the survey, although in some cases the numbers were in the hundreds, with Brazil, Colombia, and Peru all having over 35,000 respondents (MyWorld, 2015, 2018).

The top three priorities from Latin America were a good education, better job opportunities, and protection against crime and violence (see Figure 10.1). The first two were also top priorities in the global survey, but in the survey as a whole better health care came third and concerns about crime and violence were seventh (Nicolai et al., 2016: 10). This ranking reflects both the better overall health indicators in Latin America than in many other global regions, but also the levels of crime and violence which greatly affect the daily life and life chances of millions of people across the region. This is unsurprising given the estimated homicide rate in Latin America and the Caribbean is significantly higher than in any other global region; 21.3–27.3 intentional homicides per 100,000 people in 2015 compared to a global figure of 5.2–6.7 per 100,000 (United Nations, 2017: 50).

Nicolai et al. (2016) identify that the other main difference between the global priorities and those found in Latin America, are those relating to the environment; protecting forests, rivers and oceans, and action taken on climate change. The latter was the ranked the least important globally. The greater focus on environmental protection is a major difference between the MDGs and the SDGS, with a clear recognition of different ecosystems, and environmental processes, and the threats posed by human activity.

Overall the patterns are similar between South America and Central America (which includes Mexico in the survey classification), but there are some noteworthy distinctions. Both priorities relating to employment, better job opportunities and support for people who cannot work, were identified as important by a greater percentage of respondents in Central America than South America. South American respondents tended to rate freedom from discrimination and persecution, and political freedoms as more important than their Central American counterparts.

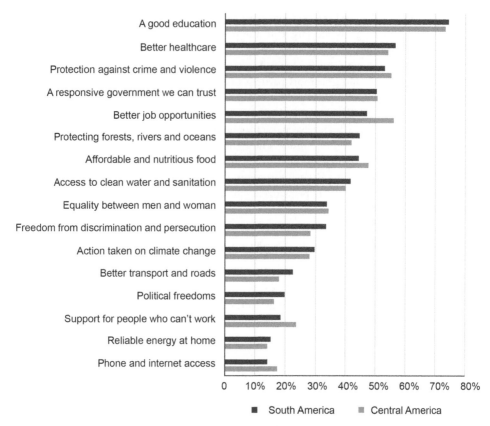

Figure 10.1 MY World Priorities (% of respondents selecting each option)

Source: Adapted from Figure 10.1 in Nicolai et al. (2016: 10).

Implementation

The ambition and scope of the SDG agenda has been a cause of both celebration and criticism (Scott and Lucci, 2015). One key note of caution has been about the mechanisms through which the targets can be achieved. As Alicia Bárcena, ECLAC Executive Director stated in the foreword to *Horizon 2030: Equality at the Centre of Sustainable Development*, there is a need to move beyond the goals and targets to consider what tools are needed to achieve them: "Without these tools, there is a risk that the Agenda will remain wishful thinking, at best implemented in an ad hoc and piecemeal fashion, contradicting the stated intention that it should be a universal, comprehensive and indivisible agenda" (ECLAC, 2016a: 9–10). This section investigates the institutional infrastructure that has been implemented in the region since the launch of the SDGs, focusing on regional cooperation, national government action, the role of city authorities, and civil society organizations. These different scales of implementation, alongside the governance structures of the United Nations and other multilateral institutions, need to be considered, even though the national scale remains key in the SDG process. In the proposal document for the SDGs, the UN Open Working Group for Sustainable Development Goals stated, "Each country has primary responsibility for its own economic and social development and the role of national policies, domestic resources and development strategies cannot be overemphasized"

(United Nations, 2014: 5). Reporting against SDG targets will also be done at the national level, although data may be later aggregated to identify regional and global patterns.

To facilitate regional cooperation around the SDGs in 2016 ECLAC established the Forum of the Countries of Latin America and the Caribbean on Sustainable Development. In doing this, ECLAC is not seeking to impose a singular set of policies and processes on its member states; something which would go against the principles of national sovereignty, and would also fail to recognize the diversity of development challenges across the region and governmental capacity to address these. In its *Horizon 2030: Equality at the Centre of Sustainable Development* document (ECLAC, 2016a), there is acknowledgement of the "special needs and particular challenges" of different groups of countries, including small island developing states (SIDS), landlocked developing countries, middle-income countries, least developed countries, and conflict/post-conflict states (p. 18). These categorizations match those used at the global level in the United Nations documentation, such as the targets for SDG 17.

In laying out a vision for achieving the SDGs in Latin America and the Caribbean, ECLAC has called for significant changes in global economic and political structures. Explicitly referencing the structuralist tradition in the region's (and particularly ECLAC's) development thought, there is a call for "a new political economy and new international and national coalitions to sustain it" (ECLAC, 2016a: 24). ECLAC explicitly acknowledges the tensions that exist between the United Nations policy-setting agenda, of which the SDGs are a very significant example, and a deregulated economic system with high levels of concentration of technology and income. For ECLAC, strong public policies, rather than a reliance on market forces, are required to deliver on the SDG Agenda (see also, Ghosh, 2015), and the 'elite multilateralism' (p. 29) which characterizes the international development system, needs to be dismantled.

ECLAC's vision for a future of sustainable development includes a significant expansion of the green economy, creating jobs and improved standards of living without destroying the environment. This is what ECLAC has called its "environmental push" to 2030 (ECLAC, 2016a). ECLAC sees this approach as a way of achieving sustainable development, but also addressing current limits to Latin American economic progress, including specialization on low-technology production and limited capacity to diversify production for export. As part of its demands for structural change in the international arena, ECLAC is calling for a "global environmental Keynesianism" based on "expansionary fiscal policies, with investments focused on technologies, goods and services linked to low-carbon production and consumption paths" (ECLAC, 2016a: 10). However, such a vision would require significant shifts in national development strategies across the globe, as well as vast additional resource at a regional level. It also remains wedded to a vision of development as growth (Moore, 2015).

ECLAC has also developed a system of annual reporting for national governments and has encouraged the region's governments to volunteer for the reviews of national policy which are conducted annually by the United Nations. In 2016, Colombia, Mexico, and Venezuela presented their voluntary national reviews, and a further 11 Latin American countries were involved in the 2017 process (Forum of the Countries of Latin America and the Caribbean on Sustainable Development, 2018: 19). Finally, ECLAC's vision favours "people-centred policies and actions, transparency and accountability" (ECLAC, 2016b: 19).

Regional cooperation through trading blocs is also identified as a route through which Latin America could make progress towards the SDGs by increasing trade and attracting foreign direct investment. Official development assistance (ODA) represents a very small percentage of finance coming into the region. This is unsurprising given the countries' status as middle-income countries in World Bank classifications. While additional ODA may be available as part of the 2015 Addis Ababa Agenda for development finance, private sector finance is going to be vital (Forum of the Countries of Latin America and the Caribbean on Sustainable Development, 2018). Regional economic cooperation faces challenges, not least because of different national development strategies that may clash, a lack of economic diversity in production, and weaknesses in infrastructure provision (Álvarez, 2016).

At a national level, Latin American governments have engaged with the SDG process in a range of ways since approving the SDG Agenda at the United Nations in September 2015. The large number of goals and targets means that there has to be a prioritization of activities, but national governments are expected to have a holistic approach to sustainable development and to incorporate the SDGs across their activities. Monitoring progress against the SDG targets also falls to national governments, raising significant questions about the data collection capacity of state institutions (Georgeson and Maslin, 2018). Of the 14 Latin American countries that have presented voluntary national reviews, 11 have set up institutional structures to implement and monitor progress (Forum of the Countries of Latin America and the Caribbean on Sustainable Development, 2018). For example, Mexico has established the Consejo Nacional de la Agenda 2030 para el Desarollo Sostenible (National Council for Agenda 2030 on Sustainable Development) and there is a Senate working group on Agenda 2030. In 2017, the Treasury published arrangements to link the National Development Plan 2013–2018 to the SDG targets (Secretaría de Hacienda y Crédito Público, 2017). This included identifying how forms of public expenditure, such as the Seguro Popular health insurance scheme and Prospera, a conditional cash transfer scheme, feed into specific targets. While such institutional structures are a welcome contribution to operationalizing the SDG Agenda, the six-year presidential cycle in Mexico, means that embedding these processes into the operation of federal, state, and local government faces significant challenges.

SDG 11 focuses on sustainable cities and communities. Given their contribution to economic growth, but also to environmental destruction including the emission of greenhouse gases, as well as the fact that urban areas are now home to over half the world's population, it is important that towns and cities have been highlighted in the SDG agenda (Simon et al., 2015; Rudd et al., 2018). Compared to other regions of the Global South, Latin America's city governments have tended to be better at implementing overarching policies for sustainable development (ECLAC and UN-Habitat, 2018; Satterthwaite, 2018), but this varies greatly across the region and the SDG Agenda places greater demands on municipal authorities, not least through the importance of inclusive urbanization, rather than just a focus on economic growth (McGranahan, Schenshul, and Singh, 2016).

Grassroots organizations were significant in administering the MY World consultation survey, but within the SDGs themselves, civil society organizations are rarely mentioned, appearing only in Target 17.17 about partnerships for sustainable development (Willis, 2016). For Satterthwaite (2018), with his focus on the SDGs relating to sustainable urban living, community-driven processes are a key way in which progress could be made. He highlights the role of networks such as Slum/Shack Dwellers International and their constituent organizations in achieving significant improvements in urban living standards for the poorest groups through their work with city authorities.

For a region with strong networks of civil society organizations, the SDGs provide a framework around which grassroots mobilization and action can be structured. While there may be some organizations which would see this as being coopted into an externally driven and imposed agenda around the dubious concept of 'development' (see Arturo Escobar's comments in Esteva and Escobar, 2017), the SDG agenda also provides opportunities to shape agendas to achieve greater equity and social justice.

Since the SDGs were launched, civil society organizations have mobilized around the Goals in two main ways. First, they have sought to raise awareness among the general population around the SDGs and second, they have pressed politicians to deliver on their commitments to achieve the targets. For example, in Peru, young people under the umbrella heading of the Millennials Movement have worked as SDG 'ambassadors' through both the MY World vote (UN SDG Action Campaign, 2015, 2017), and the promotion of the SDGs in conjunction with youth movements such as the Scouts, university student networks, and Interquorum (a non-party political, non-religious youth organization) (Red Interquorum, 2018). Similarly, in Chile, the European Union–funded Asocia 2030 brings together about 350 civil society organizations to campaign, educate, and mobilize around the SDGs, with a focus on poverty reduction, gender equality, reducing inequalities, taking action on climate, and the development of peace, justice, and strong institutions (Asocia 2030, 2018). While these engagement activities are important, and feed into Target 4.7 about learning and awareness of sustainable development, achievement of the SDGs requires the significant participation of grassroots organizations in the actual delivery of the Goals, as suggested by Satterthwaite.

Achievements and challenges

While it is still very early in the SDG period, examining the institutional commitment and mechanisms, alongside external conditions, means that there have already been attempts to predict outcomes. Nicolai et al. (2016) produced a regional scorecard for Latin America and the Caribbean based on one target for each goal, apart from SDGs 8 and 9 which are targeted at the least-developed countries, of which there is only one, Haiti, in the LAC region. This scorecard process highlights not only the limits in data availability and accuracy, but also the diversity within the region.

For the 15 targets examined, Nicolai et al. conclude that Latin America as a whole is performing well in relation to reducing inequality (see below), ending extreme poverty, providing universal access to energy, and providing universal access to sanitation. However, there is a need to reform existing processes to achieve this. More radical changes are needed to reach the targets by 2030 for halting deforestation, ending hunger, reducing maternal mortality, providing universal secondary education, ending child marriage (particularly challenging in South America), and mobilizing domestic resources for development partnerships. Based on Nicolai et al.'s analysis, to achieve most of the environmental targets as well as reducing violent deaths requires a reversal on current actions as the targets "are heading in the wrong direction" (2016: 16).

A focus on inequality has been one of the significant shifts in emphasis from the MDGs to the SDGs. The MDGs' prioritization of extreme poverty reduction meant that targets were to increase the share of the global population above a poverty line. This continues to be significant in the SDGs with SDG Target 1.1 being to end extreme poverty (people living on less than US$1.90 a day), but under the SDGs other forms of poverty are identified and there is a specific goal, SDG 10, on reducing inequality. The engagement with inequality reflects a commitment to equality as intrinsically important as part of a more just world where human rights are upheld. A key message of the SDG Agenda is that "no one will be left behind." However, reducing inequality is also targeted because of evidence that inequality hampers economic growth, social harmony, and political stability (World Bank, 2006; Wilkinson and Pickett, 2010).

According to the analysis by Nicolai et al. (2016) progress towards Target 10.1 – "By 2030 progressively achieve and sustain income growth of the bottom 40% of the population at a rate higher than the national average" – is going well in Latin America, although some policy reforms will be required. The region's countries have experienced reduced income inequality, as measured by the Gini Index, in recent decades (see Table 10.2). This has been a result of social protection schemes, significant flows of remittances, and increased employment. It must be stressed that despite declines in income inequality, the figures remain very high. Globally, countries with much more equal societies, such as the Scandinavian states, have Gini Index figures of 20–30. Latin America had six countries with figures of over 50 in 2010–2015. The only other region with such a concentration of inequality in its nations is Southern Africa (UNDP, 2016: 180–196). According to the World Bank's *Atlas of Sustainable Development Goals 2018*, in Costa Rica, Mexico, and Nicaragua income growth of the poorest 40% was slower than the average in the period 2009–2014 (World Bank, 2018: 39).

For Latin America different axes of exclusion and disadvantage have resulted in income inequality, and the concomitant challenges of access to health, education, and good standards of living. In many parts of the region, indigenous populations and peoples of African descent experience extreme forms of discrimination (Escobar, 2008). While Target 10.2 calls for the promotion of "social, political and economic inclusion of all, irrespective of age, sex, disability, race, ethnicity, origin, religion or economic or other status," ECLAC (2016a: 23) makes a particular point of highlighting the "lack of reference to indigenous peoples or Afro-descent groups" in the SDGs. Within Latin America and the Caribbean, there are an estimated 46 million indigenous people and 130 million Afro-descendant people (Forum of the Countries of Latin America and the Caribbean on Sustainable Development, 2018: 15).

The concept of *buen vivir*, roughly translated as 'living well' has been adopted as a focus of national development in Ecuador and Bolivia, drawing on indigenous concepts of living in

Table 10.2 Patterns of income inequality in Latin America 2000–2015

	Income Gini Index[a]	
	2000–2010[c]	*2010–2015*[d]
Colombia	58.5	53.5
Bolivia	57.2	48.4
Honduras	55.3	50.6
Brazil	55.0	51.5
Panama	54.9	50.7
Ecuador	54.4	45.4
Guatemala	53.7	48.7
Paraguay	53.2	51.7
Chile	52.0	50.5
Mexico	51.6	48.2
Peru	50.5	44.1
Costa Rica	48.9	48.5
Argentina	48.8	42.7
Dominican Republic	48.4	47.1
Uruguay	48.8	42.7
El Salvador	46.9	41.8
Venezuela	43.4	46.9[b]

Notes: [a] The income Gini index is a measure of income distribution, where 0 income is evenly distributed in the population and 100 is when one person has all the income. [b] Refers to period earlier than 2010

Sources: [c] UNDP (2010: 152–154) [d] UNDP (2016: 206–209)

harmony with the natural environment with a focus on a collective, rather than an individual approach to life (Acosta, 2017). While incorporating indigenous concepts into national strategies could be seen as an important process of inclusion, it is in the operation of policies that the real measure of inclusion can be seen. For indigenous populations in urban areas in both Bolivia and Ecuador, experiences of discrimination and exclusion are commonplace (Horn, 2018).

A recognition of different cosmovisions is present in both ECLAC (2016a) and United Nations (2014) documentation on SDGs and the environment where there are clear statements about how some groups refer to the natural environment as 'Mother Earth' and see nature as having rights just as humans do. While this recognition is important, there is little in the SDG environmental goals and targets that demonstrates an ability to engage with the implications of such conceptions of the world. The role of 'traditional knowledge' in addressing environmental challenges is sometimes mentioned in passing (e.g. Forum of the Countries of Latin America and the Caribbean on Sustainable Development, 2016: 56), but there is a failure to consider in depth how traditional or indigenous knowledge might inform or be incorporated into sustainable development or environmental policy (Mistry and Berardi, 2016).

It is in the achievement of the environmental goals that Latin America is currently forecast to face most challenges (Nicolai et al., 2016). Having been successful under the MDG system, the expansion to include greater environmental aspects threatens to undermine the region's record of success against international development targets. There are some signs of improvement in the protection of biodiversity and attempts to control deforestation. For example, the average proportion of each terrestrial, inland freshwater, and mountain key biodiversity area that is covered by protected areas increased in Latin America and the Caribbean from 34% in 2000 to 40% in 2010 and 42% in 2017 (United Nations, 2017: 48). The rate of deforestation is also slowing down; 51.3% of the land area of Latin America and the Caribbean was forest in 1990, 47% in 2010, and 46.4% in 2015 (United Nations, 2017: 49). This is partly a reflection of the adoption of sustainable management strategies, particularly involving local communities. Deforestation rates in protected areas of the Brazilian Amazon were a quarter of those in non-protected areas (UNEP/IUCN, 2016 in Forum of the Countries of Latin America and the Caribbean on Sustainable Development, 2018: 29). However, this does indicate that even in protected areas deforestation is occurring. Forest conservation policies can face obstacles due to lack of staff and enforcement mechanisms, trade-offs against other policy priorities such as export earnings, and poor policy development (Rosa da Conceição, Borner, and Wunder, 2015).

ECLAC's environmental push to 2030 stresses the importance of developments in carbon-neutral technology. This includes encouraging shifts to renewable energy. For most of the region, renewables contribute a relatively small share of annual energy generation, but in some countries renewable sources dominate. For example, in 2015 99% of Costa Rica's energy came from renewables and the figure was 92.8% for Uruguay (Forum of the Countries of Latin America and the Caribbean on Sustainable Development, 2018: 54). Moving away from fossil fuels requires significant investment, most likely from private sector interests, but there could also be subsidies or taxation to encourage companies and consumers to make the switch. A shift to renewables would help meet the SDGs around energy and climate action, as well as contributing to employment goals. More than two million people were employed in the renewable energy sector in 2010, but there is significant scope for expansion in the number of jobs should the move away from fossil fuels take place (albeit with a knock-on effect on jobs in the oil and gas sectors). However, making this move requires government action and finance, as well as changes in consumer behaviour. There are also challenges in the development of renewable energy generation because of its potential impact on the livelihoods and communities of rural populations as hydroelectric, wind, or solar power plants are established.

Conclusions

In this chapter I have outlined how national governments, regional organizations, and civil society in Latin America have engaged with the SDG process. The expansive nature of the SDGs and their targets means that priorities, forms of implementation, and possibilities of success will vary by region and at different scales. Flexibility to operate not just within the bounds of national sovereignty, but also as part of processes of transparency and public engagement are part of the SDG remit. As a region, Latin America has made significant strides in developing and embedding Agenda 2030 into national policies, but the effectiveness of these processes remains to be seen. The region's governments have also sought to challenge the global development institutions, seeking to query the concentration of power among a few states, large corporations, and financial institutions, and the continued reliance on market forces to drive the development process. The America First agenda of the Trump administration, and the threats of trade wars and greater protectionism across the globe, does not bode well for a form of global cooperation that could underpin SDG success. For some, the SDGs are "an attempt at measuring the world" (Sachs, 2017: 2578) and seeking to constrain what development is and should be. If the SDG agenda is to really deliver on its aim of leaving no one behind, and to address the diversity of the region's populations, then good data and measurement are required, but there also needs to be an openness to a range of voices and perspectives.

References

Acosta, A. (2017) Living well: Ideas for reinventing the future. *Third World Quarterly* 38(12): 2600–2616.

Álvarez, A. M. (2016) Retos de América Latina: Agenda para el Desarrollo Sostenible y las Negociaciones del Siglo XXI. *Revista Problemas del Desarrollo* 186(47): 9–30.

Asocia 2030. (2018) *Asocia 2030 Homepage.* http://proyectoasocia2030.cl/proyecto-asocia-2030/

Escobar, A. (1995) *Encountering Development: The Making and Unmaking of the Third World.* Princeton, NJ: Princeton University Press.

———. (2008) *Territories of Difference: Place, Movements, Life, Redes,* Durham NC: Duke University Press.

Esteva, G. and Escobar, A. (2017) 'Post-Development @ 25: On 'being stuck' and moving forward, sideways, backward and otherwise'. *Third World Quarterly* 38(12): 2259–2272.

Forum of the Countries of Latin America and the Caribbean on Sustainable Development. (2018) *Second Annual Report on Regional Progress and Challenges in Relation to the 2030 Agenda for Sustainable Development in Latin America and the Caribbean,* Santiago: ECLAC.

Georgeson, L. and Maslin, M. (2018) Putting the United Nations Sustainable Development Goals into practice: A review of implementation, monitoring, and finance. *Geo: Geography and Environment* e00049. https://doi.org/10/101002/geo2.49

Ghosh, J. (2015) Beyond the Millennium Development Goals: A Southern perspective on a Global New Deal. *Journal of International Development* 27: 320–329.

Hart, G. (2001) Development critiques in the 1990s: *culs de sac* and promising paths'. *Progress in Human Geography* 25: 649–658.

Horn, P. (2018) Indigenous people, the city and inclusive urban development policies in Latin America: Lessons from Bolivia and Ecuador. *Development Policy Review* 36: 483–501.

Instituto Nacional de Estadística y Geografía (INEGI). (2013) *The Millennium Development Goals in Mexico: Progress Report* 2013. Aguascalientes: INEGI.

McGranahan, D., Schenshul, D. and Singh, G. (2016) Inclusive urbanization: Can the 2030 Agenda be delivered without it? *Environment & Urbanization* 28(1): 13–34.

MDG Track. (2018) *Monitoring Progress Towards the Millennium Development Goals.* www.mdg.track.org (Accessed 7 July 2018).

Mistry, J. and Berardi, A. (2016) Bridging Indigenous and scientific knowledge: Local ecological knowledge must be placed at the center of environmental governance. *Science* 352(6291): 1274–1275.

Moore, H. (2015) Global prosperity and sustainable development goals. *Journal of International Development* 27: 801–815.

MyWorld 2015. (2018) *MyWorld Analytics.* http://data.myworld2015.org/ (Accessed 6 July 2018).

Nicolai, S., Bhatkal, T., Hoy, C. and Aedy, T. (2016) *Projecting Progress: The SDGs in Latin America and the Caribbean*. London: ODI.

Red Interquorum. (2018) *Red Interquorum Homepage*. www.redinterquorum.org/ (Accessed 7 July 2018).

Rosa da Conceição, H., Borner, J. and Wunder, S. (2015) Why were upscaled incentive programs for forest conservation adopted? Comparing policy choices in Brazil, Ecuador and Peru. *Ecosystem Services* 16: 243–252.

Rudd, A., Simon, D., Cardama, M., Birch, E. L. and Revi, A. (2018) The UN, the Urban Sustainable Development Goal and the New Urban Agenda'. In T. Elmqvist, X. Bai, N. Frantzeskaki, C. Griffith, D. Maddox, T. McPhearson, S. Parnell, P. Romero-Lankao, D. Simon and M. Watkins (eds) *Urban Planet: Knowledge Towards Sustainable Cities*. Cambridge: Cambridge University Press, pp. 180–196.

Sachs, W. (1992) (ed) *The Development Dictionary: A Guide to Knowledge as Power*. London: Zed Books.

———. (2017) The Sustainable Development Goals and *Laudato si*: Varieties of post-development? *Third World Quarterly* 38(12): 2573–2587.

Satterthwaite, D. (2018) Who can implement the Sustainable Development Goals in urban areas? In T. Elmqvist, X. Bai, N. Frantzeskaki, C. Griffith, D. Maddox, T. McPhearson, S. Parnell, P. Romero-Lankao, D. Simon and M. Watkins (eds) *Urban Planet: Knowledge Towards Sustainable Cities*. Cambridge: Cambridge University Press, pp. 408–410.

Scott, A. and Lucci, P. (2015) Universality and ambition in the post-2015 development agenda: A comparison of global and national targets. *Journal of International Development* 27: 752–775.

Secretaría de Hacienda y Crédito Público (SHCP). (2017) *Vinculación del Presupuesto a los Objetivos del Desarrollo Sostenible*. Mexico City: SHCP. www.gob.mx/cms/uploads/attachment/file/231527/Lineamientos_p_y_p_2018_Anexo_2_Vinculacion_ODs.pdf

Simon, D., Arfvidsson, H., Anand, G., Bazaz, A., Fenna, G., Foster, K., Jain, G., Hansson, S., Marix Evans, L., Moodley, N., Nyambuga, C., Oloko, M., Ombara, D. C., Patel, Z., Perry, B., Primo, N., Revi, A., Van Niekerk, B., Warton, A. and Wright, C. (2015) Developing and testing the Urban Sustainability Goal's targets and indicators – a five-city study. *Environment & Urbanization* 28(1): 49–63.

United Nations. (2014) *Open Working Group Proposal for Sustainable Development Goals*. New York: United Nations.

———. (2017) *The Sustainable Development Goals Report 2017*. New York: United Nations.

———. (2018) *Sustainable Development Goals*. www.un.org/sustainabledevelopment/sustainable-development-goals/

United Nations Development Programme (UNDP). (2010) *Human Development Report 2010: The Real Wealth of Nations: Pathways to Human Development*. New York: UNDP.

———. (2015) *Millennium Development Report 2015*. New York: United Nations.

———. (2016) *Human Development Report 2016: Human Development for Everyone*. New York: UNDP.

United Nations Economic Commission for Latin America and the Caribbean (ECLAC). (2016a) *Horizon 2030: Equality at the Centre of Sustainable Development*. Santiago: ECLAC.

———. (2016b) *Resolutions Adopted by the Economic Commission for Latin America and the Caribbean in the Thirty-Sixth Session*. Santiago: ECLAC.

——— and UN-Habitat. (2018) *Regional Action Plan for the Implementation of the New Urban Agenda in Latin America and the Caribbean 2016–2036*. Santiago: ECLAC.

United Nations Environment Programme (UNEP) and International Union for Conservation of Nature and Natural Resources (IUCN). (2016) *Protected Planet Report 2016: How protected areas contribute to achieving global targets for biodiversity*. Cambridge: UNEP and IUCN.

United Nations Millennium Campaign. (2015) *UN Millennium Campaign 2015 Report*. New York: United Nations.

United Nations SDG Action Campaign. (2015) *The Millennials Movement: Jóvenes Peruanos por un Mundo Mejor*, 11 August. https://sdgactioncampaign.org/2015/08/11/the-millenials-movement-jovenes-peruanos-por-un-mundo-mejor/

———. (2017) *Programa de Embajadores Péru Agenda 2030: Por el desarrollo sostenible y el mundo que queremos*, 16 March. Available at: sdgactioncampaign.org/2017/03/16/programa-de-embajadores-peru-agenda-2030-por-el-desarrollo-sostenible-y-el-mundo-que-queremos/

Wilkinson, R. and Pickett, K. (2010) *The Spirit Level: Why Equality is Better for Everyone*. London: Penguin Books.

Willis, K. (2016) Viewpoint: International development planning and the Sustainable Development Goals (SDGs). *International Development Planning Review* 38(2): 105–111.

World Bank. (2006) *World Development Report 2016: Equity and Development*. New York: World Bank.

———. (2018) *Atlas of Sustainable Development Goals 2018*. New York: World Bank.

11

THE WAR ON DRUGS IN LATIN AMERICA FROM A DEVELOPMENT PERSPECTIVE

Guadalupe Correa-Cabrera[1]

Introduction

The evolution of drug trafficking and organized crime in Latin America, as well as government strategies to fight these problems have driven unprecedented levels of violence, affected patterns of development and income distribution, and prompted structural economic changes in a number of countries. With the goal of reducing substance abuse and fighting drug trafficking in the Western Hemisphere, the United States and a number of Latin American countries have centred regional anti-narcotics policy and cooperation on supply-side efforts to combat the so-called drug cartels and Central American gangs that allegedly facilitate the drug trade. The United States has led and supported such efforts, which include the militarization of anti-drug trafficking operations in a scheme that is widely known as a "War on Drugs."

The present chapter analyzes the so-called drug wars in Latin America from a development perspective. It assesses the overall economic effects of US-driven anti-narcotics policy and the militarization of security in key countries of the region. This analysis characterizes the resulting conflicts as "modern civil wars" or economic wars related to the control of material resources in strategic zones of Latin America. This chapter also identifies the groups that seem to have benefited the most (directly or indirectly) from novel criminal schemes, governments' responses to new forms of organized crime, and the resulting brutality. The main winners (or potential winners) appear to be corporate actors in the energy sector, transnational financial companies, private security firms (including private prison companies), and the US border-security/military-industrial complex. This piece focuses on Colombia, Mexico, and Central America – the Northern Triangle countries in particular (Guatemala, Honduras, and El Salvador).

The present analysis draws primarily on two recent works: Paley (2014) and Correa-Cabrera (2017). The first book explores how US anti-drug policies in Colombia, Mexico, and Central America have served to benefit particular private economic interests in the region. The second work proposes a new theoretical framework for understanding the emerging face, structure, and economic implications of organized crime in Mexico, as well as the effects of government responses to these phenomena. Arguing that the armed conflict between transnational criminal organizations and the Mexican state resembles a civil war, key beneficiaries of this war are identified.

The transformation of organized crime in Latin America

Violence related to drug wars – also linked to the appearance of new forms of organized crime – has reached unprecedented levels in Latin America. This phenomenon has affected entire communities, led to the loss of thousands of lives, and has caused a human tragedy of considerable dimensions in several countries of this region. In Colombia, for example, a 20-year war to dismantle the drug cartels, paramilitaries, and guerrilla groups cost approximately 15,000 lives and displaced thousands of families and small farmers from their lands (Huey, 2014). In Mexico, according to some estimations, the drug war has claimed over 100,000 lives (Molloy, 2017; SESNSP, 2018).[2] During this period, more than 30,000 people have been reported missing/disappeared (RNPED, 2018) – with many of these disappearances linked to organized crime. Thousands of citizens have become internal refugees, displaced within Mexico or seeking refuge abroad. This unparalleled increase in violence has been accompanied by the widespread use of barbaric, terror-inflicting methods (Correa-Cabrera, 2017).

At the same time, drug trafficking organizations have diversified their operations and are involved in lucrative new activities, such as extortion, kidnapping, migrant smuggling, weapons smuggling, and trafficking of crude oil, natural gas, and gasoline, among others. These activities have been made possible by a new relationship that exists now between drug cartels and a new set of actors. New corruption networks have been built among criminal organizations, local police and law enforcement agencies, politicians at all levels, and federal authorities. Formal businesses, including transnational companies, have also established new connections with these criminal groups (Correa-Cabrera, 2017).

In the Central American Northern Triangle (Guatemala, Honduras, and El Salvador), the control of territory and criminal activities by transnational gangs or *pandillas*, such as *Barrio 18* and *Mara Salvatrucha* (MS-13) has extended considerably. Endemic corruption and a broken prison system have contributed to this significant expansion and to the "loss of the monopoly on violence" by the State in specific regions of these countries. The basic activities of these Central American criminal groups involve the extortion of a number of small businesses and families in gang-controlled zones. At the same time, certain factions of these gangs participate in the drug trade, have established direct connection with the most important drug cartels in the Hemisphere (such as the Sinaloa Cartel and the Zetas) with the aim of transporting drugs along the territory they control, or distributing them among the local population.

Overall, we have observed in recent decades the evolution of drug trafficking organizations and criminal gangs into truly transnational entities. A number of them have access to military equipment and high-caliber arms; and some of them receive protection from public officials or law enforcement agents. At the same time, the governments of these countries have reacted to this new power of paramilitary-style criminal organizations by increasing militarization in the fight against organized crime – supported by US military aid and within the scope of the US War on Drugs. These developments have significantly elevated the levels of violence and terror in specific regions of Latin America, in particular in certain parts of Mexico and Central America. The ensuing environment bears a resemblance at times to that of a civil war (Correa-Cabrera, 2017: 2–3).[3]

The paramilitarization of organized crime in Latin America, the growing power of Central American gangs and the expansion of their criminal model, and a more violent confrontation between criminal syndicates, as well as the militarization (and paramilitarization) of the fight against illegal actors and drug trafficking have produced situations in some regions of the Americas that can be analyzed through the framework of civil wars as presented in academic literature.

This type of approach has seldom been considered by scholars focusing on these issues; but due to the dynamics, magnitude, and type of actors involved in armed conflicts of a number of Latin American countries, a framework of this type might be useful to advance our understanding of the security crisis in some countries of the Western hemisphere (Correa-Cabrera, 2017).

Armed conflicts related to drug trafficking and other forms of transnational organized crime in Latin America have the characteristics of civil wars, and more specifically of "new" or "modern" civil wars – i.e., those that are: essentially "criminal, depoliticized, private, and predatory" (Kalyvas, 2001: 100). It seems that in these Latin American cases, economic agendas and corporate interests, rather than political ideologies, have been driving violent armed conflicts. In other words, it is likely that conflict in Mexico, Colombia, and the Northern Triangle countries related to the War on Drugs has been mainly driven by economic opportunities – more than by grievances. This view departs from classical theories on wars and armed conflicts. The present chapter analyzes such conflicts, their main causes, as well as the effect of the so-called drug wars from a development perspective (Correa-Cabrera, 2017: 126).

US-supported drug wars in Latin America

President Richard Nixon first declared the US War on Drugs in 1971, while positing drug abuse as "America's public enemy number one." This war, in the view of some, was adopted by (or imposed upon) other countries to their detriment and has continually evolved into new forms in the decades since it was introduced. According to Paley (2017: 6–7), the US drug war "criminalize[s] and dehumanize[s] communities inside and outside the United States, while also strengthening the repressive State apparatus (police, courts, prisons, army)." Within the context of the US War on Drugs, Latin American countries such as Colombia, Mexico, Guatemala, Honduras, and El Salvador have militarized their fight against the drug trade and organized crime. This strategy has been advanced in part through US-backed policy initiatives like Plan Colombia, the Merida Initiative, and the Central America Regional Security Initiative (CARSI).[4] These initiatives have promoted legal and policy reforms as well as "the militarization of aid and the steering of anti-drug money towards fostering the creation of more welcoming investment policies" (Paley, 2014: 89). In general, this framework:

> introduces (or increases) the presence of soldiers and marines in patrolling streets and neighborhoods, and militarizes the training, weaponry and composition of police forces throughout host nations. . . . Militarized policing strategies are funded by the US Department of Defense, the State Department (through USAID) and the Department of Justice, with supplemental funds from host nations, and in Central America and Colombia include the active participation of on-duty US police and soldiers.
>
> *(Paley, 2017: 17)*

The US-backed militarization of security strategies in these Latin American countries has coincided with a visible increase in the murder rate as well as with the further militarization of organized crime, and the creation or strengthening of countrywide structures of paramilitary control. As Paley (2014) observes, the militarization of criminal groups in response to state militarization of drug trafficking routes breeds a form of paramilitarism. In some countries of the region – such as Colombia and México – extra-official armed groups appear to "do the army's job," though they seemingly have no formal connnections with the military forces or government authorities.[5] Hence, an "unconventional security policy apparently has transitioned into the use of paramilitary groups" in some regions of Latin America (Correa-Cabrera, 2017: 106).

It is worth mentioning that the US-backed initiatives that promote a military strategy against drug production and trafficking in Colombia, Mexico, and Central America have not achieved the ostensible goal of weakening narco-related activities. In fact, drug trafficking in these regions has not visibly decreased. At the same time, however, non-state armed actors have been strengthened, thus "increasing extra-legal violence with no apparent effect on its stated goal of curbing drug production" (quoted in Paley, 2014: 55). Plan Colombia, for example, hasn't significantly reduced the amount of cocaine for sale in the United States, and homicide rates in the Andean country remain quite high. Regardless, Plan Colombia has been touted by authorities as a successful initiative. Overall, the US drug war in Latin America "hasn't managed to curb drug production or use but has now cost taxpayers more than a trillion dollars" (Fernández, 2017). On the contrary, some of the most visible results of the War on Drugs in Latin America are the increasing levels of violence and a sharp increase in homicide rates following an unconventional security strategy. Moreover, it has been said that "the use of the military puts civil liberties and human rights at risk, as well as the rule of law" (Machuca, 2014: 12).

Latin American drug wars: a development perspective

In practice, Latin American drug wars seem to have little to do with illicit substances and much more "to do with the transformation of the business environment" (Paley, 2014: 82). This section examines the War on Drugs in Latin America from a development perspective, focusing on the economic impacts of drug trafficking and drug-related violence.

There are contrasting views regarding the calculations of the overall impact of drug trafficking on the economy. Loret de Mola (2001), for example, estimated that the Mexican economy would decline approximately 63% if this illegal activity was completely eradicated – due to the way organized criminal activity has permeated the Mexican economy. Neilson (2004) refers to the global drugs trade as a form of counterglobalization that must be understood not as an alternative economy, but as central to the global capitalist economy. Rios (2008), on the other hand, estimates that drug trafficking generates for Mexico economic losses of nearly 4.3 billion dollars per year. With such large discrepancies in estimates, it is unclear what kind of economic impact these illicit activities have had the national economy as a whole (Correa-Cabrera, 2013). While it is frequently assumed that violence related to organized crime has negative effects on Latin American economies and economic development in general, development indicators as represented by the Human Development Index do not however necessarily show this trend (see Figure 11.1). In fact, the impact on economic development as a result of the War on Drugs appears to vary from country to country and it is not clear how the related variables are correlated quantitatively. Further, the costs of violence are unevenly distributed among different social groups. Through qualitative studies, however, it seems to be the case that this phenomenon particularly affects the most vulnerable segments of the population, such as the poor, micro- and small business owners, small land owners, and indigenous peoples. In some countries like Mexico, for example, notwithstanding the extreme levels of violence furthered by the drug war, the economy has continued to grow and to be perceived as an attractive destination for foreign investors (Correa-Cabrera, 2013).

It is extremely useful to assess the impact of the drug wars in Latin American economies and socioeconomic structures by using the analytical framework and examples provided by Paley (2014). In her work, she stresses "the importance of critical research and writing on the conflicts in Colombia, Mexico, and elsewhere in the Hemisphere that take into consideration resource extraction as a driving force behind whatever the current dominant explications of the conflicts are" (Correa-Cabrera; in Paley, 2014: 5). For Paley, the so-called drug wars aren't

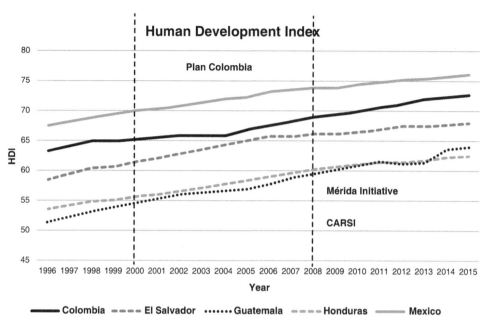

Figure 11.1 Human Development Index (HDI): Mexico, Northern Triangle, and Colombia

"about prohibition or drug policy," but instead, are wars "in which terror is used against the population at large in cities and rural areas," while "parallel to this terror and the panic it generates, policy changes are implemented which facilitate foreign direct investment and economic growth" (2014: 5). This is the development phase of the War on Drugs. Paley refers to this scheme as "drug war capitalism," which in her view is advanced as a "war on the people" and their communities.

Overall, Paley connects drug war violence to capital accumulation in Latin America. She places drug-related conflicts "into the broader context of US and transnational interests in the hemisphere" and links "anti-drug policies to the territorial and social expansion of capitalism" (2014: 33). Her analysis specifically shows "how narcotics prohibition can facilitate the expansion and maintenance of capitalism, and how the militarized enforcement of prohibition can provide an off-the-books source of funding for reactionary and paramilitary groups involved in strengthening State repression of community networks and popular resistance" (2017: 8). She argues, more specifically, that the "enforcement of prohibition strengthens States and smooths the functioning of capitalism in four ways: through policy changes introduced to support prohibition, through formal militarization, by creating cash economies to ensure the smooth operation of capitalism, and by outsourcing the costs of war, which has long been an essential part of State development within a capitalist world economy" (2017: 4–5).

The idea of drug war capitalism considers "factors other than drugs . . . as potentially influencing violence aimed at local populations in resource rich areas" (Paley, 2014: 135). Hence, drug wars in Latin America today can be seen as modern economic wars that have the dual function of extracting resources and increasing control over resource-rich territories. In Paley's (2014: 5) view, one should seriously consider resource extraction as a driving force behind any explanation of drug war-related conflicts. For example, the key role of natural resources in civil conflicts and the proliferation of armed groups has been evident in Colombia, a country rich in oil, gas, coal, cropland, gold, emeralds, and other valuable minerals (Gray, 2008). In this Latin American nation, it seems that the presence of such resources, as well as that of drugs, and other commodities has fueled violence "by eroding trust and discipline in armed groups," and has lengthened the conflict "by giving weaker groups the means to keep fighting" (Gray, 2008: 78).

Some evidence for a number of Latin American countries shows that indeed, during drug war times, internal conflicts, and militarization have been concentrated in "areas deemed important for energy projects or resource extraction." Many of these conflicts have also taken place "where there are fierce social and land conflicts related to the imposition of mega-projects" such as oil and natural gas exploration or exploitation, large-scale agriculture, hydroelectric projects, large-scale forestry, among others (Paley, 2014: 174). Drug war policies also seem to have helped the US to exercise further control and achieve their foreign policy strategic aims in the region (Correa-Cabrera, 2017).

Examples of drug war capitalism in Latin America

To show how drug war capitalism works in practice one can provide a number of examples in Latin America. The cases of Colombia, Mexico, Guatemala, Honduras, and El Salvador are quite illustrative. Corporate interests have been advanced and protected in the framework of the Latin American drug wars initiated and supported by the United States. In different countries of the region, major infrastructure has been built and privatization programs have been initiated during a period of intense violence. These actions have been implemented with the intention of increasing prosperity and creating economic alternatives that would force fewer people into the drug trade (Paley, 2017: 16).

Transnational oil and gas companies are among the biggest winners in this new context. For example, Paley (2014) notes that during the times of Plan Colombia the state oil company, Ecopetrol, was privatized, and new laws were introduced to promote foreign investment. At the same time, she observed that the army was used to protect oil pipelines belonging to transnational companies. Hence, during this time, "foreign investment in the extractive industries soared and new trade agreements were signed" (Paley, 2014: 8). Extreme violence in this country coincided with a period of increased resource exploitation. According to Gray (2008: 79) drug cultivation was not the only driving force behind violence in this Andean country, but also "licit development in the oil, mining, and agribusiness sectors." Actually, paramilitary displacement here transformed newly de-populated land for palm oil and extractive projects (Paley, 2014: 162).

The same thing has happened in Mexico in the time of the drug wars. Extreme levels of violence connected to the paramilitarization of organized crime and the militarization of security have greatly influenced the recent transformation of Mexico's energy sector. In December 2013, the Mexican Congress approved constitutional changes to open up even more Mexico's hydrocarbons industry to the participation of private transnational businesses. At the same time, Mexican states rich in hydrocarbons – such as Chihuahua, Coahuila, Nuevo León, Tamaulipas, San Luis Potosí, and Veracruz – have been militarized in the framework of the War on Drugs (see Figure 11.2). Forced displacements due to severe drug-related violence have been frequent in these territories. In this same context, the government of Mexico intends to attract massive foreign investments to tap into the country's energy resources (Correa-Cabrera; in Paley, 2014: 8). Similarly, "[m]ining projects have been among the most conflictive sites of recent capitalist expansion in Mexico, and the majority of gold and silver production in the country takes place in states with some of the highest rates of violence" (Paley, 2014: 100).

These phenomena are characteristic not only of Colombia and Mexico, but of several regions of the world. Similar dynamics can be observed and documented in other parts of the

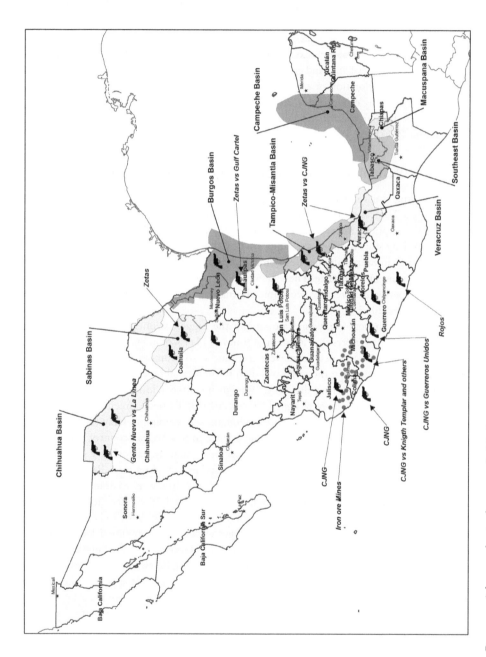

Figure 11.2 Drug war violence and natural resources in Mexico

hemisphere, such as for example, Guatemala and Honduras. Internal conflict – gang violence, or modern civil war – in these Latin American countries has been concentrated in strategic areas for energy projects and resource extraction (see Figure 11.3). In fact, criminal paramilitarization, gang violence, militarization, and other forms of paramilitarism (including mercenaries or self-defense groups collaborating with armed forces or entrepreneurs) coincide with regions that show important social and land conflicts, where some segments of the society oppose the imposition of mega-projects, particularly those related to oil and natural gas exploration or exploitation (Paley, 2014). Modern civil war and gang-related violence have had the ultimate effect of restricting social mobilization and supporting private investments in a number of extractive industries (Correa-Cabrera, 2017). At the same time, some indigenous communities and small land owners in these countries have had their lands taken away by war; these properties have been – or will potentially be – acquired by transnational corporations whose aim is to extract natural resources or develop mass tourism.

Evidence shows that in the case of Guatemala, "internal conflict has unravelled in areas deemed important for energy projects or resource extraction" (Paley, 2014: 184). Here, national security policy seems to have been influenced by extractive industries, including hydrocarbons, hydroelectric projects, and mining. In this context, some have reported "how police and army defend corporate interests against indigenous resistance, and how paramilitary groups are formed by elites looking to protect their interest" (Paley, 2014: 185). It is interesting to notice how the presence of drug traffickers or gangs in resource-rich territories of Guatemala (like Laguna del Tigre or El Petén) hasn't affected production of the most profitable industries – including biofuels, large-scale agriculture, mass tourism, hydroelectric projects, among others (Paley, 2014: 183).

Final observations: who benefits from the US-supported war on drugs in Latin America?

Similar patterns are observed in Honduras and El Salvador where anti-gang efforts and anti-drug trafficking operations have been militarized. For activists and others "involved in struggles against oil drilling, mining companies, hydroelectric projects, wind farms, large-scale forestry and the exploitation of petrified wood, the militarization [and paramilitarization] of [security] brings with it direct and deadly consequences" but benefits corporate capital and the owners of extractive industries (Paley, 2014: 212). A number of academics, journalists, and political analysts arrive to similar conclusions (see also Alvarado, 2015; Zavala, 2018).

Cockcroft (2010), for example, claims that Mexico's drug war has been an excuse for militarizing the country. In his view, militarization prevents the rise of opposition, which would eventually allow transnational companies "to gain greater control over . . . oil, minerals, uranium, water, biodiversity, and immigrant labor" (40). Similarly, for Jalife-Rahme (2014) what is really at stake in drug wars is the control of oil resources. Following this same logic, journalist Carlos Fazio maintains that the ultimate goal of this new security model would be to support transnational corporations by concentrating power in the government and eroding all sorts of regional autonomy, thus allowing these companies to access lands, and control resource extraction and energy generation (cited in Chávez, 2013). These same dynamics seem to take place in different regions of Latin America and mainly benefit the United States and the corporate sector in general. Fazio used the example of the US Chiquita Brands (successor of United Fruit Company), which used paramilitaries in Colombia to displace farmers in order to use those farming areas for its own benefit (Chavez, 2013).

Fazio also warns against the indiscriminate use of the term "failed State," a category that, he claims, comes from US think tanks and institutions like the CIA or the Pentagon, and has often

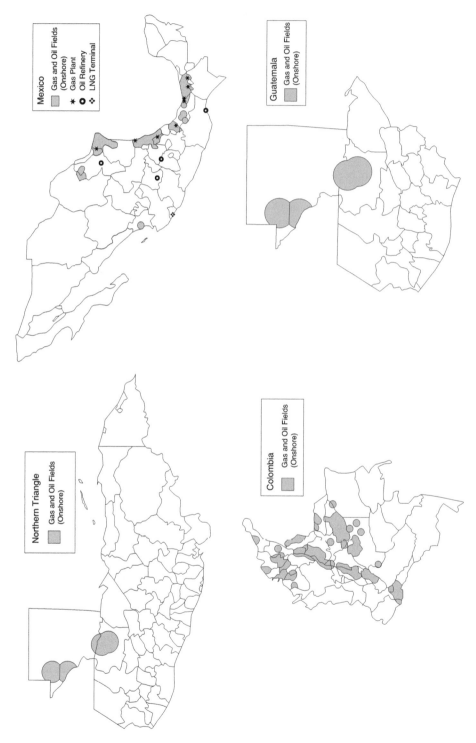

Figure 11.3 Oil and gas (Mexico, Northern Triangle, and Colombia)

served as an excuse to destabilize countries and then control strategic resources (Chávez, 2013). He notes that the concept of failed State was used in the early years of Mexico's drug war, and recalls when then Secretary of State Hillary Clinton stated the "existence of narco-insurgency in Mexico, comparing Mexico's criminal organizations with rebel armed groups." Under this logic, policies that should be implemented by the Mexican government to fight organized crime – backed by the United States – would require the usage of "counter-insurgency" tactics. Cockcroft agrees with this argument; according to him, "High officials of the US government and its armed forces blather a lot about failed States. We are told that these [S]tates are bleeding to death and only a transfusion of military intervention can save the patients" (Cockcroft, 2010: 40).

The War on Drugs has significantly transformed Latin American economies, as well as income distribution and development patterns in the region in the most recent years. These new developments have taken place at a large cost in terms of human lives. Vulnerable "rural populations continue to be displaced from their lands" and are victims of "state and non-state violence" (Paley, 2014: 78). The militarization of security in the region has served "to pacify urban and rural populations, defend transnational corporate interests against community organization and strengthen the repressive apparatus of the state" (Paley, 2017: 17). Overall, the War on Drugs has disproportionately impacted the poorest segments of Latin American societies. In fact, the primary victims of this US-supported policy are poor people, migrants, indigenous communities, and peasants. This phenomenon has contributed to enhance inequalities and has claimed hundreds of thousands of lives. Unfortunately, drug wars in Latin America have been largely ineffective.

Notwithstanding the very significant financial, material, and human resources spent over the past decades, the drug trade and organized crime in the region have not been eradicated. What is more, drug-related deaths and violence have reached extremely high levels. In fact, "one can observe today a vast drug supply which is always guaranteed in the very large US market. Simultaneously, [drug trafficking] has significantly benefited transnational banks in multi-million dollar money laundering operations" (Correa-Cabrera, 2017: 212).

On the whole, the main losers of these wars are the most vulnerable people: small land- and business owners, and those who do not have the resources to flee or defend themselves against extortion, kidnappings, and other forms of brutality by drug cartels, transnational criminal gangs or *pandillas*, paramilitaries, and government forces. Forced displacements, mass disappearances, and militarization in key regions of Latin America "have emptied strategic lands and left them available for future investments." At the same time, these modern wars have become "a big business for private actors who provide security services to the government, entrepreneurs, and even criminal groups" (Correa-Cabrera, 2017: 219).

The drug war in Latin America has had the effect of bringing financial gain for specific groups or actors (national, foreign, and transnational). Overall, the greatest beneficiaries of drug war violence seem to be transnational actors: banks, private security companies, the private prison system, and energy firms. The United States has also profited significantly from the so-called drug war through the expansion of its border-security/military-industrial complex (Correa-Cabrera, 2017: 227).

It is worth noting that forced displacements, disappearances, and depreciation of land values in key violent areas of Latin America have not halted investment in energy and commercial infrastructure, or mass tourism. Energy contractors have not stopped working, and the expansion of large investment projects continues despite the high risk posed by organized crime and the large number of disappearances. It is also interesting to observe that while drug cartels, criminal paramilitaries, and Central American gangs have affected mostly micro, small, and medium entrepreneurs and poorer families or individuals, "they have hardly touched other transnational interests" (Correa-Cabrera, 2017: 227). Armed conflicts in Latin America in the framework of the War on Drugs have promoted private security businesses, which are sometimes "part of a set of contract schemes that include consulting services, training, and the massive sale of weapons. These services can be offered to [organized crime groups] as well as to law enforcement agencies" (Correa-Cabrera, 2017: 229). The role of private security companies in drug-related armed conflicts in Latin America can be compared, to some extent, to their role in the wars in Iraq and Afghanistan. My views on these wars are the following:

> In such conflicts, . . . specialized private companies have been hired by the US government or by other transnational private businesses to protect their facilities, personnel, and interests in areas of extreme violence and confrontations. The privatization of security – including the privatization of the prison system – seems to be a worldwide trend.
>
> *(Correa-Cabrera, 2017: 229)*

Finally, among the major winners of armed conflicts in Latin America are US arms-producing companies. The militarization of security in Latin America and along the US southern border – as a response to drug cartels and gang violence – has essentially benefited arms-producing companies and private security contractors. Purchases of arms and military equipment to fight drug cartels, transnational criminal gangs, or criminal paramilitary have increased substantially in the region during the past few years. At the same time, criminal groups have responded in kind to the governments' offensive. The paramilitarization of organized crime and the growing conflicts among different criminal groups for control of key Latin American territories had already significantly increased illegal imports of arms and military equipment into Mexico, Colombia, and Central America. Hence, a vast amount of legal and illegal arms has arrived in Latin America within the context of the drug war (Correa-Cabrera, 2017: 228).

Overall, the effects of Latin American drug wars or the militarization of security in a number of countries in the region have not been positive in terms of development, equality, and social justice. What is worse, these US-backed policies seem to have failed in their quest to reduce drug consumption, violence, and insecurity in the region. Considering these unsatisfactory results, it might be advisable to look at different alternatives to solve the problems of drug abuse, drug trafficking, and organized crime in this region and the Western Hemisphere in general. It could be worth it to evaluate the possibility of legalizing or regulating the usage of certain drugs or the performance of specific drug-related activities. In this regard, it would be quite interesting to further examine the case of Bolivia that decided to end its drug war "by kicking out the DEA and legalizing coca" (Tegel, 2016). The country's policy of legal but regulated production of coca seems to have effectively reduced drug-related conflict. At the same time, other complementary policy options and community actions should be considered.[6] The actual solution to drug-related problems might require a multi-dimensional approach, but drug wars do not seem to be the most effective and appropriate option.

Notes

1 Excerpts and key arguments of this chapter appear in Guadalupe Correa-Cabrera's book entitled: *Los Zetas Inc.: Criminal Corporations, Energy, and Civil War in Mexico* (University of Texas Press, 2017). Special thanks to the publisher for granting permission to use this material. The author also thanks Kurt Birson for his invaluable help as research assistant in this project.
2 It is worth noting that the total number of deaths reported by the Mexican government are not solely attributed to the War on Drugs. It is quite difficult to calculate the exact amount of deaths caused by drug-related conflicts.

3 On Mexico's "modern civil war," see Chapter 6 of Correa-Cabrera (2017).
4 Mexico has maintained this scheme until today. In December 2017, the Mexican Congress approved the Law of Internal Security, which will formally regulate the deployment of the military in the country. This took place more than a decade after the president deployed it to allegedly fight the drug cartels (Díaz, 2017: par. 1).
5 See Saab and Taylor (2009).
6 See Neate and Platt (2010).

References

Alvarado, I. (2015) Mexico's ghost towns: Residents seeking asylum in U.S. fear returning to deadly Juárez Valley. *Al Jazeera America*, June 17. http://projects.aljazeera.com/2015/09/mexico-invisible-cartel/ (Accessed 30 April 2018).

Chávez, A. (2013) Michoacán, el laboratorio para acabar con las autonomías: Carlos Fazio. *Desinformémonos Periodismo de Abajo*, 3 November. http://desinformemonos.org/2013/11/michoacan-el-laboratorio-penista-para-acabar-con-las-autonomias-carlos-fazio/ (Accessed 30 April 2018).

Cockcroft, J. D. (2010) *Mexico's Revolution: Then and Now*. New York: Monthly Review Press.

Correa-Cabrera, G. (2013) Desarrollo empresarial, inversión extranjera y crimen organizado en México: Los efectos reales de la violencia (2006–2010). *Panorama Socioeconómico* 31(46): 29–40.

———. (2017) *Los Zetas Inc.: Criminal Corporations, Energy, and Civil War in Mexico*. Austin: University of Texas Press.

Díaz, L. (2017) Victims of Mexico military abuses shudder at new security law. *Reuters*, 15 December. www.reuters.com/article/us-mexico-violence-victims/victims-of-mexico-military-abuses-shudder-at-new-security-law-idUSKBN1E92LR (Accessed 30 April 2018).

Executive Secretariat of the National System of Public Safety. (SESNSP, 2018) *Informe de Víctimas de Homicidio, Secuestro y Extorsión 2017*. Mexico City: SEGOB.

Fernández, B. (2017) Time to declare war on the US 'war on drugs' in Latin America. *Upside Down World*, 12 October. http://upsidedownworld.org/archives/international/time-to-declare-war-on-the-us-war-on-drugs-in-latin-america/ (Accessed 30 April 2018).

Gray, V. J. (2008) The new research on civil wars: Does it help us understand the Colombian conflict? *Latin American Politics and Society* 50(3): 63–91.

Huey, D. (2014) The US war on drugs and its legacy in Latin America. *The Guardian*, 3 February. www.theguardian.com/global-development-professionals-network/2014/feb/03/us-war-on-drugs-impact-in-latin-american (Accessed 30 April 2018).

Jalife-Rahme, A. (2014) *Muerte de Pemex y Suicidio de México*. Mexico City: Orfila Valentini.

Kalyvas, S. N. (2001) New' and 'old' civil wars: A valid distinction? (research note). *World Politics* 54(October): 99–118.

Loret de Mola, C. (2001) *El Negocio: La Economía de México Atrapada por el Narcotráfico*. Mexico City: Grijalbo.

Machuca, M. F. (2014) *From Militarization to New Forms of Paramilitarism in Mexico*. Professional Report, Master in Public Policy and Management (MPPM). Brownsville: University of Texas at Brownsville.

Molloy, M. (2017) Homicide victims in Mexico 2007-November 2017. *Frontera List*, 27 December. https://fronteralist.org/2017/12/27/homicide-victims-in-mexico-2007-november-2017/ (Accessed 30 April 2018).

National Registry of Data of Missing or Disappeared Persons. (RNPED, 2018) *Database of Missing or Disappeared Persons, 2006–2017*. Mexico City: RNPED, SESNSP, SEGOB.

Neate, P. and Platt, D. (2010) *Culture Is Our Weapon: Making Music and Changing Lives in Rio de Janeiro*. New York: Penguin Books.

Neilson, B. (2004) *Free Trade in the Bermuda Triangle . . . and Other Tales of Counterglobalization*. Minneapolis: University of Minnesota Press.

Paley, D. M. (2014) *Drug War Capitalism*. Oakland, CA: AK Press.

———. (2017) State power and the enforcement of prohibition in Mexico. *Mexican Law Review* 10(1): 3–20.

Ríos, V. (2008) *Evaluating the Economic Impact of Mexico's Drug Trafficking Industry*. Working Paper. Boston, MA: Harvard University, Government Department. www.gov.harvard.edu/files/Rios2008_Mexican DrugMarket.pdf

Saab, B. Y. and Taylor, A. W. (2009) Criminality and armed groups: A comparative study of FARC and paramilitary groups in Colombia. *Studies in Conflict and Terrorism* 32(6): 455–475.

Tegel, S. (2016) Bolivia ended its drug war by kicking out the DEA and legalizing coca. *Vice News*, 21 September. https://news.vice.com/article/bolivia-ended-its-drug-war-by-kicking-out-the-dea-and-legalizing-coca (Accessed 30 April 2018).

United Nations Development Programme (UNDP). (2016) *2016 Human Development Report: Development for Everyone*. New York: UNDP.

Zavala, O. (2018) *Los Carteles No Existen: Narcotráfico y Cultura en México*. Barcelona: Malpaso Ediciones.

12

DIVERSITIES OF INTERNATIONAL AND TRANSNATIONAL MIGRATION IN AND BEYOND LATIN AMERICA

Cathy McIlwaine and Megan Ryburn

Introduction

Referred to by Durand and Massey (2010) as a series of 'new world orders,' the population flows into and out from Latin America since the 1500s have been characterized by huge diversity. Not only has the direction of flows changed over time, but migrants have moved to the continent, they have left, they have moved within and they have maintained complex transnational ties across borders. More than merely reflecting multiple patterns of mobilities, these movements have been the cornerstone of nation-building in Latin America, underpinned by expressions of intersectional power, exclusions, and inequalities (Wade, 2010). Furthermore, they act as important barometers of socio-economic, political, and cultural change in the continent and other parts of the world. This chapter traces these processes focusing on three sets of movements: first, early flows from Europe to Latin America; second, migration from Latin America to Europe and the United States (US); and third, movements within Latin America, before exploring the ways in which transnational ties link these together. The chapter also argues that these processes are often underpinned by inequalities of power manifested in multiple ways, and that the complexities, multidirectionality, and transnationality of migration within and beyond Latin America is often overlooked and simplified.

Multiple temporalities and spatialities of international migration flows: from Europe to Latin America

While an excavation of the meanings of the 'Age of Discovery' and the nature of colonial endeavours and exploitation is beyond the scope here, it is important to note that the movement of Spanish and Portuguese to what was 'invented' as the 'New World' or Latin America (Mignolo, 2005), were in effect among the first migrants to the continent. Undergirding the colonization process was slavery, which although not migration *per se*, entailed the forced movement of more than 12 million people from the African continent to Latin America from 1500s until the 1800s (Durand and Massey, 2010: 22). At the same time, relatively small numbers of migrants from Europe and the Middle East began to settle in various countries across the region. This 'mixing' of peoples came to be referred to as 'miscegenation' or '*mestizaje/mestiçagem*' which has been interpreted both positively in terms of a 'racial democracy' and negatively as rooted in racism (Wade, 2010, 2017).

Not until the second half of the 19th century did people move in large numbers to Latin America from Europe and various parts of Asia. Between this time and the 1950s, European migrants, mainly from Italy, Spain, and Portugal moved primarily to Argentina, Brazil, Cuba, Uruguay, and Chile (approximately 7.5 million people) (Durand and Massey, 2010: 22). Indeed, Italians, Spanish, and Portuguese constituted over two-thirds of migrants to Latin America between the 1870 and 1930 (Goebel, 2016: 2). More specifically, Argentina has often been identified as the country with the largest numbers of European migrants settling; between 1820 and 1932, more than 6 million moved there (followed by more than 4 million to Brazil) (Padilla and Peixoto, 2007) (see below). The scale of these movements, especially between 1880s and 1930 was 'quantitatively unprecedented' as a global mass movement of people (Goebel, 2016: 2) and arguably created one of the 'new global orders.' Although there were many more migrants moving from Europe to the US at this time rather than Latin America, in some specific countries such as Argentina and Uruguay, the ratio of arrivals to residents exceeded that of the US (ibid).

The reasons explaining international migration patterns are complex although one common factor lay in racial engineering. It is often argued that European migration was favoured by Latin American elites who sought to encourage it as a way of 'whitening' and thus, purportedly, 'improving' the population (Wade, 2017; see Bastia and Vom Hau, 2014). However, this has been challenged as too simplistic. For example, although Mexico and the Dominican Republic nurtured desires for racial 'whitening' with attempts to lure European settlers, this was never realized despite some migration of Chinese and Middle Eastern migrants to Mexico and the acceptance of 750 Jewish refugees from Nazi Germany to the Dominican Republic in 1937 (Goebel, 2016). In reality, the reasons for moving were multiple and depended on labour demand, access to land, and high salaries together with immigration policies in Latin America as well as poverty, hardship, and politics in Europe and beyond, not to mention improved transport and communication (Padilla and Peixoto, 2007).

The exact nature of these flows took different forms depending on the country and the time period. For instance, Chinese migration was concentrated in Peru where migrants moved to work on coastal plantations, as well as to the Panama Canal and sugar and banana plantations in Cuba, the Dominican Republic, and Costa Rica throughout the late 19th and early 20th centuries. The Japanese moved to Latin America after the slowdown of the main flows from Europe, moving primarily to Brazil (approximately 190,000) and Peru (approximately 20,000) (Durand and Massey, 2010: 22). This complexity was compounded first, by migrants moving within Latin America when they arrived (see below), and second, by mislabelling certain nationality groups. For instance, Germans who migrated to Argentina were really from the Lower Volga in Russia while 12,000 Irish who settled there in the 19th century were referred to as English (Goebel, 2016: 10). Furthermore, around half of all European migrants in the late 19th and early 20th centuries returned home (ibid.: 6). This was especially common among Spanish and Portuguese who returned usually when they failed economically (Padilla and Peixoto, 2007).

Therefore, while this period of migration was critically important in nation-building in Latin America, the flows were much more multifaceted and dynamic than originally thought, especially in relation to intra-regional migration. Yet they also laid the foundations for subsequent transnational ties between Latin America and the rest of the world as the predominant flows broadly changed direction in the 1950s.

Multiple temporalities and spatialities of international migration flows: from Latin America to Europe and beyond

Variations in global economic and political development within and beyond Latin America continued to fuel international migration into and out from the region after the Second World War. However, this period marked a shift from migrants arriving in Latin America towards emigration elsewhere. Not only did post-war reconstruction and subsequent economic growth in Europe mean that people were less likely to want to migrate abroad, but population growth and Import Substitution Industrialization (ISI) in many Latin American countries brought economic improvements meaning that much migration was rural-urban and closely related with urbanization processes (Pellegrino, 2000) (see below). Yet, labour demand in the US in the 1940s and 1950s still meant that Puerto Rico and Mexico provided abundant supplies of low-wage migrant labour in agriculture and services. While as US citizens Puerto Ricans could move relatively easily, albeit on neocolonial terms, Mexican migrants entered either as undocumented or through a range of guest worker programmes, including the Bracero Program in the 1940s (Calavita, 2010).

Cold War politics, US hegemony, and the rise of authoritarian regimes also played a role in shaping international migration (Pellegrino, 2000). From the 1950s to the 1990s, the US was involved in direct interference or military interventions in many Latin American countries resulting in a range of different outcomes; in Cuba, following the 1959 Revolution, the US opened its doors to Cuban refugees, while the intervention in the Dominican Republic in 1965 led to out-migration of those on the Left to the US (Portes and Rumbaut, 2006). US support for right-wing dictatorships in Argentina, Brazil, Bolivia, Guatemala, and Nicaragua also led to increased migration, much through exile, albeit varying according to country. For instance, while Nicaraguan refugees could secure permanent residence in the US, most Guatemalan and Salvadorans were only able to apply for temporary protected status (TPS) (Durand and Massey, 2010). The latter has since been rescinded in 2018 by the Trump administration for 200,000 Salvadorans residing in the US since 2001 following two earthquakes and joining withdrawals of TPS for 45,000 Haitians and Nicaraguans in 2017.

Although emigration grew significantly during this period, increasing from 1.6 million to 11 million in Latin America and the Caribbean between 1960 and 1990s (Pellegrino, 2000: 399), subsequent flows have been even higher. This was primarily linked with the imposition of neoliberal economic development models that supplanted ISI policies, and which led to widespread hardships throughout the continent as part of the 'Lost Decade.' In some cases, this led to a decline in skilled immigration, as in Venezuela where the oil industry had attracted many skilled workers from abroad (Durand and Massey, 2010), whereas in many others, people moved to escape poverty and improve their economic opportunities. Flows continued to increase to the US and other countries with significant ties to Latin America, such as Brazilians moving to Portugal and Japan (Pellegrino, 2000). This was facilitated in Japan by changes in the immigration system allowing descendants of Japanese to enter on favourable terms (Tsuda, 1999).

The turn of the millennium witnessed increased emigration, but also marked diversification. Between 2000 and 2010, 50% of Latin Americans migrating (excluding Mexico) moved to the US, 24% to other Latin American countries, and 13% to Spain (OIM, 2015: 5). Although 9/11 led to some curtailment of flows into the US and contributed to diversification of European destinations (McIlwaine, 2011), movements into North America dominated; in 2015, 25 million migrants lived in the region representing an increase from 10 million in 1990. Around half of this comprised people born in Mexico (12.5 million), constituting the largest country-to-country migration corridor in the world (IOM, 2018: 75–76). There are important populations from

Colombia, Peru, and Ecuador (more than 1.5 million), with recent growth from Paraguay, Venezuela, and Brazil linked with political and economic crises in these countries (IOM, 2017: 7).

Despite the dominance of Mexico-North American flows, 4.6 million Latin Americans resided in Europe, representing an increase from 1.1 million in 1990. While Spain was the primary destination with 1.8 million migrants, many also moved to Italy, Portugal, Sweden, the Netherlands, the UK, France, and Germany and were most likely to have moved from Colombia, Brazil, and Ecuador (IOM, 2018: 76). While early flows to Europe included Cubans for political reasons, as well as other exiles fleeing authoritarian regimes (such as Chileans moving to Sweden, Norway, and the UK), economic factors have dominated. These have included poverty, income inequality, and lack of opportunities in Latin America together with labour demand in the booming economies of southern Europe especially in care, cleaning, and construction sectors (prior to the crisis of 2008) (McIlwaine, 2011). These flows, which were largely feminized in nature, were further stimulated by socio-cultural similarities and existing family ties in Spain and Italy, together with favourable immigration legislation that encouraged migration through various bilateral regularization programmes (McIlwaine, 2012). Indeed, the gendered variations in migration patterns reflect both the gender ideologies in home countries as well as the nature of labour demand in destinations. Reflecting on migration to the US, Donato (2010) notes that feminized flows of Dominican migration compared to more masculinized flows from Mexico reflect flexible gender identities in the former and more 'traditional' ideologies in the latter.

Although it is acknowledged that much contemporary movement beyond borders of Latin America revolves around economic exigencies, political circumstances have continued to play an important role. Just as Cold War geopolitics and authoritarian rule influenced migration in the second half of the 20th century, armed conflict and everyday violence in several countries precipitated contemporary movements. For example, the Colombian armed conflict has created the world's largest population of internally displaced peoples (7.2 million as of 2016) and extensive cross-border movements to Venezuela and Ecuador (IOM, 2018: 78). It has also led to international migration, or what has been termed 'transnational displacement' underpinned by threats to personal security and fear linked to everyday as well as political and gender-based violence and erosion of economic livelihoods (McIlwaine, 2014 on Colombians in London). In addition, everyday urban violence in many Central American countries and Mexico, often linked with gang and drug violence, has led to migration and burgeoning asylum claims. In 2015, there were 250% more asylum claims in the US from El Salvador, Honduras, and Guatemala between 2013 and 2015, and an increase of 155% in claims from Mexicans in the US (IOM, 2018: 79).

These contemporary processes are dynamic and in constant flux. South America has over 10 million emigrants with a regional average of 5.4% of the total population moving. However, while absolute numbers of emigrants are growing, there has been a slowdown between 2010 and 2015 (IOM, 2017). While this is partly linked with people being less likely to move, it also relates to the return of those born or naturalized abroad. For example, in 2015, 118,598 people from the US lived in South America, especially in Brazil and Ecuador. Similarly, following the 2008 recession, many Latin Americans with Spanish nationality returned home; in Ecuador, the stock of Spanish migrants increased between 2005 and 2015 from 3,658 to 7,473 (IOM, 2017: 5). There has also been an increase in migration from Africa and Asia as Europe and North America have become increasingly difficult to access. While the numbers from African countries are small, between 2004 and 2014, Argentina, Brazil, and Chile granted almost 50,000 permanent residency permits to Chinese citizens (of whom 58% were male) (IOM, 2017: 5). Yet, for the region as a whole probably the most significant international migration dynamics relate to intra-regional flows.

Multiple temporalities and spatialities of international migration flows: within Latin America

Although intra-regional migration in Latin America has always been important, it has increased significantly recently. In South America, 70% of all migrations are within the region, now outnumbering extra-regional movements (IOM, 2017). Internal migration has also been integral to the creation of Latin American countries even if it has been overlooked (Rodríguez and Busso, 2009). Partly interrelated with internal migration across proximate borders has been rural-urban movements, with the latter being rapid and contributing to the urbanization of the continent as a whole. While 41% of the region lived in urban areas in the 1950s, this stood at 80% in 2014 – significantly higher than other regions (UN DESA, 2014). Factors motivating these processes include the dominance of ISI as an economic strategy, combined with changes in agricultural production, which also contributed to emigration (Cerutti and Bertoncello, 2003; see above). Today, internal migration is primarily urban-urban, which again feeds into international processes as most migrants moving abroad originate from cities (McIlwaine and Bunge, 2016).

This has been accompanied by shifting and diversifying intra-regional migration which has tended to be strongly feminized (Martínez Pizarro and Orrego Rivera, 2016). Although extra-continental migration was the predominant form to Latin America from the colonial period into the early 20th century, from the 1920s, intra-regional flows – especially between neighbouring countries – began to gradually increase as migration from Europe declined. Argentina has historically been one of the main receptor countries and is emblematic of these patterns (see above). In 1914, migrants from outside Latin America comprised 27% of Argentina's total population, and migrants from neighbouring countries only 2.6% (Martínez Pizarro, 2011: 101). By 2000, migrants comprised 4.2% of the total, of which 60% were from neighbouring countries, principally Paraguay, Bolivia, Chile, and Uruguay (Martínez Pizarro, 2011: 101). By 2015, this had risen to 5% (South America's highest average) (IOM, 2017: 2).

Argentina is joined by Costa Rica and Venezuela as countries that have traditionally had significant intra-regional migrant populations. In Costa Rica, migrants constituted 7.5% of the total population in 2000 (CEPAL, 2006), rising to 9% in 2015 (IOM, 2018: 76), mainly Nicaraguans. This represents the continuation of a trend that commenced in the late 19th century when the expansion of the banana plantations attracted Nicaraguan workers to Costa Rica (Fouratt and Voorend, 2017). Subsequently, Nicaraguans have gradually moved from agricultural to more urban-based employment (CEPAL, 2006). This mirrors tendencies in Argentina, where intra-regional migrants were initially involved in the agricultural sector, but have progressively moved to the cities (Bastia, 2007).

As with all types of migration across borders noted above, intra-regional migration was spurred by both economic and political factors. In the case of Nicaraguan migration to Costa Rica, for example, in addition to economic instability, the armed conflict from 1978 to 1990 also compelled people to leave. Similarly, repression and human rights abuses in the Southern Cone dictatorships of the 1960s to the 1980s led to the exile of hundreds of thousands. Whilst people fled to all continents except Antarctica (see above), certain countries within Latin America, most notably Venezuela, were important hubs for refugees from the region. For example, Chilean exiles, comprising around 200,000, or 2% of the country's 1973 population, settled in over 140 countries, but it is estimated that over 40% of them settled in Venezuela (Sznajder and Roniger, 2007).

Venezuela has also been a key destination for refugees and migrants from neighbouring Colombia seeking to escape armed conflict and everyday urban violence (McIlwaine, 2014; see above). For several decades, Colombians have constituted the largest intra-regional migrant population in Latin America. In 1990, there were approximately 600,000 intra-regional Colombian migrants, reaching 700,000 in 2000 with around 90% moving to Venezuela (CEPAL, 2006: 23). In turn, Haiti currently sends the second largest number of refugees abroad linked with disasters (IOM, 2018: 78) while Central American migration directly to Mexico and transit migration through Mexico to the US has also been very significant (ibid.).

Much transit migration has been irregular, with apprehensions moving southwards. Indeed, in 2014 and 2016, arrests of Central Americans exceeded those of Mexicans trying to cross the Mexican-US border. These have also diversified to include Haitians as well as migrants of African and Asian origins trying to cross into the US via Mexico (IOM, 2018: 79). This burgeoning transit migration has entailed high levels of exploitation as well as structural and everyday violence and death among migrants as they negotiate their way towards their dream in the US, often via smuggling routes (Vogt, 2013). The response to transit and other intra-regional movements has been increased border enforcement, focusing especially on combatting smuggling and the associated industry. For example, Costa Rica closed its borders to Cubans in 2015 and to all irregular migrants in 2016, while Mexico implemented its Southern Border Plan in 2014 to reduce irregular migrant flows from Central America (IOM, 2018).

In the southern cone, Chile has also experienced recent increases in intra-regional migration. While in 1992, there were an estimated 114,597 foreigners living there (0.9% of the total population), this more than quadrupled to 465,319 migrants (or around 2.7% of the population) in 2015 (CASEN, 2015). The majority (90%) are from other Latin American countries, but with an increasing diversity in countries of origin; previously most migrants were from Argentina and Peru, yet there are now growing numbers from Bolivia, Colombia, Venezuela, Dominican Republic, and Haiti. For example, the Colombian population in Chile has increased by 345% in the past decade (Rojas Pedemonte and Silva Dittborn, 2016) while the Bolivian population has increased markedly and concentrated in precarious employment in labour niches such as agriculture, domestic work, and wholesale garment retail (Ryburn, 2016). Other countries in the Southern Cone such as Brazil are also receiving increasing and diversified migration such as increasing Haitian migration (IOM, 2017; OIM/IPPDH, 2017).

Another important economic factor affecting intra-regional migration has been the 2008 global economic crisis (see above). Whilst impacts have been felt throughout Latin America, they have been less severe than in the US or Europe. Indeed, some migrants to the US and Spain returned briefly to their countries of origin following the fall-out from the economic crisis, only to move onwards again to an intra-regional destination. Additionally, those who previously would have considered extra-continental migration became less inclined given the crisis and the relatively stable economies of certain countries such as Chile, which has been part of the OECD since 2013 (IOM, 2017). Furthermore, restrictive immigration regimes in the US and Europe also act as a deterrent to extra-continental migration (McIlwaine, 2015a). Migration regimes in Latin America are comparatively relaxed, although certainly not entirely unrestrictive; indeed, groups of certain nationalities and/or ethnicities and socio-economic backgrounds may face particular discrimination (Ryburn, 2018 on Chile).

Since 2010, Haiti and Venezuela have also seen a sharp spike in emigration prompted by a combination of factors. In Haiti, many people left in the aftermath of the disasters – from the year 2000 to 2010, just after the earthquake, Haitian migration rose by 392%, from 64,360 to 317,054 (Martínez Pizarro and Orrego Rivera, 2016). The Dominican Republic received the vast majority of Haitians, but as indicated above, Brazil and Chile have also become important receiving countries (OIM/IPPDH, 2017). In Venezuela, intense political instability since the latter years of Hugo Chávez's presidency and following his death has prompted many to leave; an estimated 606,281 Venezuelans left the country in 2015 alone (OIM, 2015). Reversing previous migration flows, many are going to neighbouring Colombia, as well as to Argentina and Chile (IOM, 2017; Martínez Pizarro and Orrego Rivera, 2016).

The contemporary flows are indicative of the growing diversity of intra-regional migration in Latin America. Whilst this represents a new phase in Latin American migration, it is also consistent with what has been a very dynamic migration context over several centuries. The multiplicity of flows, motivations for migration, and transnational connections over time are therefore more complex than is often recognized.

Transnationality in Latin America and beyond

While the discussion above has delineated the main patterns and processes of migration into and out from Latin America, it is also important to identify the broader significance of these movements. While it is not the intention to rehearse the range of theoretical approaches that have been developed to understand international migration flows, it is important to show how some of the main conceptual framings play out in the Latin American context and how this context has shaped the theorizing in the first place. Whether this is related to maximizing earning power as in neo-classical theories, risk-sharing from a New Economics of Labour Migration (NELM) viewpoint, or household networks and strategies identified through structuration perspectives, various forms of inequalities influence why people move (see Castles, de Haas, and Miller, 2014; McIlwaine, 2015b for discussions). Yet these inequalities span source and destination countries and as a result, a relational approach is needed to understand international migration as part of a transnational system of ties, circuits, and spaces that link migrants and their lives together over space. While not exclusively, much research on transnationalism has drawn on Latin American and Caribbean examples (Basch, Glick Schiller, and Szanton, 1994).

Latin American transnationality has taken multiple forms. Analyses of political transnationality, for instance, tend to revolve around how Latino migrants have engaged in electoral and civic politics in the US, especially in terms of the mutually reinforcing nature of transnational political engagement and wider integration processes (Portes, Escobar, and Arana, 2008). However, this geographical focus has recently broadened to include the experiences of Latin Americans in European contexts. Research with Colombians in the United Kingdom (UK) and Spain has highlighted the gendered nature of political participation where working- rather than middle-class women are more active (McIlwaine and Bermudez, 2011), and the ambivalent forms of citizenship exercised in relation to external voting (McIlwaine and Bermudez, 2015). Not only does this demonstrate the gendered importance of politics in integration processes in destinations, but also the potential power of migrants in influencing politics back home, not least through remittance income.

Economic transnationalism is another core dimension of Latin American transnationality through remittance sending and their role in addressing development problems as part of the wider migration-development nexus. While sending money back home can be viewed as a key driver in reducing poverty (Acosta et al., 2008), international migrants make considerable sacrifices in order to send money home. Remittance receipts can also create social problems back home not least because of their dubious sustainability (Wills et al., 2010). Yet, remittances sent to Latin America continue to grow apace; in 2016, they grew nearly 8% compared to 2015, amounting to US$70 billion (Orozco, 2017: 3). Despite their contradictory role, they are likely to remain a key economic driver and safety net in Latin America into the future. Furthermore, while much research on remittances has been on extra-continental flows, they are also extremely important across more proximate borders (Melde et al., 2014).

Social transnationality, maintained through 'social remittances' denoting "ideas, behaviours, identities and social capital that flow from receiving- to sending-country communities" and back again (Levitt, 1998: 926), are also increasingly recognized as important and indeed, provide the mechanisms through which to tie political, economic, and cultural links together. Referring to the Dominican Republic and the US, Levitt and Lamba-Nieves (2011) show how social remittances operate individually across borders between friends and families and collectively through civil society organizations as well as through the ideas and values that migrants take with them when they move). Religious remittances are part of wider forms of social remittances and can refer to the transfer of faith as people move, institutional links between churches across borders as well as virtual linkages through, for example, transmitting church services over the internet between countries (Sheringham, 2013).

Another important dimension of this relates to gender and specifically how gendered power relations transform as women and men move (Donato, 2010; Mahler and Pessar, 2001). It is generally accepted that such transformations are contradictory, entailing some improvements in women's lives in relation to improved access to the labour market and the independence concomitant with having access to an income (Boehm, 2008). On the other hand, gender inequalities can be intensified, especially if women are undocumented or experience gender-based violence (McIlwaine, 2010). Furthermore, it is more difficult to change gender ideologies compared with gender practices, the latter being more malleable, and the outcomes for women vary in intersectional ways (ibid). Gender transformations must also be viewed within wider processes of inequality through the extension of transnational chains of emotional labour or 'global care chains' and "the international division of reproductive labour" (McIlwaine and Ryburn, 2019 for discussion). While much of this work refers to Latin Americans moving to the US and Europe in terms of women from poorer countries moving to carry out reproductive labour hitherto conducted by wealthier women who have entered the paid workforce (Hondagneu-Sotelo, 2001), increasingly it occurs within the region (Ryburn, 2016). Also important to acknowledge is that women not only migrate, often becoming transnational mothers in the process, but they are also left behind in Latin America, especially as grandmothers, elder sisters, and other female relatives with caring responsibilities for children of migrants (Herrera, 2013).

Many aspects of these transnational ties can be combined into analyses of transnational social spaces. Although it is often assumed that transnational linkages are binational, Latin American migration has become more complicated often linking multiple countries and peoples together. From the perspective of Latin Americans in London, in 2010, 36.5% had migrated via another country, with 38% moving from Spain (McIlwaine, 2012). Subsequently, research showed that 80% of Latin Americans who had migrated from Europe came from Spain through onward migration (McIlwaine and Bunge, 2016; also Mas Giralt, 2017). These mobility processes feed into the construction of multifaceted transnational spaces comprising families living across borders, remittances flowing not only back home to Latin America, but also to intermediate destinations, as well as a host of other ties through negotiations over immigration status, political engagements, and civil society organizations (McIlwaine, 2012, 2015a).

It is therefore important to acknowledge that Latin American international migration has created a diverse system of transnational spaces and multiple linkages both within the continent and beyond, often buttressed by inequalities of power. Although many wealthy professional and highly skilled Latin Americans move around the world with ease, and it is rarely the very poorest in society who move, most who move do so to enhance their social, economic, and/or political livelihoods. However, the reality of the outcomes of migrating, at least in initial stages can be extremely harsh. Again using the London example, half of Latin Americans recently arrived from Spain (who had EU citizenship) worked in low-paid, precarious cleaning jobs as their only option (usually because of English language difficulties). Their lives beyond the labour market were equally precarious with many sharing poor-quality, overcrowded accommodation with other families or individuals (McIlwaine and Bunge, 2016). These forms of precarity also prevailed (and are arguably worse) in intra-regional movements in terms of labour exploitation (McIlwaine and Ryburn, 2018), poor health provision (Gideon, 2014), and housing provision (Rojas Pedemonte and Silva Dittborn, 2016) especially when compared with their native-born counterparts. Relatedly, across the world, Latin American migrants face frequent discrimination and racism in their everyday lives (Tijoux, 2016), and even physical violence, as among Central American migrants who transit through Mexico (Vogt, 2013). These hardships are likely to increase into the future as the Trump administration in the US implements ever more draconian measures to prevent migration and to deport those already settled there, many who originate in Latin America and especially Mexico where Trump's rhetoric has been especially insidious (see above). Such hostile migration regimes have been replicated in other parts of the world, especially in European countries, and are likely to contribute to increased intra-regional migrations within the continent.

Conclusion

This chapter has examined the nature of international and transnational migration within and beyond Latin America. Through tracing a range of temporal and spatial framings around flows into, out from, and within the continent, it has also assessed some of the core socio-economic and political reasons underlying these movements at different scales. As well as arguing that international migration has been integral in nation-making processes, it also suggests that a range of different forms of transnationality have tied people and countries together in multiple ways across borders. Overall, the chapter has argued for enhanced recognition of the complex diversities of international migration within and beyond Latin America; not only have these been more diverse than first thought in looking back historically, but are also likely to diversify further into the future, especially within the region.

References

Acosta, P., Calderón, C., Fajnzylber, P. and Lopez, H. (2008) What is the impact of international remittances on poverty and inequality in Latin America? *World Development* 36: 89–114.

Basch, L., Glick Schiller, N. and Szanton Blanc, C. (1994) *Nations Unbound*. Amsterdam: Gordon and Breach.

Bastia, T. (2007) From mining to garment workshops: Bolivian migrants in Buenos Aires. *Journal of Ethnic and Migration Studies* 33(4): 655–669.

Bastia, T. and vom Hau, M. (2014) Migration, race and nationhood in Argentina. *Journal of Ethnic and Migration Studies* 40(3): 475–492.

Boehm, D. A. (2008) 'Now I am a man and a woman!' Gendered moves and migrations in a transnational Mexican community. *Latin American Perspectives* 35(1): 16–30.

Calavita, K. (2010) *Inside the State: The Bracero Program, Immigration, and the I. N. S.* New Orleans: Quid Pro Quo Books.

CASEN. (2015) *Inmigrantes: principales resultados (versión extendida)*. Santiago, Chile: Observatorio del Ministerio de Desarrollo Social.

Castles, S., de Haas, H. and Miller, M. (2014) *The Age of Migration*. 5th Ed. Basingstoke: Palgrave Macmillan.

CEPAL. (2006) *Migración Internacional*. Santiago, Chile: América Latina y el Caribe Observatorio Demográfico.

Cerutti, M. and Bertoncello, R. (2003) *Urbanisation and Internal Migration Pattern in Latin America*. Buenos Aires: Centro de Estudios de Población Argentina.

Donato, K. M. (2010) US migration from Latin America: Gendered patterns and shifts. *Annals of the American Academy of Political and Social Science* 630(1): 78–92.

Durand, J. and Massey, D. S. (2010) New world orders: Continuities and changes in Latin American migration. *Annals of the American Academy of Political and Social Science* 630(1): 20–52.

Fouratt, C. and Voorend, K. (2017) Sidestepping the state: Practice of social service commodification among Nicaraguans in Costa Rica and Nicaragua. *Journal of Latin American Studies* 'First view' October 16. https://doi.org/10.1017/S0022216X17001195 (Accessed 7 December 2017).

Gideon, J. (2014) *Gender, Globalization, and Health in a Latin American Context*. London: Palgrave Macmillan.

Goebel, M. (2016) Immigration and national identity in Latin America, 1870–1930. *Oxford Research Encyclopedia of Latin American History*. http://latinamericanhistory.oxfordre.com/view/10.1093/acrefore/9780199366439.001.0001/acrefore-9780199366439-e-288?print=pdf (Accessed 1 December 2017).

Herrera, G. (2013) Gender and international migration. *Annual Review of Sociology* 39(1): 471–489.

Hondagneu-Sotelo, P. (2001) *Doméstica: Immigrant Workers Cleaning and Caring in the Shadow of Affluence*. Berkeley, CA: University of California Press.

International Organization for Migration (IOM). (2017) *South American Migration Report No. 1. Migration Trends in South America*. Buenos Aires: IOM. https://robuenosaires.iom.int/sites/default/files/Documentos%20PDFs/Report_Migration_Trends_South_America_N1_EN.pdf (Accessed 7 December 2017).

———. (2018) *World Migration Report 2018*. Geneva: OIM.

Levitt, P. (1998) Social remittances: Migration driven local-level forms of cultural diffusion. *The International Migration Review* 32(4): 926–948.

——— and Lamba-Nieves, D. (2011) Social remittances revisited. *Journal of Ethnic and Migration Studies* 37(1): 1–22.

Mahler, S. J. and Pessar, P. R. (2001) Gendered geographies of power: Analyzing gender across transnational spaces. *Identities* 7(4): 441–459.

Martínez Pizarro, J. (2011) *Migración internacional en América Latina y el Caribe: Nuevas tendencias y nuevos enfoques*. Santiago, Chile: CEPAL.

——— and Orrego Rivera, C. (2016) *Nuevas tendencias y dinámicas migratoria en América Latina y el Caribe*. Santiago, Chile: United Nations/OIM/CEPAL.

Mas Giralt, R. (2017) Onward migration as a coping strategy? Latin Americans moving from Spain to the UK post-2008. *Population, Space and Place* 23(3): 1–12.

McIlwaine, C. (2010) Migrant machismos: Exploring gender ideologies and practices among Latin American migrants in London from a multi-scalar perspective. *Gender, Place and Culture* 17(3): 281–300.

———. (2011) Introduction: Theoretical and empirical perspectives on Latin American migration across borders. In C. McIlwaine (ed) *Cross-Border Migration Among Latin Americans: European Perspectives and Beyond*. New York: Palgrave Macmillan, pp. 1–17.

———. (2012) Constructing transnational social spaces among Latin American migrants in Europe: Perspectives from the UK. *Cambridge Journal of Regions, Economy and Society* 5(2): 271–288.

———. (2014) Everyday urban violence and transnational displacement of Colombian urban migrants to London, UK. *Environment and Urbanization* 26(2): 417–426.

———. (2015a) Legal Latins: Creating webs and practices of immigration status among Latin American migrants in London. *Journal of Ethnic and Migration Studies* 41(3): 493–511.

———. (2015b) International migration. In C. Cooper and J. Michie (eds) *Understanding All Our Futures: Why Social Sciences Matter*. Basingstoke: Palgrave Macmillan, pp. 176–191.

——— and Bermudez, A. (2011) The gendering of political and civic participation among Colombian migrants in London. *Environment and Planning A* 43: 1499–1513.

———. (2015) Ambivalent citizenship and extra-territorial voting among Colombians in London and Madrid. *Global Networks* 15(4): 385–402.

McIlwaine, C. and Bunge, D. (2016) *Towards Visibility: The Latin American Community in London*. London: Trust for London.

——— and Ryburn, M. (2019) Metropolitan mobilities: Transnational urban labour markets. In R. van Kempen and T. Schwanen (eds) *Handbook of Urban Geography*. Cheltenham: Edward Elgar.

Melde, S., Anich, R., Crush, J. and Oucho, J. (2014) Introduction: The South-South migration and development nexus. In R. Anich, J. Crush, S. Melde and J. Oucho (eds) *A New Perspective on Human Mobility in the South*. New York: Springer, pp. 1–20.

Mignolo, W. D. (2005) *The Idea of Latin America*. Malden, MA: Wiley-Blackwell.

OIM/IPPDH. (2017) *Diagnóstico regional sobre migración haitiana.* Buenos Aires: OIM/IPPDH.

Organisación Internacional para las Migraciones (OIM). (2015) *Dinámicas Migratorias en América Latina y el Caribe (ALC), y entre ALC y la Unión Europea.* Brussels: OIM.

Orozco, M. (2017) *Remittances to Latin America and the Caribbean in 2016.* Washington, DC: Inter-American Dialogue. www.thedialogue.org/wp-content/uploads/2017/02/Remittances-2016-FINAL-DRAFT. pdf (Accessed 7 December 2017).

Padilla, B. and Peixoto, J. (2007) *Latin American Immigration to Southern Europe.* Washington, DC: Migration Information Source.

Pellegrino, A. (2000) Trends in international migration in Latin America and the Caribbean. *International Social Science Journal* 52(165): 1468–2451.

Portes, A. and Rumbaut, R. G. (2006) *Immigrant America: A Portrait.* Berkeley, CA: University of California Press.

Portes, A., Escobar, C. and Arana, R. (2008) Bridging the gap: Transnational and ethnic organizations in the political incorporation of immigrants in the United States. *Ethnic and Racial Studies* 30: 1056–1090.

Rodríguez, J. and Busso, G. (2009) *Migración interna y desarrollo en América Latina entre 1980 y 2005.* Santiago, Chile: CEPAL.

Rojas Pedemonte, N. and Silva Dittborn, C. (2016) *La migración en Chile: breve reporte y caracterización.* Santiago, Chile: Obersvatorio Iberoamericano sobre Movilidad Humana, Migraciones y Desarrollo. www. extranjeria.gob.cl/media/2016/08/informe_julio_agosto_2016.pdf (Accessed 7 December 2017).

Ryburn, M. (2016) Living the Chilean dream? Bolivian migrants' incorporation in the space of economic citizenship. *Geoforum* 76: 48–58.

———. (2018) *Uncertain Citizenship: Everyday Practices of Bolivian Migrants in Chile.* Berkley: University of California Press.

Sheringham, O. (2013) *Transnational Religious Spaces: Faith and the Brazilian Migration Experience.* Basingstoke: Palgrave Macmillan.

Sznajder, M. and Roniger, L. (2007) Exile communities and their differential institutional dynamics: A comparative analysis of the Chilean and Uruguayan political diasporas. *Revista de Ciencia Política* 27(1): 43–66.

Tijoux, M. E. (2016) *Racismo en Chile: La piel como marca de la inmigración.* Santiago, Chile: Ediciones Universitaria.

Tsuda, T. (1999) The motivation to migrate: The ethnic and sociocultural constitution of the Japanese-Brazilian return-migration system. *Economic Development and Cultural Change* 48(1): 1–31.

UN DESA. (2014) *World Urbanization Prospects: The 2014 Revision.* New York: United Nations.

Vogt, W. (2013) Crossing Mexico: Structural violence and the commodification of undocumented Central American migrants. *American Ethnologist* 40(4): 764–780.

Wade, P. (2010) *Race and Ethnicity in Latin America.* London: Pluto Press.

———. (2017) Liberalism and its contradictions: Democracy and hierarchy in *mestizaje* and genomics in Latin America. *Latin American Research Review* 52(4): 623–638.

Wills, J., Datta, K., Evans, Y., Herbert, J., May, J. and McIlwaine, C. (2010) *Global Cities at Work: New Migrant Divisions of Labour.* London: Pluto.

13

REGIONAL ORGANIZATIONS AND DEVELOPMENT IN LATIN AMERICA

Andrés Malamud

Introduction

There is no established convention for the designation of "developed" and "developing" countries in the United Nations system. However, some of its agencies, as well as other international organizations, have created categories that can be used as proxies. For Latin America, the picture is grim: only Argentina and Chile are ranked "very high" in the UNDP Human Development Index (2016), only Chile and Uruguay are considered as "high-income economies" by the World Bank (2017), and no Latin American country is deemed to be an "advanced economy" by the International Monetary Fund (2017). To the dismay of nationalists, the Commonwealth of Puerto Rico is included in all three top groups. Apparently, economic integration with the US has contributed more to development than national independence and regional integration – even though Puerto Rico is less developed than the 50 US states. Low levels of development (as conventionally understood) in Latin America are due to a combination of causes, some of which are rooted at the national level while others are related to the failure of regional organizations.

In South America, notes Mazzuca (2017), income per capita is five times larger than in tropical Africa but five times smaller than in the advanced North Atlantic economies. This makes of the region neither a poster child nor a basket case as regards economic development. Mazzuca (2017: 18) points to political geography as the main reason thereof:

> Some countries in South America could have followed the economic path that Australia and New Zealand initiated in the mid-19th century. Such a path was not followed because of the way in which national boundaries were demarcated. The key legacy of the process of border demarcation was twofold: on the one hand, the creation of two territorial colossuses, Argentina and Brazil, that were dysfunctional combinations of subnational economies; on the other, the emergence of smaller countries that were not powerful enough to become the engine of development for South America as a whole. Even though some small countries originally had viable economies, as was the case of Chile and Uruguay, they were in fact hurt by the dysfunctional economic nature of their giant neighbors.

The two largest economies were dysfunctional because a highly productive core region coexisted with a larger backward periphery, which drained resources in exchange for political support. Although Mazzuca's analysis is focused on South America, he explicitly allows for its extension to the rest of Latin America, with Mexico in an equivalent position to Argentina and Brazil, and Costa Rica occupying a similar place as Chile and Uruguay. Since regional organizations can be decoded as *politically organized geography*, Mazzuca's analysis of political geography as enabler or obstruction to development is very timely – though not uncontroversial. His argument that dysfunctional national integration arrested both national and regional development opens the door to two rival hypotheses. The optimistic hypothesis posits that national dysfunctionalities can be overcome through efficient regional complementation; the pessimistic one suggests that national dysfunctionalities doom regional cooperation, as the collective cannot outdo its constitutive parts. Latin American states have apparently decided in favour of the optimistic hypothesis, and they have accordingly tried to pursue national development through regional organization at least since the mid-20th century. The paramount institution behind this goal was the UN sponsored Economic Commission for Latin America (today ECLAC), established in Santiago, Chile, in 1948.

Regional tools for development

There are three recognized sources of development: accumulation of human, physical, and social capital; investment in innovation; and political institutions. Regional organizations are instances of the last category; therefore, it is relevant to understand how political institutions promote development. They do so through three mechanisms: "government credibility, legal protection and enforcement, and public good provision. These factors are tantamount to self-restriction, restriction of others, and enabling of others respectively" (Arias, 2015: 421) – or, to simplify, democracy, rule of law, and public goods.

Each of the mentioned factors can be disaggregated for analytical purposes: democracy encompasses checks and balances, universal franchise, and competitive elections; the rule of law guarantees property rights and contract enforcement; and public goods include interstate peace and security, larger financial and trade markets, and connectivity infrastructure such as energy and transportation. Regional organizations in Latin America can be evaluated according to the degree to which they have addressed these issues.

Democracy

Initially, democracy was neither a goal of nor a condition for regional integration. Both the Latin American Free Trade Association (LAFTA) and the Central American Common Market (CACM), created in 1960, included non-democratic governments among their founding members. Moreover, neither of them mentioned the word "democracy" in their foundational treaties. Economic development and regional integration were conceived of as purely technical issues, in complete isolation from the type of government of the member states and the decision-making procedures of the newly established common institutions.

Only after the third wave of democratization, which in Latin America began in 1978, did the link between regional organizations and democracy emerge. In the early 1990s, the issue of democracy took centre stage in the three main Latin American blocs. In Mercosur, the foundational treaty was signed only after Paraguay got rid of its long-time dictator, President Stroessner, and joined previously democratized Argentina, Brazil, and Uruguay in 1991. At around the same time, between 1989 and 1990, the Andean Community was revived through the establishment of a new body, the Andean Presidential Council. The connection of the new institution with democracy was made evident when it suspended Peruvian membership in the wake of President

Fujimori's 1992 *autogolpe* – or self coup, which ironically means the opposite as Fujimori dissolved congress so that he could hold power unchecked. In Central America, pacification and democratization led to the institutionalization of presidential meetings in 1991, transforming the CACM into the Central American Integration System (SICA). By the mid-1990s, all subregional organizations in Latin America had turned from complete indifference to full commitment to democracy. The time was ripe for the next step: the development of democratic clauses.

Democratic clauses are the operative instrument of democratic conditionality. In turn, democratic conditionality is a strategy developed by some international organizations to induce candidate and/or member states to comply with their democracy standards. SICA signed the Framework Treaty on Democratic Security in 1995, which promoted democracy and the rule of law in all the member states; Mercosur adopted a democratic clause through the Ushuaia Protocol in 1998; and the Andean Community did alike through the Additional Protocol to the Cartagena Agreement in 2000. They were crowned by the Inter-American Democratic Charter, adopted on 11 September 2001 by a special session of the General Assembly of the Organization of American States held in Lima, Peru. However, several studies have questioned the effectiveness of democratic conditionalities in general and of democratic clauses in particular.

After analysing three cases in the European Union, Schimmelfennig, Engert, and Knobel (2003: 515) conclude "that the impact of democratic conditionality has been marginal, but not irrelevant." Domestic conditions, i.e. governmental cost–benefit calculations, appear to be more important for compliance than reinforcement by reward. Governments and state elites were far more decisive for effectiveness than societal conditions or transnational channels. This is coincident with the conclusion arrived at by Van der Vleuten and Ribeiro Hoffmann (2010: 737), who found out that, in the EU, Mercosur and the South African Development Community, violations of democratic principles sometimes go unsanctioned. They argue that the enforcement of a democratic clause depends on whether "intervention serves the geopolitical, domestic political or material interests of regional leading powers, or if pressure by a third-party with a matching identity increases the reputational costs of non-intervention" (Van der Vleuten and Ribeiro Hoffmann, 2010: 755). They conclude in a realist vein: in the absence of external pressure, "the interests of the regional leading power explain the behaviour of a regional organization" (Van der Vleuten and Ribeiro Hoffmann, 2010: 755). This finding anticipated why the impeachments of Paraguay's Fernando Lugo in 2012 and Brazil's Dilma Rousseff in 2016 provoked different regional reactions: while both Mercosur and UNASUR suspended Paraguay's membership, they did not even convene to discuss the events that took place in Brazil.

Closa and Palestini (2015) have taken a step further to show that, in Latin America, the adoption of democratic protection mechanisms by regional organizations has not contributed per se to democratic consolidation. Instead, the performance of those mechanisms "is tied to the interests of governments that are both their rule makers and their enforcers in concrete political crises" (Closa and Palestini, 2015: 8). Hence, governments design a democratic clause to minimize its probability to escape their discretionary control, so it ends up enforcing regime stability rather than democracy. Closa and Palestini suggest that this bias in favour of the incumbent governments is not exclusive of the organization they study, i.e. UNASUR, but structural to the link between regional organizations and democracy in Latin America.

At the regional level, several organizations have established some kind of parliament or parliamentary assembly (Malamud and De Sousa, 2007). In some countries, among which various member states of SICA, the Andean Community, and Mercosur, parliamentarians are popularly elected. However, in no case have these parliaments legislative authority. Regional decisions are made by consensus of the national executives.

Two conclusions are in order. First, Latin American regional organizations have evolved from democratic indifference to democratic protection, but their efficacy as guarantors of national democracy has been low at best. Second, as long as such thing as regional democracy exists, it is rooted exclusively at the national level.

Rule of law

Well-defined and protected property rights incentivize innovative entrepreneurship, while contract enforcement prevents parties from failing to honour their contracts (Arias, 2015: 425). The crucial agents to perform these tasks are the courts of law.

Several regional courts exist in Latin America. The Inter-American Court of Human Rights, established in 1979, belongs to the Organization of American States system and is focused on limiting state arbitrariness towards individuals rather than guaranteeing property rights and contract enforcement or adjudicating in conflicts between states. The latter functions are purportedly performed by three subregional rather than hemispheric courts: the Court of Justice of the Andean Community (founded in 1979), the Central American Court of Justice (which was in operation between 1907 and 1917 and was reestablished in 1991), and Mercosur's Permanent Review Tribunal (inaugurated in 2004). Of these multi-purpose tribunals, the most deeply scrutinized has been the Andean Court.

According to Helfer, Alter, and Guerzovich (2009: 45), "the Andean Tribunal of Justice is one of the most active international courts in a world increasingly populated by international courts and tribunals." Its agenda is dominated by disputes relating to trademarks, patents, and other intellectual property rights, which within the Andean Community are regulated at the regional rather than the national level. Helfer, Alter, and Guerzovich (2009: 8) find that the Tribunal "has contributed to building an effective rule of law for intellectual property in a region of relatively weak national legal systems." This has created economically valuable and enforceable private property rights and has increased national agencies' fidelity to the rule of law. They conclude that, as a result of its more than 1,400 rulings, "intellectual property protection in the Andean Community looks different than it does elsewhere in Latin America" (Helfer, Alter, and Guerzovich, 2009: 46), having established what they enthusiastically call "a rule-of-law island." They only temper their enthusiasm to admit that the growing ideological differences that divide the member states "have hampered efforts to adopt new regional laws and enabled the United States to negotiate with Andean countries bilaterally. In these one-on-one settings, the United States possesses far greater leverage to pressure each country to adopt policies that favour American interests" (Helfer, Alter, and Guerzovich, 2009: 47). Notably enough, the European Union adopted the same one-on-one setting when it accepted that a bi-regional agreement be signed only by Colombia and Peru in 2012, to the exclusion of Bolivia and Ecuador (the latter of which joined in 2017).

Mercosur's Tribunal has been far less active than its Andean counterpart. Indeed, it issued only six infringement proceedings and three preliminary rulings between 2005 and 2012, after which it did not produce any further rulings. Neither firms nor individuals are allowed to resort to the Tribunal, which is only accessible to the governments and courts of the member states. The most resounding case that the Tribunal had to hear regarded the suspension of Paraguay after the ousting of President Lugo, in 2012. The plaintiff claimed that the measure violated Mercosur's norms, while the defendants argued that the Tribunal had no competences to intervene, as the issue was political and not judiciable. The justices (or more accurately arbiters, as the five Tribunal members are not judges according to Mercosur regulations) settled the dispute with a decision that split the difference between the largest states . . . and the court itself. First, the Tribunal decided in favour of its own competence to adjudicate on cases such as the one at hand, as Paraguay contended. But subsequently they decided that Paraguay's allegation did not meet the necessary legal requirements, and so it was declined. Two jurists that drafted a report to be presented before the Uruguayan Council for International Relations assessed that the political will of the largest members had prevailed over Mercosur's law, signalling that the bloc lacked "a working juridical system" that provided it with "legal security, . . . institutional stability, . . . and international credibility" (Arbuet-Vignali and Vignali Giovanetti, 2012: 30). Mercosur proved to be a politically minded organization rather than a community of law.

The contrasting cases of the Andean Court and Mercosur's Tribunal depict South America as a region in which the protection of property rights and legal enforcement are not only parcelled and heterogeneous, but also probably in decline. In some cases, the reluctance of the courts to enforce regional regulations is due to the notion of their own vulnerability: as judges anticipate that national governments will not comply with the court's rulings, they prevent institutional embarrassment by not issuing any. These conditions are not auspicious for development.

As regards the degree of legal security provided by regional organizations, Arnold (2017) points to a further problem: non-incorporation. In most Latin American arrangements, joint decisions do not enjoy direct effect or direct applicability: in order to enter into force, all regulations should be internalized by every member state according to their domestic provisions. This means that no regional norm, even those adopted by unanimity, can be enforced until it has not been ratified by the last member state. In Mercosur, for example, two-thirds of the directives require incorporation, and only half of them have obtained it. Why do member states approve agreements in the regional bodies and then fail to validate them domestically? Arnold blames defective behaviour on the gap between the reward from signing agreements and the costs of implementing the ensuing policies. Whatever the cause, the result contributes to further eroding the legal predictability of the bloc.

Public goods

Regional organizations are expected to provide three main types of public goods: interstate security, enlarged markets, and physical infrastructure. The first good aims at peace, the second at scale, and the third at connectivity.

Interstate security

While Central America had a record of domestic and interstate violence until very recently, South America has not witnessed interstate wars between major powers since 1883, and altogether since 1942 (Kacowicz, 2005; Mares, 2001). These factors have led to two outcomes: first, not survival but development has become South American states' top priority; second, high politics has been conducted through diplomatic rather than military means. Given a historical low level of intra-regional exchanges, regional public goods have been usually defined on the negative, mostly as the avoidance of harmful externalities. No country in the region has been capable of either forcing or buying off its neighbours alone, so acting hegemons have traditionally been extra-regional actors. The correlation of forces among the major powers, namely Argentina and Brazil, was balanced until the 1980s, and mutual distrust as much as the scarcity of extra-regional threats prevented them from building a joint security architecture. However, the balance of power had already started to tilt towards Brazil, and by 2000s it became evident that bipolarity was no longer an apt description of the regional state of affairs (Martin, 2006; Schenoni, 2012). Only then did Brazil start to invest in the creation of a governance framework to keep the region stable and extra-regional powers away. Brazil's "low, late and soft investment in regional security governance is explained by a combination of low regional threats, insufficient national capabilities, a legalistic culture of dispute settlement, and the participation in transgovernmental networks that substitute for, or subtly underpin, interstate cooperation and regional institutions" (Malamud and Alcañiz, 2017: 18). Indeed, illegal activities have promoted regional integration more effectively than state strategies, as drug trafficking, smuggling, and corruption – as the Brazilian criminal investigation known as Car Wash[1] has unearthed – create more trans-border activities than states can regulate or sanction. It is no surprise then, as Merke claims (2017: 148), that security cooperation has concentrated on preventing regional public bads rather than providing regional public goods.

Enlarged markets[2]

Systematic thinking and political advocacy for Latin American integration set foot when Argentine economist Raúl Prebisch was appointed director of the UN Economic Commission for Latin America (today ECLAC) in 1948. In 1950 he published *The Economic Development of Latin America and its Principal Problems*, which became a cornerstone of the region's economic thinking and inspired the integration projects that were launched in the following three decades. Between 1964 and 1969, Prebisch served as the founding secretary-general of the United Nations Conference on Trade and Development (UNCTAD).

ECLAC aimed at the enlargement of national markets through the creation of a common market. The coalition of technocrats and reformist politicians led by Prebisch considered that economic cooperation was the only means to overcoming traditional dependence on primary-commodity export trade. As the heretofore model of development – import-substitution industrialization (ISI) – was reaching its limit of exhaustion within the national markets, larger markets entailing economic diversification and technological modernization were necessary to advance further development (Wionczek, 1970). Other accounts also mention the creation of the European Community as triggering the integrative efforts, on grounds that the resulting trade diversion was damaging Latin American countries that were primary-commodity exporters (Mattli, 1999).

ECLAC's drive for integration came about in two waves. The first one saw the establishment of LAFTA and CACM in 1960; the second led to the creation of the Andean Pact (later Andean Community or CAN) in 1969 and the Caribbean Community (CARICOM) in 1973. A third wave took place later, as the transitions to democracy developed from the 1980s onwards, and saw the creation of the Common Market of the South (Mercosur) and the relaunching of both the CACM and the CAN. Labelled "open regionalism," as they aimed to combine regional preference with extra-regional openness, these processes reached an early success and most of them are still in existence – whether slightly or radically changed. Yet, none achieved its initial objectives.

If Santiago de Chile-based ECLAC was the think tank where it all started, its offspring soon crossed the Andes and installed its headquarters in Buenos Aires. The Institute for Latin American Integration (INTAL) is a unit of the Inter-American Development Bank (IDB) created in 1965 as a research, consultancy, and diffusion agency. Under the aegis of IDB president Felipe Herrera, it launched a series of publications that were to become the most important vehicle of reflection on Latin American integration for decades.

Most of the INTAL production followed ECLAC's historical-structuralist approach, a framework that included a great deal of conceptual innovations: centre-periphery relations, deterioration in the terms of trade, structural imbalance of payments, structural inflation and unemployment, development planning, and regional integration (Bielschowsky, 1998). However, INTAL provided a more pluralist environment and was influenced by events that ECLAC's founders could not have foreseen such as the oil shocks, the end of the Cold War and, more recently, the emergence of China. INTAL's dealing with integration has been done through the prism of the economy, mainly focusing on trade and investment. Although not blind to politics, its grasp is chiefly technocratic.

In terms of development, the performances of regional development banks have been crucial. The role of these institutions is usually underestimated, even in Europe (Clifton, Díaz-Fuentes, and Gómez, 2017). In Latin America there are two main cases: the Inter-American Development Bank (IDB) and CAF – Development Bank of Latin America.

The IDB was established in 1959 to finance development in Latin America and the Caribbean. The idea of a development institution had been suggested at the First Pan-American Conference, in 1890, to back the creation of an inter-American system, but it was born almost 70 years later as a result of the initiative of Brazil's President Juscelino Kubitschek. With headquarters in Washington, DC, it works under the aegis of the Organization of American States (OAS). The Bank is owned by 48 sovereign states, which are its shareholders and members. Only the 26 borrowing countries are able to receive loans.

CAF is a development bank established in 1970 within the framework of the Andean Pact, though it later expanded much beyond CAN. Today it is owned by 19 countries, 17 of which are in Latin America and the Caribbean plus Spain and Portugal, as well as 13 private banks. It promotes sustainable development through credit operations, non-reimbursable resources, and support in the technical and financial structuring of projects in the public and private sectors of Latin America. Its establishing agreement was signed in Bogotá, Colombia, in 1968, but its headquarters have always been located in Caracas, Venezuela.

Although national rather than regional, there is another lending institution that has had a great role in promoting development in the region: the Brazilian National Development Bank (BNDES). According to Hochstetler (2014), the BNDES saw dramatic increases in its financial resources for lending after 2005. The bank was "internationalized" to finance exports and support foreign direct investments and other international economic activities, while at the same time keeping "many of the bank's resources inside Brazil or tightly linked to Brazilian firms" (Hochstetler, 2014: 360). While BNDES loans are not accompanied by traditional, intrusive conditionalities, "they are strongly conditioned on the use of Brazilian firms and/or products to access funding. As such, they follow the logic of Brazilian development needs as much or more as those of their recipients" (Hochstetler, 2014: 360).

Henderson and Clarkson (2016: 43) further erode the belief that, "by pooling resources to address such common international economic issues as development funding and financial crises, regional financial organizations contribute to economic stability." Instead, South American organizations have seen their effectiveness limited by the economic asymmetry of their member states. The research findings suggest that, as regards the regional giant, unilateral and bilateral financial initiatives have brought Brazil "far more significant economic and political results than have regional financial organizations" (Henderson and Clarkson, 2016: 43).

Until 2014, when Operation Car Wash revealed a regional scheme of corruption centred on national oil giant Petrobras, Brazil's approach to regional finance had looked better than its two competitors. On the one hand, the "free market-oriented project of the U.S. envisioned extension of a NAFTA-like regulatory framework hemisphere-wide, promising Latin Americans better financial services, credit, and investment in exchange for strong financial property protections and [...] reduced financial policy space for their governments" (Armijo, 2013: 95); on the other, Venezuela's vision of "Bolivarian" finance, exported to the Caribbean and upper

Andes and loosely incarnated in the Bolivarian Alliance for the Peoples of Our America (Spanish acronym ALBA), promoted "assertive state management vis-à-vis both foreign and domestic investors, populist redistribution, and increasing reliance on non-market financial transactions" (Armijo, 2013: 95). Brazil's project covered the middle ground, promising to unite South America through the "creation of continent-wide physical infrastructure and capitalist financial markets, while retaining an ongoing role for public sector banks responsive to central government priorities" (Armijo, 2013: 95). The collapse of Venezuela's economy, the ousting of the Workers' Party in Brazil, and the advent of Donald Trump led to the breakdown of the three projects.

The most recent creation as regards capital markets is the Latin American Integrated Market (Spanish acronym MILA). This program integrates the stock exchange markets of Chile, Colombia, Mexico, and Peru, the founding members being the Lima Stock Exchange, the Santiago Stock Exchange, and the Colombia Stock Exchange. MILA has become Latin America's largest stock exchange, surpassing Brazil's Bovespa. It aims to develop an integrated capital market through providing investors with a greater supply of securities and larger sources of funding. It works under the umbrella of the Pacific Alliance, a four-member but open-to-enlargement organization aimed at freeing trade and enticing extra-regional investment.

Physical infrastructure[3]

Transport costs in Latin America are generally higher than in other developing areas. The region lags behind "in terms of adequate roads and railways as well as in port and airport efficiency" (Nolte, 2017: 5). Connectivity in infrastructure, energy, and telecommunications has been promoted by two major projects: the Initiative for the Integration of the Regional Infrastructure of South America (IIRSA) and the Mesoamerican Integration and Development Project (Portales, 2017: 308).

It is commonly accepted that density, distance, and division are key to economic development. Density is understood as the spatial concentration of economic activity; it tends to reinforce itself because of home market effects, as companies concentrate in places that feature relatively large markets, reducing their unit costs and generating a surplus output that is exported. Distance is about cost and time to communicate and to transport intermediate and final goods. Division captures tariff and non-tariff barriers, for instance fluctuating currency exchange rates. Distance and division can be summarized as transport cost. Development policies aim to boost density, overcome distance, and end division. As a first step of regional development, the reduction of transport costs – or, the impact of distance and division – reinforces leading areas because economic activities concentrate there at medium transport costs. As a second step, transport costs become so low that companies can choose their location according to considerations of factor endowment. This generates economic impulse for peripheral areas.

These two-step processes start from geoeconomic nodes, defined as geographic cores of economic networks. There are two countries in Latin America whose scale and location allow them to perform the role of a geoeconomic node: Brazil for South America and Mexico for Central America. This explains the existence of the two large subregional projects.

Following an invitation by Brazil's President Fernando Henrique Cardoso, all 12 South American presidents met for the first time in Brasilia on 1 September 2000. With the presence of the IDB and CAF presidents, they agreed on a blueprint for the coordination of energy, telecommunication, and transport policies. They identified 12 development axes and founded IIRSA as a loose intergovernmental initiative, a technical forum for cooperation on regional infrastructure that coordinates investment in projects that physically interlink the South American countries. Until 2014, the IIRSA portfolio reached US$ 130 billion and a total of 544 projects. The development axes Mercosul-Chile and Peru-Brazil-Bolivia are the largest, with a respective investment share of 39 and 22% of the total IIRSA investment. The remaining axes cover the rest of South America, ranging from the southern Andes to Guyana's highlands.

As concerns energy cooperation, accomplishments have been extremely modest. Physical integration, and thus Brazil's geo-economic nodality, remains mostly rhetoric. The *Gran Gasoducto del Sur*, which Venezuelan president Hugo Chávez envisaged in 2005 as a giant network of gas pipelines across the continent, was stillborn. Brazil's pipeline network itself has only recently been interconnected nationwide. Its focus lies on the exploitation of untapped resources in the Amazon Basin and Atlantic Ocean, not on regional trade. The Bolivia-Brazil natural-gas pipeline GASBOL connects Santa Cruz in the east of Bolivia with Porto Alegre and São Paulo in Brazil; other than that, there are only minor cross-border pipelines, which link Argentina, Bolivia, Chile, and Uruguay as well as Colombia and Venezuela.

With regards to trade in electricity, the potential of further integration looks equally dim. Experts of the Latin American Energy Organization (OLADE, 2003) state that Argentina and Brazil possess a realistic transfer potential of 5,000 megawatts, which equals only 4% of Brazil's installed capacity. Argentina and Chile were expected to reach transfers of not more than 500 megawatts; so were Brazil and Uruguay. Colombia and Ecuador, as well as Ecuador and Peru, were predicted to transfer 250 megawatts bilaterally. Further publications by OLADE projected scenarios of regional cooperation but fell short of recording major accomplishments.

Palestini and Agostinis (2018) investigated why South American functional cooperation emerged in the absence of economic interdependence and market-driven demand. In their view, the phenomenon "can be explained largely by the articulation of a *regional leadership* and its effect on the convergence of *state preferences*" (Palestini and Agostinis, 2018: 46). However, there was variation in policy outcomes: while Brazil's undisputed leadership on transport infrastructure turned this policy area relatively successful, divergent state preferences and Venezuela's defiant leadership turned energy integration unsuccessful. In 2009, IIRSA was incorporated into the South American Council of Infrastructure and Planning (Spanish acronym COSIPLAN). This is one of UNASUR's 12 ministerial councils, and its creation has meant no boost whatsoever for IIRSA.

The seed of the Mesoamerican Project was sowed at around the same time as IIRSA, when Mexico's President Vicente Fox announced the Puebla-Panama Plan on 12 March 2001. After several changes and merges that took social concerns into consideration, the Mesoamerica Integration and Development Project was launched in 2008 by the presidents of Colombia, Mexico, and Central America as a technical instance for cooperation and integration in the region. The Dominican Republic was taken on board the following year.

The Project is organized around four economic axes, which constitute the vertebral, and five social axes, best defined as sideshows. Funding is provided by the IDB, CAF, and the Central American Bank for Economic Integration (BCIE). According to the Project website, its portfolio included 107 operations between 2008 and 2015: most of them (48) involved transport, followed by trade facilitation and competitiveness (25) and energy (20). The overall investment came to around $3 billion. Independent assessments of the Project's performance are difficult to find, but slight improvements in the transport and energy infrastructure of the region have been registered.

Recapitulation and prospects

The contribution of regional organizations to Latin American development has been slight at best. Partly due to the prioritization of other goals such as national sovereignty or regime stability, and partly due to resource scarcity and poor management, regional organizations have proliferated but national levels of development have remained low and, above all, heterogeneous across countries. The latter feature testifies to the negligible impact of regional organizations, as their very existence was intended to promote convergence and smooth differences. None of the three sources of development, namely democracy, rule of law, and public goods, have shown to attain a major impact.

Although some believe that regional organizations have strengthened democracy, either by locking it in at country level or by democratizing their own procedures, the reverse effect is more visible. Democratization at the national level was a key factor behind the renewal of South American integration in the 1980s. In Central America, in contrast, Dabène (2009) has made the plausible argument that regional organizations fostered national democracy in the 1990s.

As to the rule of law, regional regulations and tribunals are rarely seen by foreign investors or local agents as drivers of legal protection and contract enforcement. Indeed, investment in the region is usually committed under the umbrella of foreign courts or of the International Centre for Settlement of Investment Disputes (ICSID). Although Mercosur and the Pacific Alliance's members have enjoyed initial reputation boosts, as investors use "the company states keep" as proxy for a country's willingness to honour its sovereign debt obligations (Gray, 2013), the rollercoaster of Latin American regionalism looks like a burden rather than asset.

Concerning public goods, interstate security has occasionally been strengthened by actions taken within regional organizations. However, there were three limitations to these actions. First, they were mostly reactive and concentrated on avoiding public bads (such as the escalation of interstate disputes like the one involving Colombia and Ecuador in 2008) rather than creating public goods (positive peace). Second, they were effective to deal with conflicts among smaller states rather than large powers. Third, localized geopolitical tensions have remained unsolved, creating additional barriers to integration (see the case of Bolivia and Chile, who share several collective memberships but hold no bilateral diplomatic relations). Regional organizations have also been unable to promote enlarged regional markets. This was mostly due to extra-regional developments rather than regional malfunction: as intra-regional interdependence is low all through Latin America, extra-regional markets have more leverage for both imports and exports than neighbouring markets. Finally, physical infrastructure has received a moderate boost in both South and Central America, especially as regards transport – energy and communications lagging well behind. However, investment in waterways and land roads has not only been inchoate and segmented, it has also been focused on moving commodities across and out of the region rather than promoting intra-regional exchanges and value chains.

Should trade and investment opportunities for Latin America remain dependent on extra-regional poles such as China and the US, and should intra-regional threats remain low, regional organizations will continue to be marginal as drivers of development. However, an unexpected opportunity looms in the horizon: if illicit activities such as smuggling, drug trafficking, and transnational corruption keep growing, the Latin American states may find it necessary to deepen regional cooperation in order to face the new threats. Paradoxical as it might look, informal unlawful integration could be at the root of further formal integration.

Notes

1 Allegedly the largest in world history, Operation Car Wash has been commanded by Brazilian federal judge Sérgio Moro since 17 March 2014. It has led to the conviction of top businessmen and politicians, including former President Lula, and indirectly to the impeachment of President Dilma Rousseff.
2 This section contains fragments from Malamud (2010).
3 Unreferenced data in this section come from Scholvin and Malamud (2014).

References

Arbuet-Vignali, H. and Vignali Giovanetti, D. (2012) Laudo N° 01/2012 del T.P.R. Un vacío imposible de llenar. *Estudios del CURI* N° 08/2012 (Consejo Uruguayo para las Relaciones Internacionales), Montevideo.

Arias, L. M. (2015) Political institutions and economic development. In J. Gandhi and R. Ruiz Rufino (eds) *Routledge Handbook of Comparative Political Institutions*. London; New York: Routledge, pp. 421–440.

Armijo, L. E. (2013) Equality and regional finance in the Americas. *Latin American Politics and Society* 55(4): 95–118.

Arnold, C. (2017) Empty promises and nonincorporation in Mercosur. *International Interactions* 43(4): 643–667.

Bielschowsky, R. (1998) Evolución de las ideas de la CEPAL. *Revista de la CEPAL* – Nro. Extraordinario 21–45.

Clifton, J., Díaz-Fuentes, D. and Gómez, A. L. (2017) The European Investment Bank: Development, integration, investment? *Journal of Common Market Studies* 1–18. doi:10.1111/jcms.12614

Closa, C. and Palestini, S. (2015) *Between Democratic Protection and Self-Defense: The Case of UNASUR and Venezuela*. EUI Working Paper RSCAS 2015/93.

Dabène, O. (2009) *The Politics of Regional Integration in Latin America: Theoretical and Comparative Exploration*. New York: Palgrave Macmillan.

Gray, J. (2013) *The Company States Keep: International Economic Organization and Sovereign Risk in Emerging Markets*. Cambridge: Cambridge University Press.

Helfer, L. R., Alter, K. J. and Guerzovich, M. F. (2009) Islands of effective international adjudication: Constructing an intellectual property rule of law in the Andean community. *The American Journal of International Law* 103(1): 1–47.

Henderson, J. and Clarkson, S. (2016) International public finance and the rise of Brazil. *Latin American Research Review* 51(4): 43–61.

Hochstetler, K (2014) The Brazilian National Development Bank goes international: Innovations and limitations of BNDES' internationalization. *Global Policy* 5(3): 360–365.

International Monetary Fund (2017) *World Economic Outlook Database*. https://www.imf.org/external/pubs/ft/weo/2017/01/weodata/index.aspx

Kacowicz, A. (2005) *The Impact of Norms in International Society: The Latin American Experience, 1881–2001*. Notre Dame, IN: University of Notre Dame.

Malamud, A. (2010) Latin American regionalism and EU studies. *Journal of European Integration* 32(6): 637–657.

——— and Alcañiz, I. (2017) Managing security in a zone of peace: Brazil's soft approach to regional governance. *Revista Brasileira de Política Internacional* 60(1): e011.

——— and De Sousa, L. (2007) Regional parliaments in Europe and Latin America: Between empowerment and irrelevance. In A. Ribeiro Hoffmann and A. van der Vleuten (eds) *Closing or Widening the Gap? Legitimacy and Democracy in Regional International Organizations*. Aldershot: Ashgate, pp. 85–102.

Mares, D. (2001) *Violent Peace: Militarized Interstate Bargaining in Latin America*. New York: Columbia University Press.

Martin, F. E. (2006) *Militarist Peace in South America: Conditions for War and Peace*. New York: Palgrave Macmillan.

Mattli, W. (1999) *The Logic of Regional Integration: Europe and Beyond*. Cambridge: Cambridge University Press.

Mazzuca, S. L. (2017) Critical juncture and legacies: State formation and economic performance in Latin America. *Qualitative and Multi-Method Research* 15(1): 29–35.

Merke, F. (2017) Lo que sabemos, lo que creemos saber y lo que no sabemos de América Latina. *Pensamiento Propio* 45: 143–164.

Nolte, D. (2017) Trade: The undervalued driver of regional integration in Latin America. *GIGA Focus – Latin America* 5, September. www.giga-hamburg.de/en/publication/trade-the-undervalued-driver-of-regional-integration-in-latin-america (Accessed 2 February 2018).

OLADE. (2003) *La situación energética en América Latina*. Quito: Organización Latinoamericana de Energía.

Palestini, S. and Agostinis, G. (2018) Constructing regionalism in South America: The cases of sectoral cooperation on transport infrastructure and energy. *Journal of International Relations and Development* 21(1): 46–74.

Portales, C. (2017) Public goods and regional organizations in Latin America and the Caribbean: Identity, goals, and implementation. In A. Estevadeordal and L. W. Goodman (eds) *21st Century Cooperation:*

Regional Public Goods, Global Governance, and Sustainable Development. London; New York: Routledge, pp. 287–311.

Prebisch, R. (1950) *The Economic Development of Latin America and Its Principal Problems.* New York: United Nations.

Schenoni, L. L. (2012) Ascenso y hegemonía: pensando a las potencias emergentes desde América del Sur. *Revista Brasileira de Política Internacional* 55(1): 31–48.

Schimmelfennig, F., Engert, S. and Knobel, H. (2003) Costs, commitment and compliance: The impact of EU democratic conditionality on Latvia, Slovakia and Turkey. *Journal of Common Market Studies* 41(3): 495–518.

Scholvin, S. and Malamud, A. (2014) *Is There a Geoeconomic Node in South America? Geography, Politics and Brazil's Role in Regional Economic Integration.* ICS Working Paper 2/2014, University of Lisbon.

UNDP (2016) *Human Development Report 2016. Human Development for Everyone.* http://hdr.undp.org/en/content/human-development-report-2016-human-development-everyone

Van der Vleuten, A. and Ribeiro Hoffmann, A. (2010) Explaining the enforcement of democracy by regional organizations: Comparing EU, Mercosur and SADC. *Journal of Common Market Studies* 48(3): 737–758.

Wionczek, M. S. (1970) The rise and the decline of Latin American economic integration. *Journal of Common Market Studies* 9(1): 49–66.

World Bank (2017) *World Development Indicators 2017.* https://openknowledge.worldbank.org/handle/10986/26447

14

LATIN AMERICA AND THE UNITED STATES

Gregory Weeks

Introduction

This chapter traces the study of the relationship between Latin America and the United States. Approaches have evolved in terms of orientation, theoretical perspective, and methodology. As a political scientist, I spend most – though not all – of my space discussing scholarship in that discipline. I touch on only a minuscule fraction of the thousands of books and articles published on the subject, so by necessity I leave out many such works.[1]

The chapter concludes with a discussion with suggestions for where future research might fruitfully go. In a roughly chronological manner, the chapter shows how a predominantly pro-US policy literature became intensely critical after the US response to the Cuban Revolution. Yet after the end of the Cold War, that perspective shifted again as younger scholars questioned whether the critical approach downplayed Latin American agency. Within that discussion of changes over time, the chapter shows how US-based analyses differed from those coming from Latin America. The orientation and methodologies employed have been different. Latin America is much less concerned with security, while scholars in the United States pay too little attention to the works published in Latin America. Making these comparisons allows for a clearer sense of what avenues future work could take.

Background

Prior to World War II, few scholars paid much attention to US-Latin American relations. Historians wrote first and tended to offer up straightforward narrative history. Significantly, US scholars took for granted that the United States was a positive force in the region and that it should intervene for the benefit of Latin American countries. This came at a time when the United States intervened and even occupied multiple Latin American countries, which continued well into the 20th century. Aside from the notion that US "assistance" was a boost for the region, these works did not explore causal relationships. They were descriptive and simply assumed good intentions.

In 1899, historian John Holladay Latané gave a series of lectures published the following year as *The Diplomatic Relations of the United States and Spanish America* (Latané, 1900). His book concludes with a favourable assessment of the Monroe Doctrine. The same easy acceptance of

US benevolence is evident throughout the first textbook on the topic, by diplomatic historian Samuel Flagg Bemis (1943), who invoked "protective imperialism" as a wholly salutary outcome of US policy.

The study of US policy toward Latin America mushroomed after the Cold War began in the late 1940s and for a time scholarly works remained complimentary. The 1959 Cuban Revolution quickly shifted the approving view of US policy toward a critical stance. During the 1960s, a new generation of scholars stopped accepting the US government's public statements on foreign policy and questioned US government motives. Ironically, many of them did so while receiving federal funding for that research. Indeed, administrations deeply dedicated to fighting Communism funded fieldwork by US scholars that yielded highly condemnatory published studies. They saw the imbalance of power between the two regions not as a source of protection but rather as aggression from a domineering state.

After the Cuban Revolution: focus on power

One result of the Cuban Revolution was a spate of military coups that ushered in repressive dictatorships, with the 1964 Brazilian coup as a watershed. Conservative military leaders labelled virtually all progressive movements as subversive and toppled democratic governments, arguing they were saving the country from Communist rule. Without fail, the US government lent its support to those governments, which generated more scholarly backlash. Especially after the coup that overthrew the democratically elected government in Chile of Salvador Allende in 1973 and the outrage it generated, books and articles proliferated at a dizzying pace. As graduate students or young assistant professors, scholars came of age at a time when distrust of US foreign policy generally was high, and when evidence of chronic wrongdoing was being unearthed for the first time. Overwhelmingly, these historians and political scientists were critical of US policy, especially with regard to support (or more accurately lack thereof) for democracy and human rights.

Realism emerged as a major theoretical approach in International Relations and was used extensively to explain the relationship between the United States and Latin America. There are many variants of realist theory but in its simplest form it posits, "Self-help is necessarily the principle of action in an anarchic order" (Waltz, 1979: 111). The options of any given state are conditioned by their capacity, or power, to pursue actions based on self-interest. In the context of US-Latin American relations, this meant analysis of power imbalance. As one prominent textbook noted, it entailed "how the United States has chosen to apply and exercise its perennial predominance" (Smith, 2008: 5). Similarly, the "history of U.S.-Latin American relations has always been characterized and shaped by significant differences in military and economic capabilities and the absence of international institutions to constrain the actions of the United States" (Weeks, 2015: 2). For the most part, this meant discussing US efforts to use its hegemonic position to Latin America's disadvantage, and the ways in which governments in the region responded.

Saving the region from Communism was not the focal point for Latin American scholars. Dependency theory was a Latin American theoretical counterpoint, where US hegemony was explained as a structural outcome of global capitalism. The core, which in the 20th century meant the United States, extracted primary products from Latin America while exporting finished manufactured goods. This unequal arrangement prevented independent Latin American economic development. US policy was therefore considered part of an economic imperial project. For dependency theorists, Latin American agency was tightly constrained and conditioned by the power of US capital and the US government that supported its expansion. Andre Gunder

Frank (1986: 114), for example, argued that "metropolis-satellite relations are not limited to the imperial or international level but penetrate and structure the very economic, political, and social life of the Latin American colonies and countries." Dependency theory consumption dropped off around the 1980s in the United States, whereas in Latin America the core assumptions of the dependency approach, if not always the theory itself, have remained more analytically relevant. Notably, dependency-oriented arguments appear periodically in the statements of leftist Latin American leaders.

The Cold War context reinforced US-centered scholarly orientation. Fundamental assumptions in the literature were that the United States was politically hegemonic, economically dominant, and aggressive. Dependency theory emerged from those assumptions, as did a deluge of critical works, sometimes Marxist, often based largely on documents showing evidence of US wrongdoing in the cases of Cuba (Williams, 1962), Chile (Petras and Morley, 1975), and Central America (Pearce, 1982). These and many other works tended to view relations in terms of US empire. The analytic result, however, was often to subsume Latin American agency within US power.

Since Cuba was at the centre of the Latin American Cold War, and because Fidel and then Raúl Castro continued to rule the country long after that international conflict ended, the literature on US-Cuban relations is vast (Schoultz, 2009 offers a good overview and analysis). From many different ideological angles, most of these works examined some element of the interaction of US power and Cuban resistance. Cuba is of special interest to scholars because of its perceived strategic importance combined with fascination (or frustration, depending on your viewpoint) with Fidel Castro's ability to thumb his nose at the US government and persevere against great odds.

The same extensive scholarly treatment is true for Central America, which suffered extreme levels of violence and political instability, particularly in the 1980s. The United States fought a covert war against the Nicaraguan government, overtly funded a genocidal military dictatorship in Guatemala, and supported a repressive government and army in El Salvador. President Ronald Reagan was fixated on the region, arguing that he was promoting democracy and fighting Communism even while his allies sponsored widespread violence. Not surprisingly, the academic response was harsh (e.g. see LaFeber, 1993).

End of the Cold War

The Soviet Union's fall in 1990–1991 fundamentally shifted the study of US-Latin American relations. Within a short time frame, not only did the spectre of Soviet encroachment disappear, but the Sandinistas in Nicaragua were voted out, the Salvadoran government and rebels signed a peace treaty, the dictatorship of Augusto Pinochet left power, and the Cuban dictatorship floundered without an external patron. The absence of threat to the United States created space for researchers, especially in the United States, to question long-held assumptions and probe new areas.

Analytically, the end of the Cold War prompted many to question the primacy of power to understand the dynamics of US-Latin American relationships. Kathryn Sikkink's (2004) work on the power of human rights ideas is an important example, as she demonstrated that ideas can have their own influence independent of the power of even a hegemon. As human rights abuses became publicized, with the activism of international non-governmental organizations they received more attention both globally and in the US Congress. Over time, this meant US policy makers could not ignore their importance and policy changed accordingly. The imbalance of power between the United States and Latin America certainly mattered, but it did not necessarily determine outcomes as much as previously assumed.

The 2000s marked a new generational shift in the United States, as scholars who were not even alive at the time of the Cuban Revolution entered academia. To criticisms of US policy they added Latin American agency and questioned whether *everything* the US government did was harmful almost all the time. As Tanya Harmer (2011: 274) wrote about the Chilean coup, "And it was, in the end, other Chileans who let this happen." When Russell Crandall (2006: 4) asserted, "we must ask whether U.S. bayonets helped lead to more democracy, not less," it was not the argument of an ideologue. Even established scholars began to question the assumption that some combination of national security and economic self-interest drove US policy. In fact, other variables like Latin American lobbying mattered a lot (Grow, 2008).

In both History and Political Science, US-published studies brought Latin American agency more into the picture, with the logic that "Mononational research tends to produce mononational explanations and to ignore the role of players from countries other than those whose words are examined" (Crandall, 2006: 4). A number of authors pushed back on the power imbalance emphasis and showed how Latin American governments mattered more than generally appreciated (Long, 2015; Mora and Hey, 2003; McPherson, 2013). They were not passive actors having power exerted on them, but rather were instrumental in shaping outcomes. Even efforts during the Cold War both by the US and the Soviet Union to co-opt and control Latin American intellectuals was frustrated by their ability to define their own version of nationalism that was beholden to neither (Iber, 2015).

At the same time, the ongoing declassification of US government documents fostered rich and highly critical analyses of Cold War policy. In particular, President Bill Clinton's Chile Declassification Project released approximately 23,000 documents related to US-Chilean relations. Examples of publications using such documents include Kornbluh (2003) and Morley and McGillion (2015) on US policy toward the Pinochet dictatorship; Harmer (2011) on US policy toward Salvador Allende; McSherry (2005) on Operation Condor; LeoGrande and Kornbluh (2014) on US-Cuban relations; and the US role in the 1964 Brazilian coup (Pereira, 2018). No doubt these will continue to emerge and will lead to reassessment of the US role in Latin American political crises.

Thematically, one area that received increased attention was the US-led "drug war," which entailed securitized policies of military training, coca eradication, and interdiction. Colombia received particular attention (Crandall, 2002) but Bolivia and the rest of the Andean region did as well (Loveman, 2006). US military aid to these countries, with Plan Colombia and Operation Blast Furnace in Bolivia, were large-scale operations aimed at destroying drugs with a supply side policy focus. The thrust of most of this literature was that US policy was counterproductive and harmful, and rarely if ever considered local realities. Attacking the source, such as destroying coca plants, uprooted not only the plants but entire communities, and remained too blind to the demand side, which was the US consumer of cocaine and other narcotics.

In the past decade, the study of US policy expanded to include the themes of criminal gangs and their ties to immigration. In the 1980s, migrants from El Salvador formed gangs (most notably MS-18) in Los Angeles, which over time spread back to El Salvador and across the United States. Works include the effects of past US policy, as the US funded the violence that drove Salvadorans out of their country (García, 2006); problems of hardline (so-called *mano dura*) policies (Cruz, 2010); the rise of children migrating (Donato and Sisk, 2015); and the negative effects of deportation (Menjívar, Morris, and Rodríguez, 2017). Similar to works on narcotrafficking, they emphasized the destructive outcomes of US policy, but also Latin American policies when they adopted hardline, militarized measures.

After Hugo Chávez's election in 1998, US-Venezuelan relations received considerable attention. He came to office with a clear message of rejecting US policy and forging socialist policies to reduce poverty and income inequality. The George W. Bush administration in particular was hostile, and in 2002 applauded the coup that briefly removed Chávez from power. From then, bilateral relations were especially tense. The literature includes analyses that place the bilateral relationship in historical perspective (Kelly and Romero, 2002); of Venezuela's efforts to balance US power (Corrales, 2009); and use of oil revenue to challenge US hegemony (Clem and Maingot, 2011). Similar to studies of US-Cuban relations, the ideological bent of such works varies widely, but the vast majority reflect power imbalance, where for example Venezuela resists US policy preferences and forges international alliances and organizations that consciously exclude the United States.

That development has gone hand in hand with studies of China's rising influence in Latin America and how that affects the United States (Roett and Paz, 2008; Gallagher, 2016; Denoon, 2017). The Chinese economic presence in Latin America is considerable but still relatively new so its long-term impact on US-Latin American relations remains mostly a matter of speculation. Latin America views the Chinese influence largely through the lens of trade and economic calculation. Chinese trade with Latin America increased 22-fold between 2000 and 2015, and Foreign Direct Investment reached into the tens of billions (OECD, 2015). In the United States, the focus is often on how China may threaten US security and hegemony. At times these can veer into alarmism, and assume Latin America passivity in the face of Chinese influence, which the United State should counter.

Studies on the Barack Obama presidency have focused on his doctrine of engagement (Kassab and Rosen, 2016) and use of soft power (Weeks, 2016) but there are also critics on the left who argue that President Obama showed more substantive continuity than change with the George W. Bush presidency (e.g. Buxton, 2011). The latter criticisms became less common after 2014, when Obama thawed diplomatic relations with Cuba, but even more so after the election of Donald Trump. Trump, whose presidency is too new for publication of academic works at the time of this writing, based his initial Latin America policy on racial slurs, insults toward allies and adversaries alike, harsh criticisms of NAFTA, and partial rolling back of Obama's Cuba policy. When former Chilean President Ricardo Lagos said that Trump's election "was not good news for the world" he spoke for many Latin Americans (quoted in Romero, 2016).

Methodology

This chapter has covered the substance of the literature but we need also to consider methodology. One constant has been qualitative methods. Of course, this is to be expected for work by historians, but it is also true of political scientists, whose discipline has otherwise been moving in a decidedly quantitative direction. As Mariano Bertucci (2013) points out after an extensive study of peer-reviewed books and articles, the study of US-Latin American relations does not reflect broader trends in international relations research. This fact has been noted in Latin America as well, where the use of quantitative political science has grown (Merke and Reynoso, 2016).

There are various possible answers to this question. Authors have tended to provide considerable historical context to explain US-Latin American relations, which leads towards more historical (and thus qualitative) analyses. Nonetheless, in Political Science we might expect more mixed methods approaches, which would reflect trends in the discipline in this direction, but in general those have been the exception. General International Relations literature focuses extensively on interstate conflict and terrorism, which are much less prevalent in Latin America than elsewhere in recent years. Latin America has the lowest incidents of interstate war of any region in the world. The worst guerrilla conflicts, such as the Colombian Revolutionary Armed Forces (FARC) or the Peruvian Shining Path, were negotiated to an end and defeated militarily, respectively. Current levels of violence in Latin America are strongly tied to narcotrafficking and the wide availability of guns, neither of which has been a major topic for general IR scholars.

General treatments of US-Latin American relations are without exception qualitative and typically contain considerable historical background, complete with primary documents, to provide context. Foci have included how US policy makers denigrated Latin Americans (Schoultz, 1998); how US policy was part of a global project (Grandin, 2006; Loveman, 2010); how the US relentlessly promoted a policy of globalization (O'Brien, 2007); the extent of intervention (Livingstone, 2009); and ways in which Latin America challenges hegemony (Tulchin, 2016). The underlying assumption is that broad historical context is required to understand political and economic development. That requires careful examination of how US policy developed over time, with qualitative narratives showing the ebbs and flows of US intervention and the use of US military and economic power.

Nonetheless, a number of studies of the US impact on the Colombian conflict have been quantitative. That was possible in large part because the Colombian government and the United Nations collected coca cultivation data, while the United States provides data on foreign aid. Examples include the effects of military aid on conflict (Dube and Naidu, 2015; Jadoon, 2017); effect of aid on coca cultivation (Rouse and Arce, 2006); and the effects of Plan Colombia more generally (Banks and Sokolowski, 2009; Franz, 2016). It is worth noting, however, that there is an enormous literature on these topics and it remains overwhelmingly qualitative. The qualitative literature is also almost entirely critical, paying close attention to the human cost of the conflict.

Another area that has been more quantitative analysis is trade, which is the one area of US-Latin American relations relevant to economists, both from the US and from Latin America. It has been more conducive to quantitative methods because international institutions in particular (e.g. the World Bank and the Economic Commission for Latin America and the Caribbean, or CEPAL) have produced publicly available data for years. The US pushed Latin American countries to liberalize their economies after the debt crisis of the 1980s and the North American Free Trade Agreement (NAFTA), signed by the US, Canada, and Mexico, went into effect in 1994. Similar agreements gradually spread throughout the region for over 20 years, though the election of Donald Trump, who harshly criticized NAFTA, stalled such efforts. US political and economic influence flowed through both international institutions like the United States Agency for International Development (Scott and Steele, 2011) but also through economics training at US universities, which facilitated the flow of neoliberal thought to Latin American political elites (e.g. Biglaiser, 2002).

Trade openness receives attention (Avelino, Brown, and Hunter, 2005), as does the North American Free Trade Agreement (Caliendo and Parro, 2015; Campos-Vázquez, 2013; Hassan and Nassar, 2017) and free trade agreements in general (Baier, Bergstrand, and Vidal, 2007). The quantitative studies tend to be positive about the aggregate effects of trade. The volume of qualitative studies is much larger and tends to be more negative, focusing on those Latin Americans who have not benefited or have even become worse off because of free trade.

Within Latin America itself, the study of International Relations is relatively new. Of course, the dependency school came directly out of the region, specifically from CEPAL. The immediate concern of IR scholars was autonomy, in particular from the United States, though the exclusive focus on resisting dependence slowed by the 1990s (Tickner, 2003). In South America, Latin American IR scholarship expanded rapidly in the 2000s, heavily influenced by US approaches. As one recent study found, the three theories found most commonly in published works were liberalism, realism, and constructivism (Madeiros, Barnabé, Albuquerque, and Lima, 2016). Further, qualitative methods were predominant, but gradually being overtaken by quantitative ones.

Latin American studies of bilateral relations are overwhelmingly qualitative, unless they focus squarely on trade.

In an excellent literature review, Giacalone (2012) outlines the approaches of Foreign Policy Analysis in the larger Southern Cone countries. With variation, Latin American scholars have tended to use US theoretical models but adapt them to local realities and infuse those more with normative arguments, often founded on Marxist principles. Across all countries, autonomy was an important variable. Overall, Latin American studies of International Relations are often interdisciplinary and transnational, as one extensive review of Mexican articles showed (Cid Capetillo, 2008). US studies are not. Not surprisingly, Latin American perspectives are often highly sceptical of the motives behind US policy, such that even prominent outlets like *Foreign Affairs en Español* publish articles referring to lies as a principle of US policy toward Latin America (Boron, 2006).

Another important difference between US and Latin American scholars is the issue of security. For US-based scholars, the concept is framed primarily in terms of how the current US administration defines it. That shifted from the Cold War to the drug war, and then also to organized crime and transnational terrorism. US researchers pay little attention to how Latin American policy makers define security, and how it differs from security assumptions in the United States. Given that foreign policy decisions flow in part from calculations of opportunities and security concerns, this represents a major shortcoming.

Perhaps more significantly, Latin American governments sometimes consider US policy and actions to be part of a threat they face. Certainly, this is the case with a number of leftist governments, especially Venezuela, where the Bush administration supported the 2002 coup and the relationship has been entirely adversarial since. But some Latin American countries, for example Bolivia, also came to view the traditionally militarized US response to the drug war as a threat to its citizens. Even more recently, many governments in the region believe climate change to be a major threat to security, whereas the Trump administration declares it to be false science.

Hey (1997: 652) conducted an extensive literature review on studies of Latin American foreign policy and concludes that "[w]hen the core deems a policy area salient, it is likely that core pressure will affect Latin American foreign policy in the desired direction." In other words, US policy preferences – she looks at the market-driven "Washington Consensus" in particular – were decisive even for Latin American foreign policy decisions. She argues that the literature ignored Latin American foreign policy bureaucracies as too unprofessional and underskilled to merit inclusion into any overarching argument. Hey concludes with a call for more theory-building.

Indeed, the US-centric nature of the field poses potential obstacles to a fuller understanding of the US-Latin American relationship. To a large degree, data collection centres on the United States: interviews with US policy makers, analyses of US aid programs, public opinion data, government-generated data, archival documents, and the like. In the past, some of this stemmed from the lack of reliable data in Latin America, especially during eras of repressive military regimes, but that has changed significantly. Latin American data is now more available, but IR scholars have been slower to utilize it.

It may be that many found the language skills, time, and resources required to do extensive fieldwork in Latin America to be too onerous and expensive. US-published works contain considerable reference to interviews with US policy makers. Former cabinet secretaries, assistant secretaries, ambassadors, national security advisors, and members of Congress become rich sources of information about US foreign policy. However, the other side of "international relations," namely Latin American policy makers, gets very little attention. These can only be found in case studies generally published in the country itself (e.g. Bywaters, 2014; Fernández de

Castro, 2015; Tickner, 2016). To be fair, many (perhaps even most) analyses of Latin American foreign policy toward the United States use official documents and statements much more than interviews as well, so in general this is an underused resource.

Only recently have Latin American public opinion data become available and used to understand US-Latin American relations. These have focused on sources of "anti-Americanism" (Baker and Cupery, 2013, Azpuru and Boniface, 2015) and whether US influence is diminishing (Azpuru, 2016). They questioned assumptions about how Latin Americans perceive the United States, which in fact is more positive than conventional wisdom would suggest (perhaps at least until Donald Trump's election). These analyses were made possible by the Latin American Public Opinion Project (LAPOP) at Vanderbilt University, which has been conducting surveys in the region regularly since 2004. These works also point to a different methodological approach, which combine qualitative and quantitative methods. A mixed methods approach takes advantage of newly available data but infuses it with the context that has made qualitative work so valuable. Until recently, researchers have used LAPOP data primarily to understand Latin American domestic politics, so the extension to US-Latin American relations is a welcome, albeit nascent, development.

Future research

The study of US-Latin American relations has come a long way and has gained in nuance as time goes on, but there is still work to be done. There is tremendous variety of approach, methodology, and theoretical perspective. That diversity is essential for understanding the new era of US-Latin American relations we are in, characterized by Latin America reaching out to other parts of the world for trade and other kinds of exchange while US attention wanders or, in the case of Donald Trump, even becomes unpredictably hostile.

Especially in a (mostly) democratic era in Latin America, where so many resources are now publicly available, researchers should maintain and even expand having the Latin American perspective front and centre. In large part, this will entail increased use of sources from individual countries, both primary and secondary. Yet especially for US researchers, it requires shedding a long-held belief that US power is overwhelming to the point of losing sight of Latin American agency.

More attention should be paid to the differing definitions of security in the United States versus Latin America. Too narrow a focus on Latin American resistance to US policy obscures the fact that policy makers in the region are viewing policy choices through a lens that keeps local issues more in mind. Especially through the efforts of Hugo Chávez, new international organizations like the South American Union (UNASUR) and the Community of Latin American and Caribbean States, are defining regional priorities without US participation. Scholars should examine how these define new notions of security and how much they diffuse throughout the region and affect US-Latin American relations. At the same time, US policy makers and domestic politics should receive more attention and ideally placed in theoretical context. For US presidents, Latin America has not been a major priority, which means policy gets made in large part by lower level cabinet and National Security Council appointees.

From a methodological standpoint, the increased quantification of International Relation studies has barely touched US-Latin American relations. In and of itself, this is not a problem since different types of research questions are best suited to different methods. Some issues require qualitative case studies to explain historical context and complex interplay between different political actors. Nonetheless, the dearth of quantitative work should at least give us pause. In particular, this would be an opportune time for exploration of what mixed methods might accomplish. The study of Latin American public opinion toward the United States points one direction that this might take.

Finally, scholars should more deliberately apply theoretical perspectives from the subfield of International Relations, regardless of methodology. The wealth of largely descriptive studies offers considerable insight, but do not always build upon past literature to develop new theoretical insights. Moreover, US scholars would benefit greatly from greater attention to the literature produced in Latin America, which is heavily influenced by theories generated in the United States but which retains its own flavour and develops its own insights based upon local contexts.

Note

1 In other words, I apologize if I do not cite the publications of whoever happens to be reading this.

References

Avelino, G., Brown, D. S. and Hunter, W. (2005) The effects of capital mobility, trade openness, and democracy on social spending in Latin America, 1980–1999. *American Journal of Political Science* 49(3): 625–641.

Azpuru, D. (2016) Is U.S. influence dwindling in Latin America? Citizens' perspectives. *The Latin Americanist* 60(4): 447–472.

———— and Boniface, D. (2015) Individual-level determinants of anti-Americanism in contemporary Latin America. *Latin American Research Review* 50(3): 111–134.

Baier, S. L., Bergstrand, J. and Vidal, E. (2007) Free trade agreements in the Americas: Are the trade effects larger than anticipated? *The World Economy* 30(9): 1347–1377.

Baker, A. and Cupery, D. (2013) Anti-Americanism in Latin America: Economic exchange, foreign policy legacies, and mass attitudes toward the colossus of the North. *Latin American Research Review* 48(2): 106–130.

Banks, Catherine M. and Sokolowski, John A. (2009) From war on drugs to war against terrorism: Modeling the evolution of Colombia's counter-insurgency. *Social Science Research* 38(1) (March): 146–154.

Bemis, S. F. (1943) *The Latin American Policy of the United States*. New York: Harcourt, Brace and Company.

Bertucci, M. E. (2013) Scholarly research on U.S.-Latin American relations: Where does the field stand? *Latin American Politics and Society* 55(4): 119–142.

Biglaiser, G. (2002) The internationalization of Chicago's economics in Latin America. *Economic Development and Cultural Change* 50(2): 269–286.

Boron, A. (2006) La mentira como principio de política exterior de Estados Unidos hacia América Latina. *Foreign Affairs en Español* 6(1): 61–68.

Buxton, J. (2011) Forward into history: Understanding Obama's Latin American policy. *Latin American Perspectives* 38(4): 29–45.

Bywaters, C. C. (2014) El 'No' de Ricardo Lagos a la invasion de Ira ken 2003: El proceso de toma de decisiones de política exterior en Chile. *Estudios Internacionales* 46(177): 65–88.

Caliendo, L. and Parro, F. (2015) Estimates of the trade and welfare Effects of NAFTA. *The Review of Economic Studies* 82(1): 1–44.

Campos-Vázquez, R. M. (2013) Why did wage inequality decrease in Mexico after NAFTA? *Economía Mexicana Nueva Epoca* 22(2): 245–278.

Cid Capetillo, I. (2008) Avances y aportaciones sobre teoría de relaciones exteriores. *Relaciones Internacionales* 100: 33–50.

Clem, R. S. and Maingot, A. P. (2011) (eds) *Venezuela's Petro-Diplomacy: Hugo Chávez's Foreign Policy*. Gainesville: University Press of Florida.

Corrales, J. (2009) Using social power to balance soft power: Venezuela's foreign policy. *The Washington Quarterly* 32(4): 97–114.

Crandall, R. (2002) *Driven by Drugs: U.S. Policy Toward Colombia*. Boulder, CO: Lynne Rienner.

————. (2006) *Gunboat Democracy: U.S. Interventions in the Dominican Republic, Grenada, and Panama*. Lanham, MD: Rowman & Littlefield.

Cruz, J. M. (2010) Central American *Maras*: From youth street gangs to transnational protection rackets. *Journal of Global Crime* 11(4): 379–398.

Denoon, D. B. H. (2017) (ed) *China, the United States, and the Future of Latin America*. New York: New York University Press.

Donato, K. and Sisk, B. (2015) Children's migration to the United States from Mexico and Central America: Evidence from the Mexican and Latin American migration projects. *Journal on Migration and Human Security* 3(1): 58–79.

Dube, O. and Naidu, S. (2015) Bases, bullets, and ballots: The effect of US military aid on political conflict in Colombia. *The Journal of Politics* 77(1): 249–267.

Fernández de Castro, R. (2015) Decision making in Mexican foreign policy. In J. I. Domínguez and A. Covarrubias (eds) *Routledge Handbook of Latin America in the World*. New York: Routledge, pp. 169–179.

Franz, T. (2016) Plan Colombia: Illegal drugs, economic development and counterinsurgency – A political economy analysis of Colombia's failed war. *Development Policy Review* 34(4): 563–591.

Gallagher, K. P. (2016) *The China Triangle: Latin America's China Boom and the Fate of the Washington Consensus*. New York: Oxford University Press.

García, M. C. (2006) *Seeking Refuge: Central American Migration to Mexico, the United States, and Canada*. Berkeley, CA: University of California Press.

Giacalone, R. (2012) Latin American foreign policy analysis: External influence and internal circumstances. *Foreign Policy Analysis* 8(4): 335–354.

Grandin, G. (2006) *Empire's Workshop: Latin America, the United States, and the Rise of the New Imperialism*. New York: Metropolitan Books.

Grow, M. (2008) *U.S. Presidents and Latin American Interventions: Pursuing Regime Change in the Cold War*. Lawrence: University Press of Kansas.

Gunder Frank, A. (1986) The development of underdevelopment. In P. F. Klarén and T. J. Bossert (eds) *Promise of Development: Theories of Change in Latin America*. Boulder, CO: Westview Press, pp. 111–123.

Harmer, T. (2011) *Allende's Chile and the Inter-American Cold War*. Chapel Hill: The University of North Carolina Press.

Hassan, M. and Nassar, R. (2017) An empirical study of the relationship between Foreign Direct Investment and key macroeconomic variables in Mexico. *Journal of International Business Disciplines* 12(1): 18–30.

Hey, J. A. K. (1997) Three building blocks of a theory of Latin American foreign policy. *Third World Quarterly* 18(4): 631–657.

Iber, P. (2015) *Neither Peace Nor Freedom: The Cultural Cold War in Latin America*. Cambridge, MA: Harvard University Press.

Jadoon, A. (2017) Persuasion and predation: The effects of U.S. military aid and international development aid on civilian killings. *Studies in Conflict and Terrorism* (online early).

Kassab, H. S. and Rosen, J. D. (2016) *The Obama Doctrine in the Americas*. Lanham, MD: Lexington Books.

Kelly, J. and Romero, C. A. (2002) *The United States and Venezuela: Rethinking a Relationship*. New York: Routledge.

Kornbluh, P. (2003) *The Pinochet File: A Declassified Dossier on Atrocity and Accountability*. New York: The New Press.

LaFeber, W. (1993) *Inevitable Revolutions: The United States in Central America*. 2nd Ed. New York: W. W. Norton & Company.

Latané, J. H. (1900) *The Diplomatic Relations of the United States and Spanish America*. Baltimore: Johns Hopkins University Press.

LeoGrande, W. M. and Kornbluh, P. (2014) *Back Channel to Cuba: The Hidden History of Negotiations Between Washington and Havana*. Chapel Hill: University of North Carolina Press.

Livingstone, G. (2009) *America's Backyard: The United States and Latin America from the Monroe Doctrine to the War on Terror*. London; New York: Zed Books.

Long, T. (2015) *Latin America Confronts the United States: Asymmetry and Influence*. New York: Cambridge University Press.

Loveman, B. (2006) (ed) *Addicted to Failure: U.S. Security Policy in Latin America and the Andean Region*. Lanham, MD: Rowman & Littlefield.

———. (2010) *No Higher Law: American Foreign Policy and the Western Hemisphere Since 1776*. Chapel Hill: University of North Carolina Press.

Madeiros, M., Barnabé, I., Albuquerque, R. and Lima, R. (2016) What does the field of International Relations look like in South America? *Revista Brasileira de Política Internacional* 59(1): 1–31.

McPherson, A. (2013) *The Invaded: How Latin Americans and Their Allies Fought and Ended U.S. Occupations*. New York: Oxford University Press.

McSherry, J. P. (2005) *Predatory States: Operation Condor and Covert War in Latin America*. Lanham, MD: Rowman & Littlefield.

Menjívar, C., Morris, J. E. and Rodríguez, N. P. (2017) The ripple effects of deportations in Honduras. *Migration Studies* (in press).

Merke, F. and Reynoso, R. (2016) Dimensiones de política exterior en America Latina según juicio de expertos. *Estudios Internacionales* (Santiago) 48(185): 107–130.

Mora, F. and Hey, J. A. K. (2003) (eds) *Latin American and Caribbean Foreign Policy*. Lanham, MD: Rowman & Littlefield.

Morley, M. and McGillion, C. (2015) *Reagan and Pinochet: The Struggle Over U.S. Policy Toward Chile*. New York: Cambridge University Press.

O'Brien, T. F. (2007) *Making the Americas: The United States and Latin America From the Age of Revolutions to the Era of Globalization*. Albuquerque: University of New Mexico Press.

Organization for Economic Cooperation and Development (OECD). (2015) *Latin American Economic Outlook 2016*. Paris: OECD Publishing.

Pearce, J. (1982) *Under the Eagle: U.S. Intervention in Central America and the Caribbean*. Boston, MA: South End Press.

Pereira, A. (2018) The US role in the 1964 Coup in Brazil: A reassessment. *Bulletin of Latin American Research* 37(1): 5–17.

Petras, J. and Morley, M. (1975) *The United States and Chile: Imperialism and the Overthrow of the Allende Government*. New York: Monthly Review Press.

Roett, R. and Paz, G. (2008) (eds) *China's Expansion into the Western Hemisphere: Implications for Latin America and the United States*. Washington, DC: Brookings Institution Press.

Romero, M. C. (2016) Ricardo Lagos y triunfo de Trump en EE.UU.: 'No es una buena noticia para el mundo. *El Mercurio*, 9 November. www.emol.com/noticias/Nacional/2016/11/09/830412/Ricardo-Lagos-y-triunfo-de-Trump-en-EEUU-No-es-una-buena-noticia-para-el-mundo.html (Accessed 5 February 2018).

Rouse, S. M. and Arce, M. (2006) The drug-laden balloon: U.S. military assistance and coca production in the Central Andes. *Social Science Quarterly* 87(3): 540–557.

Schoultz, L. (1998) *Beneath the United States: A History of U.S. Policy Toward Latin America*. Cambridge, MA: Harvard University Press.

———. (2009) *That Infernal Little Cuban Republic: The United States and the Cuban Revolution*. Chapel Hill: The University of North Carolina Press.

Scott, J. M. and Steele, C. A. (2011) Sponsoring democracy: The United States and democracy aid to the developing world, 1988–2001. *International Studies Quarterly* 55(1): 47–69.

Sikkink, K. (2004) *Mixed Signals: U.S. Human Rights Policy and Latin America*. Ithaca, NY: Cornell University Press.

Smith, P. H. (2008) *Talons of the Eagle: Latin America, the United States, and the World*. New York: Oxford University Press.

Tickner, A. B. (2003) Hearing Latin American voices in International Relations studies. *International Studies Perspectives* 4: 325–350.

———. (2016) Exportación de la seguridad y política exterior de Colombia. *Análisis*, December. http://library.fes.de/pdf-files/bueros/kolumbien/12773.pdf (Accessed September 2017).

Tulchin, J. (2016) *Latin America in International Politics: Challenging U.S. Hegemony*. Boulder, CO: Lynne Rienner.

Waltz, K. N. (1979) *Theory of International Politics*. New York: McGraw-Hill.

Weeks, G. (2015) *U.S. and Latin American Relations*. Malden, MA: Wiley-Blackwell.

———. (2016) Soft power, leverage, and the Obama doctrine in Cuba. *The Latin Americanist* 60(4): 525–539.

Williams, W. A. (1962) *The United States, Cuba, and Castro: An Essay on the Dynamics of Revolution and the Dissolution of Empire*. New York: Monthly Review Press.

15

LATIN AMERICA AND CHINA

Barbara Hogenboom

Introduction

The rise of China has a great impact on Latin America's development trajectories. Economic relations with China have expanded at an unprecedented speed since the start of the 21st century. From 2000 to 2016, trade between Latin America and China increased no less than 26 times, and China became the number one trade partner for Brazil, Argentina, Chile, Peru, and Uruguay (IISCAL, 2018). China is also becoming a source of major direct investments and development loans to Latin American countries. These trends are primarily driven by China's growing demand for raw materials and the Chinese government's Go Global strategy since the 2000s. After the crisis of the 1980s and the disappointing economic performance during the 1990s, most Latin American countries greatly benefitted from the commodity boom of 2003–2013. The boom was partly driven by China's economic growth, and in this same period China became the world's leading importer of several raw materials of which Latin America holds large reserves, especially iron ore, copper ore, bauxite, and oil. With the increased interest of the Chinese government, companies, and banks in the region, the countries have become less dependent on US and European markets, capital, and policies. However, the subsequent downturn of global prices for key metals, oil, and soy has reminded the region of the risks of such commodity-based growth, whereas cheap Chinese products continue to harm Latin America's manufacturing sector.

While most Latin American governments have applauded and stimulated these new relations, the rapidly growing influence of China has been subject to several debates in society and academia. Contrary to government discourses about the benefits of these South-South relations and mutual development, there are also concerns that China's economic modernization deepens the exploitation of natural resources and the underdevelopment of Latin America, and that China may in time become another imperialist power taking advantage of the region. The commodity-focussed Chinese interests raise concerns about Latin America's increased dependency on extractive sectors, with both the risks of deindustrialization – also due to Chinese competition – and local conflicts around large mining and oil projects.

This chapter starts with an overview of the region's China-related development trends since 2000, and the current importance of China in trade, investment, and development funding. Next, I use the case of the oil sector to look into the particular nature of Chinese actors and their involvements in the region. The subsequent section reviews some other important sectors in China-Latin America relations. The chapter concludes with a discussion of the implications of China's increasing engagement with Latin America for the region's development.

Resource-centred relations

Since the 1980s China has shown impressive growth rates and rapid integration in the global economy. China increased its share of world exports from 1% in 1980 (UNCTAD, 2005: 133) to 17% in 2016 (Eurostat, 2017), and since 2009 the 'factory to the world' is the world's leading exporter. An addition to its economic restructuring, China's transformation into a central place for global production also results from its active role in international institutions and bilateral diplomacy. In the 1980s, China adapted its diplomatic relations to its economic policies of liberalization and it abandoned its political strategy of expanding Maoism to Latin America. In the 1990s, China's aim to deepen its insertion into the global economy and gain membership of the World Trade Organization (WTO) required good relations with as many countries as possible. Its new foreign policy proclaimed no hegemonism, no power politics, no arms races, and no military alliances, which in Latin America was seen as positive for cooperation, mutual confidence, and multilateralism.

From the 1990s onwards, China has strengthened relations with most Latin American countries (Fernández Jilberto and Hogenboom, 2012). From the start, Brazil has been China's most important partner in the region. The two countries developed intensive diplomatic, economic, and scientific relations, and as early as 1993, the Chinese government defined its relations with Brazil as a strategic alliance. Under President Lula da Silva (2003–2010), Brazil and China strengthened their relations and took several counter-hegemonic initiatives, such as the creation of the BRICS, an association of emerging economies together with India, Russia, and South Africa (the acronym is derived from the names of these five countries). Besides Brazil, China also established strategic partnerships with Venezuela, Argentina, and Ecuador. And it signed free trade agreements with Chile, Peru, and Costa Rica. The exception to this trend of intensified relations with the People's Republic of China concerns a number of Central American and Caribbean countries that hold diplomatic relations with Taiwan. As part of Beijing's One China policy, it denies diplomatic relations with these countries. Gradually, however, countries tend to give up their relations with Taiwan in order to establish relations with China, as Costa Rica, Panama, and the Dominican Republic have done in recent years.

The relations with China have also developed at the regional level. With respect to development-related organizations, China has been a member of the Inter-American Development Bank (IDB) since 2009, when it contributed $350 million to IDB programmes. The Latin American Development Bank CAF also holds institutional relations for collaboration with China, but not as a member country. The United Nations Economic Commission of Latin America and the Caribbean (ECLAC) has been making efforts to improve China-Latin America relations. Through its reports and activities, ECLAC is an important agenda-setting agency from the region itself, feeding and supporting regional policy-making processes. Numerous ECLAC publications deal with the relevance of China for the region, especially pointing at the benefits and potential of China as new destination for exports, a source of investments, credits, imports and technology, and a partner for mutual development. Furthermore, in 2014 China initiated the creation of a Forum with the Community of Latin American and Caribbean States (CELAC). This China-CELAC Forum serves as a platform for exchange and policy coordination with the region as whole.

Since the turn of the century and the start of the latest commodity boom (2003–2013), China has become a major destination for Latin American exports. These trade flows responded in particular to China's rising demand for primary commodities. For Latin America, this 'China effect' on global prices resulted in even greater extra revenues from commodity exports than the direct effect of increased export volumes to China (Jenkins, 2011). Starting at only $5 billion in 2000, the value of Latin American exports to China reached around $100 billion between 2011 and 2014 (ECLAC, 2015a: 36). For most countries China is either the first or second export destination. While for the region as a whole the US remains the main export market, China has become about as important as the European Union. Apart from this market share, high demand from China has made the region less vulnerable for economic ups and downs in the economies of its traditional trade partners. When Europe and North America were struck by the economic crisis starting in 2008, ongoing high performance in China helped to sustain high prices and thereby Latin America's commodity exports and economic growth.

The region's massive flows of goods to China are however far from diversified: most Latin American countries export just a few primary products, mainly copper, soy, iron ore, and oil (see Figure 15.1). For the period of 2011–2015, Latin America's exports worldwide consisted of 21% agricultural products, 31% extractive goods (mining, oil, and gas), and 46% manufactured goods. However, the region's exports to China consisted of 32% agricultural products, 55% extractives, and only 12% manufactured goods. Together, primary goods thus represent 77% of total exports to China (Ray and Gallagher, 2017).

Beyond the small share of manufactures in the region's exports to China, Latin America's industrial sector is very much harmed by the increased competition with cheap Chinese products. All countries have seen part of their national industry shrink or disappear, ranging from small family workshops of cloths or shoes to large-scale factories of textiles, electronics, and other consumer articles. Chinese industrial competition is very strong, both in the national economies and in third markets (Gallagher and Porzecanski, 2010). The latter has been most problematic for Mexico and Central America, which lost considerable market share to China in their main export market: the United States. This competition in export manufacturing includes the ability to attract foreign direct investment, and thereby the potential for future development of this sector. China's modernization and global integration negatively affects "the future 'spaces' open for the development of industrial exports in a liberalized world in which PRC is pre-empting many markets for products that developing countries can export" (Lall and Weiss, 2004:

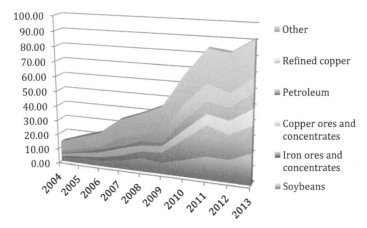

Figure 15.1   Latin American exports to China by product, 2004–2013 (US$ billion)

Source: Compiled by author using data from UN Comtrade.

23). Up to now, China's rise and the intensified economic relations with Latin America have contributed to the region's reprimarization and deindustrialization (ECLAC, 2015b).

Since the start of the 2010s, Latin America has experienced China's rise as a major source of capital, both in the form of direct investments and of loans. Already in 2004, during the first large Chinese Latin America tour by President Hu Jintao and a delegation of Chinese companies, several large investments were announced. But it took until the crisis in the US and Europe, before Chinese multinationals started to seriously invest in the region. As with the composition of Chinese imports from Latin America, a noticeable feature of Chinese investments is the disproportional share related to the primary sector. While worldwide FDI in Latin America from 2007 to 2011 went primarily to services (43%) and manufacturing (31%), followed by investment in primary goods, most of Chinese FDI in the region (almost 90%) was in the primary sector (ECLAC, 2013). From a historical point of view, China is still the new kid in town in Latin America, but at least in extractives it has been catching up fast. For instance, in a typical mining country like Peru, Chinese companies have come to account for more foreign investments than the United States or Canada.

Simultaneously, through substantial loans by Chinese banks, China has become a major lender for Latin America. From 2010 to 2015 these loans reached $ 94 billion. Almost out of nothing, the Export-Import Bank of China and especially the China Development Bank has emerged as an important source of loans. While the financial crisis forced the World Bank, the Inter-American Development Bank, and the US Export-Import Bank to limit their global activities, in 2010 Chinese loan commitments jumped to $37 billion. In effect, that year China provided more credit for Latin American development than the region's traditional channels for credits to Latin America based in Washington (Gallagher, Irwin, and Koleski, 2012). Since then Chinese development banks have continued to offer large loans to the region, as visualized in Figure 15.2. As we will see below, Venezuela, Brazil, and Ecuador have received the lion share of these Chinese loans, and most of them are somehow connected to oil. In short,

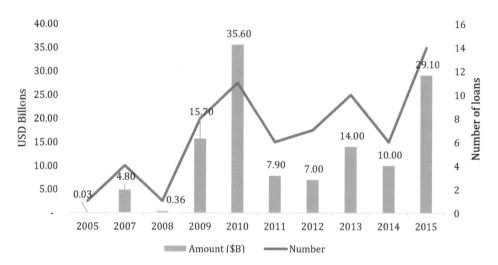

Figure 15.2   Chinese loans to Latin America, 2005–2015

Source: Compiled by author using data from Thedialogue.org (2016). Inter-American Dialogue | China-Latin America Finance Database (www.thedialogue.org/map_list/).

like Chinese imports and investments, also these Chinese loans for development revolve around Latin America's primary sector.

Chinese particularities: the case of oil relations

A short review of the oil relations between Latin America and China illustrates the importance of these rather new connections, and the ways in which Chinese actors act partly different from Western companies, governments, and banks. After gradual economic liberalization from the late 1970s onwards, China's membership of the World Trade Organization in 2001 marked the start of its rapid globalization. That same year the Chinese government announced the Go Global strategy: a comprehensive policy agenda to make its state-owned companies and banks operate abroad in order to catch up with other large economies. Furthermore, with respect to oil, China's search for energy security and its large financial reserves coincide with Latin America's resource wealth and its need for foreign capital to exploit it. Energy security is clearly a top political priority for Beijing. According to Lee (2012: 77–78), "Beijing considers not just reliable and uninterrupted but also cheap supply of energy as essential to it national and domestic political interest," and beyond economic reasons, "securing such access is also essential for mitigating risks to the survival of the regime in China." Or as a Chinese official mentioned in an interview: "oil is blood." Likewise, political regimes in Latin American oil countries are highly dependent on this commodity for exports, economic growth, and public sector revenues.

Next to metal ores and soy, oil is a key commodity in China's relations with Latin America. In 2014 China replaced the United States as the world's largest importer of crude, and Latin America accounts for around 10% of China's demand for imported oil. The region holds substantial oil reserves, especially Venezuela, Mexico, and Brazil, and to a lesser extent Ecuador, Colombia, and Argentina. As in most other parts of the world, the oil sector and oil relations of Latin America and China are characterized by a prominent role of the state and huge state-owned companies. Latin America's three largest companies are the state-owned oil companies Petrobras (Brazil), Pemex (Mexico), and PDVSA (Venezuela). And of China's state-owned oil companies, Sinopec and CNPC are even among the top five largest companies in the world. It may come as no surprise that governments of oil-producing countries like Venezuela, Argentina, Brazil, and Ecuador have aimed at gaining access to Chinese oil investments, and have been quite successful.

Predominantly four Chinese companies have started operations in Latin America: Sinopec, CNPC (globally listed as PetroChina), CNOOC, and Sinochem. These giant state-owned companies serve as important instruments for the Chinese government's energy security agenda. Buying oil in the international market and even large long-term supply contracts were deemed insufficient, and since the 1990s the government made them to globally invest and operate. As newcomers in a global environment dominated by US and European transnational companies, Chinese oil companies (COCs) initially had to settle for less profitable or more risky projects (Vermeer, 2015). On the other hand, they have also had the advantage of support of the Chinese government and development banks. Abroad COCs act as autonomous companies, through commercial branches listed on foreign stock markets, but in reality many large investments also involve government-to-government deals. Compared to companies from Western countries Chinese companies in general can more easily invest in projects that require infrastructural development and they are less restrained by short-term shareholder value maximization (Gonzalez-Vicente, 2012).

While North American and European transnational companies (TNCs) remain key foreign investors in the Latin American oil sector, companies from China are now of importance too.

In 1997, CNPC became active in Venezuela and next in Ecuador, through joint ventures with the national state-owned oil company or concessions. Later on Sinopec and Sinochem started separately in Colombia and expanded to Brazil and other countries, while CNOOC started in Argentina. Between 2003 and 2016, Chinese companies invested $25 billion in the region's oil and gas sector, about similar to Chinese mining investments (Avendano, Melguizo, and Miner, 2017). This investment largely coincided with the years of the latest global oil boom, until 2014, when the sector became much more lucrative and dynamic than before. Also in Latin America, Chinese companies partly invest in 'difficult environments,' such as countries like Venezuela and Ecuador, areas with limited infrastructure or projects involving more technical or social risks. Their limited experience with the region occasionally results in tensions, negative media coverage, and anti-Chinese sentiments. In Brazil, for instance, in the case of the 2013 auction of Libra oil fields, there were a few anti-Chinese protests (described as "absurd xenophobia" by Brazilian President Rousseff).

Besides Chinese oil companies, the Chinese government and banks are also actively involved in advancing oil relations with Latin American countries. At home, the Chinese state controls as well as aligns the energy and finance sectors by coordinating their bureaucracies and by appointing the top executives of state-owned companies and banks. Chinese state-owned banks have offered indispensable financial aid for the transnationalization of Chinese oil companies. Active support from the China Development Bank and Export and Import Bank of China point at a state-led synchronization of COCs and Chinese banks (Kong, 2010: 67–69; Kong and Gallagher, 2016). Abroad, China uses high-level diplomacy to secure oil supply and help advance the transnationalization of its companies. Simultaneously, the abovementioned state-owned Chinese banks have provided massive loans to countries with oil: Brazil, Ecuador, and especially Venezuela (see Figure 15.3). Since Venezuela and Ecuador had limited access to other sources of foreign credit, the Chinese loans were rather crucial to realize their development plans. Some of the so-called oil-backed loans are for sectors other than energy, in particular to pay for large infrastructure projects, which in turn involve big contracts for Chinese companies to build roads, railways, hydroelectric dams, or ports. Such loans are guaranteed by long-term (10–30 years) oil supply contracts to Chinese oil companies and involve multi-party arrangements between a Chinese bank, the Chinese and Latin American state-owned oil company, and

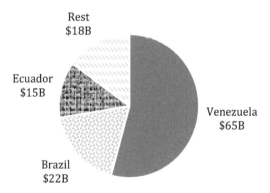

Figure 15.3 Distribution of Chinese loans to Latin America, 2007–2015 (USD Billions)

Source: Compiled by author using data from Thedialogue.org (2016). Inter-American Dialogue | China-Latin America Finance Database (www.thedialogue.org/map_list/).

the governments (Downs, 2011). Debts are thus to be paid back with oil instead of dollars, over long periods of time. As the large oil-based loans of the China Development Bank to Venezuela show, loans were regulated in such a way that the risks of the Chinese Bank were limited (Sun, 2012). In short, the Chinese government and development banks establish strong relations with Latin American governments and companies in ways that differ from US and European entities.

This Chinese way of making package deals, in the oil sector as well as other sectors, cannot be practiced throughout Latin America. Government-to-government arrangements are used in countries like Venezuela and Ecuador, where the governments are open to such deals. In contrast, in Chile and Mexico, open international bidding is required for large projects (they are both members of the OECD). Under these open market conditions, a company-to-company model has to be used and Chinese companies have to make their bid alongside Western multinationals. With countries such as Brazil, Argentina, and Peru, a mixed model is applied, involving substantial government influence as well as company-to-company arrangements. The cases of Argentina and Brazil show that such willingness is not necessarily related to the government's ideological position. While the presidencies of Macri (since 2015) and Temer (since 2016) meant a break with the progressive regimes of their predecessors and a return to neoliberal market policies, they have both continued to use package deals to attract large Chinese loans and investments. In the case of Argentina, after some renegotiations, Macri agreed to a $20 billion loan from the China Development Bank for railroad and hydroelectricity projects involving Chinese companies – a deal he had in fact criticized when campaigning for the presidential elections. These special economic relations indicate that China is increasing its economic and political leverage in Latin America.

Shortly after international oil prices dropped sharply to half their previous prices, China offered a helping hand to befriended oil countries in trouble. In 2015, Venezuela received $20 billion in Chinese loans for energy, social, and industrial projects and Ecuador received $7.5 billion for infrastructure, education, and sanitation programmes. Brazil's state-owned oil company Petrobras also received large Chinese loans (Hogenboom, 2017). Venezuela's deep crisis is of great concern to the Chinese government, banks, and oil companies (Xu, 2017). There are worries about Venezuela's overall instability and its ability to repay Chinese loans, while companies operating there have faced security problems. Nevertheless, it is in the interests of the Chinese government to maintain good relations with the nation that holds the world's largest proven oil reserves. Offering a lifeline has also been in the interest of the Chinese banks and oil companies with capital invested in Venezuela.

While Latin American governments have welcomed the fact that the Chinese government, companies, and development banks are different, some of their particularities and shady deals feed concerns in society and academia. In public opinion there are doubts about the formal win–win discourses, and whether Latin America profits as much from the loans-for-oil deals and the long-term oil supply contracts as China does. For example, a considerable share of the Latin American oil provided to Chinese companies in order to repay the loans never reaches China, but is sold in the global oil market and is most probably consumed in the United States (Jiang and Sinton, 2011). In Latin America, this profit-seeking behaviour of COCs with 'their oil' did not go well with the discourses of leftist presidents like Chávez and Correa who presented oil relations with China as being based on partnership and friendship. Distrust of Chinese oil companies also exists among Latin American companies, which hinders cooperation. Petrobras for instance seems unwilling to share its expertise with CNOOC, and its technology transfer with Sinopec remains weak despite their agreement for technological cooperation. Latin America's oil relations with China are therefore multifaceted, involving both possibilities as well as pitfalls.

Other sectors of Chinese interest

China's demand for resources and Chinese interests to investment abroad fit rather well with Latin America's resource wealth and need for capital. The region's mining sector has also been strongly affected by Chinese influences. China imports massive volumes of metals from Latin America, like copper, iron, silver, and nickel from countries like Chile, Brazil, Peru, and Cuba, respectively. Chile, for instance, is the second most important source of copper for China, which was one of the main initial Chinese interests behind establishing a free trade agreement with Chile in 2006. The region has also received increasing flows of Chinese direct investment. Already before China's Go Global policy, some smaller Chinese mining investments were made in the region, starting in Peru. Since then, Chinese mining companies have especially invested in Peru and Ecuador. Chinese mining companies are not as large as the oil companies but they are also state-owned. The Chinese government has been facilitating the international expansion of these state-owned companies, especially through offering them soft loans (Gonzalez-Vicente, 2012).

As China is a rapidly developing society with 1.4 billion inhabitants, Latin America's agricultural sector is also important. Urbanization and a growing middle class affect the volumes and patterns of consumption. Agriculture accounted for 30% of Chinese imports from the region in 2013, but the lion share (three quarters) of these goods are imports from Brazil (ECLAC, 2015a). The main product is soybeans, which are to feed China's livestock, reflecting the growing consumption of meat, especially pork and chicken. Some Chinese companies invest in this sector, too, by buying large agricultural companies.

The Chinese government and development banks have a keen interest in supporting infrastructure in the region. Through special arrangements the mutual benefits are multiplied: Chinese banks provide loans for infrastructure, Latin American governments partly contract Chinese construction companies to get the job done, usually involving local as well as Chinese labourers, and as a result Latin American governments are able to get large infrastructure projects completed. Better infrastructure in the region is also in the interest of China because it eases and lowers the price of imports from the region and serves the interests of Chinese companies operating in Latin America. Recently, Latin America has been added to China's global One Belt, One Road (OBOR) initiative for connectivity and infrastructure, launched by President Xi Jinping in 2014. A prominent country in this context is Panama, which in 2017 replaced its diplomatic relations with Taiwan for relations with China and immediately signed an agreement with China about OBOR and Chinese support for the harbour of Colon, on the Atlantic side of the Panama Canal. An important new institution for OBOR and loans for large infrastructure projects is the Asian Investment Infrastructure Bank (AIIB), an international development bank set up by China, for which Brazil, Argentina, Chile, Bolivia, Peru, Ecuador, and Venezuela are registered as prospective members. Finally, over the past few years there have been numerous media reports about the possible construction of a Chinese-funded canal between the oceans in Nicaragua. In practice, however, very little progress has been made with this plan. Next to natural, economic, and geopolitical obstacles, Nicaragua's relations with Taiwan instead of China stand in the way, and the Chinese company HKND that is supposed to finance the new canal is not state-owned but a private entity. In short, although it has been widely discussed and produced a substantial amount of opposition in Nicaragua, it is uncertain if and when a new canal will be build.

Debates and new directions

Is the rise of China positive for Latin America's development? This question continues to be the topic of heavy debates. Three interrelated key topics concern the pros and cons of China's increasing influence. The first involves the room for national decision-making over development policies; the second the question of local development and local communities; and the third the region's position in the global economy. First, the debate about China's influence on Latin American development policies partly derives from the period in which their economic relations intensified, in the 2000s, when most countries in the region shifted their development model. Presidents of the 'new left' distanced themselves from two decades of neoliberal restructuring and the Washington Consensus, and captured greater state control over strategic sectors such as oil and mining. Helped by China's rise and booming commodity markets, these reforms raised the progressive governments' budget for social expenditures, fitting their development agenda for growth with redistribution (Hogenboom, 2012). Interestingly, China itself is a country where the public sector plays an omnipresent role in the economy, not comparable with any Latin American country. How this also affects its engagement with the region is still not sufficiently understood, which causes tensions in the cross-Pacific relationship (Dussel Peters, 2015).

Moreover China emerged as a foreign partner for this new development model, through direct investment and especially loans. Positive accounts stressed that contrary to Washington-based financial institutions like IMF, World Bank, and the IDB, Chinese development banks do not pose macroeconomic policy conditions on their loans or interfere in national decision-making. And they pointed out that Chinese state-owned companies are also willing to invest in countries with other than open market models. In effect, it meant that countries opting for more profound post-neoliberal reforms were able to access a new source of foreign capital. Even though resource nationalism, such as Venezuela's oil policies, at times also went against the interests of Chinese companies (e.g. higher taxes), the emergence of China as a friendly 'rich' nation offered an alternative pathway for development to Latin America's resource-rich countries. China's rise thus constituted an enabling factor for the region's new political elites to use resource wealth for more state-led economic development and social redistribution.

On the other hand, oil-producing countries in Latin America have noted that Chinese loans involve long-term ties and some downsides. Chinese oil-related loans are not necessarily cheap since China's development banks have become more commercial and profit-seeking, often lending at commercial rates (Downs, 2011). For example, the CDB interest rates for its loans to Brazil in 2009 and to Ecuador in 2010 and 2011 were higher than those of the World Bank (Kong and Gallagher, 2016: 26). Another concern is about the implications of the large long-term oil supply contracts through which Venezuela, Brazil, and Ecuador have to pay their debts. It means that a country cannot decide to limit oil extraction, even if a new government wished to. As a consequence of low oil prices, countries are obliged to strongly expand production volumes to still make their debt payments. In Venezuela, Chinese package deals implied that oil-backed loans already came with greater participation of Chinese oil companies, and when oil prices plummeted more oil was supplied to COCs while also more foreign investment was required to raise production (Cardona Romero, 2016). So while Chinese development banks do not impose policy conditions, there are some strings attached that may negatively affect room for future development in other ways.

The involvement and physical presence of Chinese actors in the region is transformational and challenging for both sides (Ellis, 2014), and this is especially the case at the local level. The arrival of Chinese mining, oil, and infrastructure companies and workers has caused several conflicts with local communities, governments, or companies. As latecomers, Chinese companies have still limited experience with overseas direct investment, including their social and environmental impacts. The Chinese non-governmental Global Environmental Institute (2014: 196) states that around the globe "Chinese investors' misbehaviours on social and environmental issues are numerous and can include: inexperience, ignorance, complacency, sheer disregard for regulations, along with lax and patchy local law enforcement." Their report stresses the importance of sustainability principles in Chinese foreign projects. The government in Beijing recognizes the need for Chinese companies to also catch up in this area, and has designed corporate social responsibility (CSR) policies and guidelines, although most of this is still soft policy. Chinese companies themselves also want to adopt CSR practices in order to be (seen as) equals among global companies, and recently the Hong Kong Stock Exchange has made environmental, social, and governance reporting mandatory. Still, in China's extractive sectors and among large state-owned companies progress has been low, especially when it comes to transparency and accountability or multi-stakeholder consultation (Tan-Mullins and Hofman, 2014). It comes as no surprise then that the *Latin American Economic Outlook 2016* calls for greater transparency and regulation of Chinese lending to the region, particularly when it comes to environmental impacts (OECD, ECLAC, and CAF, 2015).

Various communities throughout the region have experienced the inability or unwillingness of Chinese companies to fully engage with local stakeholders. To improve these practices requires greater and coordinated efforts of the Chinese companies, development banks, and government as well as Latin American governments. And it is up to scholars and NGOs to scrutinize plans, policies, and practices from the global and national to the local scale, including stepping up studies in remote areas in which some of the Chinese companies operate. For instance, in 2016 Andes Petroleum (a joint venture of CNPC and Sinopec) signed contracts for two controversial oil projects in Ecuador's southern Amazon. The oil blocks not only lie in an untouched region with delicate ecosystems and high biodiversity, they also overlap with indigenous territory. The blocks border on the Yasuní ITT national park, where some of the largest oil reserves are located. The government of Ecuador plans to develop more oil fields in this sensitive area and Sinopec will develop six additional fields for Ecuador's state-owned oil company PetroAmazonas. Civil society groups and scholars have expressed great concern about the effects of these developments on the environment and indigenous peoples. As Ray and Chimienti (2015) argue, the new concessions are a test for both Chinese oil companies and the Ecuadorian government on a range of challenging issues: protection of the environment and indigenous territory, transparency and public accountability, and taking the opinion of the local community into account in decision-making and planning.

Latin American governments approach China as a new ally for national development, and ECLAC stresses the benefits and potential of China as a partner for regional development (ECLAC, 2012, 2015a), but critical voices from Latin American academia, NGOs, and civil society groups declare that China's expanding influence helps to sustain a commodity-based style of development that is economically, politically, and ecologically unsustainable. Already at the heights of the commodity boom, they pointed at the long-term risks of dependence on Chinese capital and commodity-based development. In their eyes, Latin America's long history of dependency on powerful foreign partners who seek access to the region's commodities is repeating itself. The players have changed, but not the rules of the game. Maristella Svampa (2013) coined the term Commodity Consensus, which criticizes the increased dependency of Latin American economies on commodities and foreign capital, including Chinese capital, under progressive regimes that claimed to be post-neoliberal. In these debates concerns about the power asymmetry in bilateral relations with China are connected to wider critiques on Latin America's neo-extractivist development model. Clearly, the implications of China's rise for sustainable development in Latin America are likely to remain a key issue for the years to come (Ray et al., 2017).

Power asymmetries between Latin American and Chinese actors represent a key challenge for Latin America's relations with China. So far, China has been in the lead. Even for Brazil or Mexico, the Chinese government, markets, companies, and banks are very powerful counterparts. And Latin American countries still mainly deal with China on an individual basis, and compete among each other for Chinese trade and capital. In order to change the unequal relation with China, and effectively foment Latin America's long-term development, both sides of the Pacific now seem to realize that Latin America has to interact more as a region with China. China took the initiative to create the China-CELAC Forum, which at its first meeting in Beijing was presented as "a new platform, new starting point and new opportunity for dialogue and cooperation" (Beijing Declaration, 2015). China's President Xi Jinping announced that he aims to increase China-Latin America trade to $500 billion and Chinese investments to $250 billion by 2025 (Beijing Declaration, 2015). The China-CELAC Cooperation Plan for 2015–2019 is directed at shifting from trade and investment in Latin America's natural resources to a new model of South-South cooperation with more consideration of the region's development needs. An expert on China-Latin America relations claims that this cooperation plan "is an unprecedented opportunity for Latin America to upgrade its industrial competitiveness and environmental protection," and "the best and balanced opportunity for Latin American economic development in 80 years" (Gallagher, 2016: 172–173). However, to materialize this opportunity requires stronger leadership by Latin American national governments as well as a joint Latin American agenda.

Moreover, critical debate, democratic decision-making, and accountability of governments will be equally indispensable. As studies of the historical underdevelopment of the region have shown (Buncker, 1985; Galeano, 1973), when it comes to large-scale resource extraction Latin American elites tend to become either greedy, or overly optimistic of the developmental effects, or both. Time and time again, governments have accommodated large extractive projects driven by foreign forces from core economies through special deals and major public investments. In most cases, foreign companies and banks as well as local economic and political elites profited more that Latin American societies. Instead of modernization, often debt accumulation and deepened dependency of commodities and foreign capital were the outcome. In light of these risks, Venezuela's current multi-crisis, after a decade of heavy Chinese involvement in its oil sector and development projects, is a warning sign for Latin America as well as China.

References

Avendano, R., Melguizo, A. and Miner, S. (2017) *Chinese FDI in Latin America: New Trends with Global Implications*. Washington, DC: Atlantic Council.

Beijing Declaration of the First Ministerial Meeting of the CELAC – China Forum. (2015) www.itamaraty.gov.br/index.php?option=com_content&view=article&id=9743:documentos-aprovados-na-i-reuniao-dos-ministros-das-relacoes-exteriores-do-foro-celac-china-pequim-8-e-9-de-janeiro-de-2017&lang=es (Accessed 10 May 2018).

Buncker, S. G. (1985) *Underdeveloping the Amazon: Extraction, Unequal Exchange, and the Failure Of the Modern State*. Urbana and Chicago: University of Illinois Press.

Cardona Romero, A. M. (2016) *China en Venezuela. Los Préstamos por Petróleo*. Bogotá: Asociación Ambiente y Sociedad.

Downs, E. S. (2011) *Inside China, Inc: China Developments Bank's Cross-Border Energy Deals*. Washington, DC: Brookings Institution Press.

Dussel Peters, E. (2015) The omnipresent role of China's public sector in its relationship with Latin American and the Caribbean. In E. Dussel Peters and A. C. Armony (eds) *Beyond Raw Material: Who Are the Actors in the Latin America and Caribbean-China Relationship?* Buenos Aires: Friedrich Ebert Stiftung, pp. 50–72.

ECLAC. (2012) *China and Latin America and the Caribbean: Building a Strategic Economic and Trade Relationship*. Santiago: ECLAC.

———. (2013) *Chinese Foreign Direct Investment in Latin America and the Caribbean China-Latin America Cross-Council Taskforce*. Santiago: ECLAC.

———. (2015a) *Latin American and the Caribbean and China: Towards a New Era in Economic Cooperation.* Santiago: ECLAC.

———. (2015b) *Latin American and the Caribbean in the World Economy 2015.* Santiago: ECLAC.

Ellis, E. (2014) *China on the Ground in Latin America: Challenges for the Chinese Impacts on the Region.* New York: Palgrave Macmillan.

Eurostat. (2017) *The EU, USA and China Account for Almost Half of World Trade in Goods.* http://ec.europa.eu/eurostat/web/products-eurostat-news/-/DDN-20170824-1?inheritRedirect=true (Accessed 10 May 2018).

Fernández Jilberto, A. and Hogenboom, B. (2012) (eds) *Latin America Facing China: South-South Relations Beyond the Washington Consensus.* Oxford; New York: Berghahn Books (CEDLA Latin America Studies Series).

Galeano, E. (1973) *Open Veins of Latin America: Five Centuries of the Pillage of a Continent.* New York: Monthly Review Press.

Gallagher, K. P. (2016) *The China Triangle: Latin America's China Boom and the Fate of the Washington Consensus.* Oxford: Oxford University Press.

———, Irwin, A. and Koleski, K. (2012) *The New Banks in Town: Chinese Finance in Latin America.* Washington, DC: Inter-American Dialogue.

——— and Myers, M. (2014) *China-Latin America Finance Database* (updated with 2015 data). Washington, DC: Inter-American Dialogue. www.thedialogue.org/map_list/

——— and Porzecanski, R. (2010) *The Dragon in the Room: China and the Future of Latin American Industrialization.* Redwood City, CA: Stanford University Press.

Global Environmental Institute. (2014) *Environmental and Social Challenges of China's Going Global.* Beijing: China Environment Press.

Gonzalez-Vicente, R. (2012) Mapping Chinese mining investment in Latin America: Politics or market? *The China Quarterly* 209: 35–58.

Hogenboom, B. (2012) Repolitisation of hydrocrabons and minerals in Latin America: Extraction for development with equity and sustainability? *Journal of Developing Societies* 28(2): 133–158.

———. (2017) Chinese influences and the governance of oil in Latin America: The cases of Venezuela, Btrzail and Ecuador. In M. Parvizi Amineh and Yang G. (eds) *Geopolitical Economy of Energy and Environment.* Leiden; Boston, MA: Brill, pp. 172–211.

IISCAL. (2018) *China América Latina Brief*, Marzo. Washington, DC: Iniciativa para las Inversiones Sustentables China – América Latina.

Jenkins, R. (2011) El "efecto China" en los precios de los productores básicos y en el valor de las exportaciones de América Latina. *Revista CEPAL* 103: 77–93.

Jiang, J. and Sinton, J. (2011) *Overseas investments by Chinese National Oil Companies: Assessing the Drivers and Impacts.* Paris: International Energy Agency.

Kong, B. (2010) *China's International Petroleum Policy.* Santa Barbara, CA: ABC CLIO.

——— and Gallagher, K. P. (2016) *The Globalization of Chinese Energy Companies: The Role of Sate Finance.* Boston, MA: Boston University.

Lall, S. and Weiss, J. (2004) *People's Republic of China's Competitive Threat to Latin America: An Analysis for 1990–2002.* ADB Institute Discussion Paper no. 14. Tokyo: Asian Development Bank Institute.

Lee, J. (2012) China's geostrategic search for oil. *The Washington Quarterly* 35(3): 75–92.

OECD Development Centre, ECLAC and CAF. (2015) *Latin American Economic Outlook 2016: Towards a New Partnership with China.* Paris: OECD.

Ray, R. and Chimienti, A. (2015) *A Line in the Equatorial Forests: Chinese Investment and the Environmental and Social Impacts of Extractive Industries in Ecuador – Latin America Economic Bulletin.* Discussion Paper 2015–2016. Boston, MA: Boston University.

——— and Gallagher, K. P. (2013) *2013 China – Latin America Economic Bulletin.* Boston, MA: Boston University.

———, Lopez, A. and Lopez, C. (2017) *China and Sustainable Development in Latin America: The Social and Environmental Dimension.* London: Anthem Press.

Sun, H. (2012) Energy cooperation between China and Latin America: The case of Venezuela. In M. Parvizi Amineh and G. Yang (eds) *Secure Oil and Alternative Energy: The Geopolitics of Energy Paths of China and the European Union.* Leiden; Boston, MA: Brill, pp. 213–244.

Svampa, M. (2013) "Consenso de los Commodities" y lenguajes de valoración en América Latina. *Nueva Sociedad* 244: 30–46.

Tan-Mullins, M. and Hofman, P.S. (2014) The shaping of Chinese Corporate Social Responsibility. *Journal of Current Chinese Affairs* 43(4): 3–18.

Thedialogue.org (2016) *Inter-American Dialogue | China-Latin America Finance Database.* www.thedialogue.org/map_list/.

UNCTAD. (2005) *World Investment Report 2004.* New York; Geneva: United Nations.

Vermeer, E. B. (2015) The global expansion of Chinese oil companies: Political demands, profitability and risks. *China Information* 29(1): 3–32.

Xu, Y. (2017) *China's Strategic Partnerships in Latin America: Case Studies of China's Oil Diplomacy in Argentina, Brazil, Mexico, and Venezuela, 1991–2015.* Lanham, MD: Lexington Books.

16

LATIN AMERICA AND THE EUROPEAN UNION

Anna Ayuso
Translated by Anna Holloway

Introduction

Historical relations between Latin America and Europe are quite complex, insofar as they combine elements of continuity and change and are affected by regional and extra-regional variables that alter the power balance and the nature of alliances. At this juncture, the interregional dynamic of EU-Latin America relations is mostly determined by elements of change that reflect the uncertainties of a shifting international context. After the great financial crisis of the first decade of the 21st century, a new era has begun. It is an era characterized by different material capacities and power imbalances that offer more alternatives in possible associations with emerging actors, and by the weakening of the traditional links derived from the colonial past. Paradoxically, however, all players are witnessing a reduction of their autonomy due to the increasing mutual dependence that globalization entails and the need to tackle common problems such as economic instability, security, climate change, or the Sustainable Development Goals (SDGs) of the UN 2030 Agenda.

The EU-LAC strategic bi-regional partnership that was launched in the 1999 Summit of Heads of State and Government in Rio de Janeiro is undergoing a process of transition, from the traditional model of interregionalism based on North-South cooperation to more horizontal and complex relations (Ayuso and Gratius, 2016), leading to tensions both within the Latin American region and within the EU. The social effects of the crisis and of the imposed austerity policies have favoured populist discourses, altering the political party system and threatening the stability of institutions weakened by scandals of corruption. After nearly a decade of economic stagnation in the EU, and of deterioration in Latin America due to the end of a cycle of high prices in raw materials, both regions are struggling to get back on the track of growth.

The situation created by the reduction in demand from China and the protectionist attitude of the US under the rule of Donald Trump – who has pulled out of the negotiations of mega-agreements such as the Transatlantic Trade and Investment Partnership (TTIP) and the Trans-Pacific Partnership (TPP) and is pursuing a more restricted North American Free Trade Agreement (NAFTA) – could create favourable conditions for the unblocking of the pending negotiations between the EU and Mercosur countries, the EU's biggest market in the region.

The two great lines that have traditionally characterized EU-LAC cooperation for the past decades, namely regional integration and social cohesion, are still partly in force but are following different strategies (Ayuso, 2008). On the one hand, the current diversity of Latin American regionalism makes it impossible to compare with the European model of integration; therefore, the interregionalist approach has had to incorporate many different actors and ways of relating (Ayuso and Gardini, 2017). In addition, the EU's incapacity to respond to the uneven effects of the economic crisis in the eurozone has created great gaps within societies that have particularly affected the more vulnerable sectors of the population. The European welfare state has been internally debilitated and can hardly be set as a model, especially at a time when Latin America is facing a new period of unpopular adjustments and reforms. The image of Europe as a normative power has also been weakened by Brexit and by its inability to create a common front of solidarity in the face of the refugee humanitarian crisis.

However, changes are taking place in the international sphere that favour the consolidation of interregionalism as a tool for multilevel governance through the creation of intermediary mechanisms (Cooper, Hughes, and de Lombaerde, 2008) which allow for partnerships of a variable geometry, capable of leading to a convergence in principles and norms (Teló, 2001). In a multipolar world, the EU and Latin America defend multilateralism as a tool for global governance and seek to create alliances. However, the fragmentation of power has made the framework of interregional and bilateral relations increasingly complex, with the involvement of more actors representing different interests (Hardacre and Smith, 2009).

The long history of EU-LAC relations offers a valuable legal, institutional, and social framework at a time when this association is confronted with the need to look for more balanced ways of adapting to the structural changes of the world order. This chapter begins by focusing on the evolution of interregional relations from the viewpoint of comparative regionalism. Then it analyzes how this transformation affects the three traditional pillars of these relations: economic association, development aid, and political dialogue. The latter is identified as the strategic element for updating the foundations of the partnership, seeking to strike a balance between the defence of shared cosmopolitan values, multilateralism, the questioning of the colonial heritage of traditional Western liberal values, and the search for alternative post-hegemonic narratives.

Interregionalism in constant reconstruction

The processes of regional integration in Latin America and Europe have evolved in parallel, albeit heterogeneously, for over half a century. They have developed and shifted in response to changes in the global context and interregional dynamics: of expansion and deepening in the EU and of a higher complexity and heterogeneity in the Latin American region. The Central American Common Market (CACM) (1958), the Andean Pact (1969), and the Latin American Free Trade Association (1960)[1] were contemporaries of the European Economic Community (EEC), whose treaty came into effect in 1958. There was a significant mirror effect which became obvious in the institutional and normative structures, but there were also obvious differences. While Europe pursued an increased supranationality, the Latin American countries fostered the creation of intergovernmental institutions, kept the internal sphere separate from the regional norms, and avoided sharing sovereignty (Ayuso, 2012).

Both regions adopted protectionist policies, albeit in different fields: the EEC protected its agricultural market through the Common Agricultural Policy (CAP) and the Customs Union, the LAC region protected industrial production following the structuralist guidelines of the UN Economic Commission for Latin America and the Caribbean (ECLAC), which advocated policies of import substitution[2] in order to develop national industries and reduce technological dependence. The external projection of these divergences and asymmetries is still an obstacle for the improvement of economic relations – one that keeps surfacing during trade negotiations (Ayuso, 2010a).

The 1980s foreign debt crisis in Latin America and the "lost decade" that followed it para-lysed regional integration and led to tensions in interregional relations with Europe, whose creditor countries aligned themselves with the policies of the International Monetary Fund (IMF). However, the processes of democratic transition in the region stimulated European soli-darity, particularly the Peace Process in Central America which led to the 1984 San José Dia-logue and the meetings of the Rio Group launched in 1987. European support for regional integration grew strong with the first agreements between the European Communities and the Latin American regional integration organizations, beginning with the Andean Pact in 1983 and continuing with the CACM in 1985.

This process intensified at the beginning of the 1990s, when a second wave of regionalism promoted by the ECLAC under the name of "open regionalism" fostered new initiatives such as the creation of the South American Common Market (Mercosur) by Brazil, Argentina, Uru-guay, and Paraguay in 1991. Likewise, the Tegucigalpa Protocol (1991) renewed the Central American Integration System (SICA), the Trujillo Protocol (1996) created the Andean Com-munity (CAN), and the Caribbean Community (CARICOM) modified its founding treaty (1992). This new regionalism aimed at enhancing Latin America's negotiating capacity, improv-ing its participation in the international market, and diversifying its relations with the rest of the world in order to reduce dependence (Mellado, 2010).

The rise of regionalism was aligned with the process of increasing globalization and liber-alization promoted by the negotiation of the Uruguay Round of the General Agreement on Trade and Tariffs (GATT) and the creation of the World Trade Organisation (WTO) in 1994. Meanwhile, the US signed the 1992 North American Free Trade Agreement (NAFTA) with Mexico and Canada and launched the 1994 Action Plan for the Americas; the goal was to nego-tiate a Free Trade Area of the Americas (FTAA) which would encompass the entire region. This liberalising tendency prevailed for a decade; following the example of Mexico, Chile signed a Free Trade Agreement (FTA) with the United States in 2003, the countries of Central America in 2004, and Colombia in 2006.

In response, the EU proposed a strategic association with LAC that they launched in the 1999 Rio de Janeiro Summit. The differential factor of this bi-regional partnership was that it aimed to consolidate a space of political agreement and interregional cooperation that would complement the establishment of a free trade zone (FTZ).[3] In this process, the promotion of Latin American regional integration took place through the bloc negotiation with existing organizations and the elaboration of strategies of interregional cooperation. Thus, a certain rivalry was created between the hemispheric models spearheaded by the US and the EU (Pollio, 2010), and advances in the map of FTAs followed a parallel course. Four regional programmes were created for the partnership schemes: one for LAC and three for CAN, Central America, and Mercosur, which were renewed every five years.

The political turn to the Left that took place in many LAC countries at the beginning of the 21st century shifted the paradigm of economic and social policies and affected Latin American regionalism. A new generation of initiatives appeared, characterized by different viewpoints such as post-liberal regionalism, which was more political and opposed economic liberalization (Sanahuja, 2010; Da Motta Veiga and Rios, 2007), or post-hegemonic regionalism which pur-sued greater autonomy from the United States (Tussie and Riggirotzi, 2012). However, these heterodox initiatives did not follow a common pattern (Van Klaveren, 2012). They include the Bolivarian Alliance for the Peoples of Our America (ALBA, 2004), promoted by Venezuela and Cuba; the Union of South American Nations (UNASUR, 2008), spearheaded by Brazil;

and the Community of Latin American States and the Caribbean (CELAC, 2010) and incorporating Mexico and the Caribbean to the regional dialogue, or the Pacific Alliance (2010) between Chile, Peru, Colombia, and Mexico, countries that were more favourable to economic liberalization. These organizations joined the already existing ones (SICA, Mercosur, CARICOM, and CAN) to create a multilateral regional structure composed by different interrelated layers, engendering ties of cooperation but also a certain rivalry between the different projects (Nolte, 2013).

All these institutions aspire to an economic and/or social integration but do not contemplate the transfer of sovereign competencies. Decisions are made consensually or unanimously at the highest level in which a single member can veto a decision (Ayuso, 2010b). This "summit diplomacy" (Rojas Aravena, 2012) depends greatly on leadership changes and corresponding ideological shifts. Material and ideological differences are expressed in extra-regional alliances and in the tendency of states to sign an FTA with the US or the EU, open up to the Asian-Pacific market, or resist.

The EU has had to adapt its regional strategy to the aforementioned changes. Although it initially took on the role of "external federator" (Sanahuja, 2007), this approach has been changing in order to adjust to the emergence of new forms of leadership, political alliances, and new spaces of regional cooperation. Thus, EU-Latin American interregionalism covers all the categories articulated by Hänggi (Gardini and Ayuso, 2015). Relations between regional groupings are preserved through the more compact blocks (CARICOM, SICA, and Mercosur), following the pattern of *pure regionalism* in harmony with the desire of the EU to foment regional integration. However, there are other categories such as *transregionalism* between regional blocks where states participate individually without holding legal personality in common, as is the case of CELAC; or *hybrid interregionalism* which refers to relations between regional organizations and individual countries and includes the political dialogue of the EU strategic partnership with Brazil and Mexico or the FTAs signed with Colombia, Peru, and Chile. These respond to a selective bilateralism (Gratius, 2008) with key regional actors in order to strengthen strategic alliances.

The multiplicity of interregional relations between LAC and the EU has led to this association being considered one of complex interregionalism (Hardacre and Smith, 2009), understood as a multilevel interregionalist strategy which, in principle, serves as an intermediary formula when the necessary conditions for a pure form of interregionalism are not in place. However, the characteristic feature seems to be that of complementarity; complementarity between different types of interregionalism which are more or less present in diverse sectors. Thus, different types of interregionalism are unevenly distributed between the three pillars of relations, according to the conversational partners involved in each of these areas.

The effects of the changing economic cycle in Latin America which began with the drop in raw material prices (especially oil) have had political consequences that favour economic openness and provide an opportunity to complete the map of interregional agreements. To this we must also add the 2016 agreement with Cuba: while it is not a free trade agreement, it does put an end to the exception of Cuba by incorporating it to the three pillars of interregional relations (Ayuso and Gratius, 2017). This panorama clashes with the protectionist and anti-Cuban stance of the Trump administration and creates a gap between the different strategies towards the Latin American region, following a period of convergence during the Obama administration.

However, other actors have increased their presence in Latin America; China, in particular, has become one of the region's main economic partners and the major external factor of economic growth, despite recent slowdown (see Hogenboom, this volume). China's heightened presence has enabled it to increase its political influence in Latin America and contributed to reducing the traditional role of the EU as a civil power committed to the defence of human rights both within the Union and abroad (Ayuso and Gratius, 2016). In what follows, we shall look into how these changes are expressed in the three major dimensions of the relation.

The future of economic interregional association

As mentioned, the bi-regional strategic partnership involved advancing towards an interregional FTZ. However, the map of agreements was hindered by the fragmentation of different processes of Latin American economic integration and the diverse levels of development in Latin American countries. The initial project involved a gradual approach, first through partnership agreements with the more economically developed countries and regional integration organizations, and then with a progressive opening that would eventually encompass the entire region (Ayuso and Foglia, 2010).

The partnership agreements incorporate all three pillars and were intended to contribute to the convergence of bi-regional economic relations, as well as to adapt to the evolution of integration projects within the region and to the level of development of countries and regional blocs by fostering a process of regulatory convergence. That is why the EU has made advances in negotiations with integration groups such as the SICA, the CAN, or Mercosur conditional on the advance of said groups in the establishment of their own Customs Unions, for which Europe offers technical and financial subregional cooperation.

As a result of the difficulties encountered in the process and of the complexity of the negotiations, the first EU-LAC partnership agreements were bilateral, with Mexico (2000) and Chile (2002) who had already signed an FTA with the US. That is, it was the hybrid, interregionalist scheme that came first, and also the one to first incorporate the element of mutual trade liberalization in strategic regional partnerships. The two countries have kept up the bilateral scheme and are now in the process of renewing these FTAs so as to adapt them to new standards and advance towards a regional regulatory confluence, maintaining certain flexibility according to the characteristics of each country (Ayuso, 2009).

Bilateral free trade agreements were also signed with Colombia and Peru in 2010. On this occasion, the initial bloc negotiations with CAN were interrupted due to internal differences within the bloc and, although cooperation and pure interregionalism remained present in the political dialogue, the latter was greatly debilitated. In 2014 a trade agreement was signed with Ecuador, expanding the map of agreements but also introducing different elements. Thus, the evolution of negotiations with CAN shows that the EU followed a strategy of adaptation in the face of the disintegration of the former after Venezuela left it to join Mercosur. Bolivia has remained in CAN while at the same time negotiating its incorporation to Mercosur, and has not signed a trade agreement with the EU (see Figure 16.1).

The subsequent incorporation of Mexico, Chile, Peru, and Colombia in the Pacific Alliance did not alter the current scheme of agreements. Although the EU has expressed its interest in strengthening ties with this organization, its low institutionality and the lack of effective competencies allow only for transregional relations with this bloc. Pure interregionalism has been kept up, more or less successfully, with the other regional blocs of economic integration.

The first agreement was signed with CARIFORUM (2008), followed by SICA (2010); both adhered to a bloc to bloc format which allowed for the conservation of pure interregionalism. Negotiations on a partnership agreement between the EU and Mercosur remained stagnant for more than 15 years, but the bloc scheme was not dissolved. They were resumed in 2010 and have not yet been completed, despite successive rounds of negotiations. However, political changes in Argentina – with the election of Mauricio Macri – as well as in Brazil – with the

---

Existing Partnership Agreements with FTZ
Mexico (1997) – under renegotiation
Chile (2002) – under renegotiation
Cariforum (2008)
Central America (2010)
Existing Trade Agreements
Peru and Colombia (2010)
Ecuador (2014)
Political Dialogue and Cooperation Agreement
Cuba (2016)
Under Negotiation
MERCOSUR

Figure 16.1   EU–Latin America agreements

removal of President Dilma Rousseff and her replacement by Michel Temer – have opened a window of opportunity for a purely interregionalist agreement, given that the new heads of state are more favourable to liberalization. The negotiation was also facilitated by the suspension of Venezuela by Mercosur due to political disagreement. Bolivia is in the process of entering Mercosur but is not participating in any trade negotiation. However, due to its low level of income, this country is one of the few Latin American states that are still being supported by the Generalised System of Preferences (GSP+).

The difficulties in sealing the EU–Mercosur deal are the result of asymmetries between the two blocs and of the opposition of the more protectionist sectors in both regions. Despite advances in the Mercosur negotiations, Mexico and Chile are also facing complications due to different regional processes that hinder the convergence of regulatory bases in a common bi-regional agenda and threaten with causing commercial distortions between the different groups and countries.

A new paradigm of cooperation for the 2030 agenda

The 1999 Rio Action Plan launched a new phase in the relations of interregional cooperation, providing a framework based on consensual agreement. However, in practice, the EU developed its partnership at three levels: bi-regional cooperation through horizontal programmes aimed at all countries within a scheme of hybrid regionalism and variable geometry; subregional cooperation with the main blocs of the region (SICA, CAN, and Mercosur), following a pure region-to-region scheme; and certain bilateral programmes with each country which fall into the category of transregionalism. To this we must add the partnership with CARICOM, subsumed within the broader bloc of the African, Caribbean, and Pacific Group of States (ACP).

In this scheme, regional integration was promoted mainly through subregional programs, while bilateral programmes were differentiated according to the level of development; the bi-regional scheme served mostly to facilitate technical cooperation and exchange between public and private institutions so as to boost interregional or triangular cooperation with exchange programmes such as AL-Invest (SMEs); ALFA (Universities); @lis (new technologies); Euro-Solar (renewable energies) or EuroSocial (social policies), amongst others.

The action plan of the Madrid 2010 summit highlighted that it was essential to attend to the needs of the citizens, for which six major thematic priorities were established.[4] In the 2013 Santiago de Chile Summit, the first to take place between the EU and the CELAC, the 2013–2015 Action Plan incorporated the issues of gender and investment, as well as of entrepreneurship for sustainable development; the 2015 Brussels Summit added Higher Education and Citizen Security. Dialogue was not the only element contemplated: planned activities and expected outcomes were established for all key areas but were only partially reflected in the 2014–2020 regional Multiannual Indicative Programme for Latin America.[5] This creates a certain disparity between Summit agreements and the strategy of the EU towards the region, leading to a problem of institutional accountability. Furthermore, European countries have bilateral agreements with the LAC region on economic, cultural, social, and scientific collaboration, which have been expanding the framework of existing relations. These intersections appear as a variant of hybrid interregionalism; while not directly connected, they interact in a way that leads to mutual completion but also to a potential rivalry between different players.

The EU's regional planning towards LAC (2014–2020) has abandoned subregional programmes with organizations that do not echo its model of integration. It has limited bloc partnerships to Central America and the Caribbean. The regional partnership with Mercosur and the Andean Community was not renewed. There were also changes in bilateral cooperation: many medium to high-income LAC countries stopped being eligible for bilateral cooperation with the EU. As of 2014, "graduated" countries (Argentina, Brazil, Chile, Colombia, Costa Rica, Ecuador, Mexico, Panama, Peru, Uruguay, and Venezuela) can only participate in regional programmes and in the horizontal lines opened up to all countries (with the exception, for the time being, of Colombia and Ecuador). Although funding of bi-regional cooperation has increased, the disappearance of many of the bilateral programmes means transitioning from a North-South cooperation scheme to more horizontal and transversal relations with an increased presence of South to South and triangular exchange programs. This type of cooperation is more common amongst medium-income countries, such as the ones prevailing in LAC, and has been promoted by many governments in the region (Sanahuja, 2015). Brazil, Chile, Argentina, or Mexico have already assumed the role of donor and seek a bilateral transfer of experiences and knowledge, albeit with limited resources (Ayllón, Ojeda and Surasky, 2014). Along the same line, the new 2017 European Global Strategy speaks of the need to promote "more innovative forms of cooperation" (EEAS, 2016).

The experience of programmes of regional cooperation such as EuroSocial should serve to promote the incorporation of the Sustainable Development Goals (SDGs) into national policies. Insofar as it is a global agenda that must be adapted to each national context, the 2030 Agenda is an opportunity to once again steer an already existing dialogue on social cohesion policies towards the incorporation of citizenship participation with parity of representation; this will foster the sharing of knowledge and know-how and, above all, promote the innovative initiatives that are needed for the transition towards a model of sustainable and socially inclusive growth. The negative impact of the economic crisis – both in Europe and the LAC region – has increased territorial inequalities and endangered the welfare state (Herman, 2014).

The elaboration of public policies that incorporate the SDGs will allow for more transparent objectives and indicators as well as for the establishment of priorities on the basis of the causes of the great inequalities in Latin America and the EU, priorities that will have to be incorporated at the different levels of interregional cooperation. To this end, as important as it may be to define goals and indicators that will measure results, it is also necessary to establish instruments of social participation at the national and local level and to align them with international monitoring.

The EU and LAC have the opportunity to turn their interregional cooperation into a substantial contribution to the agenda of global development.

Political dialogue as a strategic element of the relation

One of the pillars of the interregional relation has been dialogue at the highest level. Although summits are carried out with other regions as well, such as the dialogue with ASEAN or EU-Africa, the particularity of the EU-LAC political dialogue is that it takes place at many different levels and incorporates numerous actors (Ayuso, Mattheis, and Viilup, 2015). The EU-LAC strategic interregional partnership was Europe's answer to the US project for the liberalization of trade in the hemisphere (De Lombaerde, Kochi, and Briceño Ruiz, 2008), but it was also an outlet for the reluctance of certain Latin American countries to accept the hegemonic presence of the US – reinforced after the end of the Cold War – and its meddling in the region.

Bi-regional political dialogue prevailed over subregional dialogues through the different integration mechanisms: with Central America (1984), the Rio Group (1986), the CAN (1996), Mercosur (1995), and the Caribbean Community (2000). In addition, the EU included political dialogue in the bilateral partnership agreements signed with Mexico and Chile. These mechanisms of political dialogue led to the creation of sectorial interregional dialogues, such as the dialogues on drug trafficking, security, migrations, social cohesion, or the environment, which were later on incorporated in the agenda of cooperation of the EU-CELAC summits.

The changes taking place in Latin American integration have conditioned the development of the structure of regional dialogue through the emergence of new organizations such as UNASUR, the Pacific Alliance, and CELAC, and the opening up of EU strategic relations towards emerging powers; to these we must add the ones already established with the traditional powers (the United States, Canada, and Japan) and regional associations (such as Africa, Latin America, ASEAN, or the Mediterranean) (Gratius, 2011). The 2008 EU regional strategy for Latin America[6] steered towards the reinforcement of bilateral relations with specific countries, particularly Brazil: there was an effort to grant the latter a privileged status with the creation of a strategic association during the 2006 Portuguese presidency of the EU Council. In the following EC Communication "EU-Latin America: Global Players in Partnership" in September 2009, the Commission reiterated its will to intensify strategic dialogue with Mexico and Brazil and consolidate bilateral relations.

Political dialogue was complemented with mechanisms of consultation and association which incorporated diverse social, parliamentary, and institutional actors and started to create a multi-level contact network. The existence of this dense social network with flows of all types grants special qualities to the EU-LAC dialogue. At the institutional level, the inter-parliamentary meetings performed a qualitative leap with the creation in 2006 of the Euro-Latin American Parliamentary Assembly (EUROLAT), which meets on a regular basis. The multilevel dialogue is further enriched with the incorporation of Entrepreneurial Meetings, Meetings of the Organized Civil Society, the Academic Summit, the Meeting of Courts of Justice, or the Summit of the Peoples. The creation of the EU-LAC foundation in 2011 also follows along this line: its goal is to strengthen ties between citizens and civil organizations in LAC and Europe and thus boost economic, social, cultural, or scientific relations.

On its part, CELAC has been representing all LAC countries since 2011 and serves as an opposing party to the European institutions, creating an autonomous forum in the region which does not include two countries of the North (United States and Canada) but does comprise Cuba. The existence of a united opposing party from the Latin American side helps structure bi-regional dialogue with a presidency and a troika. The existence of a more symmetric conversational partner also allows for the linking of the CELAC thematic dialogues to those of the EU regional Action Plan.

The opening up of political dialogue towards non-governmental actors is an essential tool of the bi-regional relation, insofar as it has the potential to increment the influence of the multilateral international agenda on essential aspects that affect the administration of global public goods. This dense web of relations makes associations more substantial beyond specific conjunctures, and contributes to the creation of more horizontal relations. The challenge is to manage to transcend these relations and promote a greater democratization of global governance, linking the local to the global and using the interregional space as a platform. However, there have been no obvious results at the political level, considering the modest results of the two last summits in Santiago de Chile (2013) and Brussels (2015) (Ayuso and Gratius, 2015).

The shifting international context offers opportunities but, at the same time, poses significant obstacles. Although the lack of US attention to the LAC region can lead to an increased interest of the latter in creating ties with the EU (Malamud, 2017), the protectionist attitude of the great global power and the destabilizing effects this could have at the international level is a risk for both regions. Furthermore, the increased presence and influence of China challenges the EU to create incentives and make strategic alliance with the region more attractive. At the same time, China appears today as an actor that fosters the multilateral system. The traditional discourse on shared EU and LAC values such as democracy and human rights is still in place, but there is a different approach to their effective implementation and to the relativism displayed by other international players. Likewise, the political commitment contained in the dialogue and in the bi-regional action plans with the Agenda for global development is hindered by disagreements on the definition of shared but differentiated responsibilities.

Thus, both regions face a period of change but also of opportunity: one that requires an honest and transparent dialogue, rendered more difficult by political instability. An example of this is the postponement of the EU-CELAC Summit that should have taken place in October 2017 and has been postponed (indefinitely at the time of writing), greatly due to differences in and between both regions with regards to the political crisis in Venezuela. But it is also the result of the lack of concrete goals, including the incompletion of the agreement with Mercosur and the pending renewal of agreements with Chile and Mexico. Both EU and LAC are undergoing a process of change and face internal difficulties that are steering attention away from bi-regional relations. However, more than ever, strategic partnership can be a key element in the improvement of the respective processes of regional integration as well as of their contribution to global governance.

Notes

1 The goal of the Latin American Free Trade Association was to establish a Free Trade Area between the member states. However, 20 years later, without this goal having been accomplished, the Latin American Integration Association was created (LAIA). The LAIA essentially provides an umbrella for all preferential trade agreements signed between the countries in the region, be they bilateral or multilateral, without establishing a time frame for the progressive liberalization of the trade flow.
2 The structuralist theses on the unequal division of international labour disseminated by the ECLAC under the leadership of Raul Prebisch encouraged the adoption of developmentalist policies in many LAC countries.
3 The 1999 Rio Declaration specified "we envisage providing [. . .] equal attention to the three following strategic dimensions: a fruitful political dialogue respectful of International Law; solid economic and financial relations based on a comprehensive and balanced liberalization of trade and capital flows; and more dynamic and creative co-operation" (point 7).
4 Science, research, innovation, and technology; Sustainable development, environment, climate change, biodiversity, energy; Regional integration and interconnectivity to promote social inclusion and cohesion;

Migration; Education and employment to promote social inclusion and cohesion; The world drug problem.
5 The priorities are: The security-development nexus; Good governance, accountability, and social equity; Inclusive and sustainable growth for human development; Environmental sustainability and climate change; Higher education.
6 "The strategic association between the EU, Latin America and the Caribbean: A joint commitment".

References

Ayllón, B., Ojeda T. and Surasky, J. (2014) *Cooperación Sur-Sur: Regionalismos e Integración en América Latina.* Madrid: Los Libros de la Catarata.

Ayuso, A. (2008) Cooperación europea para la integración de América Latina: Una ecuación de múltiples incógnitas. *Revista española de Desarrollo y Cooperación* número extraordinaria V Cumbre entre la Unión Europea y América Latina y el Caribe:125–146.

———. (2009) *Estudio de viabilidad sobre el acuerdo de Asociación Global Inter-regional para la creación de una zona de Asociación Global.* Committee on Foreign Affairs of the European Parliament (EXPO/B/AFET/2008/61), Brussels, March:1–85.

———. (2010a) Tensiones entre regionalismo y bilateralismo en las negociaciones de los Acuerdos de Asociación Estratégica UE-ALC. *Revista Aportes para la Integración Latinoamericana* 22: 43–84.

———. (2010b) Integración con Equidad. Instrumentos para el tratamiento de las asimetrías en América del Sur. In M. Cienfuegos and J. A. Sanahuja (eds) *Una región en construcción: UNASUR y la Integración de América del Sur.* Barcelona: CIDOB, pp. 137–178.

———. (2012) Institucionalidad jurídica y tratamiento de las asimetrías: viejos y nuevos retos de la integración latinoamericana. In F. en Rojas Aravena (ed) *Vínculos globales en un contexto multilateral complejo.* Buenos Aires: FLACSO Secretaria-General-AECID-CIDOB, pp. 375–422.

——— and Foglia, M. (2010) Tensiones entre regionalismo y bilateralismo en las negociaciones de los acuerdos de asociación estratégica UE-ALC. *Revista Aportes para la Integración Latinoamericana Año* 16(22): 43–84.

——— and Gardini, G. L. (2017) EU-Latin American relations as a template for interregionalism. In F. Mattheis and A. Godsäter (eds) *Interregionalism across the Atlantic Space.* Cham: Springer, pp. 115–130.

——— and Gratius, S. (2015) ¿Que quiere América Latina de Europa? *Política Exterior* 166: 130–137.

———. (2016) América Latina y Europa: ¿Repetir o reinventar un ciclo? In J. A. Sanahuja (ed) *América Latina de la Bonanza a la crisis de la Globalización. Revista Pensamiento Propio* 44: 249–292.

———. (2017) "¿Nadar a contracorriente? El futuro del acuerdo de la Unión Europea con Cuba. In A. Ayuso and S. Gratius (eds) *Nueva etapa entre Cuba y la UE: Escenarios de futuro.* Barcelona: CIDOB, pp. 89–104.

———, Mattheis, F. and Viilup, E. (2015) Regional cooperation, interregionalism and governance in the Atlantic. In J. Bacaria and L. Tarragona (eds) *Atlantic Future Shaping a New Hemisphere for the 21st Century: Africa, Europe and the Americas.* Barcelona: CIDOB, pp. 117–135.

Cooper, R., Hughes, C. and De Lombaerde, P. (2008) (eds) *Regionalisation and Global Governance: The Taming of Globalisation?* London and New York: Routledge.

Da Motta Veiga, P. and Rios, S. (2007) O Regionalismo pós-liberal na América do Sul: origens, iniciativas e dilemas. *Serie Comercio Internacional* 62, Santiago de Chile: CEPAL:1–48.

De Lombaerde, P., Kochi, S. and Briceño Ruiz, J. (2008) (eds) *Del regionalismo latinoamericano a la integración interregional.* Madrid: Siglo XXI y Fundación Carolina, pp. 1–414.

EEAS. (2016) *Shared Vision, Common Action: A Stronger Europe: A Global Strategy for the EU's Foreign and Security Policy.* Brussels: European External Action Service.

Gardini, G. L. and Ayuso, A. (2015) EU-Latin America and Caribbean Inter-Regional Relations: Complexity and Change. In *Atlantic Future Scientific Paper* 24. Barcelona: CIDOB, pp. 1–24.

Gratius, S. (2008) *La Cumbre Europeo-Latinoamericana: ¿Hacia un bilateralismo selectivo?* Madrid: FRIDE, pp. 1–4.

———. (2011) The EU and the "special ten": Deepening or Widening Strategic Partnerships? *PolicyBrief* 76. Madrid: FRIDE: 1–5.

Hardacre, A. and Smith, M. (2009) The EU and the diplomacy of complex interregionalism. *The Hague Journal of Diplomacy* 4: 167–188.

Hermann, C. (2014) Crisis, structural reform and the dismantling of the European Social Model(s). *Economic and Industrial Democracy* 38(1): 51–68.

Malamud, C. (2017) ¿Por qué importa América Latina? *Informe Elcano* 22: 93.

Mellado, N. (2010) La Unión Europea y la integración sudamericana: espacio político birregional. In M. Cienfuegos and J. A. Sanahuja (eds) *Una región en construcción: UNASUR y la Integración de América del Sur.* Barcelona: CIDOB, pp. 351–388.

Nolte, D. (2013) *Latin America's New Regional Architecture: Segmented Regionalism or Cooperative Regional Governance?* Paper presented at the XXXI International Congress of the Latin American Studies Association (LASA), Washington, DC.

Pollio, E. (2010) What kind of interregionalism? The EU-Mercosur relationship within the emerging Transatlantic Triangle. In *Bruges Regional Integration and Global Governance Papers* 3. Bruges: College of Europe and United Nations University, pp. 1–29.

Sanahuja, J. A. (2007) Regionalismo e integración en América Latina: De la factura Atlántico-Pacífico a los retos de una globalización en crisis. *Pensamiento Propio* 44: 29–75.

———. (2010) La construcción de una región: Suramérica y el regionalismo posliberal. In M. Cienfuegos and J. A. Sanahuja (eds) *Una región en construcción. UNASUR y la Integración de América del Sur.* Barcelona: CIDOB, pp. 87–134.

———. (2015) *The EU and the CELAC: Reinvigorating the Strategic Partnership.* EU-LAC Foundation, Hamburg, p. 86.

Telò, M. (2001) Globalisation, new regionalism and the role of European Union. In M. Telò (ed) *European Union and New Regionalism.* Aldershot: Ashgate, pp. 1–20.

Tussie, D. and Riggirotzi, P. (2012) *The Rise of Post-Hegemonic Regionalism: The Case of Latin America.* London: Springer, p. 192.

Van Klaveren, A. (2012) América Latina en un nuevo mundo. *Revista CIDOB d'Afers Internacionals* 100: 131–150.

PART III

Political and cultural struggles and decolonial interventions

17

MORE-THAN-HUMAN POLITICS

Laura A. Ogden and Grant M. Gutierrez

Introduction: more than human politics

Latin American political movements, including the legal recognition of the rights and person-hood of nature in Ecuador and Bolivia, have inspired social and ecological activism around the globe, even as the outcomes of these efforts remain uncertain. Chapters in this volume detail the ongoing political conflicts over the use and commodification of nonhuman life, including contests over water (Boelens) and land (Ojeda), with particular emphasis on conflicts related to extractive industries, plantation forestry projects, and conservation efforts (Fletcher; Perreault; Finley-Brook; Valdivia). Rather than focus on how nonhuman life shapes political struggles over resources, in this chapter we focus attention on the ways plants, animals, and other forms of nonhuman life garner political consideration or contribute to the *reconstitution of the political*.

Expanding the political to include nonhuman entities in the polity, or the arena of political consideration, is predicated on two related propositions that we describe in this chapter (Braun and Whatmore, 2010, for a discussion of nonhuman politics). First, nonhuman entities have political standing because they are actors in the world in ways that Eurocentric political tradi-tions and ontologies have failed to recognize. Second, recognition of ontological difference, or the possibility of other worlds, is inherently a political challenge to Eurocentric and colonial ways of knowing and acting. Latin American Indigenous and social justice movements have been critical of this reconstitution of the political.

Accordingly, we digress from some of the central concerns of the more-than-human politics literature, which tends to be focused on the agentive *nature* or liveliness of nonhuman entities in the making of the world's heterogeneous assemblages of life. First, we bring an ethnographic sensibility to this task. Doing so reminds us that there is not just one "world" inhabited with multiple assemblages of life. Instead, following Mario Blaser (2014), we seek to understand how the existence of multiple worlds, or ontological realities, generates political possibilities. Second, we hold close the calls to decolonize the epistemic origin stories of contemporary scholarship on more-than-human worlds. Or, in Kim TallBear's (2011) words, "Is it too easy a comparison to say that Western thinkers are finally getting on board with something that is closer to an American Indian metaphysic?" Juanita Sundberg (2014), reflecting these concerns, asks us to resist "universal" Anglo Euro-centered thinkers and instead locate ideas within specific epis-temic communities. In response, we are particularly focused on understanding the influence of

Latin American Indigenous philosophy and activism on scholarship often labelled "more-than-human" in the humanities and social sciences.

Socio-natural assemblages and the politics of relationality

In Latin America, extractive capitalism and associated displacements from land and livelihoods have altered many forms of relational, multispecies living. During the colonial era in Latin America, the transformation of landscapes and life into resources for extractive economies was predicated on systems of coercive labour and the dispossession of Indigenous peoples from their lands. Neoliberal structural adjustment led to further privatization of water and forests and the introduction of non-state actors in management and advocacy (Galeano, 1973; Nash, 1979; Eckstein, 1989; Hecht and Cockburn, 1990; Barraclough, 1999; Olivera and Lewis, 2004; Liverman and Vilas, 2006). In the process, Indigenous and other knowledge systems that challenge binaries of nature-culture and traditional-modern became marginalized and excluded from the domain of politics (Conklin and Graham, 1995; Yashar, 1998; Roberts and Thanos, 2003). Arturo Escobar (2008: 4) has described these historically constituted processes of epistemic and material violence as "imperial globality."

Research and writing on more-than-human politics challenges imperial globality's binaries by attending to the agentive power of nonhuman beings, as well as insisting that other knowledge systems matter. Nonhuman beings include biophysical entities, such as animals and plants, as well as the ways other entities, such as rocks and rivers, animate life's processes of emergence and change. Much of the work in Anthropology and Geography has sought to understand how being human is contingent upon relations with non-human beings ("becoming"). Anthropologists often use the term "multispecies ethnography" to describe this research, while geographers tend to describe "more-than-human" relations and politics. This work builds from several related endeavours in philosophy and social theory that seek to reconsider nature and society (such as object-oriented ontologies, hybrid geographies, poststructuralist political ecology), decenter the human in ethics and theory (posthumanism), investigate science and technology, as well as experiment with alternative epistemologies (affect and nonrepresentational theory). Accordingly, these are projects that seek to understand the world as materially real, partially knowable, multicultured, and multinatured.

Anthropological research among Indigenous societies in the Americas has profoundly shaped the ways scholars appreciate the constitution of the political. Most notably, Eduardo Viveiros de Castro's (1998) theory of Amerindian perspectivism has been foundational to understandings of ontologically different worlds and what has been termed the "ontological turn" in Anthropology (Holbraad and Pedersen, 2017). As Viveiros de Castro (1998: 469) describes, Amerindian perspectivism is the "conception, common to many peoples of the continent, according to which the world is inhabited by different sorts of subjects or persons, human and non-human, that apprehend reality from distinct points of view." Instead of one world populated by multiple cultures, Amerindian perspectivism is a cosmology of multiple natures and a universal subject (Viveiros de Castro, 2012: 2). Building upon these insights, Eduardo Kohn (2013) has shown how multinaturalism exists within a broader semiotic system of human and nonhuman life in the Ecuadorian Amazon. For Kohn (2013: 14), engaging ontological difference "matters for politics," because "the tools that grow from attention to the ways the Runa relate to other kinds of beings can help think possibility and its realization differently," a point made earlier by Annemarie Mol (1999) which we discuss later in this chapter.

Understanding the human as constituted in relation to other beings and things, forces a reconsideration of the boundaries and binaries of the "environment" and "society." Ethnographies of Latin American socio-natural assemblages have been foundational to these insights. As an example, Hugh Raffles's (2002) *In Amazonia* upends the "natural history" genre by cataloguing how Indigenous communities, multinational corporations, and scientists are implicated in the making of the Amazon, a landscape of fluid nonhuman and human boundaries. Arturo Escobar's (2008) *Territories of Difference* incorporates diverse lines of evidence, from geologic to political histories, to demonstrate how Afro-Colombian coastal worlds challenge modernist ontologies of nature. As Escobar shows, this is a place of socio-natural mixing, where plants and water mediate the natural, the human, and the supernatural worlds. Ulrich Oslender (2016), building upon Escobar, uses the term "aquatic space" to describe the imbrications of land, water, and political activism in black communities of Colombia's Pacific Coast. In another example, Alex Nading's (2014) *Mosquito Trails* compellingly shows how efforts to control the dengue virus in Nicaragua produce unexpected human/mosquito intimacies that redefine the "social" through transspecies encounters. In these ethnographies, the humans and nonhumans make and remake each other through multispecies encounters, the flow of nutrients and matter, and in relation to the liveliness of animals, plants, bacteria, and other beings.

Yet these assemblages of life are composed and governed by unequal relations. More-than-human scholarship has produced critical appraisals of the symbolic and material absorption of other beings within capitalism and other arenas of socioeconomic power – including through discursive regimes, practices of governance, and contests over resources and the equitable distribution of environmental risk. For example, Dianne Rocheleau (2011) described the networks of relation that compose agroforestry projects in Zambrana, Dominican Republic, as a rooted network of plants and people. But this network is not an apolitical assemblage of relations. Instead, it is "all shot through with power, and linked to territories and larger systems," including gendered webs of power (Rocheleau, 2011: 225). In another example, Sundberg (2011) examines how a range of entities – from cats, to deserts, to thornscrub – are agentive to the politics and practices of border security in the southwestern United States. In her account, the Sonoran Desert is saturated by grief, littered with the debris of geopolitical failure, and a landscape where animals, plants, and climate challenge human exceptionalism (Sundberg, 2013).

Not surprisingly, conservation initiatives are a key arena of political asymmetries, with profound ecological, social, and ethical consequences for the people, animals, and plants implicated in these efforts. Recent scholarship has paid particular attention to the geopolitical scales of conservation politics. For example, José Martínez-Reyes (2016) carefully describes the "moral ecology" of the Mayan forests of the Sian Ka'an Biosphere Reverse, a UNESCO World Heritage Site. Here, as Martínez-Reyes describes, Mayan forest ontologies are in conflict with regional biodiversity conservation initiatives, with clashes intensified by state policy, landgrabbing, and neoliberal "green" development strategies. Similarly, Marcos Mendoza and colleagues (2017) show how binational Patagonian conservation and agroforestry initiatives have become increasingly incorporated into global networks of capitalism since the 1990s, transforming land and life in the process.

Of course, practices of living often resist geopolitical boundaries and conservation territories, as Karl Zimmerer (2000) describes. These boundary problems are well illustrated through efforts to manage both endangered and invasive species. Endangered jaguars (*Panthera onca*) for example, range from the southwestern United States to northern Argentina (Sanderson et al., 2002). Concerns about their protection have led to multiple forms of international conservation approaches, demonstrating the ways in which animal life produces new forms of geopolitics. For instance, efforts to promote a transboundary jaguar corridor between the US and Mexico include alliances among multiple government and non-governmental organizations that oversee a patchwork of private and protected lands. These efforts are challenged by the "increasing militarization and socio-political hostilities" of the region's transborder politics (King and Wilcox, 2008: 227). Unwanted plant and animal life also subvert political borders, sometimes leading to geopolitical tensions. Ogden (2018), for example, shows how Chileans perceive the spread of beavers into Chilean Tierra del Fuego, which Argentina originally introduced in 1946, as one more example of Argentine territorial incursion into Chilean Patagonia.

Social movements and ontological politics

Scholars use the term "ontological politics" to indicate the political possibilities that ontological difference presents for practices of world making. There are two, sometimes overlapping, intellectual traditions that shape this scholarship. Scholars associated with actor network theory and other strands of continental philosophy build from William James's use of the term "pluriverse" to indicate a philosophical rejection of universal claims about reality (see Ferguson, 2007). For example, Mol (2002: viii) defines ontological politics as "a politics that has to do with the way in which problems are framed, bodies are shaped, and lives are pushed and pulled into one shape or another" (see also Mol, 1999), while John Law and John Urry (2004) used the term "ontological politics" in a call for social scientists to reflect on the kinds of worlds their work produces. For Law and Urry (2004: 397), different research practices have the capacity to make multiple worlds through extended, diverse, and contested social and material relations. On the other hand, Mario Blaser (2010, 2014), Marisol de la Cadena (2010, 2015), and Arturo Escobar (2008, 2011, 2017) clearly articulate the influence of Indigenous ontologies and social movements in their ontological politics project. Here, we see the ways non-Eurocentric philosophies are shaping theories of more-than-human politics.

Escobar (2008: 15) reminds that the political ontology framework, through its articulation of alternative life and epistemic projects, constitutes a challenge to "coloniality and of the coloniality of nature." In doing so, ontological politics resist the hegemony of universal political logics that cede political authority to humans alone. This political challenge has become a key platform in Latin American social and ecological justice movements, particularly those inspired by Indigenous philosophy. Yet, claims to ontological difference and the inscription of non-humans in national politics is relatively new. In 2005, for example, Jean Jackson and Kay Warren's (2005: 553) thorough discussion of contemporary Latin American Indigenous movements analyzed the ways communal claims to territory and identity challenged Western, individual "rights based" models of democracy. Since then, Latin American social movements, drawing from Indigenous concepts and ontologies, continue to challenge Western political models, yet they often do so by insisting that nonhuman beings have rights in the collective. In other words, what counts as the communal or the collective has become more visible, even though sometimes this visibility is withdrawn from view, as de la Cadena (2015) describes for political expediency.

The most well-known examples are political claims surrounding *Pachamama*, often translated as "Mother Earth." "Pachamama" or "La Pachamama" are different expressions of an Andean sentient being important to Indigenous communities in Bolivia, Ecuador, and Peru (for a more detailed analysis see De la Cadena, 2015). As Joni Adamson (2012: 144) has argued, Pachamama grounds Indigenous political authority within "a 'cosmic spirituality linked to nature' thousands of years in the making." Throughout Latin America, Pachamama's authority is invoked in the context of anti-development activism as well as in support of conservation initiatives that attend to the livelihoods of Indigenous and rural communities. For example, women *campesinas* (peasants) identify with Pachamama as they work against mining development in Peru and Ecuador (Jenkins, 2015: 453). Pachamama was a central figure at the World People's Conference on Climate Change and the Rights of Mother Earth in Tiquipaya, Bolivia (Weinberg, 2010).

Her political vitality mobilized Ecuador to amend its constitution to protect and regenerate the Earth's "life cycles, structures, functions, and evolutionary processes" (Asamblea, 2008). In Bolivia, the political possibilities of Pachamama helped elect Evo Morales to the presidency, and to reframe environmental politics at multiple scales (Zimmerer, 2015). In many ways, Pachamama has become the goddess figure of environmental, social justice, and anti-capitalist politics in Latin America and beyond, including those associated with ideas of *buen vivir* (Acosta, 2013).

The foundations of the contemporary *Zapatismo* revolutionary movement in Chiapas, Mexico, is another well-known example of a social movement that engages ontological politics. Zapatista philosophy, sometimes referred to as *neozapatismo*, incorporates elements from leftist political struggles with Mayan cosmologies of space, time, and place (Collier, 1994; Harvey, 1998; Mignolo, 2002). The movement's emphasis on the relationality of people and environments includes an ontological shift in what these relations entail (Jung, 2008). For example, in a speech Major Ana Maria evokes forms of territoriality that blurs the boundaries of living and dead, past and present, human and more-than-human:

> Here in the highlands of the Mexican Southeast, our dead ones are alive. Our dead ones who live in the mountains know many things. Their death talked to us and we listened. The mountain talked to us and we listed. The mountain talked to us, the *macehualo*, we then common and ordinary people, we the simple people as we are called by the powerful.
>
> *(quoted in Mignolo, 2002: 253)*

Zapatista philosophy recognizes multiple realities or worlds. In some of these worlds, mountains speak. Yet, well aware of the asymmetrical ways in which different ontological realities govern lives and livelihoods in the context of colonialism and neoliberalism, Zapatistas offer the image of the *pluriverse*. Often translated as "a world where many worlds fit," the pluriverse has been incredibly generative to political activism and philosophy in Latin America and beyond (Escobar, 2017; Sundberg, 2014; Tischler, this volume). As Walter Mignolo (2013) notes, the pluriverse is a world "entangled through and by the colonial matrix of power" yet a way of "thinking and understanding that dwells in the entanglement."

The *political* in ontological politics, in other words, is rooted in the recognition of other worlds, particularly ones where non-humans have political standing. This recognition inherently challenges Eurocentric and colonial ways of knowing and acting in the world. The influence of Indigenous philosophy and Latin American social movements in shaping our theories of more-than-human politics has been profound, though often lost in our social theory origin stories (Sundberg, 2014). We owe an enormous debt to scholars such as Blaser (2010, 2014), de la Cadena (2010, 2015), and Escobar (2008, 2011, 2017) who make that connection explicit.

Conclusion

Today, the political possibilities of Pachamama feel diluted by the corporate co-option of her image on everything from yoga retreats to e-cigarettes. Still, as de la Cadena (2010: 336) notes, the presence of earth-beings in the political sphere has significantly disrupted political formations as we know them, both in Latin America and beyond. As Lidia Cano Pecharroman (2018) has recently examined, specific legal and political claims in New Zealand, Ecuador, India, and Colombia, particularly related to rivers as subjects, are converging to produce a global political discourse that seeks to protect the Rights of Nature. Social movements leading these efforts draw inspiration and political currency from each other. Some of these social movements associated with Rights of Nature efforts are led by Indigenous activists; others combine Indigenous activism and philosophy with anti-capitalist and radical politics; while other social movements draw inspiration from Indigenous cosmologies in the broadest sense, sometimes involving co-option. Understanding the complexities of how global political formations, such as those around the Rights of Nature, engage and transform historically and cultural specific, place-based ontological politics is an important direction for research and theory.

Like Rocheleau, whose quotation begins this chapter, we continue to be inspired by social movements that strive for different ways of being in the world. Contemporary environmental politics in Latin America increasingly draw from Indigenous ontologies and cosmologies in the constitution of the political (Escobar and Alvarez, 1992; Rappaport, 2005; Escobar, 2008). Latin American social movements have generated new forms of political activism in the region and have influenced scholars interested in more-than-human politics. The expanded polity enables new political subjects, particularly a shift in rights-based discourses based on identity and difference, to subjectivities that includes non-humans in the making of these claims. In doing so, this move offers a political critique of Western discourses of nature that has historically collapsed Indigeneity into nature and tradition.

References

Acosta, A. (2013) Build the Good Living- Sumak Kawsay. *Latineadefuego*. https://lalineadefuego.info/2013/01/08/construir-el-buen-vivir-sumak-kawsay-por-alberto-acosta (Accessed 28 November 2017).

Adamson, J. (2012) Indigenous literatures, multinaturalism, and *Avatar*: The emergence of Indigenous cosmopolitics. *American Literary History* 24(1): 143–162.

Asamblea Nacional Constituyente (2008) *El Universo*. 19 July. www.eluniverso.com/2008/07/24/1212/1217/E8C064BD52EF420CAECDB655555BF60C.html (Accessed 3 December 2017).

Barraclough, S. L. (1999) *Land Reform in Developing Countries: The Role of the State and Other Actors*. Geneva, Switzerland: United Nations Research Institute for Social Development.

Blaser, M. (2010) *Storytelling Globalization from the Chaco and Beyond*. Durham, NC: Duke University Press.

———. (2014) Ontology and indigeneity: On the political ontology of heterogeneous assemblages. *Cultural Geographies* 21(1): 49–58.

Braun, B. and Whatmore, S. (2010) The stuff of politics: An introduction. In B. Braun and S. Whatmore (eds) *Political Matter: Technoscience, Democracy, and Public Life*. Minneapolis: University of Minnesota Press, pp. ix–xxxviii.

Collier, G. A. (1994) *Basta! Land and the Zapatista Rebellion in Chiapas*. Oakland, CA: Inst. Food & Dev. Policy.

Conklin, B. and Graham, L. R. (1995) The shifting middle ground: Amazonian Indians and eco-Politics. *American Anthropologist* 97(4): 695–710.

de la Cadena, M. (2010) Indigenous cosmopolitics in the Andes: Conceptual reflections beyond "Politics". *Cultural Anthropology* 25(2): 334–370.

———. (2015) *Earth Beings: Ecologies of Practice Across Andean Worlds*. Durham, NC: Duke University Press.

Eckstein, S. (1989) (ed) *Power and Popular Protest: Latin American Social Movements*. Berkeley, CA and Los Angeles: University of California Press.

Escobar, A. (2008) *Territories of Difference: Place, Movements, Life, Redes*. Durham, NC: Duke University Press.

———. (2011) Sustainability: Design for the pluriverse. *Development* 54(2): 137–140.

———. (2017) *Designs for the Pluriverse: Radical Interdependence, Autonomy, and the Making of Worlds*. Durham, NC: Duke University Press.

——— and Alvárez, S. (1992) (eds) *The Making of Social Movements in Latin America: Identity, Strategy, and Democracy*. Boulder, CO: Westview Press.

Ferguson, K. (2007) *William James: Politics in the Pluriverse*. Lanham, MD: Rowman & Littlefield.

Galeano, E. (1973) *Open Veins of Latin America: Five Centuries of the Pillage of a Continent*. New York: Monthly Review Press.

Gudynas, E. (2011) Good Life: Germinating Alternatives to Development. *America Latina en Moviemento* 462(February). www.alainet.org/active/48054 (Accessed 28 December 2017).

Harvey, N. (1998) *The Chiapas Rebellion: The Struggle for Land and Democracy*. Durham, NC: Duke University Press.

Hecht, S. B. and Cockburn, A. (1990) *The Fate of the Forest: Developers, Destroyers, and Defenders of the Amazon*. Chicago: University of Chicago Press.

Holbraad, M. and Pedersen, M. (2017) Introduction: The ontological turn in Anthropology. In *Holbraad and Pedersen, the Ontological Turn: An Anthropological Exposition*. Cambridge: Cambridge University Press, pp. 1–29.

Jackson, J. E. and Warren, K. B. (2005) Indigenous movements in Latin America, 1992–2004: Controversies, ironies, new directions. *Annual Review of Anthropology* 34(5): 49–73.

Jenkins, K. (2015) Unearthing women's anti-mining activism in the Andes: Pachamama and the Mad Old Women. *Antipode* 47(2): 442–460.

Jung, C. (2008) The moral force of indigenous politics: Critical liberalism and the Zapatistas. *Public Archaeology* 7(4): 265–269.

King, B. and Wilcox, S. (2008) Peace Parks and jaguar trails: Transboundary conservation in a globalizing world. *GeoJournal* 71(4): 221–231.

Kohn, K. (2013) *How Forests Think: Toward an Anthropology Beyond the Human*. Oakland, CA: University of California Press.

Law, J. and Urry, J. (2004) Enacting the social. *Economy and Society* 33(3): 390–410.

Liverman, D. M. and Vilas, S. (2006) Neoliberalism and the environment in Latin America. *Annual Review of Environment and Resources* 31: 327–363.

Martínez-Reyes, J. E. (2016) *Moral Ecology of a Forest: The Nature Industry and Maya Post-Conservation*. Tuscon, AZ: The University of Arizona Press.

Mendoza, M., Fletcher, R., Holmes, G., Ogden, L. and Schaeffer, C. (2017) The Patagonian imaginary: Natural resources and global capitalism at the far end of the world. *Journal of Latin American Geography* 16(2): 93–116.

Mignolo, W. D. (2002) The Zapatistas's theoretical revolution: Its historical, ethical, and political consequences. *Utopian Thinking* 25(3): 245–275.

———. (2013) *On Pluriversality*. http://waltermignolo.com/on-pluriversality/

Mol, A. (1999) Ontologial politics: A word and some questions. *The Sociological Review* 47(S1): 74–89.

———. (2002) *The Body Multiple: Ontology in Medical Practice*. Durham, NC: Duke University Press.

Nading, A. (2014) *Mosquito Trails: Ecology, Health, and the Politics of Entanglement*. Oakland, CA: University of California Press.

Nash, J. (1979) *We Eat the Mines and the Mines Eat Us: Dependency and Exploitation in Bolivian Tin Mines*. New York: Columbia University Press.

Ogden, L. A (2018) The beaver diaspora: A thought experiment. *Environmental Humanities* 10(1): 63–85.

Olivera, O. and Lewis, T. (2004) *Cochabamba! Water War in Bolivia*. Cambridge: South End Press.

Oslender, U. (2016) *The Geographies of Social Movements: Afro-Colombian Mobilization and the Aquatic Space*. Durham, NC: Duke University Press.

Pecharroman, L. C. (2018) Rights of nature: Rivers that can stand in court. *Resources* 7(1): 13.

Raffles, H. (2002) *In Amazonia: A Natural History*. Princeton, NJ: Princeton University Press.

Rappaport, J. (2005) *Intercultural Utopias: Public Intellectuals, Cultural Experimentation, and Ethnic Pluralism in Colombia*. Durham, NC: Duke University Press.

Roberts, J. T. and Thanos, N. D. (2003) *Trouble in Paradise: Globalization and Environmental Crises in Latin America*. London; New York: Routledge.

Rocheleau, D. (2011) Rooted networks, webs of relation, and the power of situated science bringing the models back down to earth in Zambrana. In M. Goldman, P. Nadasdy and M. D. Turner (eds) *Knowing Nature: Conversations at the Intersection of Political Ecology and Science Studies*. Chicago: University of Chicago Press, pp. 209–226.

———. (2016) Crossing boundaries: Points of encounter with people and worlds 'Otherwise'. In W. Harcourt (ed) *The Palgrave Handbook of Gender and Development*. London: Palgrave Macmillan, pp. 276–283.

Sanderson, E. W., Redford, K. H., Chetkiewicz, C. B., Medellin, R. A., Rabinowitz, A. R., Robinson, J. G. and Taber, A. B. (2002) Planning to save a species: The Jaguar as a model. *Conservation Biology* 16(1): 58–72.

Sundberg, J. (2011) Diabolic Caminos in the Desert & Cat Fights on the Rio: A post-humanist political ecology of boundary enforcement in the United States-Mexico borderlands. *Annals of the Association of American Geographers* 101(2): 318–336.

———. (2013) Prayer & promise along the migrant trail in the Sonora Desert. *Geographical Review* 103(2): 230–233.

———. (2014) Decolonizing posthumanist geographies. *Cultural Geographies* 2(1): 33–47.

TallBear, K. (2011) Why interspecies thinking needs Indigenous standpoints. *Cultural Anthropology*, 24 April. https://culanth.org/fieldsights/260-why-interspecies-thinking-needs-indigenous-standpoints (Accessed 28 October 2017).

Viveiros de Castro, E. (1998) Cosmological deixis and Amerindian perspectivism. *Journal of the Royal Anthropological Institute* 4(3): 469–488.

———. (2012) Immanence and fear: Stranger-events and subjects in Amazonia. *HAU: Journal of Ethnographic Theory* 2(1): 27–43.

Weinberg, B. (2010) Bolivia's new water wars: Climate change and Indigenous struggle. *NACLA Report on the Americas* 43(5): 19–24.

Yashar, D. (1998) Contesting citizenship: Indigenous movements and democracy in Latin America. *Comparative Politics* 31(1): 23–42.

Zimmerer, K. S. (2000) The reworking of conservation geographies: Nonequilibrium landscapes and nature-society hybrids. *Annals of the Association of American Geographers* 90(2): 356–369.

———. (2015) Environmental governance through 'Speaking like an Indigenous State' and respatializing resources: Ethical livelihood concepts in Bolivia as versatility or verisimilitude? *Geoforum* 64: 314–324.

18

INTERCULTURAL UNIVERSITIES AND MODES OF LEARNING

Daniel Mato
Translated by Anna Holloway

Introduction

Since the end of the 20th century, as a result of the struggles of indigenous and Afro-descendant peoples in Latin America, different types of universities have been created that are often generically referred to as "intercultural universities." However, experiences in this field are very heterogeneous and a distinction must be made between those established "by" indigenous and/or Afro-descendant organizations and/or intellectuals, others that have been established through alliances between indigenous and Afro-descendant communities with "conventional" universities and/or different types of social organizations, and others founded by state agencies.

The present chapter offers an overview of this diversity of institutional arrangements, while paying special attention to the modes of learning developed within them. These modes of learning, as well as the incorporation of the knowledge and proposals of indigenous and Afro-descendant peoples in the learning process, are often linked to the origins and institutional models of the corresponding universities. It is also observed that these origins and institutional models condition the forms of relations between these universities and the communities and organizations of indigenous and Afro-descendant peoples, as well as the main axes of tension and conflict with the corresponding states.

The historical, social, and normative context

In addition, to understand the current situation of Higher Education amongst the indigenous and Afro-descendant populations of Latin America, one must consider not only their present reality but also significant aspects of their history. It is well-known that the history of America has been strongly marked by conquest and colonization involving massacres, land dispossessions, displacements, and the social and territorial reorganization of the original inhabitants of the continent, as well as by the massive importation of groups of enslaved Africans. To comprehend the significance of the creation of universities "of their own" (propias) by indigenous and Afro-descendant organizations and leaders, one must begin by highlighting that, in the process of the European colonization of America, the worldviews of both indigenous Americans and enslaved Africans came under sustained attack. One such attack was the banning of their religions and their forced conversion to Catholicism. Another was the prohibition of their languages, especially in public spaces and at school, when they were able to access formal education. The same occurred with their ancestral knowledges, particularly in relation to health care that the colonizers linked to the European idea of "witchcraft" – as with other knowledges in a range of areas. The founding of the new Republics in the 19th century did nothing to put an end to this situation. The new states preserved many of these practices and, through their education and cultural policies, developed national homogenizing imaginaries that sought to erase all differences. An important aspect of this socio-political, educational, and epistemological predicament is that even today, well into the 21st century, the bans and the exclusions continue in different forms and at all levels of the education system (Mato, 2008, 2012, 2016, 2018).

Currently, the constitutions of most Latin American countries acknowledge the right of indigenous peoples to their own language, identity, and other cultural practices. This recognition is enshrined in the constitutions of 15 Latin American countries: Argentina, Bolivia, Brazil, Colombia, Costa Rica, Ecuador, El Salvador, Guatemala, Honduras, Mexico, Nicaragua, Panama, Paraguay, Peru, and Venezuela. Furthermore, 14 Latin American countries have ratified the International Labor Organisation (ILO) Convention 169: Argentina, Bolivia, Brazil, Chile, Colombia, Costa Rica, Ecuador, Guatemala, Honduras, Mexico, Nicaragua, Paraguay, Peru, and Venezuela. The regulations of this international instrument along with national constitutions commit ratifying countries to compliance (Mato, 2012, 2018).

In the specific case of the indigenous peoples and considering the reluctance of most states to effectively commit to creating Higher Education provision for/with/by these peoples, it must be highlighted that Convention 169 explicitly establishes certain rights. I note briefly that Article 26 of the Convention establishes that "measures shall be taken to ensure that members of the peoples concerned have the opportunity to acquire *education at all levels* on at least an equal footing with the rest of the national community" (emphasis added). In addition, Article 27 specifies that all education programs and services provided to the indigenous peoples "shall be developed and implemented in cooperation with them to address their special needs, and shall incorporate their histories, their knowledge and technologies, their value systems and their further social, economic and cultural aspirations." It also indicates that "the competent authority shall ensure the training of members of these peoples and their involvement in the formulation and implementation of education programmes, with a view to the progressive transfer of responsibility for the conduct of these programmes to these peoples as appropriate." Finally, it stipulates that "*governments shall recognise the right of these peoples to establish their own educational institutions and facilities, provided that such institutions meet minimum standards established by the competent authority in consultation with these peoples. Appropriate resources shall be provided for this purpose*" (emphasis added).[1]

Furthermore, practically all countries in the region have specific laws in force that protect the rights of indigenous and Afro-descendant peoples, some of which are expressed in specific regulations for education. However, two studies conducted by the UNESCO International Institute for Higher Education in Latin America and the Caribbean (UNESCO-IESALC) concluded that, in most cases, the mentioned regulations of Convention 169 – which are binding for the states that have ratified it – as well as the corresponding constitutional and legal regulations – are rarely observed in practice within the countries concerned. According to this report, this failure results from many different factors that include: the lack of awareness of the rights of indigenous and Afro-descendant peoples among civil servants and government agencies; the prevalence of concealed forms of racism; the vested interests of transnational mining companies and large agribusiness and of the dominant economic and political sectors in the corresponding countries that are focusing on the exploitation of the territories of indigenous and Afro-descendant populations (Mato, 2012, 2018).

Indigenous organizations, states, and higher education

As a result of these historical and contemporary factors, along with international factors and struggles fought by the indigenous and Afro-descendant populations, some states, universities, and other types of Higher Education Institutions (HEIs) and private foundations have established special quota policies and scholarship programs to improve the opportunities of indigenous and Afro-descendant peoples in accessing and completing university studies in "conventional" institutions.[2] Despite these efforts, the real possibilities of indigenous and Afro-descendant individuals accessing and completing studies in "conventional" HEIs are alarmingly unequal, as a result of different factors linked to long histories of discrimination and present-day structural conditions of disadvantage (Mato, 2008, 2012, 2016,2018).

In addition, it must be highlighted that these policies and programmes are oriented towards the inclusion of individual people. So although they are a step in the right direction, they do not change the fact that the curricula and areas of research in conventional HEIs continue to exclude the histories, languages, and knowledges of indigenous and Afro-descendant peoples, as well as their demands and proposals. Furthermore, a perverse effect of these programs is that they frequently stimulate the migration of young professionals to the big cities and, in doing so, separate them from their communities. However, it must be acknowledged that these "programmes for the inclusion of individuals" have favoured the academic training of indigenous and Afro-descendant professionals and technicians who subsequently contributed to the improvement of the situation of their peoples, including the development of universities and other HEIs "of their own."

Few universities and other HEIs in Latin America incorporate the knowledges, languages, proposals, and modes of learning of indigenous or Afro-descendant populations in their curricula and consciously contribute to the promotion of equitable intercultural relations and forms of citizenship that ensure equal opportunities. However, as a result of the struggles of these peoples and of social actors in Latin America or of extra-regional players with similar aims, more than 100 special programmes have been established within "conventional" universities since the 1990s, programmes that aim at bridging this gap in different ways. Furthermore, various types of universities and other HEIs have been created that are often generically referred to as "intercultural" (Mato, 2008, 2009a, 2018).

Universities and other types of intercultural higher education institutions

From a conceptual point of view, intercultural, indigenous, and Afro-descendant universities are characterized by their interest in articulating the knowledges, modes of knowledge production, and ways of learning of different cultural traditions. Beyond this common feature, there are many differences between these institutions, a result of dissimilarities between the various indigenous and Afro-descendant peoples (and often between different communities of the same peoples), the historical relations with their respective national states, and the actors who have intervened in the construction of each one of the institutions and continue to participate in their administration.

Some have been created "by"organizations and/or indigenous and/or Afro-descendant intellectuals, others have been established through alliances between leaders of these peoples and "conventional" universities and/or different types of social organizations, and others by state agencies. There are currently many ongoing projects falling under each one of these categories. Many identify themselves as intercultural, including in their institutional denomination, others categorize themselves according to the orientation of the work they conduct. Some institutions created by indigenous or Afro-descendant leaders or organizations do not call themselves intercultural but rather use the name of their respective peoples, and those participating in them often use the term "of our own" (*propias*) to refer to the institutions and/or to the systems of education applied (Mato, 2008, 2009a, 2016, 2018).

A notable example amongst the Intercultural Higher Education Institutions (IHEIs) created by states is the system of Intercultural Universities (IUs) established in 2003 by Mexico's Secretary for Public Education (SEP) through partnerships with the local governments of each Mexican state in which the institutions were launched. Although, generally speaking, the 11 IUs established in Mexico respond to a specific "model" created by the SEP, there are significant differences between them. The most particular case is surely that of the Universidad Veracruzana Intercultural (Intercultural University of Veracruz), for it was created as part of the Universidad Veracruzana, a public autonomous university (Dietz, 2008; Hernández Loeza, 2018). In Bolivia, too, a presidential decree issued in August 2008 allowed for the creation of the so-called Universidades Indígenas Comunitarias Interculturales de Bolivia (Intercultural Indigenous Communitarian Universities of Bolivia): the Universidad Indígena Aymara (Aymara Indigenous University), the Universidad Quechua (Quechua University), and the Universidad Guaraní y de Pueblos Indígenas de Tierras Bajas (University of the Guaraní People and of the Indigenous Peoples of the Lowlands) (Choque Quispe, 2018). Peru, for its part, launched the Universidad Nacional Intercultural de la Amazonía (National Intercultural University of the Amazonia) and, more recently, three other intercultural national universities (Olivera Rodríguez, 2018).

In Brazil in 2005, the Ministry of Education launched the Programme for the Support of Indigenous Higher Education and Intercultural Teaching Degrees with the goal of financing the creation of study programmes in order to train teachers for indigenous schools. Between 2005 and 2013, five bids for funds were held that only raised money for specific study cycles; that is, it was a non-permanent solution. During this period, undergraduate degrees were established in 19 public universities and yet another two in federal institutions of education, science, and technology. These funds were used to offer courses that were not only delivered in university classrooms but also in the communities, so that students could learn in their own cultural contexts and community members would be encouraged to participate. When this cycle of special enrolments came to an end, seven of the universities involved continued to offer an indigenous endorsement on their degrees (Gomes do Nascimento, 2018).

A common feature of all IHEIs created by states – or of those created within "conventional" universities – is that indigenous organizations only play an advisory role in their management. They do not govern and they do not make decisions. However, in certain cases they have an essential role in the decision-making process. In any case, depending on their scope and location, these institutions do make it possible for some members of these communities to get a university degree. In addition, thanks to the initiative of teachers and indigenous and Afro-descendant leaders, relations are created with the communities that are not usually found in "conventional" universities. It is also common for wise and knowledgeable persons from the communities to offer some classes within the university or receive visits in the communities to share their knowledges. However, this participation is usually not duly and formally acknowledged and often takes place ad-honorem or is quite poorly remunerated, given that the regulations of the universities themselves do not allow it otherwise. All this creates tensions and conflicts that are indicative of the differences between the worldviews and future projects of the communities of indigenous peoples and of the IHEIs and/or the competent government bodies (Mato, 2016, 2008, 2018).

There are also Higher Education initiatives in many Latin American countries that are sustained through the alliance of indigenous and Afro-descendant organizations and/or leaders with other social actors, including "conventional" universities or groups of professors from "conventional" universities. In Puebla, Mexico, the Universidad Campesina Indígena en Red (Peasant and Indigenous Network University, UCIRED) was created in 1998 as a civil association by a group of professionals of the Centro de Estudios para el Desarrollo Rural (Research Centre for Rural Development), the institution that issues the degrees and postgraduate diplomas offered by UCIRED. In the state of Oaxaca, Mexico, the Instituto Superior Intercultural Ayuuk – Universidad Indígena Intercultural Ayuuk (Ayuuk Intercultural Higher Institute – Ayuuk Intercultural Indigenous University) was created in 2006, through a mechanism of collaboration between civil association Centro de Estudios Ayuuk (Ayuuk Research Centre) and the Sistema Universitario Jesuita (Jesuit University System) to which the institute pertains, as well as the Universidad Iberoamérica (Ibero-American University) and the Instituto Tecnológico y de Estudios Superiores de Occidente (Institute of Technological and Higher Education Studies of the West) (Hernández Loeza, 2018). This institution offers three undergraduate degrees and a Masters course which are officially recognized. In Peru, the Asociación Interétnica de Desarrollo de la Selva Peruana (Interethnic Association for the Development of the Peruvian Jungle, ADEISEP) has promoted two initiatives in collaboration with two neighbouring HEIs. Learning activities take place in the HEI classrooms as well as – more commonly – in community spaces. These include the Programa de Formación de Maestros Bilingües de la Amazonia Peruana (Education Program for Bilingual Teachers of the Peruvian Amazon) which has been hosting activities since 1988 in collaboration with the Instituto Superior Pedagógico Loreto (Loreto Higher Pedagogical Institute). Also, the Programa de Formación de Enfermeros Técnicos en Salud Intercultural (Education Program for Nurse Technicians in Intercultural Health) which, between 2005 and 2013, took place in collaboration with the Instituto Superior Tecnológico Público de Atalaya (Higher Public Technological Institute of Atalaya) (Olivera Rodríguez, 2018). In Colombia, the Organización Indígena de Antioquia (Indigenous Organisation of Antioquia) started working in 2005 with a team from the Universidad de Antioquia (Antioquia University) to create the Licenciatura en Pedagogía de la Madre Tierra (Degree in the Pedagogy of Mother Earth) and, in 2011, achieved its recognition by the Ministry of National Education (Mazabel Cuasquer, 2018). The training activities of this study program also take place mainly within the communities it serves.

In addition, there is a growing number of universities created and directly managed by indigenous and/or Afro-descendant organizations or leaders. In the Bolívar Province of Colombia in 2002, Afro-Colombian organization Proceso de Comunidades Negras (Process of Black Communities) established the Instituto de Educación e Investigación Manuel Zapata Olivella (Manuel Zapata Olivella Education and Research Institute), a community education institution that offers, through a convention signed with Universidad de La Guajira (La Guajira University), various academic training alternatives that respond to the demands and proposals of the region's Afro-Colombian communities (Mazabel Cuasquer, 2018). Also in Colombia, the Consejo Regional Indígena del Cauca (Indigenous Regional Council of Cauca) created the Universidad Autónoma Indígena e Intercultural (Autonomous Indigenous and Intercultural University, UAIIN) in 2003. The UAIIN is an integral part of this indigenous organization and, particularly, of its Programme for Intercultural Bilingual Education. The UAIIN seeks to "conserve and strengthen the political space of our peoples, provide orientation in the different forms of knowledge according to our needs and to what has been established in the Planes de Vida (Life Plans) and reinforce and shape our own processes of administration" (Palechor Arévalo, 2018: 170). The UAIIN offers undergraduate degrees on Community Pedagogy, Indigenous Administration and Management,[3] Indigenous Law,[4] Community Development, Indigenous Health,[5] Indigenous Communication,[6] Native Languages, and Revitalisation of Mother Earth.

In 2005 in Ecuador, after nine years of technical work and political mobilization, the Universidad Intercultural de las Nacionalidades y Pueblos Indígenas Amawtay Wasi (Amawtay Wasi Intercultural University of Indigenous Nationalities and Peoples, UINPIAW) was recognized by the National Council of Higher Education and became part of the National System of Higher Education. However, in 2013 the Council for the Evaluation, Accreditation, and Assurance of Quality in Higher Education suspended its activities following a controversial process of evaluation, arguing that the UINPIAW did not have enough professors holding a PhD and that it lacked libraries and other appropriate facilities. However, the UINPIAW was developing learning processes within the communities, making use of the knowledge of elders and of wise men and women. It provided academic training in human rights and indigenous peoples, agroecology, education, and architecture. At present, it continues its activities under the name of Pluridiversidad Amawtay Wasi (Amawtay Wasi Pluridiversity), as a community organization working in the field of research and ancestral knowledges, but it is not able to offer formal academic training or issue diplomas (Hooker, 2018; Mato, 2014; Tuaza, 2018).

However, on 15 May 2018, when President Correa's term came to an end, the Ecuadorian National Assembly passed with an absolute majority the Higher Education Reform Law (Ley Reformatoria a la Ley Orgánica de Educación Superior), in which the Universidad Intercultural de las Nacionalidades y Pueblos Indígenas "Amawtay Wasi" (Intercultural University of the Indigenous Nationalities and Peoples "Amawtay Wasi") was transformed into a " public institution of Higher Education, with a communitarian character, academic, administrative, financial and organic autonomy" (General Provision 5), whose highest authority for a transition period of three years will be the Senior Management Committee composed of representatives of the Secretariat of Higher Education, Science, Technology and Innovation of the Republic and of CONAIE (Confederation of Indigenous Nationalities of Ecuador/Confederación de Nacionalidades Indígenas del Ecuador). The law also stipulates that their activities will be publicly funded (Reform Provision 6).

In Guatemala, two indigenous community universities have been created at the initiative of Mayan academics: the Universidad Maya – Ixil (Maya – Ixil University), created in the Quiché region in 2011, and the Universidad Maya – Kaqchiquel (Maya – Kaqchiquel University), established in 2014 in the Chimaltenango and Sacatepéquez departments. They are both backed by the local indigenous authorities and are legally supported by community agreements that grant them legitimacy. Their curricula are oriented towards the strengthening of the identity of the indigenous and Afro-descendant peoples through the study and use of their languages, ancestral knowledges, and cultural practices. They are not recognized by the Guatemalan state and have signed cooperation agreements with the Universidad Evangélica de Nicaragua "Martin Luther King" in order to certify their curricula and issue diplomas. Their funding comes from the voluntary work of indigenous professionals, community support in matters of facilities and other needs, international cooperation funds, and moderate fees paid by the students. Ixil University offers the degree of Technician in Rural Development. Maya Kaqchiquel University offers a degree and teaching qualification in Teaching the Kaqchiquel Language and History, a degree in Social Research and Interculturality, General Naturopathic Medicine, Textile Art Engineering, Environmental Resources and Agricultural Engineering, Technician – Interpreter and Sworn Translator of the Kaqchiquel Language (Zúñiga, 2018).

In Nicaragua the situation is quite different. The struggles for autonomy on the Caribbean Coast led the National Assembly in 1987 to pass a regional autonomy law that allowed for the creation and recognition, at the initiative of indigenous and Afro-descendant intellectuals and leaders from the two Carribean Coast Autonomous Regions (North and South), of two very particular universities: the Universidad de las Regiones Autónomas de la Costa

Caribe Nicaragüense (University of the Autonomous Regions of the Nicaraguan Caribbean Coast, URACCAN) and the Bluefields Indian and Caribbean University (BICU). Both have a regional, community, and intercultural character and are recognized as such by the National University Council; furthermore, they are both assessed by the National Evaluation and Accreditation Council using participatory assessment criteria and modalities that are in keeping with the idea of interculturality. Created in 1992, URACCAN launched its activities in 1995. It currently has more than 8,000 students and offers degrees in Intercultural Medicine, Intercultural Nursing, Psychology in Multicultural Contexts, Sociology with a specialization in Autonomy, and Bilingual Intercultural Education, amongst others, and eight Masters degrees on subjects such as *buen vivir*/Vivir Bien Cosmovisions,[7] Intercultural Communication, Anthropology, and Intercultural Health. BICU has a similar number of students enrolled and offers 40 degrees such as Medicine, Business Administration, Accounting, Law, Nursing, and different types of Engineering, as well as Masters degrees in Strategic Administration, Environmental Studies and Sustainable Development, Project Formulation and Management, and others. Both BICU and URACCAN are public universities and, as such, are funded by the Nicaraguan state (Hooker, 2018; Zúñiga, 2018).

Finally, we must also mention the case of the Universidad Indígena Intercultural (Intercultural Indigenous University, UII), created in 2005 by the Fund for the Development of Indigenous Peoples of Latin America and the Caribbean (FILAC). The FILAC is a multilateral organization that is jointly managed by a body of representatives of governments and indigenous organizations. The UII aims to train indigenous professionals to be able to carry out – from an intercultural perspective – tasks of articulation, participation, and decision-making so that they are able in intervene in the politics, economy, and social organization of their respective countries. The UII is structured as a network of networks. Its activities are developed in collaboration with four networks, the Red de Universidades Indígenas Interculturales y Comunitarias de Abya Yala (Abya Yala Network of Indigenous, Intercultural, and Community Universities, that includes, amongst others, URACCAN, Amawtay Wasi, and UAIIN), the Red de Centros Académicos Asociados (Network of Associated Academic Centres, a group of "conventional" universities), the Red de Egresados y Egresadas de la UII (Network of UII Graduates), and the Cátedra Indígena Intercultural (Intercultural Academic Body, which unites a group of prominent indigenous leaders and professionals) (Hooker, 2018).

Tensions and conflicts between states and intercultural universities

The marked differences between the worldviews, future-making projects (proyectos de futuro), epistemologies, and modes of learning of indigenous and Afro-descendant peoples and the historical projects of Latin American states are sources of tension and conflict between intercultural universities (indigenous, Afro-descendant, community) and the government agencies that are responsible for education, higher education, and science and technology policies.

While both the states and indigenous and Afro-descendant organizations at times coincide in the use of the expression "Intercultural Education," they perceive it in very different ways. The former usually approach it through their own ethnocentrism and consider it as a resource for the "inclusion" of indigenous peoples "in" national society, without taking into account the particularities of their worldviews, knowledges, values, historical experiences, expectations, and future projects. The latter demand the acknowledgement and appreciation of these particularities, as well as their own right to be treated as equal while being able to hold on to their differences. This usually leads to tensions and conflicts in the process of designing and implementing education programs, as well as during the development of social research and outreach projects. Conflicts usually revolve around the following three axes:

i. The existence of rights that are enshrined in international instruments, national constitutions, and laws, and are not respected or enforced by the states themselves

Research conducted by UNESCO-IESALC led to the conclusion that while advances have been attained through different international instruments and the national constitutions of diverse Latin American countries, existing laws, regulations, and public policies often fail to enforce or implement these advances adequately or provide sufficient resources for their implementation. The reports showed that public servants working in pertinent government agencies are often in the dark with regards to the rights of indigenous and Afro-descendant peoples and individuals and, when they are aware of them, they do not feel compelled to respect them. Formal presentations made by some of these public servants, as well as conversations held with them, have led us to the conclusion that they are reluctant to honour the rights of indigenous and Afro-descendant peoples – established in constitutions, laws, and international conventions – and even consider they have the right to question them. These attitudes are the result of covert feelings of racism that are externalized in expressions of what we could call epistemological racism: an utter contempt for the knowledges of indigenous and Afro-descendant peoples, if not the outright denial of their existence (Mato, 2012, 2014, 2016,2018).

ii. Differences between the worldviews, epistemologies, and modes of learning of the states and of indigenous and Afro-descendant peoples

At present, there are very few communities of indigenous peoples living in isolation from the institutions, technologies, and products of "Western civilisation." On the contrary, there are interrelationships which, in some cases, go back centuries or at least many decades. Therefore, with few exceptions, it does not seem realistic to think of "pure" ancestral cultures or to approach this matter through dichotomous visions (Agrawal, 1995; Mato, 2011). The fieldwork conducted and the abundant literature on this matter rather seem to assert that different types of cultural *mestizajes* or "hybridizations" are very common, to the point that a growing number of indigenous organizations and leaders have created "universities" (which is not an idea originating from their own traditions), even if they call them "indigenous" or refer to them as "their own" ("propias") However, the development of these long-standing social processes has not led to the disappearance of the worldviews, languages, ancestral knowledges, and modes of learning and knowledge production that are characteristic of many indigenous and Afro-descendant peoples. Neither has it eradicated the differences between what is usually called *modern science*(which is nothing more than the name given to a specific mode of knowledge production) and the modes of knowledge production of different indigenous and Afro-descendant peoples (Mato, 2009b, 2011, 2014).

Modes of knowledge production are not isolated elements; they are linked to visions of the world that can "belong" either to the so-called modern societies of the West or to indigenous peoples. While there are many dissimilarities between the worldviews of the different indigenous and Afro-descendant peoples in the Americas, they generally share certain characteristics that distinguish them from Western-modern societies. Even though, after centuries of interaction, "*mestizajes*" are very common in this sphere as well, there are some obvious oppositions and conflicts between the two great currents with regards to worldviews and to the way the knowledge linked to them is produced. The Western-modern worldview is marked – or at least hegemonized – by the "humanity/nature" dichotomy, even if there are critical voices against it within Western societies. On the other hand, the ancestral worldviews of the indigenous and Afro-descendant peoples predominantly consider (even if through different expressions) that all that constitutes our world, including ourselves, belong to a same whole. Unsurprisingly, after centuries of contact, one can currently observe appropriations and articulations between the two visions; the creation of intercultural universities (indigenous, Afro-descendant, or "of their own") is an example of this. In fact, these spaces are usually a field of experimentation and elaboration on this matter and, therefore, also of tension and conflict (Mato, 2009a, 2009b, 2011, 2014).

It is important to bear in mind that the differences are not limited to abstract notions of worldviews and ways of producing knowledge but are, rather, expressed in specific practices. Accordingly, the "Western" vision has developed and put to use the idea of "natural *resources*" that must be "appropriated" or "exploited" in order to attain "*progress*," "*development*," and/or "*wellbeing*," the latter usually understood as the ownership of material goods. And this remains so even though the "West" recently "discovered" that access to "nature" and the enjoyment thereof are also part of a "quality of life" and a "goal" of "*human development*," and also that this development must be "*sustainable*." This opposition is not contemplated in the worldviews of most indigenous and Afro-descendant peoples, where "Mother Earth" is not seen as a provider of "*resources*" and, therefore, there is no discussion of "exploiting" it but, rather, of "respecting" it. This difference also gives rise to quite different categories of reflection and analysis, distinct meaning systems within them, and different ways of evaluating the opportunities or coexistence of diverse forms of human action, as well as diverging opinions on what type of knowledge should be produced, to what end, and by what means. Field observation has shown that such differences and tensions constantly arise in the sphere of intercultural universities and are often a subject of debate (Mato, 2009a, 2009b, 2011, 2014).

Furthermore, situated learning, a learning that occurs in and through practice, plays a fundamental role in the dissemination and reproduction of "traditional" knowledges. It is also crucial for its recreation and for innovative creations inspired by it, be they qualified as endogenous or as the result of the appropriation and transformation of elements from other cultures. In fact, learning in and through practice is also essential for the appropriation of knowledge and skills that originate in so-called Western-modern civilisation. While the latter might have its own merits, it has institutionalized school-based learning, in classrooms, that in many disciplines is divorced from spaces of practice.

Another significant aspect is that the modes of learning of indigenous and Afro-descendant populations are not only situated and practical, but that they are also linked to what is sometimes referred to as "oral tradition." However, special care must be taken when using and understanding this category, for it usually leads to misunderstandings of such processes. For example, training in farming techniques, environmental practices, the therapeutic uses of plants, decision-making and conflict resolution, as well as in other fields of knowledge, "oral tradition," is not limited to practices of speech (as the signifier "oral" suggests) but is conducted through action and in a specific context. It is not only a matter of words and can hardly take place within a classroom. This, of course, is obvious, but it is something that the categories "orality" and "oral tradition" conceal to a great extent. It is frequently forgotten that these are academic categories, that have resulted from the practices of societies based on writing and on their own forms of representing and explaining the practices of Other ways of life that exist in opposition to written cultures. Field observation and consultation with the actors directly involved allow us to conclude that it is not enough to invite wise men and women from the indigenous or Afro-descendant peoples to present their knowledge in university classrooms. The challenge is to leave the classroom, learn in the spaces where those sharing their knowledge live, in context. This is, in fact, what is done in some intercultural universities and, more exceptionally, in certain "conventional" universities as well (Mato, 2009a, 2009b, 2011, 2014).

iii. Rigidity of the evaluation criteria employed by the government agencies in charge of granting recognition and/or accreditation

The evaluation protocols established by government agencies to assess "conventional" universities are not appropriate for intercultural universities (indigenous, Afro-descendant, "propias") as they do not take into consideration their specificities – just as engineering degrees cannot be evaluated with the same criteria as philosophy degrees, or viceversa, and law studies cannot be assessed in the same way as studies on medicine, or viceversa. To this we must add the problem that the "evaluating peers" – who usually come from "conventional" universities – are rarely acquainted with issues such as intercultural differences and relations, intercultural collaboration, and other related subjects (Hooker, 2018; Mato, 2009a, 2009b).

A telling example is that of the limitations imposed by the protocols of evaluation which cannot assess the fact that, although intercultural universities may not have libraries with large collections, they do have the great wisdom of their elders and of other members who live in the communities in which they operate. These resources cannot substitute one another, and the process of evaluation should be able to assess which one is more appropriate for each purpose. These problems are generally linked to dominant ideas on "academic quality" that give excessive importance to publications in academic journals, especially international and "high impact" ones. This last criterion is largely self-referential, for "impact" is not measured by the social usefulness of the knowledge produced but by the number of times an academic article is cited in other academic journals. Neither do these criteria value the importance of experiences of social outreach, and how the universities themselves can benefit from them; experiences such as the creation of spaces for the training of students or the identification of subjects of research, and the development of research activities for teachers, researchers, and the students themselves (Hooker, 2018; Mato, 2011; Mato, 2009a, 2009b).

Conclusions

The intercultural, indigenous, and Afro-descendant universities that have been developing in Latin America since the 1990s constitute a dynamic and varied field that offers beneficial learning opportunities. Research has shown that intercultural, indigenous, Afro-descendant universities or universities "of their own" do not only improve indigenous and Afro-descendant access to Higher Education (as this issue is usually viewed in a reductionist way in terms of "inclusion"), but that they also develop innovative responses to important challenges facing contemporary Higher Education, particularly in matters of quality, pertinence, and relevance.

They also make a valuable contribution in that they respond to challenges of conducting research, providing services to the population, and creating productive initiatives and situated learning – practical and aimed at problem-solving – and they also expand the knowledges, worldviews, and analytical and creative capacities of the professionals and technicians who are trained through their courses.

Intercultural universities, universities "of their own," indigenous, or Afro-descendant universities are characterized by their interest in articulating the knowledges, modes of knowledge production, and modes of learning of different cultural traditions. That is why they are usually spaces of experimentation and elaboration on this issue, as well as sources of tension and conflict that arise from these differences. As noted, they work with different categories of analysis, different meanings systems, and are engaged in alternative forms of knowledge production.

Notes

1 International Labor Office, www.ilo.org/dyn/normlex/en/f?p=NORMLEXPUB:12100:0::NO::P12 100_ILO_CODE:C169#A26
2 By "conventional" I mean universities or other types of HEIs that have not been explicitly created and designed to respond to the needs, demands, and proposals of indigenous and Afro-descendant peoples.
3 Administración y Gestión Propia
4 Derecho Propio
5 Salud Propia
6 Comunicación Propia
7 Tr. Note Rooted in the worldview of the Quechua peoples of the Andes, *sumak kawsay* – or *buen vivir* in Spanish – describes a way of doing things that is community-centric, ecologically balanced, and culturally sensitive.

References

Agrawal, A. (1995) Dismantling the divide between indigenous and scientific knowledge. *Development and Change* 26(3): 413–439.
Choque Quispe, M. E. (2018) Educación Superior y pueblos indígenas y afrobolivianos: Retos y desafíos. In D. Mato (ed) *Educación Superior, Diversidad Cultural e Interculturalidad en América Latina.* Caracas: UNESCO-IESALC; Córdoba: Universidad Nacional de Córdoba.
Dietz, G. (2008) La experiencia de la Universidad Veracruzana Intercultural. In D. Mato (ed) *Diversidad Cultural e Interculturalidad en Educación Superior: Experiencias en América Latina.* Caracas: UNESCO-IESALC, pp. 359–370.
Gomes do Nascimento, R. (2018) Democratização da Educação Superior e a Diversidade Étnico-Racial no Brasil. In D. Mato (ed) *Educación Superior, Diversidad Cultural e Interculturalidad en América Latina.* Caracas: UNESCO-IESALC; Córdoba: Universidad Nacional de Córdoba.
Hernández Loeza, S. (2018) Educación superior, diversidad cultural e interculturalidad en América Latina. Estudio sobre México. In D. Mato (ed) *Educación Superior, Diversidad Cultural e Interculturalidad en América Latina.* Caracas: UNESCO-IESALC; Córdoba: Universidad Nacional de Córdoba.
Hooker, A. (2018) Universidades e Instituciones de Educación Superior Indígenas, Interculturales, Afrodescendientes y Comunitarias en América Latina. In D. Mato (ed) *Educación Superior, Diversidad Cultural e Interculturalidad en América Latina.* Caracas: UNESCO-IESALC; Córdoba: Universidad Nacional de Córdoba.
Mato, D. (2008) (ed) *Diversidad cultural e interculturalidad en Educación Superior: Experiencias en América Latina.* Caracas: Instituto Internacional de la UNESCO para la Educación Superior en América Latina y el Caribe (UNESCO-IESALC).
———. (2009a) (ed) *Instituciones Interculturales de Educación Superior en América Latina: Procesos de Construcción, Logros, Innovaciones y Desafíos.* Caracas: UNESCO-IESALC.
———. (2009b) (ed) *Educación Superior, Colaboración Intercultural y Desarrollo Sostenible/Buen Vivir. Experiencias en América Latina.* Caracas: UNESCO-IESALC.
———. (2011) There is no "universal" knowledge, intercultural collaboration is indispensable. *Social Identities* 17(3): 409–421.
———. (2012) (ed) *Educación Superior y Pueblos Indígenas y Afrodescendientes en América Latina. Normas, Políticas y Prácticas.* Caracas: UNESCO-IESALC.
———. (2014) Universidades Indígenas en América Latina: Experiencias, logros, problemas, conflictos y desafíos. *Revista Inclusión Social y Equidad en la Educación Superior* 14:17–45.
———. (2016) Indigenous people in Latin America: Movements and universities: Achievements, challenges, and intercultural conflicts. *Journal of Intercultural Studies* 37(3): 211–233.
———. (2018) (ed) *Educación Superior, Diversidad Cultural e Interculturalidad en América Latina.* Caracas: UNESCO-IESALC; Córdoba: Universidad Nacional de Córdoba.

Mazabel Cuasquer, M. (2018) Educación Superior, diversidad cultural e interculturalidad en Colombia. In D. Mato (ed) *Educación Superior, Diversidad Cultural e Interculturalidad en América Latina*. Caracas: UNESCO-IESALC; Córdoba: Universidad Nacional de Córdoba.

Olivera Rodríguez, I. (2018) Educación Superior y pueblos indígenas y afrodescendientes en el Perú: Avances y desafíos en el marco actual de las políticas. In D. Mato (ed) *Educación Superior, Diversidad Cultural e Interculturalidad en América Latina*. Caracas: UNESCO-IESALC; Córdoba: Universidad nacional de Córdoba.

Palechor Arévalo, L. (2018) La Universidad Autónoma Indígena Intercultural (UAIIN): Una apuesta a la construcción de interculturalidad. *Educación Superior y Sociedad* 20: 157–181.

Tuaza, L. A. (2018) Educación Superior y pueblos indígenas y afrodescendientes en Ecuador. In D. Mato (ed) *Educación Superior, Diversidad Cultural e Interculturalidad en América Latina*. Caracas: UNESCO-IESALC; Córdoba: Universidad Nacional de Córdoba.

Zúñiga Muñoz, X. (2018) Persistencias coloniales, aperturas y desafíos para la educación superior en Centroamérica. In D. Mato (ed) *Educación Superior, Diversidad Cultural e Interculturalidad en América Latina*. Caracas: UNESCO-IESALC; Córdoba: Universidad Nacional de Córdoba.

19

INDIGENOUS ACTIVISM IN LATIN AMERICA

Piergiorgio Di Giminiani

Introduction

For the last few decades, indigenous activism has been a major force in Latin American politics. By raising claims over collective rights and self-determination, indigenous activists have success-fully brought to public attention the limitations of existing models of nation-building in the region, built around principles of liberal equality and racial cohesion (see Van Cott, 2010). They have also contributed to an unprecedented reconfiguration of citizenship rights and the emer-gence of new forms of indigeneity that openly defy cemented racial hierarchies (see Warren and Jackson, 2003). While the impact of indigenous activism on ethnic relations and statehood are evident, it is hard to pin down the intrinsic features of this phenomenon, as it is characterized by profound geographic differences across Latin American countries. Indigenous activism in fact conflates apparently unrelated sets of practices. As a broad field of political action, indigenous activism can manifest in open confrontations between activists and police forces, as is often the case in mobilizations against mega-development projects and natural resource dispossession in indigenous territories. Activists, however, can also forge alliances with state actors and even participate in state infrastructure projects as a platform for the advancement of development and educational agendas (see Radcliffe and Webb, 2015). Activists can be ritual experts working side by side with community members in cultural revitalization programs funded by national and international organizations, as well as legal experts and intellectuals who make critical interven-tions in debates on constitutional reforms among policy-makers.

Given the vast heterogeneity of indigenous activism in Latin America, I cannot adequately cover all the manifestations of this phenomenon. Rather, in what follows I examine the three main political fields in which the impact of indigenous activism has been felt. In doing so, I highlight three major effects and corresponding movements. The first effect is the reconfigura-tion of citizenship rights, a process which entails a movement towards the state as the primary object of political contention in the quest for indigenous self-determination. The second effect is identity formation, with the self as the main point of reference. The third effect is world-making, which brings together different movements towards the notion of territory as the object of their political intentionality. The literature on indigenous activism has tended to treat these three dimensions in separation, mostly analyzed through the conceptual means of citizen-ship analysis, identity politics, and ontological approaches to politics. In this chapter, I advocate for a tightening of the connections linking these three effects, since projects of world-making triggered by indigenous activism are necessarily sustained by the articulation of identity claims and citizenship rights, and vice versa. Before advancing further in the examination of indigenous activism, it is important to clarify that this chapter examines debates in the literature and events ranging primarily from the 1980s to the present. The origins of indigenous activism in the region are deep, and can be traced all the way back to early uprisings against native enslavement during the Spanish and Portuguese colonial era (see Andrien, 2001). Furthermore, roots of present-day indigenous activism can be found in early 20th-century political organizations characterized by anti-racist demands for major inclusion in the nation, as well as in later peasant movements, among whose ranks were many indigenous leftist activists. Despite its blurred chronological boundaries, contemporary indigenous activism can be said to have surfaced in the 1980s as an intentional break from the platforms of class politics from which indigenous activists had been raising demands of social justice and redistribution for much of the 20th century (see Van Cott, 1994; Warren and Jackson, 2003; Yashar, 2005). In the indigenous social movements of the 1980s, questions of self-determination and autonomous governance gained a degree of visibility in the public arena unknown until then. As this chapter illustrates, the rise of new indigenous social movements coincided with the consolidation of neoliberalism, an ideology that in the field of ethnic politics in Latin America has paradoxically prompted considerable processes of recognition and valorization of indigenous culture, especially those more traditional aspects that did not pose a threat to current economic configurations, while favouring processes of natural resource commodification that entailed expropriation of indigenous territories (see Hale, 2006). The connections linking environmental, territorial, and indigenous struggles in the region are explored in the second half of this chapter, where I review some of the potential effects of indigenous demands in current debates on ecological destruction and the Anthropocene.

Reframing citizenship and the state

In the 1980s, indigenous activists grew increasingly dissatisfied with class-based politics of distribution, which they blamed for relegating the cultural particularities of indigenous groups to the bottom of national agendas (Yashar, 2005: 5). The spectacular indigenous protests that took place in 1992 on the occasion of the quincentenary of Columbus' first incursion into the Americas was the first time that many Latin Americans had been confronted with the extent of its dissemination. The new visibility of indigenous politics contrasted with its historical neglect in public and academic arenas, due largely to "the widespread but erroneous assumptions that indigenous citizens were not politically active, that they did not organize autonomously from the left or clientelist political parties, or that they were politically indistinct from the peasantry and popular classes" (Van Cott, 2010: 386).

The impact of indigenous activism has been widely diverse across Latin America. Differences across countries depend mainly on the particular indigenous demographics of each country, on the existence of transcommunity networks within indigenous groups (Yashar, 2005: 19), and the more general sensibilities of national populations towards indigenous historical issues. Despite such differences, demands have been generally treated by political elites as a threat to modernization processes, national cohesion, and liberal principles of property rights. The agendas of indigenous organizations encompassed historical demands for inclusion and anti-discrimination policies, in particular those affecting individual citizenship rights such as access to health and education, which were first raised by indigenous organizations in the 19th and 20th centuries (see Mallon, 2011). However, they also included more radical claims for political and legal autonomy as collective and territorially based rights (Van Cott, 2010: 388), which were met by far more sturdy criticisms in the public arena for their incompatibility with individual rights and supposed equality of citizenship status. Albeit subject to accusations of inauthenticity, the new wave of indigenous demands in the 1980s and 1990s was successful in pressuring governments to design unprecedented multicultural reforms. Pressure was exercised through mobilizations and indigenous participation in political parties, but also through alliances with transnational agencies and human rights advocacy groups which, by contributing to the legal recognition of indigenous demands as human rights issues, drew unprecedented attention to indigenous politics worldwide (Sieder, 2002: 2). International solidarity has been pivotal to the introduction of legal tools, which indigenous activists relied on for concrete advancements of local and national agendas. The best known of such legal tools is the Indigenous and Tribal Peoples Convention 169 of the International Labour Organisation, more commonly known as ILO 169, an international treaty ratified to this day by most Latin American countries, which establishes consultation mechanisms for the implementation of public and private actions affecting indigenous people (see Langer, 2003: xii).

One of the most evident achievements of indigenous activism in the 1990s is the interpolation of indigenous themes in processes of constitutional reforms. Colombia, Peru, Bolivia, and Ecuador were among the first countries in the region to consider indigenous demands in the revision of their constitutions, with the consequent recognition of special indigenous jurisdictions with the existing legal system (Sieder, 2002: 4). In Bolivia, constitutional reforms implemented under Evo Morales' government were animated by the resounding notion of plurinationalism which, in stark opposition to the traditional binomial of nation-state in Latin America, entails a collective diversification of citizenship rights (Gustafson, 2009: 989). Even in countries more reluctant to embark on extensive constitutional reforms, multicultural reforms were designed in the field of education, health, and development, usually under a banner of *interculturalidad* (*interculturalidade* in Portuguese). This term, which has come to define citizenship reforms throughout the region, is an often taken-for-granted umbrella notion for all state and civil society initiatives based on the premise that recognition of indigenous rights is best achieved through cross-cultural dialogue and acceptance of indigenous values by national populations, rather than through the institution of segregated enclaves (see Solano-Campos, 2013). Albeit possibly serving as a discursive tool for the co-option of indigenous demands as issues of national interest, *interculturalidad* has been a useful framework for indigenous organizations to compel policy makers to incorporate indigenous notions and specific concerns in any state action.

Governmental recognition of the urgency of indigenous citizenship reforms has led to profound rearrangements of indigenous-state relations in general. The empowerment of indigenous activists within the state infrastructure is well known throughout the region, and has been strategic in the advancement of indigenous agendas, even if it has often been achieved at the cost of turning militants into bureaucrats (see Radcliffe and Webb, 2015). However, although a space for collaboration with militants was opened, indigenous collective action has also been met with repressive state actions, leading to the escalation of tensions and sometimes violent confrontations between activists and police forces. In particular, demands directly affecting access, protection, and management of natural resources within and around native territories are often treated as national security risks, given that most governments in the region see their excessive extraction of primary resources as inalienable and necessary for the economic sustenance of the nation. This dichotomy is at the very core of the particular version of multiculturalism favoured by neoliberalism: an economic and political ideology based on ideals of individual accountability and laissez-faire market freedom that was dominant in Latin America in the 1980s and 1990s – the same era that saw the birth of many indigenous reforms. Under neoliberal multiculturalism, citizenship rights that can more easily be accommodated within existing economic configurations, such as rights over cultural revitalization and ethnic entrepreneurship, are encouraged; others – particularly those concerning natural resource governance, which directly threaten existing economic policies – are denied (Hale, 2006; Postero, 2007). We will come back to the limits of framing indigenous rights without proper recognition of environmental and territorial demands later in this chapter. For now, let us move to the second dimension of indigenous activism, identity formation, which relates to the internal effects of struggles over citizenship status.

Identify formation and the self

Today, indigenous activism stands as a critical reminder of the continuity of the historically deep racial hierarchies that justified cultural assimilation of indigenous people in the first place. Racial discrimination in educational and work settings has led large portions of indigenous populations to identify with national cultural values and, in some cases, to deny their indigenous status. This scenario, well known throughout the region, is a direct consequence of dominant discourses of white-European superiority, which have become operational through ideological and material processes of *blanqueamiento*, whitening (Wade, 1995: 21). Manifestations of whitening can be seen in, for instance, the celebration of European heritage in narratives about national culture, and in the once-common selective migration policies favouring white settlers in land redistribution. Whitening fantasies have coexisted with a powerful historical imaginary, that of *mestizaje*, which presupposes Latin American identity as the result of miscegenation between European and indigenous people. While the historical occurrence of extensive inter-ethnic genetic mixing is undeniable, *mestizaje* appears more of an ideology than a historical phenomenon, since it conceals the fact that mixed-race *mestizos* tend to identify more with Western values than indigenous ones which, in turn, are often romanticized as features of the past (see Alonso, 2004). By elevating the *mestizo* as the homogenous subject of Latin American identity, *mestizaje* has been functional to the consolidation of both official and popular nationalism, especially during the early post-independence stages of nation-building in the 19th century, where the corporatist principles governing indigenous enclaves during the Spanish crown were gradually being replaced by liberal models of governance (Radcliffe and Westwood, 1996: 2). For the post-independence political elite of Latin America, indigenous political and landholding structures stood as an obstacle to the application of liberal reforms of European origins, a reflection of the continuity of colonial dependence even after formal rupture with imperial powers (Quijano, 2006: 58).

The impact of indigenous activism on imaginaries of national identity is not limited to reasserting indigenous difference vis-à-vis nationalist narratives of cultural homogeneity. Indigenous activism is also responsible for the reinvigoration of indigenous belonging, which has historically been threatened by assimilatory pressures. Since the 1980s, projects of cultural, linguistic, and religious revitalization have been instrumental in encouraging indigenous populations to reconnect with their indigenous heritage (see Veber, 1998; Warren and Jackson, 2003). Central to this project is the idea of restoration, *recuperación* in Spanish, of practices and notions from the past. As shown by Rappaport in her analysis of indigenous activism among the Nasa in Colombia (2005: 25), projects of *recuperación* spearheaded by public intellectuals and ritual experts prompt an innovative and creative engagement with memory. Critical engagement with memory and colonial history has contributed to a proliferation of multiple forms of indigeneity, in turn countering essentialized and disempowering depictions of indigenous belonging in customary public discourses (De la Cadena and Starn, 2007: 12). At the same time, it has spurred a form of self-representation that has not only generally emphasized cultural cohesion, but also been instrumental in illustrating the consistency of indigenous culture to indigenous people themselves, as well as to members of the national society, despite the historical consequences of assimilation pressures (Jackson and Warren, 2005: 553). Identity formation is a process intimately intertwined with legal changes in citizenship categorization, in particular the emergence of new ethnic and racial categories built around the determination of exclusive rights. Identity, in other words, is inescapably legalizing identities, a point raised by French in framing the nature of identity as both produced by and producer of legal identification (2009). State's role in the formation of ethnic and racial identification is therefore crucial in the legalizing causes and effects of identity, but it is never a deterministic force since performative practices of identity are capable of reshaping the meanings of law itself (ibid. 14).

By helping to legitimize political demands under a banner of cultural rights, representations of indigenous cultural continuity can serve as an effective political tactic in the achievement of concrete political goals, an idea famously captured by Spivak (1985) with the expression "strategic essentialism." Essentialized self-representations by indigenous organizations have attracted much criticism from different sources, including nationalist perspectives inspired by the narrative of *mestizaje* – whereby indigenous societies have been completely incorporated into national ones – and progressive viewpoints, which have overemphasized the hybrid and dynamic character of indigenous societies in reaction to romanticized representations. Despite their differences, these critiques coalesce around an apparent misconception: that of treating representations of particular features of indigenous cosmologies and value systems, which may have been articulated in specific contexts and for specific goals, as totalizing and consensual claims of what indigenous culture is. This misconception has resulted in applying the term "essentialist" as a label to distinguish inauthentic from authentic claims, rather than viewing essentialism as a contested political field in which representations are articulated for oneself and others. For Charles Hale, the fundamental question is not whether indigenous belonging is essentialized or not, but rather how it is used and what political effects this brings out, since essentialism is inherent in all speech and action and therefore cannot be reduced to an exclusively pragmatic and self-conscious strategy (1997: 578). A focus on essentialism as a discursive process rather than a normative category can also help us to recognize how specific forms of radical alterity between indigenous and Western cultures emerge in the midst of scenarios characterized by profound identification and engagement with national culture by indigenous people themselves.

As seen so far, the restoration of indigenous social belonging for indigenous people is not simply a means, but rather a central outcome of indigenous activism with transformative effects on the individual. Accordingly, identity formation appears as an effective heuristic in examining the processes behind the emergence of new forms of self-representation in antagonism with customary colonial hierarchies. However, while it is undeniable that identity formation is a key feature of indigenous activism in the region, there are two potential shortcomings in relying on identity politics as the exclusive explicatory framework for indigenous collective action. The first concerns the risk of underestimating those structural conditions of discrimination that impede an individual's reconfiguration of their identity through discursive affirmations. Inspired by liberal ideas of self-making, such narratives presuppose that politics endows an individual with the means necessary to pass freely from one identity to another through discursive acts of self-definition (Gow, 2003). In the narrative of identity formation, therefore, little attention is paid to those non-discursive elements of daily life upon which identities are moulded in opposition to, or in line with, identity claims articulated in the political arena. The other shortcoming concerns a tendency within narratives of identity politics to presuppose a complete symbolic resignification of the objects of political claims (Di Giminiani, 2018: 10).

It is undeniable that through political action, elements of indigenous livelihood come to be endowed with new meanings; ritual practices, for instance, or landscape features like sacred sites, are often at the heart of political conflicts. Yet, to think of this process as ex novo would be to reduce all objects of indigenous politics to second-order phenomena in relation to politics; in this scenario politics would be the source of performative utterances by attaching meanings to an inert world of things waiting to be signified. Indeed, such a line of argument denies the possibility that an indigenous signification of the world is capable of motivating a political action rather than being its byproduct. Although identity formation remains an extremely useful tool in illustrating how political activism prompts the discursive ordering of multiple expressions of indigenous social belonging, by itself this process is insufficient in explaining those effects and motivations of political activism that do not appear clearly in the realm of public discourses, as in the case of daily engagement with non-humans. With this in mind, I turn to the question of how indigenous activism intervenes in wider processes of world-making, in particular by activating new forms of experiencing and imagining the territory.

World-making and the articulation of the "territory"

Many of the mobilizations that have taken place in the past decades have been triggered by natural resource extraction within indigenous territories. Although natural resource dispossession has existed for as long as colonial rule, the phenomenon has grown steadily during the past five decades following a global expansion of the natural resource market, to which Latin American countries contribute as major exporters. The proliferation of environmental conflicts in areas with indigenous populations is therefore closely associated with the commodification of natural resources, a process sustained by neoliberal governmental agendas, whereby natural resources would be best managed through a major involvement of private companies and governmental technocrats in decision-making (Perreault, 2006). The exclusion of indigenous populations from natural resource governance has led to a scenario in which activists simultaneously have to face both large transnational companies and national governments, and both in the world and in legal courts. One of the most renowned cases of indigenous opposition to global capitalism is the oil conflict in Ecuador which, since the 1990s, has seen the emergence of a large indigenous movement protesting against state and private corporations, such as Texaco, which are responsible for the contamination of extractive areas in the Ecuadorian Amazon (Sawyer, 2004).

Environmental protests led by indigenous people have acquired international visibility thanks to alliances with environmentalist advocacy groups concerned with the destruction of biodiversity spots threatened by industrial and agrarian expansion (see Ulloa, 2013). One of the earliest known cases of eco-indigenous alliance unfolded in the 1980s following news of a spiralling increase of deforestation within the Kayapó territories in the Brazilian Amazon. The international campaign in support of Kayapó demands succeeded in attracting the commitment of eco-conscious personalities such as Sting (Conklin and Graham, 1995). However, while eco-indigenous alliances may have won environmental protection for specific indigenous areas, they are inevitably characterized by conflicting interests between environmentalist and indigenous concerns. In environmental campaigns, indigenous people have often seen their role relegated to symbolic capital, useful for mobilizing politically strategic images of preservationist ideals and ecological guardianship, yet nonetheless reflecting little of their daily experiences with the environment. In conservation initiatives, indigenous hybrid environmental practices have frequently been outlawed, even when they are more essential to local livelihoods than supposedly customary practices (West, Igoe, and Brockington, 2006: 258). Subsuming indigenous environmental demands under a universal protectionist commitment ignores the possibility that such demands aim more at ensuring the continuity of complex relations between local populations and non-humans like mountains and lakes – relations that are inspired by feelings of mutual obligation – than at preserving natural resources understood as public goods or "commons" (Blaser and De la Cadena, 2017: 186).

The inadequacy of framing indigenous environmental demands in the universalizing language of global environmentalism is well known among indigenous activists, who have increasingly turned to other notions when expressing their right to maintain local forms of world-making. Among them, the notion of territory (*territorio* in Spanish and Portuguese) is certainly the most recurrent. While the significance of this notion for the protection of indigenous land is not new, the territory, as indicated by Rivera Cusicanqui, has become a major political articulator of indigenous demands (2016: 65), following a broader eco-territorial turn in Latin American political activism in the last few decades (Svampa, 2012). An emphasis on territory entails a shift from claims of citizenship rights to territorial defence and articulation, with the consequent demand for recognition of property rights over natural resources by indigenous people (Bryan, 2012). Territory is both a political construct, being the result of actions animated by a search for autonomy and self-determination, and a pre-existing entity, consisting of a complex network of place-based social relations to be defended through political actions (Escobar, 2008: 68). Under this light, territorial restoration entails efforts for the delegitimization of existing place connections inherent to indigenous cosmologies as much as performative practices of identity formation and cultural revitalization through which the meanings of the territory are stabilized for both outsides and insiders. Territorial and identity formation are therefore phenomena that cannot be thought of in separation. Performative practices of identity are also deeply imbricated with political actions aimed at the decommodification of natural resources, which in line with the recent neoliberal adjustments of natural resource market, are carried out through market means (see Prieto, 2016). Given the nature of the territory as a geographic entity, both constructed by political action and pre-dating it, indigenous activism can be understood as reactivating affective connections with the territory as much as instituting new ones. In this sense, territorial restoration involves indigenous people who live in and near their homeland as well as individuals who migrated towards urban areas as a consequence of historical land encroachment (see Di Giminiani, 2018: 147).

The emphasis placed on territorial protection and restoration by indigenous activists has drawn attention to the urgency of fostering a type of politics of self-determination that takes the question of world-making, rather than that of cultural practice, at its core. I use the term world-making to refer to a material and discursive construction of the world based on redistribution, consolidation, and erasure of ontological barriers among different entities. The potentials of world-making inherent in indigenous activism can best be grasped if framed within an approach to politics focused on ontological questions; that is, questions about what does and can exist in the world and how differences exist across different categories of being. Such an approach to politics has been defined by Blaser as political ontology, which consists of two intertwined phenomena: "On the one hand, [political ontology] refers to the power-laden negotiations involved in bringing into being the entities that make up a particular world or ontology. On the other hand, it refers to a field of study that focuses on these negotiations but also on the conflicts that ensue as different worlds or ontologies strive to sustain their own existence as they interact and mingle with each other" (2009: 11). The key ontological questions inherent in indigenous activism are two: the first concerns the preservation of those forms of life, such as agential topographic features, which are not accommodated within dominant Western narratives of nature-culture, and therefore are more easily subject to removal and depletion; the second corresponds to the legitimization of the ontological principles characterizing the lived world of indigenous groups in public debates on development and economics, where such principles are often demoted to the rank of cultural beliefs. The delegitimization of the ontological principles of indigenous lived worlds is at the very core of liberal modernity, a political project that, as argued by Escobar (2010: 9), is founded upon a dualist ontology based on the "primacy of humans over non-humans (separation of nature and culture) and of some humans over others (the colonial divide between us and them); the idea of the autonomous individual separated from community; [and] the belief in objective knowledge, reason, and science as the only valid modes of knowing." The delegitimization of the ontological principles of indigenous societies in present-day politics can unfold even in political contexts characterized by celebration of indigenous identity and structural changes in multicultural citizenship, as the case of Bolivia seems to suggest. Indigenous protests against the building of a highway, currently under construction, which cut through indigenous territories in the Bolivian area of Tipnis, drew attention to a certain ambivalence in Evo Morales' government towards indigenous self-governance. On the one hand, Morales' government enforced an *indigenista* post-liberal agenda, which helped the empowerment of the indigenous population, mostly in urban areas, through an unprecedented set of welfare reforms and self-governance policies; on the other, the economic policies of this government have shown an essential continuity with existing extractivist models, which have taken their toll mostly on indigenous people living in resource frontiers (Canessa, 2014; Rivera Cusicanqui, 2016).

By erecting a divide between truth and belief, liberal modernity serves to exclude ontological principles of indigenous lived worlds from the realm of modern, read acceptable, politics. In the liberal tradition, cosmological features of indigenous societies have been interpreted as elements of tradition detached from political affairs, as if it were possible to think of the two fields in separation in the first place (Kelly, 2011: 9). The shortcomings of such a divide become evident when we look at the ways in which political action, in particular concerning environmental and territorial rights, has allowed for a proliferation of elements of the lived world of indigenous societies within the public political arena. De la Cadena has recently illustrated the disruptive role that non-humans such as powerful mountains, known among the Quechua population as earth-beings (*tirakuna*), hold in mobilizing and directing political action in a context of growing natural resource extraction (2015: 102). The agency of earth-beings is entwined in relations of mutual obligation with humans that range from ritualistic practices to political actions in defence of these beings. The recognition of the particular agency of these beings exemplifies how indigenous activism exceeds the boundaries of liberal modernity, whereby political agency is first and foremost a human prerogative and non-humans exist as objects whose relations are rearranged by politics.

As seen so far, the potentials of world-making inherent in indigenous political action are expressed through the articulation of territorial constructs where ontological principles typical of the indigenous lived world can be reactivated and defended. However, as I argued at the beginning of this chapter, the potentials of world-making of indigenous activism cannot fully materialize unless territorial politics is sustained by struggles for the reconfiguration of citizenship rights and the discursive re-articulation of indigenous identity. In the experience of most indigenous activists, collective action is profoundly transformative. Transformation affects indigenous groups' engagement with the state through the articulation of new citizenship rights, their own sense of identity, and their experience with projects of territorial restoration, from which alternative forms of world-making can emerge. As observed during fieldwork in indigenous southern Chile (Di Giminiani, 2018), mobilizations for land restitution by the local Mapuche population have multiple effects. One of these is helping Mapuche claimants to articulate new alliances with influential non-indigenous political actors who can empower their relations with the state infrastructure. Therefore, land mobilizations were occasions of redefining their citizenship in new and critical terms. Second, mobilizations for land restitution, which necessarily entail a reflection on territorial belonging, are also part of a broader project of restoration (*recuperación*) in which cultural identity is reaffirmed through collective practices aimed at reactivating relationships with local oral history, the landscape, and indigenous ethics threatened by historical processes of assimilation. Finally, land mobilizations have compelling effects not only because, when successful, they endow individuals with more land in the logics of natural resources, but also because they prompt the emergence of new territorial configurations in which indigenous practices of place-making can be revitalized. While the success of a specific land claim might rest on the decision taken by bureaucrats and other state actors, Mapuche claimants know that victories, albeit small, can be achieved only if their actions aim simultaneously to position themselves as empowered citizens in relations with state actors, to encourage political commitment among community members through processes of identity formation, and to reconstitute a territory rather simply attain land. Significantly, these elements of indigenous politics are all equally part of what many Mapuche claimants define as their *lucha*, struggle.

Conclusion

The brief conceptual, historical, and geographical review of indigenous activism in Latin America presented in this chapter suggests that this phenomenon is best understood as a broad field of political action and opportunities, whose combined effects are capable of prompting new forms of citizenship, identity formation, and world-making. Even though indigenous demands in Latin America are increasingly subject to criticisms and attacks in line with a general shift towards more conservative and nationalistic politics, it is likely that indigenous activism will continue to be a major force in regional debates about multiculturalism and environmental governance. In particular, over the last two years, indigenous activists have emerged at the forefront of global debates on climate change. Advocates for a radical rethinking of industrial development and capitalist accumulation have looked at indigenous knowledge as a source of inspiration in debunking the ideological premises, such as belief in unrestrained growth, which have justified the industrial expansion and capital accumulation responsible for global warming (Danowski and Viveiros de Castro, 2017: 121; Kohn, 2015: 322). Indigenous activism can be a leading voice in the search for new ethical forms of inhabiting the earth in the Anthropocene epoch, a now official geological denomination referring to our present era and characterized by the transformation of humankind as a geological force on its own capable of causing irreversible transformation of earth systems. However, for the voices of indigenous activists to be truly heard, indigenous political commitment needs not to be diluted yet again in a romantic commitment to wilderness preservation, with its anti-colonial critique filtered away. Indigenous activism remains the most powerful reminder that the Anthropos in the Anthropocene is not humanity in general, but a portion of it, the industrial Western world, whose consumptive and extractive practices, put in practice first in colonial dispossession, have left all humans with a burden for the future.

References

Alonso, A. M. (2004) Conforming disconformity: "Mestizaje," hybridity, and the aesthetics of Mexican nationalism. *Cultural Anthropology* 19(4): 459–490.

Andrien, K. J. (2001) *Andean Worlds: Indigenous History, Culture, and Consciousness under Spanish Rule, 1532–1825.* Albuquerque: University of New Mexico Press.

Blaser, M. (2009) The threat of the Yrmo: The political ontology of a sustainable hunting program. *American Anthropologist* 111(1): 10–20.

———— and De la Cadena, M. (2017) The uncommons: An introduction. *Anthropologica* 59(2): 185–193.

Bryan, J. (2012) Rethinking territory: Social justice and neoliberalism in Latin America's territorial turn. *Geography Compass* 6(4): 215–226.

Canessa, A. (2014) Conflict, claim and contradiction in the new 'indigenous' state of Bolivia. *Critique of Anthropology* 34(2): 153–173.

Conklin, B. A. and Graham, L. R. (1995) The shifting middle ground: Amazonian Indians and eco-politics. *American Anthropologist* 97(4): 695–710.

Danowski, D. and Viveiros de Castro, E. B. (2017) *The Ends of the World*. Cambridge: Polity Press.

De la Cadena, M. (2015) *Earth Beings: Ecologies of Practice across Andean Worlds*. Durham, NC: Duke University Press.

———— and Starn, O. (2007) Introduction. In M. De la Cadena and O. Starn (eds) *Indigenous Experience Today*. Oxford: Berg, pp. 1–32.

Di Giminiani, P. (2018) *Sentient Lands: Indigeneity, Property and Political Imagination in Neoliberal Chile*. Tucson: University of Arizona Press.

Escobar, A. (2008) *Territories of Difference: Place, Movements, Life, Redes*. Durham, NC: Duke University Press.

————. (2010) Latin America at a crossroads: Alternative modernizations, post-liberalism, or post-development? *Cultural Studies* 24(1): 1–65.

French, J. H. (2009) *Legalizing Identities: Becoming Black or Indian in Brazil's Northeast*. Chapel Hill: University of North Carolina Press.

Gow, P. (2003) Ex-cocama: identidades em transformação na Amazônia peruana. *Mana* 9(1): 57–79.

Gustafson, B. (2009) Manipulating cartographies: Plurinationalism, autonomy, and indigenous resurgence in Bolivia. *Anthropological Quarterly* 82(4): 985–1016.

Hale, C. R. (1997) Cultural politics of identity in Latin America. *Annual Review of Anthropology* 26(1): 567–590.

————. (2006) *Más que un Indio: Racial Ambivalence and the Paradox of Neoliberal Multiculturalism in Guatemala*. Santa Fe: School of American Research Press.

Jackson, J. E. and Warren, K. B. (2005) Indigenous movements in Latin America, 1992–2004: Controversies, ironies, new directions. *Annual Review of Anthropology* 34: 549–573.

Kelly, J. (2011) *State Healthcare and Yanomami Transformations: A Symmetrical Ethnography*. Tucson: University of Arizona Press.

Kohn, E. (2015) Anthropology of ontologies. *Annual Review of Anthropology* 44: 311–327.

Langer, E. D. (2003) Introduction. In E. Langer and E. Muñoz (eds) *Contemporary Indigenous Movements in Latin America*. Wilmington: Rowman & Littlefield, pp. xi–xix.

Mallon, F. (2011) Indigenous peoples and the nation-states in Spanish America, 1780–2000. In J. C. Moya (ed) *The Oxford Handbook of Latin American History*. Oxford: Oxford University Press, pp. 281–308.

Perreault, T. (2006) From the Guerra Del Agua to the Guerra Del Gas: Resource governance, neoliberalism and popular protest in Bolivia. *Antipode* 38(1): 150–172.

Postero, N. G. (2007) *Now We Are Citizens: Indigenous Politics in Postmulticultural Bolivia*. Stanford, CA: Stanford University Press.

Prieto, M. (2016) Practicing costumbres and the decommodification of nature: The Chilean water markets and the Atacameño people. *Geoforum* 77: 28–39.

Quijano, A. (2006) El "Movimiento indígena" y las cuestiones pendientes en América Latina. *Argumentos* 19(50): 51–77.

Radcliffe, S. A. and Westwood, S. (1996) *Remaking the Nation: Place, Identity and Politics in Latin America*. London: Routledge.

———— and Webb, A. J. (2015) Subaltern bureaucrats and postcolonial rule: Indigenous professional registers of engagement with the Chilean state. *Comparative Studies in Society and History* 57(1): 248–273.

Rappaport, J. (2005) *Intercultural Utopias: Public Intellectuals, Cultural Experimentation, and Ethnic Pluralism in Colombia*. Durham, NC: Duke University Press.

Rivera Cusicanqui, S. (2016) Etnicidad estratégica, nación y (neo) colonialismo en América Latina. *Alternativa: Revista de Estudios Rurales* 3(5): 65–87.

Sawyer, S. (2004) *Crude Chronicles: Indigenous Politics, Multinational Oil, and Neoliberalism in Ecuador*. Durham, NC: Duke University Press.

Sieder, R. (2002) Introduction. In R. Sieder (ed) *Multiculturalism in Latin America: Indigenous Rights, Diversity and Democracy*. Basingstoke: Palgrave Macmillan, pp. 1–23.

Solano-Campos, A. T. (2013) Bringing Latin America's 'Interculturalidad' into the conversation. *Journal of Intercultural Studies* 34(5): 620–630.

Spivak, G. C. (1985) Strategies of vigilance: An interview with Gayatri Chakravorti Spivak. Interview by Angela McRobbie. *Block* 10: 5–9.

Svampa, M. (2012) Consenso de los commodities, giro ecoterritorial y pensamiento crítico en América Latina. *Osal* 13(32): 15–38.

Ulloa, A. (2013) *The Ecological Native: Indigenous Peoples' Movements and Eco-Governmentality in Columbia*. London: Routledge.

Van Cott, D. L. (1994) (ed) *Indigenous Peoples and Democracy in Latin America*. New York: St. Martin's Press.

———. (2010) Indigenous peoples' politics in Latin America. *Annual Review of Political Science* 13: 385–405.

Veber, H. (1998) The salt of the Montaña: Interpreting indigenous activism in the rain forest. *Cultural Anthropology* 13: 382–413.

Wade, P. (1995) *Blackness and Race Mixture: The Dynamics of Racial Identity in Colombia*. Baltimore, MD: Johns Hopkins University Press.

Warren, K. B. and Jackson, J. E. (2003) (eds) *Indigenous Movements, Self-Representation, and the State in Latin America*. Austin: University of Texas Press.

West, P., Igoe, J. and Brockington, D. (2006) Parks and peoples: The social impact of protected areas. *Annual Review of Anthropology* 35: 251–277.

Yashar, D. J. (2005) *Contesting Citizenship in Latin America: The Rise of Indigenous Movements and the Postliberal Challenge*. Cambridge: Cambridge University Press.

20

AFRO-LATINO-AMÉRICA[1]

Black and Afro-descendant rights and struggles

Deborah Bush, Shaun Bush, Kendall Cayasso-Dixon, Julie Cupples,
Charlotte Gleghorn, Kevin Glynn, George Henríquez Cayasso,
Dixie Lee Smith, Cecilia Moreno Rojas, Ramón Perea Lemos,
Raquel Ribeiro, and Zulma Valencia Casildo

Introduction

The European Conquest of what we now refer to as Latin America resulted in the mixing of three distinct continental worlds – the European, the (native) American, and the African – that themselves were composed of many different cultures, languages, spiritualities, and cosmovisions. While all three are central and foundational to historical and contemporary Latin American identities, cultural practices, and development trajectories, it is African heritages that occupy the most difficult and least prominent place. Indeed, dominant ideologies of racial mixing, *mestizaje* or *mestiçagem*, that circulate in much of the region are exclusionary, as they tend to euphemistically understand colonization processes as an encounter between "two worlds," the native American and the European, thus erasing and denying African and Afro-descendant elements. For example, José Vasconcelos' theorization of cultural hybridity through the notion of *la raza cósmica* acknowledges black presence in Mexico but does so according to an indigenist paradigm which obeys a eugenicist logic of Afro erasure (see Manrique, 2016). The reasons for this state of affairs are complex but are rooted in slavery and the forms of racial classification and hierarchy that were created in the 15th and 16th centuries by Spain and Portugal, and reworked by Northern Europeans in the 17th, 18th, and 19th centuries in order to legitimize colonialism and its multiple atrocities. Exoticizing and degrading representations of Africans as lazy, primitive, ugly, hypersexual, and prone to criminality and drug addiction received intellectual justification from all kinds of racist and Eurocentric thought, including social Darwinism and scientific racism. In particular, in the building of nationhood after independence, Latin American elites who were mostly of European descent saw their indigenous and Afro-descendant populations as obstacles to progress and looked instead to Europe for ideas, implementing policies of whitening (*blanqueamiento/branqueamento*) which encouraged European immigration. These policies were based on the racist notion that the whiter the skin colour of the nation's inhabitants, the more civilized, prosperous, and stable it would be.[2] Instead of acknowledging cultural difference and the contributions made by Afro-descendants to national development, they worked to create monolithic and monocultural states in which Africanicity would be denied, people of African descent would be assimilated, the population would become lighter-skinned over time, and Eurocentric institutions and forms of governance would be imposed. For example, Argentina developed a homogenizing national identity which invisibilized the black population and their contributions, despite the very prominent role of Afro-descendants in the country's flagship music and dance genre, tango.

While policies of whitening have been thoroughly discredited and long abandoned, the attitudes associated with them persist, generating a process of internal colonialism and endoracism that exists to this day. The legacy of such prejudices can be seen in ongoing forms of Afro-descendant exclusion and marginalization throughout the continent and the greater social mobility enjoyed by people who are lighter-skinned. While indigenous groups are subject to racism, discrimination, and dispossession of lands, they are also subject to a kind of romanticization. As a result, in some countries we see evidence of what Hooker (2005a) refers to as a dynamic of indigenous inclusion/black exclusion. This results in part from an understanding that people of African descent outside of Africa are always diasporic and therefore (according to, for example, Lawrence and Dua, 2005) are also settlers and do not possess the same rights to land and territory held by indigenous populations (although as we note below, in Colombia and Nicaragua, Afro-descendants have secured collective land titles). This view is of course highly contested and problematic, especially given the very different conditions of "settlement."[3] Some Afro-Latin Americans, particularly those who are urban-based, do, however, consciously adopt a diasporic identity (Martínez Novo and Shlossberg, 2018). At times, indigenous inclusion/black exclusion encourages those with mixed indigenous and African ancestry to claim an indigenous identity rather than an Afro one (see for example Gordon, 1998).

This chapter is written collectively by a group of Afro-descendant intellectuals, activists, and practitioners, and a number of white European and American scholars working in universities and civil society organizations in Colombia, Costa Rica, Honduras, Nicaragua, Panama, and the UK. We work for and with a range of Afro-descendant organizations including ODECO (Organización de Desarrollo Étnico Comunitario) in Honduras, AVOCENIC (Afro's Voices Center of Nicaragua) in Nicaragua, CARABANTÚ (Corporación Afrocolombiana de Desarrollo Social y Cultural) in Colombia, Centro de la Mujer Panameña (Centre of the Panamanian Woman) in Panamá, and Townbook Limón in Costa Rica. We have been collaborating around a number of key themes, but especially on Afro-descendant film and media activism, and the role that media networks and media visibility can play in the securing of political, cultural, and territorial rights, in addition to the opportunities and challenges afforded by the UN Decade for People of African Descent that runs from 2015 until 2024.[4]

This chapter provides an overview of Afro-descendant activism in Latin America with specific attention to the situation in Colombia, Costa Rica, Honduras, and Nicaragua. It outlines the forms of structural racism and invisibilization to which Afro-Latin Americans are subjected, and the ways in which activists are attempting to dismantle these forms of disadvantage and discrimination, with a particular focus on media activism. We note that a chapter of this kind, that must tend to a level of generality, cannot do justice to the degree of heterogeneity among Latin America's populations of African descent, nor to the diversity of political and cultural struggles in place. Suffice it to say that Afro identity is embraced, denied, and negotiated in a range of complex and contradictory ways, and Afro-Latin Americans are positioned differently with respect to their nation-states, suffer from geographically specific forms of disadvantage, and enjoy differing degrees of formal political recognition. So while scholarship must attend to the geographic specificities of black disadvantage, struggle, and progress, given the shared history of colonialism, slavery, and post-Second World War development, there are a number of challenges in common.

Black bodies, black thought, and racism in the Americas

Most Afro-Latin Americans are descendants of people who were enslaved by Europeans during the colonial era, or of black populations from the Caribbean islands or the US South who migrated to Latin America in search of work in agriculture, in the building of railways, or in the construction of the Panama Canal. Despite these and many other contributions to Latin American cultural and economic development, Afro-Latin Americans occupied the bottom position in dominant racial hierarchies, and were frequently subject to exploitation and coerced labour, denied citizenship or human rights, or simply ignored. The 20th century did, however, see a number of influential intellectual interventions especially by Brazilian scholars such as Manuel Raimundo Querino (1918), and Abdias do Nascimiento and Beatriz Nascimiento (1989) who challenged the dominant racial thinking of the time. The thought of Jamaican Marcus Garvey – who emphasized the importance for the fight against poverty of black organization, racial pride, education, and knowledge of black history – was extremely influential, especially in Cuba and Central America. Nevertheless, the struggle for Afro-descendant rights in Latin America and against the forms of racism, prejudice, and discrimination that constitute the principal obstacles to Afro-descendant development, citizenship, and well-being, remains rooted in a condition of invisibility that Afro-descendant intellectuals, artists, activists, and development practitioners have been seeking to overcome.

Evidence for this invisibility can be found in the ways in which white and *mestizo* Latin Americans as well as outsiders often express surprise at the presence of black and Afro-descendant populations within the continent. For example, Nicaraguan author Sergio Ramírez (2007: 9) asserts that Nicaragua's African heritage is something that people do not discuss ("de eso no se habla"). For Ramírez (2007: 9), this dominant attitude gives rise to an historical amnesia and non-admission of the "Africa that we carry within." Similarly, Creole political theorist Juliet Hooker, who has written substantially on the black struggle for recognition in Nicaragua (see Hooker, 2005a, 2005b), notes how people don't imagine that a Nicaraguan person could be black (Vílchez, 2017). In 2011, Afro-Colombian hip hop group Choc Quib Town shocked Colombians when they told Univision's Jorge Ramos that Colombia was a racist country.[5] US-based African-American scholar, Henry Louis Gates Jr, states that he was surprised to learn of the extent of the transatlantic slave trade in Latin America compared with the United States. He found it "mind-boggling" to learn that of the 11.2 million Africans who were transported as slaves to the Americas between 1502 and 1866, only 450,000 of these were taken to the US. The rest were taken to mainland Latin America, especially Brazil, and the Caribbean (Gates Jr, 2011: 2).[6] Gates' surprise reveals the need to decenter US-centric understandings of black history, as has Hooker's (2017) recent work, which focuses on black Nicaraguan influences on US intellectuals such as Frederick Douglass. Afro invisibility is also supported by historical ignorance and the teaching of official histories across the continent that erase black struggles and contributions. In particular, the historical and geopolitical significance of the Haitian slave rebellion at the end of the 18th century, which led to the creation of the first independent black republic in the world, is inadequately taught in schools in Latin America, despite its deep influence on subsequent independence struggles, the fear it instilled in slave owners in Cuba and elsewhere (see Ferrer, 2014), and the psychological impact it had on France and other colonizing countries, which spent the next decades brutally punishing Haiti for its resistance. The history of Haiti is one of several black histories in Latin America that require wider circulation; its relative erasure is indicative of the wider concealment of black resistance and power throughout the region.

The failure to come to terms with Haiti is revealed in subsequent revolutionary traditions in Latin America, that, in spite of discourses of liberation, often continued to be profoundly racist in their orientation. In Cuba, although the Independence Wars were led by black figures, and the revolution in 1959 contributed considerably to the integration of Afro-Cubans in the society (through improvements in economic status and access to health and education), racism is still "alive and well," according to Roberto Zurbano Torres. This Afro-Cuban activist and writer caused a scandal in his home country when he published an article in the *New York Times* noting persistent racial discrimination along with the underrepresentation of Afro-Cubans in the tourist industry and new privately owned businesses created in the wake of economic liberalization. According to Zurbano Torres, racism in Cuba "has been concealed and reinforced in part because it isn't talked about." With the revolution, Cuba "decreed the end of racism" to the extent that to question official narratives of racial progress became "tantamount to a counter-revolutionary act" (Zurbano Torres, 2013).

In Cuba, Brazil, and elsewhere, post-independence ideologies and discourses of racial democracy, which materialized in state-endorsed claims that racism does not exist, turned racism into a taboo subject. As Jennifer Roth-Gordon (2017) writes, Brazilians have learned to live with a set of racial contradictions: although there is ample evidence for the persistence of white superiority/black inferiority, we also find there a widely expressed pride in Brazil's racial hybridity and tolerance, which is channelled through the rhetoric of *cordialidade*. Several decades of black anti-racist activism throughout the continent do mean, however, that there has been an increase in public awareness and discussion of racism, along with the establishment of more formal practices of recognition.

There are around 150 million Afro-Latin Americans across the continent. Afro-Latin Americans refer to themselves as Afro (or Afro-Cubans, Afro-Colombians, Afro-Ecuadorians, and so on), black (*negro/pardo*), brown (*morenos*), Creoles, Garífunas, Raizales, and Palenqueros, and speak English, Spanish, Portuguese, French, Garífuna, and a number of Creole languages, such as Creole English and Palenquero (a Spanish Creole language), as well as indigenous languages. Many Afro groups, such as the Garífuna of Honduras or the Rama Kriol of Nicaragua, possess and acknowledge mixed black and indigenous ancestry. Afro identities are fluid, dynamic, and heterogeneous, and there is no fixed or universal perspective on what constitutes blackness or who can claim a black or Afro identity. Black cultures are, however, based on tangible collective histories, cultures, and modes of belonging, all of which link to connections elsewhere across the African diaspora. For example, Black Creole Nicaraguans trace their historical and familial connections to a multitude of places, including Costa Rica, England, Scotland, Jamaica, the Cayman Islands, Belize, San Andrés, Providencia, and the US South.

Brazil has the second largest black population in the world and there are large black populations elsewhere in South America, such as Colombia, Venezuela, Ecuador, and Peru, in the Latin American Caribbean islands such as Cuba, Dominican Republic, Haiti, and Puerto Rico, and in Mexico and all over Central America. They share cultural, linguistic, and religious practices, and intellectual, political, and familial affinities with Afro-descendants around the world, and in particular those of the Afro-Caribbean diaspora. Many Afro-Costa Ricans consider themselves to be Garveyites and follow the philosophies of Marcus Garvey. Garvey visited Limón in 1929 where he established a branch of the United Negro Improvement Association (UNIA), which continues to this day to be a powerful site for Afro-Costa Rican solidarity and activism. Yoruba-inspired religious traditions such as Candomblé, Umbanda, and Santería are widely practised, especially in Brazil and Cuba, but also attract significant numbers of followers in Argentina, Colombia, the Dominican Republic, Uruguay, and Venezuela. There are also many words of African origin in Spanish and Portuguese (Megenney, 1983). African influences are widespread in Latin American music, fashion, and food cultures as well as in everyday cultural values.

Afro-descendant populations in Latin America face structural racisms that are reproduced through the prejudices of policymakers, legislators, employers, broadcasters, and educators. Racism is also accompanied by specific forms of territorial stigmatization,[7] as many black populations are concentrated in certain regions, such as the Caribbean Coast of Nicaragua, the city of Limón in Costa Rica, the Chocó region of Colombia, and the Costa Chica in Mexico. The stigmatization of such sites of Afro-descendant demographic concentration works partly through mainstream media representations and narratives, and discourages both public and private investments there. As a result, many black communities are denied basic infrastructure and decent social services. Many Afro-descendants are also faced with the everyday violence of capitalist development through exploitative forms of extractivism – processes that work to exacerbate existing disadvantages.

Structural and everyday racisms and infrastructural neglect have serious material outcomes, which means that black populations are much more likely to live in poverty, be unemployed or underemployed, be incarcerated, and be victims of violence, including police brutality. They are also therefore much less likely than white or *mestizo* populations to complete high school or get university degrees, appear on the television, and own businesses. As well, Afro-Latin Americans tend to have higher rates of infant mortality and malnutrition, and lower rates of life expectancy. They must often deal with assumptions of lowly or servile status. Black lawyers are sometimes mistaken for domestic servants, and black hotel customers for waiters or chamber maids. Afro-Latin Americans with dreadlocks are often presumed to be drug addicts. They are often denied taxi rides or entry to bars and restaurants. Given the barriers that racism creates for black social mobility, many people deny their own African ancestry, even when self-reporting to censuses and other surveys. As a result, demographic data that estimate the numbers of Afro-descendants are inaccurate. This problem of invisibilization is exacerbated by the failure of states and NGOs to collect development indicators that pertain specifically to Afro-descendant populations. Consequently, Afro-descendant cosmovisions, worldviews, and development needs remain on the margins of most public policies in Latin America. There is a strong relationship between these forms of racism and the tendency observed among young Afro-descendants to neglect their language, culture, and ethnicity. Migration that results from structural racism also hinders the intergenerational transmission of cultural heritage. Structural racism therefore results in cultural loss.

Anti-racist struggle and activism

Anti-racist struggle and resistance have a 500-year history. In the colonial and postcolonial periods, enslaved Africans, maroons, and free blacks developed creative modes of resistance that have taken many forms. Slave rebellions were common and, in many places, slaves escaped from their captors and established their own maroon communities, referred to as *palenques* or *quilombos*. Slaves in captivity also developed creative forms of survival, such as *capoeira*, a martial art designed as a dance to trick the slaveowners in Brazil. Black populations played important roles in national independence struggles and fought alongside the likes of Bolívar in Venezuela and Colombia, for example. In the late 20th century, black activism became central to the new social movements that developed in the 1970s, 80s, and 90s. Many Afro-Latin American groups participated in the 2001 World Conference against Racism in Durban, South Africa, and the Latin American preconference in Santiago, Chile, in 2000. Both events helped decisively in the production of significant pan-Afro-Latin American solidarities and collective identities. As Afro-Uruguayan leader Romero Rodríguez said in Santiago, "We went in as blacks and emerged as

Afro-descendants" (cited in Baéz Lazcano, 2018). Since 2000, the term Afro-descendant, which now enjoys international and state legal recognition, sits at the centre of a counterdiscursive repertoire that mobilizes against many of the racially subordinating discourses that perform much of the work of colonization (Báez Lazcano, 2018).

There are now hundreds of Afro-descendant organizations in Latin America that have been fighting on behalf of their communities using a range of strategies, tactics, and modes of organization. Some are very large, well-established, and well-connected to the United Nations, World Bank, USAID, and other international development agencies, as well as to their respective states, while others, such as ARAAC (Regional Articulation of Afro-descendants in Latin America and the Caribbean), attempt to retain their political autonomy from such entities. Their aims are diverse, but most tend to converge around the need to make Afro-descendant populations visible and voices heard (cf., Couldry, 2010), so that discriminatory practices and attitudes can be dismantled, Afro-descendant territorial rights respected, and access to health, education, housing, and employment secured. Most seek to intervene in the formal political sphere to lobby for specific legislative or constitutional reforms that work to end discrimination and obtain recognitions.

One of the largest and most influential Afro-Latin American social movements is the Proceso de Comunidades Negras (PCN), an umbrella organization that unites a large number of grassroots organizations and NGOs. At its first assembly in Colombia in 1993, the PCN demanded rights to identity, territory, and political participation (PCN, 2012). The PCN seeks to secure self-determination and sustainable livelihoods for black communities, develop Afro-focused popular media and communication strategies, protect biodiversity and the ecosystem, and valorize (rather than demonize) blackness (see Asher, 2009). Arturo Escobar (2008) notes that the group engages in "counterwork" in the face of extreme political and ecological challenges and destructive development models. The PCN, in collaboration with the Universidad de los Andes, created the Observatory of Racial Discrimination, which provides intellectual and legal support for anti-racist initiatives and activism. It also fights against free trade agreements, extractivism, and other development policies that undermine Afro-descendant livelihoods, and uses its media resources to denounce the excessive criminalization of Afro-Colombians (see, for example, PCN, 2018). Finally, it seeks to build solidarities and collaborations with Afro-descendant movements elsewhere in the world.

The Central American Black Organization (CABO/ONECA) was founded in Dangriga, Belize, in 1995 and plays a similar role to the PCN. CABO is a transborder networking organization that seeks to strengthen and increase the visibility of the social, political, cultural, environmental, and organizational capacity of Afro-Central Americans that live in both the isthmus and the US. CABO works closely with national organizations such as ODECO in Honduras. ODECO is based in La Ceiba on the Caribbean Coast and intervenes politically to promote the placement of Afro-Honduran rights onto political agendas, improve labour rights, and expand access to health care and education. ODECO also celebrates Afro-Honduran and Garífuna cultures by promoting events such as the month-long celebration of African heritage that is held in May each year in many different countries throughout the region.

While anti-racism is central to all Afro-descendant struggles, many organizations also work with a feminist, queer, class-oriented, or intersectional focus, and seek to defend black women, domestic servants, and black LGTBIQ populations, or to promote empowering forms of Afro-aesthetics. Intersectional analyses are particularly crucial in light of the ways that class-, race-, sexual-, and gender-based forms of domination and subordination intersect, intertwine, and complexly co-constitute one another. Afro-descendant women, for instance, must simultaneously negotiate racism, sexism, and often poverty, and are frequently exposed to particularly extreme forms of violence in labour markets, at home, and on the street. Thinking intersectionally means eschewing neatly ranked and ordered hierarchies of oppression. As Betty Lozano Lerma (2010: 2, our translation) writes:

> The identity of black Colombian women is defined by the fact of being black in a discriminatory *mestizo* society; poor in a class society, and women in a patriarchal society in which their ethnically-marked features matter in a fundamental way [. . .]. For black women, neither gender, nor class, nor race/ethnicity is a central category. We are black women who have been historically impoverished, therefore the articulation of these categories without hierarchies is fundamental in order to account for the black female subject.

The intersectional entanglement of identities means that Afro-descendant women face particular challenges in negotiating and resisting domination. María Lugones' (2007) work is useful for understanding how coloniality produces intersecting identities that generate specifically gendered and racialized forms of violence (see Esguerra Muelle, this volume). Lugones produces a feminist critique of Aníbal Quijano's (2000) work on the coloniality of power and reveals how colonialism and coloniality impose a hierarchical and Eurocentric gender order wherein colonizing men and women dominate their colonized counterparts, yet where colonized women are subordinated by colonized men. These colonial hierarchies thus produce the enactment of systemic violence against black women by both white and black men, and perpetuate widespread indifference towards this violence. However, black feminists have mobilized against gender-based violence that is both endemic to many Latin American societies and much more likely to target black women. Indeed, in 2015, 50,000 women marched in Brasília in the Black Women's March to protest against racism, sexism, and genocide.[8] Nevertheless, colonial hierarchies remain institutionalized in many insidious ways and result, for example, in the frequent sexual abuse of black domestic workers by their employers (Ribeiro Corossacz, 2014); in the preference for lighter-skinned customer-facing employees in banks, hotels, and shops; and in the relative absence of diverse images of black women in advertising and television.

One form of pro-black feminist activism involves a creative and empowering embrace of what we might refer to as an Afro-aesthetics that coalesces in particular around the cultural politics of hair and fashion, and in which Afro-descendant women are dismantling racist ideologies of whitening through African-inspired clothing and natural hairstyles (see Caldwell, 2004, 2007; Pinho, 2006, Sánchez Villareal, 2017). Black women in Latin America are part of a global "natural" hair movement that is also found in Africa, Europe, and the US.[9] As Carolette Norwood (2018: 80) writes, "this diasporic movement is about resistance, personal and collective empowerment, self-affirmation, celebrating one's own natural esthetic, and establishing one's own standard of beauty from one's own standpoint."

While gender and sexuality cut across racial and ethnic modes of identification, age is also an important factor. Across the continent, populations of African descent often find themselves engaged in a generational struggle over values and cultural practices. While all cultures are dynamic, contested, and embedded in processes of globalization, there is often a sense – especially among older Afro-Latin Americans – that their culture is under threat from dominant national cultural practices and languages as well as from external influences. For example, in Nicaragua, the growing dominance of the Spanish language on the Caribbean Coast is often seen as a threat to Creole English, and older locals frequently express concern over the fact that many young black Creole Nicaraguans are starting to address one another in Spanish rather than English or Kriol, which has been an important everyday practice for the construction of collective identities and solidarities. Older Afro-descendants also extend similar concerns into the realm of popular musical tastes and styles. In Costa Rica, newer musical styles popular with Afro youth such as *reggaetón* are sometimes understood by elders as destructive of more traditional Afro-Costa Rican musical forms such as *calypso*. As we note below, organizations such as Townbook Limón are attempting, through their work, to bridge the gap between younger and older Afro-Costa Ricans.

Visual activism

Afro-descendants have long recognized that mainstream media in Latin America, as elsewhere, have been mobilized as agents of coloniality. Afro-descendant people are largely absent from entertainment television, news, and current affairs and advertising. When Afro-descendant bodies do appear on TV, it is often within the terms of a highly limited representational repertoire and an extremely narrow set of parameters and stereotypes. News anchors tend to be white or light-skinned, and *telenovelas* very often feature all white casts and storylines punctuated by only occasional appearances of Afro-descendant actors playing domestic servants or bodyguards (see *A negação do Brasil,* 2000; Nogueira Joyce, 2012). Newscasts tend not to cover Afro-led initiatives, Afro struggles for collective rights, or stories of Afro pride and success, but do deliver a steady stream of reports on Afro-Latin American crime, especially drug trafficking. There are, of course, important exceptions. Media in Latin America are no more monolithic than in other parts of the world, and are best understood to constitute a terrain of struggle, like coloniality in general. So, for instance, there has been a small number of recent *telenovelas*, including *La Mamá del Diez* (Colombia, 2017), *Déjala morir, la niña Emilia* (Colombia, 2017), and *Mister Brau* (Brazil, 2015), which feature Afro-descendant casts and explore Afro-descendant lives and storylines. Nevertheless, the politics of racial representation in Latin American media is a persistent problem. Even in 2018, after many decades of black activism in Brazil in pursuit of more inclusive practices, TV Globo came under fire from the country's Labour Prosecution Service after it failed to include *any* black actors in *Segundo Sol*, a *telenovela* set in Bahia, where 80% of the population is black (Cowie, 2018).

Within this context of media coloniality and exclusion, many Afro-Latin Americans have sought to mobilize community radio and television, film, and social media to circulate Afro-centered discourses, historical narratives, and ways of knowing; to interject Afro inflections into national and international dialogues and debates; and to promote the discussion of issues that are significant to Afro-descendant communities. They have sought to harness media to speak back to power, to circulate counternarratives and counterdiscourses that advance decolonial projects, and to allow Afro-descendant populations to see and hear themselves within the media sphere. Moreover, Afro-descendant activists, intellectuals, and mediamakers have increasingly embraced the communicative resources and affordances of the new media environment in their struggles against coloniality and invisibilization.

Townbook Limón is an example of an Afro NGO based in Costa Rica that is making use of the convergent media environment by using radio, Facebook, and YouTube along with live music performance to empower Afro youth to connect with and respect their culture. Like ODECO, whose multimedia strategy is expressed through their slogan, "we seek voices that shut up the silence," Townbook Limón aims through multiple modes of media engagement to project "the best of Limón" to themselves and the wider world. In Nicaragua, AVOCENIC (Afros' Voices Center of Nicaragua), with support from local intercultural university, URAC-CAN, began in 2012 to broadcast the first ever black TV programme in the history of Bilwi/Puerto Cabezas, entitled "Black-Creoles: Building Our Identity and Well-being Together." The show worked with the slogan: "This is who we are: Our history, our identity, our culture and our contributions" and was led by Shaun Bush and Dixie Lee, a black woman and man (both co-authors of this chapter). The programme became a recognized platform and an important space for analysis on the rights, struggles, achievements, and challenges of Afro-descendants, and on their substantial contributions to the development of the country and region.[10]

In Colombia, Carabantú has been addressing structural racism and discrimination through a wide range of initiatives that focus in particular on the importance of *etnoeducación*. Founded in Medellín in 2003, Carabantú has been working alongside other organizations in Colombia, such as the PCN, to coordinate activities such as the "Cátedra Popular Ana Fabricia Córdoba," a monthly seminar held in the city on Afro-descendant topics, and a preparation course for black students who are applying to university programmes that operate affirmative action policies. More recently, as a result of experiences organizing film screenings and talk-back sessions, Carabantú has launched the Festival Internacional de Cine Comunitario Afro Kunta Kinte (FICCA Kunta Kinte). This festival, which at the time of writing is entering its third year (2018), articulates an organic process of audience development and audiovisual recognition among communities in Medellín that have a high black population. Academic events, screenings, and workshops converge in this annual festival, which culminates with the screening of short films produced as a result of the photography and film training Carabantú offers in the *barrios* throughout the year. In these shorts, young filmmaker-participants share their experiences and community history from their situated perspective, for the world. This approach to film production illustrates how Carabantú uses cinema as an instrument for ethno-education, whereby stigmatized and marginalized communities in Medellín – communities that are characterized by their social vulnerability as a result of urban violence and the ongoing Colombian conflict – are involved in a process designed to fortify self-recognition and cultural reproduction as Afro-descendants. The film festival is generating positive results: in 2017, its academic forum was instrumental in the creation and consolidation of a new independent association, the Consejo Audiovisual Afrodescendiente WI DA MONIKONGO, aimed at galvanizing, visibilizing, and maximizing the potential of Afro-Colombian film and media practitioners working in the different regions of Colombia. WI DA MONIKONGO signifies the linguistic and cultural diversity of the group by drawing on *raizal* Creole from San Andrés islands, and Palenquero Creole, and translates roughly as *somos audiovisual*: we are audiovisual.

The multicultural turn

While there is still a lot of work to be done, Afro-descendant activism has resulted in a number of important achievements over the past two decades (see Box 20.1), particularly in terms of Afro-descendants' formal recognition as a people and as political agents. Indeed, Latin America has experienced a multicultural turn that has resulted in important new spaces of collaboration and of Afro-descendant institutionality, as well as a range of political, legal, and constitutional initiatives that have been secured. Afro-descendants are now formally recognized in a number of Latin American constitutions, such as the 1987 Nicaraguan Constitution and the 1991 Colombian Constitution. The land rights of Afro-descendants have also gained legal recognition in, for example, Laws 28 and 445 in Nicaragua, and Law 70 in Colombia. In some countries, Afro-descendants have gained important affirmative action rights.

Box 20.1 Selected key events in Afro-descendant activism and institutionality

- **1987** Nicaraguan Constitution grants recognition to black and indigenous populations and recognizes the multi-ethnic character of the Nicaraguan nation
- **1987** Law 28 "Law of Autonomy" is passed in Nicaragua, granting autonomy and collective legal and territorial rights to black and indigenous populations
- **1988** Brazilian Constitution establishes racism as a crime
- **1991** Colombian Constitution grants political recognition to Afro-Colombians
- **1992** Black Indigenous and Popular March of Resistance march to protest the celebration of the "discovery" of America
- **1992** Creation of the Network of Afro-Latin-American, Afro-Caribbean and Diasporic Women (Red de Mujeres Afrolatinoamericanas, Afrocaribeñas y de la Diáspora)
- **1992** Creation of ODECO (Organización de Desarrollo Étnico Comunitario) in La Ceiba, Honduras
- **1993** Creation of the PCN (Proceso de Comunidades Negras)
- **1993** Law 70 "Law of Black Communities," granting Afro-Colombian collective land rights and political representation, passes in Colombia
- **1995** Creation of ONECA/CABO (Organización Negra Centroamericana – Central American Black Organization) in Dangriga, Belize
- **1995** Creation of ONEGUA (Organización de Negros de Guatemala) in Livingston, Guatemala
- **1996** Creation of REMAP (Network of Afrodescendant Women of Panama – Red de Mujeres Afrodescendientes de Panamá)
- **1997** Creation of ASOMUGAGUA (La Asociación de Mujeres Garífunas Guatemaltecas)
- **1998** Creation of the Strategic Afro-descendant Alliance
- **2000** Latin American preparatory meeting for the Third World Conference against Racism, Racial Discrimination, Xenophobia and Related Intolerances, held in Santiago (Chile)
- **2001** 3rd World Conference against Racism, Racial Discrimination, Xenophobia and Related Intolerances, Durban, South Africa
- **2001** Colombia names 21 May as Día de la Afrocolombianidad (Day to celebrate Afro-Colombian culture)
- **2003** I Encounter of Afro-descendant politicians from the Americas and the Caribbean (I Encuentro de Parlamentarios Afrodescendientes de las Américas y el Caribe) held in Brasilia
- **2005** Creation of the the the Black Parliament of the Americas (Parlamento Negro de las Américas)
- **2008** Ecuadorian Constitution establish Ecuador as a plurinational state and Afro-Ecuadorians as a people
- **2010** Ecuador passes decree to create Plurinational Plan against Racism and Discrimination
- **2010** Creation of AVOCENIC (Afro's Voices Center of Nicaragua)
- **2011** Creation of Regional Articulation of Afro-descendants in Latin America and the Caribbean (ARAAC)
- **2011** Celebration of the First World Summit of Afro-descendants (Primera Cumbre Mundial de Afrodescendientes) in La Ceiba, Honduras
- **2011** UN declares International Year for People of African Descent
- **2012** First National Afro-Ecuadorian Congress held in Guayaquil

- **2012–2014** Brazil introduces affirmative action policies in higher education and the civil service
- **2013** First National Afro-Colombia Congress held in Quibdo, that convened the National Afro-Colombian Authority (ANAFRO)
- **2013** Brazil celebrates the First International Colloquium for Afrodescendant People
- **2015–2024** UN Decade for People of African Descent
- **2017** Afro-Costa Rican leader, Epsy Campbell Barr, is elected vice-president of Costa Rica

Despite these historic moves, there is a consensus among scholars of Afro-descendant rights that the path to Afro-descendant liberation is fraught with obstacles and setbacks, and that conjunctural factors mean that racism, inequality, and discrimination remain pervasive (see Hooker, 2008; Hale, 2002, 2004, 2005; Restrepo, 2018; Cupples and Glynn, 2018). Therefore, we are witnessing what Martínez Novo and Shlossberg (2018: 353) refer to as "a resurgence of inequality and racism after recognition." As Charles Hale's (2002, 2004, 2005) work has amply recognized, the multicultural turn has always been embedded in a set of contradictions, as political recognition has coincided with neoliberal forms of governance. As Hale has noted with respect to both indigenous and Afro-descendant populations, neoliberal multiculturalism engenders a narrow understanding of rights, and works to keep more radical legal and territorial demands on the margins of what is possible. Afro-descendant and indigenous leaders are admitted to the corridors of power as long as they do not challenge the territorial integrity of the nation-state nor unsettle the dominant ideologies of *mestizaje* (see Hale, 2002, 2004, 2005; this volume; Hale and Millamán, 2006). As Agustín Lao Montes (2015) writes:

> The recognition of Afro-descendants as political subjects, with their own demands and complaints, became a double-edged sword; that is to say, as it opened the way to combat racism and to promote black power, it also made way for a relative integration of their political action in the institutions of the State and agencies of international cooperation, involving such pillars of transnational capital as the World Bank and the imperial State such as USAID.

With respect to the resurgence and persistence of racism, we can note a few general trends across the continent. Though a geographically specific focus is warranted in order to explore the complexity of different contexts, this chapter seeks to crystallize some of the shared concerns and tactics which Afro-Latin American individuals and collectives are mobilizing to challenge racializing practices. In this vein, we can affirm that there are obstacles to racial equality in both the resiliently neoliberal countries and the more progressive "Pink Tide" ones (Lao Montes, 2015; Hale, this volume; Martínez Novo and Shlossberg, 2018), which reveals the racism of both the Left and the Right in Latin America. This is not surprising, as the Latin American Left has struggled to shake off its Eurocentric foundations (see Grosfoguel, 2012). These obstacles also relate to the ways in which modern/capitalist/colonial thinking continues to reproduce itself insidiously, despite the persistence and creativity of historical and contemporary resistances to it. The main obstacles include: (1) the failure of governments to implement redistributive economic policies so that cultural recognition is not accompanied by redistributions of wealth; (2) the emergence or persistence of intercultural conflicts (for example, between indigenous and Afro-descendant populations who fight over the meanings and applications of autonomy, or between rural and urban Afro-descendants whose struggles are by necessity differently framed); (3) the ways in which Afro-descendant politics are neutralized by folkloricization and exoticizing celebrations of difference (a reworking of racial democracy in the multicultural age); (4) the failure to effectively implement legislation that has been passed; (5) the dominance of *mestizo* political parties that co-opt Afro-descendant leaders and thus fragment and divide Afro communities; (6) the clashes and conflicts between state-led and grassroots initiatives; (7) the imposition of extractivist projects that lead to violence, dispossession, and displacement, such as mining initiatives in Afro-descendant communities; and (8) the criminalization and murder of Afro-descendant activists who oppose extractivist projects or other forms of harmful development.

In practice this means that post-recognition struggles are just as hard as those fought in the past. In Brazil, racism might have been recoded as a crime, but nobody ever gets convicted (Araujo, 2015). In Nicaragua, an Afro-descendant radio show on black history broadcast in Bluefields was evicted from two different stations, and the Rama Kriol community of Bangkukuk Taik had to fight both against government co-optation and for the right to refuse the construction of an interoceanic canal on their ancestral lands, about which they were never consulted (see Cupples and Glynn, 2018). In Colombia, Afro-descendant populations are still waiting for the land titles to which they are legally entitled, but are still not consulted about mining or neoliberal development projects in their territories. When they oppose such projects, they risk their lives. Despite the signing of the 2016 peace accords that led to the demobilization of the FARC and brought an end to the civil war, much of Colombia is still characterized by armed conflict and hundreds of black activists have been murdered (Medina Uribe, 2017; Diaz and Jiménez, 2018). One of Colombia's most prominent environmental and human rights defenders, Temistocles Machado from Buenaventura on Colombia's Pacific Coast, whose political leadership in defending land rights from multinationals was extraordinary, was murdered in January 2018 (Ciurlizza, 2018). Afro-Colombian activists defending their territory and ecosystem have also been internationally recognized. For example, Francia Márquez Mina was recently awarded the Goldman prize for her courageous environmental activism against illegal mining in the Cauca department (Lourdez Zimmermann, 2018), a struggle that mirrored her rising political career (*El Espectador*, 2018).

Conclusions

The challenges posed by neoliberal multiculturalism and ongoing forms of coloniality mean that Afro-descendants have to think very carefully about the ways in which they fight for justice and rights. As Hooker (2008: 280) writes:

> This reality raises important philosophical and political questions for Afro-descendant social movements in Latin America. Namely, how can Afro-descendants wage the struggle for equality? How should they frame their struggles for collective rights? Should Afrodescendants ground their claims to collective rights in arguments about the need to preserve a distinct Black culture or cultures, or should they base them instead on the need to overcome historic and present injustice? Alternatively, should they base their demands for collective rights on both kinds of normative claims? Will either of these narratives prove as persuasive to Latin American publics and elites as the notion of "indigenous rights"?

The problem remains one of worldviews and the value of non-Eurocentric and border knowledges that Afro-descendant activists often mobilize. As Arturo Escobar (2018) writes of the

Afro-Colombian activists fighting extractivism in the Cauca River valley, a different kind of world is being imagined, but this world remains unthinkable to the elites and capitalists engaged in neoliberal projects.

The proclamation by the UN of 2015–2025 as the Decade for People of African Descent is an important step in the securing of greater political visibility on national and global scales. This proclamation builds on and is inspired by the second UN Decade for Indigenous Peoples 2005–2014, which brought substantial global political visibility to indigenous groups and was welcomed and urgently required, but tended to amplify ideologies of *mestizaje*, from which Afro-descendants are excluded. There is, then, some mobilization around the UN Decade across the continent: Afro groups are collaborating around common goals and using the Decade to analyze the ongoing barriers to Afro development. The Decade, along with the UN Sustainable Development Goals, is an important mobilizing instrument through which to lobby national governments for resources and changes in policy. The network of scholars and activists concerned here has revealed large differences in how national governments are responding to and engaging with the Decade. While the Nicaraguan government has done virtually nothing, the Costa Rican government has embarked on a set of important initiatives, including state-funded advertising campaigns aimed at dismantling prejudice, such as "Yo soy afrodescendiente" (I am Afro-descendant) and "Anímate a conocerme" (Encourage yourself to get to know me).[11] Many hope the Decade will lead to the creation of a permanent forum for Afro-descendants in the UN, similar to the Permanent Forum of Indigenous Peoples, which could strategize on a global scale for the rights of people of African descent (Lao Montes, 2015). The Decade is also open to important critiques: while it might enhance visibility and recognition, it is also subject to state and supranational control and co-optation, and therefore to a degree of instrumentalization and deradicalization.

While important advances have been made, substantial challenges remain. For Afro-Latin Americans and their allies, dealing with race, racism, and anti-racism means dealing with a series of contradictions. One needs to understand that race is a social construction, but recognize racism as something horribly material and tangible. "Blackness" was constructed in the context of the slave trade and colonialism, but is now inhabited and performed in empowering ways. These involve not biological characteristics but forms of cultural citizenship and belonging, and sets of cultural practices that must be nurtured. Throughout the continent, blackness is both denied and embraced; resistance to and complicity with racial regimes are not easily separated. The most important thing to grasp, however, is that the achievement of the collective human rights of Afro-descendant people in Latin America is owning to the massive efforts of black organizations and black grassroots struggles. This undoubtedly means that the transformations for a life with identity and dignity have fallen on the shoulders of Afro-descendants, and that past struggles continue to be our/their present struggles.

Notes

1 The term Afro-Latino-América began to be used by a group of black Brazilian activists that sought to overturn the Afro invisibility implied by the term Latin America. It constituted a "paradigm-shifting concept that would eventually reverberate across the diaspora" (Reid Andrews, 2018).

2 One of the most significant defences of whitening policies can be found in *Facundo*, written by Argentinian statesman, Domingo Faustino Sarmiento (1845).

3 For a thoughtful literature review on these debates, see Dhamoon, 2015.

4 Our research has been funded by the Marsden Fund of the Royal Society of New Zealand (MAU1108) and the Arts and Humanities Research Council (RA3852). See www.afrolatin-network.hss.ed.ac.uk/ for more information.

5 The interview can be found here https://esvid.net/video/chocquibtown-entrevista-en-univision-0meXf3n6swk.html.
6 Given that some of the exploitative relationships on which slavery was based were sexual in nature, some Afro-Latin Americans are descended from slaveowners as well as slaves. In places such as Haiti and Nicaragua, a small minority of slaveowners had African ancestry.
7 See Wacquant, Slater and Pereira, 2014 for an introduction to the concept of territorial stigmatization.
8 www.2015marchamulheresnegras.com.br/.
9 Also see Kobena Mercer's (1987) influential essay on the semiotics of black hairstyles.
10 Examples of this TV programme available at: www.youtube.com/watch?v=kPUg3prYrOE www.youtube.com/watch?v=pKhCwibAezc&feature=youtu.be.
11 The video campaigns can be found here www.youtube.com/watch?v=Uvjltj4kgLQ and www.youtube.com/watch?v=Y2I8XOlSH_0.

References

A negação do Brasil (2000) [film] Directed by J. Zito Araújo. Brazil.
Araujo, A. L. (2015) The mythology of racial democracy in Brazil. *Open Democracy*, 22 June. www.opendemocracy.net/beyondslavery/ana-lucia-araujo/mythology-of-racial-democracy-in-brazil (Accessed 27 May 2018).
Asher, K. (2009) *Black and Green: Afro-Colombians, Development, Nature and the Pacific*. Durham, NC: Duke University Press.
Baéz Lazcano, C. A. (2018) Entramos Negros y salimos Afrodescendientes . . . Y aparecimos los Afro-chilenos. *ReVista: Harvard Review of Latin America* Winter. https://revista.drclas.harvard.edu/book/entramos-negros-y-salimos-afrodescendientesy-aparecimos-los-afrochilenos-0
Caldwell, L. K. (2004) "Look at Her Hair": The body politics of black womanhood in Brazil. *Transforming Anthropology* 11(2): 18–29.
———. (2007) *Negras in Brazil: Revisioning Black Women, Citizenship and the Politics of Identity*. London: Rutgers University Press.
Ciurlizza, J. (2018) Temistocles Machado adds to the long list of human rights defenders murdered in Colombia. *Democracia Abierta*, 6 March. www.opendemocracy.net/democraciaabierta/javier-ciurlizza/temistocles-machado-adds-to-long-list-of-human-rights-defenders-m (Accessed 23 June 2018).
Couldry, N. (2010) *Why Voice Matters: Culture and Politics After Neoliberalism*. London: Sage.
Cowie, S. (2018) Bahia is Brazil's blackest state – but you'd never guess it from latest TV soap. *The Guardian*, 18 May. www.theguardian.com/world/2018/may/18/brazil-segundo-sol-telenovela-white-black-cast-race?CMP=share_btn_link (Accessed 20 May 2018).
Cupples, J. and Glynn, K. (2018) *Shifting Nicaraguan Mediascapes: Authoritarianism and the Struggle for Social Justice*. Cham: Springer.
Dhamoon, R. K. (2015) A feminist approach to decolonizing anti-racism: Rethinking transnationalism, intersectionality, and settler colonialism. *Feral Feminisms* 4: 20–37.
Díaz, F. A. and Jiménez, M. (2018) Colombia's murder rate is at an all-time low but its activists keep getting killed. *The Conversation*, 6 April. https://theconversation.com/colombias-murder-rate-is-at-an-all-time-low-but-its-activists-keep-getting-killed-91602 (Accessed 25 May 2018).
El Espectador. (2018) *Francia, la líder afro que busca una curul en el Congreso*. 24 February. www.elespectador.com/elecciones-2018/noticias/politica/francia-la-lider-afro-que-busca-una-curul-en-el-congreso-articulo-741017 (Accessed 23 June 2018).
Escobar, A. (2008) *Territories of Difference: Place, Movement, Life, Redes*. Durham, NC: Duke University Press.
———. (2018) *Designs for the Pluriverse: Radical Interdependence, Autonomy, and the Making of Worlds*. Durham, NC: Duke University Press.
Faustino Sarmiento, D. (1845) *Facundo. Civilización y barbarie*. Santiago: El Progreso.
Ferrer, A. (2014) *Freedom's Mirror: Cuba and Haiti in the Age of Revolution*. Cambridge: Cambridge University Press.
Gates Jr, H. L. (2011) *Black in Latin America*. New York: New York University Press.
Gordon, E. T. (1998) *Disparate Diasporas: Identity and Politics in an African Nicaraguan Community*. Austin: University of Texas Press.
Grosfoguel, R. (2012) Decolonizing Western uni-versalisms: Decolonial pluriversalism from Aimé Césaire to the Zapatistas. *Transmodernity: Journal of Peripheral Cultural Production of the Luso-Hispanic World* 1(3): 88–104.

Hale, C. R. (2002) Does multiculturalism menace? Governance, cultural rights and the politics of identity in Guatemala. *Journal of Latin American Studies* 34(3): 485–524.

———. (2004) Rethinking indigenous politics in the era of the "indio permitido" *NACLA Report on the Americas* September/October: 16–20.

———. (2005) Neoliberal multiculturalism: The remaking of cultural rights and racial dominance in Central America. *PoLAR: Political and Legal Anthropology Review* 28(1): 1–28.

——— and Millamán, R. (2006) Cultural agency and political struggle in the era of the Indio Permitido. In D. Sommer (ed) *Cultural Agency in the Americas*. Durham, NC: Duke University Press, pp. 281–304.

Hooker, J. (2005a) Indigenous inclusion/black exclusion: Race, ethnicity and multicultural citizenship in Latin America. *Journal of Latin American Studies* 37(2): 285–310.

———. (2005b) "Beloved enemies": Race and official mestizo nationalism in Nicaragua. *Latin American Research Review* 40(3): 14–39.

———. (2008) Afro-descendant struggles for collective rights in Latin America: Between race and culture. *Souls* 10(3): 279–291.

———. (2017) *Theorizing Race in the Americas: Douglass, Sarmiento, Du Bois, and Vasconcelos*. Oxford: Oxford University Press.

Lao-Montes, A. (2015) Afro-Latin-American social movements. *Latin America in Movement Online*, 23 February. www.alainet.org/en/active/81004 (Accessed 25 May 2018).

Lawrence, B. and Dua, E. (2005) Decolonizing antiracism. *Social Justice* 32(4): 120–143.

Lourdez Zimmermann, M. (2018) El "nobel ambiental" que ganó Francia Márquez por su lucha contra la minería ilegal. *Semana Sostenible*, 23 April. http://sostenibilidad.semana.com/impacto/articulo/francia-marquez-gana-premio-goldman-por-su-lucha-contra-la-mineria-ilegal/40870 (Accessed 3 June 2018).

Lozano Lerma, B. R. (2010) Mujeres negras (sirvientas, putas, matronas): Una aproximación a la mujer negra de Colombia. *Temas de Nuestra América: Revista de Estudios Latinoamericanos* 26(49): 1–22.

Lugones, M. (2007) Heterosexualism and the colonial /modern gender system. *Hypatia* 22(1): 186–209.

Manrique, L. (2016) Dreaming of a cosmic race: José Vasconcelos and the politics of race in Mexico, 1920s – 1930s. *Cogent Arts and Humanities* 3: 1–13.

Martínez Novo, C. and Shlossberg, P. (2018) Introduction: Lasting and resurgent racism after recognition in Latin America. *Cultural Studies* 32(3): 349–363.

Medina Uribe, P. (2017) Who is killing Colombia's black human rights activists? *Okay Africa*, 2 February. www.okayafrica.com/january-least-four-black-community-leaders-murdered-colombia/ (Accessed 25 May 2018).

Megenney, W. W. (1983) Common words of African origin used in Latin America. *Hispania* 66(1): 1–10.

Mercer, K. (1987) Black hair/style politics. *New Formations* 3: 33–54.

Nascimiento, A. and Nascimiento, E. L. (1989) *Brazil: Mixture or Massacre? Essays in the Genocide of a Black People*. Dover: Majority Press.

Nogueira Joyce, S. (2012) *Brazilian Telenovelas and the Myth of Racial Democracy*. Lexington: Rowman and Littlefield.

Norwood, C. R. (2018) Decolonizing my hair, unshackling my curls: An autoethnography on what makes my natural hair journey a Black feminist statement. *International Feminist Journal of Politics* 20(1): 69–84.

PCN. (2012) *Defeating Invisibility: A Challenge for Afro-Descendant Women in Colombia*. Tumaco and Buenaventura: Afro-descendant Women Human Rights Defenders Project. www.afrocolombians.org/pdfs/Defeating%20Invisibility.pdf

———. (2018) Somos dignidad, somos pueblo libre. ¡El pueblo negro no se rinde carajo! #¡SarayTuliaMarisLibres! *Proceso de Comunidades Negras*, 11 May. https://renacientes.net/blog/2018/05/11/somos-dignidad-somos-pueblo-libre-el-pueblo-negro-no-se-rinde-carajo1-saraytuliamarislibres/ (Accessed 24 May 2018).

Quijano, A. (2000) Coloniality of power and Eurocentrism in Latin America. *Nepantla: Views from the South* 1(3): 533–580.

Pinho, P. (2006) Afro-aesthetics in Brazil. In S. Nuttall (ed) *Beautiful/Ugly: African and Diaspora Aesthetics*. Durham, NC: Duke University Press, pp. 266–289.

Raimundo Querino, M. (1918) *O Colono Prêto como Fator da Civilização Brasileira* (The African Contribution to Brazilian Civilization). Bahia : Imprensa Oficial do Estado.

Ramírez, S. (2007) *Tambor Olvidado*. San José: Aguilar.

Reid Andrews, G. (2018) Afro-Latin America by the numbers: The politics of the census. *ReVista: Harvard Review of Latin America*, Winter. https://revista.drclas.harvard.edu/book/afro-latin-america-numbers-politics-census (Accessed 26 May 2018).

Restrepo, E. (2017) Afrodescendientes y minería: Tradicionalidades, conflictos y luchas en el Norte del Cauca, Colombia. *Vibrant: Virtual Brazilian Anthropology* 14(2): 1–15.

———. (2018) Talks and disputes of racism in Colombia after multiculturalism. *Cultural Studies* 32(3): 460–476.

Ribeiro Corossacz, V. (2014) Abusos sexuais no emprego doméstico no Rio de Janeiro: A imbricação das relações de classe, gênero e "raça." *Temporalis* 14(28): 299–324.

Roth-Gordon, J. (2017) *Race and the Brazilian Body: Blackness, Whiteness, and Everyday Language in Rio de Janeiro*. Oakland, CA: University of California Press.

Sánchez Villareal, F. (2017) Cirle Tatis: la youtuber negra que se dejó el afro para combatir el racismo. *VICE*, 29 September. www.vice.com/es_co/article/qvjeym/cirle-tatis-youtube-negra-afro-combate-racismo (Accessed 23 June 2018).

Vílchez, D. (2017) Juliet Hooker: "No pensamos que el nicaragüense puede ser negro". *Confidencial/Niú*, 17 July. https://niu.com.ni/juliet-hooker-no-pensamos-que-el-nicaraguense-puede-ser-negro/ (Accessed 16 May 2018).

Wacquant, L., Slater, T. and Borges Pereira, V. (2014) Territorial stigmatization in action. *Environment and Planning A* 46(6): 1270–1280.

Wade, P. (1997) *Race and Ethnicity in Latin America*. London: Pluto Press.

Zurbano, R. (2013) For blacks in Cuba, the revolution hasn't begun. *New York Times*, 23 March. www.nytimes.com/2013/03/24/opinion/sunday/for-blacks-in-cuba-the-revolution-hasnt-begun.html (Accessed 16 May 2018).

21

ZAPATISMO

Reinventing revolution[1]

Sergio Tischler
Translated by Anna Holloway

Introduction

The Zapatista movement is amongst the most significant political expressions of the anti-capitalist movement in the world today. In fact, it might, in many ways, be *the* most significant one. Its timing took us by surprise: no one in their right minds expected that, after the fall of the Berlin Wall and the peace talks – particularly between guerrillas and governments at war in Central American countries – a guerrilla movement could suddenly emerge in Mexico, challenging both the implementation of the North-American Free Trade Agreement, a spearhead of neoliberal politics, and the traditional approach of classic guerrilla warfare. In their own style, the Zapatistas said they were an "absurd" guerrilla, for they went against the current of a world globalized by capital. That is, they went against a hegemonic temporality that apparently had no significant fissures and could not be challenged.

It was not long before the world realized that this "absurd" was, in fact, the emergence of the *extraordinary* in history, this temporality that obstinately storms in from time to time as an expression of social antagonism and class struggle, challenging the hegemonic time of domination and opening up thresholds. To Zapatismo we owe not only the example it has set, its political astuteness, and sense of humour, but also a fresh outlook on revolution, a new vision on the transformation of the world. In what follows, I wish to explain what is extraordinary in Zapatismo and why it is so important for the anti-capitalist movement.

The emergence of the extraordinary

On the 1 January 1994, the Zapatista Army of National Liberation (Ejército Zapatista de Liberación Nacional, EZLN) took San Cristóbal de las Casas in the state of Chiapas, shaking Mexico to its foundations. The images and discourses of this indigenous rebellion in the mountains of Chiapas quickly turned into a breath of hope and a reference point for the struggles of the oppressed in other parts of the world.

Why this resonance? Perhaps because it deeply touched the chords of our repressed desire for change in a world that appeared to be closed and devoid of alternatives; alternatives which are, however, very much alive in our collective memory. And it made them vibrate. It could be said that, in a sense, we had been waiting for them; we somehow realized that "they" were "us." García de León captured this moment with the following words:

> Many different "never befores" – still not fully understood – came together in a medley of circumstances and parallel incidents unleashed by the dynamic of revolt: never before had a peasant uprising questioned the all-encompassing rule of the state party; never before had this occurred with the consensus of all the participating communities; never before had a movement of this type produced so much written material and so many political alternatives for the entire nation; never before had a popular movement defied the system in such a way without being radically exterminated in a bath of blood and impunity; never before had the battlefield so clearly unfolded in the arena of language itself. Never before had an armed initiative so unmistakably unveiled peaceful solutions to the end-of-the-century Mexican dilemma, not only in the continuation of politics through war but also in its opposite: the continuation of war in the field of politics and of the impact on civil society. Never before, in our lifetime, had words so clearly signified what they were meant to signify, revealing the wooden and hypocritical nature of official discourse.
>
> *(García de León, 1994: 12–13)*

And he continues:

> we started to realise that the revolt in fact sprang from within ourselves, that it covered our entire social territory and that while we thought Mexico's indigenous peoples were paying the price for necessary progress – excluded, until now, from the alleged benefits handed out by the state-benefactor or by the new politics of "social liberalism" – they were actually carrying our own maladies on their shoulders, the crimes of an entire society devoid of democracy and justice. That is why the call of the jungle so deeply penetrated the hearts of Mexicans in the entire country. That is why their hidden face appeared before us as a mirror, a mirror in which to contemplate our own imprisoned face.
>
> *(García de León, 1994: 14)*

For some, the armed uprising of the EZLN was a cry for justice by the country's poor and oppressed against an unfair and openly corrupt system of national power. For others, more distanced from Mexican everyday life, this action symbolised the cry of humanity oppressed by capital. The truth is that, with the armed taking of San Cristobal de las Casas and the actions that followed, the rebel army delivered a material and symbolic blow to the system of domination and turned into a moral-ethical beacon for the world's struggles and hopes for transformation. In other words, the *time of the extraordinary* was making an appearance, severing the threads of the tightly knit fabric of domination in contemporary history, and the dignity embodied by the Zapatistas emerged as a revolutionary subject (Holloway, 2000).

We believe it is precisely this centrality of the emergence of the extraordinary that should be the starting point in the discussion of the importance of Zapatismo for the anti-capitalist movement. In other words, the immediate resonance of Zapatismo cannot be understood only on the basis of the fairness of its particular demands but also of how it talked about old issues in a new and innovative way, giving them a different meaning.

And it is revolution we are talking about. Zapatismo added to the indigenous element of revolution – expressed through demands for land and self-government – that of national power.

Furthermore, it did so in a novel way that interrogated the discursive orthodoxy of the historical Left. The Zapatistas more or less claimed that what mattered the most in revolution was not taking power, but rather dissolving the pillars of domination that result in vertical relationships of power between the leaders and the led, between those who are "above" and those who are "below." Hegemony and homogeneity had no place in this idea of revolution. They also explained that the EZLN as an army was destined to disappear because no army is democratic, not even an insurgent army with revolutionary credentials. And all these ideas were embedded in a praxis guided by the following Zapatista principles:

1   To serve, not serve oneself.
2   To represent, not replace.
3   To construct, not destroy.
4   To lead by obeying.
5   To propose, not impose.
6   To convince, not conquer.
7   To work from below and not seek to rise.

The new idea of the revolutionary subject

Another surprising aspect of the Zapatista idea of revolution is that of the revolutionary subject. Far from appealing to a homogeneous figure of emancipation which, in the historical process, becomes universal through its political realization in the state – as presented in the classic orthodox idea of revolution – Zapatismo speaks of a world of "below and to the left" in which "many worlds fit." The idea of the revolutionary subject is that of multiple subjects expressing the multiplicity of anti-capitalist struggles, each one of which entails a particularity that cannot be subsumed in a vanguard or universal abstract. This spectrum includes the struggles of indigenous peoples and peasants, of women, of all types of urban workers, of the unemployed, of gays and lesbians; of all those who, in one way or another, are dominated by the multiple heads of what the Zapatistas call "the capitalist Hydra" (EZLN, 2016) and take on the challenge of standing against this power with a critical and belligerent attitude.

The idea of the subject is the idea of multiple anti-capitalist struggles in dialogue and "mutual recognition" (Gunn, 2015) that agree on specific moments of struggle but mostly aim at creating an anti-capitalist, polyphonic We. Against the idea of a homogeneous subject emerges what Benjamin (2002[1968]) calls a "constellation" of struggles and subjects. In this context, politics is perceived as a continuous process of coming to agreement in which dialogue plays a central role: a dialogue which, to be revolutionary, must overcome the traditional aspect of political communication that separates the elites – those who do the talking – from the rank and file. This implies a different time. A time that goes against what Max Weber (2005) highlighted as one of the political attributes of the "small group," characteristic of the elites and of the instrumental action of modern politics that consists in fast-paced decisions claiming autonomy from the masses. If ordinary people are to decide on their destiny – the Zapatistas seem to be telling us – this cannot be done within a homogeneous time which lies at the core of a politics that disregards the particularities of struggles and becomes the time of those "who are above," of the elites and the vanguard.

Importantly, in the Zapatista idea of politics the construction of the *we* also involves the subversion of the patriarchy by women. The voices of women are not simply one more component of this we; their colours illuminate anti-capitalist struggle, a struggle that is at the same time understood as anti-patriarchal, anti-sexist, and anti-classificatory. In this sense, the anti-capitalism of zapatismo does not separate the economic from the political and the cultural but, rather, understands these expressions as parts of a complex relation of "multiple heads," bound together by a dehumanizing social logic of exploitation and domination. The patriarchy is one of those "heads."[2]

In other words, the resonance of zapatismo is due not only to the fairness of its struggle and the bravery of its actions. Time as a critical category plays a crucial role in this matter, for Zapatismo created a rupture in the time of capitalist domination that seemed unbreakable, which, thanks to of a dialectic of *discontinuity* with the revolutionary tradition, gave rise to the time of revolution and hope, that appeared with a new face, a new image.

Discontinuity and revolutionary tradition

When we speak of discontinuity in Zapatismo we think of a double rupture: that of the continuity of capital's domination as a system of rule, but also that of the classical cannon of the revolutionary subject, namely, the one that emerged from the Marxist-Leninist codification of historical experience in the Bolshevik revolution in Russia, modified in certain aspects by the Cuban Revolution. I will refer to this aspect further on.

The taking of power by the Bolsheviks consecrated an idea of revolution which revolved around the vanguard party – considered the organized expression of the true consciousness of the proletariat – and the historical necessity to transform this vanguard into a state supported by the proletarian masses and allied with the poor peasantry, in the specific case of rural and backward Russia. Therefore, the taking of state power by the vanguard party is considered the fundamental condition for the transformations that will presumably lead to a socialist society, the precursor of communism, in agreement with the interests of the proletariat which would objectively express the general interest of a society in the process of emancipation. All this on the basis of the theoretical backbone of Lenin's canonical work *What is to be Done?* It is through this codification of the Bolshevik experience that the state-centered dogma of social revolution came into being.

The Cuban revolution changed the idea of Latin America in the 1960s: it brought to the forefront the reality of a socialist revolution in a part of the world that was considered to be "underdeveloped," and that – according to the classical cannon – had not developed its productive forces sufficiently and did not have a proletariat with enough political presence to truly contemplate a historical change of such magnitude. Before this event, the struggles led by communist parties took place in the context of a democratic and anti-oligarchic revolution that cleared the path for modern capitalism, considered an agent for the progressive development of the productive forces of society (see Caballero, 2002). The reality of social revolution posited by the Cuban experience came together with a new organizational form that was not the classic Leninist party of the Third International, but a new historical figure that would play a crucial role in the continent's revolutionary struggles: the rural guerrilla.

The figure of the guerrilla dramatically changed the idea of the revolutionary vanguard and the perception of the peasantry as a subject of social change. The characteristic centrality of the urban proletariat as the revolutionary subject in the Leninist doctrine was replaced by that of the peasantry as the driving force of socialist revolution in the predominantly rural populations of Latin American countries. The mode of organization would no longer be a party focused socially and politically on urban centres. The topography of revolution changed. The jungles and the mountains now appeared as the setting for a new possible world. Urban struggles continued to matter but the drive for change would be generated in the jungles and the mountains.

However, the new idea of revolution remained loyal to the classical Leninist canon as far as the state, the vanguard, and other issues were concerned. Perhaps it was not possible to think of revolution outside the *state form*[3] because the historical constellation within which it took place viewed the state as a central political category. The figure of the Soviet socialist state as a successful form of social change still illuminated the struggles of the people and was considered – not without objective reasons – a condition for the success of what was understood as anti-imperialist struggles. In a way, one could think of the guerrilla model that applied to many Latin American experiences as an adaptation of Leninism – and not so of "Marxism-Leninism" – to the Latin American context and struggles. It was a moment of creativity compared to the mechanical model of the communist parties that followed the theoretical blueprints of the Third International.

In other words, the classical canon of the revolutionary subject understood politics as part of the *state form*, and revolutionary organization – be it party or guerrilla – as an anticipation of a new historical synthesis called socialism, which expressed the true identity of the state and the subaltern classes.

Unlike politics as the monopoly of vertical time in the *state form*, Zapatismo privileges and deploys the horizontal time of politics, putting the question of autonomy in the Zapatista territories and the development of a *we* that is constituted by the set of struggles from "below and to the left" in the national and international sphere at the centre of anti-capitalist struggle. This is a "world in which many worlds fit," whose signature document is probably the 2005 Sixth Declaration of the Lacandon Jungle (EZLN, 2005).

This perspective seems to update past community resistances of the indigenous peoples, particularly in Chiapas, and profoundly empathizes with the experiences of self-determination in the revolutions of the 20th century, defeated in the context of the triumph of the state-centered experience of the socialist revolution. The revolution from below, expressed particularly through the example of the soviets of workers and peasants in Russia, is part of a history of state synthesis based on the revolutionary subject understood as vanguard and hegemony, categories rejected by the Zapatistas, partly – we assume – because they realized this type of experience did not lead to social emancipation.

In this sense, Zapatismo can be considered a creative response to the crisis of the classical canon of the revolutionary subject, reminding us of Rosa Luxemburg's (2002) critique of Lenin on the issue of the revolutionary vanguard, structured so as to vertically lead the labour movement and the revolutionary process in general.

Autonomy and horizontal time

Some of the abovementioned are linked to the fact that autonomy is a fundamental political category in Zapatismo. Let us go back in time. The Zapatista uprising in January 1994 shocked and awakened Mexican society. This was expressed through an intense social mobilization that forced the government to stop the more open and brutal actions of counter-insurgent war in the territories that had been occupied by the armed rebels. One of the most significant results of the Zapatista resonance was precisely this, in the sense that it not only showed that their struggle had at once become a national ethical-moral point of reference, but also that it was a mirror where a society fed up with corruption saw its own reflection. It was as if this society, which had put up with the electoral fraud of the PRI (Institutional Revolutionary Party) that brought Carlos Salinas de Gortari to power, was saying: if there is anything decent in national politics, it is Zapatismo. Their voice, the voice of the indigenous and the oppressed which has for centuries been silenced, can never again be ignored.

The government was forced to enter a process of negotiation with the insurgents. The official outcome of this process of negotiation, which the Zapatistas opened up to society by inviting democratic personalities, social leaders, and critical academics to participate in the dialogue with the government, were the so-called San Andres Accords. Established political parties were excluded from the invitation: this set the tone for Zapatista politics, which rejects democracy "from above" that rules over those who are "below," as well as the established political cal parties who represent it. The question of indigenous autonomy was placed at the centre of the negotiations.

Once approved by the participants, the Accords were forwarded to Mexican Congress. The latter passed a law that undermined the critical profile of indigenous autonomy, one that the government representatives in the dialogue had already been forced to accept. This was perceived by a significant part of Mexican society as betrayal. However, the government had in fact remained true to its *raison d'être*: to prevent the development of an autonomy that defied the liberal model of political self-determination. In doing so, it revealed the contradictions and political and social limits of this model, determined by the web of domination which it serves.

However, Zapatismo did not remain paralysed within the web of Mexican institutional politics. In 2003 it went against the grain and formally initiated the process of creating autonomous governments in the recovered territories. The "Caracoles"[4] and the "Juntas de Buen Gobierno" (Councils of Good Government) were established, launching one of the most creative facets of Zapatismo. Generally speaking, Zapatista autonomy can be understood as self-government which responds to community forms of social organization and is guided by the idea of direct democracy, a result of their critique of representative democracy as the form of domination of capital. Autonomy is a series of practices creating a government based on gender parity and including, amongst others, the rotation of political offices and the elimination of "specialists" in the administration; thus, governing becomes truly a collective experience. The government is also in charge of monitoring the implementation of collective agreements reached through consensus in Zapatista territories; of administrating an economy of common areas destined to support the material reproduction of the self-government; and of overseeing an autonomous health and education system, amongst others.

Nevertheless, we must always be alert before the danger of perceiving autonomy as a fixed and finished phenomenon that takes place in a territory liberated and emancipated from capital and the state. Its deployment cannot be understood in linear terms. Autonomy must be understood as a category of struggle, therefore open and ongoing. Or, if you prefer, as a *figure-process* oriented towards the emancipation of the relations of domination and a radical, non-instrumental change in how we relate with nature. At its core lies the desire to transform social relations horizontally, on the basis of collective decisions and practices that work towards the elimination of the "above" and the "below" of domination expressed in multiple ways, among them is patriarchal domination; the Women's Revolutionary Law[5] was created to attend to this state of affairs.

Importantly, Zapatista autonomy cannot be understood in reductive and narrow terms as nothing more than the self-government that is practised in the recovered territories; it involves a complex relation between the local, the national, and the international. The Zapatista image of the *caracol* expresses the inward and outward movement that autonomy entails. The Zapatistas are well aware of the fact that the government's strategy is to reinforce the enclosure of the occupied territories and cause the seed of autonomy to wither. Thus, one of the tactics of the insurgents has been to break the enclosure imposed by the state and capital. From the moment of their first armed public appearance, they have learned to move with great creativity – and some contradictions – in the sphere of national politics. Without the national and international political presence of Zapatismo, based on the ethical-moral force of their existence, the Mexican government would have most probably managed to tighten the political and military siege and attack the rebels more openly.

In August 2013, the Zapatistas organized an event called "The Small Zapatista School" (La Escuelita Zapatista). On this occasion, the Zapatistas showed the world their achievements but also the problems facing the autonomy process in their territories. The first cycle, which was broadcast through videoconference to different parts of the country, was attended by 1,700 participant-students. In line with their anti-patriarchal stance, each one of the attendees was accompanied by a Zapatista *compañero/compañera* called Votanes and Votanas, who took care of the participants, explained to each one personally aspects of the books that had been given to them and, generally speaking, described the experience of autonomy, their achievements, problems, advances, and setbacks.

It was not the renowned figures of the EZLN who were at the forefront of this event but ordinary people of all ages and in the massive collective meetings that took place every day in the auditorium of the Indigenous Centre for Integral Learning Fray Bartolome de las Casas (CIDECI) in San Cristóbal de las Casas, it was young people, the new generations of young men and women who had been educated in the autonomous system of education of the Zapatistas, who took the lead.

To all of us who had the privilege of attending as students, they expressed that Zapatista autonomy – as a figure of anti-capitalist struggle – is an arduous and difficult but attainable process. That it is not possible to replicate their experience in other places for there are irreducible particularities that subvert any attempt at a general model or recipe, and that each one of us, in our own way, can contribute to the construction of a collective *we* "from below and to the left"; that what is needed is organization and the will to change. They showed us that autonomy is also a language and that a new language is needed so that social change becomes thinkable; that this language challenges us to think differently, it is a kind of anti-grammar that goes against the tide and puts the grammar of power and hegemony in crisis (see Tischler, 2014). Probably one of the most accomplished works on the political aesthetics of Zapatismo is the essay "Between light and shadow" by Marcos (Subcomandante Marcos, 2014).[6]

Years ago, I visited the Oventic *caracol* and spent time with *compañeros* and *compañeras* from the Zapatista system of education. They explained to me their concept of time, to which I often recur to exemplify the complex process of Zapatista autonomy and the knowledge that this practice encompasses. These young people told me the Zapatistas have *three times*: exact time, just time, and necessary time. Precise time is linked to commerce and the clock; in general terms, to the figure of the commodity and capital. Just time has to do with the reproduction of the community fabric and the relation with the land; it points towards a non-instrumental social relation between people and nature. And necessary time, the time of the revolution understood in horizontal, anti-capitalist terms. Importantly, just time cannot be subordinated to exact time, because within Zapatismo, just as in other indigenous community experiences, it can be redefined, deepened, and expanded by necessary time. This points towards a national anti-capitalist perspective and speaks of the need, for the time being, for the existence of EZLN as an army (Tischler, 2012). It has also led to a revolutionary selection of collective memory and tradition, as well as to the fight against negative elements of memory and tradition, such as the patriarchy.

In a seemingly simple language, the *compañeros-as* told me that those times exist simultaneously as expressions of social relations: they complement and contradict each other at the same time. In other words, autonomy can be understood only within a complex, non-linear temporality, where horizontality is a constant struggle, a practice and a horizon that necessarily transcend the local sphere. In certain ways, Zapatista autonomy updates the debate between the populist Russians and Marx – where the former claimed that the peasant community (*mir*) was a central factor in the abolition of capitalism in Russia – as well as the pervasive ideas of Mariátegui on the need to understand the indigenous community (*ayllu*) as one of the historical forces of communism in Peru.

A detotalizing perspective

Autonomy, and Zapatista politics in general, seems to point towards what we call a dialectic of *detotalization* (Tischler, 2012). We cannot go into details regarding its theoretical implications, but we shall say that all grammar of revolutionary power has been linked to the category of totality, understood in positive terms. Lenin and Gramsci thought of revolution from this perspective, but it was probably Lukacs (1971) who was its main exponent.

For him, following along the lines of Hegel's positive dialectic, revolution would be the realization of totality, accomplishing the unity between subject and object that had been torn apart by capital. Benjamin (2002) questions this dialectic in his critique of the idea of revolution as progress in linear and homogeneous time. Benjamin's idea of constellation contains this critique.

However, it was Adorno (1990) who presented the conceptual and material mechanism of totality as a relation of domination in more detail. In this mechanism, universality is the result of a process of identification-homogenization in which particularities are stripped of their qualities in a relation of subordination. This, according to our perspective, refers to the law of value and the rule of concrete labour over abstract labour: totality is a category of capital (see Tischler, 2012). Therefore, it must be understood in terms of a negative dialectic that sets out to overcome said category and achieve social emancipation. Amongst other things, this perspective allows us to break the totalizing coating of the concept of class struggle, and this is crucial if we wish to understand the subject on the basis of the centrality of antagonism and struggle and not in the static, sociological terms of the primacy of structural positioning (see Nasioka, 2017).

We believe it is this perspective that we find in the Zapatista critique of the vanguard and of the idea of the homogeneous revolutionary subject. In this sense, we can also understand the category of autonomy as a critical category; that is, in terms of a detotalizing category.

However, it could be argued that the abovementioned is in no way related to the Zapatista language: this is true if we opt for a superficial approach. The counter-argument is that Zapatismo, in adding a critical tension to all tradition belonging to the classical revolutionary canon, urges revolutionary thought to look at these matters critically. In this sense, theoretical issues that might seem quite remote are in fact part of the new constellation of anti-capitalist struggles, a constellation where Zapatismo has had a crucial resonance. If we are to be consistent with Zapatismo, this aspect of critique is also part of their history and cannot be ignored.

Threshold

At the time of writing, Zapatismo and the National Indigenous Congress (Congreso Nacional Indigena, CNI) decided to fight for the inclusion of a female indigenous candidate in the 2018 electoral process: *compañera* María de Jesús Patricio, better known as Marichuy. Generally speaking, it seems that the goal was to mobilize the "below and to the left" of indigenous and peasant Mexico with a consensus candidate that would make the whole country aware of the struggles of rural communities against the violence of capital and would unveil the complicit silence of the systemic parties. At the same time, there was the intention to activate urban networks in solidarity with the struggle of the peasants and the Zapatistas.

However, the symbolic dimension of the candidacy must also be stressed. Marichuy represents those who are lowest amongst Mexico's underdogs. The dark-skinned complexion of the indigenous woman represented by Marichuy is the face of the insubordination that grows in the "below." And in this insubordination indigenous women play a crucial role, for anti-capitalist struggle is at the same time an anti-patriarchal one.

One of the risks of this initiative was that the CNI might become trapped in the web of institutional politics. However, the anti-systemic and anti-capitalist demands of the movement go beyond the prevailing institutionality and portray the movement as an answer to what the Zapatistas call the capitalist "storm." In their words, it is a way of introducing the "agenda of those who are below" in order to produce *cracks* in the "agenda of those who are above."

The precedent of this initiative was the 2006 Other Campaign (Otra Campaña). On that occasion, Zapatismo erupted in Mexican electoral time with the goal, amongst others, of revealing the constitutive fallacy of procedural democracy and presidential campaigns, and of creating a big front of struggles from "below and to the left."

Delegate Zero, the name of Subcomandante Marcos in the Other Campaign, spearheaded the process. It is well known that the elections were fraud-ridden and the candidate of the institutional Left, Manuel López Obrador, was deprived of a probable triumph. Amidst the climate of frustration produced by the fraud, the followers of López Obrador accused Subcomandante Marcos of having contributed to the dispossession suffered by their candidate, given that Zapatismo had openly confronted the person they considered would almost certainly win. The goal was to highlight the differences and antagonisms between two political projects perceived as left-wing, as opposed to the clearly neoliberal programs that dominate the Mexican political spectrum. An issue that has not yet been sufficiently clarified is the impact that this confrontation had on the image of Zapatismo, particularly in the urban sphere.

However, the conflicts and contradictions between the two currents revealed something more profound and still very much present: the confrontation between a long-lasting, national-popular trend led by López Obrador, with a memory that is largely institutional and a project that remains within the boundaries of the policies of capital, and a new one, that of the Zapatistas, anti-capitalist and critical of bourgeois institutionality. The narrative of López Obrador appeals to certain institutional forms that emerged from the Mexican revolution and includes popular demands but, in practice, offers nothing more than a public administration committed to combating corruption as a way of favouring capitalist accumulation. On the contrary, Zapatismo reclaims the unfulfilled popular dreams of the Mexican revolution in terms of a new anti-capitalist democracy constructed from below and focused on autonomy, with the aforementioned consequences.

In any case, it must be stressed once again that Zapatismo cannot be reduced to conjunctural politics or to the effort put into the ongoing elaboration of autonomy in the Councils of Good Government. As I have tried to show in this chapter, Zapatismo's *now-time* rejects all linear, mechanic, and reductionist interpretations. It is a complex dialectic between the local, the national, and the international. The initiatives that are inscribed in this dimension can be seen as a "thunderbolt" that sheds light on the "state of exception" in which we live so as to turn it into a true state of exception (Benjamin, 2002[1968]).[7] This "thunderbolt" represents the untimely and sudden character of Zapatismo, as well as the extraordinary historical density of their *now-time*, the expression of a new revolutionary time that allows us to imagine the possibility of a new anti-capitalist constellation.

As stated at the beginning of this chapter, Zapatismo represents the *extraordinary* in current political history, opening up a new perspective in the constellation of today's anti-capitalist struggles. Its detotalizing idea of revolution involves an anti-vanguard/anti-hegemonic revolutionary subject and the central role of horizontality in social relations and government. This becomes blatantly clear in the experience of autonomy, which emblematically has the face of a woman in reference to the head-on fight against the patriarchy.

However, there are no certainties in struggle. The only certainty is the need to struggle. The contributions of Zapatismo do not guarantee anything that can be measured as "success," if this is to be interpreted in terms of power, instrumental rationality, or linearity. Zapatismo is something else: it is the struggle to create a new world from the unfolding of a new revolutionary constellation. This is the path on which its vision and its actions are set.

Notes

1 Translated by Anna Holloway.
2 The literature behind the ideas in this chapter emerge from the writings and communiqués of the EZLN, especially the work of former Subcomandante Marcos, now Subcomandante Galeano. Readers seeking more detailed accounts of this work should consult the five volumes of documents published by Ediciones Era (see EZLN, 2003; Monsivais, 1995, 1997, 2003; Poniatowska and Monsivais 1994) and also the publications found on enlacezapatista.ezln.org.mx.
3 On the issue of the state as a form of social relations, see Holloway (2002).
4 Caracoles are autonomous Zapatista government entities.
5 The Women's Revolutionary Law includes a series of laws that emerged *from below*, as a demand of women themselves in the struggle against patriarchal domination in the Zapatista communities. The law consists of ten points:

First – Women, regardless of their race, creed, colour, or political affiliation, have the right to participate in the revolutionary struggle in any way that their desire and capacity determine.

Second – Women have the right to work and receive a fair salary.

Third – Women have the right to decide the number of children they bear and care for.

Fourth – Women have the right to participate in the matters of the community and have charge if they are freely and democratically elected.

Fifth – Women and their children have the right to PRIMARY CARE in health and nutrition matters.

Sixth – Women have the right to education.

Seventh–Women have the right to choose their partner and not be forced to marry.

Eighth – No woman shall be beaten or physically mistreated by relatives or strangers. Rape and attempted rape will be severely punished.

Ninth – Women can occupy positions of leadership in the organization and hold military ranks in the revolutionary armed forces.

Tenth – Women will have all the rights and obligations specified by the revolutionary laws and regulations.
6 In this document, written in homage to assassinated Zapatista comrade Galeano, the political aesthetic of the Zapatistas is clearly expressed as going against the logic of vanguard and individualism that had characterized left-wing politics.
7 For the Zapatista version, see SubGaleano (2015).

References

Adorno, T. (1990) *Negative Dialectics*. London: Routledge.

Benjamin, W. (2002[1968]) Thesis on the philosophy of history. In W. Benjamin and H. Arendt (eds) *Illuminations*. New York: Schocken, pp. 253–264.

Caballero, M. (2002) *Latin America and the Comintern*. New York: Cambridge University Press.

EZLN. (2003) *EZLN: Documentos y comunicados, Tomo 4*. México: Ediciones Era.

———. (2005) Sixth declaration of the Lacandón Jungle. *Enlace Zapatista*. http://enlacezapatista.ezln.org.mx/sdsl-en/ (Accessed 6 May 2018).

———. (2016) *Critical Thought in the Face of the Capitalist Hydra: Contributions by the Sixth Commission of the EZLN*. Durham, NC: PaperBoat Press.

García de León, A. (1994) Prólogo a EZLN. In E. Poniatowska and C. Monsivais (eds) *EZLN: Documentos y comunicados, Tomo 1*. México: Ediciones Era, pp. 11–15.

Gunn, R. (2015) *Lo que usted siempre quiso saber sobre Hegel y no se atrevió a preguntar*. Buenos Aires: Herramienta Ediciones; BUAP.

Holloway, J. (2000) El zapatismo y las ciencias sociales en América Latina. *Revista Chiapas* 10: 171–176.

———. (2002) *Change the World Without Taking Power: The Meaning of Revolution Today*. London: Pluto Press.

Lukács, G. (1971) *History and Class Consciousness*. Cambridge, MA: MIT Press.

Luxemburg, R. (2002) *Leninism or Marxism, or: Organizational Questions of Russian Social Democracy*. Montreal: Kersplebedeb.

Monsivais, C. (1995) *EZLN: Documentos y comunicados, Tomo 2*. México: Ediciones Era.

———. (1997) *EZLN: Documentos y comunicados, Tomo 3*. México: Ediciones Era.

———. (2003) *EZLN: Documentos y comunicados, Tomo 5*. México: Ediciones Era.

Nasioka, K. (2017) *Ciudades en insurrección: Oaxaca 2007/ Atenas 2008*. México: Cátedra Jorge Alonso 2014 Award.

Poniatowska, E. and Monsivais, C. (1994) *EZLN: Documentos y comunicados, Tomo 1*. México: Ediciones Era.

Subcomandante Insurgente Marcos (2014) Between light and shadow. *Enlace Zapatista*, 27 May. http://enlacezapatista.ezln.org.mx/2014/05/27/between-light-and-shadow/ (Accessed 6 May 2018).

SupGaleano. (2015) The storm, the sentinel and night-watch syndrome. *Enlace Zapatista*, 4 April. http://enlacezapatista.ezln.org.mx/2015/04/04/the-storm-the-sentinel-and-night-watch-syndrome/ (Accessed 6 May 2018).

Tischler, S. (2012) Revolution and detotalization. *Journal of Classical Sociology* 12(2): 267–280.

———. (2014) No es lo mismo resistir para sobrevivir que resistir para transformar el mundo. La Escuelita Zapatista: ¿Desafío epistemológico? *Herramienta* 55.

Weber, M. (2005) *Economy and Society*. Berkeley, CA: University of California Press.

22

COUNTER-MAPPING DEVELOPMENT

Joe Bryan

Introduction

Counter-mapping's promise for development lies with its claims to make the invisible visible, producing a new, more accurate understanding of space through popular participation. Maps play an instrumental role in this approach, documenting the existence of indigenous territories, customary use, informal settlements, and other entities previously not found on state-issued maps. The overall goal is one of challenging state sovereignty, forcing a reconciliation with its constitutive exclusions by way of securing a place on the map. On those terms, counter-mapping has had considerable success across Latin America. No more visible measure of that success exists than the massive transfer of state-owned lands to various forms of community ownership as part of a region-wide "territorial turn" (Offen, 2003; Bryan, 2012). At the same time, the spatial expansion and intensification of capitalism continues unchecked by these new forms of territoriality, raising the question of what, if anything, counter-mapping counters?

Addressing that question requires understanding how maps work, particularly with regard to their role in shaping dominant forms of power and economy that counter-mapping challenges. Any success attributed to counter-maps is often more about their ability to work as maps than it is about countering the dominance of capitalism and state power. That paradox forms the core of the widely discussed "dilemmas" of counter-mapping (Wainwright and Bryan, 2009; Hodgson and Schroeder, 2002; Hale, 2005; Peluso, 1995). Social movements and scholars have increasingly addressed these dilemmas by shifting attention away from maps as representations, and towards the importance of struggles over the understanding of space that informs map making and use. Collectively, these approaches to mapping draw out how dominant understandings of space have been historically produced through colonialism and capitalism, upending appearances that space is somehow ahistorical or politically neutral. Counter-mapping demonstrates the fundamental incompleteness of the production of space as a constitutive aspect of hegemony, making visible the ways in which familiar forms of power and economy expand and intensify (Craib, 2004). At the same time, counter-mapping opens up an important site for challenging those processes through the production of new geographies rather than the reconfiguration of existing ones (Counter Cartographies Collective, Dalton, and Mason-Deese, 2012). That effort resonates with and critiques calls for post-development, providing a means of advancing in the concrete towards new forms of power, authority, and space.

Whose maps, what space?

Though the origins of counter-mapping are multiple and contested, they can generally be traced back to the end of the Cold War and the rise of neoliberal policies in Latin America. Both of those historical moments introduced new ways of seeing space in terms of the expansion of capitalist markets and democratization of political representation. Counter-mapping took hold as a response to both, challenging the expansion of capitalist development into historical frontiers and the racial and ethnic hierarchies constitutive of nations. Aided by increased access to cartographic technologies such as GPS units and GIS software, marginalized communities mapped themselves back on the official cartography of Latin America. At its most basic, a counter-map was any kind of map that simply showed what was excluded or erased from state-issued maps, equating absence with political marginalization. Over the course of the 1990s the approach flourished with indigenous peoples and Afro-descendent communities, particularly those living in forested frontiers. Both used maps to claim rights to territories established by their customary use and occupancy of land and resources.

Their maps challenged the colonial foundations of state claims to territorial sovereignty. They also critiqued demands for agrarian reform championed by movements on the Left calling for transfer of lands to those who worked them. In the forests of the Amazon basin, Guyana shield, and Caribbean coast of Central America, customary uses went well beyond agriculture, including activities such as hunting and fishing that were as invisible to Left demands for agrarian reform as they were to state officials advocating frontier settlement and resource concessions. Maps of indigenous and Afro-descendent territories challenged both, breaking with the emphasis on the nation-state as development's main unit of analysis. This last point was made clear by geographer Bernard Nietschmann's (1995) oft-repeated claim that more indigenous land would be defended and reclaimed with maps than with guns.

Counter-mappers like Nietschmann were not the only ones using maps in this way. By the late 1990s, Latin America was awash in projects variously termed "ethnocartography," "participatory mapping," and "community mapping" (Herlihy and Knapp, 2003; Bryan and Wood, 2015). Many of these projects shared an emphasis on community participation, producing maps of territories defined by customary use and occupancy for the goal of advancing recognition of indigenous and Afro-descendant territories. In each setting, there was shared sense of the necessity of mapping. If communities did not map themselves, they would almost surely have to abide by the exclusions and authority of maps made by states. It was a matter of map or be mapped (Bryan, 2009; Offen, 2009).

The cartographic imperative concealed sharp political divergences between the attitude mapping projects adopted towards state power and capitalist development. In Colombia, state-backed mapping projects set out to advance demarcation and titling of indigenous and Afro-descendant lands as a means of countering their violent dispossession by paramilitaries and guerilla groups. Mapping was considered a step towards a new, community-oriented approach to development backed by the World Bank (Asher, 2009; Offen, 2003). The World Bank supported similar efforts to document "land tenure" in indigenous and Afro-descendent communities in Central America as part of broader efforts at reforming property rights (Gordon, Gurdian, and Hale, 2003). That work complemented legal efforts to win recognition of indigenous peoples' rights to land as property as a matter of protecting their basic human rights (Wainwright and Bryan, 2009; Hale, 2006). In the Amazon Basin and Central America, non-governmental organizations backed mapping projects aimed at improving conservation strategies through recognition of indigenous land rights as blueprint for sustainable development (Smith et al., 2003; Stocks, 2003; Stevens, 1997). All of these projects could claim a degree of novelty, even as their political content was debated (Salamanca and Espina, 2012; Sletto et al., 2013). Did they amount to a democratization of space (Herlihy and Knapp, 2003)? Or were they the hallmark of neo-liberal multiculturalism, offering recognition of indigenous and Afro-descendent land rights in exchange for acquiescence to free market policies (Hale, 2005)?

Both lines of questioning underscore the efficacy of mapping. As much as they differed in their application and intent, all of these projects produced maps that worked as maps. The novelty of their content – place names, land use, non-state territories – was recognizable to judges, state officials, and World Bank staff, among others, due to their adherence and mastery of carto-graphic conventions. Claims about the legitimacy of community maps hinged on the accuracy of the information conveyed, locating customary use and occupancy with the level of precision afforded by GPS units. As innocuous as georeferencing can appear, its standards are defined by an abstract understanding of space as something measurable and quantifiable that underpins cartography. This understanding of space is not without history nor geography. It has coevolved with colonialism and capitalism, facilitating the application of political economic and racial log-ics to understanding and organizing space. The growing literature on the history of cartography in Latin America illustrates as much, showing the subsumption of indigenous spaces to the logics of colonial authority, state sovereignty, and property.

The trouble with much of this literature is that even as it acknowledges the range and diver-sity of this process, it retains a singular focus on maps as representations of an already, always existing concept of space (Dym and Offen, 2011). Community mapping, including counter-mapping, makes a similar move. In treating communities and territories as objects of mapping and recognition, they tend to elide how both are historically formed. Their recognition as com-munal properties and territories in the present moment draws on that history at the same time that it completes its elision. What goes missing is the role that cartography plays in primitive accumulation, converting land and resources into property and establishing it as a geographical category that comes into existence "dripping from head to foot, from every pore, with blood and dirt," to borrow Marx's line (Marx, 1990: 926). That physical violence is backed by the structural and symbolic violence of limiting representation of space to cartography, treating its boundaries and categories as the basis for law and inclusion in society. There can no "develop-ment," broadly construed, without this abstract understanding of space (Lefebvre, 1991; see also Coronil, 2000). Related arguments outline the role of maps in producing historically and geo-graphically contingent distinctions between "the West and the Rest," Third and First Worlds, Global South and North, and so on (Hall, 1992; Lewis and Wigen, 1997). Accordingly, any challenge made from within that space will find itself circumscribed, a point widely discussed in terms of the dilemmas of counter-mapping in Latin America and beyond (Bryan and Wood, 2015; Wainwright and Bryan, 2009; Mollett, 2013; Peluso, 1995; Hodgson and Schroeder, 2002). Not unlike counter-hegemonic processes, the horizons of political possibility for counter-mapping are limited by state and capital (Escobar, 2010). The trouble with maps is not maps themselves but the space of which they are part and parcel.

The link between the proliferation of mapping and neoliberalism demonstrates the impor-tance of the production of space as a critical component of capitalist accumulation. Neo-liberalism's spatial qualities are readily apprehended through its expansion of markets and intensification of capital accumulation. The former can be understood in terms of the spread or expansion of the kind of abstract space described above, combining the destruction of previ-ously recognized forms of state property associated with "national lands" and frontiers. Claims for recognition of indigenous territories have played a part in this transition, predicated on the idea of dispelling, once and for all, the colonial fiction of state or national lands (Gordon, Gurdian, and Hale, 2003). The dismantling of agrarian reform and communal lands in places like Mexico illustrates another example of this expansion. Related processes of intensification manifest as the proliferation of sites of capital accumulation within abstract space, through everything from genetically modified seeds to micro credit schemes. Such efforts are often grasped as the fragmentation of previously established social wholes, particularly by social movements geared towards defending hard-earned supports from states. And yet, they can also be viewed as the deepening of the very same capitalist forms that produced that perceived social whole to begin with.

Indeed, many of the entities made visible through counter-mapping, beginning with the very idea of a sedentary, ethnically defined community, bear the indelible traces of colonialism and capitalism in their form. Concentrated settlements in the Amazon are as much an artefact of missionization, the reduction system, and rubber tapping as they are shaped by customary use (Porto-Gonçalves, 2001). National territories of the Mapuche and Miskito make reference to their prior recognition as part of systems of Spanish and British colonial rule, respectively (Bryan, 2011b; Nahuelpan Moreno et al., 2012). The very category of communal lands in Mexico belies a similar history of using land ownership as a means of managing labour markets while clearing up space for industrial agriculture (Morton, 2013; Rus, 1994). None of these are meant to undermine the importance of contemporary efforts to defend those same spaces. Rather, such details are essential to confronting the question of space that underlies those struggles and their use of maps.

Community mapping projects across Latin America demonstrate this point. In Colombia's Choco region, extensive mapping and titling of black lands has done precious little to stop dispossession by paramilitary groups and oil palm plantations (Asher, 2009). After initial reluctance to recognize indigenous and black lands, the Nicaraguan state has titled nearly a third of the national territory to community ownership. Nonetheless, titles have done little to stop massive invasion of lands by settlers from other parts of the country and ensuing deforestation (Miranda Aburto, 2016). A similar process has occurred in Honduras, where titling of indigenous lands in particular has gone hand-in-hand with their forcible incorporation into narco-economies with catastrophic effects for both forests and people (McSweeney et al., 2014). In other countries like Bolivia and Ecuador, recognition of indigenous territories has served a pretext for opening up lands for extractivism, insisting that with ownership of land and resources comes the obligation to make use both for the benefit of the nation – without the option of saying no (Anthias and Radcliffe, 2015; Escobar, 2010; Postero, 2017).

In more than one case, the response to the ongoing threat of dispossession has been the production of another map, multiplying the production of abstract space rather than opposing it. There is perhaps no better example of this predicament than the controversy surrounding the American Geographical Society's Bowman Expeditions program. The program's first expedition was to Mexico, tasked with mapping communal lands held by indigenous communities in the Huasteca Potosina in San Luís Potosí and the Sierra Juárez in Oaxaca. Led by geographer Peter Herlihy, the project's goal was to create maps of communal lands that could be used to evaluate the impact of a Mexican program aimed at dividing those lands into individually owned parcels (Herlihy, 2008). That goal resonated with communities, several of whom cooperated with the project out of their own sense of its importance for protecting communal lands. The project imploded, however, following the communities' discovery that the entire project was funded by the US Army as part of an effort to develop new counter-insurgency tactics (Bryan and Wood, 2015; Wainwright, 2013). Expedition reports implied that all of the data gathered by the Expedition – place names, archaeological sites, trails, water, and so on – were turned over to the US Army. By contrast, communities were given paper maps containing a small fraction of the information they had helped produce (Cruz, 2010). Despite the controversy, the American

Geographical Society has continued the program with support from the US military, sending expeditions to Colombia, Honduras, Panama, and the Lesser Antilles. Collectively the expeditions reinforce arguments equating the absence of private property rights with weak state presence, poverty, and drug trafficking. Mapping plays an instrumental role in creating property rights with the ostensible goal of improving security through economic integration, recycling Cold War development narratives for a new era.

From maps to mapping

Struggles over space can seem no less abstract than the categories through which space is understood. How does one go about making a different world, advancing a general analysis of abstract space through concrete action? This question is partially addressed through innovations in the uses of maps and understanding of their relationship to power. That move requires abandoning any idea of maps as instruments or tools that can be wielded for good or bad purposes, and replacing it with a more nuanced attention to their role in producing spaces. A growing number of social movements have addressed this question through renewed emphasis on mapping as a practice that extends beyond maps to include they ways they are produced, read, and used. In Brazil and Colombia, this approach is often referred to as "social cartography" by way of distinguishing it from other, competing terms such as "ethnocartography," "participatory mapping," and "counter-mapping" (Almeida, 2013; Restrepo, Velasco, and Preciado, 1999). The difference is more than a matter of semantics. It underscores the importance of putting social processes before the map, inverting instrumentalist uses of maps to fit black and indigenous territories, among others, into state maps (Wainwright, 2008; Sletto, 2009; Bryan 2011a).

Social cartography's emphasis on mapping raises any number of interesting points. Among them is a significant reconsideration of power that both builds on and addresses the shortcomings of counter-mapping. One innovation involves the use of mapping as a means of discovery and analysis built through social movements. A number of research collectives have taken this approach to mapping patterns of dispossession and capital accumulation. Their work is perhaps less "community-based" than earlier versions of counter-mapping envisioned. Their strength lies instead with producing maps that are accessible and circulate socially, with the open possibility of creating new forms of alliance and struggle.

One approach has simply been to aggregate maps of indigenous and black territories, by way of drawing out common aspects of struggle among places that might otherwise seem disparate and distant. A recent effort to map Guaraní communities demonstrates this approach. Initially launched as a strategy for organizing communities displaced by the soy boom in Brazil, Argentina, and Paraguay, the project has since expanded to include communities in Bolivia and Uruguay as well (Centro de Trabalho Indigenista, 2008; Campahna Guaraní, 2016). The resulting map has been used in turn to elicit trans-border collaboration between Guaraní organizations, producing a set of political and spatial connections previously fractured by states (EMGC, 2016).

While significant, the sorts of connections made by the Guaraní mapping project posits a weak form of politics vulnerable to appropriation by other interests. To wit, the US-based World Resources Institute has coordinated a parallel effort to map Guaraní communities in Paraguay, using many of the same techniques and some of the field partners from the broader effort. Funded by the US Agency for International Development, the WRI project contends than mapping indigenous land rights is a key step towards opening up space for the expansion of Paraguay's booming cattle industry through improved "supply chain management" and avoidance of conflicts with Guarani communities. The project is modelled on the Indonesia government's "One Map" project, and seeks to "cartographically 'unify' Paraguay's mapping systems by integrating them" (Veit and Sarsfield, 2017: 40). How Guaraní communities respond to this sort of cartographic primitive accumulation remains to be seen.

Other efforts have used maps to document the drivers of displacement and dispossession that affect various poor and rural populations. The Argentina-based group Iconoclasistas have done this through their mapping of the soy complex, documenting its serial dispossessions of land and bodies through consolidation of property, agrochemical use, and displacement of labour. Iconoclasistas have used interest in their maps to train people in how to produce their own maps as means of analysing common problems and responding collectively to the spatial dynamics of capital accumulation (Iconoclasistas, 2013). The GeoComunes collective in Mexico has taken a similar approach, compiling maps of state-issued mining and energy concessions across the country.[1] Human Rights Everywhere/Geoactivismo in Colombia has also focused on mapping concessions alongside compiled maps from black and indigenous communities.[2] Collectively these efforts chart the line between capital accumulation and an outside that it cannot tolerate, drawing attention to the zero point at which capitalist social relations are spatially created in land and bodies (De Angelis, 2017; Federici, 2012).

Body mapping and corporeal cartographies expand understandings of dispossession into the realm of unpaid labour and social reproduction (Colectivo Miradas Críticas, 2017; Mignorance, 2017). Typically, body mapping efforts have been used to document the bodily effects of violence. Much like efforts to map land as property, they risk complicity in the production of individualized victimhood that has long been one of neoliberalism's key sites of subject formation. However, there are growing numbers of corporeal cartography projects that seek to connect bodies with the situations where such violence occurs as demonstrated by the work of the GeoBrujas collective in Mexico. Their work productively analyzes how capitalist accumulation operates in and through the production of differences of gender, race, and class. They also counter the idea of the individual body as the object of analysis, resituating it within social relations of reproduction and capital accumulation.

Those efforts resonate with a growing number of mapping projects that attempt a more rigorous engagement with different spatialities, challenging abstract understandings of space as homogenous for everyone, everywhere, and at all times. Some of this work follows the "ontological turn" in the social sciences, directing attention to indigenous understandings of space (and time). That effort usefully breaks with the dominant understanding of abstract space as a unified thing. Instead of understanding it through measurement and representation, indigenous ontologies are immersed in the experience of being in a mesh of social and more-than-human relations (Blaser, 2014; De la Cadena, 2015; Oslender, 2016). Many of these efforts disrupt the dominance of maps as means of understanding and seeing space. A design painted on a ceramic bowl may contain profound spatial and historical information with reciprocal effects on bodily comportment and social interaction. As tempting as it may be to translate that information on to maps, it has to be made clear that any such effort places greater value on passing knowledge from one setting to another than it does on the fidelity to meaning and form of the original. That difference contrasts with counter-mapping practices that aim to transpose alternative ontologies onto dominant understandings of space through cartography in what amounts to a kind of unidirectional or monolingual translation that leaves hierarchies of knowledge and power intact. At the same time, the emphasis on ontology doubles down on the importance of difference as a means of structuring knowledge and space with mixed effects. If there is one spatial aspect of capitalism that rises above the rest, it is its repeated capacity to colonize difference as a vehicle for capitalist accumulation (Bessire and Bond, 2014). The nuances of alternative ontologies scarcely need to be understood in order for capitalism to exploit them.

An atlas produced by a group of Mapuche researchers makes headway on that task in Chile (Melin, Mansilla, and Royo, 2017). The atlas uses a series of historical maps to show how Wallmapu – the living space of the Mapuche people – has been colonized by national cartographies of Chile and Argentina through ethnic differentiation, military invasion, and agrarian reform. Through these practices, Wallmapu has been reduced to a Mapuche territory differentially claimed and inhabited by a heterogenous array of Mapuche social movements and communities. The historical maps, many lifted from the colonial archive, are therefore presented alongside maps produced by the authors of the atlas in conjunction with Mapuche communities. The overall attempt is one of decolonizing the very idea of Mapuche territory through making Wallmapu visible again. Part of that task requires dismantling notions of "the Mapuche" as occupying a homogenous, ethnically defined territory. In its place, the atlas cultivates an appreciation of the differences in location and experience of Mapuche communities, producing Wallmapu as space of dialogue and exchange that slips away from state constructions of Mapuche identity and territory. Far from being the projection of a singular spatial rationale, the maps join together different languages and knowledges. To read them requires an immersion in both with an eye towards how they arrange different knowledges and space/times. The state and community maps work dialectically to develop this analysis, underscoring the authors' intent to produce a new space rather than reclaim and defend an old one. Their effort puts forth an idea of mapping as a practice of decolonization through the production of a new space.

The Wallmapu atlas resonates with efforts by social movements to produce new territorialities, pluralizing understandings of space through proliferation. Zapatista communities' refusal of maps and state recognition are one of more discussed examples of this trend, directing their efforts instead towards producing the spaces necessary for reproduction of a collective form of life (Reyes and Kaufman, 2015; Reyes, 2015). Theirs is a geography that attempts a break with abstract understandings of space as comprised of so many pieces that fit together like a jigsaw puzzle, occupying the cracks and fissures in state geographies while opening and producing others (Marcos, 1997, 2018). The Zapatistas' approach has been critiqued for its inability to organize alliances equal in social and spatial reach to the capitalist totality they confront (Wilson, 2014; Harvey, 2017). And yet, there are any number of efforts across Latin America involved in similar kinds of efforts, ranging from rubber tapping communities in the Amazon to "slum" dwellers (Porto-Gonçalves, 2001; Zibechi, 2012). These efforts share an effort to rethink politics and space, rejecting the emphasis on state power that binds legal recognition with counter-hegemonic movements from the left (Reyes, 2012).

Conclusion

Mapping, counter and otherwise, will not lead to any immediate "counter-hegemonic" alternative to development. But that should not define its limits. Rather, mapping provides a means of critically challenging the abstract notion of space historically essential to colonization and capitalist accumulation. The stakes are admittedly high. Every new resource concession, privatization, and frontier makes the defence of territory all the more urgent and pressing. At the same time, it reinforces the dominance of an understanding of space constituted through dispossession. The challenge of mapping is to chart the contours of that process, recognizing its constitutive exclusions and displacements to produce a space in which indigenous peoples, Afro-descendants, women, the poor can only every have a narrowly constrained part. All of these processes can be grasped cartographically, though their complexity exceeds both maps and the understanding of space that they rely on and produce. If counter-mapping is of any use, it lies with its ability to use those excesses to challenge the primacy of maps as a means of understanding space, and the regime of power that accompanies it.

Notes

1 *GeoActivismo*. https://geographiando.net (Accessed 21 April 2018).
2 *GeoComunes*. http://geocomunes.org (Accessed 21 April 2018).

References

Websites

Almeida, A. W. B. D. (2013) Nova Cartografia Social: territorialidades específicas e politização da consciência das fronteiras. In A. W. B. D. Almeida and E. D. A. Farias Jr. (eds) *Nova Cartografia Social: Povos e Comunidades Tradicionais*. Manaus, BR: UEA Ediciones, pp. 157–173.

Anthias, P. and Radcliffe, S. A. (2015) The ethno-environmental fix and its limits: Indigenous land titling and the production of not-quite-neoliberal natures in Bolivia. *Geoforum* 64: 257–269.

Asher, K. (2009) *Black and Green: Afro-Colombians, Development, and Nature in the Pacific Lowlands*. Durham, NC: Duke University Press.

Bessire, L. and Bond, D. (2014) Ontological anthropology and the deferral of critique. *American Ethnologist* 41(3): 440–456.

Blaser, M. (2014) Ontology and indigeneity: On the political ontology of heterogeneous assemblages. *Culutral Geographies* 21(1): 49–58.

Bryan, J. (2009) Where would we be without them? Knowledge, space, and power in indigenous politics. *Futures* 41(1): 24–32.

———. (2011a) Walking the line: Participatory mapping, indigenous rights, and neoliberalism. *Geoforum* 42: 40–50.

———. (2011b) Cartografías del colonialismo: mapeo indígena en Nicaragua. *Boletín de la Asociación para el Fomento de los Estudios Históricos en Centroamérica*, 48. http://afehc-historia-centroamericana.org/index.php?action=fi_aff&id=2586 (Accessed 21 April 2018).

———. (2012) Rethinking territory: Social justice and neoliberalism in Latin America's Territorial turn. *Geography Compass* 6(4): 215–226.

——— and Wood, D. (2015) *Weaponizing Maps: Indigenous Peoples and Counterinsurgency in the Americas*. New York: Guilford Press.

Campahna Guaraní. (2016) *Mapa Guaraní Continental*. www.socioambiental.org/pt-br/mapas/mapa-guarani-continental-2016 (Accessed 21 April 2018).

Centro de Trabalho Indigenista. (2008) *Mapa Guaraní Retã*. http://guarani.roguata.com/map (Accessed 23 April 2018).

Colectivo de Miradas Críticas del Territorio desde el Feminismo. (2017) *Mapeando el Cuerpo-Territorio: Guía Metodológica para Mujeres que Defienden sus Territorios*. Quito, Ecuador: Territorio y feminismo, Instituto de Estudios Ecologistas del Tercer Mundo, y CLACSO.

Coronil, F. (2000) Towards a critique of globalcentrism: Speculations on capitalism's nature. *Public Culture* 12(2): 351–374.

Counter Cartographies Collective, Dalton, C. and Mason-Deese, L. (2012) Counter (mapping) Actions: Mapping as militant research. *ACME: An International e-Journal for Critical Geographies* 11(3): 439–466.

Craib, R. (2004) *Cartographic Mexico: A History of State Fixations and Fugitive Landscapes*. Durham, NC: Duke University Press.

Cruz, M. (2010) A living space: The relationship between land and property in the community. *Political Geography* 29(8): 420–421.

De Angelis, M. (2017) *Omnia sunt communia: On the commons and the Transformation to Postcapitalism*. London: Zed Books.

De la Cadena, M. (2015) *Earth beings: Ecologies of Practice across Andean worlds*. Durham, NC: Duke University Press.

Dym, J. and Offen, K. (2011) (eds) *Mapping Latin America: A Cartographic Reader*. Chicago: University of Chicago Press.

EMGC (Equipo Mapa Guaraní Continental). (2016) *Cuaderno Mapa Guaraní Continental: Pueblos Guaraníes en Argentina, Bolivia, Brasil, y Paraguay*. Equipo Mapa Guaraní Continental, Campo Grande, MS.

Escobar, A. (2010) Latin America at a crossroads. *Cultural Studies* 24(1): 1–65.

Federici, S. (2012) *Revolution at Point Zero: Housework, Reproduction, and Feminist Struggle*. Oakland, CA: AK Press.

Gordon, E., Gurdián, G. and Hale, C. (2003) Rights, resources, and the social memory of struggle: Reflections on a study of Indigenous and Black community land rights on Nicaragua's Atlantic Coast. *Human Organization* 62(4): 369–381.

Hale, C. R. (2005) Neoliberal Multiculturalism: The remaking of cultural rights and racial dominance in Central America. *Political and Legal Anthropology Review* 28(1): 10–28.

———. (2006) Activist research v. cultural critique: Indigenous land rights and the contradictions of politically engaged anthropology. *Cultural Anthropology* 21(1): 96–120.

Hall, S. (1992) The West and the rest: Discourse and power. In S. Hall and B. Gieben (eds) *The Formations of Modernity: Understanding Modern Societies*. Cambridge: Polity Press, pp. 275–332.

Harvey, D. (2017) Listen, Anarchist! A personal response to Simon Springer's 'Why a radical geography must be anarchist'. *Dialogues in Human Geography* 7(3): 233–250.

Herlihy, P. H. and Knapp, G. (2003) Maps of, by, and for the peoples of Latin America. *Human Organization* 62(4): 303–314.

Herlihy, P. H. et al. (2008) A digital geography of indigenous Mexico: Prototype for the American geographical society's Bowman expeditions. *The Geographical Review* 98(3): 395–415.

Hodgson, D. L. and Schroeder, R. A. (2002) Dilemmas of counter-mapping community resources in Tanzania. *Development and Change* 33(1): 79–100.

Iconoclasistas. (2013) *Manual de Mapeo Colectivo: Recursos Cartográficos Críticos para Procesos Territoriales de Creación Colaborativa*. Buenos Aires: Tinta Limón.

Lefebvre, H. (1991) *The Production of Space*. Oxford: Wiley-Blackwell.

Lewis, M. W. and Wigen, K. E. (1997) *The myth of continents; a critique of metageography*. Berkeley, CA: University of California Press.

Marcos, S. I. (1997) *The Seven Loose Pieces of the Global Jigsaw Puzzle: Neoliberalism as a Puzzle, the Useless Global Unity Which Fragments and Destroys Nations*. www.elkilombo.org/documents/sevenpiecesmarcos.html (Accessed 21 April 2018).

———. (2018) *The Zapatistas' Dignified Rage: Final public speeches of Subcommander Marcos*. Chico, CA: AK Press.

Marx, K. (1990) *Capital, Volume I*. London: Penguin Books.

McSweeney, K. et al. (2014) Drug policy as conservation policy: Narco-deforestation. *Science* 343(6170): 489–490.

Melin, M., Mansilla, P. and Royo, M. (2017) *Mapu Chillkantukun Zugu: Descolonizando el mapa del Wallmapu, construyendo cartografía cultural en territorio Mapuche*. Temuco: Pu Lof Editores.

Mignorance, F. (2017) *Las geografías del cuerpo*. https://geoactivismo.org/las-geografias-del-cuerpo/ (Accessed 21 April 2018).

Miranda Aburto, W. (2016) El infierno de los miskitos. *Confidencial*, 7 June, online. https://confidencial.com.ni/infierno-los-miskitos/ (Accessed 21 April 2018).

Mollett, S. (2013) Mapping deception: The Politics of mapping Miskito and Garifuna space in Honduras. *Annals of the Association of American Geographers* 103(5): 1227–1241.

Morton, A. D. (2013) *Revolution and State in Modern Mexico: The Political Economy of Uneven Development*. Lanham, MD: Rowman & Littlefield.

Nahuelpan Moreno, H. et al. (2012) (eds) *Ta Iñ Fijke Xipa Rakizuameluwün: Historia, Colonialismo y Resistencia desde el País Mapuche*. Temuco: Ediciones Comunidad de Historia Mapuche.

Nietschmann, B. Q. (1995) Defending the Miskito reefs with maps and GPS: Mapping with sail, scuba and satellite. *Cultural Survival Quarterly* 18(4): 34–37.

Offen, K. (2003) The territorial turn: Making black territories in Pacific Colombia. *Journal of Latin American Geography* 2(1): 43–73.

———. (2009) O mapeas o te mapean: Mapeo indígena y negro en América Latina. *Tabula Rasa* 10: 163–189.

Oslender, U. (2016) *The Geographies of Social Movements: Afro-Colombian Mobilization and the Aquatic Space*. Durham, NC: Duke University Press.

Peluso, N. L. (1995) Whose woods are these? Counter-mapping forest territories in Kalimantan, Indonesia. *Antipode* 27(4): 383–406.

Porto-Gonçalves, C. W. (2001) *Geo-grafías: Movimientos Sociales, Nuevas Territorialidades y Sustentabilidad*. México, DF: Siglo XXI.

Postero, N. (2017) *The Indigenous State: Race, Politics, and Performance in Plurinational Bolivia*. Oakland, CA: University of California Press.

Restrepo, G., Velasco, A. and Preciado, J. C. (1999) *Cartografía social*. Tunja, CO: Universidad Pedagógica y Tecnológica de Colombia.

Reyes, A. (2012) Revolutions in the revolutions: A post-counterhegemonic moment for Latin America? *South Atlantic Quarterly* 111(1): 1–27.

———. (2015) Zapatismo: Other geographies circa "the end of the world". *Environment and Planning D: Society and Space* 33(3): 408–424.

——— and Kaufman, M. (2015) Sovereignty, indigeneity, territory: Zapatista autonomy & the new practices of decolonization. In F. Luisetti, J. Pickles and W. Kaiser (eds) *The Anomie of the Earth: Philosophy, Politics, and Autonomy in Europe and the Americas*. Durham, NC: Duke University Press, pp. 44–68.

Rus, J. (1994) The comunidad revolucionaria institucional: The subversion of native government in Highland Chiapas. In G. Joseph and D. Nugent (eds) *Everyday Forms of State Formation: Revolution and the Negotiation of Rule in Modern Mexico*. Durham, NC: Duke University Press, pp. 265–300.

Salamanca, C. and Espina, R. (2012) *Mapas y Derechos: Experiencias y Aprendizajes en América Latina*. Rosario, Argentina: Editorial de la Universidad Nacional de Rosario.

Sletto, B. I. (2009) "We Drew What We Imagined": Participatory mapping, performance, and the arts of landscape making. *Current Anthropology* 50(4): 443–476.

———, Bryan, J., Torrado, M., Hale, C. and Barry, D. (2013) Territorialidad, mapeo participativo y política sobre recursos naturales: La experiencia de América Latina. *Cuadernos de Geografía: Revista Colombiana de Geografía* 22(2): 193–209.

Smith, R. C. et al. (2003) Mapping the past and the future: Geomatics and indigenous territories in the Peruvian Amazon. *Human Organization* 62(4): 357–368.

Stevens, S. (1997) (ed) *Conservation Through Cultural Survival: Indigenous Peoples and Protected Areas*. Covello, CA: Island Press.

Stocks, A. (2003) Mapping dreams in Nicaragua's Bosawas reserve. *Human Organization* 62(4): 344–356.

Veit, P. and Sarsfield, R. (2017) *Land Rights, Beef Commodity Chains, and Deforestation in the Paraguayan Chaco*. Washington, DC: USAID Global Tenure and Climate Change Program.

Wainwright, J. (2008) *Decolonizing Development: Colonial Power and the Maya*. Malden, MA: Wiley-Blackwell.

———. (2013) *Geopiracy: Oaxaca, Militant Empiricism, and Geographical Thought*. New York: Palgrave Pivot.

——— and Bryan, J. (2009) Cartography, territory, property: postcolonial reflections on indigenous counter-mapping in Nicaragua and Belize. *Cultural Geographies* 16(2): 153–178.

Wilson, J. (2014) The violence of abstract space: Contested regional developments in southern Mexico. *International Journal of Urban and Regional Research* 38(2): 516–538.

Zibechi, R. (2012) *Territories in Resistance: A Cartography of Latin American Social Movements*. Oakland, CA: AK Press.

PART IV

Gender and sexuality, cultural politics and policy

23

GENDER, POVERTY, AND ANTI-POVERTY POLICY

Cautions and concerns in a context of multiple feminizations and 'patriarchal pushback'

Sarah Bradshaw, Sylvia Chant, and Brian Linneker

Introduction

This chapter considers how women have been incorporated into development policy in Latin America in recent decades with particular reference to Conditional Cash Transfer (CCT) programmes, which have increasingly been targeted to, and through, female beneficiaries. Drawing on existing data and case study evidence, our discussion critically examines the 'engendering' of development policy and the feminization of poverty-reduction programmes, and their success (or otherwise) in helping to forge pathways out of poverty for women and girls. Given mixed outcomes in respect of alleviating gender-differentiated burdens of income and asset poverty, we highlight a number of shortfalls and exclusions. In particular we argue that anti-poverty efforts targeting women have been more efficiency- than equality-driven, and in many cases have done relatively little to address gendered rights and responsibilities within households and communities. The delivery of economic resources through women does not seem to have redressed persistent gendered disparities in time, voice, mobility, or security, let alone income poverty.

Beyond trends relating to the feminization of poverty alleviation, there have been other discernible feminizations in Latin America since the 20th century, including of urban populations and household headship, of educational attainment, and employment/labour markets, all buttressed by a feminization of policy initiatives, family law, and human rights, and by a feminization of politics and protest, both formal and informal (see Chant with Craske, 2003; Chant and McIlwaine, 2016). One less than sanguine example of feminization, however, is 'femicide' or the killing of women for being women (see Staudt, 2008; Sweet and Ortíz Escalante, 2010). That this has occurred at a time of falling poverty in general, if not for women, leads us to question how different 'feminizations' interact and how this might help us understand gendered experiences of poverty and privation in contemporary Latin America.

In addition to stressing the need to adopt a more multidimensional approach to poverty, we underline the importance of complementary interventions, not least relating to gender-based violence (GBV) and violence against women and girls (VAWG). For policy interventions to make real advances in gender equality we propose that a broader stance on gendered privations, which takes into account intersecting sites of oppression and resistance among men as well as women, might aid in achieving this goal.

Measuring and understanding gendered poverty

Despite advances in multidimensional poverty indicators (see Alkire et al., 2013), and attempts to establish measures of individual deprivation (Wisor et al., 2014), household income continues to be the benchmark in international arenas, as in Target 1.1 of the headline Sustainable Development Goal (SDG) to eradicate poverty. Yet income is not the sole or even principal factor in determining well-being among women and girls, with other privations arguably being of equal (or greater) importance, such as 'time poverty,' 'asset poverty,' and 'power poverty,' all of which interrelate with one another to some degree (see Chant, 2007, 2008). While women might increasingly engage in income-generating or 'productive' activities, potentially reducing their income poverty, the fact that they generally have to combine productive work with reproductive tasks in their homes and communities means that their time poverty becomes greater. Women's heavy workloads and paltry opportunities for rest and recuperation can have negative implications for their health, and in turn rebound negatively on income poverty (ibid.; see also Chant, 2016b; Gammage, 2010; Noh and Kyo-Seong, 2015). Due to gendered societal norms, it may be inordinately difficult for women to translate their own income into voice and agency in the home such that they end-up continuing to experience power poverty marked, inter alia, by limited control over household decision-making and household assets such as land and housing (Bradshaw, 2002, 2013; Chant, 1997b, 2007; Kabeer, 2003: 198; Murphy, 2015: 77).

Power poverty impacts not only on people's capacity to access, own, or control tangible assets such as household income, but also on more intangible assets such as mobility and security. Women and girls experience differing poverties in variegated ways, which are often critically contingent on how gender intersects with class, 'race,' sexuality, age, and so on. Yet in many parts of Latin America, not to mention across the globe, several fathers, husbands, brothers, and sons curb the mobility of 'their' womenfolk by attempting to keep them 'off the streets' and out of public places, and confining them primarily to the socially acceptable space of the home and immediate neighbourhood (see Chant and McIlwaine, 2016). Some men exert control over women's lives through policing their bodies and sexualities, with violence, and the threat of violence, often being crucial mechanisms. Gendered power imbalances in intimate personal relationships are further institutionalized through governance regimes and legal systems, education, and religion. Patriarchal structures exist at every level of society, and households may be as much sites of oppression as of solidarity for many women (Chant, 2007, 2016a).

While the majority of households have at their core a heterosexual couple, other household forms exist and are growing in number, including female-headed households (FHHs). FHHs are a fluid and diverse group, varying in respect of their composition, age, access to support from ex-partners and the state, as well as in the catalysts underpinning their formation (Chant, 1997a, 2016a). Although 'male-deficit' household arrangements (Barrow, 2015) may be associated with income poverty, in that globally women earn less than men and are clearly bereft of a co-resident male partner's wage, for some women, heading their own domestic units may yield non-monetary benefits. This includes greater control over assets and life decisions, and more tranquility and security, especially if women have extricated themselves from violent spousal relationships. Beyond this, and notwithstanding that some FHHs are at an above-average risk of privation, especially when they comprise a lone woman and young dependent children, a number of studies reveal little difference in income poverty between female and male-headed households (MHHs) (Bradshaw, 2002; Chant, 2007), and in Latin America, specifically, there continues to be a very uneven picture (Liu, Esteve, and Treviño, 2017).

Mixed evidence as to the extent to which FHHs are economically poorer than MHHs is riddled with definitional and data-related issues (see Bradshaw, Chant, and Linneker, 2017; Chant, 2003a, 2003b, 2007, 2016a). It is also clouded by how the feminization of household headship interacts with the feminization of poverty and the feminization of policy. Official statistics point to growing proportions of self-declared FHHs in Latin America generally. This conceivably owes to increased ability, confidence, and strategic instrumentality to self-report (Chant and McIlwaine, 2016: 75; Liu, Esteve, and Treviño, 2017), as well as to a range of gender equality, and female-oriented anti-poverty, interventions which allow women the possibility of exiting injurious domestic arrangements (Chant, 2009: 26–27). The question as to whether feminized anti-poverty programmes contribute to patterns of feminized poverty is interesting in this context.

'Feminization of poverty'

Most Latin American countries have risen to the ranks of 'middle income' status over recent decades, yet inequalities persist. In the World Income Inequality Database for 2017, Latin America stood out as having national Gini coefficients consistently in the top two highest categories (UNU-WIDER, 2017). Inequality is also apparent in the realm of gender, notwithstanding marked variations. For example, the Global Gender Gap Report (GGGR) for 2016 placed Nicaragua (an albeit questionable) tenth out of the 144 countries for which the index was calculated, while Guatemala came 105th (see WEF, 2016). Yet, while the situation of women across the region varies, their relative poverty is often assumed as a generic given.

Since the Fourth World Conference on Women in Beijing in 1995 it has become commonplace to accept that poverty has a 'female face' and that women are "70% of the world's poor, and rising" (Chant, 2016b: 1–2; see also UNDP, 1995). This suggests not only that poverty is feminized but also that there is a 'feminization of poverty' – or an increase in numbers of women relative to men among the poor over time. A 'feminized' or 'feminizing' poverty has also typically been linked with the 'feminization' of household headship, with FHHs resolutely constructed as the 'poorest of the poor.' These assumptions have remained remarkably persistent on account of relentless repetition in scholarly and policy circles, as well as in popular media (see Chant, 2008: 16, 2016b: 2; also Bradshaw, Chant, and Linneker, 2017; Wisor et al., 2014). As recently as 2016, for example, the Deputy Director of UN Women suggested that sustainable development is not possible if the "feminisation of poverty" continues (Puri, 2016), even if the previous year UN Women's *Progress of the World's Women 2015–2016* stated that "it is unknown how many of those living in poverty are women and girls" (UN Women, 2015: 45).

Analysing existing studies which have sought to measure gendered poverty across time and space (notably by Medeiros and Costa, 2006, 2008 and ECLAC, 2014), suggests that while there had been a 'de-feminization' of poverty in many Latin American countries in the 1990s/early 2000s, some nations have recently witnessed a 're-feminization' of poverty (Bradshaw, Chant, and Linneker, 2017). Our analysis of existing data which compares FHHs with MHHs suggests some relatively robust trends for six countries, pointing to poverty undergoing (re)feminization in Chile, Costa Rica, and Mexico, and de-feminization in Bolivia, Brazil, and Venezuela.

However, these trends might have as much to do with how we measure poverty as actual poverty (Bradshaw, Chant, and Linneker, 2017). In part this is because poverty measures continue to adopt the household as the unit of assessment and comparisons are accordingly between FHHs and MHHs rather than between women and men (ibid.).

Heteronormative assumptions pertaining to what constitutes a household mean that non-normative, same-sex, households are often rendered invisible, while disaggregating data by sex occludes trans people and/or those with fluid gender identities. It is interesting that while lone mother/female-headed households are often still not socially accepted, they have become a readily adopted evaluative category for comparing men's and women's relative poverty, despite scholarly protestations to the contrary (see Chant, 1997a, 1997b, 2003a, 2007; Kabeer, 2003: 81).

Feminized international policy discourse

From the late 20th century major international development agencies have promoted gender mainstreaming and the 'engendering' of development. World Bank research in the early 2000s, for example, highlighted that societies with higher levels of gender discrimination tend to experience slower economic growth (World Bank, 2001; World Bank GDG, 2003). When women and girls have systematically lower access to education and health this translates into 'less than optimal' levels of labour market participation and entrepreneurship. At the same time better-educated women have lower fertility as well as lower infant and child mortality rates. Thus, for the World Bank, equality in education can help increase a country's economic growth, and curtail population expansion, thereby bringing about per capita economic benefits. The World Bank also recognizes the efficiency gains from channelling resources through women, as epitomized in its flagship 2012 *World Development Report on Gender Equality and Development* (World Bank, 2011). Herein, 'smart economics' justifications for 'investing in women' remain alive and well beneath the veneer of promoting gender equality as a fundamental human right (see Chant, 2012; Chant and Sweetman, 2012).

Women have become a key target for social safety nets and other welfare programmes, most notably the CCTs favoured by international development banks. Latin America is home to two of the earliest cash transfer programmes – *Progresa/Oportunidades* (formerly *Solidaridad*) in Mexico and *Bolsa Família* in Brazil. These interventions, designed and promoted by their respective national governments, provide cash transfers which aim to reduce the poverty of households now and in the future, through investing in the health and education of children, eliminating child labour, and enhancing adult participation and productivity in labour markets (see Buvinič, Das Gupta, and Casabonne, 2009; Johnston, 2016; Lavinas, 2016). Women are the 'beneficiaries' of monetary disbursements and are also responsible for compliance with future-proofing anti-poverty measures. Since investing in women is seen to be one of the most efficient routes to ensuring wider development aims, there has been a "generalised bid to alleviate poverty primarily, or even exclusively, through women," effectively producing a "feminisation of poverty alleviation" (Chant, 2003a: 27, 2008; Roy, 2002). Indeed, in Latin America the percentage of women participating in poverty reduction programmes has historically been higher than the percentage of women identified as poor (ECLAC, 2004). Yet dealing with poverty is "arguably as onerous and exploitative as suffering poverty" (see Chant, 2016b: 19), and while UN Women (2015: 45) suggest that part of the reason for the reported general global decline in poverty are "new social policies" which have included women (see also GEOLAC, 2013), it is perhaps no surprise that gender-differentiated poverty shows signs of persisting, if not deepening, in several parts of Latin America.

The feminization of poverty alleviation

CCTs have been described as the most important innovation in social policy in Latin America over the past 15 years (Levy, 2015). Following Mexico's *Progresa/Oportunidades*, and Brazil's *Bolsa*

Familia, similar schemes began to emerge across the region (Bradshaw, 2008; Feitosa de Britto, 2007; Jenson and Nagels, 2018; Johnston, 2016; Molyneux and Thomson, 2011; Tabbush, 2010). As previously intimated, CCTs have a twofold aim – to address immediate material deprivation, and to tackle the intergenerational transmission of poverty through building children's capabilities (see González de la Rocha and Escobar Latapí, 2016; Johnston, 2016; Lavinas, 2016; Molyneux, 2006, 2007). Although cash transfers are paid to women who have the responsibility to ensure that programme conditions are fulfilled, they are not necessarily aimed at improving the situation of women or reducing women's poverty *per se* (see Chant, 2016b). Indeed, available data suggest that while CCTs have reduced the gap between women's income and poverty thresholds, the contribution is small, at 12% of the extreme poverty line, and 7% of the general poverty line (GEOLAC, 2013: 57; see also Lavinas, 2016). This owes arguably to the fact that CCTs tend to focus more on the 'empowering,' rather than the income poverty-reducing, nature of cash transfers for women. The suggestion that providing women with money of 'their own' allows them to make decisions over what is purchased for the household (Adato and Hoddinott, 2010) conceivably reduces feminized 'power poverty.' However, do CCT programmes which rely on women's socially constructed altruistic behaviour as mothers not only limit economic well-being gains for women themselves, but also obstruct pathways out of feminized 'power,' as well as 'income' and 'asset' poverty?

Given the extensive coverage and longevity of the Mexican and Brazilian CCTs, they might be expected to have had an impact on national level poverty, and, as they target women, gendered poverty too. Yet from our own analysis of available data, there appears to have been a feminization of income poverty in Mexico over time, compared with a recent de-feminization in Brazil (see Bradshaw, Chant, and Linneker, 2017, 2018). In Mexico, the priority of CCTs in their various incarnations has been to build human capital, while the main objective in Brazil has been to transfer resources to poor households. Although measures which target resources to women with the aim of strengthening the human capital of children reduce general income poverty, and reveal quite significant advances for boys, and often especially girls, in terms of health and education outcomes (see González de la Rocha, 2008; González de la Rocha and Escobar Latapí, 2016), they appear to do little to reduce the income poverty of women. Indeed, detailed micro-level intra-domestic studies suggest that when women bring income into the home male partners may retain more of their earnings for discretionary personal use (Bradshaw, 2002, 2013; Chant, 2007, 2016b). Moreover, when women's income contributions are not 'earned,' and are construed as part of women's mothering role, male partners may not value these resources and thus inhibit gains in women's personal agency (Bradshaw, 2008). While women's income, asset, and power poverty might not be reduced by CCTs, it is also the case that their time poverty can increase. CCTs demand that young children are taken for health checks and that women participate in 'voluntary' community tasks, often at quite considerable distance from their homes, thereby requiring major temporal investments and trade-offs (see Molyneux, 2006, 2007). Underlying assumptions that women have 'spare' time or are not 'working', despite their frequent involvement in income-generating ventures, negates the reality that housework is also work, and typically time-consuming and arduous, especially in poor neighbourhoods (Chant, 2014; Chant and McIlwaine, 2016).

While the requirements of CCT participation impact upon women's available time for other activities it should also be recognized that CCT disbursements are seldom sufficient to guarantee a viable living (Chant, 2016b; Lavinas, 2016: 2). In order to meet programme conditions many women, particularly female heads, either have to reduce or curtail involvement in income-generating activities and thus risk aggravated monetary poverty, or juggle income-generating activities with CCT demands and increase their time poverty. Indeed, evidence suggests that female CCT 'beneficiaries' spend fewer hours in the labour market and more doing unpaid domestic work and caregiving than women who do not receive cash transfers (GEOLAC, 2013: 62). This points to some women, such as female heads, being effectively 'priced out' of the anti-poverty programmes which target them, with the need to find cash meaning that they do not have sufficient time to comply with the conditions (see Bradshaw, Chant, and Linneker, 2017; Feitosa de Britto, 2007; Hernández Pérez, 2012). This has led to the suggestion that "the capacity of the CCTs for transforming the lives of poor women through the transfer of monetary income (one step forward) is more than neutralized by the consolidation of their caregiver role, which has multiple negative implications (two steps back)" (GEOLAC, 2013: 61).

The gendered, maternalist stereotypes on which CCTs are predicated may also entrench a variety of pernicious inequalities between women and men, with Johnston (2016: 112) alerting that: "It is worrying that . . . cash transfers may have cemented unequal gendered norms over the responsibility for household welfare" (see also Chant, 2014; Escobar Latapí and González de la Rocha, 2009; Molyneux, 2006, 2007; Molyneux and Thomson, 2011; Tabbush, 2010). Yet the very existence of policies and programmes aimed at women may be enough to influence popular perceptions that they are benefiting, and more so than men (Chant, 2009; Milanich, 2017). However, while the policy focuses on women and girls may create perceptions of positive change relative to men and boys, this may not reflect reality. Perceptions of positive change may also provoke a negative response, or male 'backlash,' which appears to manifest itself in GBV in various guises.

Feminizations and femicides

Notwithstanding greater reportage of VAWG over time, especially in cities, there seems to have been a rise in extreme and public acts of VAWG in most Latin American nations in recent years, including acid attacks (Colombia), fatal rapes (Argentina, Brazil), and more generally increased assaults on women in public spaces, frequently taking the form of sexualized attacks involving groups of men who sometimes record and circulate these online. The rising number of women dying from violent acts has been such that the concept of 'femicide' has evolved to express this trend.

Femicide is a crime involving the violent and deliberate killing of women on grounds of their gender. Although globally men and boys still account for the majority of the people who suffer violent deaths, women and girls are 16% of violent fatalities, and during 2010–2015 there were 64,000 women killed violently each year (Widmer and Pavesi, 2016: 1). Since femicide has only just begun to feature explicitly in criminal codes in the region (GEOLAC, 2015), longitudinal statistics are hard to come by and are prone to underestimate the extent of the problem (Gabor, 2016). None the less, estimates suggest that over half the countries with very high femicide rates at a world scale are in Latin America (UN Women, 2016; Waiselfisz, 2015; Widmer and Pavesi, 2016). Mexico is ranked sixth globally for crimes against women (Waiselfisz, 2015), and it is conjectured that husbands, boyfriends, or family members are implicated in at least 60% of femicides (UN Women, 2013). Despite significant attempts by the Mexican government to address the issue, including legal recognition of femicide as a crime, there is still an estimated impunity rate of 95% (Gabor, 2016: 1). In 2015, seven women on average in Mexico were killed each day as a result of GBV, a rate 15 times higher than the world average (ibid.), and rising, with an estimated increase in femicide of 46% since 2013 (Waiselfisz, 2015).

Among possible drivers of VAWG in Latin American countries are the legacy of armed and civil conflict and the rise in organized crime and gang formation (especially male gangs), which may not only account for high levels of generalized violence, but also GBV, including male-on-male violence (Jones and Rodgers, 2009; Moser, 2004, McIlwaine, 2013; Wilding and Pearson, 2013). Indeed, data point to nations with the highest rates of violent death for men and boys also having the highest rates for women and girls (Widmer and Pavesi, 2016: 2). A link between urbanization, poverty, and inequality has also been implicated (Choup, 2016). In urban slums, for instance, men may compensate for their loss of ability to fulfil socially prescribed roles of protector and provider by (re)asserting authority and power through GBV and VAWG. Recent trends suggest that urban men may be responding 'collectively' to their perceived loss of power through ever more aggressive and public displays of violence (Chant and McIlwaine, 2016; Wilding, 2010; Wilding and Pearson, 2013). This is in spite of, and perhaps in part because of, ever more vocal and visible resistance to VAWG by regionally strong and internationally aligned grassroots feminist solidarity movements such as '*Ni Una Menos*' which, through strategic alliances and inspired use of social media, have attempted to drive home that aggression against women and girls should, and will, not be tolerated.

In the 1990s a purported 'crisis of masculinity' was identified as men's public and private patriarchy was threatened through female strides in education, employment, politics, and policy interventions (see Chant, 2009; Chant with Craske, 2003). As previously intimated, the 'over-achievement' of women and girls may well have fuelled male anxieties around erosion of their key means of controlling women's mobility, bodies, and sexuality. Reactions to masculine 'role failure' in many instances have seen some men distancing themselves from shared daily life and obligations to families, households, and offspring, and often intensifying recourse to multiple relationships (polygyny) with invidious impacts on younger generations (Chant, 2009). Women's assumed efficiency in delivering policy outcomes and the political capital accruing to the explicit targeting of women in the name of gender equality, bolstered by the MDGs and SDGs from 2000 onwards, has arguably reinforced men's routine exclusion from social and anti-poverty policies. Recognition of women's rights has also led to a perceived penalization and pathologization of men in 'pro-female legislation' such as in the realm of 'responsible paternity,' which has conceivably exacerbated men's sense of alienation (see Chant, 2007, 2009; Johnston, 2016; Milanich, 2017). Although, as we have seen, when targeted by policymakers women do not always benefit, perceptions of relative gain may be enough to disenfranchise men to the extent that visceral male power is (re-)asserted. In the absence of 'alternative' gender scripts for men, particularly as 'hands-on' husbands and fathers, an 'aggravated machismo' appears to have emerged, with male violence being key to perpetuating masculine power over women and girls.

In her analysis of post-2011 'Arab Spring' VAWG in North Africa and the Middle East, Kandiyoti (2013) questions whether (re-)assertions of male power should be seen as "patriarchy in action" or "patriarchy in crisis," positing that "masculinist restoration" comes into play "at the point when patriarchy-as-usual is no longer fully secure, and requires higher levels of coercion to ensure its reproduction" (ibid.). Kandiyoti's notion that recourse to (public and collective) violence (or the condoning of violence) may represent more than a "resurgence of traditionalism," leads us to question whether we are witnessing an intensification of (traditional) expressions of masculinity, or new forms of asserting patriarchal control. Should we regard current exaggerated and violent expressions of masculinity in Latin America as 'abnormal' or as an intensification of the 'normal,' or, what Bradshaw, Linneker and Overton (2017), have referred to as "supernormal patriarchy"? That the feminization of policy, including anti-poverty policy, seems to have provoked an epidemic of intensified GBV may suggest a 'patriarchal pushback' against perceived, if not real, advances for women and girls. Rather than 'men in crisis,' given the feminizing trend in violent deaths in Latin America, as well as entrenching poverty, would a 'crisis for women' be a more appropriate concept?

Conclusions

While data to support a general trend toward a feminization of poverty in Latin America remain lacking, there has been a decided feminization of poverty alleviation programmes, and of wider initiatives to address gendered disparities in employment, education, heath, and political representation. Such interventions while, on the surface, well-intentioned and politically aligned with global development objectives, may be as much about facilitating economic growth as improving women's well-being.

Anti-poverty programmes such as CCTs may bring very few personal gains for women in terms of their ability to reduce income, asset or power privation. They may, also, simultaneously, increase female time poverty, despite claims by national governments and international development banks that CCTs are highly successful and 'empowering' for women, and are clearly construed at the grassroots by some men as exclusionary and hostile during an epoch in which their power and privileges, along with their customary roles as protectors and providers, have been undermined. The outcome of such programmes may accordingly be a male 'backlash' against women, individually and collectively.

While in the 1990s a purported 'crisis of masculinity' began to enter scholarly and policy discourse, in more recent years the manifestation of male crisis in acts of violence against women suggests that a new and rather different 'crisis for women' is presenting itself, both in degree and kind. The notion of 'patriarchal pushback' attempts to capture the new, more collective and public, retaliation of men as a group against women as a group in Latin America and beyond. It is encapsulated in visible and visceral acts of extreme violence against women which seek to limit the latter's mobility and autonomy through provoking fear on the street (*'calle'*) as well as in the home (*'casa'*). Notwithstanding mobilizations among women to protest such extreme acts, and legal and policy initiatives to tackle femicide, when the underlying reasons for rising GBV are not addressed, such violence may become the 'new normal.' This makes it imperative to tackle the range of issues at play in a holistic fashion, and to insist that policies which progress the feminization of poverty alleviation, education, employment, politics, and public life, do so for equality rather than efficiency reasons. Grassroots initiatives launched by NGOs such as Promundo and Puntos de Encuentro that seek to bring men on board in the struggle for gender egalitarianism alongside other forms of equality, may herald a way forward (see Chant, 2016b; Promundo, 2016; Solórzano et al., 2008).

Although 'patriarchal pushback' might be in part a product of policies ostensibly aimed at reducing feminized poverty, we also need to question whether the recent rise in VAWG is playing a part in re-feminizing poverty in Latin America through making women and girls fearful of capitalizing on their newfound capabilities. Without due interrogation and action, we could forfeit the achievements scored over several years of feminist struggle, rather than defend and extend women's and girls' rights.

Acknowledgements

Thanks are due to an Urban, Development and Planning Cluster Seed Fund Grant from LSE's Department of Geography and Environment for supporting our research.

References

Adato, M. and Hoddinott, J. (2010) (eds) *Conditional Cash Transfers in Latin America*. Washington, DC and Baltimore: International Food Policy Research Institute and John Hopkins University Press.

Alkire, S., Meinzen-Dick, R., Peterman, A., Quisumbing, A., Seymour, G. and Vaz, A. (2013) The women's empowerment in agriculture index. *World Development* 52(1): 71–91.

Barrow, C. (2015) Caribbean kinship research: From pathology to structure to negotiated family processes. In A. Coles, L. Gray and J. Momsen (eds) *The Routledge Handbook of Gender and Development*. London: Routledge, pp. 215–224.

Bradshaw, S. (2002) *Gendered Poverties and Power Relations*. Managua: Fundación Puntos de Encuentro.

———. (2008) From structural adjustment to social adjustment: A gendered analysis of onditional cash transfer programmes in Mexico and Nicaragua. *Global Social Policy* 8(1): 188–207.

———. (2013) Women's decision-making in rural and urban households in Nicaragua: The influence of income and ideology. *Environment and Urbanization* 25(1): 81–94.

———, Chant, S. and Linneker, B. (2017) Gender and poverty: What we know, don't know, and need to know for agenda 2030. *Gender, Place and Culture* 24(12): 1667–1688.

———. (2018) Challenges and changes in gendered poverty: The feminisation, de-feminisation and re-feminisation of poverty in Latin America. *Feminist Economics* (in press).

Bradshaw, S., Linneker, B. and Overton, L. (2017) Extractive industries as sites of supernormal profits and supernormal patriarchy? *Gender and Development* 25(3): 439–454.

Buvinič, M., Das Gupta, M. and Casabonne, M. (2009) Gender, poverty and demography: An overview. *The World Bank Economic Review* 23(3): 347–369.

Chant, S. (1997a) *Women-headed Households: Diversity and Dynamics in the Developing World*. Houndmills and Basingstoke: Palgrave Macmillan.

———. (1997b) Women-headed households: Poorest of the poor? Perspectives from Mexico, Costa Rica and the Philippines. *IDS Bulletin* 28(3): 26–48.

———. (2003a) *Female Household Headship and the Feminisation of Poverty: Facts, Fictions and Forward Strategies*, Gender Institute Working Paper, New Series, Issue 9, London School of Economics and Political Science. http://eprints.lse.ac.uk/574/

———. (2003b) *The Engendering of Poverty Analysis in Developing Regions: Progress Since the United Nations Decade for Women, and Priorities for the Future*, Gender Institute Working Paper, New Series, Issue 11, London School of Economics and Political Science. http://eprints.lse.ac.uk/573/

———. (2007) *Gender, Generation and Poverty: Exploring the 'Feminisation of Poverty' in Africa, Asia and Latin America*. Cheltenham: Edward Elgar.

———. (2008) The "feminisation of poverty" and the "feminisation" of anti-poverty programmes: Room for revision? *Journal of Development Studies* 44(2): 165–197.

———. (2009) The feminisation of poverty in Costa Rica: To what extent a conundrum? *Bulletin of Latin American Research* 28(1): 19–44.

———. (2012) The disappearing of "Smart Economics"? The world development report 2012 on Gender Equality: Some concerns about the preparatory process and the prospects for paradigm change. *Global Social Policy* 12(2): 198–218.

———. (2014) 'Exploring the 'feminisation of poverty' in relation to women's work and home-based enterprise in slums of the Global South. *International Journal of Gender and Entrepreneurship* 6(3): 296–316.

———. (2016a) Female household headship as an asset? Interrogating the intersections of urbanisation, gender and domestic transformations. In C. Moser (ed) *Gender, Asset Accumulation and Just Cities: Pathways to Transformation*. London: Routledge, pp. 21–39.

———. (2016b) Women, girls and world poverty: Equality, empowerment or essentialism? *International Development Planning Review* 38(1): 1–24.

——— with Craske, N. (2003) *Gender in Latin America*. London; New Brunswick, NJ: Latin America Bureau and Rutgers University Press.

Chant, S. and McIlwaine, C. (2016) *Cities, Slums and Gender in the Global South: Towards a Feminised Urban Future*. London: Routledge.

Chant, S. and Sweetman, C. (2012) Fixing women or fixing the world? "Smart Economics", efficiency approaches and gender equality in development. *Gender and Development* 20(3): 517–529.

Choup, A. (2016) Beyond domestic violence survivor services: Refocusing on inequality in the fight against Gender-Based Violence in the Americas. *Bulletin of Latin American Research* 35(4): 452–466.

Economic Commission for Latin America and the Caribbean (ECLAC). (2004) *Roads Towards Gender Equity in Latin America and the Caribbean*. Santiago de Chile: ECLAC.

———. (2014) *Social Panorama of Latin America 2014*. Santiago de Chile: ECLAC. www.cepal.org/en/publications/37626-social-panorama-latin-america-2014

Escobar Latapí, A. and González de la Rocha, M. (2009) Girls, mothers and poverty reduction in Mexico: Evaluating Progresa-Oportunidades. In S. Razavi (ed) *The Gendered Impacts of Liberalisation: Towards Embedded Liberalism?* New York; Geneva: Routledge; UNRISD, pp. 435–468.

Feitosa de Britto, T. (2007) *The Challenges of El Salvador's Conditional Cash Transfer Programme, Red Solidaria*, Country Study 9, Cash Transfer Research Programme, International Poverty Centre, Brasilia. www.ipc-undp.org/pub/IPCCountryStudy9.pdf

Gabor, M. (2016) *Femicide: Not One More*, Council on Hemispheric Affairs, 24 October. www.coha.org/femicide-not-one-more/

Gammage, S. (2010) Time pressed and time poor: Unpaid household work in Guatemala. *Feminist Economics* 16(3): 79–112.

Gender Equality Observatory of Latin America and the Caribbean (GEOLAC). (2013) *Annual Report 2012 – A Look at Grants, Support and Burdens for Women*. United Nations Publications, ISBN: 978-92-1-121832-9, Santiago, Chile. https://www.cepal.org/publicaciones/xml/5/50235/GenderEquality Observatory.pdf. Santiago de Chile: GEOLAC.

———. (2015) Femicide or feminicide as a specific type of crime in national legislations in Latin America: An on-going process. *Notes for Equality*, 17 July. https://oig.cepal.org/sites/default/files/noteforequality_17_0.pdf

González de la Rocha, M. (2008) Life after Oportunidades: Rural program impact after 10 years of implementation. In Secretaría de Desarrollo Social (SDS) (ed) *External Evaluation of Oportunidades 2008. 1997–2007: 10 Years of Intervention in Rural Areas, Volume 1*. Mexico DF: SD, Coordinación Nacional del Programa de Desarrollo Humano Oportunidades, pp. 128–199.

——— and Escobar Latapí, A. (2016) Indigenous girls in rural Mexico. A success story? *Girlhood Studies* 9(2): 65–81.

Hernández Pérez, A. (2012) *Gender and the Constraints to Co-Responsibility in the Context of Seasonal Migration: The Case of Mixtec Women and the Oportunidades Programme in Oaxaca, Mexico*. Mimeo: Department of Geography and Environment, London School of Economics.

Jenson, J. and Nagels, N. (2018) Social policy instruments in motion: Conditional Cash Transfer instruments from Mexico to Peru. *Social Policy and Administration* 52(1): 323–342.

Johnston, D. (2016) Cost-cutting, co-production and cash transfers: Neoliberal policy, health and gender. In J. Gideon (ed) *Handbook on Gender and Health*. Elgar: Cheltenham, pp. 98–116.

Jones, G. A. and Rodgers, D. (2009) (eds) *Youth Violence in Latin America: Gangs and Juvenile Justice in Perspective*. New York: Palgrave Macmillan.

Kabeer, N. (2003) *Gender Mainstreaming in Poverty Eradication and the Millennium Development Goals: A Handbook for Policy-makers and Other Stakeholders*. London: Commonwealth Secretariat.

Kandiyoti, D. (2013) Fear and fury: Women and post-revolutionary violence. *50:50 Inclusive Democracy*. www.opendemocracy.net/5050/deniz-kandiyoti/fear-and-fury-women-and-post-revolutionary-violence

Lavinas, L. (2016) *NOPOOR Policy Brief No.13: Conditional Cash Transfers: Pros and Cons*. Brasilia: International Poverty Centre. http://nopoor.eu/publication/nopoor-policy-brief-nr-13-conditional-cash-transfers-pros-and-cons

Levy, S. (2015) Is social policy in Latin America heading in the right direction? Beyond Conditional Cash Transfer Programs. *Brookings Op-Ed*, 21 May. www.brookings.edu/opinions/is-social-policy-in-latin-america-heading-in-the-right-direction-beyond-conditional-cash-transfer-programs/

Liu, C., Esteve, A. and Treviño, R. (2017) Female-headed households and living conditions in Latin America. *World Development* 90(2): 311–328.

McIlwaine, C. (2013) Urbanisation and gender-based violence: Exploring the paradoxes in the Global South. *Environment and Urbanization* 25(1): 65–79.

Medeiros, M. and Costa, J. (2006) *Poverty Among Women in Latin America: Feminization or Over-representation?* Working Paper 20, International Poverty Centre, Brasilia. www.ipc-undp.org/pub/IPCWorking Paper20.pdf

———. (2008) Is there a feminisation of poverty in Latin America? *World Development* 36(1): 115–127.

Milanich, N. (2017) Daddy issues: "Responsible Paternity" as public policy in Latin America. *World Policy Journal* 34(3): 8–14.

Molyneux, M. (2006) Mothers at the service of the New Poverty Agenda: Progresa/ Oportunidades, Mexico's Conditional Transfer Programme. *Journal of Social Policy and Administration* 40(4): 425–449.

———. (2007) *Change and Continuity in Social Protection in Latin America: Mothers at the Service of the State*. Gender and Development Paper 1 (Geneva: United Nations Research Institute for Social Development). www.unrisd.org (Accessed 31 July 2016).

——— and Thomson, M. (2011) Cash transfers, gender equity and women's empowerment in Peru, Ecuador and Bolivia. *Gender and Development* 19(2): 195–212.

Moser, C. (2004) Urban violence and insecurity: An introductory roadmap, editor's introduction. *Environ-ment and Urbanization* 16(2): 3–16.

Murphy, S. (2015) Glass ceilings and iron bars: Women, gender and poverty in the post-2015 agenda. *Global Justice: Theory, Practice, Rhetoric* 8(1): 74–96.

Noh, H. and Kyo-Seong, K. (2015) Revisiting the "feminisation of poverty" in Korea: Focused on time use and time poverty. *Asia Pacific Journal of Social Work and Development* 25(2): 96–110.

Promundo. (2016) *Promundo's Annual Report 2015* (Rio de Janeiro: Promundo). http://promundoglobal. org/wp-content/uploads/2016/08/Promundo-Annual-Report-2015-Final.pdf

Puri, L. (2016) *Sustainable Development Is Not Possible If Feminization of Poverty Continues*. Remarks by UN Women Deputy Executive Director Lakshmi Puri at the Opening Ceremony of the 2016 W20 Meet-ing on 24 May. www.unwomen.org/en/news/stories/2016/5/lakshmi-puri-speech-at-the-opening-ceremony-of-the-2016-w20-meeting (Accessed 9 January 2017).

Roy, A. (2002) *Against the Feminisation of Policy*, Comparative Urban Studies Project Policy Brief, Wood-row Wilson International Center for Scholars, Washington, DC. www.wilsoncenter.org/topics/pubs/ urbanbrief01.pdf

Solórzano, I., Bank, A., Peña, R., Espinoza, H., Ellsberg, M. and Pulerwitz, J. (2008) *Catalyzing Individual and Social Change Around Gender, Sexuality, and HIV: Impact Evaluation of Puntos de Encuentro's Communication Strategy in Nicaragua*, Horizons Final Report. Washington, DC: Population Council.

Staudt, K. (2008) *Violence and Activism at the Border: Gender, Fear and Everyday Life in Ciudad Juárez*. Austin: University of Texas Press.

Sweet, E. and Ortiz Escalante, S. (2010) Planning responds to gender violence: Evidence from Spain, Mexico and the United States. *Urban Studies* 47(10): 2129–2147.

Tabbush, C. (2010) Latin American women's protection after adjustment: A feminist critique of Condi-tional Cash Transfers in Chile and Argentina. *Oxford Development Studies* 38(4): 437–451.

United Nations Development Programme (UNDP). (1995) *Human Development Report 1995*. Oxford: Oxford University Press.

UN Women (UNW). (2015) *Progress of the World's Women 2015–2016: Transforming Economies, Realising Rights*. New York: UN Women. http://progress.unwomen.org/en/2015/pdf/UNW_progressreport.pdf

———. (2016) *Femicide in Latin America*. www.unwomen.org/en/news/stories/2013/4/femicide-in-latin-america

UNU-WIDER. (2017) *WIID – World Income Inequality Database* (Helsinki: UNU-WIDER). www.wider. unu.edu/project/wiid-world-income-inequality-database (Accessed 14 November 2017).

Waiselfisz, J. J. (2015) *Mapa Da Violência 2015: Homicídio De Mulheres No Brasil*. Brasilia: FLACSO. www. mapadaviolencia.org.br/pdf2015/MapaViolencia_2015_mulheres.pdf (Accessed 7 November 2017).

Widmer, M. and Pavesi, I. (2016) A gendered analysis of violent deaths. *Small Arms Survey: Research Notes* 63, November. www.smallarmssurvey.org/fileadmin/docs/H-Research_Notes/SAS-Research-Note-14.pdf

Wilding, P. (2010) "New Violence": Silencing women's experiences in the favelas of Brazil. *Journal of Latin American Studies* 42: 719–747.

——— and Pearson, R. (2013) Gender and violence in Maré, Rio de Janeiro: A tale of two cities?. In L. Peake and M. Rieker (eds) *Rethinking Feminist Interventions into the Urban*. London: Routledge, pp. 159–176.

Wisor, S., Bessell, S., Castillo, F., Crawford, J., Donaghue, K., Hunt, J., Jaggar, A., Liu, A. and Pogge, T. (2014) *The Individual Deprivation Measure: A Gender-Sensitive Approach to Poverty Measurement*. Mel-bourne: International Women's Development Agency Inc. www.iwda.org.au/assets/files/IDM-Report-16.02.15_FINAL.pdf

World Bank. (2001) *Social Protection Strategy: From Safety Net to Springboard*. Washington DC: World Bank.

———. (2011) *World Development Report 2012: Gender Equality and Development*. Washington DC: World Bank. http://econ.worldbank.org/WBSITE/EXTERNAL/EXTDEC/EXTRESEARCH/EXT WDRS/EXTWDR2012/0,,contentMDK:23004468~pagePK:64167689~piPK:64167673~theSite PK:7778063,00.html

World Bank Gender and Development Group (GDG). (2003) *Gender Equality and the Millennium Develop-ment Goals*. Washington, DC: World Bank GDG.

World Economic Forum (WEF). (2016) *Global Gender Gap Report 2016* (Geneva: WEF). http://reports. weforum.org/global-gender-gap-report-2016/ (Accessed 30 August 2017).

24

GENDER, HEALTH, AND RELIGION IN A NEOLIBERAL CONTEXT

Reflections from the Chilean case

Jasmine Gideon and Gabriela Alvarez Minte

Introduction

Today, the majority of Latin American health systems are constituted of an active private sector alongside the public sector. While some attention has been given to the gendered impact of privatization in health across the region, mainly in lower-income counties, little attention has been given to the impact of privatization in middle-income countries. The small body of existing research on the gendered implications of health sector reform in Latin America has argued that questions of gender equity have not been adequately addressed (Ewig, 2010; Gideon, 2014).

Drawing on the case of Chile, discussion in this chapter reflects on some of the gendered impacts of privatization in the health system. In Chile, privatization was first initiated under the military dictatorship of General Pinochet, a government which espoused very conservative views of women and reproductive rights (Pieper-Mooney, 2009). At the same time gendered historical legacies have clearly shaped patterns of access and usage within the health system. Debates around the role of the private sector in health have highlighted the importance of accountability structures if health inequalities are to be effectively addressed by private providers (Montagu and Goodman, 2016). This chapter considers the gendered nature of accountability structures in the private sector in health in Chile and asks what this can potentially contribute to promoting gender equality and gender justice in health across the region.

The first part of the chapter outlines what is meant by the private sector in health care, asking which actors are included in this definition and how the private sector is organized in Latin America. The chapter then provides an overview of the process of privatization across the region before focusing more specifically on the structure of the Chilean health system and the role of the private sector within it. Specific emphasis is given to the embedding of gendered and highly conservative historical legacies which, as the chapter argues, have continued to reinforce women's primary role in the health system as being shaped around biological reproduction. While these legacies do frame policy processes and women's access to services in the public sector, the economic and political power of significant private actors in the health sector has enabled these actors to circumvent regulation and legislation that has among other objectives, sought to improve gender equity in health. One of the most significant implications of this is that powerful business elites in the health sector have played a significant role in constraining women's access to their sexual and reproductive rights.

Contextualizing the private sector in health care

There is considerable variation in the composition, size, and nature of the private sector in health across Latin America. Private sector actors can include local-level drug peddlers and individual clinical practitioners at one end of the spectrum, to corporate hospital chains and international private insurers at the other (Mackintosh et al., 2016: 596). While the majority of private health providers tend to be "for profit," others may be operated by religious, charitable, or similar organizations. Within much of Latin America the private health care sector tends to provide high quality care to the wealthier population sectors within stratified health systems where low-income users rely on lower quality public provision (Mackintosh et al., 2016). Pharmacies are also an integral part of the private sector and the cost of purchasing medications is frequently borne by households rather than the state and these out of pocket payments can have a very detrimental effect on low-income households (Dintrans, 2018). Moreover, effective regulation of pharmacies and drug stores can be particularly problematic (Montagu and Goodman, 2016). Indeed, as the Chilean case demonstrates, pharmacies can be very powerful actors in the health sector and exert a strong influence over policy outcomes (Casas, 2004).

Yet despite the growing prominence of private provision in the health sector, much is still unknown about their operation and contribution (Horton and Clark, 2016; Peters, Mirchandani, and Hansen, 2004). As critics have noted inadequate regulation, lack of access for the poor, increased risk of inappropriate treatment that maximizes provider profit and reliance on public sector trained staff are all often characteristic features of the private sector (Wadge et al., 2017: 1). Moreover, evidence of these issues can be found across Latin America, particularly when viewed through a "gender lens"[1]. For example, considerable concern has been expressed over the high levels of caesarean sections performed on women in the region which can be particularly high within the private sector (Heredia-Pi et al., 2014; Murray and Elston, 2005).

Montagu and Goodman (2016) contend that it is important to think more broadly about what regulation constitutes beyond a focus solely on statutory rules laid down by governments and self-regulation implemented by relevant professional bodies. Self-regulation can be problematic because it is often assumed that these professional bodies are well placed to assess performance and are able to control providers yet, as is evident in the Chilean case, this is not always the reality. Moreover, they tend to place more emphasis on providing leadership and protection to the medical community with minimum disciplining of members (Montagu and Goodman, 2016: 615). As is discussed in more detail below, regulation of the private health insurance companies in Chile, the ISAPRES (*Instituciones de Salud Previsional or* private health insurance institutions), has not always been effective and many of the companies have found ways to circumvent regulations imposed on them (Martinez-Gutinez and Cuadrado, 2017).

Privatization of health care services in Latin America

By the early 1980s the crisis of the hegemonic development model signalled an end to the Import Substitution Industrialization (ISI) model of industrialization and scarce economic resources led policy makers to question the universalist principles that had previously informed welfare provision. Although health indicators in the region had improved overall since the 1960s, these improvements masked deep-rooted inequalities and the lack of access and coverage faced by large numbers of the population. The Pan American Health Organization (PAHO) estimated that in the early 1990s, around 130 million people in Latin America and the Caribbean had no access at all to formal health care (Abel and Lloyd-Sherlock, 2000: 1). Health systems across the region were highly centralized and characterized by high levels of inefficiency and inequitable allocation of resources among population groups and regions, resulting in poor quality care (Mesa-Lago, 2008). Within the region there was a broad agreement around the need for social sector reform, and that if it was to be made more efficient, it had to be brought into closer alignment with the market and pluralized service delivery (Molyneux, 2008: 779). Internationally debates around the need for an increased emphasis on "the social" were also gaining ground, particularly in light of evidence from UNICEF on the "human cost" of economic reforms (Cornia, Jolly, and Stewart, 1987). In response, with the exception of Cuba, Latin American health sectors were subject to a series of reform programs during the 1980s and 1990s (Mesa Lago, 2008). These reforms typically emphasized the marketization of health provision and introduced programs of privatization and liberalization and the targeting of services in order to increase efficiency and improve equity in health outcomes. By 2005 nearly all the countries in the region had introduced some type of health care reform which, broadly speaking, were organized around five central pillars:

- the separation between financing, provision, and regulation of services;
- the strengthening of prepayment schemes to grant a stable market and to establish health funds management for profit;
- the pricing of medical activities that turns them into commodities and the definition of health plans or packages and their costs;
- the conversion of public clinics and hospitals into autonomous enterprises;
- the introduction of users' freedom of choice of health fund manager and/or service provider (Mesa Lago, 2007).

While these reforms may be broadly categorized as neoliberal, it is also worth noting that such generic and totalizing terms can mask the complex realities of individual country cases (Molyneux, 2008).

Yet following the implementation of reforms across the region there is little evidence that equity improved. Indeed, critics argued that the exclusion of marginalized groups was further exacerbated with private sector institutions more accountable to transnational corporations than health users (Iriart, Merhy, and Waitzkin, 2001). This has led to the creation of new commercial interests in the health system, which have limited reform efforts in several parts of the region, and particularly in Chile (Ewig and Palmucci, 2012; Pribble and Huber, 2010).

More recently growing attention has been given to what have been termed "post neoliberal"reforms in Latin America and their implications for social policy. While considerable debate remains over the validity of the term, Ruckert and colleagues (2017:1584) have argued that

> the notion of post-neoliberalism remains useful if we understand it not as a complete break with neoliberalism, but rather as a tendency to break with certain aspects of neoliberal policy prescriptions, without representing a set of strict policies or a clearly identifiable policy regime.

With regard to social policy, post-neoliberal states have been characterized by increased social spending and particular emphasis placed on conditional cash transfers and attention to human development. Grugel and Riggirozzi (2018) contend that policies centred around rights-based approaches to welfare provision are also a central element of post-neoliberal social policy which conceptually offer a clear break with the more narrowly framed entitlements in neoliberal approaches.

Within the health sector new approaches to health care financing can be identified across Latin America which emphasize the right to health rather than the individual risk approach that prevailed in the 1990s (Chapman, Forman, and Lamprea, 2017). In Chile, the Plan AUGE (Universal Access with Explicit Guarantees/*Acceso Universal con Garantías Explícitas*), introduced in 2002, was intended as way of expanding universal health coverage and securing the financial protection of all households in relation to health care. Nevertheless, critics argue that this has only had limited success, particularly as the Plan AUGE only covers 80 specific health conditions (Nuñez and Chi, 2013). Moreover, critics have also suggested that the selection of the 80 health conditions has been highly technocratic and does not reflect any publicly stated preferences (Chapman, Forman, and Lamprea, 2017), again pointing to concerns from a gender perspective as women's voices are predominantly excluded from this process (Gideon, 2014).

Moreover, users registered in private health insurance company plans (the ISAPRES) are still given privileged access to better quality care within the Plan AUGE more rapidly than public sector users (Koch, Pedraza, and Schmid, 2017). Given that women are more likely to be in the public sector rather than the private sector, this has important gendered implications. Moreover, a wide body of literature highlights the burden of out of pocket payments (for example the cost of buying medicines and other pharmaceutical products) on poorer households and that this can be a major contributor to health inequalities. Chile has a particularly high burdens of out-of-pocket payments (Dintrans, 2018) and several recent Chilean studies demonstrate that on poorer households and that low-income elderly users have been particularly negatively impacted (Cid and Prieto, 2012; Koch, Pedraza, and Schmid, 2017). Again given the higher number of older users, particularly women, in the public sector, this is likely to have clear gendered implications.

The Chilean health sector

Chile has a mixed health insurance system comprised of the public sector – *Fondo Nacional de Salud* (FONASA – National Health Fund) – and the private sector *Instituciones de Salud Previsional* (ISAPRES – private health insurance institutions). Around 76% of the population is covered by FONASA while 19% is covered by the ISAPRES (ISAPRES de Chile, 2016) – the remainder are primarily covered by other private health providers or a specific system for the armed forces, only a very small percentage of the population lack any health coverage. The ISAPRES have tended to engage in risk selection, only offering health care coverage to "healthy" individuals and this has meant that FONASA is left to cover the greater proportion of the "risky" population, including women of reproductive age, the elderly, the poor, and the sick (Bitrán, 2013). Legislation governs the benefits that must be offered to FONASA users (Ley 18, 469) and while ISAPRES plans must offer the same benefits they are not obligated to include more than this in any specific package.

As argued above, following privatization within the health sector private-for-profit corporations now play an active role in the Chilean health system and several of the ISAPRES companies are controlled by subsidiaries of multinational corporations outside of Chile. Ownership of the ISAPRES is not always easy to trace, for example the US-based company Aetna purchased interests in some ISAPRES companies during the 1990s but these were later sold to the Dutch company ING Group NV (Waitzkin, 2011). In 2015 profits made by the ISAPRES companies in Chile reached around US$60 million while profits for the past decade surpassed $US860 million (Rotarou and Sakellariou, 2017: 500). At the same time recent revelations have shown that some ISAPRES companies also constituted part of powerful business conglomerates making significant political donations over the past ten years, highlighting the powerful grip of the ISAPRES on the political, economic, and social sphere in Chile (Rotarou and Sakellariou, 2017: 500).

Privatization of the health sector in Chile

Chile was the first country in Latin America to implement a series of structural reforms within the health sector under the military dictatorship of General Pinochet. The presence of the military government, which quelled any opposition to the proposed reforms, facilitated the imposition of "laboratory-like conditions" in which to restructure the Chilean health system and introduce an element of free market competition (Miranda, Scarpaci, and Irarrazaval, 1995).

The privatization process in Chile was domestically driven by the influence of the neoliberal thought of the Chicago school of economics, though the Chilean technocrats – or Chicago Boys – appointed in government (Valdés, 1995). A group of neoliberal economists, educated at the *Universidad Católica*, and then in the USA, primarily but not exclusively at the Department of Economics of the University of Chicago. This group was behind the neoliberal policies and privatization policies implemented by the regime. A second factor was the conservative ideology that also spearheaded most of the state reform process during the late 1970s and 1980s (Huneeus, 2000). During the authoritarian regime, a conservative sector influenced policy decisions and the conservative agenda was a fundamental part of the dictatorship's idea of the nation, with a conservative view of the family, motherhood, and in the belief in a natural order that regulates gender relations. Moreover, the power of the *Junta* was key to reinforcing a patriarchal and maternalistic view in policymaking, in particular in the 1980 Constitution and the 1974 National Development plan (Cristi and Ruiz, 2014; Huneeus, 2000; Pollack, 1999; Moulián and Vergara, 1981). The regime emphasized sexual differences and the control of women through domesticity and connected patriotic ideals with family values. Pinochet, using the image of himself as the nation's father, requested the cooperation of "patriotic mothers" to carry this idea forward: the persistence of discriminatory practices and gender stereotypes were basic to the regime's idea of nationhood (Pieper-Mooney, 2009; Araujo, 2009). Family planning, access to contraception, and availability of abortion were affected as the *Junta* made efforts to eliminate the previous government's progressive policies that had a holistic approach to maternal and child health and wanted to ensure access to birth control in state-controlled clinics (Pieper-Mooney, 2009).

The combination of this religious traditional ideology of the family and traditional gender relations and with neo-liberal ideas and policies united in a "conservative synthesis" spearheaded the principle of a subsidiary state, respectful of the autonomy of intermediate bodies – such as the family – and the need for defending the principle of authority, order, and discipline, and a strong anticommunist sentiment (Valdivia Ortiz de Zárate, 2006: 61). The regime started a process of "conservative modernization," neo-liberal policies in the labour market and a conservative religious discourse (Oyarzún, 2005). This was not only a discursive tool, but implied a dismantling of the access to sexual and reproductive rights of the previous decades. For example, there is anecdotal evidence of intrauterine devices (IUD) being removed from women without their consent (Jiles and Rojas, 1992; Jiles, 1994; Pieper-Mooney, 2009; Casas and Vivaldi, 2014), research on contraception stagnated and access to reproductive health suffered (Jiles, 1994), public health institutions stopped advertising family planning services (Casas, 2014) and the regime controlled all institutions working on family planning, and they were mandated to report on initiatives, in particular if they worked with international organizations (ODEPLAN, 1979). The regime also policed young peoples' sexuality. They banned pregnant students from attending regular daytime classes, forcing them to go to evening schools (Montecino and Rossetti, 1990). In addition, women were used to absorb the negative consequences of the decrease in access to services by providing unpaid care work within their families (Pieper-Mooney, 2009). Finally, in 1989, the regime criminalized abortion in any circumstance.

A number of studies of Latin American social policy have looked at the importance of historical legacies in the development of health systems and highlighted the ways in which many of these deeply embedded historical legacies remain unchallenged in subsequent reform processes (Haggard and Kaufman, 2008; Huber and Stephens, 2001). In addition, some research has specifically highlighted the gendered nature of these historical legacies across the region and how these continue to shape both the design and implementation of health policy, often resulting in negative outcomes for women (Ewig, 2010; Gideon, 2014). Moreover, these studies have also drawn specific attention to the ways in which gender inequalities are also cross cut by inequalities of age, class, race, and ethnicity, highlighting the need for more intersectoral approaches to understanding the impact of health sector reforms (Gkiouleka et al., 2018).

The next section of the chapter highlights how the "conservative synthesis" in Chile has continued to influence the design and delivery of health policies and how, at the same time, the political and economic strength of the private health insurance companies has allowed them to resist and undermine attempts to curb their power through increased regulation. As the next section demonstrates, the outcomes of these processes have in effect meant that women continue to be discriminated against by the ISAPRES companies and perhaps more significantly, women continue to be denied access to their sexual and reproductive rights.

The ISAPRES and the lack of accountability to women

As argued above economists and technocrats have played a central role in designing policy reforms in Chile and much of Latin America. Research has also shown how part of the explanation for the gender bias in much of Latin American public policy can be attributed to their prominent role (Montecinos, 2009). Both professional groups tend to favour theoretical approaches inherently opposed to feminist approaches, which push for a restructuring of the economic and social spheres and challenge existing norms and values (Gideon, 2014). Moreover, in the Chilean case, the strong conservative influence has also sought to maintain the status quo, especially in the context of the post-democratic transitions across the region.

Within the health sector, gender-blind technocratic planning is evident in the disparities occurring within the ISAPRES, notably in terms of access to services and levels of care. Women have been particularly discriminated against within the ISAPRES, especially the prohibitive cost of plans for women which Michelle Bachelet campaigned to end during her presidency (Gideon and Alvarez, 2016). Higher rates of charging were justified by the ISAPRES in terms of women's tendency to take more sick leave than men (OPS/ OMS, 2002: 19), yet such calculations fail to account for differences in gender roles. Analysis by the regulatory body of the ISAPRES, the *Superintendencia de ISAPRES* showed that although ISAPRES spend more on women during their reproductive years, they also spend considerably more on men of 69 years or more (Muñoz, 2004). The gendered notion of risk that is built into the design of the ISAPRES points to the gender-blindness of economists and technocrats responsible for health planning and the ways in which gendered rules are set. Although since 2010 a series of regulatory changes have sought to eliminate these gender differences in rates of charging for plans and sought to prevent the sale of plans to women which excluded services relating to pregnancy and birth, evidence continued to emerge of the ongoing selling of these discriminatory plans to women (Gideon and Alvarez, 2016).

The private health insurance system also raises concerns regarding women's entitlements to health as the majority of women registered in the ISAPRES are included as "dependents" – generally as part of their husband's plan rather than as a plan holder themselves. In 2016 just over 45%of ISAPRES beneficiaries were women but of those only around 36%were registered as the plan holder (Sánchez, 2017: 8). If women are primarily accessing private health care services through their husbands' entitlements, this suggests a more regressive approach to health care access and not one in line with wider commitments of the Chilean government to gender equality, for example through support of the Sustainable Development Goals which promote gender equality as well as the right to health (ECLAC, 2016).

Women's lack of access to sexual reproductive rights

The "conservative synthesis" in Chile also left a series of institutional legacies behind that have allowed conservative sectors to ensure their principles are protected not only by authoritarian legacies, but also by decentralization and privatization policies that limit the state's ability to promote redistributive policies to ensure equal access that still have an effect during the democracy. The macro level policies that privatized the health care system, as well as the decentralization and privatization of education, all influenced the reach of policies ensuring universal access to services or information in particular in access to emergency contraception (EC) and access to universal sex education that had clear gendered effects. These legacies and the authoritarian enclaves (Siavelis, 2009) protected conservative interests and the values of the conservative elite, including the conservative business with interest in the private health sector, and have made the conservative influence effective in maintaining the institutional status quo and resisted policies on access to sex education, emergency contraception, and abortion, even in a society where social change was imminent. Current conservative advocates defend conservative ideas by invoking the principles established during the dictatorship, for example by private clinics rejecting services that are against their core believes, in particular in the case of voluntary sterilization and abortion (El Diario Financiero, 2012; Urquieta, 2015). This shows how the limited capacity of the state to act and regulate certain realms has also given space to conservative private business to influence and implement heath policies, limiting the state's capacity to design and implement them. The socioeconomic framework, and efforts to privatize key services, can obstruct the access of the general population by curtailing state expenditure and therefore blocking access to sexual and reproductive services. The neoliberal framework and its emphasis on the protection of private "liberties" in Chile is linked to conservative ideas about the patriarchal family, that has historically limited peoples access to sexual and reproductive health and rights (Casas, 2008; Casas and Herrera, 2012; Casas and Ahumada, 2009).

During the 2010s, the government attempted to ensure universal distribution of emergency contraception (EC) in Chile but were successfully challenged in this process by a conservative section of the economic elite, most significantly some of the owners of several high-street pharmacy chains. Out of the three main pharmacy chains involved in this challenge, Salcobrand, Fasa, and Cruz Verde, two were linked to the conservative Catholic group Opus Dei (Salcobrand's CEO and main stakeholder is the Yarur family while the Harding family have ownership of Cruz Verde; both families are connected to Opus Dei). All three chains stopped the selling of EC in their pharmacies, and made the available compounds disappear from their counters. Salcobrand publicly argued that the decision not to distribute EC was based on conscientious objection, as they argued there were fears the EC could be micro abortive, and therefore against the moral beliefs of the leadership of the company. This made the EC pill disappear from the market, effectively blocking the government's attempts to ensure that the pill was sold to anyone requesting it (Bobadilla, 2007; Casas, 2008; La Nación, 2007; Alvarez Minte, 2017). These conservative business elites also have commercial interest in private health care provision and have been active in rejecting services that are against their core believes, in particular in the case of voluntary sterilization and abortion. (El Diario Financiero, 2012; Urquieta, 2015). More recently, since the de-criminalization on abortion in three cases, the constitutional tribunal has guaranteed conscientious objection to institutions and as such, private clinics have already stated that they will not perform abortions, even when they are guaranteed by law (Tribunal Constitucional, 2017).

This again shows how the limited capacity of the state to act and regulate certain realms has given space to conservative private businesses to influence and implement heath policies, limiting the state's capacity to design and implement them. The socioeconomic framework, and efforts to privatize key services, can obstruct the access of the general population by curtailing state expenditure and therefore blocking access to sexual and reproductive services. The neoliberal framework and its emphasis on the protection of private "liberties" in Chile is linked to conservative ideas about the patriarchal family, that has historically limited peoples access to sexual and reproductive health and rights (Casas, 2008; Casas and Herrera, 2012; Casas and Ahumada, 2009).

Conclusions

The chapter has drawn attention to some of the gendered impacts that accompanied the growth of the private sector in health in Chile. While there are often implicit assumptions that private is "better" than public, the chapter points to areas of concern which reveal a clear gender bias in the operation of the private sector. The discussion highlights some of the ways in which women's health rights and entitlements, particularly in relation to sexual and reproductive health, can be constrained where the private sector is particularly powerful and is able to resist attempts to regulate it. While in the Chilean case the specific nature of the institutional legacies and political context has enabled conservative private business elites to play an active role in the health sector, the case does underline the importance of effective regulatory structures if any real inroads are to be made in achieving gender-equitable outcomes in health.

Note

1 See Gideon (2014) for a full discussion of looking at the health sector through a gender lens.

References

Abel, C. and Lloyd-Sherlock, P. (2000) Health policy in Latin America: Themes, trends and challenges. In P. Lloyd-Sherlock (ed) *Healthcare Reform and Poverty in Latin America*. London:ILAS, University of London, pp. 143–162.

Alvarez Minte, G. (2017) *The Conservative Resistance Against Women's Bodily Integrity in Latin America: The Case of Chile*. Doctoral thesis, submitted to Dept of Geography, Birkbeck, University of London.

Araujo, K. (2009) Estado, Sujeto y sexualidad en el Chile postdictatorial. *Nomadías* 9: 11–39.

Bitrán, R. (2013) *Explicit Health Guarantees for Chileans: The AUGE Benefits Package*. Universal Health Coverage Studies Series, No. 21, World Bank, Washington, DC.

Bobadilla, S. (2007) Venta de La Píldora En Farmacias. *La Tercera*. www.jesus.cl/iglesia/paso_iglesia/recortes/recorte.php?id=7711 (Accessed 2 May 2018).

Casas, L. (2004) La batalla de la píldora. El acceso a la anticoncepción de emergencia en América Latina. *Revista de Derecho Y Humanidades* 10: 183–208.

———. (2008) La saga de la anticoncepción de emergencia en Chile: Avances y desafios. *Serie Documentos Electrónicos, Programa Género Y Equidad FLACSO* 2.

Casas, L. (2014) Women and reproduction: From control to autonomy – the case of Chile. *Journal of Gender, Social Policy & the Law* 1(12): 427–452.

———— and Ahumada, C. (2009) Teenage sexuality and rights in Chile: From denial to punishment. *Reproductive Health Matters* 17(34): 88–98. doi:10.1016/S0968-8080(09)34471-7.

Casas, L. and Herrera, T. (2012) Maternity protection vs. maternity rights for working women in Chile: A historical review. *Reproductive Health Matters* 20(40). Reproductive Health Matters: 139–147.

Casas, L. and Vivaldi, L. (2014) Aborto: Una clara oportunidad para legislar. *Comunicación Y Medios* 30: 241–254.

Chapman, A.R., Forman, L. and Lamprea, E. (2017) Evaluating essential health packages from a human rights perspective. *Journal of Human Rights*16(2): 142–159.

Cid, C. and Prieto, L. (2012) El gasto de bolsillo en salud: el caso de Chile, 1997 y2007. *Revista Panamericana de Salud Pública* 31(4):310–316.

Cornia, A., Jolly, R. and Stewart, F. (1987) *Adjustment with a Human Face, Vol. 1.* Oxford: Clarendon Press.

Cristi, R. and Ruiz, C. (2014) Conservative thought in twentieth century Chile. *Canadian Journal of Latin American and Caribbean Studies* 15(30):27–66.

Diario Financiero, El. (2012) Clínicas Que No Harán Abortos Concentran Más de 220 Camas. *El Diario Financiero*, February.

Dintrans, P.V. (2018) Out-of-pocket health expenditure differences in Chile: Insurance performance or selection? *Health Policy* 122(2): 184–191.

ECLAC. (2016) *Report of the Fifty-Third Meeting of the Presiding Officers of the Regional Conference on Women in Latin America and the Caribbean*, Santiago, 26–28 January. www.cepal.org/sites/default/files/events/files/report_of_the_fifty-third_meeting_of_the_presiding_officers_of_the_regional_conference_on_women_in_latin_america_and_the_caribbean.pdf (Accessed 12 February 2018).

Ewig, C. (2010) *Second-Wave Neoliberalism: Gender, Race and Health Sector Reform in Peru.* University Park, PA: Pennsylvania State University Press.

———— and Palmucci, G. A. (2012) Inequality and the politics of social policy implementation: Gender, age and Chile's 2004 health reforms. *World Development* 40(12): 2490–2504.

Gideon, J. (2014) *Gender, Globalisation and Health: Issues and Challenges in a Latin American Context.* Basingstoke and New York: Palgrave Macmillan.

———— and Alvarez Minte, G. (2016) Institutional constraints to engendering the health sector in Bachelet's Chile. In G.Waylen (ed) *Gender, Institutions, and Change in Bachelet's Chile.* New York: Palgrave Macmillan, pp. 147–169.

Gkiouleka, A., Huijts, T., Beckfield, J. and Bambra, C. (2018) Understanding the micro and macro politics of health: Inequalities, intersectionality & institutions – A research agenda. *Social Science & Medicine* 200: 92–98.

Grugel, J. and Riggirozzi, P. (2018) New directions in welfare: Rights-based social policies in post-neoliberal Latin America. *Third World Quarterly* 39(3): 527–543.

Haggard, S. and Kaufman, R. (2008) *Development, Democracy and Welfare States: Latin America, East Asia and Eastern Europe.* Princeton, NJ: Princeton University Press.

Heredia-Pi, I., Servan-Mori, E.E., Wirtz, V.J., Avila-Burgos, L. and Lozano, R. (2014) Obstetric care and method of delivery in Mexico: Results from the 2012 National Health and Nutrition Survey. *PLoS One* 9(8): e104166. doi:10.1371/journal.pone.0104166

Horton, R. and Clark, S. (2016) The perils and possibilities of the private health sector. *The Lancet* 388(10044): 540–541.

Huber, E. and Stephens, J. (2001) *The Development and Crisis of the Welfare State: Parties and Policies in Global Markets.* Chicago: University of Chicago Press.

Huneeus, C. (2000) Technocrats and politicians in an authoritarian regime. The 'ODEPLAN Boys' and the 'Gremialists' in Pinochet's Chile *Journal of Latin American Studies* 32(2): 461–501.

Iriart, C., Merhy, E. E. and Waitzkin, H. (2001) Managed care in Latin America: The new common sense in health policy reform. *Social Science and Medicine*52:1243–1253.

ISAPRES de Chile. (2016) *ISAPRES, 1981–2016: 35 Years Supporting Chile's Private Health System.* www.isapre.cl/PDF/35YEARSIsapres.pdf (Accessed 30 April 2018).

Jiles, X. (1994) Historia de Las Políticas de Regulación de La Fecundidad En Chile. In T. Valdés and M. Busto (eds) *Sexualidad Y Reprodución; Hacia Una Construcción de Derechos.* Santiago, Chile: CORSAPS/FLACSO, pp. 129–136.

———— and Rojas, C. (1992) *De La Miel a Los Implantes. Historia de Las Políticas de Regulación de La Fecundidad En Chile.* Santiago, Chile: CORSAPS.

Koch, K.J., Pedraza, C.C. and Schmid, A. (2017) Out-of-pocket expenditure and financial protection in the Chilean health care system – a systematic review. *Health Policy* 121(5): 481–494.

Mackintosh, M., Channon, A., Karan, A., Selvaraj, S., Cavagnero, E. and Zhao, H. (2016) What is the private sector? Understanding private provision in the health systems of low-income and middle-income countries. *The Lancet* 388(10044): 596–605.

Martinez-Gutierrez, M.S. and Cuadrado, C. (2017) Health policy in the concertación era (1990–2010): Reforms the Chilean way. *Social Science & Medicine*182:117–126.

Mesa-Lago, C. (2007) Social security in Latin America: Pension and health care reforms in the last quarter century. *Latin American Research Review* 42(2): 181–201.

———. (2008) *Reassembling Social Security: A Survey of Pension and Health Care Reforms in Latin America*. Oxford: Oxford University Press.

Miranda, E. J., Scarpaci, L. and Irarrazaval, I. (1995) A decade of HMOs in Chile: Market behavior, consumer choice and the state. *Health and Place* 1:51–9.

Molyneux, M. (2008) 'The 'neoliberal turn' and the new social policy in Latin America: How neoliberal? how new? *Development and Change* 39(5):775–97.

Montagu, D. and Goodman, C. (2016) Prohibit, constrain, encourage, or purchase: How should we engage with the private health-care sector? *The Lancet* 388(10044): 613–621.

Montecino, S. and Rossetti, J. (1990) (eds) *Tramas Para Un Nuevo Destino: Propuestas de La Concertación de Mujeres Por La Democracia*. Santiago Chile: Arancibia Hnos.

Montecinos, V. (2009) Economics: The Chilean story. In V. Montecinos and J. Markoff (eds) *Economists in the Americas*. Cheltenham; Northampton, MA: Edward Elgar.

Moulián, T. and Vergara, P. (1981) Estado, ideología y políticas económicas En Chile: 1973–1978. *Revista Mexicana de Sociología* 43(2): 845–903.

Muñoz, M. S. (2004) Reforma de salud y equidad de género contenidos. *Superintendencia de las ISPARES*. www.supersalud.gob.cl/documentacion/666/articles-4013_recurso_1.pdf (Accessed 12 February 2018).

Murray, S. and Elston, M. A. (2005) The promotion of private health insurance and its implications for the social organisation of healthcare: A case study of private sector obstetric practice in Chile. *Sociology of Health & Illness* 27(6): 701–721.

Nación, L. (2007) *La trastienda de la batalla por la pastilla del día después: Dorando la píldora*. Santiago, Chile: La Nación.

Nuñez, A. and Chi, C. (2013) Equity in health care utilization in Chile. *International Journal for Equity in Health* 12:58. https://equityhealthj.biomedcentral.com/articles/10.1186/1475-9276-12-58 (Accessed 30 April 2018).

ODEPLAN. (1979)*Política Poblacional Aprobada Por Su Excelencia El Presidente de La República Y Publicada En El Plan Nacional Indicativo de Desarrollo (1978–1983)*. Santiago de Chile: Presidencia de la República.

OPS/OMS. (2002) *Discriminación de las mujeres en el Sistema de Instituciones de Salud Previsional*. Regulación y perspectiva de género en la reforma. Género, Equidad y Reforma de la Salud en Chile. Working Paper No. 1. OPS/OMS, Santiago, Chile.

Oyarzún, K. (2005) Ideologema de La Familia: Género, Vida Privada Y Trabajo En Chile, 2000–2003. In X. Valdés and T. Valdés(eds) *Familia Y Vida Privada*. Santiago Chile: FLACSO Chile, pp. 277–310.

Peters, D.H., Mirchandani, G.G. and Hansen, P.M. (2004) Strategies for engaging the private sector in sexual and reproductive health: How effective are they? *Health Policy and Planning*19(suppl_1): i5–i21.

Pieper-Mooney, J. (2009) *The Politics of Motherhood: Maternity and Women's Rights in Twentieth-Century Chile*. Pittsburgh, PA: University of Pittsburgh Press.

Pollack, M. (1999) *The New Right in Chile, 1973–1997*. London: Palgrave Macmillan.

Pribble, J. and Huber, E. (2010) *Social Policy and Redistribution Under Left Governments in Chile and Uruguay*. Colegio Carlo Alberto, Working Paper no. 177. www.carloalberto.org/assets/workingpapers/no.177.pdf

Rotarou, E.S. and Sakellariou, D. (2017) Neoliberal reforms in health systems and the construction of long-lasting inequalities in health care: A case study from Chile. *Health Policy* 121(5): 495–503.

Ruckert, A., Macdonald, L. and Proulx, K.R. (2017) Post-neoliberalism in Latin America: A conceptual review. *Third World Quarterly* 38(7): 1583–1602.

Sánchez, M. (2017) *Análisis estadístico del sistema Isapre con enfoque de género – año 2016, Documento de Trabajo, Departamento de Estudios y Desarrollo, Superintendencia de Salud*. Chile. www.supersalud.gob.cl/documentacion/666/articles-16434_recurso_1.pdf (Accessed 12 February 2018).

Siavelis, P. (2009) Enclaves de La Transición Y Democracia Chilena. *Revista de Ciencia Política* 29(1): 3–22.

Urquieta, C. (2015) La elite conservadora tras las clínicas de la Cámara de la Construcción que se oponen a la ley de aborto. *El Mostrador*, 16 February.

Tribunal Constitucional. (2017) *Sentencia Tribunal Constitucional, Rol 3729–3717 Respecto Del Proyecto de Ley Que Regula La Despenalización de La Interrupción Voluntaria Del Embarazo En Tres Causales*. Tribunal Constitucional, Santiago, Chile.

Valdés, J. G. (1995) *Pinochet's Economists: The Chicago School in Chile*. Cambridge: Cambridge University Press.

Valdivia Ortiz de Zárate, V. (2006) Lecciones de Una Evolución: Jaime Guzmán Y Los Gremialistas, 1973–1980. In V. Valdivia Ortiz de Zárate, R. Alvarez and J. Pinto (eds) *Su Revolución Contra Nuestra Revolución: La Pugna Marxista-Gremalista En Los Ochenta*. Santiago, Chile: Lom Ediciones, pp. 49–99.

Wadge, H., Roy, R., Sripathy, A., Fontana, G., Marti, J. and Darzi, A. (2017) How to harness the private sector for universal health coverage. *The Lancet* 390(10090): e19–e20.

Waitzkin, H. (2011) *Medicine and Public Health at the End of Empire*. Oxford: Routledge.

25

MEN AND MASCULINITIES IN DEVELOPMENT

Matthew Gutmann

Introduction

Despite decades of effort on-the-ground and in academic scholarship to incorporate men and masculinities into work related to Gender and Development, for a range of reasons gender is still associated by practitioners and various publics overwhelmingly with women. This is true in development work on issues ranging from microfinancing to family welfare, in government programs related to educational equity, as well as in academic settings, where gender studies are usually synonymous with women's studies. An additional problem is that men are seldom clamouring for inclusion in gender-based programs. A central goal of the powerful feminist movements that emerged throughout the world in the 1960s and 1970s was that societies, governments, households, and communities needed to pay attention to basic issues of gender inequalities and directly and immediately take steps to address these concerns. That men are far more often than women in dominant positions socially, politically, and within families and neighbourhoods means that men are often loath to give up their privileges and tackle gender imbalances. This poses additional obstacles to involving men in development efforts.

For both these reasons – because gender is still too often equated with women only, and because men are not as a group demanding changes in gender relations – the struggle for gender equality is widely regarded as a pressing issue for women alone. This situation pertains to matters of health, migration, and governance among many other issues. But as we will see, and as difficult as the real challenges are, many efforts in development work are hobbled without men's inclusion. The most complex part is to determine how to involve men without losing sight of priorities focused on structural, cultural, and ideological gender vulnerabilities and the marginalization of women.

Unsurprisingly, when it comes to men and masculinities in development, there has been more interest in gender-based violence (GBV) and men who have sex with men (MSM) than all the other topics combined, from mental health to armed conflict to migrant remittances to university enrolments to gender justice. In part these objectives reflect the social struggles that gave rise to a focus on men and masculinities: feminist movements and public health campaigns against HIV/AIDS. In addition, however, attentiveness to violence and sexual health also reflects inattention to structural inequalities that have proved especially intransigent, including the gendered qualities of political and economic power in societies across the globe.

Drawing on recent scholarship and practice, this chapter takes stock of successes and ongoing obstacles to involving men in development efforts in the region, in the context of the move in Latin America and the rest of the world from a period up until the 1990s, when the branch of development work was referred to as Women in Development (WID), to the professed inclusion of men in development work after the mid-1990s, that was accompanied by a nominal change of the overarching term for this branch of development work to Gender and Development (GAD). For background readings on men and development, see studies by Chant and Gutmann (2000); Cornwall, Karioris, and Lindisfarne (2016); and Paulson (2016).

The main thematic issues addressed here include best practices related to achieving gender equality, and ongoing debates to boost the participation of men based primarily on narrowly conceived interests (for example, men's self-care, sexual dysfunction, testicular and prostate cancers) in contrast to programmes that promote shared goals (for example, birth spacing and timing of children, mutual decision-making in couples). For example, in addressing issues of reproductive health and gender-based violence, whether and how to involve men is of central concern. To ignore and/or exclude men when they are so centrally involved makes little sense, but reasonable concerns about women losing leadership and control over these matters likewise are sensible and indeed imperative. In addition, and directly related, we will examine how women's autonomy simultaneously serves as catalyst and challenge to changing men's relationships at the household level and more broadly in society.

The chapter is organized around key topics related to men, masculinities, and development in Latin America: health, including reproductive health and sexuality, gender-based violence, and life expectancy; migration, including displacements, refugees, and remittances; and governance and citizenship, including armed conflict and gender justice. And this chapter asks three questions that are often used to frame discussions of men and masculinities in development:

One, Who needs men?
Two, Cui bono, or, Who stands to benefit from gender equality?
And, three, What does power have to do with men and masculinities?

Who needs men?

Beyond the obvious fact that men have always been in leadership positions in most development agencies and programs worldwide, arguments against including men in development initiatives on the ground are legion. Major government efforts in Latin America in recent years are based in part on the explicit premise that men, at least working-class men, are not to be trusted. In the programs Bolsa Família in Brazil and PROGRESA-Oportunidades-Prospera in Mexico, for example, poor women, and more specifically poor mothers, are "preferentially" given funds to support their families, with various requirements involving education, health, and nutrition of their children. As the "agents of change," women and not men, mothers and not fathers, receive the funds, all of which has been heralded as innovative and a key reason for the success of these programs. In this instance, in other words, men are not only unnecessary and not part of the solution, they are part of the problem that these "conditional cash transfer" programs seek to overcome.[1]

Among NGO personnel working in the development field, Chant and Gutmann (2002: 270) list among the reasons that men are found on the periphery of gender and development the need to protect for women and children the few resources available, concerns that men might hijack funds and leadership, lack of understanding that men, too, are engendered and engendering, and of vital concern the fact that, with the exception of gay men and HIV/AIDS, there has never been any significant interest and involvement by men in gender and development issues.

The transforming sociodemographics of gender in Latin America in the 21st century add additional complexity to matters of gender equality. Women across the continent have achieved parity in enrolment through university-level education. Further, in many Latin American countries there are more women than men in higher education; as fertility and maternal mortality rates plummeted, tens of millions of women were able to secure paid work outside the home (see Camou and Maubrigades, 2017). At the same time, the gender pay gap remains enormous, with men out-earning women in virtually all positions.

As women's employment numbers have risen and their ability to establish themselves more autonomously, financially and in other ways, has strengthened, gender-based violence has if anything increased for women across class lines. The experience in Peru reported by Mitchell is common, "when women enter the workforce and begin to gain economic independence from their partners they reported that the levels of violence often increased" (2013: 101). And, finally, although ten countries in Latin America have had a woman as head of state, and although the percentage of women in national legislatures in greater in several Latin American countries than others like the United States, on community, municipal, regional, and national levels men continue to dominate political life in Latin America.

Cui bono, or, who stands to benefit from gender equality?

A central enigma of gender and development is that what happens to women can and does affect the men in their lives, both negatively and positively. To take one example, gender-based violence in Latin America usually involves men who are well known to the women who are the victims. Since the beginning of studies of men and masculinities in development, there has been a tension evident in scholarship and grassroots efforts between highlighting what masculinities scholar Raewyn Connell (1987) has termed the challenge to many men of losing the "patriarchal dividend," on the one hand, and on the other appealing to a set of unique and universal "men's needs," for example, men's self-care, sexual dysfunction, testicular and prostate cancers (see also Connell, 2011).

So, do men have more to lose or gain by gender equality, and what does this mean for gender and development? Linked to this question are implicit assumptions about whether men are even able to change, or whether they are in too many ways "hard-wired," especially with respect to sexuality and aggression. These two issues in turn directly impinge on the underlying question of why men do not express more enthusiasm for gender and development projects, and if and how this could change.

Two problems

One problem in examining men and masculinities in development in Latin America is methodological: much of what we know about gender and development is based on self-reporting, and when there is a discrepancy between women and men, the tendency in Latin America as elsewhere is to believe information provided by women when it is at odds with reports from men. In most instances this seems exceedingly reasonable. The trouble is that in one survey after another on men and masculinities in development in Latin America the flimsiness of the data is made abundantly clear. Assumptions seem to follow on conjecture, opinion is built on partial experience. The fact remains that we have far too few empirical studies of men and masculinities in development in Latin America, and this hampers our ability to make useful generalizations.[2] Sensible conjecture and subjective familiarity may be fine starting points, but they are insufficient.

The next problem in talking about men and development is one of orientation: development and development studies are about studying down, focusing on the relatively powerless. In looking at men and masculinities, this inherent orientation leads to rather blatant bias, with the implication, for instance, that every sort of gender-based problem is worse the further down the social ladder you descend. Nowhere has this been shown; it is nonetheless widely believed. In other words, a gender problem can appear to be a problem for a particular class of men, and that class rarely involves studying up.

Gender, health, and development in Latin America

Reproductive health

Development in the area of reproductive health provides good illustrations of the perennial issue of structural factors and individual choice, as well as the fact that gender and development in Latin America still largely operates within a binary framework.[3] Throughout Latin America, beginning in the late 1960s and especially in subsequent decades, international development agencies and foundations, as well as local governments, developed painstaking, comprehensive, and widespread campaigns to foster the spread of contraceptive adoption, birth planning, and birth spacing. Catholicism in this respect seemed to have little impact on most women: as soon as they gained access they availed themselves of new birth control methods, and birth rates plummeted accordingly.

At the same time as these organizations and organizers were achieving great success in promoting family planning for women and improving women's health in general, a curious and unintentional outcome of these planned parenthood campaigns was that men were essentially planned out of the equation. The clearest indication of the accomplishments of these birth control campaigns was the dramatic fall in fertility rates in the region, from just under six births per 1,000 women in 1965 to just over two in 2017. The clearest indication that men were not central to these campaigns is evident in every document produced in this period by governments, NGOs, and foundations (see Gutmann, 2007).

At best, this simply reflected a focus that did not include men. At worst, it reflected beliefs about men and reproduction – in particular that men as a group were irresponsible sexual partners, more interested in sexual conquest and siring many children than anything else – that made them sexually unreliable and even deceitful. The policy in family planning campaigns to ignore men not only lumped them into a rather homogenous category, but worse, was ultimately premised on a conviction that men were sexually motivated and governed by their underlying biology (see Gutmann, 2007). Ochoa-Marin and Vásquez-Salazar (2012; see also Viveros, 2002) similarly call attention to the challenge in Colombia faced by government agencies mandated to provide reproductive health care; they are aimed almost exclusively at women and integrating men has proved difficult in both programing and implementation. These practices of dismissing men from reproductive health work in development continues in nefarious ways, such as in Mexico through what it called there the Oferta Sistemática, or, the Standard Offer.

What this program entails is the following: whenever a woman between 15 and 49 years old appears at a government health clinic or hospital in need of assistance – recurring migraines or bronchitis or skin lesions or any other problem or symptoms – medical personnel are instructed to talk with the woman about birth control. Do you use something? Have you had any problems with your method? If you don't use contraception, why not? Unless a man accompanies a woman to the medical appointment, he is not privy to such a line of inquiry, much less the target of these questions. This program highlights again that contraception is exclusively a female concern.

It was not always this way. Though numbers are difficult to come by, we can ask what, if anything, men did or did not do before the advent of the birth control pill and the widespread availability of modern birth control for women. In countries throughout Latin America men used condoms, and they practised rhythm methods as best they could. They also achieved high marks from their wives on the manliness scale when they proved expert at the self-control required for withdrawal to work. Only years later, when widespread assumptions about men's inherent reproductive irresponsibility began to be questioned, did scholars bother to inquire about men's actual sexual practices, desires, and feelings in the past, and to not take them quite as much for granted, as if all men in Latin America were the same and had been since at least the Conquest. It turned out men in Latin America, not all but a lot of them, demonstrated self-control, love, and a desire to space out children by learning when to pull out (see, for example, Hirsch and Nathanson, 2001).

By neglecting the deep and abiding concerns of many men regarding birth spacing and timing of children, and mutual decision-making in couples, the stereotype of men and reproductive irresponsibility is spread despite all evidence to the contrary for many if certainly not all men in Latin America (see Castro, 2000; Gutmann, 2007). While not always viewed through the prism of development, studies on and work with MSM have direct relevance here, including the benefits of queering gender and development more broadly. At the same time, because of its fixation on women, health concerns of HIV/AIDS have often been categorized as separate from the field of gender and development in Latin America. This is unfortunate, not the least because the success of countries like Brazil in addressing HIV/AIDS could serve as a model in many respects for the engagement of social movements with government policy. It can also contribute to a problem noted by Welsh (2014: 44) in which "the models and dynamics of patriarchal power" are reproduced within LGBT organizations in Nicaragua, resulting in the struggle for the elimination of all forms of homophobia being unnecessarily held back.

Gender-based violence

In addition to reproduction, perhaps nowhere is the disadvantage of excluding men from development work clearer than with respect to GBV. It is also no accident that more feminist organizations of and for men in Latin America have been formed around this issue than any other.[4] These initiatives operate within a range of models as to the causes and solutions to GBV, from individual psychologies that development agencies must identify and therapeutically eradicate to indictments of highly militarized states in which violence permeates the social pores.

There are even theories of why men in Latin America behave as they do that are linked to biological presumptions, for instance, in the invocation of nonhuman primate categories of alpha and beta males in a recent collection on men and development published by the World Bank, *The Other Half of Gender* (Bannon and Correia, 2006: 5). To the extent that men's violence, migratory patterns, sexuality, and everything else are attributed to their biology, remedial efforts are crippled from the start. Among many other studies promoting work with men on GBV, Vargas Urías (2012) shows the perils of placing the blame for this pandemic on natural causes, insisting rather on the need to change what men do, not what they supposedly are by virtue of their bodies. One more controversial question relating to GBV in Latin America is whether in a sense all violence is gender-based, that is, including male-on-male violence, or whether this waters down the usual connotation and emphasis of GBV on women as victims and men as perpetrators. Work in Colombia and other postconflict zones is especially significant in this respect (on development and gender in postconflict Colombia, see Gómez Alcaraz and García Suárez, 2006; Theidon, 2009).

Life expectancy

In 2017, the average life expectancy for women in Latin America and the Caribbean was 79 years old; for men, it was 73. These numbers varied widely by country. In El Salvador the figure was nine years apart, 78 years for women, 69 for men. (By contrast, the comparative figures for the United Kingdom that same year were 83 and 79, respectively.) Straddling the line between ideas that naturalize male behaviour and those that view cultural factors as predominant is the formulation advanced by some development specialists in the region that masculinity itself is a *factor de riesgo* (de Keijzer, 1997; Amuchástegui, 2006), and that only by appreciating male violence, for example, in this way can it be confined.

Few deny that violence is involved in the markedly growing gap in life expectancies for men and women in Latin America.[5] Although certainly not male-only problems, men are involved more than women in vehicular crashes, alcohol consumption and abuse, work-related injuries, suicide (four men for every woman), and homicide: men are nine times more likely to be murder victims as well as the murderers in Latin America.

Men, migration, and development in Latin America

For young men born in rural areas in Latin America in the 21st century, the choice is often between migration and the military; men make up 93–96% of enlisted ranks in Latin American armed forces. Migration for men in the region, as for women, is voluntary, forced, documented, undocumented, contracted, and trafficked. Men on the whole migrate more and over greater distances than women. (On gender and migration in Mexico, see Barrera Bassols and Oehmichen Bazán, 2000.)

Attention to migration by men, displacement of men, and refugee status claimed for men is extensively documented in many ways, yet gender is not usually one of them. This is a product of the narrow view of gender as being important to understand for women but not men on the part of researchers and organizers, and it reflects the language migrant and refugee men use describe their experiences and circumstances. In addition, although the impact of men's migration on the women left behind in villages and other communities is a major focus of attention, the gendered impact on men is more alleged than verified. We are only beginning to understand the gendered aspects of displacement, such as the three million people forced to flee their homes in the decades-long conflict in Colombia (see Theidon, 2009 for a start).

In many parts of Latin America migrant men are the primary vectors for the transmission of HIV, and understanding how they themselves first get infected and how and whom they infect cannot be understood or addressed except in relation to their masculinities and sexualities related to migration patterns. More specifically, in contemporary Latin America, AIDS is a disease of migration and modernity. Forced to migrate for economic and/or political reasons – too often the two are pitted against each other, whereas for questions of men and masculinities they are of a piece – self-identified heterosexual migrant men may have sex for the first time with other men, returning home to infect their wives and girlfriends. What is more, and directly relevant to matters of development and enforced underdevelopment, while they may be infected in the metropoles, they are forced to migrate yet again, this time back to their home countries, where medical care may be deficient but it is more available and less expensive than in the inhospitable lands where they toiled (see Gutmann, 2010).

In 2016, remittances to Latin America, mainly sent by men who had migrated for work outside the region, surpassed $70 billion. $27 billion went to Mexico and over $10 billion to Central America (Orozco, 2017). The crucial role of remittances economically in Latin America is widely acknowledged and discussed. The gendered aspects are less considered. To the extent that *mantener a la familia* is a codified rubric for exemplary masculine comportment, and insofar as opportunities to make a living and support that family financially are circumscribed in rural and more periodically urban areas of the region, then remittances are directly linked in the minds and lives of men and women in Latin America to the worthiness and manhood of tens of millions of migrant men displaced as a result of economic, political, and armed turmoil.

Men, governance, and citizenship in Latin America

In Mexico City, São Paulo, and Rio de Janeiro at rush hour women have the option to ride on women (and child) only subway cars. In each of these cities there has been noisy opposition to this kind of *sólo mujeres* gender segregation, usually by young feminists. That may seem surprising, but it shouldn't. No one is challenging the severe problem much less trauma of sexual assault. A 2014 report names Bogotá as having the most dangerous metro system for women; Mexico City's is number two. At the same time, what separation of women from men, or the option for women to separate, says about men reveals underlying perceptions about men, male violence, and what public policy can do to address the problem of sexual assault.

Specifically, young feminists have argued, to separate means to capitulate, if not to men per se, then to male assaults on women, as if to say, that's just the way men are, that's just what men do. If that is the foundational estimation then corralling men is advisable. If this is, instead, certain men exerting male prerogative, then it is the male privilege that must be challenged, stopped, and punished.

Sólo mujeres means that only women are allowed to enter certain cars. Every time a woman gets on the subway at rush hour she is reminded that she can receive certain protections from men clawing at her, making rude comments, and threatening her. Simultaneously the program means that men are forbidden from entering those same cars. Every time a man gets on the subway at rush hour he is visually provoked to think about the reason for the separation of men from women, think about what actions could have led to isolating men from women, and whether it is women being split up from men or the other way around. This is an example of public policy that provides a challenge to men, masculinities, and development – there is widespread social debate and no readily apparent solution much less quick fix to the problem of sexual assault on metros in Latin America.

Another example of governance and citizenship in Latin America provides a clearer, salutary, and replicable example of a social problem with a notable solution: AIDS in Brazil. As Richard Parker has documented, popular mobilization in Brazil in the 1980s and 1990s demonstrated "the key role that civil society organizations and activist initiatives . . . played in the development of both government and intergovernmental agency responses to the epidemic" (2011: 22). Through organizations like the ABIA (Brazilian Interdisciplinary AIDS Association, founded in 1986), Brazil's institutional response to the AIDS crisis, with all its defects, became a model for the world, both in terms of providing relatively low-cost care, for the number of people afflicted who were served, and the fact that most people suffering from HIV/AIDS at that time were young, gay men who were taking matters into their own hands and achieving remarkable successes.[6]

In addition, this experience in Brazil highlights competing priorities often found more broadly among development NGOs, namely whether to highlight service provision or political pressure on existing governmental institutions as the primary goal of NGO efforts. The successes in Brazil's HIV/AIDS work hold lessons far more broadly, including with respect to key regions of the Global South handling the crisis with far more speed, care, and coverage than many countries in the Global North. The lessons are also significant for men and masculinities in development in Latin America, because caring for a stigmatized male population was made the centerpiece of AIDS work, unapologetically and with a dogged determination.

Another social experiment also in Brazil had more mixed results. The first state-sponsored Women's Police Station (delegacia da mulher) was opened in 1985 with the aim of providing more competent, compassionate, and consistent support for women, including around issues of gender-based violence. Underfunded from the beginning, the results have been uneven, with ample evidence that other police used these women's stations as an excuse to pay less attention to GBV. Nonetheless, reporting of GBV went up, if prosecutions did not always follow.[7]

The case of militarized masculinities arising through the decades-long armed conflict in Colombia is both unique in Latin America for its longevity and illustrative of similar zones of civil war and militarism in the continent. Kimberly Theidon (2009) has done some of the best thinking of connections between weapons, masculinities, and violence in Colombia in particular. One of the key contradictions she identifies is that the "cultural and political economy of militarized masculinity" reveals "how little access former combatants have to civilian symbols of masculine prestige," such as education, legal income, and decent housing (2009: 5, 18). In a context in which alternatives to going to war seemed imaginary, young men in Colombia joined military units of every kind for a range of often incongruous reasons. To escape poverty, look for adventure, avenge a murdered relative, and achieve social mobility were all common motivations. Ex-combatants emerged at the ceasefire, however, with only one marketable skill: "prowess with a weapon and combat experience" (Theidon, 2009: 16).

The Disarmament, Demobilization, and Reintegration (DDR) program in Colombia was little concerned with what should happen to men (and women) reared on military masculinities. Following demobilization and what was previously a highly militarized daily routine, health care personnel reported a noticeable increase in household GBV among ex-combatants. Either because this problem was viewed as tertiary to other more pressing concerns or not, the effect was to extend and strengthen stereotypical relationships between ex-combatants and the women in their families, representing both a lost opportunity and an exacerbation of a deeply problematic social crisis.[8]

Conclusion

Theidon's (2009: 33) point that, "Addressing violent masculinities should be a key concern when 'adding gender'" to DDR interventions could apply to adding gender in development more generally: leaving men and masculinities out of the mix may seem a prudent course, but it too often perpetuates more problems than it avoids. The issue is less conceptual than political. In the sphere of development work, men have been identified as engendered and engendering for decades, at least since major international conferences in 1994 in Cairo and the next year in Beijing.[9]

The obstacle to truly adding men in development in Latin America, as elsewhere, is rooted in gender-based inequalities and tacit assumptions about the inability of men to change and be changed. In case after case, however, it is the women who are the object of development initiatives who have insisted that change was possible in how their men treated them and their children.

Introducing a collection surveying "development policy and discourse on men and boys in relation to gender equality," Edström, Das, and Dolan write that even "some donor representatives are articulating a need to integrate and anchor the engagement of men and boys in future development goals, as well as making suggestions for how these can take a more structural direction than has been the case to date" (2014: 8).

The most effective ways of involving men centrally in development is not through narrow appeals to their supposedly unique interests and concerns but instead to highlight shared experiences and goals. Women's relative autonomy in many parts of development work has been a significant achievement, and it is one that can serve as both catalyst and challenge to changing men's relationships at the level of households and society overall. And it is striking that a generational shift may be taking place in gender and development, with young women insisting on the participation of young men in development-related activities, and young men volunteering to get involved alongside the young women.[10]

In health, migration, and governance, men are hardly innocent bystanders. Nor are they avoidable. Worthy efforts have been made to achieve gender equality without the participation of men, but it's a gauntlet that has been run with limited success. Gender and Development must not only involve men but be characterized by men alongside women at the core of its mission, just as this gender integration can never lose its focus on inequality, power, and justice.

Notes

1 See discussions about faulty assumptions about men leading to twisted logic when building development programs throughout the authoritative volume on men and development, by Cornwall, Edström, and Greig (2011). See also chapters in the related volume by Cornwall, Karioris, and Lindisfarne (2016).

2 For a detailed examination of issues of self-reporting, discrepant reporting, and decision-making in households, see Becker, Fonseca-Becker, and Schenck-Yglesias (2006).

3 See Stern et al. (2003) for a sophisticated study of adolescents, masculinities, and sexualities.

4 For example, CORIAC in Mexico; Projeto H in Brazil; Puntos de Encuentro in Nicaragua. These are in addition to organizations whose mandate may be broader but who have GBV as a major focus, like Instituto Papai in Brazil and Salud y Género in Mexico; and larger NGOs who maintain a GBV emphasis as well, like International Planned Parenthood, Ford Foundation, and the Pan-American Health Organization. For a synopsis history of one NGO working with men that began in Brazil, Promundo, see Barker et al. (2011).

5 See discussion in Paulson 2016; this study is the best and most comprehensive to date on men and masculinities in development in Latin America.

6 For caregiving by men more generally, see Barker (2014).

7 On Women's Police Stations in Brazil, see Hautzinger (2007); Paulson (2016); Morrison, Ellsberg, and Bott (2004).

8 On masculinity and violence in Colombia, see also Gómez Alcaraz and García Suárez (2006).

9 For a good summary of the significance of Cairo and Beijing for men and masculinities in development, see Wanner and Wadham (2015).

10 This point is made briefly in Sweetman (2013: 8) in a stellar overview of working with men on gender equality. This phenomenon of young men volunteering in development work is also evident in work currently underway between Brown University, International Planned Parenthood Federation, and Mexfam in Mexico City.

References

Amuchástegui, A. (2006) ¿Masculinidad (es)?: Los riesgos de una categoría en construcción. In G. Careaga and S. Cruz Sierra (eds) *Debates sobre masculinidades: Poder, desarrollo, políticas públicas y ciudadanía.* Mexico City: Universidad Nacional Autónoma de México, pp. 159–181.

Bannon, I., and Correia, M. C. (2006) (eds) *The Other Half of Gender: Men's Issues in Development.* Washington, DC: World Bank.

Barker, G. (2014) A radical agenda for men's caregiving. *IDS Bulletin* 45(1): 85–90.

Barker, G., Nascimento, M., Ricardo, C., Olinger, M. and Segundo, M. (2011) Masculinities, social exclusion and prospects for change: Reflections from Promundo's work in Rio de Janeiro, Brazil. In A. Cornwall, J. Edström and A. Greig (eds) *Men and Development: Politicizing Masculinities*. London: Zed Books, pp. 170–184.

Barrera Bassols, D. and Oehmichen Bazán, C. (2000) (eds) *Migración y relaciones de género en México*. Mexico City: Grupo Interdisciplinario sobre Mujer, Trabajo y Pobreza, A.C. and Universidad Nacional Autónoma de México.

Becker, S., Fonseca-Becker, F. and Schenck-Yglesias, C. (2006) Husbands' and wives' reports of women's decision-making power in western Guatemala and their effects on preventative health behaviors. *Social Science and Medicine* 62: 2313–2326.

Camou, M. M. and Maubrigades, S. (2017) The lingering face of gender inequality in Latin America. In L. Bértola and J. Williamson (eds) *Has Latin American Inequality Changed Direction? Looking Over the Long Run*. Cham: Springer, Open Access.

Castro, R. (2000) *La vida en la adversidad: El significado de la salud y la reproducción en la pobreza*. Cuernavaca: Universidad Nacional Autónoma de México, Centro Regional de Investigaciones Multidisciplinarias.

Chant, S. and Gutmann, M. (2000) *Mainstreaming Men into Gender and Development: Debates, Reflections, and Experiences*. Oxford: Oxfam.

———. (2002) "Men-streaming" gender? Questions for gender and development policy in the 21st century. *Progress in Development Studies* 2(4): 269–282.

Connell, R. (1987) *Gender and Power: Society, the Person and Sexual Politics*. Berkeley, CA: University of California Press.

———. (2011) Organized powers: Masculinities, managers and violence. In A. Cornwall, J. Edström and A. Greig (eds) *Men and Development: Politicizing Masculinities*. London: Zed Books, pp. 85–97.

Cornwall, A., Edström, J. and Greig, A. (2011) (eds) *Men and Development: Politicizing Masculinities*. London: Zed Books.

———, Karioris, F. G. and Lindisfarne, N. (2016) (eds) *Masculinities Under Neoliberalism*. London: Zed Books.

De Keijzer, B. (1997) El varón como factor de riesgo: Masculinidad, salud mental y salud reproductiva. In E. Tuñón (ed) *Género y salud en el sureste de México*. Villa Hermosa: ECOSUR/UJAD, pp. 67–81.

Edström, J., Das, A. and Dolan, C. (2014) Introduction: Undressing patriarchy and masculinities to re-politicise gender. *IDS Bulletin* 45(1): 1–10.

Gómez Alcaraz, F. H. and García Suárez, C. I. (2006) Masculinity and violence in Colombia: Deconstructing the conventional way of becoming a man. In I. Bannon and M. C. Correia (eds) *The Other Half of Gender: Men's Issues in Development*. Washington, DC: World Bank, pp. 93–110.

Gutmann, M. (2007) *Fixing Men: Sex, Birth Control, and AIDS in Mexico*. Berkeley, CA: University of California Press.

———. (2010) New labyrinths of solitude: Lonesome Mexican migrant men and AIDS. In S. Chant (ed) *The International Handbook on Gender and Poverty: Concepts, Research, Policy*. London: Edward Elgar, pp. 321–326.

Hautzinger, S. (2007) *Violence in the City of Women: Gender and Battering in Bahia*. Berkeley, CA: University of California Press.

Hirsch, J. S. and Nathanson, C. A. (2001) Some traditional methods are more modern than others: Rhythm, withdrawal and the changing meanings of sexual intimacy in Mexican companionate marriage. *Culture, Health and Sexuality* 3(4): 413–428.

Mitchell, R. (2013) Domestic violence prevention through the constructing violence-free masculinities programme: An experience from peru. *Gender and Development* 21(1): 97–109.

Morrison, A. R., Ellsberg, M. and Bott, S. (2004) *Addressing Gender-Based Violence in the Latin American and Caribbean Region: A Critical Review of Interventions*. Washington, DC: World Bank.

Ochoa-Marin, S. C. and Vásquez-Salazar, E. A. (2012) Salud sexual y reproductiva en hombres. *Revista Salud Pública* 14(1): 15–27.

Orozco, M. (2017) Remittances to Latin America and the Caribbean in 2016. *The Dialogue*. www.thedialogue.org/wp-content/uploads/2017/02/Remittances-2016-FINAL-DRAFT.pdf (Accessed 28 January 2018).

Parker, R. (2011) Grassroots activism, civil society mobilization, and the politics of the global HIV/AIDS epidemic. *Brown Journal of World Affairs* 17(2): 21–37.

Paulson, S. (2016) *Masculinities and Femininities in Latin America's Uneven Development*. New York: Routledge.

Stern, C., Fuentes-Zurita, C., Lozano-Treviño, L. R. and Resoo, F. (2003) Masculinidad y salud sexual y reproductiva: Un estudio de caso con adolescentes de la Ciudad de México. *Salud Pública de México* 45: S34–S43.

Sweetman, C. (2013) Introduction: Working with men on gender equality. *Gender and Development* 21(1): 1–13.

Theidon, K. (2009) Reconstructing masculinities: The disarmament, demobilization, and reintegration of former combatants in Colombia. *Human Rights Quarterly* 31(1): 1–34.

Vargas Urías, M. A. (2012) Análisis de la construcción y reproducción de la masculinidad en la trata de personas: un enfoque revisionista con propuestas para impulsar el trabajo con hombres. *Género y Salud en Cifras* 10(2): 34–44.

Viveros, M. (2002) *De quebradores y cumplidores: Sobre hombres, masculinidades y relaciones de género en Colombia.* Bogotá: Universidad Nacional de Colombia.

Wanner, T. and Wadham, B. (2015) Men and masculinities in international development: 'Men-streaming' gender and development? *Development Policy Review* 33(1): 15–32.

Welsh, P. (2014) Homophobia and patriarchy in Nicaragua: A few ideas to start a debate. *IDS Bulletin* 45(1): 39–45.

26

LGBTQ SEXUALITIES AND SOCIAL MOVEMENTS

Florence E. Babb

Introduction

In recent decades, LGBTQ (lesbian, gay, bisexual, transgender, and queer) sexualities and social movements have become increasingly visible throughout Latin America. While research reveals nonheteronormative sexualities and gender identities in earlier periods, emerging cultures and politics around these identities in more recent times have led to greater public awareness. In part, this awareness reflects transnational currents in LGBTQ mobilization, which many times lend support to Latin Americans who are finding a voice and self-expression. Nonetheless, there is growing knowledge that diverse sexualities and gender identities have local histories that do not simply mirror those in the Global North. We now see broad appeals for recognition and social justice across the region, for acknowledgement that sexual rights are human rights, and, in some cases, for rethinking the terms of development so that LGBTQ populations may achieve their right to social inclusion, and to lives free of prejudice and precarity.

In what follows, I trace some broad currents in research on same-sex sexuality and emergent LGBTQ populations seeking social justice, and then I turn briefly to the case of Nicaragua, where I have conducted research on the subject, to illustrate the shifting terrain of sexual politics in that Central American nation and beyond. I contend that a more capacious approach to Latin American development is critical to understanding these human rights questions and how they are implicated in contemporary struggles.[1]

Past and present nonheteronormative sexualities and gender identities in Latin America

A small body of research has explored the history of diverse sexualities in the Latin American region from precolonial to postcolonial times. Scholars examining the impact of the conquest and colonialism in the lives of indigenous peoples have discussed the changes brought about in sexuality and marriage during the colonial period and the sometimes unintended consequences of clashing interests in the postcolonial formation of nation-states. Trexler (1995) moves beyond Eurocentric approaches to understanding gender, power, and indigenous sexuality at the time of conquest in the Americas and discovers evidence of homosexuality and cross-dressing, or transvestism, challenging Eurocentric views of native lives and sexualities. Like Trexler, Horswell (2005) finds support for third-gender nonconforming persons among indigenous Americans in the past. Others have traced a gender continuum, drawing on the evidence for pre-Hispanic transgender individuals and for their transgression's prohibition by the Spanish, for example in Peru 450 years ago (Campuzano, 2008). *Travestis* have nonetheless endured through the present in many parts of Latin America (Kulick, 1998; Prieur, 1998).

More research has considered colonialism's impact on men's than on women's same-sex practices in Latin America (Sigal, 2003) and specifically among indigenous peoples, such as Yucatecans (Sigal, 2000). While much of the literature has been concerned with male experience, in some cases authors have recognized gender relations more broadly as conditioning masculinity. We know that there was generally greater fluidity in gender relations and identity before European influence grew more pronounced, but further research is needed on both women's and men's same-sex sexuality from precolonial to postcolonial times.

Like the historical research, ethnographic research on contemporary cultures in Latin America has also tended to emphasize men's rather than women's lives. This is interesting in itself, emerging a couple of decades after a rich feminist scholarship on women and gender began focusing on the region. Much of the new writing on same-sex relations was influenced by this feminist work, yet the (largely male) authors generally have concentrated on what was becoming most strikingly visible: distinct male sexual cultures and practices in urban Latin America. From early efforts to document gay men's everyday life experiences (Carrier, 1995, on Mexico; Lumsden, 1996, on Cuba; Parker, 1999, on Brazil), scholars began to address other urgent questions, including the men's reckoning with HIV/AIDS (Carrillo, 2002, on Mexico; Padilla, 2007, on the Dominican Republic); their emergent gay identities and mobilization (Green, 1999, on Brazil); and their struggles against both homophobia and racism (Allen, 2011, on Cuba).

The ethnographic literature also offers detailed accounts of the lives of travesti and transgender individuals and communities, including extraordinary efforts of cross-dressing men and transgender women who alter their bodies and dress to better conform to their gender identities (Kulick, 1998, on Brazil; Prieur, 1998, on Mexico; Lancaster, 1997, on Nicaragua). For many of these individuals, employment opportunities are limited to select areas of work, such as hairdressing and sex work, meaning that they face economic challenges and health risks, as well as stereotyping. Marcia Ochoa's (2014) recent work brings contestants in Venezuela's beauty pageants into the same discussion with transgender women (*transformistas*) in an analysis of modernity and the performance of femininity in the national culture. In many cases, despite their sexual (and, often, racial and class) marginalization in the wider society, these individuals find supportive communities that enable them to resist further exploitation.

Scholarship on LGBTQ populations in Latin America has debated key issues, such as whether men who have sex with other men are engaging in "indigenous" or local sexual practices that should not be burdened with meanings emerging from organized gay movements to the North (Lancaster, 1992). For some, it is more suitable to use local terminology and understandings, for example, describing passive and active male partners (referring to penetrative sex), and recognizing that in certain cases the active partner may be viewed as heterosexual, or bisexual, having sex with both women and men, rather than as gay. Indeed, cultural differences in LGBTQ practices and identities must be reckoned with, as José Quiroga (2000) shows in the case of Cuba and elsewhere in Latin America (and Latino America), where invisibility may be used as a strategy in opposition to the transnational imperative to "come out." Others, however, have countered with the increasing evidence of the transnational circulation of LGBTQ identities and cultural politics

Figure 26.1   Men performing as women in Carnaval in Huaraz, Peru
Source: Photo by author.

that have been at least selectively appropriated and embraced throughout the region, arguing that it would be a mistake to overlook these currents of change (Parker, 1999; Babb, 2003). We will see that the rise of LGBTQ movements in Latin America can only be attributed to wide-ranging combinations of local, national, and transnational factors.

As noted, attention to women among sexually minoritized groups has been seriously under-represented, in part due to women's lower public visibility in many cases, but this has been changing over the past two decades (Balderston and Guy, 1997; Green and Babb, 2002). Like some of the research on men, a few studies of women have discovered same-sex practices that challenge the globalized models of LGBTQ cultures and identities. Gloria Wekker (2006) makes an important contribution with her intimate monograph on working class women's *mati* sexual culture (which favours having both female and male partners over marriage) in the Afro-Surinamese community, where their women-centered families coexist with their heterosexual relationships; she shows that the emphasis for these women is on sexual practices and not on identities.

In contrast, other scholars and activists have focused on self-identified lesbians and their experiences, including a book-length study by Nicaraguan psychologist Mary Bolt González (1996), whose pioneering research into discrimination and self-esteem was based on in-depth interviews with women in her country; she found that intersecting inequalities surrounding gender and sexual orientation coupled to make these women's lives particularly challenging, even if lesbians often passed under the radar more readily than gay men in Nicaraguan society. Cymene Howe (2013) also focuses on Nicaraguan lesbians in her monograph, taking an ethnographic approach in order to consider women in the context of the revolutionary culture that both suppressed and gave rise to LGBTQ political organizing; she examines everyday life, national developments, and the media as a force for mobilizing around sexual rights struggles. More will be said later of Nicaraguan sexual cultures and politics.

Queering development in Latin America

Before considering what it would mean to define Latin American development in more inclusive terms that would do justice to LGBTQ populations, it is important to say a word about the suitability of the word "queer" in relation to sexualities in the region. Using the terminology of "queer" in place of LGBT or same-sex sexuality has been somewhat controversial in both the Global North and the Global South, because it may be viewed as homogenizing diverse sexual identities or as imposing an external episteme (from the North) in places where there are other ways of knowing sexual experience. I use it as an analytical term to suggest an overarching rubric that recognizes the fluidity of sexualities that are understood to be always in formation, and that counters the tendency to think in taxonomic, static, and binary ways about sexual identities and practices. Moreover, I find the term increasingly suitable in those Latin American contexts where self-identified LGBTQ individuals, including academics and activists, are using the term. When I turn to the Nicaraguan case, I will mention the recent formation of the arts and activist group Operación Queer as one example of locally defined queer politics (that also engages with a transnational queer politics).[2]

While feminists have for decades offered critiques of development initiatives that overlook gender or treat women only in instrumentalist ways, such critical approaches are only slowly advancing in relation to heteronormative development programs. Amy Lind and her collaborators have offered powerful critiques of projects that give inadequate attention to the lives, families, and communities of those whose sexual and gender identities position them in the social margins. One pioneering intervention called for rethinking Western development's assumptions about the sort of families and households that are served by development initiatives (Lind and Share, 2003). When such development plans are introduced, they have much potential for unintended consequences, such as targeting only normative, heterosexual nuclear families and overlooking same-sex couples and their children (as well as single-parent households, and households without children). Likewise, women may be targeted as biological reproducers, neglecting the needs of lesbians and others who may or may not be parents. When sexually nonconforming individuals are left out of development processes, they may not receive needed resources or health care, or, as significantly, social recognition. This may contribute to their further marginalization and lower self-esteem. At the same time, their exclusion can mean that development projects are deprived of the positive input and energy that LGBTQ individuals and communities might contribute.

When, on the other hand, initiatives have been directed to HIV-positive men, for example, reductionist development programs may see the men as little more than their HIV status, and, if they are LGBTQ-identified, little more than their sexuality. This might be compared to mainstream development's logic concerning women in general, viewing them fundamentally as mothers and socializers of the next generation, keepers of tradition, and vehicles for implementing development goals. This tendency within development projects to reduce individuals to their bodies and to their capacity to work and thereby to help alleviate poverty, has deleterious effects for heterosexuals too, of course, but it can have harsher consequences for those whose bodies are deemed transgressive. Queer-identified women are less often identified in this way, yet their neglect can mean inadequate attention to their health care or to their economic challenges (as both women and queer) – nor is it likely that lesbians' activism around issues affecting others will be recognized (e.g., supporting the rights of men with HIV/AIDS or the rights of heterosexual women for reproductive rights and freedom from domestic violence).

In an anthology on development and sexual rights, Lind (2010: 1) writes that "For too long, people who do not fit within socially prescribed sexual and gender roles in their societies have been seen as irrelevant to or 'outside' the project of development." This insight led her to bring together queer theorists who would challenge the field of development to reimagine its policies and practices, and indeed its notions of who is entitled to full citizenship rights. In this way, contributors to the anthology argue, new sexual subjects would find a place in the development narrative and in visions for a more just future. Instead of being viewed as irrelevant or even antithetical to the broad project of national development, LGBTQ individuals would be seen as productive and valued members of society. Critical and activist scholarship like this has posed urgent questions and called for rethinking development objectives. Of course, it remains to be seen whether under conditions of neoliberal globalization in so much of Latin America, queering development can make real gains. What seems clear is that critiques and proposals must be accompanied by grassroots activism if they are to have a significant impact.

The project of bringing together a critical feminist and queer perspective on development is also taken up in the work of Andrea Cornwall, Sonia Corrêa, and Susie Jolly (2008) and contributors to a volume on "development with a body," which reconsiders sexuality, human rights, and development. Several areas taken up are worth discussing here, as they sometimes fall outside of discussions of development that focus largely on vulnerabilities and victimization, especially where minoritized populations are at centre. While homophobia and violence against LGBTQ individuals are urgent social problems to address, far less attention has been devoted to the right of these individuals to full and satisfying sexual lives. Indeed, sexual pleasure and everyone's right to it is not frequently on the development agenda whether it concerns heterosexuality or other sexualities. Yet these scholar-activists want to be sure that a politics of sexual rights takes up pleasure as well as risk in the lives of LGBTQ people.

In a different yet related way, sex workers (both heterosexual and LGBTQ) have staked a claim for the right to gain their livelihood in the ways they choose. In this case, the issue is generally one of workers' rights rather than a right to pleasure, extended to those who employ their bodies in order to earn a living. The right to provide sexual services to others as a form of employment can be especially important where work is scarce and sex work often pays better than other jobs in the informal economic sector. It is crucial to distinguish between sex trafficking and sex work because while the former can involve individuals in critical danger with almost no control over their lives, sex workers often pride themselves on their independence and professionalism as they set the terms of their transactions with clients.

Gay, bisexual, and, in some cases, straight-identified men attract an international gay clientele as they perform sex work, satisfying their partners and wives that their line of work can contribute to higher earnings and at times continuing support through remittances sent to the Dominican Republic or Cuba (Padilla, 2007; Allen, 2011; Hodge, 2001). In certain areas, sex workers have organized to demand better conditions and to resist campaigns to remove them from city streets. Most attention to activist sex workers, however, has focused on women who seek male clients, as in a case of Brazilian sex workers who formed a fashion design company to diversify and sell to a wide public; more connection to queer sex tourism would challenge assumptions about masculinity and family in Latin America (Brennan, 2016). More research is needed on LGBTQ sex workers in the context of the growing and globalized movement for sex worker rights, and for countering both police violence and the perception that sex workers ought to be "rescued." Instead, sex workers are deserving of recognition and justice as full citizens, and as change agents around development issues such as containing the spread of HIV in Latin America.

LGBTQ social movements

By now it should be clear that the growing visibility of gay men, lesbians, transgender, and queer individuals has come about because more are coming out of the shadows, which in itself is a political act of courage in contexts characterized by homophobia or, at best, a heteronormative expectation that those who are sexually different will conceal their orientation. In the last few decades, some in LGBTQ communities have gone further to form organizations and social movements to address their social exclusion and marginalization. Continental gatherings have been held to strategize and form alliances and while struggles are often still in the fledgling stage, it is now possible to point to some notable successes. Indeed, some writers have emphasized that while LGBTQ people in the Latin American region continue to experience serious human rights violations, they have also seen progressive legislation, in some cases going beyond that found in nations in the Global North.

For example, same-sex marriage is now legal in Argentina, Brazil, Uruguay, Colombia, and in several states of Mexico including Mexico City; and same-sex civil unions are legal in Chile and Ecuador. In 14 Latin American nations, laws now prohibit workplace discrimination based on sexual orientation, and adoption is afforded to same-sex couples in a number of countries. Transgender individuals have made gains in Cuba and Bolivia. Arguably, legislation has outpaced social attitudes in some regions (Brochetto, 2017). Catholicism, Evangelicalism, and cultures of machismo still present barriers to change, though there are indications of growing tolerance in the region. Among some LGBTQ rights groups there is continuing debate over whether to give priority to such legislative issues as support for same-sex marriage, or instead, to such issues as anti-LGBTQ violence – but they agree that sexual rights are a key and urgent issue of our time.

Omar Encarnación (2016), and Javier Corrales and Mario Pecheny (2010), take a broad regional perspective and argue for a decentering of our views on the emergence of LGBTQ movements in Latin America. Recognizing early forms of resistance to European-imposed heteronormativity and the local roots of recent struggles over sexual rights, these analysts do not overlook transnational currents; yet they contend that tropes of gay movements everywhere being similarly inspired by landmark events like the Stonewall uprising in New York City in 1969 are misguided. We need to examine histories and contemporary politics of gay, lesbian, and trans-identified movements where they occur in order to correct such short-sighted approaches (Green, 1999, on Brazil; Lumsden, 1996, on Cuba; Howe, 2013, on Nicaragua). Fortunately, scholarship is advancing on the subject.

As Corrales (2015) notes, left-leaning and economically developing nations have not necessarily been in the forefront of change, even given the presence of LGBTQ movements. Examples include Nicaragua, where progress has been slow, and Peru, where the government has stalled on support for progressive legislation. In Nicaragua, the Sandinista government has sometimes given lip service to LGBTQ organizations in order to appear "modern" and also as an evident effort to drive a wedge between these groups and feminist organizations (which have been critical of the government). And in Peru, when the first "out" national politician Carlos Bruce introduced a bill to support same-sex civil unions, it was defeated by Congress and seems unlikely to move forward soon (the Catholic Church has opposed same-sex unions and a 2013 poll showed that 65% of Peruvians oppose them); the same congress removed sexual orientation and gender identity as protected categories under a recent hate crime bill. Peru has for decades had a strong feminist movement and made strides in relation to gender equality, but LGBTQ rights have gained less ground. Several nations have been especially slow to embrace sexual rights agendas, for example in the English-speaking Caribbean, and in some cases conservative groups have found support from transnational homophobic interests, such as faith-based organizations in the United States. Such challenges have meant that some rights organizations have embraced more conservative politics in order to try to win acceptance, potentially compromising their more far-reaching objectives (Corrales, 2015).

Through their activism, gay men and lesbians are seeking the right of sexual expression, the right to express their desires and affections free of social prejudice, and to have their relationships recognized, while transgendered individuals seek to achieve recognition of their gender identity and their full citizenship rights (Corrales and Pecheny, 2010: 5). Obstacles remain daunting, resulting in uneven or stalled development in efforts to attain LGBTQ rights. Some of these obstacles can be summarized: the dilemma of being a small minority in a region where many still find comfort in remaining in the "closet" rather than becoming more visible as activists; the difficulty of finding political allies when even those on the left may be wary about adding sexual rights to their agendas, which tend to favour social class issues; cultural resistance by families in areas where young adults often reside longer at home before establishing themselves independently (due to economic or other reasons); backlash or complacency once initial gains have been made and the broader public no longer sees urgency to LGBTQ concerns; the continued strength of conservative religious interests that block policy and other change.

Some analysts have suggested that until Latin America reaches higher levels of economic development the region may not see more widespread support for LGBTQ rights. It may be true that where basic needs of food, shelter, and health care are wanting, individuals may have little incentive or even energy to devote to political organizing for sexual rights. Nonetheless, the sexual lives and pleasures of the poor are no less compelling than those of higher social classes in their societies. Indeed, the desire to be with loved ones is felt as keenly among the poor as the wealthy, and the social support that would be afforded by the right of LGBTQ people to form families and households of their choosing may be critically important to those with fewer advantages in life. Rather than being faulted for not "coming out" and becoming more visible, such individuals should be better understood and supported. Political analysts may regret the loss of more LGBTQ individuals standing up to be counted, but this must be weighed against the social costs for such women and men in coming out under adverse conditions.

Regardless of the obstacles to LGBTQ organizing, those who have followed developments most closely describe the current moment in positive terms. For Javier Corrales (2015b), "Outside of the North Atlantic, no region in the world has undergone more progress in expanding LGBT legal rights than Latin America. And for Omar Encarnación (2016), "Latin America, famous around the world as a bastion of Catholicism and machismo, has in recent years emerged as the gay rights leader of the Global South. More surprising yet, several Latin American nations are international gay rights trendsetters." Real gains for sexual rights *have* been achieved in the region and as societies gradually become more tolerant and inclusive, there will be still more opportunities for LGBTQ movements to gain ground. As they do, it will be important for their internal politics to reflect their expectations of the wider society: to recognize social differences in the plural, and the complex intersections of sexuality with race, gender, class, and other social vectors, that can either enrich or divide – but with adequate attention can produce strong coalitions on which to build programs for social transformation.

Nicaraguan currents of change

While LGBTQ movements in such nations as Argentina, Brazil, and Mexico may be better known, the small Central American nation of Nicaragua has drawn the attention of activists and scholars who have noted the postrevolutionary development of a growing community of individuals calling for recognition and rights in a society that historically sought to stifle mobilization around sexual difference.[3] The deeper history of nonheteronormative sexualities and gender identities in Nicaragua is only beginning to be told,[4] but recent research has examined the stirrings of a more public LGBTQ presence that came to the fore by the late 1980s. Only after efforts by the Sandinista government to quell organizing around sexual rights near the end of the decade of revolutionary governance (1979–1990), following the 1990 electoral loss of the Sandinistas and the subsequent neoliberal turn, were both feminist and LGBTQ activism to take off in more autonomous directions. Freed of the demands of a verticalist Sandinista political party, the decade of the 1990s saw the florescence of numerous activist groups, NGOs, publications, and social venues catering to a sexually diverse clientele.

By this time, several NGOs that were known for their progressive work in confronting the spread of HIV/AIDS and in promoting sexual health and education could more freely serve a diverse and often queer public. Lesbians were frequently in the forefront of activism, to the point that gay men sometimes bristled at their own secondary participation. NGOs receiving international support were occasionally criticized for being overly influenced by global political currents, and the transnational aspects of some organizing activities were apparent. For example, *jornadas*, periods of activity devoted to showing gay pride and countering sexual prejudice, were held in late June, around the anniversary of the Stonewall uprising, when marches and other events take place in cities around the world. Nevertheless, there was much that was distinctly Nicaraguan in the coming together of feminist, left, and queer activists in several cities around the country. During the 1990s, political organizing was galvanized by efforts to challenge the country's sodomy law, which criminalized same-sex sexuality of both gay men and lesbians, as one of the most repressive in the hemisphere. At the same time, there were occasions for celebration, such as an annual Miss Gay performance and competition, which became well enough established that it was held in Managua's elegant national theatre.

With Daniel Ortega's reelection as President in 2006, and his governance through the present, there was a return to state-led development and an increasingly authoritarian rule. Self-styled as a left-oriented populist, Ortega's government came down hard on members of the left, notably feminists and others departing from the Sandinista party line, who took issue with antidemocratic practices. During this time, LGBTQ groups continued to gain visibility and sometimes even found expressed support from Ortega, likely springing from his desire to seem modern when it would not be politically costly to do so – or to divert attention from other antiprogressive actions ("pink washing"). In 2008, the oppressive sodomy law was repealed, making same-sex sexual activity legal, as part of the unveiling of a new penal code. Trans and other LGBTQ Nicaraguans seized the moment and in more recent years began holding *plantones*, monthly demonstrations, on one of Managua's principal traffic circles. The visibility may have gained them some cultural capital if not necessarily further legal or political rights.

While discrimination based on sexual orientation is banned in the areas of employment and access to health care, same-sex couples are ineligible for the legal benefits that opposite-sex married couples enjoy. Indeed, in 2014 Nicaragua's Congress adopted a new Family Code restricting marriage and adoption to heterosexual couples. The strong influence of the Catholic Church coupled with social conservatism clearly stood in the way of extending these rights to LGBTQ Nicaraguans. A Pew Research Center survey in 2013 showed that 77% of Nicaraguans opposed same-sex marriage, with just 16% in favour, and others expressing no opinion. Despite NGOs' initiatives to educate a broad public and a hugely popular television series, "Sexto Sentido," with sympathetic gay and lesbian characters, attitudes have been slow to change.

Elsewhere I have suggested that the broad-based participation of young men and women in the Nicaraguan revolution in the late 1970s and in the health and literacy brigades that followed provided opportunities to leave home and family and discover new desires that in some cases led to the expression of LGBTQ identities and sexual orientations. In this way, Nicaraguans' path to organizing around sexual rights was distinct from that in much of the Global North, where urban societies offered various pathways to youthful independence and to sexual self-discovery.

In 1989, a decade after the revolutionary victory, gay and lesbian-identified Nicaraguans marched openly for the first time in the July 19 march honouring the tenth anniversary of the Sandinista revolution. In 1991, Nicaraguans held their first Gay Pride event and the following year they launched a Campaign for Sexuality without Prejudices in response to the Chamorro government's reactivation of the nation's draconian sodomy law. From that point, a number of LGBTQ groups emerged and began joining together in coalition for the yearly jornadas. While women played a prominent role as organizers and activists in such groups as Nosotras, Grupo Safo, Entre Amigas, and Grupo Lesbiana por la Visibilidad, they have continued to face exclusionary practices in the wider society in the 21st century.

To fast forward to 2013, it is notable that a collective of artists, academics, and activists joined together that year in Managua as Operación Queer in order to give more visibility to gender and sexual difference. The collective's impact has been both cultural and political as it takes up questions of the body and identity and of forms of exclusion relating not only to gender and sexuality but to social class, ethnicity, age, ability, and aesthetics. Those in the collective have addressed whether there is something like a community among "queer" Nicaraguans, understood to be diverse and fluid in its formation. Intellectually and culturally sophisticated, Operación Queer has significant political potential as well. As one indication of its ability to push LGBTQ Nicaraguans further in their analysis, Operación Queer has addressed its own elitism and takes a highly self-critical approach; the collective recognizes, for example, the lived experience of a trans sex worker in Managua who may be unfamiliar with queer theory and yet bravely transgress norms of sexual performance on a daily basis.[5]

We can learn from the experience of this Central American nation. The achievements of LGBTQ groups in Nicaragua are in some ways exceptional, drawing on mobilization strategies honed during a decade of revolutionary government and then taking off in more autonomous directions. Activists found a political space under new conditions of post-1990 neoliberalism, and then deepened their resolve under the contradictory terms of the unholy pact formed by Sandinistas and the Catholic Church in the post-2007 period. Now that it is better understood that same-sex and queer sexualities have a deep history in Nicaragua, it is unnecessary to shun association with transnational currents of LGBTQ activism as not truly "Nica." Feminist and queer Nicaraguans have had to navigate an unstable political ground during the last decade, but to some degree the nation has become more tolerant and has come to recognize sexual and gender identities as diverse and not "unnatural." This is a definite step forward in granting greater dignity to sexual minorities in Nicaragua, as elsewhere in Latin America, and it demonstrates what can be achieved with or without state support. As a result of their struggles, many in the LGBTQ community have undergone personal transformation, and that in itself is no small accomplishment. Nonetheless, they face numerous challenges ahead as they strive to overcome the historical legacy of a still-heteronormative postrevolutionary society. As such, their recent experience shines a light on the sort of advances and the limitations that we have seen throughout the region.

Notes

1 Although it is beyond the scope of this chapter, queer literature and film in Latin America have also contributed meaningfully to Latin American development and social mobilization. Queer writers and filmmakers have struggled to gain visibility and recognition and their cultural production would be the subject of another chapter.
2 A rich discussion of the saliency of queer theory and politics in Latin America may be found in several contributions to the Ecuadorian journal *Íconos*, including Cordero Velásquez, 2011. See also Babb, 2014.

3 See Babb, 2001, 2003, 2018 for more detailed discussion of the emergence of LGBTQ culture and politics in Nicaragua. In addition to my own field research, sources I have found useful include Bolt 1996; Thayer, 1997; Howe 2013, and Kampwirth, 2014. To date, the discussion of gender and sexual rights in Nicaragua has focused mainly on the country's dominant *mestizo* population, and more attention should be directed to minoritized indigenous and Afro-descendant peoples.

4 See the forthcoming work of V. González-Rivera and K. Kampwirth, *One Hundred Years of LGBT History in Nicaragua* (manuscript in progress). See also González-Rivera, 2014.

5 For discussion of Operación Queer I consulted various websites including www.bienalcentroamericana.com/2016/08/08/operacion-queercochona/.

References

Allen, J. S. (2011) ¡*Venceremos? The Erotics of Black Self-Making in Cuba.* Durham, NC: Duke University Press.

Babb, F. E. (2001) *After Revolution: Mapping Gender and Cultural Politics in Neoliberal Nicaragua.* Austin: University of Texas Press.

———. (2003) Out in Nicaragua: Local and transnational desires after the revolution. *Cultural Anthropology* 18(3): 304–328.

———. (2014) Sexualities in Latin America and the Caribbean. In B. Vinson (ed) *Oxford Bibliographies in Latin American Studies.* New York: Oxford University Press.

———. (2019) Nicaraguan legacies: Advances and setbacks in feminist and LGBTQ activism. In H. Francis (ed) *A Nicaraguan Exceptionalism? Debating the Legacy of the Sandinista Revolution.* London: Institute of Latin American Studies (ILAS), University of London. In press.

Balderston, D. and Guy, D. J. (1997) (eds) *Sex and Sexuality in Latin America.* New York: New York University Press.

Bolt González, M. (1996) *Sencillamente diferentes . . .: La autoestima de las mujeres lesbianas en los sectores urbanos de Nicaragua.* Managua, Nicaragua: Fundación Xochiquetzal.

Brennan, D. (2016) Sex worker activism and labor. In M. Gutmann and J. Lesser (eds) *Global Latin America into the Twenty-First Century.* Berkeley, CA: University of California Press, pp. 240–250.

Brochetto, M. (2017) The perplexing narrative about being gay in Latin America. *CNN*, 4 March. https://edition.cnn.com/2017/02/26/americas/lgbt-rights-in-the-americas/index.html (Accessed 25 September 2018).

Campuzano, G. (2008) Gender, identity and *travesti* rights in Peru. In A. Cornwall, S. Corrêa and S. Jolly (eds) *Development with a Body: Sexuality, Human Rights and Development.* New York: Zed Books.

Carrier, J. (1995) *De los Otros: Intimacy and Homosexuality among Mexican Men.* New York: Columbia University Press.

Carrillo, H. (2002) *The Night Is Young: Sexuality in Mexico in the Time of AIDS.* Chicago: University of Chicago Press.

Cordero Velásquez, T. (2011) Comentarios desde el sur. *Iconos: Revista de Ciencias Sociales* 40: 129–135.

Cornwall, A., Corrêa, S. and Jolly, S. (2008) (eds) *Development with a Body: Sexuality, Human Rights and Development.* New York: Zed Books.

Corrales, J. (2015a) The politics of LGBT rights in Latin America and the Caribbean: Research agendas. *European Review of Latin American and Caribbean Studies* 100: 53–62.

——— (2015b) *LGBT Rights and Representation in Latin America and the Caribbean: The Influence of Structure, Movements, Institutions, and Culture.* Chapel Hill: University of North Carolina.

——— and Pecheny, M. (2010) (eds) *The Politics of Sexuality in Latin America: A Reader on Lesbian, Gay, Bisexual, and Transgender Rights.* Pittsburgh, PA: University of Pittsburgh Press, 2010.

Encarnación, O. G. (2016) *Out in the Periphery: Latin America's Gay Rights Revolution.* New York: Oxford University Press.

González-Rivera, V. (2014) The alligator woman's tale: Remembering Nicaragua's 'first self-declared lesbian.' *Journal of Lesbian Studies* 18(1): 75–87.

Green, J. N. (1999) *Beyond Carnival: Male Homosexuality in Twentieth-Century Brazil.* Chicago: University of Chicago Press.

——— and Babb, F. E. (2002) (eds) Special Issue: Gender, Sexuality, and Same-Sex Desire in Latin America. *Latin American Perspectives* 29(2).

Hodge, D. (2001) Colonization of the Cuban body: The growth of male sex work in Havana. *NACLA Report on the Americas* 34(5): 20–24.

Horswell, M. J. (2005) *Decolonizing the Sodomite: Queer Tropes of Sexuality in Colonial Andean Culture*. Austin: University of Texas Press.

Howe, C. (2013) *Intimate Activism: The Struggle for Sexual Rights in Postrevolutionary Nicaragua*. Durham, NC: Duke University Press.

Kampwirth, K. (2014) Organizing the Hombre Nuevo Gay: LGBT politics and the second Sandinista revolution. *Bulletin of Latin American Research* 33(3): 319–333.

Kulick, D. (1998) *Travesti: Sex, Gender, and Culture among Brazilian Transgendered Prostitutes*. Chicago: University of Chicago Press.

Lancaster, R. N. (1992) *Life Is Hard: Machismo, Danger, and the Intimacy of Power in Nicaragua*. Berkeley, CA: University of California Press.

———. (1997) Guto's performance: Notes on the transvestism of everyday life. In D. Balderston and D. J. Guy (eds) *Sex and Sexuality in Latin America*. New York: New York University Press, pp. 9–32.

Lind, A. (2010) (ed) *Development, Sexual Rights and Global Governance*. New York: Routledge.

——— and Share, J. (2003) Queering development: Institutionalized heterosexuality in development theory, practice and politics in Latin America. In K. K. Bhavnani, J. Foran, and P. A. Kurian (eds) *Feminist Futures: Re-Imagining Women, Culture and Development*. New York: Zed Books, pp. 55–73.

Lumsden, I. (1996) *Machos Maricones and Gays: Cuba and Homosexuality*. Philadelphia, PA: Temple University Press.

Ochoa, M. (2014) *Queen for a Day: Transformistas, Beauty Queens, and the Performance of Femininity in Venezuela*. Durham, NC: Duke University Press.

Padilla, M. (2007) *Caribbean Pleasure Industry: Tourism, Sexuality, and AIDS in the Dominican Republic*. Chicago: University of Chicago Press.

Parker, R. (1999) *Beneath the Equator: Cultures of Desire, Male Homosexuality, and Emerging Gay Communities in Brazil*. New York: Routledge.

Prieur, A. (1998) *Mema's House, Mexico City: On Transvestites, Queens, and Machos*. Chicago: University of Chicago Press.

Quiroga, J. (2000) *Tropics of Desire: Interventions from Queer Latino America*. New York: New York University Press.

Sigal, P. (2000) *From Moon Goddesses to Virgins: The Colonization of Yucatecan Maya Sexual Desire*. Austin: University of Texas Press.

———. (2003) (ed) *Infamous Desire: Male Homosexuality in Colonial Latin America*. Chicago: University of Chicago Press.

Thayer, M. (1997) Identity, revolution, and democracy: Lesbian movements in Central America. *Social Problems* 44(3): 386–406.

Trexler, R. C. (1995) *Sex and Conquest: Gendered Violence, Political Order, and the European Conquest of the Americas*. Ithaca, NY: Cornell University Press.

Wekker, G. (2006) *The Politics of Passion: Women's Sexual Culture in the Afro-Surinamese Diaspora*. New York: Columbia University Press.

PART V

Labour and campesino movements

27

RURAL SOCIAL MOVEMENTS

Conflicts over the countryside

Anthony Bebbington[1]

Introduction

In 2016, at least 200 land and environmental, defenders were murdered – the dead-liest year on record. Not only is this trend growing, it's spreading – killings were dispersed across 24 countries, compared to 16 in 2015.

(Global Witness, 2017: 6)

Defenders of the Earth, the Global Witness 2017 report on the global killing and criminalization of land and environmental defenders, makes for sobering reading. The follow year's report, *At What Cost*, is worse still:

In 2017, Global Witness documented 207 killings of land and environmental defend-ers – ordinary people murdered for defending their forests, rivers and homes against destructive industries. This is six more murders than in 2016, making it the worst year on record.

(Global Witness, 2018: 9)

These are appalling figures and for Latin America they are worse still. By far the most dangerous continent in which to be an environmental defender is Latin America, with 110 of 185 killings in 2015, 122 of 200 killings in 2016, and 121 of 207 killings in 2017. Within Latin America, the most killings in 2017 were in Brazil (57), followed by Colombia (24). However, on a per capita basis, the most dangerous countries in which to be an environmental defender remain Nicaragua and Honduras. Killings were linked to conflicts over the control of land and natural resources: in 2016 these conflicts were most frequently related to the presence of mining, oil, and gas industries, though also to logging and agroindustry, while in 2017 killings linked to agroindustry increased markedly.

Not all those killed are the classic faces of social movements. They include government park rangers, lawyers, and journalists working on environment, human rights, corruption, investment, and dispossession, as well as community and indigenous leaders. However, in a broader con-ceptualization, those killed are very much part of movements. They are involved in collective arguments over narratives about development and resource control, legitimate rights and rights holders, and the values towards which socio-economic and political organization should aspire.

These Global Witness documents frame the discussion in this chapter. They are vitally important in their own right and what they report on should always be a core concern for those doing political ecologies and political economies of rural change in Latin America. Furthermore, they make apparent issues that underlie the argument of this chapter. First, the Latin American countryside continues to be the venue of conflicts and struggles that will define the future face and feel of the region. Second, the countryside continues to be of central importance to any reflection on the region, notwithstanding the increasing urbanization of Latin America and of Latin American Studies. Third, the countryside is not just a rural question (it probably never was). The killings reported by Global Witness are related to global and national flows of commodities and finance, flows that articulate the rural and the urban always. In this sense, movements that emerge to contest them are not best understood as purely rural social movements. While they are movements that may have a concern for transformations happening in the countryside, they are addressing processes that exceed *lo rural* and that will almost certainly have to have urban components if they are to stand any chance of being successful in their own terms.

In the light of these observations, the remainder of this chapter is organized as follows. I begin with a reflection on the meanings of social movements and follow this with a discussion of what it might mean to place the word rural in front of this term. The third section considers the historical nature and role of rural social movements in the region, while the fourth discusses and reflects on some contemporary movement processes in the Latin American countryside. The closing section draws out broader observations, taking the text back to the very deliberately chosen subtitle of the chapter, "Conflicts over the countryside." This subtitle is meant to convey the idea that there are many conflicts spread across Latin American countrysides, that these are conflicts over what that countryside should be for, who it should be for, and what its future should be, and that these conflicts are not just rural in nature but instead about country in the fullest sense of the term.

Rural ... social movements

Social movements

The definition of what constitutes a social movement is contested, both politically and analytically. Politically, the contestation generally has much to do with arguments over legitimacy and significance. When Ecuador's then president Rafael Correa talked of anti-mining protests as *cuatro pelagatos* (four nobodies) (Moore, 2009), while activists and participants saw socio-environmental movements, what was at stake was the legitimacy of particular forms of socio-political action. In a similar vein, when critics characterize mobilizations over rural issues as the work of once-upon-a-time Trotskyists turned green, or of misguided urbanites, they are trying to argue that such mobilizations are neither a movement nor legitimate. Such political arguments over legitimacy and what is a real movement can also occur within social movements, as some actors argue that others are not really part of the movement because they have the wrong residential, class, racial, ethnic, gender, employment, or other background.

Parallel to such political disputes, the conceptual challenge of defining a movement is also fraught. Where some writers see movements, others see only organizations, networks, or mobilizations. Where some see organic processes, others see only a series of events. And then there is the argument as to whether, in order to be conceptualized as a social movement, a phenomenon has to carry within it seeds of an alternative way of organizing society, or whether a completely interest-based process can also be a movement.

Against the backdrop of such debates, I take social movement to refer to a process of mobilization that is sustained across time and space, rather than as a specific organization. Thus, while formal organizations can be *part of* social movements, movements are more than organizations and also include the more nebulous, uncoordinated, and cyclical forms of collective action, popular protest, and networks that serve to link organized and dispersed actors in processes of social mobilization (Tilly, 1985). They are politically and/or socially directed collectives of usually several networks and organizations aiming to change elements of the political, economic and social system (Ballard et al., 2005: 617).[2]

One way of characterizing movements is to ask whether they are primarily concerned with issues of accumulation, social reproduction, or identity. In this three-way distinction, the first type of movement would be one that emerges in response to dynamics of capital accumulation. If we follow Harvey (2003) and distinguish between accumulation by exploitation and accumulation by dispossession, this gives us a further means of distinguishing between sub-types of movement. Thus, a movement that emerges around prices, wages, and value distribution within commodity chains would be a response to accumulation by exploitation. Conversely, processes of accumulation through dispossession of land, water, natural resources, territory, and other assets are more likely to induce emergence of movements around territory, ethnicity, environment, justice or place (c.f. Escobar, 2008).

Other movements emerge around the distribution and provision of services and assets that are collectively consumed and provided by the state. Such services might include shelter, basic infrastructure, education, or health, and more specifically for rural areas, bilingual education or irrigation system provision. A third type of movement has roots in structured relationships of prejudice based on identity. These movements can be understood as challenging the terms of recognition (Lucero, 2008) under which certain identity-based groups are subject to disadvantage as a consequence of the ways in which they are viewed and governed by other, more powerful groups. Indigenous or Afro-Latin American movements, often with deep roots in rural communities, would be examples of this type of movement. Comparing movements in terms of the domains in which they are active, we might say that the first type contests relationships linked to production, the second contests relationships linked to the state and collective consumption, whilst the third contests relationships within society. The boundaries are, however, fluid.

In addition to their broad orientation, social movements can be distinguished by scale and strategy. While some movements are localized, others operate at national and transnational scales, either as specific movements or as federations and confederations of movements. Thus, Vía Campesina might be deemed a transnational movement, the Mesa Nacional Contra la Minería in El Salvador a national level movement that articulated a series of national and sub-national actors, and the Asociación de Pueblos Guaraní in Bolivia a subnational movement, albeit one that sometimes seeks to impinge on national debates, laws, and policies. Across these different scales, some movements adopt strategies that are oppositional, others engage in strategies that emphasize negotiation, others focus considerable effort on delivering services to their members, and others combine elements of each of these strategies. There is, then, a lot going on within the category of social movement. Adding the qualifier rural does not necessarily simplify this diversity.

What makes a rural social movement rural?

What does it mean to place the word rural in front of the term social movement? Here I note three possibilities.

First, might this mean that the participants in the movement are residents of rural areas? There are a number of difficulties here. One is the simple challenge of determining where the division between rural and urban lies: is it defined by settlement size, population density, land use? The phenomenon of circular migration between the countryside and the city also complicates classifications by residence. In addition, such a classification could not easily deal with movements in which urban dwelling professionals or lawyers, for instance, played an important role alongside, say, farmers or indigenous community members. A definition based on the residence of members can also be used against movements whose processes involve such urbanites by critics who seek to dismiss them as fake movements on the grounds that they are not pure rural.

A second possibility might be that a rural social movement is one whose issue is rural. The issues most frequently associated with the murders of environmental defenders reported by Global Witness might qualify here: logging, agroindustry, land dispossession, dam building, indigenous territorial loss, extractive industries. However, these are also issues around which urban populations mobilize and for which they feel sufficient concern as to render them urban issues also. Extractive industry or agroindustry conflicts can mobilize city residents concerned about water resources, environmental quality, forest cover, food quality, or models of development. Indeed, it might be argued that movements are more likely to have an impact on such issues when they can enrol people living in different places, or even countries, and located at different points along the value chain or global production network of which the activity is a part. The implication seems to be that activities that drive dispossession that is experienced in the countryside are not only rural issues, and can induce sustained mobilization and concern from people living in quite different places who see these as *their* issues too.

A third possibility is that to call something a rural social movement simply means that it is addressing a phenomenon whose primary *venue* is in the countryside. Indeed, this framing is one way to argue for the continued importance of rural studies: namely that the countryside is a key venue of some of the most pressing issues of our times.[3] It is a critical venue for climate change (because of land cover issues), biodiversity (because of forest cover issues), human rights (as per the data reported by Global Witness), models of development (because of the resurgence of extractivist thinking), and water security (because of the relations between land cover and regional water availability and quality). These are not rural issues, and those who care about them do not only live in rural areas, but they have to be addressed in great measure through the countryside.

This only gets us so far, because these issues can still mobilize different constituencies and raise different socio-political challenges. Here the distinctions introduced earlier may be helpful for giving further focus. Thus, rural social movements (addressing issues with rural venues) may address these as issues of accumulation, of collective consumption, or of identity politics; they may address them at a range of scales; and their strategies may range from the confrontational (e.g. open protest and litigation), to the conciliatory (e.g. collaboration with the state to protect forests), through to the self-directed (as in the case of movements of private conservation areas).

20th-century rural social movements

The Latin American countryside has long been a venue of rebellion and protest (Stern, 1987). This tradition of mobilization is an important precursor to more recent social movements insofar as memories and iconic figures are often invoked, and contemporary issues are framed as parts of longer standing injustices related to dependency, sovereignty, and race. Open protest is an ever-present part of political repertoires in the region, and this long history has put Rebellion in the Veins, in Dunkerley's (1984) memorable turn of phrase. Such protest and rebellion have done much to shape contemporary institutions and generative ideas in Latin America. Laying a discussion of the long history of rural protest on one side for lack of space, this section briefly explores this relationship for the 20th century.

The Latin American countryside was both venue and incubator of some of the more momentous political transformations of the last century. Revolutions in Mexico, Bolivia, Cuba, and Nicaragua had important rural roots, as did the Zapatista transformation of politics in Chiapas and Mexico. While these processes cannot easily be framed as rural social movements, nor were they entirely distinct from the phenomenon: they articulated rural residents, together with urban, intellectual, and international actors; they were concerned with injustices that were felt with particular acuteness in the countryside, even when their agendas reached well beyond rural issues; and they were cases of sustained, issue based mobilization as opposed to isolated protest events. These processes also transformed both national and rural institutions.

While most 20th-century rural social mobilization has been at a less-than-revolutionary scale, it has played important roles in transforming the face of Latin America. For instance, the wave of land reform legislation that affected much of the region from the 1960s to the early 1980s was in considerable measure a consequence of rural social movements. The US together with elites in the region worried that the increasingly assertive and organized peasant protest over access to, and the distribution of, land that characterized the 1950s and 60s could culminate in political transitions similar to that of the Cuban Revolution. Land reform was identified as a mechanism that might defuse this protest (de Janvry, 1981; Grindle, 1986). Initial reforms were modest in scale and intent, and in several countries sustained protest following these reforms induced a second generation of reforms that deepened land redistribution, as in Chile, Peru, and Ecuador.

A second set of transformations relates to the recognition of indigenous territories. While this is still an incomplete and imperfect reform, it remains the case that in just 40 years the status of indigenous rights over territory has been vastly elevated in juridical form and public debate alike. This change owes everything to the emergence of indigenous movements, and lowland/ Amazonian and Central American movements in particular. Key moments in this process were the indigenous uprising in Ecuador in June 1990, the indigenous March for Dignity and Territory from the Bolivian lowlands to La Paz in September 1990, and the continent-wide counter-quincentenary activities of 1992. National phenomena such as these, coupled with subnational processes of organization and alliances with NGOs, lawyers, and others, placed the plurinational character of Latin America ever more firmly in public debate and opened up space for legislative progress on territorial rights, prior consultation, and bilingual education.

A third example is the case of the rubber tappers movement in Brazil, one of whose early leaders was Chico Mendes, murdered in 1988. The rubber tappers contested the hugely unequal patterns of land ownership in the Brazilian Amazon, and the association of land concentration with deforestation. They made the case, on the basis of their own activities, for an alternative mode of forest governance in which rubber tappers would combine the protection of forest with the pursuit of livelihood through the creation of extractive reserves (Hecht and Cockburn, 2011; Anderson, 1990; Brown and Rosendo, 2000). The rubber tappers' model of extractive reserves became, arguably, the most visible example of grassroots, market-oriented forest governance in Latin America and went on to inform many experiments in community-based forestry.

These three examples are illustrative not only because they suggest different ways in which rural social movements have shifted policy and thinking in the region, but also because in each case (land reform, indigenous territory and rights, and community-based forest governance), periods of gains have been followed by counter-reforms. The land redistributions of the 60s and

70s were followed by a slow but sure re-concentration of land ownership and a steady privatization of collective forms of property; efforts to stall indigenous land titling, or to roll-back land rights and protections, have been legion; and pressure on forest cover and areas of forest used and managed by forest dwellers has intensified greatly with the deepening of an extractivist model of development combining mining, hydrocarbons, infrastructure, and agroindustry. The implication is social mobilization may have induced change in the region, but absent sustained pressure and a profound change in ways of thinking and forms of valuation of people, diversity, and nature, such gains are always fragile.

These examples also speak to a narrative about changing modes of rural social mobilization. The argument is made that as rural social organizations slowly emerged in the period from the 1930s, they were broadly class-based in nature, and based on the identity of the peasant or peasant worker. This class-based orientation reflected the role of communist and socialist parties in the origin of these organizations, and the general dominance of peasant, as opposed to indigenous, thinking within the left as it sought to make sense of agrarian transitions. The class-based nature of these movements was also reflected in their demands for land and improved working conditions, as opposed to territory or cultural rights. From the 1980s onwards, however, it is argued that indigeneity increasingly became the frame through which rural questions were approached, and around which populations organized. This frame originated from organizing processes among Amazonian groups and the Maya in Guatemala, among others. With time, it also began to influence how existing class-based rural social movement organizations thought of themselves, as well as inspiring new processes of reaffirmation of identity, such as those that began to occur, for instance, among Mapuche peoples in Chile.

While this sequentialist narrative oversimplifies greatly, it does help to describe a genuine shift in the second half of the 20th century: namely the steady rise of race and ethnicity as identities around which rural mobilization was organized. While class-based demands and identities have not gone away, ethnic and racial claims are far more prominent in rural protest, national debate, and movements' own self-identification than was the case in the 1960s.

Finally, the three illustrations draw attention to the extent to which rural social movements have come into being and been sustained by actors whose origins and aspirations are always more than rural, more than agrarian, and more than local. While people living in rural areas are consistently at the core of these processes, it is hard to find movements where priests, lawyers, NGOs, political activists, and intellectuals have not also had a part to play, and where national and international resources have not also been mobilized. Many of these patterns have continued into processes of mobilization in the early 21st century, albeit with some twists.

Contemporary conflicts over the countryside

Since the turn of the century the macro-political economy of the Latin American countryside has been broadly framed by three phenomena that have deepened over the last two decades. First is the commodities consensus (Svampa, 2015), the increasingly convergent view among political and economic elites that national development strategy should hinge around the promotion of export-oriented primary industries: mining, hydrocarbons, agroindustry, and, in some countries, industrial forestry. This consensus has itself been driven by the super-cycle of high mineral prices between, approximately, 2000 and 2012 and the significant increase in Chinese interest in Latin America's primary industries. This has led to rapid growth of exports to China and increased Chinese investment in Latin American resources industries and more recently large-scale infrastructure and civil engineering companies (Gallagher, 2016; Ray et al., 2017). Second is the increasing importance of the narco-economy in the countryside. While the political, economic, and cultural significance of drug production and distribution dates back at least to the 1980s, its presence has become more widespread as cartels extended their presence in rural areas to control production, protect distribution channels, and launder profits in other land and natural resource-based investments. A source of rural violence, the narco-economy has also been identified as a driver of land cover change in Mesoamerica (McSweeney et al., 2014).

Third, has been the introduction of new asymmetries of power into the countryside. Asymmetric relations are hardly a new phenomenon, with deep roots in the *encomienda, hacienda, gamonalismo*, and church, among other institutions. However, the ascendance of primary industrial and narco sectors has brought new elites into the countryside. Transnational and national capitalist enterprises are far more present in rural areas than was the case 30 years ago. Meanwhile, some parts of the smaller scale subnational capitalist class have succeeded in new agricultural, mining, and service sector investments and become increasingly powerful. This is reflected in the increasing leverage of a body like the National Society for Mining, Petroleum and Energy in Peru, or of networks of initially medium-scale farmers in Brazil who took advantage of increasing Chinese demand for soybean and beef, acquiring large swathes of land and nurturing their own political grouping, the Bancada Ruralista, through which they have been able to secure control of governorships, seats in Congress, and Ministerial positions.

These three phenomena have had important effects in framing the terrain of rural social movements, especially those addressing issues of accumulation and of identity. First has been a relatively rapid expansion of the resource extraction frontier. This has brought extractive industry and agroindustrial investments and elites into the territories of rural populations, the forests of the Amazon, Chaco or Cerrado, and protected areas. These incursions create the conditions for conflict as elites seek access to and control over resources that have historically been used by others. Second has been an uptick in rural violence reflecting the significant possibilities for rent capture that these new frontiers have created. Increasing numbers of actors seem prepared to use force and violence to secure these rents. Third has been a palpable transnationalization of resource control which has stirred narratives of resource nationalism and resource sovereignty alongside more general arguments about dependency and imperialism that have resonance in some social movements. Fourth has been the increased assertiveness and power of these newly arrived actors in territorial and environmental governance, reducing the relative power of historical and traditional authorities.

To argue that, as the political economy of rural and agrarian development changes, so rural social movements change in lock step, would be too reactive an interpretation of social movements. It would minimize the extent to which movements are proactive in pursuing political, material, and ideational aspirations, while at the same time downplaying the degree to which social movement organization is subject to path dependency. Indeed, organizations once created do not easily cease to exist, leadership once ascendant does not quickly walk away from authority, and political objectives once crafted continue to circulate as organizing ideas. Many of the movement organizations of the latter 20th century continue to exist in the 2010s. Campesino organizations of the 1960s and 1970s still have offices and greater or lesser resonance with some populations (the CCP of Peru, FENOC [now FENOCIN] of Ecuador, the CSUTCB of Bolivia, FEDECOAG of Guatemala, and so on), while indigenous movement organizations are also entering their fourth decade (CONAIE in Ecuador, CIDOB, much debilitated, in Bolivia, AIDESEP in Peru), and the Proceso de Comunidades Negras (PCN) of Colombia celebrates a quarter of a century in 2018.

That said, this new context *has* gone a long way in framing movement dynamics, either eliciting new movements or leading existing movements to modify their agendas or become involved in conflicts of a type that they had not previously engaged. Most of the organizations noted in the previous paragraph now deal much more explicitly with questions of extractive industry, forest loss, free prior informed consent, and transnational enterprise than was the case in the 20th century. Meanwhile, new movements have emerged in response to the conse-quences of the commodities consensus. The bulk of these new movements are subnational in scope, taking forms such as the Frentes de Defensa against extractives that emerged across Peru, but others have been national, such as the Mesa Nacional Contra la Minería in El Salvador, or CONACAMI again in Peru. Some of these movements focus on issues of accumulation by dispossession, some focus on identity politics, and some (though fewer) also focus on politics of redistribution (e.g. the allocation and use of tax revenues from resource extraction).

Conflict and movement emergence around this new context has also underlain some of the most iconic and politically consequential conflicts of the last two decades. An argument can be made that this new context and the alignment of social movements behind it, helped drive the rise to presidential power of Evo Morales and the Movimiento al Socialismo in Bolivia. While Morales already had a strong base in the cocalero movement and the national peasant and colonist confederations in Bolivia, two of the important precursors to his election were the Water War of Cochabamba, and then the Gas War two years later (Perreault, 2006). Each of these mass mobilizations were at their core motivated by ideas of resource nationalism and sovereignty made acute by the increasing presence of transnational capital, and the capacity of this capital to control these resources. Morales' elevation as a national leader through the second of these mobilizations in particular, and the alignment behind him of what was to become his government of social movements, were largely artefacts of increasingly angry resource national-ist sentiments.

As a second conflict, the Bagua conflict (or *Baguazo*) in Peru, which led to 34 deaths, mostly of policemen, occurred at a large-scale mobilization triggered by government decrees and manoeuvres that would have facilitated extractive industry and large-scale private investment in areas that indigenous populations deemed their collective or community territories. That conflict dominated national debate for a year or more subsequently and was ostensibly a trigger for national legislation in 2011 on prior consultation for indigenous communities (though the subsequent roll out of this law has left activists very disappointed). Also in Peru, the conflict over Cerro Quilish between 1998 and 2004 became both a national and Latin America-wide example of what social mobilization against extractive industry could achieve because it led the mining company, Minera Yanacocha, to withdraw from a mine expansion in the face of coordinated rural and urban mobilization over the impact of the expansion on water resources. In many respects, the more recent failure of the same company to initiate a new mine, Minas Conga, has been a replay of the Quilish mobilizations. Interestingly, the 2017 legislation of the Government of El Salvador to ban mining in the country was also made on the basis of concerns about water security under conditions of climate change, and again in response to a decade of sustained social mobilization around this issue.

The extractivist macro-economy has, then, become one of the primary venues for rural social mobilization over the last two decades. That mobilization has in turn had consequences of political significance, sometimes stretching beyond the country in question and the specific issue of resource extraction. However, these observations merit several qualifications of which I note three. First, there are many more instances in which movements do not emerge, or the response to movement is repressive. In these instances, the reason may have something to do with the techniques of those elites who will gain from extractivism. On the one hand the presence of new resource extraction companies comes accompanied, often, by instruments of corpo-rate social responsibility, community development, and local employment generation. Together these instruments can create sufficient incentives to diffuse potential mobilization in the face of loss of land, resources, governing authority, or territorial control. In other instances, the threat or use of violence can disarticulate mobilization. The sources of such violence are always opaque, and may involve companies, local traffickers who will gain from resource extraction, or increasingly in Mexico and Central America, narco-interests with links to resource extraction. High-profile deaths, such as the murder of Berta Cáceres in Honduras in relation to a dam itself linked to broader resource extraction, are just the most visible examples of such violence.

The second caveat is that socially organized, movement-like responses to the extractive economy are not always or only ones of contentious politics. A clear example of this is the Alianza Mesoamericana de Pueblos y Bosques across Central America and Mexico. This alliance of indigenous, peasant, and forest using peoples reflects a response to the pressures on their territories resulting from the rapid growth of investment in oil palm, sugar cane, mining, hydrocarbons, logging, and hydroelectric dams. In this instance, however, the response combines negotiation with government authorities over policy, some litigation, and above all efforts to develop community-based alternatives for the commercial management of forests and forest territories. The argument is that such responses are more viable in the long term than protest alone, because movements' bases need livelihood opportunities, and because the organizations can engage with government more effectively if they also come armed with demonstrable experience of alternative, productivist ways of managing forest lands.

A third and final caveat is that such movement responses are not necessarily progressive in a political sense – or, at least, they have features that would be criticized by certain readings of progressive politics. I note two examples. First is the gender politics of such movements. While there are clear exceptions, the internal dynamics of many movements are both masculinist and male-dominated, and visits to leadership offices will encounter spaces dominated by men and symbols of maleness. Reflecting this, the primary concerns of such movements frequently focus on issues of concern to younger and middle-aged men, rather than issues that might be prioritized by women, children, or the elderly. The second example is the tendency towards authoritarian intolerance within movements. A particularly visible example of this might be the approach of the Morales government of social movements to dissent and critique. This is an approach that has closed NGOs and research centres, has criticized critical scholars, leaving them cautious of what they say publicly, and has consciously weakened and divided indigenous movements that have been critical of national leadership. On a smaller, more everyday scale are the many examples of intolerance in the ongoing functioning of movements. The implication is that while rural social movements *might be* democratizing (Fox, 1990), there is absolutely no reason to presume that this is actually the case.

Final observations

The Latin American countryside is the venue of some of the most critical contestations for the future face of Latin American societies, and, indeed, of global society. Whether or not the larger part of Amazonian, Chaqueño, and Cerrado primary forest is cleared for soybean, meat, oil palm, mineral, and hydrocarbon production will have clear implications for future relationships between climate, water resources, and quality of life that stretch well beyond the countryside. Clearing the forest is likely to render many Latin American cities hotter, drier, and more dangerous to health: the near cataclysmic water crisis in Sao Paulo in 2016 is an indicator of what is to come if deforestation proceeds apace. The fate of the forest (Hecht and Cockburn, 2011) will also be the fate of those who live by the grace of that forest. What happens to rural land cover in the future will depend on the status, rights and presence of indigenous, quilombola and cabo-clo peoples in Latin American society, and on how far the region will accept that plurinational territorialities can coexist. The future of these forests will have immense implications for global climatic regulation and biodiversity.

How conflicts over the countryside work out will, as Global Witness's reports imply, say a great deal about the status of human rights, protest, and alterity in future Latin Americas. The outcomes of these conflicts will also have implications for the extent to which Latin American economies continue to be narrowly resource-dependent. If the resource extraction option for rural Latin America wins out, the implications will be national, not just rural. As discussed earlier, the rise of the Bancada Ruralista in Brazil is *both* a consequence of earlier periods of soybean expansion in Mato Grosso *and* a driver of a model of development that combines resource extraction and infrastructure across the Amazon, and channels public and private investment towards these sectors. Any deepening of an extractive economy in the countryside will also deepen political structures that fix a political economy that channels resources and orients incentives towards further resource extraction.

The significance of rural social movements must, therefore, be read through their place within, and their implications for, the political settlements that guide broader economic and political relationships in Latin America. While rural social movements can be celebrated for their alternativeness, their prefigurative politics, their seeds of progressivism, or their anticipation of post-capitalism (Gibson-Graham, 2006), what really matters is whether their collective arguments win out or not. Winning or losing determines the future of political settlements, and from those settlements flow many other outcomes (Bebbington et al., 2018). The stakes are very high – clearly so, otherwise there would not be so much political protection for the killing of movement leaders and activists.

The either/or's laid out in the preceding paragraphs are doubtless framed too starkly, and might be criticized for being alarmist, melodramatic, and lacking in ethnographic or tactical political nuance. However, some of the material discussed in this chapter, and the data with which Global Witness confronts us, suggest that these either/or's are a valid and important lens through which to read the place of rural social movements. What rural social movements *are*, but more importantly what they *achieve*, will have immense implications for what Latin America will *mean* and will *be* in the future.

Notes

1 I acknowledge the financial support of an Australian Research Council Australia Laureate Fellowship (FL160100072) and a Climate and Land Use Alliance grant (Contract # 1607–55271) that allowed me to prepare this chapter. I am grateful to countless friends and colleagues whose ideas, work, and conversations have influenced my understanding of social movements in general, and in Latin America in particular. There are too many such people to mention everyone here, but I do note that this chapter is particularly influenced by the ideas of, and my collaborations with, Tom Carroll, Fernando Eguren, Denise Humphreys Bebbington, Diana Mitlin, and Pablo Ospina.
2 Here I draw from Bebbington, Mitlin et al. (2010).
3 I am indebted to Fernando Eguren for this observation.

References

Anderson, A. (1990) (ed) *Alternatives to Deforestation: Steps Toward Sustainable Use of the Amazon Rain Forest*. New York: Columbia University Press.

Ballard, R., Habib, A., Valodia, I. and Zuern, L. (2005) Globalization, marginalization and contemporary social movements in South Africa. *African Affairs* 104(417): 615–634.

Bebbington, A., Abdualai, A-G., Bebbington, D. H., Hinfelaar, M. and Sanborn, C. (2018) *Governing Extractive Industries: Politics, Histories, Ideas*. Oxford: Oxford University Press.

————, Mitlin, D., Mogaladi, J., Scurrah, M. and Bielich, C. (2010) Decentring poverty, reworking government: Movements and states in the government of poverty *Journal of Development Studies* 46(7): 1304–1326.

Brown, K. and Rosendo, S. (2000) Environmentalists, rubber tappers and empowerment: The politics and economics of extractive reserves. *Development and Change* 31(1): 201–227.

De Janvry, A. (1981) *The Agrarian Question and Reformism in Latin America*. Baltimore: Johns Hopkins University Press.

Dunkerley, J. (1984) *Rebellion in the Veins*. London: Verso.

Escobar, A. (2008) *Territories of Difference: Place, Movements, Life, Redes*. Durham, NC: Duke University Press.

Fox, J. (1990) (ed) *The Challenge of Rural Democratization: Perspectives from Latin America and the Philippines*. London: Frank Cass.

Gallagher, K. (2016) *The China Triangle: Latin America's China Boom and the Fate of the Washington Consensus*. New York: Oxford University Press.

Gibson-Graham, J. K. (2006) *A Postcapitalist Politics*. Minneapolis: University of Minnesota Press.

Global Witness. (2016) *On Dangerous Ground*. London: Global Witness.

————. (2017) *Defenders of the Earth*. London: Global Witness.

————. (2018) *At What Cost?* London: Global Witness.

Grindle, M. (1986) *State and Countryside: Development Policy and Agrarian Politics in Latin America*. Baltimore: Johns Hopkins University Press.

Harvey, D. (2003) *The New Imperialism*. Oxford: Oxford University Press.

Hecht, S. and Cockburn, A. (2011) *The Fate of the Forest. Developers, Destroyers, and Defenders of the Amazon*. Updated Ed. Chicago: University of Chicago Press.

Lucero, A. (2008) Indigenous political voice and the struggle for recognition in Ecuador and Bolivia. In A. Bebbington, A. Dani, A. de Haan and M. Walton (eds) *Institutional Pathways to Equity: Addressing Inequality Traps*. Washington, DC: World Bank, pp. 139–168.

McSweeney, K., Nielsen, E. A., Taylor, M. J., Wrathall, D. J., Pearson, Z., Wang, O. and Plumb, S. T. (2014) Drug policy as conservation policy: Narco-deforestation. *Science* 343: 489–490.

Moore, J. (2009) Ecuador: Mining protests marginalized, but growing. *Upside Down World*, 21 January.

Perreault, T. (2006) From the *Guerra del Agua* to the *Guerra del Gas*: Resource governance, neoliberalism and popular protest in Bolivia. *Antipode* 38(1): 150–172.

Ray, R., Gallagher, K., López, A. and Sanborn, C. (2017) (eds) *China and Sustainable Development in Latin America*. London: Anthem Press.

Stern, S. (1987) (ed) *Resistance, Rebellion, and Consciousness in the Andean Peasant World, 18th to 20th Centuries*. Madison: University of Wisconsin Press.

Svampa, M. (2015) Commodities consensus: Neo-extractivism and enclosure of the commons in Latin America. *South Atlantic Quarterly* 114(1): 65–82.

Tilly, C. (1985) Models and realities of popular action. *Social Research* 52(4): 717–747.

28

LABOUR MOVEMENTS

*Maurizio Atzeni, Rodolfo Elbert, Clara Marticorena,
Jerónimo Montero Bressán, and Julia Soul*

Introduction

The cycles of conflicts and struggles and the forms of workers' collective resistance registered in the last few decades in Latin America reveal the heterogeneous and uneven ways in which the working classes are made and remade. Place-specific socio-political contexts and their effects on employment relations (McGrath-Champ, Herod, and Rannie, 2010), the organizational power of workers within the structure of the labour process and capitalist production (Hyman, 2006; Darlington, 2014), and the role of the state in mediating between different class interests (Panitch and Albo, 2014, Panitch and Chibber 2013) are all factors that highly influence the dynamics of conflict and the emergence of labour movements.

This notwithstanding, in the country-specific analysis presented in the following pages we have followed Silver's (2003, 2014) comparative historical analysis of world labour unrest as a general frame of reference to explain "the making and remaking" of labour movements. The vantage point in Silver's approach rests on establishing transhistorical and transnational connections between patterns of labour movement struggles and the historical development of the capitalist mode of production. While Marxist scholars argue that the internal contradictions of capital force capitalists to identify new strategies for improving profitability and reducing costs when they face crises of over accumulation (Harvey, 1982, 2014), Silver argues that we should also consider the dynamics of the continuous tensions between the capitalist drive for profitability and workers' resistance within the workplace. In so doing, she identifies four main strategies deployed by capitalists. The first involves the spatial relocation of productive activities to low-wage countries, a process that has given rise to globalized production chains (Mezzadri, 2016). The second strategy seeks to reduce the dependence of capital on living labour by introducing new labour saving technology, such as automation or by outsourcing digitalized work processes (Woodcock, 2018). The third and fourth strategies are derived from the first two and are typical of companies in core economies: shifting production to services and new industrial (more value-added) sectors, and moving from industrial production to financial speculation.

For our purposes, Silver's approach is important in two senses: (a) the continuous existence of labour unrest is rooted in structural dynamics of development common to different countries and producing similar patterns in terms of organizational forms and types of conflict; and (b) both workplace and social movements struggles represent forms of class resistance. In

Latin America, this has been theorized as the relationship between state-defined development trajectories and different types of working-class politics (Collier and Handlin, 2009: 48). The incorporation of the labour movement into the political system of middle-income countries in Latin America occurred in the first half of the 20th century during the first phase of import-substitution industrialization (ISI). In this period, working-class politics were channelled through strong unions and labour based parties (Collier and Collier, 1991), which remained at the centre of the political arena until the final crisis of the second ISI phase in the 1970s and 1980s.

In the final decades of the 20th century, the implementation of neoliberal policies has represented the most important systemic factor shaping the social reality of Latin America. The liberalization of labour markets and labour relations reforms, processes of privatization, deregulation, the reduction of welfare programmes, in the context of the transnationalization and financialization of the economy, have created widespread poverty, unemployment, and informality, further increasing class divides (Portes and Hoffman, 2003). Social opposition and mobilization against these policies have moved away from the strong union-labour party link, producing in many countries the growth of new community-based organizations representing the interests of the poorest workers (for instance *cocaleros* in Bolivia, *Sem Terra* in Brazil, *Piqueteros* in Argentina) and new political alliances and movements openly opposing neoliberal policies (Collier and Handlin, 2009). The anti-neoliberal struggles of the 1990s and early 2000s prepared the ground for the political turn of many Latin American governments to a populist centre-left, with anti-imperialist tones, what has been called the Pink Tide (see for instance Chodor, 2014; Grigera, 2017; Enríquez and Page, this volume). The context of commodities boom and economic growth in which these reformist governments emerged, favoured redistributive policies, the creation of employment, the improvement of state social provisions, the extension of basic labour rights, and the reduction of poverty. In various cases, this favourable context also helped to revitalize trade unions in the formal sector and led to the emergence of new grassroots workplace-based organizations and the consolidation of welfare social programmes. Despite these positive changes, the structural conditions inherited from previous decades and the continuous dependence of Latin American economies on an extractive export-oriented model, have limited the consolidation of reformism in the region, giving rise to a new conservative turn which is accompanied by pro-market reforms and anti-labour policies in countries like Chile, Mexico, Argentina, and Brazil. This new cycle represents for the working classes a regressive change that remains to be analyzed.

Changes in economic models and political systems, and transformations of production and distribution processes have cyclically appeared in the development of capital as ways to restore accumulation. However, we can also expect that "new agencies and sites of struggle emerge along with new demands and forms of struggle, reflecting the shifting terrain on which labour-capital relations develop" (Silver, 2003: 19).

In the following pages we aim to briefly sketch how these broad range of transformations have impacted on working conditions and on the agencies of workers in contexts with different political trajectories: stable neoliberalism (Mexico and Chile); stable popular reformism (Ecuador, Bolivia, and Venezuela); conservative shift (Argentina and Brazil).

Stable neoliberalism: Mexico and Chile

Unlike the other cases explored in this chapter, Mexico and Chile have followed a stable neoliberal path along recent years. The significant protests against neoliberalism in these countries (especially in Chile between 2011 and 2013) did not influence the course of institutional politics to the extent they did in the other countries. The fact that the economic situation did not reach the point of a deep crisis that undermined the legitimacy of neoliberal policies may partly explain this continuity. In the case of Chile, the stable legacy of neoliberalism can also be traced to the consequence of state-led political violence in the 1970s and early 1980s. Until the Pinochet dictatorship, the Chilean labour movement was one of the most radical of the region, becoming one of the main supports of Allende's democratic pathway to socialism through "cordones industriales," massive factory take-overs and land occupations (Winn, 1986; Gaudichaud, 2004). Such radicalism was violently confronted with massive incarcerations, torture, and repression during the 1973–1990 dictatorship, which was successful in producing labour de-mobilization and the imposition of a stable neoliberal consensus that exists until today (Winn, 2004).

In recent years, both countries have deepened their economic dependence on the export sector. The Chilean economy is mostly dependent on the export of primary commodities, and like Mexico, shows the growth of a service sector based on precarious low-paid jobs (Narbona, Páez, and Tonelli, 2011). The steady economic growth in the past decades has been accompanied by growing social inequality and a weak social safety net (Torche, 2005; Bank Muñoz, 2017: 32). In Mexico, since the mid-1980s, and most importantly with the start of the North American Free Trade Agreement in January 1994, governments have sought to take advantage of the country's privileged location and its large workforce of more than 50 million workers in order to attract foreign investment for industrial export markets. Cutting down labour costs has since then become the obsession of governments, to the detriment of the internal market. While unemployment is low (3.46%), the minimum salary is amongst the lowest in the world, and the second lowest in Latin America (in 2017 it was US$4.2 a day according to statistics from the Instituto Nacional de Estadística y Geografía (INEGI)[1], only above that of Nicaragua). Furthermore, there exists a clear segmentation between those working in manufacturing industries and those who work providing services for the internal market. While the average salary for the former is US$4 per hour, salaries in the rest of the economy drag down the average to US$1.93 per hour. It is in the latter that unregistered labour (which affects 56.5% of the total workforce) is concentrated.

Labour legislation reflects the continuation of a neoliberal path, especially in Chile, where the basic foundations of the neoliberal regime imposed through violent repression during the dictatorship have not changed in significant ways. Indeed, the cornerstone of Chilean labour legislation is a modified version of the highly restrictive labour code that the military imposed through violence. This labour code determines de-centralized negotiations between labour and capital at the company level, and severely undermines the right to strike by allowing employers to hire strike replacements. These two basic features – decentralized bargaining and restrictions on the right to strike – persist even after the last reform attempt pursued in 2015–2016 by President Bachelet (Perez, 2017) and have resulted in a labour market that shows low unemployment levels but high levels of job precariousness. In Mexico, where bargain contracts are also negotiated on a firm-basis, the reform of the Federal Law of Work in 2012 meant further flexibilization, such as the extension of periods of probation, the legalization of outsourcing, and an increase in wages paid by the hour. Despite these changes, the reform was supported by the main union confederations.

Just as in other parts of the world, Latin American unions face falling rates of unionization and strong social fragmentation. This creates a challenging scenario that forces them to coordinate demands of different economic sectors and groups of workers. In Mexico, the weak representation of trade unions has paved the way for growing labour flexibilization since the mid-1980s and a 68.8% drop in the minimum wage between 1982 and 2016 (De la Garza, 2014). Not only is the ratio of unionization extremely low (14.5% in 2010 according to De la Garza, 2012), but also "union simulation" is extensive. This can be seen for instance in the so-called "protection contracts," signed secretly between employers and "ghost unions" or in the more explicit corporatist practices of many formal trade unions, who have historically tended to present themselves as guarantors of the social peace between labour and capital. These practices of "underground" negotiation have a long history, and neither the creation of a new union confederation in 1997 (UNT), nor the change in the governing party in 2000 (after over seven decades of PRI hegemony) seem to have had significant effects on their implementation (Bensusán, 2016; De la Garza, 2012).

Perspectives on workers' organization in both countries are starkly different, with the Chilean labour movement facing a better situation. In this country, there is a growing grassroots union movement that in recent years has led struggles for better salaries and collective bargaining rights and against workplace precariousness. In the export industries, sub-contracted workers in the mines and the forestry industry have been at the forefront of the movement (Durán-Palma and Lopez, 2009; Perez Valenzuela, 2017), whereas in the service sector it was dock workers and retail workers who developed the most innovative and democratic strategies (Bank Munoz, 2017; Santibáñez and Gaudichaud, 2017).

The strength and power of the grassroots movement lies in the renovation of the union movement from below. However, workers still face major difficulties when trying to upscale demands to the national level. During the wave of protests initiated in the late 2000s (Donoso and Von Bulow, 2016: 4), the main national struggles were oriented towards broader social rights, such as struggles for indigenous rights, rural land and urban housing, a public and free university system, and a public and universal pension system. Despite the fact that these struggles were not restricted to labour rights, workers were involved and unions did have a say, such as in the struggle for a universal pension fund (Pérez, 2017: 284–286). In the near future, the grassroots democratic experience of workers in strategic sectors of the economy and the above-mentioned struggles to guarantee "common goods" (education, land, nature) might provide the horizon for a renewed labour movement that is able to upscale the struggle through its relations with other mobilized sectors of society. In Mexico, there are growing experiences of organization amongst street vendors into so-called civic associations, who struggle for the legalization of their activity and for better working conditions for workers in street markets (Tilly et al., 2014; Morales Hernández, 2013; Swanson, this volume). However, informal workers – who account for over half the workforce of the country – are largely fragmented and unorganized. Real wages continue a downward slide, and there are no signs of change in this respect.

Rise and decline of stable popular reformism: Bolivia, Ecuador, and Venezuela

The victory of Hugo Chávez in Venezuela in 1998 inaugurated an era of popular reformist governments in many Latin American countries. Together with Ecuador and Bolivia, these countries are still under reformist governments – despite the recent signs of conservatism given by president Lenin Moreno in Ecuador – and have escaped so far the general shift to the right in the region.

The current political cycle in Bolivia started in 2005 with the electoral victory of Evo Morales and the *Movimiento al Socialismo*. Morales – the first indigenous man to achieve the presidency – is supported by an indigenous, peasant, working-class popular social basis forged during the anti-neoliberal struggles of the early 2000s, the so called *guerra del agua* and *guerra del gas*, against the privatization of water and for the nationalization of gas respectively. These were social and political turning points. During the election campaign, grassroots organizations supporting Morales had two key demands: the formation of a Constituent Assembly to change the constitution, and the nationalization of oil and gas reserves; demands that were indeed taken forward by Morales. Since Morales assumed the presidency however, peasant organizations and unions of formally employed workers have played different roles at least initially. The former were part of the so-called "*Pacto de Unidad*" signed at the end of 2003, which expressed the indigenous core of the anti-neoliberal movement (Soruco Sologuren, 2009; Chavez León, Mokrani, and Uriona Crespo, 2010). Meanwhile, the workers' confederation, COB, claimed to have a "political independent" role but by 2008 was incorporated – in a subordinated manner – to the block of social organizations officially supporting the government. Since then, COB has taken part in the formulation of laws and official propositions (Fornillo, 2011; CEDLA, 2016).

In Ecuador, when assessing the conditions of the working class in the ten years that correspond to Correa's double presidential mandate, it is clear that despite the government's progressive discourse, few improvements have been made. After initial support from social movements, the government marginalized these and particularly CONAIE, the most important indigenous based organization that led the anti IMF and anti-neoliberalism struggles in the 1990s and early 2000s (Dávalos, 2014). Towards the end of his presidency, Correa entered in open conflict with organizations struggling against his policies, accusing them of preparing the ground for a *Golpe Blando* (soft coup) by the conservative Right.

In the new Bolivarian Republic of Venezuela, the victory of Hugo Chávez in the 1998 elections and the sanctioning of a new Constitution a year later expressed a significant transformation in class relations. The failure of the 2002 coup further radicalized Chavez' project. While the "Bolivarian Revolution" and "Socialism of the 21st Century" sought to improve the living standards and working conditions of the working classes, this process has been subject to a number of contradictions and, although inequality has decreased, it still remains high. Workers, peasants, and indigenous movements became the social bases of the Bolivarian Revolution. In 2003 the government created Misiones, oriented to fight poverty, illiteracy, health problems, and housing – among others – throughout Venezuela. This reinforced social and grassroots organization, although critics questioned state control over social organizations. Also, indigenous movements achieved the recognition of their rights through the 1999 Constitution and the approval of the Organic Law of Indigenous Peoples and Communities in 2005. However, formal recognition of indigenous territories is still minimal (Angosto Ferrández, 2017: 182).

In terms of economic policies, the governments of the region showed little intention to foster structural changes. All three basically worked to redistribute the wealth generated from the export of a few extractive commodities taking advantage of their high prices. This helped to build electoral consensus, the consensus of the commodities as Svampa (2015) put it, but at the same time deepened their dependence on these products, making the *extractive neo-developmentalist model* (Feliz, 2015) unsustainable in the long term. In Bolivia, the exploitation of natural resources (gas, oil, and mining) continues to be the main source of income. The key strategy of Morales' economic policies is to increase the appropriation of state surplus in order to support industrialization policies. The increase of hydrocarbons and mining rent and royalties, as well as the setting of tougher conditions for foreign investments, has been the main result of the nationalizations. However, from 2010 onwards the fall in the prices of commodities in the context of the international crisis forced the government to ease foreign investment conditions at the expense of state surplus appropriation.

In Ecuador, the neo-extractivist economic model based on the export of oil, fish and tropical fruits such as bananas from extensive plantations has basically kept historical patterns of labour conditions unaltered. Informality and underemployment still count for about 60% of the total labour force and this has immediate consequences not just in terms of the protection of labour rights but also in terms of salaries. Although minimum wages for formal workers went from

US$170 to US$375 in ten years (2007–2017), the average wage only increased from US$314.8 to US$326.2 due to the preponderance of informality, even after several years of sharp economic growth (Bayas, 2017).

In Venezuela, the deeper dependence on oil exports and the lack of plans to diversify the economic structure of the country are dramatic. This is why the international economic crisis and the fall in the price of oil increased the economic and political contradictions of the Bolivarian Revolution. This situation is leaving the government without resources to contain the political crisis, and strengthened the growth of opposition to Maduro's government.

Progressive governments in these countries have showed a number of contradictions in dealing with unemployment, workers' rights, and their relationship with the unions and the labour movement in general. In Bolivia, changes introduced by government policies have improved the conditions of particular groups of workers, limiting fixed-term contracts and recognizing historical demands related to the pension system. Additionally, there has been a continuous increase in the minimum wage since 2007. However, labour informality prevails, with 71% of the workforce in that condition, mainly because of the extent of "non-waged" forms of labour subsumption – family or community labour, individual contracts, home labour, and so on. Since 2000, the share of waged workers increased from 30% to 40% in 2011. However, subcontracting is also rife and it is the main anti-union strategy of companies in strategic economic sectors. There is indeed a broad agreement about the continuities of neoliberal features in the labour market as much as in labour relations. The high unemployment rate is understood as the main reason for the continuity of labour precariousness and general low wages as persisting features of Bolivian labour market. The government responds to it through programmes of low-wage temporary contracts for unskilled jobs (LO/FTF, 2016).

The Venezuelan government increased the minimum wage and the 2012 Labour Law prohibited outsourcing (Bonilla, 2011; Lucena e Iranzo, 2016). However, these advances were more the result of government strategic priority in building what has been called the Socialism of the 21st century than of trade union actions. Initially, Chávez's government promoted the organization of cooperatives and the nationalization of companies in oil, iron and steel, financial, electrical, and food sectors as well as workers' control. In many companies, workers took the initiative to establish workers' councils and in 2006 they formed the Revolutionary Front for Workers in Co-Management and Employed Enterprises (FRETECO). Workers' control, however, has had contradictory outcomes. Government support has enabled unproductive companies to recover, but the amount of worker control has actually been quite limited. Occupations also gave rise to experiences of articulation and solidarity between workers and peasants, as in the case of the state-owned socialist company Pedro Camejo, which provides specialized machinery and transport services for the agricultural sector. This company, created by Chávez in 2007, was occupied by workers in 2013 in opposition to mismanagement with the backing of peasant population and the local mayor (Azzellini, 2017). In Bolivia, workers' self-organization processes continue at the territorial level, joining together waged workers, self-employed groups, and community organizations as was the case of the "Gasolinazo" conflict at the end of 2010 (Quiroga et al., 2012)

In Ecuador, despite some improvements in poverty and inequality indicators during Correa's first mandate, and a left-wing discourse promoting social rights – such as those encoded in the *buen vivir* communitarian policies and sanctioned in the new 2008 Constitution – Correa's policies towards labour have been plagued with attacks on workers and social movement organizations (Gaussen, 2016). In addition to the unchanged nature of the structural conditions of the labour market, Correa criminalized social protest and has worked to control and co-opt popular organizations (Santillana and Webber, 2015). In this context, the government arrested indigenous leaders, made changes to the criminal code to punish social protest as act of terrorism and sabotage, limited the right to strike, and passed a decree banning all political actions by social organizations. As a consequence of these actions, historical trade unions such as UNE (*Unión Nacional de Educadores*, the teachers' union) were dissolved. In parallel, the creation of new pro-government social organizations with financial support from the government (such as *Red Agraria*, *Red de Maestros*, *ParlamentoLaboral*, and the *Central Unitaria de Trabajadores*) was actively promoted, therefore "cloning" social organizations in an attempt to dispute the representativeness of the historical unions (Daza and Santillana, 2015).

These contradictory approaches generated a number of reactions and struggles from the workers and their historical and/or newly created organizations. In all three countries, public sector workers have been among the most strongly organized. In Bolivia, for example, the Miners' Federation successfully fought for workers' control of the tin mine of Huanuni. As a result, the mine is now managed by a tripartite entity formed by the local union, the now state-owned company, and the government. In other mining areas, however, the government has supported private investment (CEDLA, 2016; Fornillo, 2009).

In Ecuador, popular marches and protests led by CONAIE and trade union organizations intensified in the first years of Correa's second term. A 2015 march involving students and teachers opposing the educational reform paralysed highways, took control over symbolic public spaces, and practically blocked the circulation of people and commodities in the cities, covering over half the country's territory (Daza and Santillana, 2015).

In Venezuela, throughout the whole period of popular reformism there has been an increase in labour conflicts, not only for economic reasons, but also due to inter-union disputes (Lucena e Iranzo, 2016). The union landscape has become more complex, with the creation of new pro-government and independent unions and growing opposition to the Bolivarian process from traditional unions with corporatist orientations – such as the Confederation of Workers of Venezuela (CTV).

Under the MAS government, Bolivia became one of the fastest-growing countries in Latin America. Indigenous populations became a new political subject that is today at the centre of the political scene. There is a strengthening political alliance between indigenous groups and unions in opposition to the conservative elites that control economic resources, as well as a large portion of the labour force. This reality sets a new historically unique political and ideological background for the current discussions about Evo Morales' reelection and about the perspectives of Bolivian workers.

In Ecuador, the election of Lenín Moreno – Correa's former vice-president – and his concessions to some of the demands of workers and social movements, in a break away from Correismo, have generally been welcomed. He recently reverted some of the most restrictive policies previously introduced by Correa to control social organizations. However, these measures might be just a superficial change to keep the framework of the neo-extractivist model unaltered. Moreno's recent signs of conservatism, and the popular support he received in February's plebiscite, are worrying in this respect.

In Venezuela, opposition unions (CTV, CGT, CODESA, and UNT) together with the Democratic Unity Table (MUD), an opposition political party created in 2012, have deepened their confrontation with the government. At the time of writing, the economic and political crisis was continuing to deteriorate. US interventionism has complicated domestic issues, forcing the government to become increasingly authoritarian and repressive in the attempt to defend Maduro and with him the conquests that workers, peasants, and indigenous people gained during *Chavismo*. Plunged into a deep crisis, the Bolivarian Revolution does not seem to find its way forward.

Conservative shift: Argentina and Brazil

After over a decade of progressive governments, Argentina and Brazil have shifted to the right in the last two years, with far-reaching consequences for workers and the labour movement in general. Progress in many social indicators during the ruling of the Workers' Party (PT) in Brazil (2003–2017) and *Kirchnerismo* in Argentina (2003–2015) was evident, although flexibility and informal and precarious labour remained high.

With the triumph of Luiz Inazio "Lula" da Silva in the 2003 presidential election, the *Partidos dos Trabalhadores* (Workers' Party) came to power. The PT was the political arm of a broad array of social movements and organizations forged in the confrontation against the military government and during the democratic (and neoliberal) regimes. The period was abruptly interrupted in mid-2016 by an intervention by congress that some political groups have characterized as a *parliamentary coup*. The vice-president – an entrepreneur belonging to the party PMDB – was appointed president and has since then set a clear path against workers' rights. In Argentina, *Kirchnerismo* (a fraction of the Peronist Party that could be identified as the center-left fraction of the party) came to power after the deep social, economic, and political crisis that culminated in massive social mobilizations at the end of 2001. In this context, *Kirchnerismo* applied policies aimed to boost the internal market through the improvement of employment indicators and a recovery – albeit slow – of real wages. In November 2015, the victory of a coalition between the right-wing PRO party and the more conservative arm of the historical *Partido Radical* brought Macri to power, and entailed a substantive shift in the state approach to labour relations (Elbert, 2016).

Improvements in workers' rights were significant during this progressive reformist period. In Brazil, leaders and intellectuals linked to social movements such as the *Movimento sem terra* (MST, landless movement), or the trade unions federated in the *Central Unica do Trabalhadores* (CUT) among others, were appointed to ministries and to state-owned companies. Workers' rights were also strengthened with the creation of the Labour Relations Council in 2010; the implementation of the National Plan to Eradicate Slave Labour, launched in 2003 by the previous government jointly with the ILO (Cardoso and Gindin, 2009; Figueira and Neide, 2017); and an increase in the minimum wage.

In Argentina, the government used the context of high commodity prices – especially soybean – to increase its revenues through taxing commodity exports, redistributing these resources in favour of a light industrialization through subsidies, a policy that proved successful in boosting employment and that led some to deem the Kirchnerist administrations as *neo-developmentalist* (Feliz, 2015). The state recovered a central place in the channelling, institutionalization, control, and regulation of labour conflicts, through collective bargaining and an active labour policy (Etchemendy and Collier, 2007; Palomino, 2008). After a confrontation with the rural bourgeoisie in 2008 and the defeat in the legislative elections of 2009, Cristina Fernández de Kirchner (2007–2015) launched a series of redistributive policies such as the *Assignación Unica Universal* (a child benefit scheme for low-income families), and nationalized the pension funds privatized in the 1990s, claiming to be deepening the "Model of Growth with Social Inclusion." However, the impact of the world economic crisis, the lack of investment, and the increase in inflation led to an economic retrenchment and a stagnation of jobs creation, which were contained by state policies through fiscal expansion until Macri came to power.

These social advances responded also to processes of revitalization of the labour movement since the crisis of neoliberalism in the late 1990s and early 2000s, and it is fair to argue that they were aimed at containing the rising discontent with neoliberal policies. During the period, the more favourable political and economic context also played a key role in strengthening the bargaining power of the main unions. During Lula's first term union density increased, though more among public sector than private sector workers. Union success in organizing on the shopfloor – through delegates or factory committees – remained otherwise low. Collective bargaining continued being much more important for manufacturing unions than for rural workers – which account for almost half the unionized workers but accounted for just 9% of agreements. As early as 2004 a group of radicalized workers and unions left the CUT in search of a more confrontational relation with the PT government (Boito, 2007, Boito and Marcelino 2011; Leher et al., 2010).

In Argentina, a number of studies (Etchemendy and Collier, 2007; Atzeni and Ghigliani, 2008, Palomino, 2008; Senén González and Del Bono, 2013) have highlighted the processes of union revitalization, expressed in the strengthening of the union leadership and the reorganization of workplace politics, both of which were historical traits of the labour movement in Argentina (Atzeni and Ghigliani, 2008 2013; Varela, 2015; Elbert, 2017; D'Urso and Longo, 2017; Marticorena, 2017). The active role of the state in regulating labour formed the material basis of an alliance with the main workers' confederation, the General Confederation of Labour (CGT), until early in Cristina Fernandez' second period in office, when the CGT broke up and the alliance with the government ran into crisis (Marticorena, 2015).

Some authors point to certain neoliberal continuities in the economic policies of the PT (Boito, 2007; Flynn, 2007) and the Kirchner governments (Azpiazu y Schorr, 2010; Piva, 2015). This is especially true in Brazil, where neoliberal monetary policies to control deficit and inflation and privatizations were very important during Lula's first term. A further continuity with neoliberalism in both countries is of course the deepening of labour flexibility and informality. The latter particularly remains high and there is the sensation that in the context of economic growth that has characterized the period, more improvements could have been made. Conflicts in Brazil are mainly connected to wages but also with two key regulations favouring flexibilization: the bank of hours (a more flexible device to compute working days according to management needs) and the profits participation bonus (PLR) – a bonus set over a share of annual companies' profits that is claimed by unions at company level. Both of them became key issues in collective bargaining, mainly in the manufacturing sector (Cardoso and Gindin, 2009). Today, demands on PLR are one of the main issues in company-level strikes, whereas public workers (mainly from the health and education sectors) led the bargaining conflicts at the federal level. Struggles of rural workers' movements led by the MST against the dramatic reduction of lands under Agrarian Reform as much as against the advance of agribusiness through land-grabbing processes, are also a key arena of conflict. Indeed, since 2000, agribusinesses have become the main target of rural workers' movements struggles, but official support to these has kept unaltered along the PT's terms and Temer's administration.

In Argentina, despite a GDP growth rate of around 8% between 2003 and 2007, the recovery of wages after the crisis was slow in relation to the increase in labour productivity. Precarious work remained high and labour conditions and legislation maintained the flexible contractual forms introduced in the 1990s (Marticorena, 2014). The lack of structural changes and the continuity of a sharp income inequality in the context of the economic stagnation of the last years put in evidence government's economic policies limitations. The strong dependence on commodity exports and the support of the government for agribusinesses and large mining and oil corporations – to the detriment of the protection of the environment and of indigenous people's rights – highlight the importance of extractivist activities for the sustainability of the government's economic plans.

Today, workers from Brazil and Argentina are facing a general offensive against their rights and living conditions, which involves the increase of unemployment and informality as much as the decrease in real wages, the implementation of regressive labour policies (through a broad Labour Reform Law in Brazil and through the promotion of regressive agreements in Argentina), the reform of the pension systems, and mounting repression. In both countries the union movement is slowly forging a confrontational dynamic that seems not to be strong enough to influence power relations. In Argentina, co-optation and official political alignments seem to prevail among the historical union leadership, whereas in Brazil unions seem to have lost the power of mobilization they once had.

Conclusions

Political developments in Latin America in the last two decades provide rich documentation to analyze the relationships between macro-politics and the labour movement. The crisis of neoliberalism and the more recent shift to the right in the region, have taken different countries through a variety of political paths. Here we have grouped these countries into three categories: stable neoliberalism, recent conservative shift, and stable popular reformism. In looking at these events through a country-specific approach in seven countries, we have seen important differences and substantial commonalities between them.

While in Mexico and Chile the lack of a deep economic crisis seems to have limited struggles against neoliberal policies, in Argentina, Brazil, Ecuador, Venezuela, and Bolivia the Pink Tide was a response to the broad popular opposition to policies of privatization, deregulation, flexibilization, and financialization. Populist centre-left-wing governments came to power by threading alliances with either unions or indigenous and peasants' organizations. However, despite some "concessions," in most countries these alliances did not last long.

Perhaps the most important commonality to all of these countries is the continuity of the main macroeconomic policies. Governments have sought to take advantage of the high commodity prices in order to redistribute the wealth originated in mining and rural activities, without seeking to transform the role of these countries in the international division of labour. This appears to be the key element that explains the current crisis of popular reformism in the region, since the world economic crisis and the plummeting of the prices of commodities have led to much slower rates of growth, or even to plain stagnation, limiting state resources for welfare policies.

Labour precarity and flexibilization also continued to be high despite improvements enjoyed by formal workers. While Mexico and Chile introduced new labour legislation deepening flexibilization – as did Brazil – the other countries did not modify the laws of flexibilization inherited from the neoliberal (and even authoritarian) administrations. An exception is Venezuela, which virtually banned subcontracting in its 2012 labour law reform. In Mexico, Ecuador, and Bolivia informal workers account for over half of the workforce, and little – if any – progress to protect them was made in the last two decades.

In terms of trade unions and labour movements, even in Argentina and Brazil, where union revitalization has been important, state control of the unions seems to have deepened. In Venezuela and Ecuador, Chávez and Correa even promoted the formation of new pro-government unions.

The rise of grassroots workplace unions and of informal workers' organizations has been important in Argentina, whereas in Chile they seem to pose the biggest threat to governments, in a general radicalization of the labour movement since 2011. Nevertheless, current perspectives for the labour movement and the working classes in the region seem grim in a context of economic stagnation, political shift to the right, and a new attack of the White House on these countries through tighter financial policies and intervention into internal politics.

Note

1 http://www.inegi.org.mx/

References

Angosto Ferrández, L. F. (2017) Indigenous peoples, social movements, and the legacy of Hugo Chávez's Governments. *Latin American Perspectives* 44(1): 180–198.

Atzeni, M. and Ghigliani, P. (2008) Nature and limits of trade unions' mobilisations in contemporary Argentina. In *Labour Again Publications*. Amsterdam: International Institute of Social History. www.iisg.nl/labouragain/documents/atzeni-ghigliani.pdf

———. (2013) The re-emergence of workplace based organisation as the new expression of conflict in Argentina. In G. Gall (ed) *New Forms and Expressions of Conflict at Work*. Basingstoke: Palgrave Macmillan, pp. 66–85.

Azpiazu, D. and Schorr M. (2010) La industria argentina en la postconvertibilidad. In D. Azpiazu, M. Schorr and V. Basualdo (eds) *La industria y el sindicalismo de base en la Argentina*. Buenos Aires: Cara o Ceca, pp. 29–59.

Azzellini, D. (2017) Class struggle in the Bolivarian process: Workers' control and workers' councils. *Latin American Perspectives* 44(1): 126–139.

Bank Muñoz, C. (2017) *Building Power from Below: Chilean Workers take on Walmart*. Ithaca, NY: Cornell University Press.

Bayas, T. (2017) Post commodity boom: Ecuador and Labour market conditions. *Council on Hemispheric Affairs*. www.coha.org/wp-content/uploads/2017/09/Tomas-Bayas-Market-Labor.pdf

Bensusán, G. (2016) Organizing workers in Argentina, Brazil, Chile and Mexico: The authoritarian-corporatist legacy and old institutional designs in a new context. *Theoretical Inquiries in Law* 17: 131–161.

Boito, A. (2007) Class relations in Brazil's new neoliberal phase. *Latin American Perpsectives* 34(5): 115–131.

——— and Marcelino, P. (2011) Decline in unionism? An analysis of new wave of strikes in Brazil. *Latin American Perspectives* 41(5): 94–109.

Bonilla, J. (2011) *El movimiento sindical venezolano frente a la situación socio-laboral: desafíos y propuestas*. Caracas: ILDIS.

Cardoso, A. and Gindin, J. (2009) *Industrial Relations and Collective Bargaining: Argentina, Brazil and Mexico Compared*. ILO Industrial and Employment Relations Department (DIALOGUE). Working Paper No. 5.

CEDLA. (2016) *Una década y media en el mundo del trabajo: Compendio del Boletín Alerta Laboral (2000–2016)*. La Paz: Bolivia.

Chavez Leon, P., Mokrani Chavez, D. and Uriona Crespo, P. (2010) Una década de movimientos sociales en Bolivia. *OSAL* 28: 71–93.

Chodor, T. (2014) *Neoliberal Hegemony and the Pink Tide in Latin America: Breaking Up With TINA?* Basingstoke: Palgrave Macmillan.

Collier, R. B. and Collier, D. (1991) *Shaping the Political Arena: Critical Junctures, the Labor Movement, and Regime Dynamics in Latin America*. Princeton, NJ: Princeton University Press.

——— and Handlin, S. (2009) Introduction: Popular representation in the interest arena. In R. B. Collier and S. Handlin (eds) *Reorganizing Popular Politics: Participation and the New Interest Regime in Latin America*. University Park, PA: Pennsylvania State University Press, pp. 3–31.

Darlington, R. (2014) The role of trade unions in building resistance: Theoretical, historical and comparative perspectives. In M. Atenzi (ed) *Workers and Labour in a Globalized Capitalism: Contemporary Themes and Theoretical Issues*. Basingstoke: Palgrave Macmillan, pp. 111–138.

Dávalos, P. (2014) *Siete ensayos sobre el posneoliberalismo en el Ecuador*. Bogota: Ediciones desde abajo.

Daza, E. and Santillana, A. (2015) Movilizaciones en Ecuador: cambio de ciclo y perspectivas criticas, *La linea del fuego*. https://lalineadefuego.info/2015/09/22/movilizaciones-en-ecuador-cambio-de-ciclo-y-perspectivas-criticas-por-esteban-daza1-y-alejandra-santillana2/

De la Garza Toledo, E. (2012) *La situación del trabajo en México, 2012: El trabajo en la crisis*. México, DF: Universidad Autónoma Metropolitana (Iztapalapa).

Donoso, S. and Von Bulow, M. (2016) *Social Movements in Chile: Organization Trajectories, and Political Consequences*. New York: Palgrave Macmillan.

———. (2016) Introduction: Social movements in contemporary Chile. In S. Donoso and M. Von Bulow (eds) *Social movements in Chile: Organization Trajectories, and Political Consequences*. New York: Palgrave Macmillan, pp. 3–27.

Durán-Palma, F. and López, D. (2009) Contract labour mobilisation in Chile's copper mining and forestry sectors. *Employee Relations* 31(3): 245–263.

D'Urso, L. and Longo, J. (2017) Radical political unionism as a strategy for revitalization in Argentina. *Latin American Perspectives*, online first. http://journals.sagepub.com/doi/full/10.1177/0094582X17736042

Elbert, R. (2016) Labour politics and the return of neoliberalism in Argentina. *Global Dialogue Magazine* International Sociological Association 6(3). http://globaldialogue.isa-sociology.org/labor-politics-and-the-return-of-neoliberalism-in-argentina/ (Accessed 18 October 2018).

———. (2017) Union organizing after the collapse of neoliberalism in Argentina: The place of community in the revitalization of the labor movement. *Critical Sociology* 43(1).

Etchemendy, S. and Collier, R. (2007) Down but not out: Union resurgence and segmented neocorporatism in Argentina (2003–2007). *Politics and Society* 35(3): 363–401.

Feliz, M. (2015) Neo-developmentalism, accumulation by dispossession and international rent. Argentina, 2003–2013. *International Critical Thought* 4: 499–509.

Figueira Rezende, R. and Neide, E. (2017) Slavery in today's Brazil: Law and public policy. *Latin American Perspectives* 44(6): 77–89.

Flynn, M. (2007) Between subimperialism and globalization: A case study in the internationalization of Brazilian capital. *Latin American Perspectives* 34(6): 9–27.

Fornillo, B. (2011) El regreso de la Patria Minera en Bolivia. En sindicalismo revolucionario durante el primer gobierno del MAS. In P. Abal Medina, B. Fornillo and G. Wyczyker (eds) *La forma sindical en Latinoamérica: Miradas contemporáneas.* Buenos Aires: Nueva Trilce, pp. 19–36.

Gaudichaud, F. (2004) *Poder popular y cordones industriales.* Santiago de Chile: LOM.

Gaussen, P. (2016) El fin del trabajo o el trabajo como fin? Proceso Constituyente y reforma laborales en el Ecuador de la revolución ciudadana (2007–2013). *Revista Latinoamericana de Derecho Social* 23: 31–55.

Grigera, J. (2017) Populism in Latin America: Old and new populisms in Argentina and Brazil. *International Political Science Review* 38(4): 441–455.

Harvey, D. (1982) *Limits to Capital.* London: Verso.

———. (2014) *Seventeen Contradictions and the End of Capitalism.* London: Profile Books.

Hyman, R. (2006) Marxist thought and the analysis of work. In M. Korczynski, R. Hodson and P. Edwards (eds) *Social Theory at Work.* Oxford: Oxford University Press, pp. 26–55.

Leher, R., Coutinho da Trindade, A., Botelho Lima, J. y Costa, R. (2010) Os rumos das lutas sociais no período 2000–2010 *OSAL* 28: 49–70.

LO/FTF. (2016) *Labor Market Profile Bolivia.* Danish Trade Council for International Development and Cooperation, Copenhagen.

Lucena, H. and Iranzo, C. (2016) Venezuela: los estudios laborales (1993–2014). In E. De la Garza Toledo (ed) *Los estudios laborales en América Latina.* Anthropos: UNAM, pp. 179–208.

Marticorena, C. (2014) *Trabajo y negociación colectiva. Los trabajadores en la industria argentina, de los '90 a la posconvertibilidad.* Buenos Aires: Imago Mundi.

———. (2015) Avances en el estudio de la relación entre sindicalismo y kirchnerismo. Revista *Socio-histórica. Cuadernos del CISH* 36. Universidad Nacional de La Plata. Facultad de Humanidades y Ciencias de la Educación. Centro de Investigaciones Socio Históricas. www.sociohistorica.fahce.unlp.edu.ar/article/view/SH2015n36a04

———. (2017) Contribución al debate sobre la organización de base en la Argentina reciente a partir de la dinámica sindical en el sector químico. *Revista Conflicto Social* 10(18): 224–257. http://publicaciones.sociales.uba.ar/index.php/CS/article/view/2659/2288

McGrath-Champ, S., Herod, A. and Rannie, A. (2010) *Handbook of Employment and Society: Working Space.* Cheltenham: Edward Elgar.

Mezzadri, A. (2016) *Sweatshop Regimes in the Indian Garment Industry.* Cambridge: Cambridge University Press.

Morales Hernández, C. (2013) Organización y negociación en los Tianguis mexicanos. In A. Orsatti and I. Escalona (eds) *Experiencias sindicales de formalización mediante organización sindical y diálogo social en América Latina y el Caribe.* Sao Paulo: Trade Union Confederation of the Americas, pp. 72–74.

Narbona, K., Páez, A. and Tonelli, P. (2011) *Precariedad laboral y modelo productivo en Chile.* Santiago: Fundación SOL.

Palomino, H. (2008) La instalación de un nuevo régimen de empleo en Argentina: de la precarización a la regulación. *Revista Latinoamericana de Estudios del Trabajo* 13(19): 121–144.

Panitch, L. and Albo, G. (2014) *Transforming Classes: Socialist Register 2015.* London: Merlin Press.

——— and Chibber, V. (2013) *Registering Class: Socialist Register 2014.* London: Merlin Press.

Pérez, P. (2017) *Business, Workers, and the Class Politics of Labor Reforms in Chile, 1973–2016*. San Diego: University of California.

Piva, A. (2015) *Economía y política en la Argentina Kirchnerista*. Buenos Aires: Batalla de Ideas.

Shade, L. (2015). Sustainable development or sacrifice zone? Politics below the surface in post-neoliberal Ecuador. *The Extractive Industries and Society* 2(4): 775–784.

Portes, A. and Hoffman, K. (2003) Latin American class structures: Their composition and change during the Neoliberal Era. *Latin American Research Review* 38(1): 41–82.

Quiroga, M. S., León C., Meneses, O., Pacheco, H. and Ríos, P. (2012) *Perfiles de la conflictividad social en Bolivia (2009–2011) Análisis multifactorial y perspectivas*. La Paz: UNIR.

Santibáñez, C. and Gaudichaud, F. (2017) Los obreros portuarios y la idea de "posición estratégica" en la posdictadura chilena (2003–2014). In J. Ponce, C. Santibáñez and J. Pinto (eds) *Trabajadores y trabajadoras. Procesos y acción sindical en el neoliberalismo chileno, 1979–2017*. Santigo de Chile: America en Movimiento.

Santillana, A. and Webber, J. R. (2015) Cracks in correismo? *La linea del fuego*. https://lalineadefuego. info/2015/08/14/cracks-in-correismo-by-alejandra-santillana-ortiz-jeffrey-r-webber/

Senén González, C. and Del Bono, A. (2013) *La revitalización sindical en Argentina: Alcances y perspectivas*. Buenos Aires: Prometeo; UNLAM.

Silver, B. J. (2003) *Forces of Labor: Workers' Movements and Globalization Since 1870*. Cambridge: Cambridge University Press.

———. (2014) Theorising the working class in twenty-first century global capitalism. In M. Atenzi (ed) *Workers and Labour in a Globalized Capitalism: Contemporary Themes and Theoretical Issues*. Basingstoke: Palgrave Macmillan, pp. 111–138.

Soruco Sologuren, X. (2009) Estado Plurinacional – Pueblo. Una construcción inédita en Bolivia, Revista *OSAL* 26: 19–33.

Svampa, M. (2015); Commodities Consensus: Neoextractivism and enclosure of the commons in Latin America. *South Atlantic Quarterly* 114(1): 65–82.

Tilly, C., De la Garza Toledo, E., Sarmiento, H. and Gayosso Ramírez, J. L. (2014) Los trabajadores que se organizan en la plaza: Contra-movimiento de una fuente inesperada. *Revista de Economía Crítica* 18: 160–180.

Torche, F. (2005) Unequal but fluid: Social mobility in Chile in comparative perspective. *American Sociological Review* 70(3): 422–450.

Varela, P. (2015) *La disputa por la dignidad obrera: Sindicalismo de base fabril en la zona norte del conurbano bonaerense 2003–2014*. Buenos Aires: Imago Mundi.

Winn, P. (1986) *Weavers of Revolution: The Yarur Workers and Chile's Road to Socialism*. New York: Oxford University Press.

———. (2004) *Victims of the Chilean Miracle: Workers and Neoliberalism in the Pinochet era, 1973–2002*. Durham, NC: Duke University Press.

Woodcock, J. (2018) Digital labour and workers' organisation. In M. Atenzi and I. Ness (eds) *Global Perspectives on Workers and Labour Organisations*. New York: Springer Nature, pp. 157–173.

29

LABOUR, UNIONS, AND MEGA-EVENTS[1]

Maurício Rombaldi
Translated by Brooke Parkin

Introduction

The signs from the current global political conjuncture are that global capitalism is taking significant steps to attack openly the regulation of labour. In various countries, this attack has taken place through proposals to alter the organs of the state and regulatory statutes with a view to reducing social rights of working people. In recent years, the most significant expression of this movement has been the labour reform initiatives tried out in various European and Latin American countries, which confirm that these initiatives are not individual cases, but the consequence of a wave of liberal governments whose aim is to offer greater freedom to capital by challenging rights considered to constrain the free circulation and accumulation of capital. Within this framework, unions have played a fundamental role in curbing this movement to roll-back rights by establishing limits, although insufficient, to the actions of capital and the subsequent imposition of degrading working conditions. This chapter deals with this last point and, in particular, with the union strategies for regulating labour relations alongside economic enterprises that take shape during sporting mega-events.

Unions in Brazil have taken a number of globally visible actions in this respect. These actions have resulted from the possibilities created by the presence, within the country, of a series of transnational corporations (TNCs) and large sporting events, both of which have been the objects of global activist agendas. In recent years in Brazil, international coordination amongst workers has moved beyond the traditional networks of metal workers in German firms.[2] Initiatives with different scopes have gradually been established in companies in different sectors like commerce, private security, and forestry.[3] International coordination amongst workers in the aviation and construction sectors has also gained momentum, profiting from the window of opportunity presented by the preparations for events like the 2014 World Cup and 2016 Olympic Games.

With regards to these events, the case of the *Fédération Internationale de Football Association* (FIFA) is exemplary of how sports have become a vehicle for capitalism to penetrate national economies, through new product development and opening markets to sponsors. Whilst FIFA carries out activities analogous to a TNC,[4] its influence is amplified beyond the sporting events. The organization imposes demands on the host country of the World Cup, that include demands of infrastructure and legal guarantees to protect their sponsors, without making explicit reference to social protection for workers involved in the preparations for the football (McKinley, 2011: 16).

Union strategies in sporting events

These sports competitions have become opportunities to scale up union strategies because they use large numbers of workers who gain prominence through the public attention around, and during the preparations for, the events. Moreover, the need to regulate work through negotiations with entities that are not traditional national unions has converged with the expertise accumulated by global union federations (GUF). Learning from the strategies developed to negotiate with organizations like FIFA, the International Olympic Committee and local organizers of the sports events, as well as local unions coordinating actions together – an exercise they are rarely disposed to do – has, principally, enabled links to develop between national and international agendas, consequently, strengthening connections between different levels of workers' organizations.

Some international union initiatives were developed in this way in the first decade of the 21st century. The *Play Fair* campaign was conceived of to coordinate and unify global union organizations and NGOs taking action to improve the working conditions at events like the Olympics and the World Cup. Made up of the International Trade Union Confederation (ITUC), the global union federation IndustriALL that represents, amongst others, textile workers, the Building and Wood Workers International (BWI), and the NGO *Clean Clothes Campaign*, the first edition of the campaign sought to establish minimum social standards for workers involved in preparations for the sporting events.

Directed at the 2012 Olympics in London, the campaign resulted in an agreement between British unions and the organizing committee of the Games, a say in the procedure for contracting third-party companies – by means of implementing a best practice protocol for outsourcing – and denouncements of the working conditions that the Chinese labourers who made the mascots were subject to, as well as of the social audits considered fraudulent and ineffective.[5]

In the same vein, *Play Fair* was present during the preparations for the 2010 World Cup in South Africa, supporting an action developed by BWI federation. BWI's campaign, entitled *Campaign for Decent Work Towards and Beyond 2010*, mediated the relationship between unions from different political positions, produced dossiers on working conditions, and provided support in national negotiations with FIFA. In this regard, the federation held meetings in Zurich between South African union members from the construction sector and representatives from FIFA, including its president at the time, Joseph Blatter. These actions resulted in the inclusion of local unions in the stadium inspections, a 39% increase in enrolment in unions in the construction sector between 2006 and 2009, and a 12% salary readjustment after a national strike in July 2009.

The case of Brazil

In Brazil, sporting mega-events were embedded in a developmental project adopted by the governments of Lula da Silva and his successor, Dilma Rousseff. One of the economic policies that was foundational to this project was the Plano de Aceleração de Crescimento (Growth Acceleration Programme), which underpinned a number of large-scale works. The construction sector took on a decisive role nationally, as a representation of what was possible from federal investment in infrastructure and, at the same time, significantly increasing the number of jobs, but this was not consensual (Veras, 2014). The 2014 World Cup and the 2016 Olympics were conceived in this setting full of expectations about the relationship between investment and increasing jobs. During the preparations for the 2014 World Cup, the Ministry for Sports suggested that it expected that the mega-event would be influential in creating close to 330,000 permanent jobs, between 2009 and 2014, and another 380,000 temporary jobs during the event itself (Ministério do Esporte, 2010). In the construction sector, the increase in job openings was related to planned works at 12 airports, six ports, and another 44 transit projects (Ministério do Esporte, 2014). Furthermore, 12 stadiums were built and refurbished[6] specifically for the World Cup, at a calculated total of USD $2.5 billion (ibid).

The contrast between the positive expectations and the reality of the preparations proved to be a complex setting in which to consolidate best practices at work. A significant portion of works was marked by delays accompanied by public pressure to finish them within the time frame stipulated by FIFA. By May 2012, it was estimated that the public bidding process had only been completed for 25% of transport projects (Amora, 2012), and that 41% of works for the World Cup had not even started (UOL, 2012). Although problematic for working conditions in the construction sector, this opened up positive possibilities for union negotiations.

On 31 March and 1 April 2011, in Rio de Janeiro, BWI launched the *Campaign for Decent Work Before and After 2014*. Despite the large quantity of unions existing in Brazil, due to particularities of union structures, up until its launch there were only a few Brazilian unions with institutional memberships through affiliations to BWI.[7] Data from the Ministry of Labour found that, in 2014, 10,813 unions existed in the country, suggesting an intensely fragmented space of organized unions (Cardoso, 2015: 494), but with the potential to scale up the actions of the global federation within the country. In fact, by the end of the campaign in 2014, GUF had 25 affiliated Brazilian union organizations, up from only five in 2011.[8]One of the principal actions that made up the strategy was the coordination, by BWI, of 16 unions, six state federations, and two Brazilian confederations,[9] many of them with different political orientations. GUF gathered and distributed information related to strikes and local agreements, produced campaign materials, and promoted national and international meetings between entities from unions, the organizers of the World Cup, and representatives from government, the International Labour Organization (ILO), and FIFA.

The organizations participating in the campaign who, on the whole, represented workers from host cities of the World Cup,[10] showed a significant lack of coordination between unions and confederations from the construction sector. Moreover, the fragmentation in representation was seen in the division created by affiliations to four central unions – Central Única dos Trabalhadores (CUT), Força Sindical, União Geral dos Trabalhadores (UGT), and Nova Central Sindical – and in the construction sector's non-existent experience at negotiating nationally for issues like an agreement on a basic salary, a recurring demand amongst the leadership. At this juncture, as a foreign organization BWI served as a bridge between the Brazilian organizations and they discussed "only what was possible to reach a consensus on (. . .) the non-negotiable differences were put aside," according to the vice-president of the Latin American committee of BWI at the time of the campaign, Edison Bernandes.

Given the divisions between unions, during the campaign Brazilians went on to develop a platform of joint actions that resulted in a *manifesto*, symbolizing an initial consensus over their demands from the construction sector. Later on, a *common national agenda* was established based on this material. It was an unedited document that unified demands and proposed a national agreement between workers and business owners to secure a minimum national wage, benefits, and the guaranteed right in each workplace to union or worker committee membership, amongst other things.[11]

BWI, together with the host union, designed the actions of campaigns in the host cities of the World Cup and experimented by uniting ideologically opposed unions. They organized visits to stadiums and other works in the cities, as well as meetings with representatives from government, the local organizers of the World Cup, the press, and other bodies. These actions were complemented concurrently by publicizing news on the BWI websites and by actions taken abroad by European unions that did not just share the results of the campaign but helped developed projects related to the initiative.

The 6th of March 2012 in Brasilia marked a landmark in national discourse, when the demands of construction workers, from the national agenda, were presented to the National Confederation of Industry (NCI), an organization of employers, with the message "workers should also be part of the 2014 World Cup." Even though the agenda has not resulted in a national agreement, the initiative laid the foundations by defining the content of the negotiations established in various regions of the country. For the leaders, rather than bring demands together into one negotiation, the unified agenda served to support demands established locally.

Labour negotiations in the construction sector were characterized by intense conflict in the period between 2011 and 2014. During this period, there were 28 strikes at stadiums being constructed or refurbished,[12] the majority of them in the first two years. The reason being that in the latter two years of work there was less demand for labourers, either because some of the stadiums were already finished by June 2013,[13] or because the unfinished ones were reaching the final stages when there is less manual labour needed. According to the report from the Inter-Union Department of Statistics and Socio-Economic Studies (DIEESE), a little more than half of the standstills were related to labour conditions or to the violation of rights already established in agreements or legislation (DIEESE, 2015).

Despite the reference provided by the national agenda, the different mobilizations in the stadiums were not coordinated with each other. One of many explanations for the lack of coordination is that there were different dates for the annual negotiations – established beforehand by region – between unions and companies, which meant that local mobilizations had different timings and, also, that unforeseen conflicts emerged. This last point refers to the shutdowns that occurred as a result of poor work conditions, caused by being offered spoiled food (Maracanã), accidents at work (Maracanã/Arena Amazônia), poor hygiene conditions in changing rooms and lack of overtime pay (Mineirão), or lack of adequate work wear (Arena Grêmio). Furthermore, the nine fatal accidents at the Pantanal, Corinthians, Grêmio, Amazônia arenas and the Mineirão and Mané Garrincha stadiums should be emphasized as they are no less representative of the labour conditions experienced during the preparations for the World Cup.

According to the DIEESE report, the strikes for new concessions or to broaden the already existing ones were linked, principally, to introducing, maintaining, or improving meal allowances, acquiring health care plans, and demanding salary adjustments. Concerning these demands, the campaign gave a sense of unity. In addition to demanding equal salaries and rights between different regions of the country, in March 2012, just before presenting the national agenda to NCI, there was talk in the national press of the possibility of a general strike in the construction sector. Even though an agreement had not been reached nationally, it is worth noting that during the preparations for the World Cup, negotiations by unions in the construction sector in the 12 host cities had been successful. As indicated by DIEESE's (ibid) system for monitoring salaries, the lower wage level for all agreements between 2009 and 2013 was readjusted to values above inflation, as measured by INPC-IBGE,[14] and above the adjustment in the national minimum wage. According to the study, beyond the wage level, the salaries of the other labourers involved in the works on stadiums were readjusted to exceed inflation. The increase in real wages above this minimum level was greater than the average real increases registered by SAS-DIEESE for all professional categories, and for categories in the construction and furniture sectors in the country. Even though salary adjustments have varied at different works, they have achieved a variety of concessions, like increases in meal allowances, in overtime pay, in transport vouchers, in health insurance, in monthly bonuses, and in the share of profits and earnings. Therefore, it can be said that, in general, the mobilizations were successful. They not only enabled improvements in salaries but also improvements in labour conditions.

In contradiction to the highly lucrative possibilities that come with organizing the World Cup and the pressure to deliver the sites, the mobilizations achieved positive results for labour negotiations. On the international level, despite the efforts of BWI to talk to FIFA, no agreements were reached to include Brazilian unions in stadium inspections, as had been the case in South Africa. Similarly, more than three years after the World Cup, no national agreement to regulate salary levels or other demands from the common agenda has been put into effect.

The 2016 Olympic Games in Rio de Janeiro established, in the same way as the World Cup preparations did, a challenging environment for unions and workers. On the one hand, there was international pressure to complete works in good time for the tournament to begin. On the other hand, workers and their unions were interested in getting decent work conditions. The calculated cost of 37.5 billion Reais (TCU, 2016)[15] for work on the Olympic park and urban travel (a metro line, motorways, increasing and modernizing the Transbrasil bus system [BRT] and the refurbishment of the Olympic stadium amongst other things), at the same time as countless strikes for improvements in salaries intensified even more the contradictions in the preparations for the Games.

In 2014, following the end of the campaign set up by unions for the World Cup, a subsequent homonymous initiative was established for 2016. The *Campaign for Decent Work Before and After 2016* was developed in conjunction with SITRAICP, a union affiliated with BWI, and represented close to 15,000 construction workers in the state of Rio de Janeiro. They drew up a series of materials and documents designed to promote safe practices on the Olympic construction sites, in negotiations over the use of timber certified by a socio-environmental stamp,[16] in jointly pressuring local authorities to establish negotiations with the unions in Rio de Janeiro, the *carioca* unions, as well as in the search to influence public opinion over accidents and conflicts in the workplace. In addition, BWI, based on the experience of the preparations for the last sporting event, anticipated a complex situation in Brazil in terms of work safety and so produced a safety code for construction works in Rio de Janeiro. It was based on research into the accidents in the construction sector during the works leading up to 2014. This established a constant dialogue between the organizing committee of the Games, which eventually resulted in the committee adopting these suggestions for protocol into their work safety policy. Nevertheless, it is worth emphasizing that, despite all these efforts, there were countless workplace accidents during the construction of the Olympic facilities.[17] Eleven deaths and the three very serious accidents were counted, on top of more than 1,600 infractions identified by inspectors, which resulted in works being interrupted.

The data for accidents are not unique to the preparations in Brazil, but point to a recurring scenario in the construction sector globally. They reveal a contrast between the organizer's search for, and peddling of, an "Olympic spirit" or *fair play* in the sports, and the poor work conditions experienced by those who prepare the mega-events. Sixty deaths of construction workers were counted at the Sochi Winter Olympics in 2012 (BWI, 2014), two more died during the preparations for the 2018 Winter Olympics in Korea (BWI, 2018), and the preparations for the 2022 World Cup in Qatar are proving even less encouraging, over a thousand workers have already lost their lives (Russell, 2016).

A significant part of the labour problems identified at construction sites in Rio de Janeiro arose because the municipal government did not have to consider health and safety principles when contracting firms to work on the Olympic sites. Moreover, the idiosyncrasy of *carioca* trade unionism differentiated its practices from the ones developed collectively by the collaboration of national unions, during the preparations for the 2014 World Cup. For example, according to the president of SITRAICP, Nilson Duarte Costa, during the preparations for 2014 there were deals agreed collectively with companies so that the trade unionists had free access to workplaces. During the works for 2016 this sort of agreement was more difficult because of the variety of works and their geographical spread (a large part of them were outsourced).[18]

In the years before the sporting events, scattered works and the difficulty unions had in brokering solutions to the problems encountered at workplaces were significant factors in intensifying a situation that was already very common in the Brazilian construction sector. Considering the large number of shutdowns experienced in general in the construction sector in Brazil in 2012, and especially at the stadiums in the three years that preceded the World Cup, the period from 2014–2016 in Rio de Janeiro was marked by 15 shutdowns,[19] which, in total, involved close to 37,000 workers and resulted in accumulated wage increases of close to 23%[20] in the period.

Global achievements and obstacles

On the international level, the different actions of BWI's global strategy for sporting mega-events have given signs of achieving positive results. Among them, in Russia, BWI, the Russian Construction Workers Union (RBWU), and the FIFA Local Organising Committee signed a cooperation agreement (BWI, 2016) to guarantee decent and safe working conditions during the construction and renovation of the stadiums for the 2018 World Cup in the country. In the same vein, in November 2017, the trade union signed an agreement with the construction company QDVC in Qatar, which planned human rights regulation for the workplace, accommodation, and recruitment of workers involved in the preparations for the 2022 World Cup (BWI, 2017).

However, in spite of the success of international union strategies, as seen at the World Cup, and the experience accumulated during actions targeting previous sporting mega-events, during the preparations for the Olympics, there were difficulties in coordinating the international agenda and the agenda developed by the *carioca* unions, as well as the Brazilian unions from different states and with different political affiliations. BWI and unions from Rio de Janeiro struggled to coordinate international strategies with ones undertaken locally.

There are two converging interpretations of the obstacles to developing international union actions together with organizations from Rio de Janeiro. On the one hand, Costa (2005) argues that, when analysing the internationalization of CUT, the priority given to the need for everyday order, locally, can act as a brake on the internationalization of local organizations. In the case of Rio de Janeiro, highly regulated labour relations that provide for annual negotiations over salaries and other demands, as well as the abrupt increase in the number of works and workers involved, could have favoured prioritizing traditional actions and left a smaller opening for more international-style actions. On the other hand, Veras (2014) analyzed the contradictions that exist in the development of large-scale works in the country and focused especially on shutdowns, like the ones at the Santo Antônio and Jirau hydroelectric plants in the state of Rondônia in 2011, which had close to 38,000 workers. According to Veras, these mobilizations, which shocked the unions, emphasized the disconnection between the union leadership and the represented workers. Above all the analysis highlights a series of difficulties encountered by unions in a context of abrupt growth in the number of workers on large projects.

In the same vein, in the case of the Olympics in Rio, in contrast to the experience of unions in the preparations for 2014, the difficulties in getting national support from construction unions outside of Rio were noticeable. With the exception of seminars organized by BWI in the host city of the Games, where unions from various political tendencies and regions of Brazil were present, there were no significant actions of solidarity or actions to engage with the campaign by Brazilian construction unions outside of the state of Rio de Janeiro. In terms of the permanence of experiences from the 2014 World Cup, this suggests there are obstacles to continuing the levels of unity and the consensus reached during the formation of the national common agenda from the previous campaign.

A similar important point is the fact that the campaign, driven by the international organization and supported by SITRAICP, in certain moments, gave signs of having not been incorporated organically by local unions. Even if the union developed its actions in cooperation with BWI, when participating in events and meetings together with Rio2016 and local authorities, as well as contributing by organizing delegations to visit the construction sites, a series of actions developed by SITRAICP during the period of the campaign maintained the same standards for negotiations and mobilizations as traditionally used. Practices like dialogue with companies in the sector, opening annual negotiations, and the strikes that follow from this process did not change much in character and, therefore, the union did not classify them as part of an international strategy, but as a result of the local labour relations.

Therefore, from the results of the campaigns implemented by international unions for 2014 and 2016, it is possible to state that the possibilities of success of international union strategies depend on the details of national and local unionism. In addition to the wage gains, the campaign directed at the 2014 World Cup constituted an unprecedented negotiation in which unions acted in a unified manner and institutionalized their internationalization by significantly increasing the number of BWI memberships. In the case of the campaign directed at the Games in Rio de Janeiro, it demonstrated that the *status quo* of unions in the construction sector, replete with local disagreements and difficulty bringing together an international agenda with a national one, is far from being effectively transformed.

From the perspective of the unions, according to an interview with the BWI's global campaign coordinator, Carlos Antonio Q. Añonuevo, the results achieved by the initiatives in 2014 and 2016 prove that "international campaigns need to be designed according to the local negotiation and mobilization capacities. This influences directly, which objectives are defined for each activity and what results can be expected." Amongst the examples that support this statement is the campaign developed during the preparations for the 2022 World Cup in Qatar, in which the GUF sought to foster the development of organizations capable of representing local workers. If a large part of the manual labour used is immigrant labour, certain strategies would move to organize workers in the communities that developed during the workers' migration to the country. Therefore, it might be possible to claim that the possibilities for success for international campaigns are different, which would make any comparisons that try to assess the success of campaigns relative. So, in Qatar the international dimension of the campaign is more significant, given that there are no grassroots unions. In Brazil, that there were unions capable of establishing salary negotiations and maintaining their own structures, contributed to the formation of agendas related to issues of health and safety, and other more specific points on organizations in the workplace.

Conclusion

This chapter sought to demonstrate how certain union strategies have had successes and failures in terms of regulating labour. Above all, despite their limitations, they have constituted potential avenues for regulating labour on a global scale, and there have been some agreements established recently, whether between BWI and companies involved in the preparations for the 2022 World Cup in Qatar or with workers in the aviation sector at LATAM (Feller and Conrow, 2017), to add to the efforts to open negotiations with Qatar Airways. The quick additional observations one can draw from these cases are that there is a new protagonist in international activism and that sporting mega-events have become a laboratory for unions from which they can learn valuable lessons.

Notes

1 This chapter is the culmination of a series of academic articles previously published by the author on the theme of sporting mega-events and the regulation of labour relations. For earlier works, see: Rombaldi (2017) andCottle and Rombaldi (2013).

2 Workers' representatives are expected, by German legislation on co-management, to be included on the boards of directors and to participate in any of the firm's strategic decisions. This has been crucial to workers at German TNCs, spread across different countries, sharing information as well as establishing international networks of solidarity. These are made up of, but not limited to, the networks of workers' unions at firms like *Volkswagen*, *Daimler-Benz*, and BASF.

3 Wallmart, Prosegur, and Arauco are firms operating in commerce, private security, and forestry respectively, and have all been the subject of international union actions to form networks. The first two firms have been the subject of campaigns promoted by UNI Global Union, and the latter by Building and Wood Workers International (BWI).

4 McKinley (2011) sets out one of the arguments for how FIFA is a TNC, which focuses on the intimate connection between the sports marketing firm, International Sports and Leisure(ISL), and FIFA. ISL is a key piece in the close relationships established with large sponsors and communication firms. Moreover, it received the naming rights to the World Cup and took on a central role commercially in defining contracts related to the football.

5 Information was obtained from the February 2012 agreement between the Local Organising Committee for the London Games and the Trade Union Congress / Play Fair, and the reports produced by ITUC (2012).

6 In fact, during the period 14 stadiums were constructed or refurbished. In addition to the 12 planned for the 2014 World Cup, another two had work done on them privately. They were the Arena Grêmio, in Porto Alegre, and the Arena Palmeiras, in São Paulo.

7 Even today, the structure of unions in Brazil follows the same model that was adopted and consolidated by the labour laws of 1943 during the government of Getúlio Vargas. In this system, unions are legalised in a state register that abides by the principles of *uniqueness* and *monopoly of representation*. This means that all unions are intimately linked to recognition from the state, which establishes geographical limits to its areas of operation and the category of workers represented (Martins Rodrigues, 2009).

8 CONTICOM, FETICOM-SP, FETICOM-RS, SINDPRESP, and Sindicato Solidariedade-São Caetano/SP.

9 Unions participating in the campaign: STICC-POA, SINTRACON-CTBA, SINTRAPAV-PR, SINDECREP-SP, Sindicato Solidariedade – São Caetano/SP, SINDPRESP-SP, SINTRAPAV-SP, SINDECREP-RJ, SITRAICP-RJ, SINTRACONST-ES, SITRAMONTI-MG, STICMB-DF, SINTRAICCCM-MT, SINTEPAV-BA, SINTEPAV-CE, STICONTEST-AM, SINTRACOM-SBC. State federations: FETICOM-SP, FETICOM RS, FETRACONSPAR, FETRACONMAG-ES, FETIEMT, FSCM-CUT. National Federations and confederations: FENATRACOP, CONTICOM/CUT.

10 BWI's initial strategic priority was cooperation between the unions of the host city. However, over the course of the campaign, unions from other regions joined the proposed actions.

11 As per the original document *Unified National Agenda*.

12 Although not part of the World Cup, the Arena Grêmio, in Porto Alegre, and Arena Palmeiras, in São Paulo, are counted here as work was done on them at the same time as the other stadiums. They were also planned and financed through loans from an investment package received for works for the World Cup.

13 In addition to the Arena Grêmio, inaugurated at the end of 2012, six stadiums were finished by June 2013 for the Confederations Cup. They were the Maracanã, Fonte Nova, Arena Pernambuco, Castelão, Mineirão, and Mané Garrincha.

14 National Consumer Price Index, calculated by the Brazilian Institute of Geography and Statistics.

15 Of this amount, close to R$ 24 billion was spent on construction works in urban infrastructure, public transport, mobility, and sanitation (TCU, 2016).

16 Amongst their platforms for action, the BWI supported an initiative to promote tripartite dialogue between workers, governments, and firms. Out of this the BWI policies for certifying wood with the FSC's socio-environmental stamps emerged.

17 According to data presented by the fiscal auditor of labour, Elaine Castilho, at the closing ceremony of the BWI campaign for the Olympics in Brazil, in June 2016.

18 According to a 2014 dossier by CUT (Central Única dos Trabalhadores,2014) focused on outsourcing and development, in Brazil outsourcing is characterized by longer working days, lower pay, higher turnover in jobs, and a larger number of accidents.

19 Two of them were state-wide strikes at the beginning of 2014 and 2015, lasting six and three days respectively, as a result of negotiations over salaries.

20 According to data obtained at the collective labour conventions of SITRAICP in the years 2014, 2015, and 2016. I used documents from the trade union. There are no academic references or available online refences.

References

Amora, D. (2012) Copa tem só 25% de obras de mobilidade urbana licitadas, diz TCU. *Folha de S. Paulo*, 2 May. www1.folha.uol.com.br/esporte/1084662-copa-tem-so-25-de-obras-de-mobilidade-urbana-licitadas-diz-tcu.shtml (Accessed 24 September 2013).

BWI. (2014) *Decent Work, Fair Play for All Teams*. Carouge: Building and Wood Workers International.

———. (2016) *BWI and RBWU Sign Cooperation Agreement with FIFA for Russia 2018*. www.bwint.org/es_ES/cms/prioridades-10/sindicato-11/campanas-deportivas-21/noticias-22/russia-bwi-and-rbwu-sign-cooperation-agreement-with-fifa-for-russia-2018-2419

———. (2017) *Construction Company to Sign Agreement with BWI in Qatar*. www.bwint.org/cms/news-72/construction-company-to-sign-agreement-with-bwi-in-qatar-854 (Accessed 10 December 2017).

———. (2018) *One Worker Killed and Two Injured at PyeongChang, Site of the 2018 Winter Olympics*. https://odoo.bwint.org/cms/news-72/one-worker-killed-and-two-injured-at-pyeongchang-site-of-the-2018-winter-olympics-728 (Accessed 6 June 2018).

Cardoso, A. (2015) Dimensões da crise do sindicalismo brasileiro. *RevistaCaderno CRH* 18(75): 493–510.

Costa, H. A. (2005) O sindicalismo, a política internacional e a CUT. *RevistaLua Nova* 64: 129–152.

Cottle, E. and Rombaldi, M. (2013) Les leçons de la Coupe du monde de foot de la FIFA en Afrique du Sud, le Brésil et l'héritage des syndicats. In Collectif CETIM (ed) *La Coupe est pleine! Les désastres économiques et soci aux des grands événements sportifs*. Geneve: CETIM, pp. 62–95.

Central Única dos Trabalhadores. (2014) *Terceirização eDesenvolvimento: Uma conta que nãofecha*. São Paulo: Central Única dos Trabalhadores. https://cut.org.br/system/uploads/ck/files/Dossie-Terceirizacao-e-Desenvolvimento.pdf (Accessed 6 June 2018).

DIEESE. (2015) *Indicadores da Agenda de Trabalho Decente*. São Paulo: DIEESE. www.dieese.org.br/anu ario/2015/sistPubLivreto7TabalhoDecente.pdf

Feller, D. and Conrow, T. (2017) El poder de los sindicatos de aviación en América del Sur: la red sindical LATAM ITF. *Nueva Sociedad* 271b: 165–187.

ITUC. (2012) *Jogando com os direitos dos trabalhadores: um informe sobre a produção de artigos promocionais para os Jogos Olímpicos de Londres 2012*. www.ituc-csi.org/IMG/pdf/play_fair_pt_final.pdf

Martins Rodrigues, L. (2009) *Partidos e Sindicatos: Escritos de Sociologia Política*. Rio de Janeiro: Centro Edel-stein de Pesquisas Sociais.

McKinley, D. T. (2011) Fifa sports-accumulationcomplex. In E. Cottle (ed) *South Africa's World Cup: A Leg-acy for Whom?*Durban: UniversityKwaZulu Natal Press, pp. 13–39.

Ministério do Esporte (2010) *Impactos Econômicos da Realização da Copa de 2014 no Brasil*. Brasília: Minis-tério do Esporte.

———. (2014) *Balanço Final para as Ações da Copa do Mundo Fifa Brasil 2014*. Brasília: Ministério do Esporte.

Rombaldi, M. (2016) Diferentes ritmos da internacionalização sindical brasileira: uma análise dos setores metalúrgico e de telecomunicações. *Caderno CRH*, online 29: 535–552.

———. (2017) Campañas por trabajo decente en megaeventos deportivos en Brasil: Estrategias sindicales innovadoras en el sector de la construcción. *Nueva Sociedad* 271b: 165–187.

Russell, S. (2016) Family still waiting to hear truth of Qatar metro worker's death. *Playfair Qatar*, 15 March. www.playfairqatar.org.uk/category/uncategorized/ (Accessed 8 September 2016).

TCU. (2016) *O TCU e as Olimpíadas de 2016*, 4ª Ed. Brasília: Tribunal de Contas da União.

UOL. (2012) 40,6% das obras da Copa do Mundo de 2014 ainda não começaram, segundo governo federal. *UOL Copa*, 23 May. https://copadomundo.uol.com.br/noticias/redacao/2012/05/23/41-das-obras-da-copa-do-mundo-de-2014-ainda-nao-comecaram-segundo-governo-federal.htm?cmpid=copiaecola (Accessed 23 September 2013).

Veras, R. (2014) Brasil em obras, peões em luta, sindicatos surpreendidos. *Revista Crítica de Ciências Sociais* 10: 111–136.

30

STREET VENDORS

Kate Swanson

Introduction

In cities across Latin America, street vendors are an ever-present part of urban life. At busy intersections, drivers receive competing sales pitches from newspaper vendors, fruit sellers, window washers, and more. In public parks and plazas, boys and men ply their trades as shoe shiners. On buses, young people pitch sad stories of both hardship and resilience in order to secure donations from captive audiences. In popular tourist districts, children sell flowers, arts, candies, and more to travellers who appear to have cash to spare. While street vendors have a long history in Latin American cities, their presence is also connected to high rates of poverty and inequality across the region. In fact, more often than not, those who work on the streets hail from the poorest *favelas* and urban *barrios*. Moreover, many are from marginalized and racialized groups, since those on the margins often have limited opportunities to get ahead in the more formalized urban economy.

In this chapter, I review current literature on the urban informal sector and street vendors in Latin America. I begin by outlining key characteristics of the urban informal sector, before exploring the lives of those who work on the streets. I then explore rising conflicts over access to public space in the region. While some cities have adopted a punitive approach to regulating city streets, other have embraced a more inclusive "right to the city" approach. I conclude by urging planners and scholars to explore alternative futures for truly inclusive Latin American cities.

The informal economy

Street commerce has a long history in Latin America. Before supermarkets and shopping malls took hold, urban shoppers purchased the majority of their perishable goods and petty commodities in public markets, or from *ambulantes* – mobile street vendors carrying their supplies in hand. This form of commerce isn't unique to Latin America. In fact, around the world street vendors and public markets have long been the urban norm. For instance, in Victorian-era London, street commerce was a regular feature of everyday life, as evocatively captured in the writings of Charles Dickens. As cities in industrialized nations developed, street commerce gradually lost ground to formal markets as city planners worked to modernize and regulate public spaces.

In poorer nations where gaps between the rich and poor remain high, street commerce remains a regular feature of urban life. In the 1970s scholars adopted a new term to describe this unregulated form of entrepreneurial capitalism. The International Labour Organization (ILO) popularized the term "informal economy" to describe an unregulated labour sector operating outside the control of the state. Work in the urban informal economy was characterized as small-scale and family-based, with low entry barriers in terms of skill, capital, and organization (Portes and Haller, 2005). While the informal economy describes work that operates outside of waged, formal labour (such as microenterprises, small unregulated industries, as well as more hidden forms of labour including domestic work, day labourers, illegal trade, among others), the most visible manifestation of the urban informal sector is the ubiquitous street vendor.

Many perceive the urban informal sector in a largely negative light and as a sector in need of municipal and state regulation. Some argue that it is a form of underemployment where workers lack labour protections and are at risk of exploitation (ILO, 2018). Others argue that the informal sector presents an income-earning alternative for the poor in an otherwise over-regulated state economy (de Soto, 1989). Feminist scholars have argued that the urban informal sector offers women the freedom to combine entrepreneurial work with childcare responsibilities. As self-employed labourers, women can bring their young children with them to their place of work, particularly if they work on the streets or in public markets (Swanson, 2010). It also gives women a chance to earn much needed income and gain more power in both their households and communities (Hays-Mitchell, 1999; Radcliffe, 1999). In a region shaped by profound racial hierarchies, the urban informal sector also offers income-earning opportunities for racialized groups, including Indigenous peoples and Afro-descendants. Given high levels of racism barring entry into the formal market economy, street commerce is often one of the most viable options for marginalized workers (Swanson, 2013).

Today, estimates suggest that close to 50% of Latin Americans participate in the informal labour economy. Participation rates are highest in Central America. For instance, informal labour participation rates in both Guatemala and Honduras are above 70% (Gonzalez, 2015). Perhaps not surprisingly, these nations also have some of the highest inequality indices in the region (World Bank, 2018). In these regions, the informal economy remains the norm and the vast majority of day-to-day business transactions pass under the radar of state and municipal regulations. Yet, while the unregulated informal and regulated formal sector are often constructed in opposition to one another, in reality these sectors are much more fluid. As noted by Goldstein (2016: 21), there is "perpetual slippage between the formal and the informal" as workers and consumers move between so-called "under the table" opportunities and those subject to taxation and regulation.

Street workers

Many of those who work on the streets are first or second generation rural migrants. This is especially the case in nations with high Indigenous populations, such as Bolivia, Ecuador, and Guatemala. As described elsewhere (see Ruttenberg, Enríquez, and Page, this volume), in the 1980s much of Latin America underwent profound change through Structural Adjustment Programs (SAPs). Across the region, nations accepted IMF financial aid in exchange for sweeping cuts to education, social spending, health care, infrastructure, agricultural subsidies, and removal of trade barriers. While this increased wealth for some, many experienced rising inequality and declining opportunities. Small-scale rural farmers were no exception. A shift to export-led agriculture, rising competition from outside markets, and an increasingly cash-driven consumer economy meant that small-scale farmers began struggling to make ends meet (Popke and

Torres, 2013). As a result, many were forced to migrate to cities in order to find income, most often in the informal sector. In the city, they joined thousands of other new migrants, many of whom began building makeshift homes in the urban periphery. As a result, barrios and favelas swelled as former rural dwellers sought better opportunities in rapidly growing cities (Davis, 2006). This resulted in a dramatic rural-urban shift: in 1960, 50% of Latin America's population lived in cities (Fay, 2005). Today, 75% of Latin Americans live in cities (Carr, Lopez, and Bilsborrow, 2009), making it the most urbanized region in the developing world (UN, 2016).

These mass rural-urban migrations are complicated by regional racial-spatial divides. Much of the region is dominated by the ideology of *mestizaje*, which assumes historical racial mixing of Spaniards, Africans, and Indigenous peoples (Wade, 2002). However, mestizaje is not only about physical whiteness, but also about discursive whiteness. This ideology encourages individuals to gradually evolve from "primitive" Indianness into more "civilized" states of being – states that eventually become incompatible with Indigenous ways (Bonnett, 2000; de la Cadena, 2000). According to dominant geographical imaginaries, Indigenous peoples and Afro-descendants are perceived to belong in "traditional" rural spaces, whereas white-*mestizos* are associated with "modern" urban spaces (Radcliffe and Westwood, 1996; Rahier, 1998). Whites and Indians are, in fact, often constructed in an oppositional binary; modernity and urban progress are associated with whiteness whereas backwardness and rural decay are associated with Indianness (Swanson, 2007a). In Mexico City, Martínez speaks to middle-class racialized anxieties concerning street vendors in the historical centre. She argues that the "specter of the Indian" permeates depictions of urban street vendors as "amoral, menacing, backward, and incommensurable others" (Martínez, 2016: 541). Moreover, they are perceived as a threat to urban renewal and a cosmopolitan modernity. These types of imaginaries are further informed by a long-standing hygienic discourse that is used to legitimize efforts to sanitize and cleanse public spaces of undesirable elements, particularly Indigenous and Afro-descendant bodies (Colloredo-Mansfeld, 1998).

Malena's story is perhaps typical of many street vendors. She grew up in a Quichua-speaking village in the high Andes. In the 1990s, many from her village began migrating to the city in search of better opportunities. Since the profits from agriculture were so low, they required a source of income to pay for education, buy goods, and help pull their families out of extreme poverty. Malena began working on the streets of Quito at the age of seven. She would make the six-hour trip to the city on weekends and school holidays to earn income for her family. By the age of 14 (when I first met her), Malena spent many of her days both begging and selling candies to tourists and locals. These were hard days, where she endured rejection and condescending remarks from urban residents who would say things such as, "go get a job," or "get off the streets" (Swanson, 2010). At the age of 19, Malena's mother died, leaving her to care full time for her three younger siblings, the youngest of whom was only four. While she managed to find work for several years in an Indigenous-run microlending cooperative, by age 28, she was back on the streets again. This time, she moved to the city permanently in the hopes of making ends meet. With only a 6th grade education and few marketable skills, employment options for women like Malena remain limited. Yet, she was acutely aware of the injustices of her situation. Even as a teenager, she was astounded and angered by the profound wealth disparities in the city. She spoke to the racism she experienced and expressed a deep cynicism regarding municipal efforts to "clean up" the city and push street vendors back to their villages. And no matter how hard Malena worked and strove to get ahead, it was very difficult for her to improve her lot in life.

Across Latin America, stories like Malena's are replicated en masse. As cities grow, the poor continue to seek out better opportunities wherever they can. In fact, transnational migration is on the rise within the continent, as residents struggle to find work to support their families. For instance, deeply entrenched poverty exacerbated by a series of environmental disasters has recently pushed thousands of Haitians to build new lives in Brazil, Chile, and Mexico. Life for the urban poor can be gruelling as they struggle to make ends meet amidst vast social and structural inequalities. Geographers and other academics have documented the life and struggles of street vendors and informal sector workers throughout the region (Bromley and Mackie, 2009a; Crossa, 2009; Davis, 2013; Donovan, 2008; Galvis, 2014; Mackie, Swanson, and Goode, 2017; Müller, 2016; Parizeau, 2015; Steel, 2012). Others have also focused explicitly on the lives of children on the streets (Aitken et al., 2006; Aufseeser, 2018; Bromley and Mackie, 2009b; Swanson, 2007b). Children and young people are involved in many types of work, but more visible activities include: shining shoes, guarding cars, working in markets, begging, and selling candies, handicrafts, flowers, and other small goods. Because they work on the streets, they are often perceived as "street children," regardless of their situations. Yet, young people have a range of experiences and connections to the streets. Much like the artificial boundary created between the formal and informal sector, children often live much more fluid lives. Some work on the streets after school and return home each night; some live on the streets full time; others move between these categories depending upon their situations regarding work, home, and school. In Ecuador, I came to know a lively and very capable nine-year-old Indigenous boy who divided his time attending school in the mornings and selling candies and begging from tourists in the afternoons. At the end of the day, he'd return to his family's rented apartment for the night. He maintained his school and work schedule mostly independently, despite his young age.

There are numerous misconceptions regarding those who work on the streets. Women are often perceived as "bad mothers" who are exploiting their innocent children. Able-bodied men are particularly maligned on the streets, which may be why day labour construction work is often a more attractive option for men (Swanson, 2007b). While there are certainly exploitative situations on the streets, the reality is that most people working on the streets are merely trying to improve their situations – even children. In an ideal world, young children wouldn't have to work on the streets; however, the reality is that many need cash income in order to survive in an otherwise unequal world.

The right to the city

As rising precarity pushed increasing numbers of street vendors onto the streets and plazas of Latin America's cities, middle class and elite anxieties regarding rising urban disorder began to grow (Becker and Muller, 2013; Martínez, 2016). Urban image has become more and more important around the world, as urban planners struggle to enhance their city's global competitiveness and attract international investment. In the industrialized north, New York City's urban revitalization project of the late 1980s and early 1990s became widely cited as a successful model for urban change. Led by then Mayor Rudy Giuliani and Police Commissioner William Bratton, New York City embarked upon a zero tolerance campaign to tackle so-called quality of life crimes, which included: squeegee windshield cleaners, panhandling, street artists, street vendors, among others (Smith, 1996). According to Mayor Giuliani, these street-based activities were indicative of "a city out of control" (Smith, 1998: 3). Zero tolerance policing quickly caught on and began to spread around the world as a way to improve the urban image, attract financial investment, and increase tourism (Mitchell, 2011; Smith, 2002). Since the 1990s, Giuliani and Bratton have been hired as crime and security consultants, or cited as the inspiration for policing strategies in nations across Latin American, including: Mexico (Becker and Müller, 2013; Crossa, 2009; Davis, 2013; Mountz and Curran, 2009), Brazil (Goode, Swanson, and Aiken, 2013; Wacquant, 2003), Ecuador (Swanson, 2007a), Peru (Aufseeser, 2018), Venezuela (Andrews and Bratton, 2008), Chile (DePalma, 2002), Argentina (Dammert and Malone, 2006), Honduras (Rodgers, Muggah, and Stevenson, 2009), El Salvador (Hume, 2007; LaSusa, 2015; Zilberg, 2007), Guatemala (Rodgers, Muggah, and Stevenson, 2009), and the Dominican Republic (Howard, 2009), to name a few (Swanson, 2013).

For street vendors, this has been bad news. In Ecuador, for instance, Mayor Jaime Nebot of Guayaquil hired Bratton as a crime and security consultant in 2002. While Bratton suggested an overhaul of the city's anti-crime strategies, the policing reforms were minimal, perhaps due to lack of funds, political will, or both. Instead, much of the effort focused on pushing street vendors out of the key tourist districts, thus worsening their ability to earn income. In Mexico City, the municipality set out to "rescue" the historical centre from street vendors, by turning it into a gentrified middle-class neighbourhood (Becker and Müller, 2013; Crossa, 2009; Davis, 2013; Martínez, 2016; Mountz and Curran, 2009). In this instance, Mexican billionaire Carlos Slim contracted Giuliani Partners at a price of $4.3 million dollars. Yet, as Mountz and Curran (2009) astutely note, what Giuliani brought to Mexico City was the illusion of control, rather than real change. The removal of street vendors did little to drop crime rates, but did succeed in increasing property values. In Brazil, Rio de Janeiro contracted Giuliani Partners to help the city prepare for the 2014 FIFA World Cup and 2016 Olympic Games. Thereafter they implemented a strategy called *Choque de Ordem*, or Shock of Order, which targeted street vendors, pamphleteers, and other urban informal sector workers for removal. According to a spokesperson for the Department of Public Order, raids on street vendors were designed to "combat visual pollution" (Soifer, 2009), which was perceived as a catalyst for public insecurity and crime (Forte, 2011). Yet, while cleansing the streets of "visual pollution," Shock of Order policing ended up pushing the urban poor into ever more precarious positions. Activists resoundingly criticized the municipality for pursuing hygienist urban cleansing policies, which the municipality fervently denied (Schmidt and Robaina, 2017).

These examples are merely three instances reflective of larger efforts to implement zero tolerance – also known as *mano dura*, or iron fist policing – throughout the region (see also Rodgers, this volume). Social organizations and street vendor federations have actively fought against these crackdowns; moreover, the effectiveness of these policies has been widely disputed. As in Rio de Janeiro, some argue that these approaches are merely state-sanctioned social cleansing policies designed to create "safer feeling" public spaces and attract global capital investment (Swanson, 2013). Meanwhile, the lives and well-being of the city's urban poor are made more difficult as they struggle to make ends meet within the context of rising inequality, displacement, and marginality.

Yet, not every Latin American city has embraced this punitive approach. In fact, some cities have explicitly adopted a "right to the city" approach. According to Harvey, the right to the city is a collective cry for an alternative urban life. He states that everyone has the right to the city, including the dispossessed – who have the "right to change the world, to change life, and to reinvent the city more after their hearts' desire" (Harvey, 2012: 25). Some municipalities and nations have adopted this approach in their municipal policies and even their national constitutions. Bogotá, for instance, has been heralded as a progressive model for modern cities (Beckett and Godoy, 2010). Over the last 20 years, city administrators have emphasized the right to the city approach in an effort to promote more inclusionary urban space (Galvis, 2014). Rather than criminalize street vendors, Bogotá chose to work with the urban poor to provide job training and alternative employment options. In other words, marginalized groups were targets of inclusive social policies, rather than exclusionary and punitive ones. Moreover, municipal planners worked to re-envision and re-invest in the city's social and educational infrastructure in order to build a more inclusive urban citizenship (Beckett and Godoy, 2010).

Ecuador has also embraced the Right to the City discourse. In 2008, the nation revised its constitution under President Rafael Correa. Article 31 is especially progressive. It states:

> Persons have the right to fully enjoy the city and its public spaces, on the basis of principles of sustainability, social justice, respect for different urban cultures and a balance between the urban and rural sectors. Exercising the right to the city is based on the democratic management of the city, with respect to the social and environmental function of property and the city and with the full exercise of citizenship.
>
> *(Republic of Ecuador, 2008)*

Constitutional changes also prevent authorities from permanently seizing street vendors' commercial goods, thus granting them some legal protection (Art. 329). Yet, several years earlier, many of Quito's street vendors had already been cleared from the city's historical centre in an effort to "recover" public space (Kingsman, 2006; Middleton, 2003; Swanson, 2007a). This sector remains off limits to street vendors, and those who violate this rule are subject to substantial fines (Garcia, 2017). Much like in Mexico City, this cleansing of the historical centre has resulted in rising property values and gentrification (Burgos-Vigna, 2017). It would seem that despite constitutional change, not everyone has an equal right to the city.

Although politicians might embrace progressive policies, implementation is often a different story. In Bogotá, for instance, Galvis (2014, 2017) argues that the city's inclusionary policies are merely rhetoric masking ongoing exclusion. In fact, he argues that while Bogotá is often heralded as a model city for progressive urbanism, street vendors, sex workers, and homeless populations continue to experience harassment and exclusion in public spaces. He states, "other ways of mobilizing urban life as a means to promote urban equality must be judged not in terms of their abstract embrace of equality, but rather on the way they treat those whose lives and livelihoods depend on realizing their right to circulate in and appropriate public space" (Galvis, 2017: 97). He argues that ongoing exclusion in Bogotá is based upon long-standing social, racial, and class hierarchies that continue to marginalize the city's poor. Galvis' work in Bogota brings into question debates regarding more supportive and possibly post-revanchist urban policies that purport to embrace inclusion, rather than exclusion (DeVerteuil, 2012; Mackie, Swanson, and Goode, 2017). While policies may be intentionally progressive, exclusion will continue as long as social, racial, and class hierarchies remain intact.

Future directions

Given the long history of street vending in Latin America and the fact that it is a practice that is deeply entrenched in regional norms and practices, it will be especially difficult to eliminate in the long run, despite the wishes of municipal planners and developers. Moreover, street vending is often a strategy reserved for the poorest and most destitute of workers. Many are first- or second-generation migrants, who have been forced from their lands – or even countries – in an effort to improve their socio-economic situations. As stated by a street vendor in Quito, "Each vendor has a backstory, whether we're Ecuadorian or from another country. What we're looking for is a way to get ahead honestly, but we need authorities to support us, rather than persecute us" (Merizalde, 2017). While perhaps stating the obvious, recognizing street vending as a survival strategy rather than a criminal offence would help encourage municipalities to develop viable strategies to both aid street workers and enhance urban development. Yet, inequality remains a fundamental problem throughout the Americas. And as inequality continues to rise, more and more workers are pushed onto the streets, given limited other options.

While perhaps a monumental task, states must address ongoing and rampant inequality in the region. Given a choice, many of those working in the urban informal sector would prefer to find employment in more stable and less precarious positions. Moreover, we must take Galvis' critiques seriously. While cities may adopt inclusionary policies, how are those policies enacted in practice? How do these policies affect the lives of the city's urban poor? Do policies allow residents to envision alternative urban futures that truly provide a right to city for all? Providing a universal right to the city is a laudable goal, yet notoriously difficult to enact. As stated by Meneses-Reyes and Caballero-Juárez (2014: 381), "Given judicial precedents, one can conclude that selling on the streets is no less than a right that any citizen may exercise in a public space." Nevertheless, policies and practices don't always align, particularly in the context of the most marginalized urban residents.

References

Aitken, S., López Estrada, S., Jennings, J., and Aguirre, L. M. (2006) Reproducing life and labor: Global processes and working children in Tijuana, Mexico. *Childhood* 13: 365–387.

Andrews, W. and Bratton, W. (2008) Crime and politics in Caracas. *City Journal* 18(4). www.city-journal.org/html/crime-and-politics-caracas-13133.html (Accessed 10 January 2018).

Aufseeser, D. (2018) Challenging conceptions of young people as urban blight: Street children and youth's ambiguous relationship with urban revitalization in Lima, Peru. *Environment and Planning A: Economy and Space* 50(2): 310–326.

Becker, A. and Müller, M. M. (2013) The securitization of urban space and the "rescue" of downtown Mexico City: Vision and practice. *Latin American Perspectives* 40: 77–94.

Beckett, K. and Godoy, A. (2010) A tale of two cities: A comparative analysis of quality of life initiatives in New York and Bogotá. *Urban Studies* 47: 277–301.

Bonnett, A. (2000) *White Identities: Historical and International Perspectives.* Harlow: Pearson Education Limited.

Bromley, R. and Mackie, P. (2009a) Displacement and the new spaces for informal trade in the Latin American city centre. *Urban Studies* 46: 1485–1506.

———. (2009b) Child experiences as street traders in Peru: Contributing to a reappraisal for working children. *Children's Geographies* 7: 141–158.

Burgos-Vigna, D. (2017) Quito, a world heritage city or a city to live in? Right to the city and right to heritage in the 'Good Living State'. *City* 21(5): 550–567.

Carr, D., Lopez, A. C. and Bilsborrow, R. (2009) The population, agriculture, and environment nexus in Latin America: Country-level evidence from the latter half of the 20th century. *Population and Environment* 30: 222–246.

Colloredo-Mansfeld, R. (1998) "Dirty Indians", radical indígenas, and the political economy of social difference in modern Ecuador. *Bulletin of Latin American Research* 17: 185–205.

Crossa, V. (2009) Resisting the entrepreneurial city: Street vendors' struggles in Mexico City's historic center. *International Journal of Urban and Regional Research* 33: 43–63.

Dammert, L. and Malone, M. F. T. (2006) Does it take a village? Policing strategies and fear of crime in Latin America. *Latin American Politics and Society* 48: 27–51.

Davis, D. (2013) Zero tolerance policing, stealth real estate development, and the transformation of public space: Evidence from Mexico City. *Latin American Perspectives* 40: 53–76.

Davis, M. (2006) *Planet of Slums.* New York: Verso.

DeVerteuil, G. (2012) Does the punitive need the supportive? A sympathetic critique of current grammars of urban injustice. *Antipode* 46: 874–893.

Donovan, M. G. (2008) Informal cities and the contestation of public space: The case of Bogotá's street vendors, 1988–2003. *Urban Studies* 45: 29–51.

de la Cadena M. (2000) *Indigenous Mestizos. The Politics of Race and Culture in Cuzco, Peru 1919–1991.* Durham, NC: Duke University Press.

DePalma, A. (2002) The Americas court a group that changed New York. *The New York Times,* 11 November.

de Soto, H. (1989) *The Other Path: The Invisible Revolution in the Third World.* New York: Harper & Row.

Fay, M. (2005) (ed) *The Urban Poor in Latin America*. Washington, DC: World Bank.

Forte, J. (2011) UOP Shock and Order in Ipanema. *The Rio Times*, 27 December.

Galvis, J. P. (2014) Remaking equality. Community governance and the politics of exclusion in Bogotá's public spaces. *International Journal of Urban and Regional Research* 38: 1458–1475.

———. (2017) Planning for urban life: Equality, order, and exclusion in Bogotá's lively public spaces. *Journal of Latin American Geography* 16: 83–105.

Garcia, A. (2017) Los informales se concentran en 20 sectores de Quito. *El Comercio*, 18 de junio. www.elcomercio.com/actualidad/comercioinformal-control-quito-sectores-ventas.html (Accessed 9 January 2018).

Goldstein, D. M. (2016) *Owners of the Sidewalk: Security and Survival in the Informal City*. Durham, NC: Duke University Press.

Gonzalez, E. (2015) Weekly chart: Latin America's informal economy. *Americas Society/Council of the Americas*. 2 April. www.as-coa.org/articles/weekly-chart-latin-americas-informal-economy (Accessed 9 January 2018).

Goode, R. J., Swanson, K. and Aiken, S. C. (2013) From God to men: Media and the turbulent fight for Rio's favelas. In G. H. Curti, J. Craine and S. Aitken (eds) *The Fight to Stay Put: Social Lessons Through Media Imaginings of Urban Transformation and Change*. Stuttgart: Franz Steiner Verlag, pp. 161–180.

Harvey, D. (2012) *Rebel Cities: From the Right to the City to the Urban Revolution*. New York: Verso.

Hays-Mitchell, M. (1999) From survivor to entrepreneur: Gendered dimensions of microenterprise development in Peru. *Environment and Planning A* 31: 251–71.

Howard, D. (2009) Urban violence, crime and the threat to 'democratic security' in the Dominican Republic. In D. McGregor, D. Barker and D. Dodman (eds) *Global Change and Caribbean Vulnerability: Environment, Economy and Society at Risk*. Kingston: University of West Indies Press, pp. 298–316.

Hume, M. (2007) Mano Dura: El Salvador responds to gangs. *Development in Practice* 17: 739–751.

International Labour Organization (ILO). (2018) *Informal Economy*. www.ilo.org/global/topics/employment-promotion/informal-economy/lang – en/index.htm (Accessed 9 January 2018).

Kingsman, G. E. (2006) *La Ciudad y los Otros, Quito 1860–1940: Higienismo, Ornato y Policía*. Quito: FLACSO-Universidad Rovira e Virgili.

LaSusa, M. (2015) Guiliani in Rio. *NACLA*. 26 October. https://nacla.org/news/2015/10/26/giuliani-rio (Accessed 9 January 2018).

Mackie, P., Swanson, K. and Goode, R. (2017) Reclaiming space: Street trading and revanchism in Latin America. In A. Brown (ed) *Rebel Streets, Informal Economies and the Law*. London: Routledge, pp. 63–76.

Martínez, A. L. (2016) "You cannot be here": The urban poor and the specter of the Indian in neoliberal Mexico City. *The Journal of Latin American and Caribbean Anthropology* 21: 539–559.

Meneses-Reyes, R. and Caballero-Juárez, J. A. (2014) The right to work on the street: Public space and constitutional rights. *Planning Theory* 13(4): 370–386.

Merizalde, M. B. (2017) El control a la informalidad en Quito rebasa a los policías municipales. *El Comercio*. 16 June. www.elcomercio.com/actualidad/control-ventas-informales-policia-metropolitana.html (Accessed 9 January 2018).

Middleton, A. (2003) Informal traders and planner in the regeneration of historic city centres: The case of Quito, Ecuador. *Progress in Planning* 59: 71–123.

Mitchell, K. (2011) Zero tolerance, imperialism, dispossession. *ACME* 10: 293–312.

Mountz, A. and Curran, W. (2009) Policing in drag: Giuliani goes global with the illusion of control. *Geoforum* 40: 1033–1040.

Müller, M. M. (2016) *The Punitive City: Privatised Policing and Protection in Neoliberal Mexico*. London: Zed Books.

Parizeau, K. (2015) Re-Representing the city: Waste and public space in Buenos Aires, Argentina in the late 2000s. *Environment and Planning A: Economy and Space* 47: 284–99.

Popke, J. and Torres, R. M. (2013) Neoliberalization, transnational migration, and the varied landscape of economic subjectivity in the Totonacapan region of Veracruz. *Annals of the Association of American Geographers* 103: 211–229.

Portes, A. and Haller, W. (2005) The informal economy. In N. J. Smelser and R. Swedberg (eds) *The Handbook of Economic Sociology*, 2nd Ed. Princeton, NJ: Princeton University Press, pp. 403–427.

Radcliffe, S. (1999) Latina labour: Restructuring of work and renegotiations of gender relations in contemporary Latin America. *Environment and Planning A* 31: 196–208.

——— and Westwood, S. (1996) *Remaking the Nation: Place, Identity and Politics in Latin America*. London: Routledge.

Rahier, J. M. (1998) Blackness, the racial/spatial order, migrations, and Miss Ecuador 1995–1996. *American Anthropologist* 100: 421–430.

Republic of Ecuador. (2008) *Constitution of the Republic of Ecuador*. English translation from Political Database of the Americas. http://pdba.georgetown.edu/Constitutions/Ecuador/english08.html (Accessed 9 January 2018).

Rodgers, D., Muggah, R. and Stevenson, C. (2009) *Gangs of Central America: Causes, costs, and interventions*. Small Arms Survey, Occasional Paper 23.

Schmidt, K. and Medeiros Robaina, I. M. (2017) Beyond removal: Critically engaging in research on geographies of homelessness in the city of Rio de Janeiro. *Journal of Latin American Geography* 16: 93–116.

Smith, N. (1996) *The New Urban Frontier: Gentrification and the Revanchist City*. New York: Routledge.

———. (1998) Giuliani time: The revanchist 1990s. *Social Text* 57: 1–20.

———. (2002) New globalism, new urbanism: Gentrification as global urban strategy. *Antipode* 34: 428–450.

Soifer, R. (2009) Rowdy Rio's mayor tries Giuliani's get-tough tactics. *McClatchy Newspapers*, 27 April.

Steel, G. (2012) Whose paradise? Itinerant street vendors' individual and collective practices of political agency in the tourist street of Cusco, Peru. *International Journal of Urban and Regional Research* 36(5): 1007–1021.

Swanson, K. (2007a) Revanchist urbanism heads south: The regulation of indigenous beggars and street vendors in Ecuador. *Antipode* 39: 708–728.

———. (2007b) "Bad mothers" and "delinquent children": Unravelling anti-begging rhetoric in the Ecuadorian Andes. *Gender, Place and Culture: A Journal of Feminist Geography* 14: 703–720.

———. (2010) *Begging as a Path to Progress: Indigenous Women and Children and the Struggle for Ecuador's Urban Spaces*. Athens: University of Georgia Press.

———. (2013) Zero tolerance in Latin America: Punitive paradox in urban policy mobilities. *Urban Geography* 34: 972–988.

United Nations. (2016) *The World's Cities in 2016*. Department of Economic and Social Affairs, Population Division, Data Booklet (ST/ESA/ SER.A/392). www.un.org/en/development/desa/population/publications/pdf/urbanization/the_worlds_cities_in_2016_data_booklet.pdf (Accessed 8 January 2018).

Wacquant, L. (2003) Toward a dictatorship over the poor? Notes on the penalization of poverty in Brazil. *Punishment and Society* 5: 197–205.

Wade, P. (2002) *Race, Nature and Culture: An Anthropological Perspective*. Sterling: Pluto Press.

World Bank. (2018) *Gini Index (World Bank estimate)*. https://data.worldbank.org/indicator/SI.POV.GINI?end=2011&start=2011&view=map (Accessed 9 January 2018).

Zilberg, E. (2007) Refugee gang youth: Zero tolerance and the security state in contemporary U.S.-Salvadoran Relations. In S. Alladi Venkatesh and R. Kassimir (eds) *Youth, Globalization, and the Law*. Stanford, CA: Stanford University Press, pp. 61–89.

31

MAQUILA LABOUR

Jennifer Bickham Mendez

Introduction: export-oriented production, maquila factories, and the new international division of labour

The maquiladora sector in Latin America and the Caribbean dates back as far as Puerto Rico's 1947 Operation Bootstrap initiative, taking root 20 years later in the form of Mexico's 1965 Border Industrialization Program. Maquiladoras (or "maquilas" colloquially) are assembly factories that produce light manufactured goods designated for export to foreign markets. Maquilas are emblematic of export-led development strategies which gathered steam in the Caribbean and Mexico in the 1980s, galvanized by special US tariff programs that promoted the relocation of labor-intensive phases of manufacturing abroad by providing duty-free access to US markets (Safa, 1994). The International Monetary Foundation (IMF) and the World Bank further reinforced this development strategy in Latin America by touting it as a means to attract foreign investment and alleviate the debt crisis.

During the Cold War the US government promoted export manufacturing as a mechanism to ensure political stability – known as "security through development" – in Central America and the Caribbean (Deere and Antrobus, 1990: 154). For example, the Reagan-era Caribbean Basin Initiative of 1983 provided special duty-free access to US markets and exempted 24 countries in Central America and the Caribbean from quotas on clothing assembled from cloth made and cut in the US. Central American governments soon followed suit with the passage of a series of laws that enabled companies to bring in raw materials exempt from customs and duties, allowing firms that located production in these countries to take full advantage of this program (Pine, 2008). Meanwhile, more developed countries contended with "deindustrialization," as manufacturing jobs shifted offshore.

Key to export-oriented development strategies and accompanying global economic restructuring is the establishment of export production zones (EPZs). In the 1980s and 1990s, often-times as a condition for receiving continued aid from USAID or as part of Structural Adjustment Packages required by the IMF, governments in Mexico and the Caribbean, and later Central America, set aside special zones to attract foreign investment and diversify exports. As incentives to transnational corporations to locate production within their borders, governments offered tax holidays, freedom from import duties, special rates for utilities, and even exemption from minimum wage requirements and unrestricted profit repatriation (Safa, 1995). According to an

International Labor Organization (ILO) publication, between 1986 and 2006 the number of EPZ's worldwide increased from 176 to 3500, employing an estimated 66 million workers (ILO, 2014: iii). By 2015 *The Economist* (2015) reported the number to have risen to 4,500.

Deemed "vehicles of globalization" (see McCallum, 2011: 1), EPZ's growing proliferation throughout the region in the 1990s was also indicative of the spread of neoliberal political-economic agenda. Neoliberalism advocates the mobility of capital and the promotion of free trade as well as the reduction of the role of the state in economic decisions. To achieve these ends, neoliberalism prescribes the privatization of state-owned enterprises and the drastic reduction of the public sector and social services (Robinson, 2003). Thus, in the 1990s, EPZs and export-oriented production more generally, along with the dramatic down-sizing of the public sector, were hailed as solutions to Latin America's development "problems." Latin American countries could in theory establish a comparative advantage in the global economy by specializing in the production of labor-intensive goods, thereby creating jobs for swelling ranks of desperate, unemployed workers.

Gender, labour conditions, and women's well-being

The promise of export-oriented production as a development strategy, however, is undermined by the so-called "race to the bottom," since within the international division of labour that emerged alongside this production scheme, a country's comparative advantage lies in its ability to provide the lowest possible production costs – that is, the cheapest labour possible (Brecher, Costello, and Smith, 2000). And it is here that gender emerges as a crucial component of maquila production, as firms explicitly target young women as a preferred workforce. The notion of women as "supplemental" earners helped bolster the rise of the maquila industry, which gained strength at a time of declining job opportunities for men (Safa, 1995). Although there is variability across national contexts, the ILO estimates that between 70 and 90% of workers in EPZs worldwide are female (Milberg and Amengual, 2008). While reliable statistics are limited, the feminization of this employment sector in Latin America has been consistently documented. For example, in Honduras between 1991 and 2006 close to seven out of 10 maquila workers were women (De Hoyos, Bussolo, and Núñez, 2012).

The feminization of the maquiladora labour force reflects a broader trend over the last three decades of an upsurge in women's economic activity in Latin America and across the world, which has only recently levelled off following the global economic crisis of 2008 (Otobe, 2015). Indeed, according to the ILO (2016), in Latin America and the Caribbean, female participation in the labour force increased from 44.5% in 1995 to 52.6 percent in 2015 (6). The restructuring of the labour process and deskilling of jobs involved in export-oriented production facilitate the integration of a vulnerable, devalued workforce within labour markets that were already gender-segregated. Women's economic need exacerbated by structural adjustment programs and declining male employment propel them into the work-force in labour markets that offer few alternatives to low-wage labour in maquila factories (Tiano, 2006; Fussell, 2000). Indeed, wages within these factories are often higher than comparable employment sectors (McCallum, 2011: 4).

Gender ideologies about the differences between men and women and the kind of work for which each is suited undergird the international division of labour within a global economy. The trope of women as "cheap" and "docile" labour, culturally predisposed to be patient and submissive to male authority is central to a logic that enables the incorporation of systems of gender inequalities into the flexible accumulation strategies of neoliberal capitalism (Cravey, 1998). Research conducted in the 1980s and 1990s paints a picture of women assembly workers as largely occupying the bottom of the production hierarchy, labouring long hours under oner-ous, exploitative work conditions, usually earning below-subsistence wages. As economist Linda Lim (1983: 78) noted in the 1980s, "it is the comparative *disadvantage* of women [emphasis in original] in the wage labour market that gives them a comparative advantage vis-a-vis men in the occupations and industries where they [women] are concentrated." Gender and race ideologies also become the basis for labour control within maquila factories (Hossfeld, 1990). Researchers have documented harsh work conditions, meagre wages, and human rights violations, enabled by the devaluation of women's work and a disciplining of women workers as docile employees with "nimble fingers" and an affinity for manually dexterous work (Cravey, 1998; Tiano, 1994: 208–209; Safa, 1990). Research has exposed speed-ups and firings for pregnancy, pregnancy test-ing as a condition of employment, forced overtime (often without additional pay) within EPZs in Mexico and Central America (Tuttle, 2012; McCallum, 2011; Bickham Mendez, 2005; Peña, 1997), as well as the denial of the right to the freedom of association (Gopalakrishnan, 2007; Rodriguez-Garavito, 2005).

The stimulation of urban migration resulting from the creation of thousands of jobs in maquila factories has also exacerbated poor living conditions and elevated housing costs in bor-der towns along the US-Mexico border and in the *colonias* that sprout outside EPZs in countries like Honduras (Pine, 2008). In some of these residential areas violence against women – par-ticularly, femicides – have reached "pandemic" levels (Matloff, 2015). The most notorious case is that of Ciudad Juárez, a city that became known as the female murder capital of the world in the early 2000s and where since 1993 some 1,500 women, many of them maquila workers, have been disappeared or murdered with their cases still unresolved (Amnesty International, 2003). Researchers who have investigated the Juárez case have highlighted a correlation between trade liberalization, the development of the maquiladora industry, and femicide (Weissman, 2005; Wright, 1999).

Studies published in the 1980 and 1990s ignited an intense debate among researchers about the advantages and disadvantages of maquiladora industries for women's well-being (Lim, 1990; Tiano, 1990; Fernandez-Kelly, 1983). Early feminist research argued that women's gender sub-ordination in the home and family worked in concert with state and work-place policies to reinforce capitalist and patriarchal control over female labour and cement women's super-exploitation. Scholars documented how employers in assembly plants used cultural imagery of women as docile, manually dexterous, and patient to develop an ideology that legitimated women's employment but also defined it as always, already temporary (Carrillo and Hernandez, 1985; Fernandez-Kelly, 1983). Paradoxically, despite the gender ideology of women as tempo-rary, supplemental earners, economic conditions in countries where maquila production grew in importance required them to engage in wage-generating activities in order to meet basic household needs (Tiano, 1994: 104–105). At the same time, other research demonstrated that increased labour participation among women in many cases enabled them to gain bargaining power in the home and challenge the "myth" of the male breadwinner (Safa, 1995).

One critique of the early literature that made claims about women's super-exploitation was that it relied on data collected in the 1970s when maquila-based industrialization was in its ini-tial phases. Researchers who challenged the exploitation thesis highlighted over-generalizations about work conditions across social contexts that implicitly compared wage-levels and work conditions to reference groups in wealthier countries, generating a stereotypical image of women workers as oppressed victims without agency (see Kabeer, 2004). Other researchers cited the heavy reliance on studies of Mexico and a general lack of attention to the ways in which the maquila industry has evolved over time and in ways specific to geography and history (Domínguez et al., 2010; Ver Beek, 2001). In Puerto Rico and Mexico maquila production began as early as the 1950s and 1960s respectively, whereas export-oriented industrialization only became widespread in Central America in the 1980s (with the exception of Nicaragua where the maquila sector did not gain strength until the early 1990s). In Mexico during the debt crisis of the 1980s the demand for maquila workers soared, and factory managers turned to the hire of young men (Salzinger, 1997). As the 1980s continued, facing competition from Central America, the maquila sector in Mexico shifted to more capital-intensive and technically sophisticated production, which corresponded with a significant defeminization of the industry with the proportion of women operatives declining considerably in the 1990s (Fussell, 2000; Tiano, 1994: 24). In addition, as jobs in the maquila began to decline, employers in Mexico also shifted from hiring young, single women to hiring those who were married.

Despite these debates, there is general consensus that labour conditions in maquila factories are substandard in terms of safety and human and labour rights and that, although pay in maquila factories may be higher than other available employment options, wages remain consistently insufficient to cover basic needs and fail to provide benefits necessary for survival, such as adequate health care (Tuttle, 2012; Bandy and Bickham Mendez, 2003). Indeed, female maquila employees often complete not only a "second shift" of unpaid labour at home, but a "triple shift" of economic activity in the informal sector to supplement wages (Ward and Pyle, 1995).

Labour organizing

Efforts to address issues such as low wages, substandard working conditions, and human and labour rights violations have met with formidable challenges, but also defy notions of women workers as a compliant, unorganizable sector. Unionization efforts in the countries where maquilas are located have faced considerable obstacles in the face of highly mobile production arrangements and repressive actions on the part of state-supported management. In the case of countries, like Mexico and Nicaragua, where a strong tradition of trade unionism had existed before export-oriented industrialization took root, unions lost considerable ground as a result of a shift to flexible production strategies and the ability of firms to relocate production in the event of labour unrest (Bandy and Bickham Mendez, 2003).

Despite continuing obstacles, unions have been able to establish a presence in maquila factories and even achieve some victories in an ongoing struggle for workers' rights. For example, in Mexico shortly after the 1985 earthquake in which many garment workers were killed, the Nineteenth of September National Union of Garment workers emerged as an independent union of garment workers comprised solely of women (Tirado, 1994). Their demands included the right to maternity leave and freedom from sexual harassment as well as the suspension of Mexico's payment to the external debt. The union was successful in obtaining compensation for 8,000 women workers who had lost their employment during the earthquake (Tirado, 1994: 108).

Likewise, maquila factories in Matamoros, Mexico, have enjoyed a significant union presence, since the 1970s, and women workers have been active participants in the Trade Union of Laborers and Industrial Workers and of the Maquiladora Industry (SJOIIM) (Domínguez and Quintero, 2012). And while women have not tended to occupy the highest positions within the union leadership, they were important supporters of strikes in the 1970s and 1980s and key actors in achieving collective bargaining agreements that included clauses such as the protection of maternity leave and of pregnant women from firings (Dominguez and Quintero, 2012). Some Central American unions report high numbers of active female membership, and research has documented cases in which unions seem to be integrating a gender perspective into their organizing efforts (Dominguez et al., 2010).

Such victories are frequently short-lived, however, and struggles for safe and fair labour conditions are ongoing. More recently, especially in Central America and the Caribbean where most maquila factories are devoted to apparel production, organizing efforts have also faced challenges resulting from a shifting regulatory and economic landscape that has affected the ability of firms to remain competitive in the global economy, accentuating the threat of factory closure. Of particular note is the "shock" of the phase-out of textile quotas and preferential tariffs under the Multi-fiber Arrangement, which took place between 1995 and 2005 and which facilitated the expansion of low-cost suppliers in China, threatening the competitiveness of firms in Central America (Gereffi and Bair, 2010). The maquila industry was further impacted by the global economic recession of 2008, which also resulted in the closure of numerous factories in Central America (REDCAM, 2010).

Beyond the challenges presented by management's resistance and conditions in the global economy, another significant issue pertains to gender dynamics within unions. Industrial unions in countries where maquilas are located have been largely male-dominated and have been reluctant to take-up the concerns of women workers, which include reproductive rights and health, childcare, and sexual harassment. In some cases in unions that have been able to establish an active presence in maquilas, women organizers have experienced paternalism, gendered divisions of labour in activism, and resentment of women's public voice and leadership (Bandy and Bickham Mendez, 2003; Needleman, 1998: 153–154); Indeed, in Central America in the 1990s hierarchical decision-making and cultures of machismo on the part of male unionists resulted in women organizers' leaving unions to form their own autonomous spaces "by and for women" (Bickham Mendez, 2005, 2002).

Women's activism and organizing strategies

Indeed, women's activism surrounding issues related to the maquilas has extended well beyond traditional unions to community organizations, grassroots associations, and NGOs. Women in communities where factories are located have organized to address issues of health, education, as well as violence against women. For example, along the US-Mexican border networks of community-based women's organizations have promoted economic justice and human rights using a variety of tactics, including legislative and legal strategies as well as direct action (Téllez and Sanidad, 2015). In many cases community-based women's groups have secured essential services in their *colonias*, educated women workers about their labour rights, and provided support and training about gender violence and sexual abuse (Bergareche, 2006; Coronado, 2006; Bickham Mendez, 2002). Women's activism has also taken forms that transcend rigid definitions of NGOs and labour and community organizations (Domínguez and Quintero, 2012). For example, in El Salvador, Guatemala, and Nicaragua women's organizations have either collaborated with unions or have incorporated maquila workers or ex-maquila workers directly into their leadership (Domínguez and Quintero, 2012; Bickham Mendez, 2005).

In addition to organizing at the community level, women workers' organizations have formed coalitions that span national borders. For example, in the 1990s women workers on both sides of the US-Mexican border and in Canada joined together in the struggle for social justice in the face of the North American Free Trade Agreement (NAFTA) (Tuttle, 2012; Bandy, 2004; Domínguez, 2002). Central American women's labour organizations formed regional coalitions to seek to improve conditions in maquila factories in Nicaragua, El Salvador, Guatemala, and Honduras (Bickham Mendez, 2002). And by the early 2000s women from Central America, Mexico, and Asia began to come together to hold dialogues and compare strategies for fighting abuses against women workers (Bickham Mendez, 2005).

Transnational organizing efforts

Transnational efforts have also involved collaborations with labour organizations and other groups in the Global North to address violations of workers' rights, especially in a context in which states have been reluctant or lack the capacity to enforce existing labour laws. Such cross-border organizing efforts raise the question of who should be held accountable for enforcing fair labour standards in maquila factories. For example, since 1989, various groups on both sides of the US-Mexico border – including a number of US-based unions – have organized as a transnational network known as the Coalición para Justicia en las Maquilas (CJM) (Tuttle, 2012; Bandy, 2004). This coalition or its organizational members have turned to filing complaints with NAFTA's National Administrative Offices (NAOs), charging violations of the North American Agreement on Labor Cooperation (NAALC) (Bandy, 2004). In some cases, these coalitional efforts were able to achieve precedent-setting rulings and settlements that resulted in environmental clean-up, the recuperation of workers' jobs and severance pay, and even the establishment of independent unions (Tuttle, 2012: 193; Bandy, 2004: 416). At the same time, these successes have been modest in the face of repressive actions on the part of factory owners. And the NAALC's record of implementation and effectiveness has been extremely weak (Tuttle, 2012; Domínguez et al., 2010).

Beginning in the mid-1990s, Central America women workers' organizations from four countries joined together to form a regional coalition in an effort to pressure maquila factories across the region to improve conditions and agree to uphold workers' human rights and comply with local labour laws (Bickham Mendez, 2002). Supported financially by international NGOs, the primary strategy of this coalition was to launch regional campaigns for factory owners to sign a "code of ethics." These efforts were successful in achieving some legislative goals such as the institution of minimum wage standards (Bickham Mendez, 2005). The 2000s saw massive closures of maquila factories in Honduras, El Salvador, Guatemala, and Nicaragua, and in the 2010's this network continued to hold regional meetings to raise awareness about the effects of the global economic crisis on the lives and well-being of women workers (REDCAM, 2010).

Other transnational efforts have included corporate campaigns waged by transnational advocacy networks (TANs) to pressure firms to adopt corporate codes of conduct agreeing to adhere to fair labour standards. And in the 1990s transnational campaigns attracted considerable media attention by targeting well-known brands such as the GAP and Kathie Lee Gifford's line of clothing for Wal-Mart. While TAN's were successful in pressuring a number of companies and firms to adopt codes of conduct or allow a union presence, a number of cases also highlighted contradictions within campaigns and their limitations for achieving sustained improvements in working conditions and the protection of workers' rights. For example, in 1995 international pressure waged against the GAP resulted in the successful establishment of a union in Mandarin International, a factory that operated in El Salvador. Although the independent trade union survived for some time, it was never able to achieve a collective contract. Worse still, after media attention faded, the factory closed its doors, leaving workers jobless (Armbruster-Sandoval, 2005: 75–80).

Corporate codes of conduct sparked contentious debates in the 1990s and early 2000s (Armbruster-Sandoval, 2005). Critics note that once a code is signed, ongoing monitoring may not occur or involves commercial monitors whose effectiveness is questionable. Alternatively, changes may only be put in place just before an inspection by third-party monitors. In plants with hundreds of employees, workers may be unaware of the existence of codes. And researchers, like Barrientos (2007), point out that such codes tend to be more effective at addressing outcomes, like wage-levels, onerous work hours, and health standards but less so when it comes to collective bargaining rights and freedom from discrimination (Domínguez and Quintero, 2012). Other researchers argue that different kinds of campaigns have a differential impact on enforcing labour rights, citing the potential of codes of conduct for checking the power of transnational corporations when campaigns successfully deploy a combination of external pressures from TAN's and the local empowerment of NGOs, democratic unions, and labour support organizations (See Domínguez et al., 2010; Rodríguez-Garavito, 2005: 224).

On the other hand, deep-seated asymmetries characterize relations *within* transnational networks that form to pressure maquila factories to improve standards, and transnational allies may not hold the same vision and goals as groups on the ground. Tensions that emerge within cross-border networks reveal how power differentials that relate to gender, along with relations of race, ethnicity, class, and nation can "fracture the space of transnational civil society and constrain opposition to neoliberalism" (Bandy and Bickham Mendez, 2003: 174).

Opportunities for future research

A review of the literature on export-oriented production reveals the need for renewed scholarly attention to labour conditions and organizing efforts in maquila factories in the region as well as numerous opportunities for future study. Indeed, there is a paucity of high quality, comparable data on working conditions in EPZs worldwide (Milberg and Amengual, 2008). Research from the first few years of the 2000s suggest that the Central American Free Trade Agreement (CAFTA) has affected labour conditions as well as levels of production of maquila factories in Central America and the Dominican Republic, but there is a lack of research that documents working conditions and labour organizing in the context of this free trade agreement. Likewise, we know little about the ongoing impact of the global economic crisis on the maquila industry and workers. The US presidential election of Donald Trump in 2016 will also no doubt have an impact on maquila production in Latin America, calling for new research in this area. The Trump administration has advocated for the US to exit from NAFTA. The renegotiation of this trade agreement, if it occurs, will almost certainly exert a dramatic effect on maquila production and labour conditions in Mexico.

Additionally, a recent trend within the international division of labour of today's global economy that calls out for further investigation is the embodied labour of service work, which in the 21st century has "gone global" through the international off-shoring of call centres. While the labour of the maquila worker is largely invisible to the consumer who purchases a product assembled far from his or her homeland, in the case of international call centres, consumer and labourer interact directly through voice-based service encounters. Firms use satellite technology to divert calls across national lines and back to consumers, often without any indication that the call involves an international connection. Researchers have noted that it is through such transnational interactions – a significant aspect of the "product" delivered to consumers – that understandings of nationhood and citizenship are both constructed and expressed (Mirchandani and Poster, 2016). While off-shoring of both call centres and data entry has existed since the beginning of the 2000s in India (Poster, 2007) and the English-speaking Caribbean (Freeman, 2000), it is only more recently that US firms have begun to locate call centres in Latin America. And notably, these centres are now situated in many of the same countries that have been sites with a significant presence of EPZ's – for example, in the Dominican Republic (Rodkey, 2016), Mexico (Anderson, 2013), El Salvador (Rivas, 2016), and Guatemala (Meoño Artiga, 2016). In the case of international call centres, by off-shoring the production process, companies take advantage not only of inexpensive labour costs, but of an English-speaking workforce that is intimately familiar with US culture.

This new employment sector brings together diverse groups of workers. On the one hand the labour force often consists of bilingual, upper-class university students. However, increasingly, recently deported individuals who have lived in the United States for the majority of their lives have emerged as a significant sector of the labour force in these centres (Meoño Artiga, 2016). With limited opportunities to integrate into either the society or the labour market of their countries of origin, these "repurposed" workers provide call services for a considerably lower wage than would be paid to a US-based labour force. While the maquila industry has relied on women as the preferred labour force, deportees who work in call centres in Latin America are often male – forcibly returned migrants whose criminalization and often racialization follows them back to their country of origin.

As the 21st century progresses and continues to be characterized by permeable borders for capital, but increasingly fortified, closed territorial lines for labourers, it will become increasingly important to investigate work conditions within a shifting international division of labour. In particular, the recent and growing political dominance of neoconservativism, blended with ethnic-nationalism, and trade protectionism in the US and Europe will no doubt exert a considerable impact on globalized production schemes and labour in Latin America and is deserving of further scholarly attention.

References

Amnesty International. (2003) *Intolerable Killings: Ten Years of Abductions and Murders of Women in Ciudad Juarez and Chihuahua.* http://web.amnesty.org/library/index/engamr410262003 (Accessed 29 April 2018).

Anderson, J. (2013) From U.S. immigration detention center to transnational call center. *Voices of Mexico* (95): 87–91.

Armbruster-Sandoval, R. (2005) *Globalization and Cross-Border Labor Solidarity in the Americas: The Anti-Sweatshop Movement and the Struggle for Social Justice.* New York: Routledge.

Bandy, J. (2004) Paradoxes of transnational civil society: The Coalition for Justice in the Maquiladoras and the challenges of coalition. *Social Problems* 51(3): 410–431.

———— and Bickham Mendez, J. (2003) A place of their own? Women organizers negotiating the local and transnational in the maquilas of Nicaragua and Northern Mexico. *Mobilization* 8(2): 173–188.

Barrientos, S. (2007) Do workers benefit from ethical trade? Assessing codes of labor practice in global production systems. *Third World Quarterly* 28(4): 713–729.

Bergareche, A. (2006) The roots of autonomy through work participation in the Northern Mexico border region. In D. J. Mattingly and E. R. Hansen (eds) *Women and Change at the U.S. – Mexico Border: Mobility, Labor, and Activism.* Tucson, AZ: University of Arizona Press, pp. 91–102.

Bickham Mendez, J. (2002) Creating alternatives from a gender perspective: Central American women's transnational organizing for maquila workers' rights. In N. A. Naples and M. Desai (eds) *Women's Activism and Globalization: Linking Local Struggles and Transnational Politics.* New York: Routledge, pp. 121–141.

————. (2005) *From the Revolution to the Maquiladoras: Gender, Labor and Globalization in Nicaragua.* Durham, NC: Duke University Press.

Brecher, J., Costello, T. and Smith, B. (2000) *Globalization from Below: The Power of Solidarity.* Cambridge, MA: South End Press.

Carrillo, J. and Hernández, A. (1985) *Mujeres Fronterizas en la Industria Maquiladora.* Tijuana: Centro de Estudios Fronterizos del Norte de México.

Coronado, I. (2006) Styles, strategies and issues of women leaders at the border. In D. Mattingly and E. Hansen (eds) *Women and Change at the U.S.–Mexico Border: Mobility, Labor and Activism.* Tucson: University of Arizona Press, pp. 142–158.

Cravey, A. J. (1998) *Women and Work in Mexico's Maquiladoras.* Boulder, CO: Rowman & Littlefield.

Deere, C. D. and Antrobus, P. (1990) *In the Shadows of the Sun: Caribbean Development Alternatives and U.S. Policy.* Boulder, CO: Westview Press.

De Hoyos, R. E., Bussolo, M. and Núñez, O. (2012) Exports, gender wage gaps, and poverty in Honduras. *Oxford Development Studies* 40(4): 533–551.

Domínguez, E. R. (2002) Continental transnational activism and women workers' networks within NAFTA. *International Feminist Journal of Politics* 4(2): 216–39.

———, Icaza, R., Quintero, C., Lopez, S. and Stenman, A. (2010) Women workers in the maquiladoras and the debate on global labor standards. *Feminist Economics* 16(4): 185–209.

Domínguez, E. R. and Quintero, C. (2012) Labor organizing among women workers in maquiladoras: Crossing the border of gender and class in the cases of Matamoros, Mexico and San Marcos, El Salvador. *Bodies and Borders in Latin America*. Serie HAINA VIII:77–95.

The Economist (2015) Special economic zones: Not so special, 4 April.

Fernandez-Kelly, M.P. (1983) *For We Are Sold: I and My People*. Albany: SUNY Press.

Freeman, C. (2000) *High Tech and High Heels in the Global Economy: Women, Work, and Pink Collar Identities in the Caribbean*. Durham, NC: Duke University Press.

Fussell, E. (2000) Making labor flexible: The recomposition of Tijuana's maquiladora female labor force. *Feminist Economics* 6(3): 59–79.

Gereffi, G. and Bair, J. (2010) *Strengthening Nicaragua's Position in the Textile-Apparel Value Chain: Upgrading in the Context of the CAFTA-DR Region*. Durham, NC: Center on Globalization, Governance & Competitiveness (CGGC).

Gopalakrishnan, R. (2007) *Freedom of Association and Collective Bargaining in Export Processing Zones: Role of the ILO Supervisory Bodies*. Geneva: International Labor Organization.

International Labor Organization. (2014) *Trade Union Manual on Export-Processing Zones*. Geneva: ACTRAVE Bureau of Workers' Activities. www.ilo.org/wcmsp5/groups/public/@ed_dialogue/@actrav/documents/publication/wcms_324632.pdf (Accessed 10 May 2018).

———. (2016) *Women at Work: Trends 2016*. Geneva: International Labor Organization. www.ilo.org/wcmsp5/groups/public/–dgreports/–dcomm/–publ/documents/publication/wcms457317.pdf (Accessed 23 April 2018).

Kabeer, N. (2004) Globalization, labor standards, and women's rights: Dilemmas of collective (in)action in an interdependent world. *Feminist Economics* 10(1): 3–35.

Lim, L. Y. C. (1983) Capitalism, imperialism, and patriarchy: The dilemma of third-world women workers in multinational factories. In J. C. Nash and M. P. Fernández-Kelly (eds) *Women, Men, and the International Division of Labor*. Albany, NY: SUNY Press, pp. 70–92.

———. (1990) Women's work in export factories: The politics of a cause. In I. Tinker (ed) *Persistent Inequalities*. New York: Oxford University Press, pp. 101–119.

Matloff, J. (2015) Six women murdered each day as femicide in Mexico nears a pandemic. *Aljazeera America*. 4 January. http://america.aljazeera.com/multimedia/2015/1/mexico-s-pandemicfemicides.html (Accessed 15 May 2018).

McCallum, J. K. (2011) *Export Processing zones: Comparative Data from China, Honduras, Nicaragua and South Africa*. Working Paper No. 21. Geneva: International Labor Organization.

Meoño Artiga, L. P. (2016) Transnational 'homies' and the urban middle class: Enactments of class, nation and modernity in Guatemalan call centres. In K. Mirchandani and W. Poster (eds) *Borders in Service: Enactments of Nationhood in Transnational Call Centres*. Toronto, ON: University of Toronto Press, pp. 152–180.

Milberg, W. and Amengual, M. (2008) *Economic Development and Working Conditions in Export Processing Zones: A Survey of Trends*. Geneva: International Labor Organization.

Mirchandani, K. and Poster, W. R. (2016) (eds) *Borders in Service: Enactments of Nationhood in Transnational Call Centres*. Toronto, ON: University of Toronto Press.

Otobe, N. (2015) *Export-Led Development, Employment and Gender in the Era of Globalization*. Employment Policy Department Employment Working Paper No. 197. Geneva: International Labor Organization. www.ilo.org/wcmsp5/groups/public/-edemp/documents/publication/wcms452341.pdf (Accessed 21 April 2018).

Peña, D. (1997) The Terror of the Machine: Technology, Work, Gender, and Ecology on the U.S.-Mexico Border. Austin, TX: Texas University Press.

Pine, A. (2008) *Working Hard, Drinking Hard: On Violence and Survival in Honduras*. Berkeley, CA: University of California Press.

Poster, W. R. (2007) Who's on the line? Indian call center agents pose as Americans for U.S.-outsourced firms. *Industrial Relations: A Journal of Economy and Society* 46: 271–304.

REDCAM (Red Centroamericana de Mujeres en Solidaridad con las Trabajadoras de las Maquilas) (2010) Impacto de la crisis económica en la vida de las mujeres. *Boletina* 2: 3–10. www.mec.org.ni/?mcportfolio=boletin-redcam-2010#11 (Accessed 15 May 2018).

Rivas, C. (2016) El Salvador works: The creation and negotiation of a national brand and the transnational imaginary. In K. Mirchandani and W. R. Poster (eds) *Borders in Service: Enactments of Nationhood in Transnational Call Centres*. Toronto, ON: University of Toronto Press, pp. 35–58.

Robinson, W. I. (2003) *Transnational Conflicts: Central America, Social Change and Globalization*. New York: Verso.

Rodkey, E. (2016) Disposable labor, repurposed: Outsourcing deportees in the call center industry. *Anthropology of Work Review*. 27(1): 34–43.

Rodríguez-Garavito, C. A. (2005) Global governance and labor rights: Codes of conduct and anti-sweatshop struggles in global apparel factories in Mexico and Guatemala. *Politics and Society* 33(2): 203–333.

Safa, H. I. (1990) Women and industrialization in the Caribbean. In S. Stichter and J. L. Parpart (eds) *Women, Employment, and the Family in the International Division of Labor*. Philadelphia, PA: Temple University Press, pp. 72–97.

———. (1994) *The Myth of the Male Breadwinner: Women and Industrialization in the Caribbean*. Boulder, CO: Westview Press.

———. (1995) Economic restructuring and gender subordination. *Latin American Perspectives* 22(2): 32–50.

———. (1998) Free markets and the marriage market: Structural adjustment, gender relations, and working conditions among Dominican women workers. *Environment and Planning* 31: 291–304.

Salzinger, L. (1997) From high heels to swathed bodies: Gendered meanings under production in Mexico's export-processing industry. *Feminist Studies* 23(3):549–574.

Téllez, M. and Sanidad, C. (2015) "Giving wings to our dreams": Binational activism and workers' rights struggles in the San Diego – Tijuana border region. In N. Naples and J. Bickham Mendez (eds) *Border Politics: Social Movements, Collective Identities and Globalization*. New York: New York University Press, pp. 323–356.

Tiano, S. (1990) "Maquiladora women: A new category of workers?" In K. Ward (ed) *Women Workers and Global Restructuring*. Ithaca, NY: ILR Press, 193–294.

———. (1994) *Patriarchy on the Line: Labor, Gender and Ideology in the Mexican Maquila Industry*. Philadelphia, PA: Temple University Press

———. (2006) The changing composition of the maquiladora workforce along the US–Mexico Border. In D. Mattingly and E. Hansen (eds) *Women and Change at the U.S. – Mexico Border: Mobility, Labor and Activism*. Tucson: University of Arizona Press, pp. 73–90.

Tirado, S. (1994) Weaving dreams, constructing realities: The Nineteenth of September National Union of Garment Workers in Mexico. In S. Rowbotham and S. Mitter (eds) *Dignity and Daily Bread: New Forms of Economic Organizing Among Poor Women in the Third World and First*. New York: Routledge, pp. 100–113.

Tuttle, C. (2012) *Mexican Women in American Factories: Free Trade And Exploitation on the Border*. Austin: University of Texas Press.

Ver Beek, K. (2001) Maquiladoras: exploitation or emancipation? an overview of the situation of maquiladora workers in Honduras. *World Development* 29(9): 1553–1567.

Ward, K. B. and Pyle, J. L. (1995) Gender, industrialization, transnational corporations, and development: An overview of trends and patterns. In C. E. Bose and E. Acosta-Belén (eds) *Women in the Latin American Development Process*. Philadelphia, PA: Temple University Press, pp. 37–64.

Weissman, D. M. (2005) The political economy of violence: Towards an understanding of gender-based murders of Ciudad Juarez. *North Carolina Journal of International Law and Commercial Regulation* 30(4): 795–866.

Wright, M. (1999) The dialectics of still life: Murder, women, and maquiladoras. *Public Culture* 11: 453–474.

32

FAIRTRADE CERTIFICATION IN LATIN AMERICA

Challenges and prospects for fostering development

Laura T. Raynolds and Nefratiri Weeks

Introduction

Originating in the Latin American coffee sector, fair trade has emerged in recent decades as an important initiative seeking to harness global markets to foster empowerment-based development. This movement strives to challenge exploitative trade relations rooted in the history of colonialism and promote social justice and environmental sustainability. Fair trade seeks to foster the wellbeing and empowerment of producers in the Global South through the provision of higher prices, stable market links, better working conditions, and community development resources. In the Global North, fair trade seeks to promote responsible consumption and provide shoppers with socially and environmentally friendly products. The non-governmental organization (NGO) Fairtrade International now certifies 20 commodities with global sales worth US$ 8 billion a year (FTI, 2017). Nearly two million producers participate in Fairtrade commodity networks, with Latin America and the Caribbean remaining a central locus of production (FTI, 2017).[1]

This chapter analyzes Fairtrade International certification and its efforts to foster development in Latin America and the Caribbean. As we demonstrate Fairtrade can promote wellbeing and empowerment in the short-term by guaranteeing favourable commodity prices to farmers and more decent work for hired labourers, as well as a social premium to support community development. Perhaps more significantly, Fairtrade can foster development in the long term through the capacity building support it provides to producer groups. While Latin America exports a range of certified products, coffee remains the focus of Fairtrade production in the region. Recent collaborations between Fairtrade and fair credit institutions and the newly established Fairtrade Access Fund have played key roles in supporting cooperatives in the region, helping to fuel long-term development goals. Although Fairtrade creates important avenues for fostering development, it cannot, on its own, lift impoverished farmers and workers out of poverty, satisfy all the commercial needs of producers, or provide the myriad social services often lacking in rural areas of Latin America.

Fair trade institutions and markets

Fair trade challenges historically unequal international trade relations and seeks to foster more equitable relations by linking marginalized producers in the Global South with progressive consumers in the Global North. This initiative works to transform North/South trade, creating more egalitarian commodity networks particularly for colonial-based agro-exports, like coffee, cocoa, tea, and sugar, which historically forged global divides. Fair trade fosters fairness in the Global South, by alleviating poverty and empowering farmers and workers, and in the Global North, by fuelling ethical consumption and the availability of socially responsible products. While fair trade's critique is akin to that of other social justice movements, what accounts for its success is the linking of visionary goals with practical engagements in creating ethical markets (Raynolds and Bennett, 2015; Raynolds, Murray, and Wilkinson, 2007).

Given its critique of the exploitative nature of capitalist relations, fair trade represents a Polanyian countermovement of social protection (Raynolds, 2002). In the words of its major proponents (FINE, 2001):

> Fair Trade is a trading partnership, based on dialogue, transparency and respect, that seeks greater equity in international trade. It contributes to sustainable development by offering better trading conditions to, and securing the rights of, marginalized producers and workers – especially in the South. Fair Trade organisations (backed by consumers) are engaged actively in supporting producers, awareness raising and in campaigning for changes in the rules and practice of conventional international trade.

As suggested here, fair trade seeks to fuel sustainable development by appreciating the full social and ecological value of products and redefining international trade around partnership principles (Raynolds, 2000).

A key contradiction for fair trade is that it operates "in and against the market," working through market channels to create new commodity networks for items produced under more favourable social and ecological conditions and simultaneously working against the conventional market forces that create and uphold global inequalities (Raynolds, 2000, 2002). The ongoing contestation between movement principles and market imperatives has since the outset generated substantial controversy. The nature and dynamics of fair trade's institutional growth reflects the negotiated outcomes of these contestations (Raynolds, 2012a). Fair trade ideas and practices have been institutionalized in two key ways: in the original yet now smaller fair trade organization model and the younger yet now dominant Fairtrade certification model.

NGOs and faith-based development groups established fair trade, purchasing handicrafts from producers in the Global South at favourable prices to sell to ethically minded Northern consumers. These pioneers created direct trade channels, cutting out intermediaries to increase producer returns. By infusing commodities with information regarding the people and places involved in production, fair trade worked to "humanize" economic transactions and "shorten the distance" between producers and consumers (Raynolds, 2000). Solidarity groups working in Mexico and Europe extended this direct trade model to coffee. Informed by political economy views of the role of unequal exchange in fuelling global poverty, these groups sought to expand fair trade in major agro-export arenas (Renard, 1999, 2003).

What is now the dominant fair trade model was established in the late 1980s based on the certification and labelling of coffee, and soon thereafter other food products, for sale in mainstream supermarkets (Raynolds, Murray, and Wilkinson, 2007). Fairtrade International (previously called Fairtrade Labelling Organizations International) was founded to coordinate this certification system, which now includes a set of distinct entities. Fairtrade International establishes standards for producers, who must be organized into democratic associations, uphold International Labor Organization standards, and promote ecological practices, and buyers/importers, who must pay minimum prices and a social premium and offer producer credit and long-term contracts. An autonomous NGO, FLO-Cert, audits for compliance. Twenty-two National Labeling Initiative affiliates promote sales of labelled products in market countries. Three regional Fairtrade Producer Networks represent and support participating producers in Latin America, Africa, and Asia (FTI, 2017).

By translating movement principles into formal bureaucratic institutions and rational rules and positioning products in mainstream markets, certification has spurred fair trade's commercial growth (Raynolds, 2017b). While some criticize Fairtrade for appropriating cultural symbols for commercial purposes, what Varul (2008) calls "consuming the campesino," certification has bred market success. Fairtrade certified commodity sales are now worth US$ 8 billion a year (FTI, 2017). Over 85% of consumers in the United Kingdom, Switzerland, and the Netherlands report familiarity with the Fairtrade label (FLO and Globespan, 2011). As noted in Table 32.1, the United Kingdom has the world's largest Fairtrade market, with retail sales near US$ 3 billion. Germany, Switzerland, France, Sweden, and the Netherlands also have large and well-developed markets. Fairtrade certified products are readily available in European supermarkets, speciality stores, universities, and workplaces. The US market share of Fairtrade products has recently fallen due to Fairtrade International's National Labeling Initiative affiliate, Fair Trade USA's, resignation from the international system and its replacement by Fairtrade America. Despite this reorganization, the United States had the third largest Fairtrade International sales in 2014, with earnings valued at US$ 766 million.

Fairtrade International certifies 20 commodities: coffee, tea, bananas, cocoa, sugar, honey, citrus, pineapples, mangoes, apples, grapes, rice, quinoa, spices, fruit juices, flowers, wine, cotton, sports balls, and gold. The vast majority of Fairtrade revenues are derived from coffee, tea, bananas, cocoa, and sugar. As noted in Table 32.2, these key products have all seen tremendous growth in recent years. Coffee, the first labelled commodity, remains the most valuable product

Table 32.1 Fairtrade International certified sales value in lead countries (US $1,000,000)[a]

	2004	2006	2008	2010	2012	2014
Europe						
UK	256	514	1297	1782	2449	2762
Germany	72	138	313	451	685	1104
France	87	209	376	402	444	491
Switzerland	169	179	249	292	401	511
Netherlands	43	52	90	158	239	286
Sweden	7	20	107	144	230	401
North America						
USA	267	627	1116	1243	1436[b]	766[c]
Canada	22	68	189	259	235	301
Pacific						
Australia/ NZ	1	9	27	167	242	281
Total[d]	$1034	$2039	$4262	$5727	$6155	$8311

Notes
a Euros converted to dollars using the US Federal Reserve average annual exchange rate.
b Figure is for 2011, the last year Fair Trade USA is included as the Fairtrade International affiliate.
c Figure includes only Fairtrade International sales labelled by its new affiliate, Fairtrade America, not by Fair Trade USA.
d Total includes countries not listed on the chart.

Sources: Compiled by the authors from FLO (2006, 2007, 2009, 2012, 2014, 2015).

Table 32.2 Top Fairtrade International labelled commodities by volume (metric tonnes)

	2004	2006	2008	2010	2012[a]	2014[a]
Bananas	80,640	135,763	299,205	294,447	331,980	439,474
Sugar	1,960	7,159	56,990	126,810	158,986	196,361
Coffee	24,222	52,064	65,808	87,576	77,429[b]	93,154[b]
Cocoa	4,201	7,913	10,299	35,179	40,559	65,086
Tea	1,965	3,883	11,467	12,356	11,869	11,030
Total[c]	126,160	217,628	505,152	585,772	655,068	840,653

Notes

a This column includes only Fairtrade International sales labelled by its new affiliate, Fairtrade America, not by Fair Trade USA.

b Although figures for prior years are for roasted coffee, this is for green coffee beans.

c Includes other labelled commodities measured by weight (e.g. cotton and quinoa), but not those measured by item or other volume measurements (e.g. flowers, sports balls, fruit juice, wine, and beer).

Sources: Compiled by the authors from FLO (2006, 2007, 2009, 2012, 2014, 2015).

and the backbone of the Fairtrade certification system. Rising Fairtrade coffee consumption is linked to the rise in speciality coffee markets. Cocoa and tea are well-established Fairtrade products with substantial markets particularly in Europe. Fairtrade sugar markets have recently surged with growing supermarket availability in many countries. Although bananas were incorporated into Fairtrade more recently than other major commodities, this product is now the most important Fairtrade product, measured by volume, and the second most important in value (Raynolds and Greenfield, 2015).

Recent organizational changes in the Fairtrade International system reflect the tensions between market and movement priorities. Prioritizing market principles, Fair Trade USA left the international system to maximize sales by certifying farmers who are not democratically organized and plantations in sectors like coffee that Fairtrade limits to small producers (Raynolds, 2012a). The new Fairtrade International affiliate, Fairtrade America, rebalances market and movement priorities. While these institutional changes have been disruptive, they have created openings for reasserting the movement's equity principles. Significantly, Fairtrade Producer Networks have recently gained equal representation with Labelling Initiatives in the governance of Fairtrade International (Raynolds, 2017a).

Fairtrade certified production and development in Latin America

Fairtrade certification was conceived by its Mexican coffee cooperative and European solidarity group founders as an avenue for enhancing development through "trade not aid." Fairtrade emerged during the late 1980s when neoliberal and structural adjustment policies were adopted across Latin America and the Caribbean, heightening competition and undercutting state support for agricultural producers. The 1989 drop in world prices aggravated this situation in coffee. The Mexican coffee cooperatives that co-founded Fairtrade saw this new initiative as a way to survive the market collapse and chart their own development (Mutersbaugh, 2002; Renard, 1999). While movement founders envisioned Fairtrade within the context of solidarity and partnership, its operations were anchored within colonial-based trade relations based on shipments of agro-exports from the Global South for consumption in the Global North. The inherent contradictions in this strategy of working "in and against" the market have shaped Fairtrade's trajectory, controversies, and development implications (Raynolds, 2002, 2012a).

Fairtrade agro-export certification has grown dramatically over the past 25 years. While production has expanded across the Global South, Latin America and the Caribbean remain a central locus of production since it is the world's major supplier of high quality coffee as well as other key agro-exports. The region continues to supply most of the world's Fairtrade certified coffee and bananas as well as much of its sugar and cocoa. Over half of Fairtrade International's affiliated producer organizations are in Latin America. There are 647 Fairtrade affiliated groups in 24 countries in the region producing certified agro-export commodities (FTI, 2016: 155–160). Although Mexican coffee associations dominated Fairtrade production in the continent, and indeed the world, until 2000 (Raynolds, 2009), certification in coffee and other products has spread substantially in recent years. As outlined in Table 32.3, Peru now leads the region in Fairtrade sales with annual revenues of US$ 234 million, followed by Colombia, the Dominican Republic, Mexico, and Honduras (in that order). Certified commodities represent an important segment of the export portfolios of several countries in the region, particularly for smaller economies like the Dominican Republic and Honduras (Raynolds, 2008). Peru (with 150) and Colombia (with 112) have the most Fairtrade certified producer organizations in the region and the world (FTI, 2016: 39).

Table 32.4 depicts the characteristics of Fairtrade producer associations in Latin America and the Caribbean's key certified agro-exports. There are 330 Fairtrade certified coffee organizations, 120 certified banana associations, and numerous groups cultivating cocoa, sugar, and flowers. Together these organizations form the Fairtrade Producer Network, the Coordinadora

Table 32.3 Characteristics of the top five Fairtrade producer countries (by sales) in Latin America and the Caribbean

	Fairtrade Sales Revenue (US $1,000ᵃ)	*Certified Producer Organizations*	*Participating Farmers & Workers*
Peru	233,800	150	65,400
Colombia	119,800	112	57,800
Dominican Republic	111,700	42	21,500
Mexico	69,100	63	38,600
Honduras	68,600	25	Not Available

Note
a Amounts converted using the 2015 Euro to Dollar exchange rate: 1.3297.

Source: Compiled by the authors from FTI (2016).

Table 32.4 Fairtrade International production in Latin America and the Caribbean (2014)

	Coffee	*Bananas*	*Cocoa*	*Sugar*	*Flowers*	*Total*ᵃ
Farmer Coops	330	69	58	46	–	572
Farmers	210,200	11,600	36,400	31,600	–	316,100
Worker Organizations	–	51	–	–	11	75
Workers	–	5,800	–	–	2,300	12,600
Total Participating Groups	330	120	58	46	11	647
Total Participating Farmers & Workers	210,200	17,400	36,400	31,600	2,300	328,700

Note
a Total includes additional products like fresh vegetables, dried fruit, and fruit juices.

Source: Compiled by the authors from FTI (2016).

Latinoamericana y del Caribe de Pequeños Productores y Trabajadores de Comercio Justo (CLAC), a strong advocate for social justice in the region and for peasant interests within Fairtrade (Renard, 2015). Fairtrade production originally involved only peasant cooperatives, and cooperatives still account for the majority of certified groups in Latin America. Yet despite strong opposition from CLAC and peasant producers, plantations have been integrated in most Fairtrade commodities to help supply growing markets and extend benefits to workers (Raynolds, 2017a). Large enterprises are excluded from Fairtrade coffee, cocoa, and sugar where smallholders are able to meet global demand, but in bananas and flowers, plantations now outnumber and out-produce cooperatives (Raynolds, 2017a). Fairtrade production in Latin America has grown in both sectors, with 572 certified cooperatives representing 316,100 farmers and 75 plantations employing 12,600 workers operating in the region (FTI, 2016: 155–160).

Fairtrade's development impacts vary by commodity and type of producer organization (Raynolds and Bennett, 2015).[2] Latin America leads the world in Fairtrade coffee production, with 75% of all exports sourced primarily from Colombia, Peru, and Mexico. The majority of the region's producer associations and farmers are engaged in coffee (FTI, 2016: 72–79). There are fewer Fairtrade certified cocoa cooperatives and farmers, located largely in the Dominican Republic and Peru, which face many of the same conditions as their coffee counterparts (FTI, 2016: 89–101).

Operating within the market, Fairtrade International seeks to spur small farmer development in two key ways. (1) Fairtrade requires that buyers pay minimum prices that are intended to guarantee producers a decent return and shore up prices when markets collapse, as they often do in agro-exports. A meta-analysis of Fairtrade impacts finds that peasant producers receive a clear income benefit from these price guarantees (Nelson and Pound, 2009), yet gains are often modest, and some question whether Fairtrade prices can provide Latin American coffee producers a decent standard of living (Jaffee, 2007; Renard, 2015). The weaknesses in Fairtrade's price interventions reflect the challenges of working within the market (Raynolds, 2009), and scholars concur that Fairtrade on its own cannot pull producers out of poverty (Raynolds and Bennett, 2015). (2) Fairtrade International seeks to fuel producer development by providing more favourable and stable market conditions. Research finds that Fairtrade pioneers and mission-driven buyers promote relational and civic ties with producers, building partnerships through personal interactions, technical support, and long-term commitments (Raynolds, 2009; Renard, 2003). Yet as Fairtrade markets have grown and become mainstream, corporate buyers have reasserted industrial and commercial conventions, using their buyer-power to fuel price competition and certification to foster bureaucratic control (Mutersbaugh, 2005; Raynolds, 2009; Renard, 2015).

Operating beyond the market, Fairtrade International requires that buyers pay a social premium that allows cooperatives to invest in development. In 2014 Latin American and Caribbean cooperatives received US$ 86.3 million in Fairtrade premiums (FTI, 2016: 157), which according to certification rules were allocated through a vote of coop members. In coffee, 44% of the Fairtrade premium was invested in coop infrastructure, 24% in farmer payments, and 22% in social services (FTI, 2016: 74). Certified producers reap substantial benefits from Fairtrade premiums (Nelson and Pound, 2009). Fairtrade premiums are particularly important for female (Smith, 2015) and indigenous producers (Lyon, 2015), who typically have very limited access to resources. Yet the strength of development impacts depends on the effectiveness of cooperative management (Raynolds, Murray, and Taylor, 2004). While strong producer associations use Fairtrade premiums to enhance individual and collective empowerment, weaker organizations find it hard to address the myriad needs of local producers (Bacon, 2005; Jaffee, 2007; Mutersbaugh, 2005). Cooperatives have invested substantially in coffee quality improvements, including organic certification, yet this maintains producer reliance on volatile export earnings (Bacon,

2005; Bacon, Rice, and Maryanski, 2015). In short, while Fairtrade premium programs can bolster local social services and production capacity they cannot take the place of government investments in rural areas.

In Latin America and the Caribbean, there are far fewer Fairtrade certified plantations and workers, than cooperatives and farmers, yet hired labour enterprises play a significant role in bananas and flowers (see Table 32.4). Both large and small producers export Fairtrade bananas, with 69 cooperatives representing 11,600 farmers and 51 plantations employing 5,800 workers located primarily in Colombia, the Dominican Republic, and Peru (FTI, 2016: 82–84). Due to their capital-intensive nature, Fairtrade flowers are produced only by large enterprises, with 11 certified plantations predominantly in Ecuador employing 2,300 workers (FTI, 2016: 133–136).

Fairtrade International hired labour certification standards shape firm participation and worker benefits. Certification rules uphold and in some cases exceed International Labor Organization guidelines related to workplace standards, worker protections, and labour rights. Fairtrade helps ensure that workers receive legally mandated benefits and may raise standards in key areas like worker health and safety, wages, and benefits (Raynolds, 2012b). As in farmer cooperatives, the Fairtrade social premium in large enterprises goes beyond the market in supporting capacity building and social programs. Plantations in Latin America and the Caribbean received US$ 9.3 million in Fairtrade premiums in 2014 (FTI, 2016: 157). Certification standards require that premiums be democratically managed by workers, with funds being allocated primarily to worker housing, educational and health services, and low interest loans (FTI, 2016: 82–84). While research suggests that certification can bring direct benefits to workers through improved work conditions and social programs, Fairtrade's most important development impact in the plantation sector may be in fostering worker's collective representation (Frundt, 2009; Raynolds, 2012b, 2016; Riisgaard, 2015).[3]

Linking Fairtrade and fair credit to enhance development impacts

Fairtrade International fosters capacity building by facilitating farmer and cooperative access to short and long-term credit. The availability of small farmer credit in Latin America and across the Global South is limited by the lack of infrastructure, geographical isolation, and volatility of agricultural markets. In the short-term, Fairtrade standards mandate that buyers provide pre-financing, a form of trade finance, to producer organizations. This short-term credit is critical since key crops such as coffee, cocoa, and sugar are slow to mature, leaving farmers to pay substantial upfront costs for inputs such as seeds and fertilizers (FTI, 2015). Access to short-term finance allows cooperative producer organizations to pay members before the entire crop is sold. Fairtrade promotes a voluntary best practice for pre-finance, encouraging buyers to supply pre-finance at no interest and facilitate or provide other forms of producer credit.

In Latin America, *coyotes*, or local intermediaries, are often the only source of credit for farmers producing coffee and other products. *Coyotes* extend credit using the crop as collateral in exchange for purchasing the crop at below market prices. Often they are the only providers of agricultural credit for small farmers and the only buyers for the farmers' crop, perpetuating relationships of exploitation between farmers and *coyotes* (Tedeschi and Carlson, 2013). The *coyotes'* monopoly power allows them to demand exorbitant interest rates, often reaching 100% per annum (Jaffee, 2007). Fairtrade International's entrance into commodity chains dominated by such oligopsonistic intermediaries decreases the market power of intermediaries, increasing the prices received by Fairtrade producers as well as by uncertified growers in the region (Podhorsky, 2015).

Numerous studies document how Fairtrade International certification increases credit access and fairer financial terms (Bacon et al., 2008; Mendez et al., 2010; Murray, Raynolds, and Taylor, 2003; Ruben, Fort, and Zúñiga-Arias, 2009). Research in Latin America finds that producers identify the provision of credit as the second most important benefit received from Fairtrade certification, second only to price guarantees (Raynolds, 2009). In rural Latin America, Fairtrade increases producers' direct access to credit via the certification's pre-finance requirement (Murray, Raynolds, and Taylor, 2003), with credit access significantly alleviating the vulnerability of farmers to environmental crises and market volatility (Bacon, 2005). Most studies find that Fairtrade producers have greater access to credit than their non-certified counterparts do, and that Fairtrade reduces farmer vulnerability through income and credit diversification (Nelson and Pound, 2009: 12). Despite the evidence that Fairtrade International certification facilitates credit access, Raynolds (2009) reports variability in credit and contract conditions, with mission-driven buyers providing substantially better financing, contracts, and organizational support to Fairtrade certified farmers in Latin America than mainstream buyers who minimize their obligations. These differential practices clearly shape development impacts in the region (see also Renard, 2015).

Fairtrade cooperatives in Latin America have insufficient access to credit, particularly to long-term financing that may be provided by mission-oriented buyers, but is not mandated under certification guidelines (Bacon et al., 2010). There is some evidence that Fairtrade certification bolsters the prestige of farmers in the region, facilitating their access to credit under more favourable terms from other lending organizations (Murray, Raynolds, and Taylor, 2003: 7). Yet outside credit is insufficient, and "Fairtrade needs to be supplemented by, and coordinated with, other development policies and initiatives in order to raise rural livelihoods to a more sustainable level" (Nelson and Pound, 2009: 10). Confirming producers' need for additional financing, a recent survey of Fairtrade farmers found that 91% had unfulfilled finance needs, with 65% reporting a need for investment credit, 29% for seasonal input finance, and most for "long-term finance for production improvements" (FTI, 2015).

Fairtrade International seeks to address this credit gap by working with ethical credit institutions to provide access particularly to long-term financing. Fairtrade has collaborated with several ethical lending and investment institutions, linking certified farmers with credit offered under fair conditions. The Fairtrade certification system was one of the "early drivers" for financing smallholder agriculture, causing ethical agricultural finance to develop on a "parallel path" with Fairtrade (Larrea, Minteuan, and Potts, 2013: 20). Given their shared values, there is a natural partnership between Fairtrade and ethical finance (Palmisano, 2015). Ethical and responsible investment, like Fairtrade, attempts to work within the market while also challenging the traditional exploitative practices of speculative and market finance which maximize profits at the expense of human and environmental wellbeing (Bridge, Murtagh, and O'Neill, 2013; Wood and Hagerman, 2010). Since certification supports better farm and cooperative management and more stable commodity markets it reduces the financial risks born by lending institutions. The synergies between Fairtrade certification and credit institutions is particularly apparent for ethically minded financiers wishing to invest in sustainable social and environmental enterprises.

Because of the inherently risky nature of providing credit to farmers in the Global South, agricultural smallholder finance institutions often work collaboratively. Ethical lenders typically work with NGOs, development banks, and conventional market actors and often coordinate with other ethically motivated initiatives. Several ethical finance institutions focus particularly on financing Fairtrade producer organizations and buyers of their products. Seven ethical finance institutions account for 90% of lending to all small farmer organizations in the Global

Table 32.5 Top seven recipients of the Fairtrade Access Fund in Latin America

Nation	Sector	Loan to Date (US $1,000)
Nicaragua	Coffee	3,378
Bolivia	Chestnuts	3,207
Ecuador	Coffee	2,263
Honduras	Coffee/Cocoa	2,238
Peru	Savings/Credit	2,004
Guatemala	Sugar	1,218
Colombia	Coffee/Cocoa/Fruit	604

Source: Compiled by the authors from Incofin (2016).

South: Alterfin, Oikocredit, Rabobank Rural Fund, responsAbility, Root Capital, Triodos, and Shared Interest (Carroll et al., 2012).

These ethical finance institutions support rural development by providing access to micro-finance, education investment, and financing for rural producer groups in Latin America and other regions. Socially oriented lenders mitigate some of the risks of working with small farmers in isolated rural areas by lending to producer groups selling Fairtrade certified coffee and other products which have relatively secure markets (Weeks, 2017). In addition to providing short-term trade finance to Fairtrade farmers, many ethical finance organizations provide access to the long-term credit that is critical for rural development but is rarely available in Latin America.

In addition to facilitating credit relations with ethical lenders, Fairtrade International has created a specific fund to provide long-term producer loans, the Fair Trade Access Fund, in collaboration with the Grameen Foundation, Incofin Investment, and KfW Development Bank (Weeks, 2017). As of December 2016, the Fair Trade Access Fund had disbursed over US$ 21 million to farmers in Latin America and Africa (Incofin, 2016). Seven of the Fund's top ten recipients are in Nicaragua, Bolivia, Ecuador, Honduras, Peru, Guatemala, and Colombia. As noted in Table 32.5, many of these loans support Fairtrade coffee producers, in keeping with coffee's prominence in Latin America's certified export portfolio. Yet loans also support producers of cocoa, sugar, fruit, and products like chestnuts that represent a diversification of traditional Latin American agro-exports. While this new loan fund cannot satisfy the long-term credit needs of all Fairtrade producers in the region, it provides an important example of ways in which socially oriented institutions can collaborate to enhance development impacts.

Conclusions

Fairtrade International certification promotes producer wellbeing and empowerment in Latin America and across the Global South by supporting both short and long-term development. In the short-term, certification provides producers with higher prices for their commodities, more stable market links, better working conditions, and social premiums to invest in community development. These benefits have been shown to promote capacity building in rural com-munities in Latin America, thus facilitating more long-term economic and social development. Fairtrade plays a critical role in facilitating access to credit for rural producers, by both mandat-ing access to pre-financing and reducing the financial risks for lenders often hesitant to work in rural areas of the Global South. Fairtrade International helps link certified producer groups with ethical finance institutions, increasing the access of rural producers to trade finance and long-term credit. In Latin America, the Fair Trade Access Fund represents an important new source of long-term financing helping to bolster the development of certified producers, their cooperatives, and their communities.

Fairtrade International supports the sustainable production efforts of producers in the Global South as well as the sustainable consumption efforts of consumers in the Global North. Yet Fairtrade's strategy of operating "in and against the market" creates a number of important challenges. Fairtrade certification in Latin America works largely within traditional agro-export sectors, like coffee, seeking to reduce producers' traditional market vulnerability by supporting sustainable development, close buyer-supplier relations, and income diversification. Although Fairtrade's bureaucratic growth has helped fuel more stable markets and generate ties with conventional and ethical financial institutions, some scholars and activists are concerned that the initiative's market success may threaten its movement principles. We argue that Fairtrade certification in Latin America has helped bolster rural capacity building and economic development by stabilizing markets, diversifying income and credit sources, and supporting social development beyond the market. Yet Fairtrade is not a panacea and must work with other promising initiatives to realize its vision of empowering farmers and workers, fuelling broad-based development, and creating a more just world.

Notes

1 "Fair trade" refers to multiple initiatives pursuing a common vision; "Fairtrade" to the certification system governed by Fairtrade International.
2 Although most scholars agree that Fairtrade has been largely beneficial in Latin America, positive impacts have been less clear in Africa and Asia due to the preponderance of plantation enterprises, capital-intensive products, and buyer-led commodity chains. For a critical view, see Cramer et al. (2014).
3 Brown (2013) argues that Fairtrade is not up to dealing with colonial legacies and contemporary labour needs.

References

Bacon, C. (2005) Confronting the coffee crisis: Can fair trade, organic, and specialty coffees reduce small-scale farmer vulnerability in northern Nicaragua? *World Development* 33: 497–511.
———. (2010) Who decides what is fair in fair trade? The agri-environmental governance of standards, access, and price. *Journal of Peasant Studies* 37: 111–147.
Bacon, C., Méndez V., Gómez, M., Stuart, D. and Flores, S. (2008) Are sustainable coffee certifications enough to secure farmer livelihoods? The Millennium Development Goals and Nicaragua's fair trade cooperatives. *Globalizations* 5(2): 259–274.
Bacon, C., Rice, R. and Maryanski, H. (2015) Fair trade coffee and environmental sustainability in Latin America. In L. Raynolds and E. Bennett (eds) *Handbook of Research on Fair Trade*. Northampton MA: Edward Elgar Publishing, pp. 388–404.
Bridge, S., Murtagh, B. and O'Neill, K. (2013) *Understanding the Social Economy and the Third Sector.* London: Palgrave Macmillan.
Brown, S. (2013) One hundred years of labor control: Violence, militancy, and the Fairtrade banana commodity chain in Colombia. *Environment and Planning A* 45(11): 2572–2591.
Carroll, T., Stern, A., Zook, D., Funes, R., Rastegar, A. and Lien, Y. (2012) *Catalyzing Smallholder Agricultural Finance.* New York: Dalberg Global Development Advisors.
Cramer, C., Johnston, D., Oya, C. and Sender, J. (2014) *Fairtrade, Employment and Poverty Reduction in Ethiopia and Uganda.* London: School for African and Oriental Studies.
FINE (FLO, IFAT, News!, EFTA). (2001) *Fair Trade Definition.* Bonn: FLO.
FLO (Fairtrade Labelling Organizations International). (2006) *Annual Report 2005/6.* FLO, Bonn.
———. (2007) *Annual Report 2006/7.* FLO, Bonn.
———. (2009) *Annual Report 2008/9.* FLO, Bonn.
———. (2012) *Annual Report 2011/2012.* FLO, Bonn.
———. (2014) *Annual Report 2013/2014.* FLO, Bonn.

———. (2015) *Annual Report 2014/2015*. FLO, Bonn.

FLO (Fairtrade International) and Globespan. (2011) *Shopping Choices Can Make a Positive Difference to Farmers and Workers in Developing Countries*. www.fairtrade.net/fileadmin/user_upload/content/2009/news/releases_statements/2011_Consumer_Media_Release_Fairtrade_Consumer_Survey__2_.pdf

Frundt, H. (2009) *Fair Bananas: Farmers, Workers, and Consumers Strive to Change an Industry*. Tucson: University of Arizona Press.

FTI (Fairtrade International). (2015) *Access to Finance*. www.fairtrade.net/programmes/access-to-finance.html

———. (2016) *Monitoring the Scope and Benefits of Fairtrade – Seventh Edition*. Bonn: Fairtrade International.

———. (2017) *Monitoring the Scope and Benefits of Fairtrade – Eighth Edition*. Bonn: Fairtrade International.

Incofin. (2016) *Fairtrade Access Fund Report Q2 2016*. Incofin Investment Management, Belgium. www.incofinim.com/en/node/3950/html

Jaffee, D. (2007) *Brewing Justice: Fair Trade Coffee, Sustainability, and Survival*. Berkeley, CA: University of California Press.

Larrea, C., Minteuan, S. and Potts, J. (2013) *Investing for Change: An Analysis of the Impacts of Agricultural Investment from Select FAST Social Lenders*. www.alterfin.be/sites/default/files/files/FAST 2013 Impact Report_0_1.pdf

Lyon, S. (2015) Fair trade and indigenous communities in Latin America. In L. Raynolds and E. Bennett (eds) *Handbook of Research on Fair Trade*. Cheltenham: Edward Elgar, pp. 422–440.

Mendez, V., Bacon, C., Olson, M., Petchers, S., Herrador, D., Carranza, C., Trujillo, L., Guadarrama-Zugasti, C., Cordon, A. and Mendoza, A. (2010) Effects of fair trade and organic certifications on small-scale coffee farmer households in Central America and Mexico. *Renewable Agriculture and Food Systems* 25(3): 236–51.

Murray, D., Raynolds, L. and Taylor P. (2003) *One Cup at a Time: Poverty Alleviation and Fair Trade in Latin America*. Fairtrade Research Group, Fort Collins: Colorado State University.

Mutersbaugh, T. (2002) The number is the beast: A political economy of organic coffee certification and producer unionism. *Environment and Planning A* 7: 1165–1184.

———. (2005) Just-in-space: Certified rural products, labor of quality, and regulatory spaces. *Journal of Rural Studies* 21: 389–402.

Nelson, V. and Pound, B. (2009) *The Last Ten Years: A Comprehensive Review of the Literature on the Impact of Fairtrade*. Natural Resources Institute (NRI), London: University of Greenwich.

Palmisano, I. (2015) *Fair Trade and Its Interrelations with Ethical Finance: Thematic Guide 2*. The Price Project: COPADE. http://ideasfactorybg.org/wp-content/uploads/2015/02/FairTradeanditsinterrelationswithEthicalFinance.pdf?iframe=true&width=100%&height=100%

Podhorsky, A. (2015) A positive analysis of Fairtrade certification. *Journal of Development Economics* 116: 169–185.

Raynolds, L. (2000) Re-embedding global agriculture: The international organic and fair trade movements. *Agriculture and Human Values* 17: 297–309.

———. (2002) Consumer/producer links in fair trade coffee networks. *Sociologia Ruralis* 42: 404–424.

———. (2008) The organic agro-export boom in the Dominican Republic: Maintaining tradition or fostering transformation? *Latin American Research Review* 43(1): 161–184.

———. (2009) Mainstreaming fair trade coffee: From partnership to traceability. *World Development* 37: 1083–1093.

———. (2012a) Fair trade: Social regulation in global food markets. *Journal of Rural Studies* 28(3): 276–287.

———. (2012b) Fair trade flowers: Global certification, environmental sustainability, and labor standards. *Rural Sociology* 77(4): 493–519.

———. (2017a) Fairtrade labour certification: The contested incorporation of plantations and workers. *Third World Quarterly* 38(7): 1473–1492.

———. (2017b) Bridging north/south divides through consumer driven networks. In M. Keller, B. Halkier, T. Wilska and M. Truninger (eds) *Routledge Handbook on Consumption*. New York: Routledge, pp. 167–178.

Raynolds, L. and Bennett, E. (2015) (eds) *Handbook of Research on Fair Trade*. Cheltenham: Edward Elgar Publishing.

Raynolds, L. and Greenfield, N. (2015) Fair trade: Movement and markets. In L. Raynolds and E. Bennett (eds) *Handbook of Research on Fair Trade*. Cheltenham: Edward Elgar, pp. 24–44.

Raynolds, L., Murray, D. and Taylor, P. (2004) Fair trade coffee: Building producer capacity via global networks. *Journal of International Development* 16: 1109–1121.

Raynolds, L., Murray, D. and Wilkinson, J. (2007) (eds) *Fair Trade: The Challenges of Transforming Globaliza-tion*. London: Routledge.

Renard, M. C. (1999) The interstices of globalization: The example of fair coffee. *Sociologia Ruralis* 39: 484–500.

———. (2003) Fair trade: Quality, market and conventions. *Journal of Rural Studies* 19: 87–96.

———. (2015) Fair trade and small farmer cooperatives in Latin America. In L. Raynolds and E. Bennett (eds) *Handbook of Research on Fair Trade*. Cheltenham: Edward Elgar, pp. 475–490.

Riisgaard, L. (2015) Fairtrade certification, conventions and labor. In L. Raynolds and E. Bennett (eds) *Handbook of Research on Fair Trade*. Northampton, MA: Edward Elgar, pp. 120–138.

Ruben, R., Fort, R. and Zúñiga-Arias, G. (2009) Measuring the impact of fair trade on development. *Development in Practice* 19(6): 777–788.

Smith, S. (2015) Fair trade and women's empowerment. In L. Raynolds and E. Bennett (eds) *Handbook of Research on Fair Trade*. Cheltenham: Edward Elgar, pp. 405–421.

Tedeschi, G. and Carlson, J. (2013) Beyond the subsidy: Coyotes, credit and fair trade coffee. *Journal of International Development* 25(2013): 456–473.

Varul, M. (2008) Consuming the campesino – Fair trade marketing between recognition and romantic commodification. *Cultural Studies* 22(5): 654–679.

Weeks, N. (2017) *The Fairtrade Access Fund: Does Linking Ethical Investment with Fairtrade Certification Enhance Credit Outcomes for Small Farmers?* MA thesis, Colorado State University, ProQuest Dissertations Publishing, Ann Arbor, MI.

Wood, D. and Hagerman, L. (2010) Mission investing and the philanthropic toolbox. *Policy and Society* 29(3): 257–268.

PART VI

Land, resources, and environmental struggles

33

DEVELOPMENT AND NATURE

Modes of appropriation and Latin American extractivisms

Eduardo Gudynas
Translated by Anna Holloway

One of the more well-known aspects of Latin American development strategies has been their dependence on natural resources. This condition has a long history that begins with the Spanish and Portuguese colony and persists under the different Republics until today. There have been undeniable efforts by countries to promote their own industrialization, but this dependence has been kept up through successive increases and decreases in the extraction of gold, sodium nitrate, rubber, cocoa, wool, soya, and oil, amongst others. This shows that the different types of development applied in the continent have always been tightly interwoven with Nature.

However, despite the key role that Nature plays, debates on development have found it difficult to approach this relation. The environment has frequently been viewed as a set of resources to be exploited and as a factor of production (generically referred to as "the earth"), or simply reduced to an open system that is external to the economy. These and other positions were brandished by very different theoretical schools and political stances, both conventional and heterodox. It became clear that there was a great divide between the different ways of understanding development and its ecological foundation.

This problematic became even greater at the beginning of the 21st century, due to the developmentalist boost caused by the high prices and increased demand for raw materials. So-called extractivisms – such as the exportation of minerals, hydrocarbons, and agri-food products – were on the rise, turning into key elements of development strategies.

Development studies, including many critical approaches, faced enormous difficulties in analysing this first stage, where the extraction of natural resources takes place. Within the classic theoretical and analytical frameworks, these extractivisms were reduced to simplistic interactions of society with Nature. This failed to address the diversity of social situations involved: from the well-known transnational corporations and the farmers who turned to illegal mining, to the fact that the same extractivist strategies of development were adopted by both conservative and progressive governments. Conventional economic approaches were not appropriate either, for their emphasis on economic valuation led them to ignore other type of values.

In order to overcome these difficulties in the analysis of extractivisms by critical development studies, researchers turned to the new concept of modes of appropriation, inspired by the idea of modes of production but involving a redefinition of both. In this chapter, we define this concept and present a brief example of how it applies to extractivisms. This introduction does not intend to be an exhaustive account of the issue, but rather a proposal that this idea be used as a tool for better understanding the interactions of development through the lens of ecology.

The idea of modes of production

The idea of modes of production has played a significant role in the analysis of development and capitalism. Originally formulated by Karl Marx, it probably appears for the first time in *The German Ideology* (co-authored with Friedrich Engels) and in other texts, such as *Capital*. Marx does not offer a specific definition, but we can argue that he understands it as the modes of productive and economic organization of a society and as a means for historical analysis.

The concept became very popular in the 1960s and 1970s. It appeared in Marxist discussions (as Althusser and Balibar's *Reading Capital* shows, 1979; also see Resch, 1992), was disseminated through manuals and texts in Latin America, and became relevant in discussions on underdevelopment and dependence, fuelling a significant amount of literature (for example, Fioravanti, 1972; de la Peña, 1978; also see the summaries on its use in development debates in Foster-Carter, 1978; Ruccio and Simon, 1986). Latin American contributions were also made to these discussions; prime examples are the works of Ernesto Laclau, Carlos Sempat Assadourian, Ciro Flammarion Santana Cardoso, and others, published in the influential journal *Cuadernos de Pasado y Presente* under the coordination of Juan Carlos Garavaglia (1973).

As it is conventionally understand, mode of production (MP) is a highly abstract concept that includes, on the one hand, the so-called productive forces such as human labour, resources like land, technologies and so on, and, on the other, the relations of production, referring to those who work and to those who appropriate part of this labour through exploitation, understood as social classes. As tools of historical interpretation, MPs constitute a social whole that has persisted for a long time. Thus, a broad range of MPs were defined throughout history (primitive communism, Asiatic, old, feudal, capitalist, and socialist).

Another approach was the one formulated by Eric Wolf (1982) from the perspective of world systems. He used the term to refer to specific sets of social relations within which the labour of extracting energy from Nature takes place through tools, knowledges, and organization. His perspective is also highly abstract (he describes three modes that are defined as kinship, tributary, and capitalist).

The use of the concept and its leading role in debates on development in general, and on Latin American development more specifically, started to fade away during the 1980s and almost disappeared in the 1990s. Some authors continued to use it within a very broad time frame (for example, Hume, 2007) or as a critique of capitalist development (Richards, 2001). Although it persists in dictionaries on Marxism (Duménil, Löwy, and Renault, 2014), the idea languished as time went by due, amongst other factors, to the fall of really existing socialism, analytical abstraction, emphasis on historiography, and to its focusing mainly on discussions on capital and the state (see Graeber, 2006). It became replaced by other concepts.

Nature, extraction, and production

A first look at the classic Marxist perceptions of MPs reveals elements that are very useful in understanding recent strategies of development in Latin America. This perspective adds a key component to them, as it interprets it as the labour through which humans exploit Nature.

A noteworthy example of this line of thought is the research conducted by Stephen Bunker on extractivism in the Amazon (for example, Bunker, 1984). His scholarly contributions did not receive the attention they deserved at the time of publication, although they were pioneering in their insistence that a new theoretical framework was needed for the analysis of the appropriation of natural resources. According to him, concepts pertaining to industrialized countries were insufficient or inadequate, while new categories were needed to incorporate environmental aspects such as the appropriation, use, and destruction of matter and energy, which cannot be calculated only in terms of labour or capital, as classic approaches did. This led Bunker to propose the concept of "mode of extraction," inspired by that of "modes of production" but drawing a clear line between the two.

Bunker's "mode of extraction" described systemic connections between very different phenomena, from labour organization, systems of ownership, or infrastructure, to ideology and beliefs. He acknowledged that his new concept was parallel to the classic idea of MPs, but presented the latter in a broader sense (understood as the relation between social, legal, political, and commercial aspects). He immediately made it clear that he did not embrace orthodox Marxist opinions that argued that modes of production reproduced themselves and, therefore, capitalism could expand indefinitely. This condition, Bunker claimed, is ecologically impossible. These and other warnings are correct and should be kept in mind.

Latin American extractivisms are very diverse, in terms of the different kinds that exist – such as open-cast mega mining, monocultures, or oil drilling in the Amazon – but these categories are also internally diverse. An orthodox use of MPs would consider them all as part of a capitalist mode and lose sight of all this diversity, which in turn would affect the search for alternatives. Analysis based exclusively on economic factors is also insufficient, for it often overlooks social and political dynamics. Furthermore, classic approaches within social sciences that focused, for example, on social classes failed to grasp the diversity of intervening actors, such as indigenous, Afro-descendant, peasant, displaced, or proletarian populations and so on.

However, the concept of "modes" offers valuable lessons, such as the consideration of the productive forces (natural resources, technology, etc.), and the relations established between them (taking into account factors like the role of capital or the state). The employment of this perspective in great "systems," such as capitalism, and large time frames partly explains the decline of the concept. However, if applied at a much smaller scale, it becomes more specific and acquires a greater potential for analysis and, therefore, becomes more useful.

Bunker was essentially right in his disagreement with conventional concepts and in the need for a new category to describe the appropriation of natural resources in the Amazon. While the concept of MPs can be of some use, new elements such as the ones already mentioned must be introduced, and especially those relating to the appropriation of Nature.

There are other lessons to be learnt from extractivisms. Strictly speaking, they are neither a "productive" sector nor an "industry," as their defenders argue. To consider the extraction and exportation of iron, for example, as "production" is a crude distortion, for nothing is being "produced": it is being extracted (and, therefore, amounts to a net loss of natural heritage). Neither does it make sense to qualify it as an "extractive industry," for no manufacturing process is involved. It is the export of commodities or raw materials. The characterizations "production" and "industry" undoubtedly aim to legitimize these activities socially and politically in the eyes of the public and to place extractivisms within industrialist imaginaries. The reproduction of these terms by academics reveals, intentions aside, a somewhat simplistic approach to extractivisms and particularly to their ecological and political connotations.

Finally, extractivisms represent a mode that is always shaped by ecological factors. This includes the location of land or deposits, the amount of available resources, whether they are renewable or not, the environmental impact of the removal of the resources and its consequences and so on. This environmental dimension was never fully incorporated in the idea of MPs, as we discussed earlier on.

All this explains the need to posit a separate concept, different from that of MPs, for the analysis of this first phase of interacting with Nature. This is how the idea of Modes of Appropriation (MAs) came into being. It is a category inspired by MPs, but one that must be different because of the particularities of the first step in the appropriation of what we call natural resources, a particular interaction with the environment that should be analyzed in more detail.

Defining the modes of appropriation

Modes of appropriation describe different ways of organizing the appropriation of natural resources (such as matter, energy, or ecological processes) for the satisfaction of human goals in each social and environmental context. Appropriation refers to the direct extraction of resources (through removing minerals, for example) but also to indirect extraction (such as crop harvesting). Their geographic scale is limited to specific locations and regions within countries, and their temporal scale equals more or less a year.

MAs articulate with MPs where the following stages in the transformation of raw materials take place. From a development perspective, the first stages involve the appropriation of raw materials, whereas the second ones involve manufacturing processes and their organizational dynamics, such as in the fabrication and commercialization of goods.

Examples of MAs include hunting and gathering in the Amazon forests, peasant farming in the Andes, open-cast mega-mines in Chile, and GMO monocultures in Argentina, Brazil, or Uruguay. It is not only about the physical act of removing something from the environment; many additional elements come into play, including understandings on what a resource is and is not, how resources are valued (economically, ecologically, aesthetically, spiritually, and so on), the labour and capital that goes into these practices, the institutionalized frameworks sustaining the appropriation (such as laws of access and property), the social relations deployed during the appropriation (the role of workers, the companies, and the state), and the channels of distribution and of accessing the modes of production. Therefore, MAs express different ways of obtaining matter and energy from the environment and also different ways of handling and transforming them in order to feed them into the following steps of other stages of production.

MAs are always anchored to specific locations, as they depend on specific resources existing in each place (such as mineral deposits, oil fields, or agricultural land) and are therefore defined and limited by ecological contexts. MPs focus on transformation, they are not tied to a specific location (for industries can settle in different places) and are, therefore, determined mostly by social factors.

It is important to highlight this particularity of the ecological limitation of MAs, given that they are essentially an interaction with Nature and cannot be socially regulated. MPs, on the contrary, have to deal with processes that occur mainly amongst humans and are, therefore, more flexible. For example, there cannot be a collective decision on creating an oil bed in a specific location, neither can depleted natural resources be recovered through political consensus. This is a fundamental difference between the two modes.

This perspective clearly refers to smaller scales compared to the conventional approach. There is no capitalist mode here, but rather many different modes of appropriation and production. Bunker's idea of extractivist modes would be one specific case amongst the modes of appropriation.

As mentioned previously, conventional approaches tend to embrace rigid schemes (based on social class, ethnicity, etc.) and cannot encompass the enormous diversity of actors and structures organized for the appropriation of natural resources. MAs, on the other hand, force analysts to consider this diversity for they take multiple dimensions into account.

By way of example to orient our thinking, we can say that the main components that characterize these modes are the following: ecological (type of natural resource appropriated, such as an extracted mineral or cultivated plot of land; geographical location; ecological context; environmental impacts of the appropriation, etc.); territorial (geographical spaces affected; social delimitation of the territories; concessions imposed, etc.); technological (use of technologies of appropriation; capacity to lessen or remedy impact; biotechnology, etc.); regimes of access and ownership; social (actors conducting the appropriation, local communities, business agents, etc.); capital (investments, profitability, surplus and related disputes, the role of enterprises, demand in global markets, etc.); normative (legal framework of the appropriations, compliance and audit, etc.); political (role of supporting political groups, discourses of political legitimization, etc.); state-related (performance of local, regional, and national governments, taxation and redistribution tools, etc.).

This list is only an example, but it illustrates how MAs must not be enclosed within purely economic descriptions and must include many other elements, such as social and ecological ones. This approach can be considered as "Marxian," but it is not limited to conventional approaches – be they Marxist or neoclassical – for it includes a detailed account of environmental factors and holds a place for non-material, symbolic relations (as occurs with conceptions and sensitivities on Nature).

Finally, we should at least mention that Marx also used the concept of appropriation in many texts, albeit in a different sense than the one proposed here. While Marx did not define it with precision, he did discuss it in relation to the modes of production and mostly to property, and then returned to the idea of a mode of capitalist appropriation (see, for example, Dussel, 1985).

The modes of appropriation of extractivisms

Let us go back to extractivisms in South America in order to illustrate the application of MAs. Conventional approaches either described them in broad categories – such as extractivisms performed by transnational or state-owned companies – or defined them all, in one way or another, as capitalist; at the same time, they had enormous difficulties in dealing with the proliferation of other practices and other actors. Through the use of MAs, researchers can acknowledge this diversity and begin to analyze it. Table 33.1 offers a very summarized presentation of the various modes involved in different mining extractivisms.[1]

It illustrates how appropriations are organized in different ways, with the intervention of different actors, the involvement of different dynamics of administration of capital and labour, and varying regimes of property and access to resources. Also, many MAs can compete within the same geographical space, as occurs in the disputes between legal and illegal miners, or between miners and farmers.

The concept of MAs as an instrument of analysis allows for the detailed dissection of each one of the types presented in Table 33.1. This can be briefly illustrated with the case of MAs in Bolivia's mining cooperatives, a case that is difficult to analyze from a conventional perspective because it does not correspond either to large-scale corporate mining nor to illegal, informal mining.

These activities increased notably during the MAS (Movimiento al Socialismo) government, going from 911 in 2006 (the year that Evo Morales assumed the presidency) to 1,630 in 2013 (registered in the National Federation of Mining Cooperatives, FENCOMIN) and employing an estimate of 120,000 miners. They have become the second largest mining conglomerate in Bolivia, with operations spreading across 611,000 hectares; they have outpaced state-owned COMIBOL (that had 329,000 hectares) and are second only to conventional private companies (that operate on properties of slightly over 1 million hectares).[2]

Table 33.1 Modes of appropriation in mining extractivisms

Traditional or old-school mining. Performed by individuals, families, or small groups of different origins (traditional, rural, indigenous, Afro-descendant, migrant or displaced, etc.). Intensive use of human labour and limited access to technology. Low capital investment and financial dependence on local traders and intermediaries who buy and resell the extracted mineral and can provide materials and machinery. Those involved in traditional or old-school mining are often trapped in conditions of poverty.

Illegal or informal mining. Performed by individuals or groups, also from very different backgrounds and in many cases originating from traditional practices. Limited access to capital, limited coordination in the access to inputs, technology (such as dredges), or political representation. They are involved in illegal networks for materials, machinery, and sales, and are victims of violence in the hands of these networks and also of the security forces. Some types of mineral resources, such as gold, can be obtained using simple technologies. These undertakings can involve thousands of people and cover vast extensions of land, with practices of increasing intensity and under terrible sanitary and environmental conditions. Most live in poverty (examples in Valencia, 2015).

Mining cooperatives. Individuals who are formally organized as cooperatives of different sizes. This allows for better conditions of access to capital and technologies and also provides marketing advantages. They extract more minerals and often operate as conventional companies in that they prioritize profit and outsource the social and environmental impact (the particularities of this are analyzed in more detail within the text).

Domestic private mining companies. More capital availability and access to technology than in the previous cases. They have access to large-scale and medium-scale technology, although they don't always follow up with adequate maintenance (examples in Torres, 2007). Working conditions tend to be poor and there are varying levels of unionization. On many occasions these companies establish partnerships with larger mining companies.

Domestic state-owned or mixed-ownership mining companies. State-owned or mixed-ownership (with the private sector) companies, but that are controlled by the state through ownership or funding. Examples of the former are COMIBOL (Bolivian Mining Corporation) or CODELCO (National Copper Corporation) in Chile; of the latter, Vale mining company in Brazil. They have more access to capital and can make significant investments. They make a more intensive use of technology, have more employees who are unionized to differing degrees, and outsource many of their activities. Their environmental and social performance is questioned and, as a result, they too are in conflict with local communities. They have direct access to global trade networks or employ intermediaries.

Transnational mining corporations. Large corporations with high capital availability and an intensive use of technology. Many display standard labour relations (less so in Chinese companies) and outsource a variety of activities. Large enterprises include schemes of corporate social responsibility, but their environmental and social performance is almost always questionable (see, for example, De Echave, 2011, for the case of Peru). They lobby governments and organize their own trade networks.

They are organized legally as cooperatives. Some are made up of a few partners who work directly, make little use of technology, and depend on intermediaries. Others, however, are medium-sized, have access to machinery, are moderately staffed, and have a greater commercial capacity. Finally, there are large cooperatives with great capital availability and access to equipment, that even have their own processing plants and a high participation of salaried workers, who work according to unequal and hierarchical labour relations. They resemble a corporate MA.

Their social performance is poor, as their employees receive poor salaries and are exposed to inadequate health and safety regimes. Social insurance is very limited (only 16% of cooperative members are enrolled in a pension fund). Their environmental performance is also deficient.

For example, land surveying for the Oruro and Potosí departments showed that 78% of the cooperatives lack environmental documents. There are constant complaints for non-compliance on this issue and various cases of tension and conflict with communities caused by water and soil contamination.

The profile of the cooperative members is varied. Some come from families with a mining tradition and others present themselves as community members previously involved in farming. Despite this fact, cooperatives are often in conflict with local communities or miners from formal companies, be they private or state-owned, and even with the government itself (for a relevant case, see Jiménez and Campanini, 2012).

This cannot be described only through the categories of identity, or of belonging to a community or a specific class. There are ruptures and confrontations between different groups within the same extractivist MAs and against other modes, such as that of farmers. Conflicts revolve around access to mining resources, to the surplus created, and to trade networks, and against other members of the community because of the effects on society and the environment.

This heterodox situation is repeated in the political arena. Federation FENCOMIN presents itself as a "trade union," even though its members are cooperatives and its practices resemble those of a chamber of commerce. It puts pressure on the government, places its own persons of trust in key (even ministerial) positions and, in exchange, offers electoral support to the MAS party. FENCOMIN insists it represents a stage in development that goes beyond the extractivisms of foreign companies. It adopts a nationalist approach to natural resources and defines itself as "classist and revolutionary, anti-oligarchic and anti-imperialist."[3] Despite this discourse, FENCOMIN sells a big part of the minerals it extracts to large foreign companies.

A focus on the components and the relations existing within this MA reveals a dynamic whereby the popular sectors, including peasants and indigenous inhabitants, organize themselves in forms that are increasingly entrepreneurial, distancing themselves both from the environmental commitments towards Mother Earth that are discussed in Bolivia and from the solidarity of cooperativism. They present themselves as anti-imperialist, but are integrated in the trade networks of the global markets. It is as if they were "borrowing" histories, symbols, and imaginaries – such as cooperativism, syndicalism, community, progressivism, and so on – to disguise an enterprise that walks in the opposite direction – and focuses on maximizing the extraction of resources and economic profit at the expense of social and environmental sustainability.

Characteristics and dynamics of the extractivist modes of appropriation

This short list can also serve as an example to illustrate the characteristics that extractivist MAs have in common (including contributions such as Bunker's, 1984). We must begin by stressing that they are all anchored to specific locations, as they depend on resources such as minerals, hydrocarbons, or soil fertility and, therefore, cannot be moved. This makes MAs organize themselves as enclaves, both ecological (for they are necessarily located where the resource is) and economic (with a predominance of external ties, rather than local or regional). These modes usually create limited economic ties with their surroundings, including hired staff, outsourcing, food and health provision services, housing, and so on, and therefore do not construct productive regional connections. In fact, some actually destroy other local productive practices.

Enclaves of appropriation follow different dynamics. In the case of mining and oil drilling, for example, they are itinerant: they arrive at an exploitation site, appropriate the resource and abandon the site when it runs dry, in order to "jump" to a different location. In agriculture or forestry, the connection lasts much longer.

In these modes, Nature is conceived as a set of "resources." Perceptions and sensitivities relating to the environment impose its fragmentation and commercialization; certain elements are qualified as "resources," identified, separated, and extracted, while other elements are disposed of. In some cases, as occurs with open-cast mega mining, the quantities removed and the land surface involved are huge and can thus be qualified as "ecological amputations." This is actively downplayed or concealed, and the MA operates within the political sphere and that of experts whose aim is to make the loss of natural heritage and the subsequent impact on the environment (such as soil or water contamination) tolerable.

Appropriation targets "resources," identified as such on the basis of their economic value, demand, and assignation of property rights. Economic valuation has very significant repercussions, for it not only reinforces utilitarian stances but also overshadows other kind of values (such as ecological, cultural, or religious ones that are defended mostly by local communities). The extractivist MA privileges one type of valuation and, with it, imposes a rationality that pursues profitability, efficiency, and competitiveness in the appropriation of natural resources. In other words, there is a commodification of social life and of the relation to Nature.

Conventional attitudes towards development accept and reproduce a valuation that only acknowledges, for example, the final resources exported (income from exportation) and ignores or excludes the economic cost of the impact on society and the environment. Thus, these MAs follow a rationality that presents itself as essentially economic but conceals the fact that its prism is distorted. Protests by local communities and conflicts that arise because of this impact are concealed or ignored, repressed or criminalized. These modes organize themselves economically and socially in order to outsource their social and environmental effects.

In any case, the economic value of a place usually decreases as the appropriation proceeds. In the example of non-renewable resources, it decreases at the pace of the extraction of the mineral or oil; in the case of renewable resources, a similar process can be unleashed by the loss of fertility. However, from the viewpoint of conventional development, success lies in the reduction of this heritage by extracting, for example, as many minerals or as much oil as possible. This dynamic is almost the complete opposite of what is observed in the industry's traditional MPs, where the value of a location increases with time. This is what occurs, for example, with industrial parks: one industry attracts others and the arrival of new enterprises lowers the cost of infrastructures as all industries share the same location. On the contrary, extractivist MAs act alone and their lifespan depends on the rate of depletion of natural resources.

Many of the traditional analyses of natural resources by development policies speak of the dichotomy between private and public or state-owned properties. However, a quick glance at the MAs in Table 33.1 reveals a more complex situation involving several ownership regimes (private, mixed, state-owned, cooperatives, and so on). Furthermore, a distinction must be made between owning a resource and having access to it, in the sense formulated by Ribot and Peluso (2003). For example, regardless of the ownership of mineral or oil resources, access almost always ends up in the hands of transnational corporations. It is becoming more and more common to come across strategies where extraction is in the hands of the state or of a mixed enterprise, but the technology and commercialization rights belong to transnational companies. This, in turn, explains many of the disputes taking place within extractivist MAs in relation to access to resources (as occurs in Bolivia).[4] It is also known that, beyond the ownership of each extractivist enclave, its overall production and insertion in the market is often controlled by transnationalized corporate actors. Be it through ownership or access, MAs impose a reterritorialization; this is the case, for example, with mining concessions or oil blocks that frequently conflict with pre-existing territories (such as rural or indigenous ones).

The proportion of labour and capital on the value of the appropriated natural resources is low compared to the one recorded in the MPs of the stage of industrialization. Anyhow, in mining or oil-drilling MAs, the most significant investment in labour and capital takes place at the initial stages with the construction of plants, platforms, and so on and, even so, the investment is much less than the profit made during the useful life of the mine site.[5]

There are other significant differences between extractivist MAs and the MPs that use their resources. Although the enclaves can be in very different locations (regarding their social and ecological characteristics) they provide similar resources within one same type (commodities). On the contrary, although manufacturing MPs can be grouped (as in an industrial park) they tend to have different final products. In extractivist MAs, costs are usually inflexible in the sense that any increase in the volume of extracted resources equals an increase in costs and, therefore, a higher demand for capital; unlike many manufacturing MPs, where an increase in production can lead to a reduction of cost per unit.

MAs can lead to significant population shifts, such as a massive influx of workers during the construction phase of a mining enterprise. However, once this initial stage is over, the number of workers plummets; many will move to other locations and others will remain in the area, often living in conditions of poverty. There are also population shifts due to the displacement of local communities when their territory is invaded by extractivisms.

Finally, MAs connect and articulate with MPs. They are not isolated from each other but rather overlap, through flows of matter, energy, and capital. Different industrial MPs, for example, depend on the supply of raw materials provided by mining and farming extractivisms. As a result, MPs turn into factors that determine the structure and dynamic of MAs. Their need for raw materials will determine which modes are considered necessary and profitable, what natural resources should be looked for, and what flows of investment follow. In the case of certain minerals, hydrocarbons, or some types of agri-food, access and commercialization are in the hands of just a few companies, who turn into oligopolies of natural resources.

Analyses and alternatives in development studies

What is commonly understood as interactions between society and Nature (or the environment) has been approached in many different ways. Although this is a problematic that exceeds the scope of the present chapter, we must remember that there is great diversity of opinion. Biological ecology, for example, examines it as a distortion of the ecosystem caused by humans. The efforts of human ecology at the beginning of the 20th century extended ecological dynamics, such as competition, to the social world. At the same time, the inverse also applies, as the social sciences have rendered these interpretations more complex, such as in the idea of Nature as a social construction. Ecological economics or ecological Marxism have, to a greater or lesser extent, used ideas such as metabolism in matter and energy, labour and value (see, for example, Foster, 2004; Burkett, 2014).

Beyond these efforts and others, the prevailing approach in development studies, environmental management, and other disciplines is still limited. Social perspectives still resist the incorporation of environmental concerns, and environmental approaches hardly consider social affairs and often ignore power relations. All this problematic is particularly obvious in the analyses of extractivisms. Furthermore, binary analysis on the basis of the opposition between private and state-owned companies, or development and underdevelopment, is inadequate and incapable of grasping the diversity and complexities of Latin America.

This is even more so in the specific case of extractivisms, for they have been fostered by governments both conservative and progressive, albeit in different ways and under different legitimizing discourses. There has been a mix of conceptual and ideological confusions, and it seems that a new Left has to be necessarily extractivist as the only way out of what is, once again, considered as underdevelopment. Natural resources are once again being extracted, but with the involvement of different political and social structures, different power relations at stake, and different discourses; however, everything flows into the same channels of global trade.

The concept of MAs presented in this chapter seeks to solve some of the limitations existing in the field of development studies, a necessary task in Latin America where development strategies continue to be highly dependent on natural resources. This creates environmental issues and conflicts with citizens and, furthermore, is not capable of tackling problems such as poverty or subordination to globalization. MAs are also an essential approach if we want to avoid the aforementioned traps and confusions and try out alternatives that adjust better to each one of the contexts observed in the continent.

It is clear that when governments, be they progressive or conservative, assert that extractivism is the only solution, they limit themselves to discussing different forms of organizing this specific type of MA. In view of this, MAs play a decisive role not only in describing these situations, but also in enhancing reflection and proposing alternatives. To claim that options within extractivisms are limited to passing from a private to a state-owned model is utterly insufficient. Analysis through the lens of MAs shows that different options will be needed for each mode, given that solutions for mining cooperatives will be different, for example, from those proposed for poor and excluded miners in the Amazon.

These alternatives coincide in that they look for solutions outside extractivist modes of appropriation. In other words, alternatives to development demand another type of interaction between society and Nature from the outset. This articulation is not only a relation based on flows, let us say of matter or energy; it is also expressed in social relations, symbols, beliefs, and affections. Here, too, this new concept can be of use, for it can incorporate sensibilities. Thus, the idea of modes of appropriation does not only aspire to a better description of the problems of development in Latin America; it also wants to contribute to a radical change in how people relate to each other and to Nature.

Notes

1 The definition of extractivisms is based on Gudynas (2018), understood as a specific type of extraction of natural resources characterized by its high volume or intensity, where half or more of the matter extracted is exported to global markets in the form of raw materials. The table is based on information gathered through seminars and workshops, consultations with qualified informants, fieldwork, and relevant literature, especially in Argentina, Bolivia, Brazil, Chile, Colombia, Ecuador, Peru, and Uruguay.
2 Data based on interviews in the city of Cochabamba; other references in Michard, 2008; Espinoza Morales, 2010; Ferrufino et al., 2011; Francescone and Díaz. 2013; Gandarillas, 2013; Poveda, 2014.
3 Resolutions of the XI National Congress of FENCOMIN in 2011, cited in Poveda, 2014.
4 For example, in Potosi (Bolivia) mining company Manquiri, a subsidiary of US transnational Coeur D'Alene, had signed contracts with seven mining cooperatives for the provision of resources (Gandarillas, 2013).
5 During the period of high commodity prices, profitability in the mining sector was estimated at 37.1% per year, largely above, for example, industrial MPs (estimated at 6.5%) (De Echave, 2011).

References

Althusser, L. and Balibar, E. (1979) *Reading Capital.* London: Verso.
Bunker, S. G. (1984) Modes of extraction, unequal exchange, and the progressive underdevelopment of an extreme periphery: The Brazilian Amazon, 1600–1980. *American Journal of Sociology* 89: 1017–1064.
Burkett, P. (2014) *Marx and Nature: A Red and Green Perspective.* Chicago: Haymarket.

De Echave, J. (2011) La minería peruana y los escenarios de transición. In A. Alayza and E. Gudynas (eds) *Transiciones: Post extractivismo y alternativas al extractivismo en el Perú*. Lima: CEPES, RedGE and CLAES, pp. 61–91.

De la Peña, S. (1978) *El modo de producción capitalista: Teoría y método de investigación*. México: Siglo XX.

Duménil, G., Löwy, M. and Renault, E. (2014) *Las 100 palabras del marxismo*. Madrid: Akal.

Dussel, E. (1985) *La producción teórica de Marx: Un comentario a los Grundisse*. México: Siglo XXI.

Espinoza Morales, J. (2010) *Minería boliviana. Su realidad*. La Paz: Plural.

Ferrufino, G. R. R., Eróstegui, T. and Gavincha, L. M. (2011) *Potosí. El cerro nuestro de cada día*. La Paz: Labor.

Fioravanti, E. (1972) *El concepto de modo de producción*. Barcelona: Península.

Foster, J. B. (2004) *La ecología de Marx*. Barcelona: El Viejo Topo.

Foster-Carter, A. (1978) The modes of production controversy. *New Left Review* 107: 47–77.

Francescone, K. and Díaz, V. (2013) Entre socios, patrones y peones. *PetroPress*, CEDIB, Cochabamba 30: 32–41.

Gandarillas, G. M. (2013) Empleo y derechos laborales en las actividades extractivas. *PetroPress*, CEDIB, Cochabamba 30: 4–7.

Garavaglia, J. C. (1973) (ed) *Modos de producción en América Latina*. Cuadernos de Pasado y Presente No 40. Córdoba: Ediciones Pasado y Presente.

Graeber, D. (2006) Turning modes of production inside out: Or, why capitalism is a transformation of slavery. *Critique of Anthropology* 26: 61–85.

Gudynas, E. (2018) Extractivism: Tendencies and consequences. In R. Munck and R. Delagdo Wise (eds) *Reframing Latin American Development*. Oxon: Routledge, pp. 61–76.

Hume, D. (2007) Modes of production. In P. Robbins (ed) *Encyclopedia of Environment and Society*. Thousand Oaks: Sage, pp. 1154–1155.

Jiménez G. and Campanini, O. (2012) Mallku Khota. *PetroPress*, CEDIB, Cochabamba 29: 24–37.

Michard, J. (2008) *Cooperativas mineras en Bolivia: Formas de organización, producción y comercialización*. Cochabamba: CEDIB.

Poveda, A. P. (2014) *Formas de producción de las cooperativas mineras en Bolivia*. La Paz: CEDLA.

Resch, R. P. (1992) *Althusser and the renewal of Marxist theory*. Berkeley, CA: University of California Press.

Ribot, J. C. and Peluso, N. L. (2003) A theory of access. *Rural Sociology* 68: 153–181.

Richards, A. (2001) *Development and modes of production in Marxian economics*. London: Routledge.

Ruccio, D. F. and Simon, L. H (1986) Methodological aspects of a Marxian approach to development: An analysis of the modes of production school. *World Development* 14: 211–222.

Torres, C. V. (2007) *Minería artesanal y a gran escala en el Perú: el caso del oro*. Lima: CooperAcción.

Wolf, E. (1982) *Europe and the People without History*. Berkeley, CA: University of California Press.

Valencia, L. (2015) (ed) *Las rutas del oro ilegal: Estudios de caso en cinco países*. Lima: Sociedad Peruana de Derecho Ambiental.

34

LANDGRABBING IN LATIN AMERICA

Sedimented landscapes of dispossession

Diana Ojeda

Introduction

The concept of landgrabbing gained attention in international policy-making, academic, and social movement circles after the financial crisis of 2007–2008 generated an increased demand for food and agrofuels by countries in the Global North (FAO, 2012; Grain, 2008). The "global rush for land," exacerbated by neoliberal policies and speculative trends, has translated into massive transactions of land, often involving foreign capital, in different countries in the Global South (Borras and Franco, 2012). These new dynamics have made evident the need to understand the multiscalar forces behind dramatic shifts in land use, tenure, and ownership; as well as their socio-environmental consequences. This seems even more urgent as its effects over lived ecologies have proven devastating.

Despite the fact that transactions over 1,000 ha have not been the norm, except for Argentina and Brazil (Borras et al., 2012a), multiple studies have consistently reported an increased dispossession of rural populations in Latin America and the Caribbean in relation to the entrenchment of extractivism: monocultures such as soy (Argentina, Brazil, Bolivia, Uruguay, and Paraguay), oil palm (Colombia, Ecuador, Peru, Guatemala, and Honduras), sugar cane (Brazil, Colombia, Guatemala, Guyana, and Trinidad and Tobago), fruits (Chile, Peru, Panama, Costa Rica, and Dominican Republic), forestry (Chile, Ecuador, Bolivia, Mexico, and Nicaragua), mining (Peru, Bolivia, and Colombia), and tourism (Costa Rica, Honduras, Nicaragua, and Haiti). These various forms of landgrabbing demand geographically specific modes of analysis within and beyond national borders.

This chapter suggests an approach to landgrabbing in Latin America and the Caribbean that brings the *sedimented landscapes of dispossession* into the analysis. It provides an introduction to contemporary landgrabbing, highlighting the importance of studying sustained, everyday struggles over meaning, use, and control of land and water, among other key elements for life sustenance. I understand landscapes as disputed and unfinished political projects that materialize in concrete assemblages of nature and society. In that sense, the sedimented landscapes of dispossession refer to the spatial aggregate of historical process of inequality, exclusion, death, and suffering (Ojeda et al., 2015: 109; see Moore, 2005). I argue that paying attention to these particular histories and materializations in space allows for a better understanding of landgrabbing's complex, multiscalar, and multidimensional character.

Landgrabbing: a contested view

Landgrabbing in the region does not necessarily follow "the dominant model of land grabs by single purchase" (Rocheleau, 2015: 696–697). This model often reduces landgrabbing to large transactions of land involving foreign capital, which result in the displacement of local populations. On the contrary, a large part of landgrabbing in Latin America can be characterized by smaller scale and less evident forms of landgrabbing which have systematically transformed land use and control, like those in Guatemala (Devine, 2016), Honduras (León, 2017), and Costa Rica (van Noorloos, 2014). And some of these processes have not necessarily altered the established property regime, like in Argentina (Goldfarb and Zoomers, 2014) and Mexico (Vásquez García, 2018). The dispossession of local populations has occurred as well without their displacement, as it has become evident in my own work in Colombia (Ojeda et al., 2015). Reducing landgrabbing to a set of hectares taken does not allow us to understand the different mechanisms through which local and communal use of land is being largely restricted (De Schutter, 2011; Edelman, 2013). It also disregards its localized dynamics and associated forms of resistance. As Edelman and León (2013: 1697) note, "the lack of historical perspective in many studies of land grabbing leads researchers to ignore or underestimate the extent to which pre-existing social relations shape rural spaces in which contemporary land deals occur" (see also Mollett, 2016).

Processes and dynamics of landgrabbing in Latin America and the Caribbean, and elsewhere, include diverse forms and mechanisms through which uneven geographies of land definition, access, use, and control have been violently forged and maintained, while negotiated and resisted. These sedimented landscapes of dispossession evidence complex processes of localized production of difference, which cannot be reduced to the workings of capital, and have historically given shape to deeply ingrained exclusions based on class, race, and gender, among other forms of differentiation. Considering these particular histories and geographies in the region challenges attempts to clearly define contemporary landgrabbing.

Jun Borras et al. (2012b: 405) suggest a definition of contemporary landgrabbing as "the capturing of control of relatively vast tracts of land and other natural resources through a variety of mechanisms and forms involving large-scale capital that often shifts resource use to that of extraction, whether for international or domestic purposes." They note too that this has occurred "as capital's response to the convergence of food, energy and financial crises, climate change mitigation imperatives and demands for resources from newer hubs of global capital." With their definition, they highlight three elements of landgrabbing in the region that push the definition beyond the "dominant model of land grabs": i) landgrabbing is essentially control grabbing; ii) its scale is determined by both the scale of land acquisitions and the scale of capital involved, and iii) they occur because of and within capital accumulation crises (Borras et al., 2012b: 404–405).

Nevertheless, this definition falls short in the comprehension of particular dynamics in the region. It leaves out the question of what is it that is being grabbed (along with control), and centres solely on the workings of capital, which can never be considered without conjoining processes of differentiation and power that include, but are not limited to, imperialism, militarization, racism, and misogyny. In that direction, Dianne Rocheleau (2015) points to the need to understand territories in their processual and discontinuous character: they are made and remade constantly and cannot be reduced to a delimited area. Her study in Mexico suggests a definition of landgrabbing as "the deployment of networked and dispersed power to unmake and remake territories across scales" (Rocheleau, 2015: 695). Such a view of power considers the multiplicity of actors involved in the constant reconfiguration of territory, including traditional political powers, new elites, multinational corporations, the state, NGOs, and armed actors. It opens the possibility of understanding organized, as well as everyday, less visible forms of resistance.

Moreover, it hints to the fact that rural communities, social movements, and environmental leaders in Latin America and the Caribbean largely define their struggle against landgrabbing as the *defence of territory* (CRAADT, 2017; De la Cadena, 2010; Escobar, 2008). This includes the defence of land, but also water, soil, food, and air, as well as other elements necessary for life sustenance, in its material and symbolic dimension. In more significant terms, women environmental leaders frame the long-standing struggle against extractivism as the *care of territory* (Márquez, 2015), situating it as part of a continuum of body and territory: the *territorio-cuerpo* (Cabnal, 2010). Their view also states the necessity to study the multi-layered histories and their materializations through and in space, in their multiple scales.

Landgrabbing's sedimented histories in Latin America and the Caribbean

The history of Latin America and the Caribbean is a violent history of landgrabbing, enclosure, and dispossession related, for the most part, to extractivism – that is, to the exploitation of nature (turned into resources) for extra-local needs (Bebbington, Bornschlegl, and Johnson, 2013; Svampa and Viale, 2014; Sauer and Almeida, 2011; Wolford, 2010). The invasion and imposition of colonial rule in the region over 500 years ago had dramatic effects on the configuration of access and control over soil, water, timber, animals, and minerals, among other life-sustaining commons. The production of a new socio-natural order gave shape to classed, racialized, and gendered socio-environmental inequalities that, despite their historical and geographic transformations, remain in force (Cárdenas, 2012; Devine, 2016; Mollet, 2016, see also Göbel, Góngora-Mera, and Ulloa, 2014; Gras, 2013).

Silvia Rivera Cusicanqui's (1993: 58) definition of the "colonial horizon of long duration" explains well the violence based on the production of a matrix of domination that continually recreates dispossession, exclusion, and discipline (see also Rivera Cusicanqui, 2010a, 2010b). According to her, this colonial horizon is characterized by "very deep and ubiquitous structural phenomena, that range from everyday behaviour and micro-power spheres, to global society's structure and organization of state and political power" (1993: 58, author's translation). As noted, this multiscalar character of landgrabbing, both in spatial and temporal terms, is one of its main characteristics.

While the imposition of the liberal order is usually understood by official history as a transition to civilization and citizenship, authors such as Silvia Rivera Cusicanqui (2010a) and Fernando Coronil (2000) analyze the constitutive relation between colonialism and capitalism, as well as the continuities between colonial power and state power, particularly in relation to the appropriation of land and labour by republican elites. Rivera Cusicanqui (2010a) further illustrates how the imposition of a masculine and lettered territoriality implied the production of boundaries, property, and parental authority in detriment of indigenous peoples in Bolivia, particularly of women. Moreover, as Silvia Federici's work shows (2004), the expropriation, devaluation, and criminalization of women played a fundamental role in the making of *capitalist patriarchy*, and thus the story of landgrabbing in the Americas cannot be told without taking into account the rampant attacks to social reproduction and the waging of war against women.

Development, as the intertwined practices of capitalist expansion and the production of the Third World (Hart, 2001), is central to this process. From its long roots in colonial domination and the plantation economy, to its close relationship with imperialism and militarization, the implementation of the development project needs to be told in relation to the concentration of land use, access, and control. It requires too the constant epistemic work of reducing land to commodity, property, area, and resource, none of which can account for the complex material and symbolic ecologies that land encompasses. Imagined geographies of empty, otherwise-productive lands are functional to landgrabbing, as development practices and their colonial core remain a central mechanism of dispossession of local populations (Escobar, 1995; Ojeda et al., 2015; Velásquez Runk, 2015).

During the last decades of the 19th century and the first ones of the 20th century, various countries in the region implemented an agro-exporting model. State intervention concentrated on commercial production to satisfy foreign markets in Europe and the United States, including beef, cereals, sugar, coffee, and bananas. By the mid-century, different state projects promoted the industrialization of food production and, in the following decades, the Green Revolution marked the pace of vast political, economic, and socio-environmental transformations (McMichael, 1997). As Jacobo Grajales (2011: 789) notes a "highly inequitable agrarian economy" was crafted and consolidated in the name of economic development, giving form at the same time to "a certain political opportunity structure favourable to land spoiling and to the conversion of these spurious capitals into legal ones." Later on, in the 1990s, with neoliberal "structural adjustment policies," the agricultural sector was radically transformed by the liberalization of land markets and the reduction of trade barriers (Deere and de León, 2001; Zoomers and van de Haar, 2004). Landgrabbing, land concentration, and the foreignization of landed property rapidly gained ground in the region (Borras et al., 2012a, 2012b). Countries such as Argentina, Bolivia, Ecuador, Brazil, and Uruguay that conformed the "pink-tide" in Latin America because of their Left-wing orientation during the early 2000s, arguably went under a "post-neoliberal" phase; nonetheless, they too experienced an increase in extractivist ventures associated with large-scale agroindustry and mining (Petras, 2012).

Contemporary landgrabbing in the region: key elements

The resulting increase of privatization and enclosure, particularly in relation to the expansion of monocultures for food, feed, and fuel, has brought upon radical transformations in the ways in which land is understood, distributed, and managed in Latin America. These transformations have restricted local possibilities to sustain life. And while landgrabbing is not a new phenomenon, "the sophistication of mechanisms and the increased complexity of the actors and society-environment relations have not made land appropriation, exclusion, and control any less violent" (Peluso and Lund, 2011: 677).

Latin America presents a generalized landscape of vast inequality and land concentration. Even if lesser in scale compared to Africa and Asia, contemporary landgrabbing in the region has had similar effects over local populations. The sometimes gradual, less evident from a macroeconomic perspective, appropriation of land for extractivist purposes has dramatically changed rural lives and landscapes. The large implementation of flex crops (which can be used as food, feed, or fuel depending on markets values) such as soy and oil palm has intensified conflicts around farmlands, but also around small plots, gardens, forests, marshes, state lands, reservations, river banks, and bypaths, among other key and interconnected spaces. Many of these monocrops, unlike others historically implemented in the region such as tobacco, coffee, and bananas use very little and occasional workforce and thus can be defined as "labor-displacing ventures" (Li, 2011).

In different countries in the region, other forms of extractivism have accompanied this vast agroindustrial expansion, including mining and hydrocarbon concessions, cattle-ranching, drug trafficking, and projects of climate change mitigation, conservation, and nature tourism. And while mining ventures have been largely explored in the region, narco land grabs and greengrabbing dynamics are not often studied in their landgrabbing dimension. Authors such as Ballvé (2012) and McSweeney et al. (2017) have brought attention to agrarian change associated with drug trafficking including control grab and forest loss, which affect large regions in Colombia and Brazil. Moreover, as Grandia (2012: 237) suggests, there is the need to consider the "power assemblages emerging from narco/cattle/industrial/military land grabbing occurring in Petén," in Guatemala, and one can argue, in other countries such as Honduras and Nicaragua.

Likewise, greengrabbing needs further attention. Roughly defined as landgrabbing carried out in the name of nature (Fairhead, Leach, and Scoones, 2012), greengrabbing can result from dynamics of exclusion, subordination, and removal resulting from projects of biodiversity conservation, nature tourism, and climate change mitigation. While there is growing research on landgrabbing in relation to the expansion of agrofuels in Latin America and the Caribbean (e.g. Cárdenas, 2012; Johnson, 2014; Manzi, 2013), not enough studies have focused on conservation strategies (Ojeda, 2012; Devine, 2016; Grandia, 2012; Rocheleau, 2015) or carbon deposits (Ojeda, 2014; Osborne, 2011; Pskowski, 2013). This perhaps responds to a persistent division between agrarian studies and environmental studies, and the difficulty to think of land rights and biodiversity conservation as opposite or even exclusionary processes (Mollett and Kepe, 2018).

In their study based on 2011 data, Borras et al. find that "land grabbing is taking place, albeit unevenly, between and within countries in Latin America and the Caribbean" and "[c]urrent conditions and trends in land deals point towards further expansion and at a faster pace in the near future" (2014: 22). Among the particular characteristics of landgrabbing in the region, they point to:

> (i) the significance of private lands transacted, (ii) the critical role played by domestic elites as key investors, (iii) the significance of intra-regional companies, (iv) the minimal scope of land deals (private or public) with the Gulf States, China, South Korea and India, and (v) the fact that land grabbing in this region occurs in countries that do not fit the usual profile of 'fragile' or 'weak' states.
>
> *(Borras et al., 2014: 22)*

Some aspects need additional consideration. On the one hand, as the authors point out, contemporary landgrabbing in the region concentrates in private lands. Nevertheless, property regimes are unclear and private, state, and communal forms of tenure often overlap. Also, the concentration of land control in few hands does not necessarily implicate purchase. Lease, contract farming, and the capture of different stages of the value chain are important mechanisms of dispossession in the region. This makes it difficult to obtain reliable quantitative data on the real extent of landgrabbing, but also puts forward the need to overcome the generalized notion that the title guarantees ownership. Colonial property regimes involving the fixation and titling of land have been understood by liberal academic and development circles as a guarantee of tenure, and even as a way out of poverty (de Soto, 2000). On the contrary, titling has historically been a landgrabbing strategy (Grajales, 2011; Hetherington, 2009; Ybarra, 2009 see also Blomley, 2005 and Mansfield, 2008 for a critique). In countries like Colombia (Grajales, 2011) and Mexico (Rocheleau, 2015) there is also a close relationship between armed actors and property rights enforcing institutions.

On the other hand, the region's specific histories and geographies prove wrong the assumption that state absence or weakness facilitates landgrabbing. On the contrary, the Latin American and Caribbean case suggests that landgrabbing often occurs in contexts of transparency, accountability, legality, and legitimacy. The state plays a key role in facilitating and maintaining landgrabbing both by state sanction of illegal and direct modes of expropriation, and by its role in maintaining a deeply uneven classed, racialized, and gendered order. This is clearly the case of Colombia (Camargo and Ojeda, 2017; Grajales, 2013; Vélez-Torres, 2014) and of most countries in the region including Bolivia (Rivera Cusicanqui, 2010a) and Guatemala (Cabnal, 2010), where different forms of violence – direct, structural, symbolic, and slow – coalesce. This requires too to go beyond the happy marriage between capitalism and state power, to look at relations of differentiation, subordination, and exclusion manifested in the conservation, peace-building, and aid industries as well (Camargo and Ojeda, 2017).

Conclusions

This chapter has highlighted key elements in the comprehension of contemporary, yet deeply rooted histories and geographies of landgrabbing in Latin America and the Caribbean. Seeking to de-center the debate on large-scale land grabs, I have argued for the need to pay attention to the sedimented landscapes of dispossession in the region. While the rapid expansion of the industrial production of flex crops like soy, oil palm, and timber has had major socio-environmental effects that we are just starting to understand, there is the need to study other mechanisms of dispossession that work at lesser-scales, such as green grabs and narco land grabs. Similarly, there is the increased need to understand state-sanctioned violence and its coalescence with development, including the reification of exclusionary property regimes and the constant reproduction of imagined geographies of misused lands.

It requires too to expand our notion of land to grasp on-the-ground socio-environmental relations in all their complexity. As Dianne Rocheleau (2015: 715) suggests, "[e]nvironmental justice with respect to land and territory requires that we consider more than just economic rights to land and legally recognized property. Land and environmental policy needs to recognize and respect networked cultures and ecologies and their construction of, and roots in, multiple territories" (Rocheleau, 2015: 715). This makes necessary to account for land and water, and many other intertwined socioenvironments, but more broadly to the symbolic and material ecologies that comprise them. After all, what is being grabbed is life in itself.

This is thus a call for a careful analysis of historical processes of differentiation, exclusion, and dispossession and their materializations in space. This requires paying attention to land concentration and the role of financial capital, but perhaps more importantly to inequality, subordination, death, and suffering so profoundly shaped by race and gender, as they are continually recreated in an everyday scale. It requires too to fully account for strategies of defence and care of territory, taking seriously different forms of dispossession and their long-lasting effects on multiple temporary and spatial scales.

References

Ballvé, T. (2012) Everyday state formation: Territory, decentralization, and the NarcoLandGrab in Colombia. *Environment and Planning D: Society and Space* 30(4): 603–622.

Bebbington, A., Bornschlegl, T. and Johnson, A. (2013) Political economies of extractive industry: From documenting complexity to informing current debates. *Development and Change. Virtual Issue* 2: 1–16.

Blomley, N. (2005) Remember property? *Progress in Human Geography* 29(2): 126–29.

Borras, S. and Franco, J. (2012) Global land grabbing and trajectories of agrarian change: A preliminary analysis. *Journal of Agrarian Change* 12(1): 34–59.

———, Gómez, S., Kay, C. and Spoor, M. (2012a) Land grabbing in Latin America and the Caribbean. *The Journal of Peasant Studies* 39(3–4): 845–872.

————, Kay, C. and Spoor, M. (2014) Land grabbing in Latin America and the Caribbean viewed from broader international perspectives. In S. Gómez (ed) *The Land Market in Latin America and the Caribbean*. Santiago, Chile: FAO, pp. 21–53.

Borras, S., Kay, C., Gómez, S. and Wilkinson, J. (2012b) Land grabbing and global capitalist accumulation: Key features in Latin America. *Canadian Journal of Development Studies/Revue canadienne d'études du développement* 33(4): 402–416.

Cabnal, L. (2010) Acercamiento a la construcción de la propuesta de pensamiento epistémico de las mujeres indígenas feministas comunitarias de Abya Yala. In *Feminismos diversos: el feminismo comunitario*. Madrid: Acsur Las Segovias, pp. 11–25.

Camargo, A. and Ojeda, D. (2017) Ambivalent desires: State formation and dispossession in the face of climate crisis. *Political Geography* 60(1): 57–65.

Cárdenas, R. (2012) Green multiculturalism: Articulations of ethnic and environmental politics in a Colombian 'Black Community'. *The Journal of Peasant Studies* 39(2): 309–333.

Coronil, F. (2000) Naturaleza del poscolonialismo: del eurocentrismo al globocentrismo. In E. Lander (ed) *La colonialidad del saber: eurocentrismo y ciencias sociales: Perspectivas Latinoamericanas*. Buenos Aires: CLACSO.

CRAADT – Concejo Regional de Autoridades Agrarias de las regiones Montaña Costa Chica del Estado de Guerrero, en Defensa del Territorio contra la Minería y la Reserva de la Biosfera. (2017) *La defensa del territorio es la defensa de la vida*. Declaration of the national meeting against the mining extractive model, Minaltepec, Guerrero, Mexico, 8 October. https://mx.boell.org/es/2017/10/11/la-defensa-del-territorio-es-la-defensa-de-la-vida

De la Cadena, M. (2010) Indigenous cosmopolitics in the Andes: Conceptual reflections beyond 'politics'. *Cultural Anthropology* 25(2): 334–370.

De Schutter, O. (2011) Forum on global land grabbing: How not to think land grabbing: Three critiques of large-scale investments in farmland. *Journal of Peasant Studies* 38(2): 249–279.

de Soto, H. (2000) *The Mystery of Capitalism: Why Capitalism Triumphs in The West and Fails Everywhere Else*. New York: Basic Books.

Deere, C. D. and de León, M. (2001) Who owns the land? Gender and land-titling programmes in Latin America. *Journal of Agrarian Change* 1(3): 440–467.

Devine, J. (2016) Community forest concessionaires: Resisting green grabs and producing political subjects in Guatemala. *The Journal of Peasant Studies* 45(3): 565–584.

Edelman, M. and León, A. (2013) Cycles of land grabbing in Central America: An argument for history and a case study in the Bajo Aguán, Honduras. *Third World Quarterly* 34(9): 1697–1722.

Escobar, A. (1995) *Encountering Development: The Making and Unmaking of the Third World*. Princeton, NJ: Princeton University Press.

————. (2008) *Territories of Difference: Movement, Life, Redes*. Durham, NC: Duke University Press.

FAO. (2012) *Dinámicas del mercado de la tierra en América Latina y el Caribe: concentración y extranjerización*. Rome: FAO.

Fairhead, J., Leach, M. and Scoones, I. (2012) Green grabbing: A new appropriation of nature? *The Journal of Peasant Studies* 39(2): 237–261.

Federici, S. (2004) *Caliban and the Witch: Women, the Body and Primitive Accumulation*. New York: Autonomia.

Grajales, J. (2011) The rifle and the title: Paramilitary violence, land grab and land control in Colombia. *The Journal of Peasant Studies* 38(4): 771–792.

————. (2013) State involvement, land grabbing and counter insurgency in Colombia. *Development and Change* 44(2): 211–232.

Grandia, L. (2012) *Enclosed: Conservation, Cattle, and Commerce Among the Q'eqchi' Maya Lowlanders*. Seattle: University of Washington Press.

Gras, C. (2013) *Agronegocios en el Cono Sur. Actores sociales, desigualdades y entrelazamientos transregionales*. DesiguALdades.net Working Paper Series 50. Berlin: desiguALdades.net

Göbel, B., Góngora-Mera, M. and Ulloa, A. (2014) (eds) *Desigualdades socioambientales en América Latina*. Bogotá: Universidad Nacional de Colombia, Ibero-Amerikanisches Institut.

Goldfarb, L. and Zoomers, A. (2014) The rapid expansion of genetically modified soy production into the Chaco region of Argentina. In M. Kaag and A. Zoomers (eds) *The Global Land Grab: Beyond the Hype*. London; New York: Zed Books, pp. 73–95.

Hart, G. (2001) Development critiques in the 1990s: Culs de sac and promising paths. *Progress in Human Geography* 25(4): 649–658.

Hetherington, K. (2009) Privatizing the private in rural Paraguay: Precarious lots and the materiality of rights. *American Ethnologist* 36(2): 224–241.

Johnson, A. (2014) Ecuador's national interpretation of the Roundtable on Sustainable Palm Oil (RSPO): Green-grabbing through green certification? *Journal of Latin American Geography* 13: 183–204.

León, A. (2017) Domesticando el despojo: palma africana, acaparamiento de tierras y género en el Bajo Aguán, Honduras. *Revista Colombiana de Antropología* 53(1): 151–185.

Li, Tania. 2011. Centering labor in the land grab debate. *The Journal of Peasant Studies* 38(2): 281–298.

Mansfield, B. (2008) (ed) *Privatization: Property and the Remaking of Nature-Society Relations*. London: Wiley-Blackwell.

Manzi, M. (2013) *Agrarian Social Movements and the Making of Agrodiesel Moral Territories in Northeast Brazil*. Ph.D dissertation, Clark University.

Márquez, F. (2015) *A las mujeres que cuidan de sus territorios como a sus hijas y sus hijos. A las cuidadoras y los cuidadores de la Vida Digna, Sencilla, Solidaria*. Open letter to the Government and the country. Bogotá, April 24. http://mujeresnegrascaminan.com/nos-levantamos-de-la-mesa/

McMichael, P. (1997) Rethinking globalization: The agrarian question revisited. *Review of International Political Economy* 4(4): 630–662.

McSweeney, K., Richani, N., Pearson, Z., Devine, J. and Wrathall, D. (2017) Why do narcos invest in rural land? *Journal of Latin American Geography* 16(2): 3–29.

Mollett, S. (2016) The power to plunder: Rethinking land grabbing in Latin America. *Antipode* 48(2): 412–32.

———— and Kepe, T. (2018) (eds) *Land Rights, Biodiversity Conservation and Justice: Rethinking parks and people*. New York; London: Routledge.

Moore, D. (2005) *Suffering for Territory: Race, Place and Power in Zimbabwe*. Durham, NC: Duke University Press.

Ojeda, D. (2012) Green pretexts: Ecotourism, neoliberal conservation and land grabbing in Tayrona National Natural Park, Colombia. *The Journal of Peasant Studies* 39(2): 357–375.

————. (2014) Descarbonización por despojo: desigualdades socioecológicas y las geografías del cambio climático. In b. Göbel, m. Góngora-Mera and A. Ulloa (eds) *Desigualdades socioambientales en América Latina*. Bogotá: Universidad Nacional de Colombia, Ibero-Amerikanisches Institut.

————, Petzl, J., Quiroga, C., Rodríguez, A. C. and Rojas, J. (2015) Paisajes del despojo cotidiano: Acaparamiento de tierra y agua en Montes de María, Colombia. *Revista de Estudios Sociales* 54: 107–119.

Osborne, T. (2011) Carbon forestry and agrarian change: Access and land control in a Mexican rainforest. *Journal of Peasant Studies* 38(4): 859–883.

Peluso, N. and Lund, C. (2011) New frontiers of land control: Introduction. *Journal of Peasant Studies* 38(4): 667–681.

Petras, J. (2012) Extractive capitalism and the divisions in the Latin American progressive camp. *Global Research*, 3 May.

Pskowski, M. (2013) Is this the future we want? The green economy vs. climate justice. *Different Takes* 79. http://hdl.handle.net/10009/933

Rivera Cusicanqui, S. (1993) La raíz: colonizadores y colonizados. In S. Rivera Cusicanqui and R. Barrios (eds) *Violencias encubiertas en Bolivia*. La Paz: CIPCA/Aruwiyiri, pp. 25–139.

————. (2010a) *Violencias (re)encubiertas en Bolivia*. La Paz: Ediciones La Mirada Salvaje.

————. (2010b) *Ch'ixinakax utxiwa: Reflexión sobre prácticas y discursos descolonizadores*. Buenos Aires: Ediciones Tinta Limón.

Rocheleau, D. (2015) Networked, rooted and territorial: Greengrabbing and resistance in Chiapas. *The Journal of Peasant Studies* 42(3–4): 695–723.

Sauer, S. and Almeida, W. (2011) *Terras e territórios na Amazônia: demandas, desafios e perspectivas*. Brasilia: UNB.

Svampa, M. and Viale, E. (2014) *Maldesarrollo: La Argentina del extractivismo y el despojo*. Buenos Aires: Katz.

van Noorloos, F. (2014) Transnational land investment in Costa Rica: Tracing residential tourism and its implications for development. In M. Kaag and A. Zoomers (eds) *The Global Land Grab: Beyond the Hype*. London; New York: Zed Books, pp. 86–99.

Vásquez García, V. (2018) Land grabbing in Mexico: Extent, scale, purpose and novelty. *Revista Mexicana de Ciencias Forestales* 8(44): 1–22.

Velásquez Runk, J. (2015) Creating wild Darién: Centuries of Darién's imaginative geography and its lasting effects. *Journal of Latin American Geography* 14(3): 127–156.

Vélez-Torres, I. (2014) Governmental extractivism in Colombia: Legislation, securitization and the local settings of mining control. *Political Geography* 38: 68–78.

Wolford, W. (2010) *Contemporary Land Grabs in Latin America.* Paper presented at the TNI-ICASFIAN Side Event to the 13th Session of the FAO World Food Security Council, October, Rome.

Ybarra, M. (2009) Violent visions of an ownership society: The land administration project in Petén, Guatemala. *Land Use Policy* 26(1): 44–54.

Zoomers, A. and van de Haar, G. (2004) (eds) *Current Land Policy in Latin America: Regulating Land Tenure Under Neo-Liberalism.* Amsterdam: Royal Tropical Institute.

35

PROTECTED AREAS AND BIODIVERSITY CONSERVATION

Robert Fletcher

Introduction

This chapter reviews the rise of and challenges faced by protected areas (PAs) aimed at biodiversity conservation throughout Latin America over the past half century in particular. It charts a similar process throughout the region whereby a global campaign championed by international environmental non-governmental organizations (NGOs) and financial institutions (IFIs) helped to inspire and fund the establishment of nationwide systems of PAs. While these PAs were initially administered predominantly in classic "fortress" fashion, in recent decades this approach has been complemented by introduction of a community-based conservation (CBC) strategy that seeks to enlist local residents as stakeholders and decision makers, introducing a series of market-based instruments (MBIs) including ecotourism and payment for environmental services (PES) to generate revenue to support this. More recently, this approach has been intensified by the rise of "post-neoliberal" politics in a number of societies that pursues a better integration of environmental and developmental concerns. Yet this has been challenged by the expansion of raw material extraction driven in large part by expanding trade relations with East Asia and elsewhere. As a result, protected areas have become key sites of renewed contestation between forces of conservation and extraction. The chapter discusses these developments and their implications for the future of biodiversity conservation in the region.

Latin America is of course an enormous, complex space containing countless continental, national, and local sociocultural variations as well as diverse landscapes and ecosystems, from Mexico's Sonoran desert to the lowland rainforests of Central America and the Amazon Basin to the Andean páramo and the Patagonian steppes. Hence this short chapter can provide only a cursory overview of the most common overarching dynamics occurring within the region. The discussion is based on my long-term research over the past two decades. Most of this has taken place in Costa Rica, where I lived and worked for six years, as well as Chile, where I did my master and doctoral thesis study, but I have conducted shorter research periods in Colombia and Pero as well. I supplement all of this direct empirical experience with a survey of research conducted by others throughout the region.

The rise of conservation

In the post WWII period, conservationists around the world united within the International Union for the Conservation of Nature (IUCN) to promote PA creation and management in societies worldwide. The aim eventually became to convince all governments to set aside a substantial portion (ideally at least 10%) of their territory (both terrestrial and marine) as formally protected terrain (see West, Igoe, and Brockington, 2006).[1] Through this campaign, the global conservation "estate" expanded exponentially in subsequent years, growing from a total area of 1 million km² in 1962, located "mainly in North America and colonial Africa" (Borgerhoff Mulder and Coppolillo, 2005: 28), to between 20.3 and 21.5 million km² worldwide by the end of the century (West, Igoe, and Brockington, 2006).[2]

These PAs are of different types and designations. The IUCN distinguishes six general categories of PA based on the level of protection and degree of human resource use allowed, as depicted in Table 35.1.

Within Latin America, this campaign for PA creation built on conservation efforts already underway by domestic actors. The region's first National Park – Desierto de los Leones – was established in 1917 by Mexico (Simonian, 1995), followed by Chile (1926), Argentina (1934), Brazil and Venezuela (both 1937), then Bolivia (1939). In this way, PAs developed slowly and sporadically throughout the region until the big global post-war push. As elsewhere, this push spurred a dramatic wave of PA expansion to include Ecuador (1959), Colombia (1960), Peru (1961), Costa Rica (1971), Dominica (1975), Panama (1976), and Belize (1986) as well as numerous other nations and territories. At time of writing, the IUCN's authoritative World Database of Protected Areas (WDPA) registers 7,791 PAs of various types throughout the region, as detailed in Table 35.2.

Table 35.1 IUCN protected area management categories

Category 1a:	Strict Nature Reserve: protected area managed mainly for science.
Category 1b:	Wilderness Area: protected area managed mainly for wilderness protection.
Category II:	National Park: protected area managed mainly for ecosystem protection and recreation.
Category III:	National Monument: protected area managed mainly for conservation of specific natural features.
Category IV:	Habitat/Species Management Area: protected area managed mainly for conservation through management intervention.
Category V:	Protected Landscape/Seascape: protected area managed mainly for landscape/seascape conservation and recreation.
Category VI:	Managed Resource Protected Area: protected area managed mainly for the sustainable use of natural ecosystems.

Source: IUCN, 1994.

Table 35.2 Protected areas of Latin America and the Caribbean (nations and territories)

Country/Territory	Total # PAs	Total Area PAs (km²)		Total Area PAs (%)	
		Terrestrial	Marine	Terrestrial	Marine
Anguilla (UK)	11	5.65	22.58	6.6	0.02
Antigua and Barbuda	16	84.54	196.92	18.57	0.18
Argentina	385	247,718.54	43,542.05	8.89	4.02
Aruba	3	35.83	0.02	18.92	0.0

Country/Territory	Total # PAs	Total Area PAs (km²)		Total Area PAs (%)	
		Terrestrial	Marine	Terrestrial	Marine
Bahamas	54	4,930.16	47,355.96	36.63	7.92
Barbados	9	5.64	10.82	1.27	0.01
Belize	120	8,401.66	3,653.60	37.68	10.08
Bermuda (UK)	28	1.51	0.25	2.08	0.0
Bolivia	167	336,405.70	0.0	30.87	0.0
Bonaire, St. Eustatius & Sabo (NL)	13	91.58	2,755.53	28.37	10.97
Brazil	2,197	2,468,491.64	61,881.39	28.94	1.68
British Virgin Islands (UK)	88	16.00	3.31	9.11	0.0
Cayman Islands (UK)	58	31.13	92.90	10.76	0.08
Chile	187	139,814.59	460,645.23	18.40	12.60
Colombia	621	162,123.47	76,391.62	14.16	10.45
Costa Rica	187	14,252.97	4,801.57	27.60	0.83
Cuba	226	18,480.57	15,818.75	16.55	4.32
Curaçao	13	69.73	10.43	15.46	0.03
Dominica	10	168.47	30.27	21.99	0.11
Dominican Republic	93	11,167.40	24,598.92	23.02	9.08
Ecuador	76	51,695.53	141,05	20.03	13.06
El Salvador	168	1,805.57	664.80	8.78	0.71
Grenada	49	34.73	11.00	9.3	0.04
Falkland Islands (UK)	33	61.10	52.09	0.49	0.01
French Guyana (FR)	38	43,937.55	1,362.33	52.91	1.0
Guadeloupe (FR)	85	1,170.30	90,957.92	69.72	99.91
Guatemala	259	34,896.27	955.83	31.57	0.81
Guyana	5	18,453.60	17.40	8.74	0.01
Haiti	8	534.08	0.0	1.95	0.0
Honduras	105	32,220.47	9,200.59	28.44	4.18
Jamaica	140	1,760.20	1,860.00	15.92	0.75
Martinique (FR)	61	925.95	47,590.85	80.52	99.89
Montserrat (UK)	1	11.18	0.0003	11.11	0.0
Mexico	1,192	284,976.78	715,466.56	14.5	21.78
Nicaragua	95	48,105.35	6,660.35	37.23	2.97
Panama	95	15,772.99	5,593.12	20.89	1.68
Peru	244	276,192.16	4,031.28	21.27	0.48
Puerto Rico (US)	83	657.22	3,077.93	7.27	1.75
St. Barthélemy	5	5.11	4,243.89	20.36	98.29
St. Kitts and Nevis	21	8.98	17.44	3.32	0.17
St. Lucia	42	116.65	34.01	18.75	0.22
Sint Maarten	2	0.0001	43.28	0.0	8.7
St. Martin	24	7.61	1,030.54	12.77	96.43
St. Vincent & Grenadines	55	91.90	80.36	22.42	0.22
Suriname	22	21,425.70	1,980.93	14.52	1.54
Trinidad and Tobago	44	1,594.83	37.07	30.59	0.05
Turks and Caicos Islands (UK)	34	451.74	149.82	44.37	0.1
US Virgin Islands (US)	39	51.82	306.28	13.79	0.85
Uruguay	29	6,150.09	931.55	3.45	0.72
Venezuela	251	496,701.20	16,499.94	54.14	3.49
TOTAL:	**7791**	**4,752,113.44**	**1,654,669.28**	**(a)21.10**	**(a)10.64**

Source: World Database on Protected Areas (www.protectedplanet.net); as of 13/11/2017.

As throughout the world, creation of marine PAs has lagged far behind terrestrial ones. In IUCN's ideal vision, each country would protect as much of its waters as its land, but in practice very few nations anywhere have achieved this. This is apparent in Latin America, where, as Table 35.2 shows, nations have protected an average of only around 10% of their marine area as opposed to more than 20% of their land. And given that the most expansive marine protection has occurred in the smallest – mostly island – territories, these figures are far more skewed when one considers the total relative area of land and water protection, respectively.

Conservation governance

Initially, Latin America's PAs were predominant managed in conformance with the widely practiced approach that World Wildlife Fund (WWF) calls "Paradigm 1" conservation[2] and Brockington (2002) terms the "fortress" strategy. Modelled on the US National Park system, this approach calls for the national states to enforce strict boundaries and terms of use and impose sanctions for their violation (hence it is often also called a "fences and fines" strategy). As in many places, however, the widespread social impacts of this approach, including often violent displacement and marginalization of local people (see especially Brockington, 2002; Dowie, 2009) led to a growing critique and shift in focus within the region towards CBC. CBC calls for integration of PAs within integrated-conservation-and-development projects (ICDPs) – WWF's "Paradigm 2" conservation[3] – the main aim of which is to deliver sustainable income-generating opportunities to members of park-adjacent communities and thereby encourage the latter to refrain from extracting resources from within the PA itself (Borgerhoff Mulder and Coppolillo, 2005).

In recent years, however, CBC, along with the ICDPs it promotes, have been questioned as well, due in large part to the fact that despite many years of experimentation and refinement clear-cut examples of success remain scarce (Wells and McShane, 2004; West, 2006). This has prompted a variety of responses, including calls for renewed protectionism (e.g., Wuerthner, Crist, and Butler, 2015) as well as expansion to a total landscape approach (Sayer, 2009). Probably the most widespread response has been growth of so-called "neoliberal conservation" promoting MBIs intended to commodify *in situ* natural resources for economic return (see Büscher, Dressler, and Fletcher, 2014). While the term "neoliberalism" has been employed in diverse ways by different commentators (see Flew, 2012), most precisely defined it refers to the political-economic philosophy promoting interrelated principles of decentralization, deregulation (or more commonly reregulation from states to non-state actors), marketization, privatization, and commodification (Castree, 2010). This philosophy, advanced in the post-war period in specific challenge to the rise of the welfare state, gained prominence in the 1980s with its adoption as the basis of the Reagan and Thatcher administrations in the US and UK, respectively, and subsequent promotion throughout the lower-income world by IFIs including the World Bank and International Monetary Fund (IMF) as a component of "structural adjustment" programs (SAPs) (Harvey, 2005).

Within conservation practice, neoliberalization is identified in a variety of trends, including the growing prominence and power of non-state actors including big international environmental NGOs such as Conservation International (CI) and the Nature Conservancy (TNC) (Chapin, 2004); increasing partnership among these BINGOs, corporations, and IFIs (Levine, 2002) to generate funding; creation of markets for trade in natural resources; privatization of resource control within such markets; commodification of resources to facilitate their trading; and devolution of resource control to NGOs and local community members. The common logic of neoliberal conservation holds that for resources to be conserved they must be economically valuable relative to other (extractive) uses.

The rise of ICDPs in the 1990s, of course, was part of this neoliberalization, in that they are generally specifically designed to: (1) decentralize and devolve governance from states to local community members, with NGOs commonly serving as intermediaries; and (2) support conservation by providing alternative livelihoods to local people living within PA buffer zones, primarily through increased integration within larger commodity markets. Increasingly, however, such indirect marketization is being superseded by the approach WWF labels "Paradigm 3" conservation,[4] which goes further to actually harness resources within the PA itself as a source of revenue in order to demonstrate the value of resources left *in situ* and thus encourage their preservation via direct economic incentives – a strategy that WWF calls "directly linking community benefits to conservation"[5] and West (2006) "conservation-as-development."

In this respect, a wide variety of prevalent conservation mechanisms, including ecotourism, bioprospecting, and payment for environmental services (PES), are considered aspects of neoliberal conservation (Büscher, Dressler, and Fletcher, 2014). Ecotourism is probably the most widespread of these. Commonly defined as "[r]esponsible travel to natural areas that conserves the environment and improves the well-being of local people" (Honey, 2008: 6), ecotourism is often grounded in a "stakeholder theory" asserting that "people will protect what they receive value from" (2008: 14). Partly due to its popularity as a conservation mechanism, ecotourism has grown dramatically in the past several decades to become perhaps "the most rapidly expanding sector of the tourism industry" (Honey, 2008: 6), which rivals oil production for status of world's largest industry.

Another aspect of neoliberal conservation has been the rise of privately owned and managed protected areas (PPAs) in addition to state-managed ones. Accurate numbers are difficult to find due to ambiguity in PPAs' definition and absence of a specific official database (see Holmes, 2013), but the WDPA records 14,296 official PPAs worldwide (Bingham et al., 2017: 18). Within Latin America, many PPAs are organized into national and regional networks. Again, hard numbers are scarce; Holmes (2015) documents 312 PPAs in Chile covering 2.12% of the nation's surface while in Costa Rica Langholz (1999) estimated some time ago that PPAs encompassed 63,832 hectares or 1.2% of the land. In some places PPAs can overlap with state-managed PAs: in Costa Rica they can be registered in the National Wildlife Refuge category (Langholz, Lassoie, and Schelhas, 2000); in Chile they can be recognized as National Sanctuaries; and in Mexico a specific modality, the Voluntary Conservation Use Area, has been created for them (Bingham et al., 2017).

Focus on Costa Rica

Costa Rica, where most of my own research has been conducted, illustrates more general dynamics of conservation development in the region. Despite its tiny size, the country has long been widely heralded as a world leader in conservation, leading Evans (1999) to brand it "The Green Republic." International interest in the country was inspired by documentation of its tremendous biodiversity: commonly cited figures claim 4–5% of the world's total contained in only 0.035% of land mass, which significant rates of endemism (see Honey, 2008: 160). This is due to the country's location at the "transcontinental meeting point" between North and South America (Evans, 1999: 3), bringing together both ecosystems and making it the "only place in the world that is 'both interoceanic and intercontinental'" (Hall, 1984, cited in Evans, 1999: 3).

To protect this impessive biodiversity, Costa Rica has developed a national system of PAs that now officially encompasses 27.60% of its land (though less than 1% of its waters) – among the highest rates of protection in the world (although not in the region) and still far exceeding even the Aichi Targets' ambitious aims. As Evans (1999: 7) relates, "The terms 'model,' 'example,' 'beacon,' 'showcase,' 'prototype,' 'the ideal,' and 'wave of the future' have all been used repeatedly to describe Costa Rica's national park system."

This PA network developed piecemeal over a number of decades in response to threats posed by agricultural expansion. In the 1950s a full 50% of Costa Rica's territory was forested (Evans, 1999). Subsequent diffusion of the chainsaw, however, dramatically transformed this picture, allowing the agricultural frontier to expand much more quickly. This was encouraged by a national forestry policy decreeing that new land could be claimed by "improving" (i.e., clearing) it. In the 1970s, deforestation accelerated due to the growth of the global fast food industry's demand for beef. By the end of the 1980s, Costa Rica had become the foremost supplier to the North American market while its deforestation rate was among the highest in the world.

Meanwhile, concern for environmental protection in the face of such dynamics has been building. In the 1960s, the alarming pace and impacts of deforestation were increasingly recognized by the biologists, mostly European and North American, who were studying the country's biodiversity (Evans, 1999). They and others began to campaign for protection of the remaining rainforest, appealing to private landholders as well as the national government. While initially facing "nearly total indifference to the problem of environmental protection" (Boza, in Steinberg, 2001: 3), both efforts eventually found sympathetic audiences. Campaigning on the private front resulted in establishment, in 1973, of the famous Monteverde Cloud Forest Reserve, owned by an immigrant community of Quakers from the United States and managed by the Tropical Sciences Center (TSC). Meanwhile, political lobbying led to a new 1969 Forestry Law that established a National Park Service, under the direction of Mario Boza and Alvaro Ugalde, that set about expropriating representative parcels of forest throughout the country's numerous eco-zones and placing them under state protection.

All of this changed dramatically following Costa Rica's implication in the 1980s debt crisis. Over the next decade and a half the country underwent three distinct rounds of structural adjustment overseen by the World Bank and IMF as conditionality for new loans (Edelman, 1999). In addition to mandating substantial cutbacks in state spending as well as the privatization of governing institutions in a variety of sectors, SAPs encouraged Costa Rica to relax the strong system of import tariffs erected during the Great Depression to protect the domestic economy from foreign competition. They also prescribed introduction of a number of so-called non-traditional exports, including textile production as well as pineapple and ornamental plant cultivation, to enhance international competitiveness, as well as provision of tax breaks and relaxation of legal barriers to external land ownership to encourage foreign direct investment (FDI).

In the environmental realm, neoliberalization had dramatic ramifications. As Steinberg (2001: 75) describes, "The Park Service staff was reduced by a fifth and in real terms the equipment and maintenance budget fell by an estimated 80 percent or more between 1980 and 1990." This decreased funding for park management led to a "governance gap" increasingly filled by NGOs, both domestic and international, grown fat on funding from IFIs and private donors as an ostensibly more efficient and flexible alternative to unwieldy state institutions (Levine, 2002). This allowed these organizations to implement independent conservation policies increasingly espousing a variety of market-based strategies, including several well-known debt-for-nature swaps negotiated with the Costa Rican state (Edelman, 1999). This quickly developed into a de facto division of labour of sorts between outside interests and the Costa Rican government, with the state providing predominantly fortress-style protection within its national parks, while international NGOs took responsibility for most of the CBC work to support this protection. Meanwhile, liberalized foreign ownership regulation led to a wave of land acquisition by external actors, such that by the early 1990s approximately 80% of the country's beachfront property lay in foreign hands (Honey, 2008: 164). This spurred the development of the extensive network of

PPAs noted earlier, owned both by NGOs and discrete individuals who began to take an active role in the nation's conservation (Langholz, 1999). Many of these new land owners invested in tourism development, helping to spur an international industry as yet another non-traditional "export" of sorts that rapidly replaced the mostly domestic tourism existing prior (Honey, 2008).

In the midst of this reform, in 1994, the formally centralized National Park Service was replaced by a decentralized National System of Protected Areas (SINAC), which organized the country into 11 dispersed "conservation areas" encompassing the numerous national parks and other protected areas previously administered by the Park Service (Evans, 1999). The new conservation strategy promoted under the SINAC system to replace the state-centered fortress strategy dominant in the previous era embodied this same tension, dividing each conservation area into three distinct land-use categories: "*Core areas* subject to absolute protection"; "*Buffer zones*, or multiple use areas"; and "*Intensive extraction zones*" (Brandon, 2004: 301).

By this time, however, the state-based conservation campaign had begun to wane, with most of the land capable of expropriation already acquired and funds for existing park maintenance – let alone new park creation – increasingly scarce. Despite these enclosure efforts, Costa Rica's forest continued to disappear at an impressive rate, such that by 1990 the percentage of total forested land had diminished to under 25% total (Evans, 1999). Notwithstanding continued issues of illegal logging and other forms of encroachment in the national parks, most of the remaining deforestation at this time was occurring on private land. Hence, mechanisms were needed to discourage deforestation on these lands in the absence of formal state regulation. In 1997, this led to establishment of an innovative nationwide PES program. Titled *Pago por Servicios Ambientales* (PSA), Costa Rica's program is commonly heralded as an exemplar of best practices, having "pioneered the nation-wide PES scheme in the developing world" (Daniels et al., 2010: 2116). The program recognizes four distinct ecosystem services that preserved forests are seen to provide: (1) carbon sequestration; (2) clean water; (3) biodiversity conservation; and (4) scenic beauty. The program provides direct payments to the owners of forest parcels for the services their land provides, intended to cover the opportunity costs of refraining from timber extraction or other alternate land uses.

From its inception, the program has been explicitly promoted as a market-based mechanism by many supporters (see Heindrichs, 1997). Both the program and law that founded it were instituted as part of the conditionality attendant to a third SAP, which receipt of the nation's third major World Bank loan was attached (Daniels et al., 2010). In forest management, the 1996 law proclaimed that "the current promotional (subsidy-based) system must be replaced by new, creative mechanisms to revive the forestry sector" (quoted in Heindrichs, 1997: 28). PSA has proven quite popular over its lifetime, such that by 2008 a total of 668,369 hectares had been officially enrolled, the majority (89%) under existing forest protection (Daniels et al., 2010: 2118). Moreover, the program continues to receive at least five times the number of applications that it is able to fund (Sierra and Russman, 2006). The principle source of program financing is a state tax on fossil fuel use, but significant funds are also provided by grants and loans from international actors including the World Bank and the Norwegian government (Daniels et al., 2010).

In large part due to its profusion of PAs, Costa Rica has also become "ecotourism's poster child" (Honey, 2008: 160). This reputation was enhanced by the National Tourism Bureau's (ICT) wildly successful slogan claiming that the country contains "No Artificial Ingredients." Currently, Costa Rica receives more than two million international visitors per year[6] and at least 60% of these are "nature bound" (Honey, 2008: 164). By 1993, tourism had become Costa Rica's largest economic sector, generating approximately "20 to 22 percent of Costa Rican's foreign exchange earnings and 7 to 8 percent of its GDP" (Honey, 2008: 163–164) and remained so until recently edged out by high-tech manufacturing.

At the same time that it is celebrated for its conservation achievements, however, Costa Rica is also frequently criticized for diametrically opposed behaviour. Thus Evans (1999: xii) highlights a "grand contradiction" represented by "Costa Rica's development of extraordinary national parks simultaneous to massive deforestation in unprotected areas." Responding to Boza's (1993) lauding of the country's environmental achievements, in 1994 Robert Hunter published a scathing critique asking "Is Costa Rica truly conservation-minded?" and answering his own question by describing "an almost anticonservation trend" in the country (Hunter, 1994: 595). Hunter cited a litany of environmental abuses, including extensive forest clearing for banana production in the northeast Sarapiquí region, "[d]ry season burning in the Santa Rosa National Park," and "gold mining in the Corcovado National Park" (ibid.). He concluded by admonishing Costa Rica for "adopting a two-face policy on environmental and conservation matters . . . aided and abetted by those who seek personal advantage while hiding behind a mask of care and concern for nature" (ibid.). Similar critiques have been levelled consistently in subsequent years (Evans, 1999; Vandermeer and Perfecto, 2005).

Like most nations around the world, Costa Rica's government, under pressure from numerous international interests, is concerned to support both development and conservation simultaneously. Thus Costa Rica, as most elsewhere in the tropics, has become "characterized by a mosaic of large plantation-type agriculture (some of it intensive and some extensive, including pastures) interspersed with medium and small farms and forest fragments" (Perfecto and Vandermeer, 2008: 175).

The extractive imperative

This dynamic points to a more widespread tension between conservation and extractive industry throughout the region, whose place in the global economy, despite decades of efforts to generate domestic industry via import substitution (ISI) measures, has remained primarily as an exporter of raw materials for processing elsewhere (Miller, 2007). Following widespread neoliberalization via SAPs in the 1990s, during which ISI was largely abandoned in favour of export-led development, such extraction increased dramatically, placing increasing pressure on natural resources at the same time as they were gaining growing attention from international conservationists as potential offsets for destructive activity elsewhere (Robinson, 2008). Costa Rica exemplifies this dynamic, as previously described. Such pressures have intensified greatly in recent decades due in large part to a dramatic increase in FDI in resource extraction in the region by representatives of East Asian countries and others in quest of raw materials to fuel their own economic expansion (Arsel et al., 2016).

The problems posed by this dynamic have been complicated in recent years by the rise of so-called "post-neoliberal" regimes. Within Latin America, such regimes are commonly divided into two general camps: a more radical one comprising Chavez' Venezuela, Morales' Bolivia, and arguably, Correa's Ecuador, on the one hand, and another encompassing more mainstream regimes such as Kirchner's Argentina, Lula's Brazil, Bachelet's Chile, Ortega's Nicaragua, and Tambo's Peru, on the other (see especially Yates and Bakker, 2014). Despite their many differences, these various regimes have generally all advocated a return to direct state management of the economy, including re-nationalization of important industries, as the basis of robust new redistributive social programs, a move sometimes framed in explicit challenge to a neoliberal development model. Yet in order to fund expanded social programs, most regimes have usually been forced to increase raw material extraction as their main income-generation strategy, exacerbating natural resource exploitation and thereby augmenting pressures on conservation. As Escobar (2010: 47) thus observes, for instance, despite Bolivia's impressive social gains under the Morales regime, "Overall the development model is such that it continues to wreak havoc on the natural environment due to its dependence on accumulation fueled by the exploitation of natural resources (e.g. hydrocarbons, soy, sugar cane, African oil palm)." Arsel et al. (2016) document the widespread occurrence of this dynamic, which they term the "extractive imperative," throughout the region.

Ecuador provides a prime example of this dilemma. In 2006, then President Rafael Correa was elected on a platform including an explicit critique of neoliberalism, which he claimed to reject in pursuit of a novel model of development between "Marx and markets" (Arsel, 2012; see also Shade, 2015). To enact this transformation Correa instituted a variety of reforms, including near-total nationalization of the oil and gas industries and enshrinement of the "Rights of Nature" within a renovated constitution. Yet the Ecuadorian economy remained heavily dependent upon extraction of raw natural resources, particularly fossil fuel drilling and mineral mining, for sale on the global market (Shade, 2015).

Hence the regime encountered great difficulties in supporting conservation. Correa's rhetoric included strong language critiquing commodification of nature resources in promoting nature's intrinsic value and inherent right to exist (Rival, 2010), and certain of his advisers took this further to advocate "bio-socialism" as an ostensibly sustainable economic strategy (see Wilson and Bayón, 2017). Yet the regime's environmental practices remained decidedly at odds with this rhetoric. Essentially, as in Costa Rica, Ecuador sought to support conservation and extraction simultaneously, expanding raw material exploitation while also pursuing conservation finance mechanisms (Wilson and Bayón, 2017). In theory at least, this process was intended to progressively shift the emphasis from extraction to conservation by using revenue from the former to fund development of effective mechanisms for the latter (Shade, 2015; Wilson and Bayón, 2017). In practice, however, mechanisms for conservation finance largely failed to develop, forcing continued reliance on extractive expansion to meet development objectives.

This reliance is exemplified by the rise and demise of the infamous Yasuní-ITT initiative. Originally proposed by domestic environmental NGOs, the initiative was subsequently adopted by the Correa administration to establish a mechanism whereby previously planned oil extraction in a small portion (called the Ishpingo-Tambococha-Tiputini, or ITT, sector) of the massive (9,823 km^2) Yasuní National Park, an important biodiversity hotspot, would be foregone in exchange for the voluntary mobilization of funds by the international community to cover the opportunity costs of this decision. When this funding proved elusive, however, the initiative was repackaged as a more specific carbon sequestration mechanism whereby refundable CO_2 offset certificates would be provided in proportion with payments received, thereby individualizing the financing from the originally envisioned collective international effort (Rival, 2010). Yet even this revamped initiative proved unappealing to potential investors, with only a small fraction of the requested funds committed and even less actually delivered, and hence the initiative was subsequently abandoned and oil drilling in the ITT sector initiated (Monbiot, 2014).

Spaces of hope

While all of this presents formidable obstacles, the outlook for conservation is not entirely dire. There exist a variety of established and emerging initiatives throughout the region that offer promising examples of a potential that may be scaled up in the future. While interaction between conservationists and indigenous people has long been conflictual in many places (Conklin and Graham, 1995; Dowie, 2009), for instance, there are several cases in which more positive relations have been forged in pursuit of new conservation spaces that seek to collapse the nature-culture divide commonly grounding PA development (West et al., 2006) while promoting indigenous self-determination. This is part of a larger global movement to promote Indigenous and Community Conservation Areas organized partly within IUCN (see www.iccaconsortium. org/). In several of these instances, ecotourism development has been used effectively to both support indigenous rights and resource conservation. A common example is a long-standing partnership between CI-Brazil and members of the Kayapó nation, via which University of Maryland personnel and students engage in collaborative research and visitation (Zimmerman et al., 2001; Chernela and Zanotti, 2014). In Peru, another long-standing partnership between a private tour operator and a community containing indigenous members has proven fruitful for both parties as well (see Stronza, 2005).

Beyond this, there is a variety of other innovative initiatives in development throughout the region. In Colombia, for example, a growing movement around creation of Zonas de Reserva Campesina (Peasant Reserve Zones) seeks to balance conservation and sustainable farming for local consumption (ILSA, 2012). There are also various efforts to redesign existing PES programs to emphasize less-conditional support for community conservation efforts over resource commodification and marketization (Shapiro-Garza, 2013; Bakkegaard and Wunder, 2014; Van Hecken, Bastiaensen, and Huybrechs, 2015). How these and other progressive initiatives fare in the face of an expanding extractive imperative is thus a vital question for future research and action.

Conclusion

This overview of issues surrounding biodiversity protection via protected area creation within Latin American has demonstrated a mixed record of success and social fall-out. While it is true that much biodiversity throughout the region would likely have disappeared long ago if not for its aggressive protection within national parks and other PAs (Wuerthner, Crist, and Butler, 2015), it is also apparent that such efforts are likely to be difficult to sustain in the face of intensified pressure on various fronts (Büscher and Fletcher, 2018). Initiatives to compensate via CBC have been challenging to realize, as have efforts to build on all this to promote neoliberalization via MBIs. Consequently, at present, the long-standing opposition between conservation and development in the region remains strong despite ongoing efforts to reconcile the two priorities.

Notes

1 This mandate was increased to 17% at the 10th Conference of the Parties (CPO-10) of the Convention of Biological Diversity (CBD) held in Nagoya, Japan, in 2010 via the Aichi Targets (see www.cbd.int/sp/targets/rationale/target-11/; accessed 3/27/13).
2 www.worldwildlife.org/bsp/bcn/about/paradigm.htm; accessed 6/24/2010.
3 www.worldwildlife.org/bsp/bcn/about/paradigm.htm; accessed 6/24/2010.
4 www.worldwildlife.org/bsp/bcn/about/paradigm.htm; accessed 6/24/2010.
5 www.worldwildlife.org/bsp/bcn/about/paradigm.htm; accessed 6/24/2010.
6 The Costa Rican Tourism Bureau recorded 2,343,213 international arrivals in 2012 (http://www.visit costarica.com/ict/backoffice/treeDoc/files/Llegadas_internacionales_2012.pdf; accessed 4/2/2014).

References

Arsel, M. (2012) Between 'Marx and markets'? The state, the 'left turn' and nature in Ecuador. *Tijdschrift voor Economische en Sociale Geografie* 103(2): 150–163.
———, Hogenboom, B. and Pellegrini, L. (2016) The extractive imperative and the boom in environmental conflicts at the end of the progressive cycle in Latin America. *The Extractive Industries and Society* 3(4): 877–879.

Bakkegaard, R. Y. and Wunder, S. (2014) Bolsa Floresta, Brazil. In CIFOR (eds) *REDD+ on the Ground*. Bogor, Indonesia: CIFOR, pp. 51–67.

Bingham, H., Fitzsimons, J. A., Redford, K. H., Mitchell, B. A., Bezuary-Creel, J. and Cumming, T. L. (2017) Privately protected areas: Advances and challenges in guidance, policy and documentation. *Parks* 23(1): 13–28.

Borgerhoff Mulder, M. and Coppolillo, P. (2005) *Conservation*. Princeton, NJ: Princeton University Press.

Boza, M. (1993) Conservation in action: Past, present and future of the national parks system in Costa Rica. *Conservation Biology* 12(6): 239–247.

Brandon, K. (2004) The policy context for conservation in Costa Rica: Model or muddle? In G. W. Frankie, A. Mata and S. B. Vinson (eds) *Biodiversity Conservation in Costa Rica*. Berkeley, CA: University of California Press, pp. 299–310.

Brockington, D. (2002) *Fortress Conservation: The Preservation of the Mkomazi Game Reserve, Tanzania*. Oxford: James Currey.

Büscher, B., Dressler, W. and Fletcher, R. (2014) (eds0 *Nature^{TM} Inc.: Environmental Conservation in the Neoliberal Age*. Tucson, AZ: University of Arizona Press.

———— and Fletcher, R. (2018) Under Pressure: Conceptualizing Political Ecologies of 'Green Wars'. *Conservation & Society*, introduction to special issue on "Political Ecologies of 'Green Wars'".

Castree, N. (2010) Neoliberalism and the biophysical environment: A synthesis and evaluation of the research. *Environment and Society: Advances in Research* 1: 5–45.

Chapin, M. (2004) A challenge to conservationists. *WorldWatch* 17(6): 17–31.

Chernela, J. and Zanotti, L. (2014) Limits to knowledge: Partnering between indigenous peoples and NGOs in the Eastern Amazon of Brazil. *Conservation and Society* 12(3): 306–317.

Conklin, B. A. and Graham, L. R. (1995) The shifting middle ground: Amazonian Indians and eco-politics. *American Anthropologist* 97(4): 695–710.

Daniels, A., Esposito, V., Bagstad, K. J. Moulaert, A. and Rodriguez, C. M. (2010) Understanding the impacts of Costa Rica's PES: Are we asking all the right questions? *Ecological Economics* 69: 2116–2126.

Dowie, M. (2009) *Conservation Refugees: The Hundred-Year Conflict Between Global Conservation and Native Peoples*. Boston, MA: MIT Press.

Edelman, M. (1999) *Peasants Against Globalization: Rural Social Movements in Costa Rica*. Stanford, CA: Stanford University Press.

Escobar, A. (2010) Latin America at a Crossroads. *Cultural Studies* 24(1): 1–65.

Evans, S. (1999) *The Green Republic: A Conservation History of Costa Rica*. Austin, TX: University of Texas Press.

Flew, T. (2012) Michel Foucault's *The Birth of Biopolitics* and contemporary neo-liberalism debates. *Thesis Eleven* 108(1): 44–65.

Hall, C. (1984) *Costa Rica: A Geographical Interpretation in Historical Perspective*. Boulder, CO: Westview Press.

Harvey, D. (2005) *A Brief History of Neoliberalism*. Oxford: Oxford University Press.

Heindrichs, T. (1997) *Innovative Financing Instruments in the Forestry and Nature Conservation Sector of Costa Rica*. Eschborn, Germany: Deutsche Gesellschaft für Technische Zusammenarbeit (GTZ) GmbH.

Holmes, G. (2013) *What Role Do Private Protected Areas Have in Conserving Global Biodiversity?* SRI Papers #46, Sustainability Research Institute, University of Leeds, UK.

————. (2015) Markets, nature, neoliberalism, and conservation through private protected areas in southern Chile. *Environment and Planning A* 47(4): 850–866.

Honey, M. (2008) *Ecotourism and Sustainable Development: Who Owns Paradise?* 2nd Ed. Washington, DC: Island Press.

Hunter, J. R. (1994) Is Costa Rica truly conservation-minded? *Conservation Biology* 8(2): 592–595.

Instituto Latinoamericano para una Sociedad y un Derecho Alternativos (ILSA). (2012) *Zonas de Reserva Campesina: Elementos Introductorios y de Debate*. Bogotá, Colombia: ILSA.

International Union for the Conservation of Nature (IUCN). (1994) *Guidelines for Protected Area Management Categories*. Gland and Cambridge: IUCN.

Langholz, J. A. (1999) *Conservation Cowboys: Privately Owned Parks and the Protection of Biodiversity*. Ph.D dissertation, Ithaca, NY: Cornell University Press.

————, Lassoie, J. P. and Schelhas, J. (2000) Incentives for biological conservation: Costa Rica's private wildlife refuge program. *Conservation Biology* 14(6): 1735–1743.

Levine, A. (2002) Convergence or convenience? International conservation NGOs and development assistance in Tanzania. *World Development* 30(6): 1043–1055.

Miller, S. W. (2007) *An Environmental History of Latin America*. Cambridge: Cambridge University Press.

Monbiot, G. (2014) It's simple. If we can't change our economic system, our number's up. *The Guardian*, 27 May, online. www.theguardian.com/commentisfree/2014/may/27/if-we-cant-change-economic-system-our-number-is-up (Accessed 26 February 2018).

Perfecto, I. and J. Vendermeer (2008) Biodiversity conservation in tropical agroecosystems: A new conservation paradigm. *Annals of the New York Academy of Sciences* 1134: 173–200.

Rival, L. (2010) Ecuador's Yasuní-ITT initiative: The old and new values of petroleum. *Ecological Economics* 70(2): 358–365.

Robinson, W. I. (2008) *Latin America and Global Capitalism: A Critical Globalization Perspective.* Baltimore: Johns Hopkins University Press.

Sayer, J. (2009) Reconciling conservation and development: Are landscapes the answer? *Biotropica* 41(6): 649–652.

Shade, L. (2015) Sustainable development or sacrifice zone? Politics below the surface in post-neoliberal Ecuador. *The Extractive Industries and Society* 2(4): 775–784.

Shapiro-Garza, E. (2013) Contesting market-based conservation: Payments for ecosystem services as a surface of engagement for rural social movements in Mexico. *Human Geography* 6(1): 134–150.

Sierra, R. and Russman, E. (2006) On the efficiency of environmental service payments: A forest conservation assessment in the Osa Peninsula, Costa Rica. *Ecological Economics* 59: 131–141.

Simonian, L. (1995) *Defending the Land of the Jaguar: A History of Conservation in Mexico.* Austin, TX: University of Texas Press.

Steinberg, P. F. (2001) *Environmental Leadership in Developing Countries: Transnational Relations and Biodiversity Policy in Costa Rica and Bolivia.* Cambridge, MA: MIT Press.

Stronza, A. (2005) Hosts and hosts: The anthropology of community-based ecotourism in the Peruvian Amazon. *Annals of Anthropological Practice* 23(1): 170–190.

Vandermeer, J. and Perfecto, I. (2005) *Breakfast of Biodiversity: The Political Ecology of Rainforest Destruction.* 2nd Ed. Oakland, CA: Food First Books.

Van Hecken, G., Bastiaensen, G. J. and Huybrechs, F. (2015) What's in a name? Epistemic perspectives and Payments for Ecosystem Services policies in Nicaragua. *Geoforum* 63: 55–66.

Wells, M. P. and McShane, T. O. (2004) Integrating protected area management with local needs and aspirations. *Ambio* 33(8): 513–519.

West, P. (2006) *Conservation Is Our Government Now: The Politics of Ecology in Papua New Guinea.* Durham, NC: Duke University Press.

———, Igoe, J. and Brockington, D. (2006) Parks and peoples: The social impact of protected areas. *Annual Review of Anthropology* 35: 251–277.

Wilson, J. and Bayón, M. (2017) The nature of post-neoliberalism: Building bio-socialism in the Ecuadorian Amazon. *Geoforum* 81: 55–65.

Wuerthner, G., Crist, E. and Butler, T. (2015) (eds) *Protecting the Wild. Parks and Wilderness, the Foundation for Conservation.* London: Island Press.

Yates, J. and Bakker, K. (2014) Debating the "post-neoliberal turn" in Latin America. *Progress in Human Geography* 38(1): 62–90.

Zimmerman, B., Peres, C. A., Malcolm, J. R. and Turner, T. (2001) Conservation and development alliances with the Kayapó of south-eastern Amazonia, a tropical forest indigenous people. *Environmental Conservation* 28(1). 10–22.

36

MINING AND DEVELOPMENT IN LATIN AMERICA

Tom Perreault

Introduction

The history of mining in Latin America is, to a considerable degree, the history of Latin America itself. If mining activity is not ubiquitous across the continent, it is certainly widespread, and in the Andean region it plays a dominant role in local and national economies. Pre-Hispanic cultures mined gold, silver, and other metals for use in ceremonial and decorative items. But it was with the arrival of the Portuguese and especially the Spanish that mining activities came to dominate and re-orient regional economies. Since the 1540s, when the Spanish established their first mines at Potosí, mining has shaped the course of Latin American history and established patterns of socially and spatially uneven development. It has been estimated that between 1492 and 1810, roughly 1,685 metric tonnes of gold and nearly 86,000 metric tonnes of silver were shipped out of what is now Latin America – a current value of roughly US$210 billion (Bebbington and Bury, 2013). The historical experience of colonial and neocolonial exploitation animates popular imaginaries of Latin America and its place in the global capitalist system, and in turn informs political programs on both right and left.

As with most resource economies, mining is marked by periods of boom and bust, which are experienced historically and spatially as regions temporarily surge in population and wealth, only to decline once the boom is over. The latest mining boom was the so-called 'super-cycle' from the early 1990s through the first decade of the 21st century, which had its greatest impact in the Andean countries but helped shape economies, landscapes, and politics in Brazil, Central America, and Mexico as well. The wealth unearthed by mining makes an array of social programs and infrastructure investments possible, generates employment, and spurs industry. But mining also frequently sparks or exacerbates an array of social ills, including violent conflict, environmental degradation, and economic inequality. As Bebbington (2012a) notes, development in Latin America – and particularly in the Andean region – cannot be understood without consideration of extractive industries. The core question, then, is whether resource extraction can produce forms of development that are more socially equitable and environmentally sustainable than has historically been the case. In what follows, I briefly consider mining's importance during the colonial era and Republican period, and then discuss the more recent experience with the mining 'super-cycle.' I then consider the potential and limitations of mining to foster socially equitable and environmentally sustainable forms of development. Considerations here include economic dependency and the so-called 'resource curse'; policies of Corporate Social Responsibility (CSR) and Free, Prior and Informed Consent (FPIC); and the relationship between mining and water contamination.

Historical legacies and contemporary processes

Mining has been a central feature of Latin American societies and landscapes since well before the Conquest. When the Spanish arrived in the Andes and what is now central Mexico, they encountered cultures with highly developed mining and metallurgical skill, which fueled the conquerors' dreams of wealth. The history of the Spaniards' search for gold and silver is well documented and is deservedly notorious (Galeano, 1973; Robins, 2011). Andean societies were (re)organized to serve mining centres according to the *mita* system, a process begun under the Incas and continued and intensified by the Spanish during the regime of Viceroy Francisco de Toledo. Under the mita system, indigenous communities were required to send a prescribed number of male corvée labourers (*mitayos*) to work in the mines each year (Brown, 2012). The mines at Potosí – high on the Altiplano and far from the sea – were supplied with mercury extracted at Huancavelica, maize grown in Cochabamba, coca from the *yungas* (semi-tropical valleys), and labour from across the entire region. Vast trade networks supplied the mines and connected distant and remote corners of the Andean region. The mines of the Andes were the Crown's mint; in the case of Potosí this was literally the case, as the Casa de la Moneda produced the bullion that flowed to Spain, and from there fueled the development of mercantilism and nascent capitalism in Europe and Asia (Marichal, 2006). Little of the wealth extracted from the mines at Potosí, Oruro, Huancavelica, Cerro de Pasco, and elsewhere remained in those regions, however. Instead, the mining economy fostered dependent capitalism, based on the extraction of minerals (and later hydrocarbons and other primary materials), and their export to metropolitan centres in Europe, North America, and most recently, Asia. What *has* remained in the Andes is the environmental and social legacy of mining. Potosí, Oruro, Huancavelica, and Cerro de Pasco are today among the most polluted and impoverished places in the Andean region. This simple fact raises important questions regarding the current strategy – held in common by all the Andean republics across the political spectrum – that emphasizes resource extraction as a development strategy (Bebbington and Humphreys Bebbington, 2011).

For its part, Mexico's mining economy emerged in the first years of the 18th century, as that of Peru (including Alto Peru – what is now Bolivia) was declining. San Luís Potosí was so named in the hope that its mines would prove as rich as those of its Andean namesake. Much of Latin America's mining economy was in decline by the early 19th century, when the independent republics emerged following the revolutionary decades of 1810s and 1820s. Hard-rock mining in Mexico and the Andean region revived somewhat during the late 19th and early 20th centuries, in large part owing to the infusion of foreign capital from the US and Great Britain. For instance, during this period Peru's mining economy was dominated by the US-owned Cerro de Pasco mine, while in Chile the famous Chuquicamata mine was operated by a series of Chilean and British firms before being purchased in the early 1900s by the Guggenheim brothers, who in turn sold it to the US-based Anaconda Copper Company in 1923 (Orihuela and Thorp, 2012). By the early 20th century, Bolivia's mining economy had largely shifted from silver (mostly from central Potosí) to tin (mostly from Oruro and northern Potosí), and the three tin barons, Simón I. Patiño, Carlos Víctor Aramayo, and Mauricio Hochschild, together controlled the large majority of the country's mining economy. The mid-20th century saw a wave of nationalizations in the mining industry, exemplified by Bolivia's COMIBOL (Mining Corporation of Bolivia), established following the country's Social Revolution in 1952 (Nash, 1993), and Chile's CODELCO (National Copper Corporation), established by the government of Salvador Allende. CODELCO proved to be so profitable that it was maintained by the neoliberal regime of Augusto Pinochet and today continues to be an engine of Chile's economic growth.

Supersized: political economies of modern mining

Following the so-called "lost decade" of the 1980s, marked by debt crisis, hyperinflation, civil wars, and economic contraction, the 1990s brought a period of relative stability and economic growth, albeit one that was unevenly experienced within and between Latin American countries. This period saw the initiation of the so-called 'super cycle' in resource extraction that has had far-reaching effects in Latin America, particularly in the mining regions of the Andean countries. As Bebbington and Bury (2013: 42) report, between 1996 and 2007, Latin American countries received US$91 billion in investments in the mining sector (and roughly US$2 *trillion* in the hydrocarbons sector). Mining investments were largely concentrated in five countries: Chile (by far the largest recipient), Colombia, Brazil, Mexico, and Peru.

The 'super-cycle' brought with it several interrelated social, economic, and territorial transformations. The increase in mining investment led inevitably to a numerical and spatial expansion in mining concessions. In Latin America, as in most of the world, states retain subsurface resource rights, constraining the ability of affected communities to exercise political control over their territories (Bebbington and Bury, 2013; Bebbington et al., 2008). New mining investments have often been in the form of mega-mines such as Peru's Yanacocha (Bury, 2004; Sosa, 2014), which have radically transformed landscapes through the enclosure of pasture, the building of roads, water lines, and other infrastructure, and the reshaping of mountainsides through the creation of massive open pits and tailings piles. The extractive boom has led to a massive increase in the concessions granted for mining and hydrocarbons development. To be sure, because concessions are areas permitted for exploration, rather than of actual extraction, maps showing the areal extent of concessions can exaggerate the potential impact of mining. Nevertheless, the sheer scale of the territory currently granted in mining and hydrocarbons concessions should give us pause. For example, in Colombia, the area under mining concessions more than quadrupled between 2002 and 2009 (from 10,500 km^2 to 47,700 km^2) (Bebbington, 2012a). In Peru, active mining claims account for roughly 14,000 km^2, some 11% of the total area of the country (Bebbington and Bury, 2009). Meanwhile, roughly 72% of the Peruvian Amazon (approximately 490,000 km^2) and 65% of the Ecuadorian Amazon (approximately 52,300 km^2) are under concession for hydrocarbons exploration, and by some estimates 55% percent of Bolivia's national territory is considered to have potential for hydrocarbons production (Bebbington, 2009). As Bebbington (2012b) notes, these statistics, and the concession maps that accompany them, expose the geographies of uncertainty, risk, and social vulnerability that confront those people living within, and downstream from, concessions.

The mining frontier continues to expand spatially and technologically. Lithium mining on the Altiplano salt flats represents one such example. While lithium deposits in Chile and Argentina are exploited by transnational firms (or national firms with transnational capital), in Bolivia, with some of the largest lithium deposits in the world, the mineral has become a symbol of anti-colonial development. The government of Evo Morales is determined to control all aspects of lithium extraction with the goal of eventually producing lithium batteries for electric cars. The creation of added value, it is thought, would help Bolivia overcome the historical legacies of dependency (Revette, 2017).

All too often, these transformations, together with rural populations' understanding that they must bear a disproportionate share of the burdens of mining while the benefits overwhelmingly accrue to distant others (in national capitals or overseas), have led to social conflict. According to Peru's Ombudsman's office (Defensoría del Pueblo), roughly 50% of all social conflicts in the country are associated with extractive activities (Bebbington, 2012b). These conflicts most often revolve around the enclosure of commons resources (water or pasture), deterioration in environmental quality (most frequently water contamination), or access to benefits such as employment or community-based development programs (Bebbington and Humphreys Bebbington, 2011; Himley, 2013, 2014). The recent upsurge in mining activity in Guatemala and Honduras has led to a wave of social conflicts that are interwoven with forms of violence widespread throughout northern Central America (Dougherty, 2017). Territorial conflicts have erupted in Colombia as well, in many cases centred on the tense relationship between mining activities and campesino or indigenous communities (e.g. Sánchez García, 2014).

Mining conflicts are perhaps most acute in Peru, where the National Confederation of Communities Affected by Mining (CONACAMI) is the largest, best known, and most militant anti-mining organization in the Americas. Mine-related conflicts are particularly common in areas like Cajamarca and Ancash (home to the massive Yanacocha and Pierina gold mines, respectively), where mining represents a potential threat to agrarian livelihoods (Bury, 2004; Himley, 2014; Sosa, 2014). But mining protests are diverse and reflect local histories and political ecologies. In the Peruvian community of Choropampa, near Yanacocha, protesters were not opposed to mining per se, but rather demanded compensation and environmental remediation following a devastating mercury spill (Li, 2017). Similar concerns were expressed by activists in Arequipa, where mining within the city limits threatened water quality for urban residents (Roca Servat, 2014). In Ecuador, with little historical experience with mining, residents of the Intag valley mobilize environmentalist discourses in opposition to mining and evoke the country's progressive Constitution, even as the government of former president Rafael Correa sought to criminalize anti-mining protest (Billo, 2017; Davidov, 2013). In Bolivia, where mining has broad-based support even among communities most affected by mining (Perreault, 2017a), the most active and militant actors involved in mine-related protest have been mining cooperatives. Through their alliances with the government of Evo Morales, as well as their willingness to use violence, mining cooperatives have promoted and enforced pro-mining policies (Marston and Perreault, 2017).

While local and regional protests have proliferated, the extraction super-cycle has ushered in a new period of prosperity in much of Latin America, particularly in urban areas, and has facilitated new social programs, infrastructure development, and the expansion of the middle class. In the Left-leaning 'Pink Tide' countries of Venezuela, Ecuador, Bolivia, and Argentina (under the Kirchners), this took the form of 'neo-extractivism' and 'post-neoliberalism,' rooted in redistributive programs funded by resource rents (Perreault, 2017b). In their commitment to resource extraction, there is little difference between the putatively 'post-neoliberal' governments of Bolivia and Ecuador and the neoliberal governments of Peru and Colombia. Indeed, as Bebbington (2012a, 2012b) has demonstrated, the rhetoric of their respective presidents is virtually indistinguishable in condemning indigenous communities and environmental activists for opposing mining and hydrocarbons projects.

According to Castillo (2013), the mining industry has changed dramatically in three interrelated ways since the outset of the 20th century. First, during the past half-century, mining has transformed from a labor-intensive to capital-intensive activity. Whereas the large-scale mines of the colonial and Republican eras employed thousands of workers, in both below- and above-ground operations, modern mines are highly mechanized, and use massive machinery to remove ore and enormous trucks to transport it. Open pit mines, which operate on a scale previously unimaginable and which permit the exploitation of low-grade ores, require relatively few workers. Such operations reduce the need for labour (and minimize the risk of labour unrest), in favour of capital-intensive machinery and techniques. These include open pit mining, cyanide heap-leaching and associated techniques to separate minerals from ore, and the use of massive equipment to move and process rock. These technological advances permitted capital-rich transnational firms to profitably extract minerals in densities as low as one gram per tonne of ore (Bridge, 2015).

The highly mechanized and large-scale nature of modern mining operations favours large, highly capitalized firms. This fact points to the second transformation in Andean mining: the dominance of transnational capital. In the 1980s and early 1990s, roughly 12% of international mining investment flowed to Latin America. By 2010, Latin America accounted for fully 33% (one-third) of global mining investment (Bebbington, 2009; Bridge, 2004b). Foreign participation in extractive industries in the Andean region was facilitated through the 1980s, 1990s, and early 2000s by a wave of neoliberal reforms designed to attract international investment. Reforms included various bilateral trade agreements and laws intended to restructure property rights, taxes, and royalty regimes. Mining laws were reformed with the aim of fostering the growth of private sector investment and accumulation. Despite the fact that the mining sectors remain diverse across the continent, with a mix of national and transnational capital, and very small-scale independent mines (including mining cooperatives and artisanal operations), the relatively few, very large-scale mines operated by transnational firms account for the large majority of ore produced, value in minerals exported, and royalty payments transferred to national treasuries (Bebbington, 2012a). Transnational investment in extractive industries may have been facilitated by neoliberal reforms, but it has been driven in large part by rapid economic expansion in China and India, and, for a time, in Brazil. Economic growth in these countries permitted sustained high prices for primary materials on world commodities markets, which has led to the rapid expansion of the mining and hydrocarbons sectors throughout the Andes, including in states such as Ecuador and Argentina, that have little history of mining (Billo, 2017; Bridge, 2004b).

This brings us to the third transformation in Andean mining: its enclave character. Recent investment patterns in the sector – transnational firms opening highly mechanized, capital-intensive mines, frequently in places with little or no history of extraction, and exporting processed minerals overseas – have fostered conditions in which the economic linkages between mining operations and their transnational parent corporations (and other transnational firms) tend to be far better established than the connections between the mining centre and the national and (particularly) the local economies. This lack of backward and forward economic linkages is characteristic of the classic 'enclave economy,' which is economically and socially isolated – in some cases literally walled off – from the surrounding community (Bridge, 2015). Highly skilled workers are often from capital cities or abroad, while everything from machinery to construction materials to food are obtained from distant locations. Investment capital also originates from outside the region, and profits are repatriated abroad or invested in the national capital. Little wonder, then, that local residents are often sceptical (if not outright hostile) to the opening of mining and hydrocarbons operations (Himley, 2010).

Can mining foster equitable and sustainable development?

Possibilities for equity? The resource curse and corporate social responsibility

The extraction and export of raw materials such as minerals or hydrocarbons represent an important – in many cases the dominant – source of government revenue for many countries in Latin America and is commonly viewed by politicians and popular sectors alike as a path toward economic development (e.g. Coronil, 1997). However, as has been documented by Bridge (2004a, 2015), extractive industries tend to create few well-paid jobs in regions where extraction takes place, while frequently resulting in lasting environmental damage and social disruption. Moreover, resource extraction seldom leads to long-term economic growth, and in many countries has depressed economic activity in non-extractive sectors (Sachs and Warner, 2001). While this is far from inevitable – resource-dependent but wealthy countries such as Norway, Canada, and Australia provide strong counter-examples – the so-called 'curse' of natural resource dependence has been widely recognized by scholars and politicians alike (Ross, 1999, 2012).

From an economic perspective, the 'resource curse' (or 'Dutch disease,' named for distortions in the Dutch economy following the discovery of North Sea oil) is expressed in inflated currencies, a concentration of talent in extractive sectors, and under-investment and concomitant slow growth in non-extractive sectors (Sachs and Warner, 2001). From a political perspective, the resource curse is manifest in social conflict, official corruption, and bloated, inefficient bureaucracies (Ross, 1999, 2012). While the more deterministic characteristics of the resource curse literature have been widely discredited (e.g. Le Billon, 2013), it is undeniable that numerous political, economic, and social dysfunctions remain stubbornly persistent in regions of resource extraction and ultimately serve to undermine many of the benefits that might otherwise derive from extractive industries. The same problem was diagnosed by a previous generation of analysts as characteristic of Latin America's dependent form of capitalist development (Frank, 1967; Galeano, 1973). Indeed, dependency thinking in populist form remains widespread in Andean countries, and animates both popular imaginaries and official policies regarding resource extraction (Perreault, 2017a).

Given the pervasive economic and political distortions that so often accompany mining, we may reasonably inquire as to the conditions by which it might promote meaningful development. As Bebbington (2012a) notes, the proponents of extractive industries assert that mining projects may catalyze development in three broad ways. The first of these is via multiplier effects as a result of direct investment by extractive firms in infrastructure and the direct creation of employment in mining and associated ancillary industries such as transportation, refining, and construction. Such forward and backward linkages are often limited, however, by the enclave character of extractive industries (see above) and the fact that ore is often minimally processed before export, while refining and finishing are often conducted in metropolitan centres in Asia, Europe, or North America (Bridge, 2015).

Second, advocates of extractive industries also point to Corporate Social Responsibility (CSR) programs as another source of direct development effects. CSR programs have become widespread among transnational mining and hydrocarbons firms as 'best practice' programs for social engagement and gaining 'social license,' or local approval for extractive activities (Himley, 2013). CSR programs are typically oriented toward the construction of community infrastructure such as roads, electricity, water supplies, or school facilities, but may entail services such as agricultural extension, health care, or funding for microenterprises (Bebbington, 2012a). Such programs have been widely criticized for coopting local officials or community leaders, and for providing inadequate compensation for what in many instances are irreversible changes in livelihoods and local environments (Billo, 2012). Moreover, CSR programs in many cases serve to depoliticize community-mine conflicts, treating deeply political struggles over land rights and livelihood opportunities as mere technical problems to be solved through the right mix of technology and policy (cf. Li, 2007). Such programs also tend to facilitate the expansion of transnational firms into rural and often remote communities, while initiating social programs that are temporary by design. In what may be considered another form of direct investment, in the 'neo-extractivist' states of Ecuador, Bolivia, and Venezuela, resource rents have been used to support cash transfer programs destined for targeted populations: the elderly, the poor, pregnant women, or children.

If multiplier effects and CSR programs represent the direct engagement of mining firms in economic and social development, then the taxes and royalties paid by those firms represent their indirect contribution, and the third way mining may influence development. For many if not most Latin American countries, the taxes and royalties paid by extractive industries represent a substantial percentage (even overwhelming majority) of national revenues. The effectiveness of resource rents in fostering development, however, depends on the institutional structures of redistribution and decentralization, which vary greatly between countries (Bebbington, 2012a).

Possibilities for sustainability? Mining, water, and participation

Of mining's many environmental effects, few are as pronounced, widespread, or persistent as its impacts on water (Bebbington and Williams, 2008; Perreault, 2013). Indeed, the geographies of water and mining are in many ways co-produced: mining cannot occur without water; and mining activities in turn shape water quality, availability, access, and use (Budds and Hinojosa, 2012). Little wonder, then, that struggles over mining often revolve less around the minerals being extracted than around the real or potential impacts on water quality, quantity, or access for competing and downstream users (Budds, 2010). Hard-rock mining requires large quantities of water in the processing of ore (to separate minerals from crushed rock by means of gravity or slurries containing cyanide, acids, or chemicals). Water pumped from below the water table is often acidified through contact with oxidizing rock, pumped into containment ponds or simply into nearby waterways, and may negatively affect surface- and groundwater quality. Water used in ore processing becomes laced with heavy metals and may carry cyanide or arsenic, commonly used in modern heap-leaching processes. When mine tailings (excavated rock) are exposed to water and air, oxidation of metal sulphides acidifies the rock and surrounding soil, leading to acidic mine drainage, when water – either precipitation or water pumped out of the mine – flows through the tailings. Acid drainage affects downstream riparian systems, with impacts that last decades or even centuries after a mine is abandoned (Montoya et al., 2010). As a result, mining significantly impacts both water *quantity*, as large-scale withdrawals are made upstream from mine sites, and *quality*, as wastewater is released (intentionally or not) into waterways downstream from mines and processing sites (Bridge, 2004a). The demand for water on the part of mine operators can also dramatically influence water *access* for competing users, and the legal structures shaping water rights. Smallholder irrigators have historically had poorly defined rights to water as compared to mining operations, whose rights to water are more clearly stipulated under the countries' mining codes (Perreault, 2008).

Mining activity can affect rural peoples and environments in a number of other ways as well. As Himley (2010) has demonstrated, mining can limit access to lands used for grazing or other uses. While rural residents often support road construction as a way to improve local transportation, road construction for mining can involve right-of-way through communal or privately owned land, which can lead to conflicts within communities or between community members and mining firms. Moreover, heavy use of mountain roads can cause erosion and excessive dust, and occasionally results in spills of toxic materials such as mercury, as happened at Choropampa, Peru, in 2000 (Li, 2017).

As with the difficulties in fostering equitable development, the challenges of promoting sustainable forms of development in extractive regions (or with revenues derived from extractive industries) are vast. As an industry based on the extraction of non-renewable resources, mining is, by definition, unsustainable. As mines exhaust the profitably extracted minerals in one site, they inevitably expand their operations in search of new deposits. Mining is thus inherently expansionary, both in terms of the extractive frontier and the amount of waste material produced (Bridge, 2000). In addition to the massive amounts of water consumed in extracting and processing ore, the large expanses of land permanently transformed through extractive activities, and the vast quantities of waste material (and associated acid drainage) produced, countries in Latin America also often face problems associated with poorly regulated small-scale, artisanal, and cooperative mining. This is particularly the case in areas of alluvial gold mining in the Amazon basin, where widespread use of hydraulic mining (using high-pressure hoses to dislodge alluvial deposits from river banks) and mercury amalgamation have destroyed huge swaths of lowland forests and riparian ecosystems. Institutional frameworks are crucial in the implementation of environmental safeguards, but often prove woefully insufficient. Environmental laws and the ministries that enforce them are often seen as subordinate to economic interests, a condition that frequently leads to poor enforcement. Industries have begun to implement processes of consultation, in accordance with international standards of Free, Prior and Informed Consent (FPIC) (Riofrancos, 2017). Consultation mechanisms have been implemented throughout Latin America to varying effect (Walter and Urkidi, 2017). Such mechanisms are designed primarily to promote the participation of indigenous communities in resource governance decisions, and have been promoted under the auspices of the International Labor Organization's Convention 169 on the Rights of Indigenous and Tribal Peoples (ILO 169) and the UN's Declaration on the Rights of Indigenous Peoples (UNDRIP). While the language of FPIC draws on international human rights law, it has often been adopted by Latin American countries in a much weaker form that emphasizes consultation (*consulta*) rather than the more stringent standard of consent. Furthermore, in many instances, the outcomes of consultation processes are non-binding if a given project is declared by the government to be in the national interest (Perreault, 2015). Thus, while standards of FPIC and associated consultation mechanisms would in theory provide a means to achieving environmentally sustainable and socially equitable development – or at a minimum, gaining social licence for extractive projects – in practice the outcomes are quite mixed (Schilling-Vacaflor, 2013).

Conclusion

Since the earliest days of the Conquest, mining has shaped patterns of wealth and poverty, population booms and economic decline, domination and dispossession. Mining and the associated export of primary commodities (along with the export of agricultural commodities from plantations) also helped structure Latin America's relationship to metropolitan centres in Europe and North America (and, more recently, in Asia), as well as its position in the global capitalist system. While modern mining techniques – particularly those used in open pit and other large-scale mining operations – are vastly different from those in use during the colonial period, they both reflect and reproduce similar patterns of spatially uneven and dependent development, as well as geographies of dispossession and environmental risk (Himley, 2014). And yet, precisely because of these well-established political and economic relations, countries such as Chile, Peru, Bolivia, Ecuador, and Colombia have only increased their dependence on the revenues derived from mining and other extractive industries. This dependence represents a powerful material constraint on attempts to seek alternatives to resource extraction (Arsel, Hogenboom, and Pellegrini, 2016; Kohl and Farthing, 2012). There is thus little doubt that mining will continue to be a central feature of Latin American societies into the foreseeable future, notwithstanding the growing calls for a post-extractivist economy (e.g. Acosta, 2009; Gudynas, 2011). The crucial question for scholars and policy makers, then, is how mining might be made to foster more socially equitable and environmentally sustainable forms of development than have historically been the case. Mining's proponents point to the industry's multiplier effects in the creation of employment and ancillary industries (so-called 'forward and backward linkages'). They also point to the taxes and royalties paid by mining firms, to corporate social responsibility (CSR) programs through which firms work directly in communities affected by mining activities, and to direct cash transfer programs funded by extractive activities. However, as Bebbington (2012a) notes, these will only produce durable, equitable, and widespread development if accompanied by institutional frameworks that allows for meaningful democratic participation in the governance of natural resources and the redistribution of mining revenues. Where they are employed, CSR programs should be used for long-term development rather than short-term objectives (such as garnering local support or defusing conflict). Moreover, environmental safeguards must be in place to minimize environmental degradation and to undertake environmental remediation in contaminated areas (a daunting task in many regions, and one that is beyond the capacity of most local or even central governments to finance and manage). Finally, consultation mechanisms, which have come into widespread use, must be conducted according to the standard of Free, Prior and Informed Consent (FPIC), and informed by internationally recognized principles of human rights (Perreault, 2015; Riofrancos, 2017). Consultation mechanisms must be designed and implemented to foster meaningful participation, the results of which must be respected by governments and mining firms. The potential thus exists for mining to foster more sustained and equitable forms of development. This is far from guaranteed, however, and will require deeper, more transparent and democratic participation in the governance of mining and natural resources. Whether and where these institutional frameworks will take hold remains an open question.

References

Acosta, A. (2009) *La Maldición de la Abundancia.* Quito: Abya Yala.

Arsel, M., Hogenboom, B. and Pellegrini, L. (2016) The extractive imperative in Latin America. *The Extractive Industries and Society* 3(4): 880–887.

Bebbington, A. (2009) The new extraction: Rewriting the political ecology of the Andes? *NACLA Report on the Americas* 42(5): 12–20.

———. (2012a) Extractive industries, socioenvironmental conflicts and political economic transformations in Andean America. In A. Bebbington (ed) *Social Conflict, Economic Development and Extractive Industry.* London: Routeldge, pp. 3–26.

———. (2012b) Underground political ecologies: The second annual lecture of the cultural and political ecology specialty group of the Association of American Geographers. *Geoforum* 43: 1152–1162.

——— and Bury, J. T. 2009. Institutional challenges for mining and sustainability in Peru. *Proceedings of the National Academy of Sciences of the United States of America* 106(41): 17296–17301.

———. (2013) New geographies of extractive industries in Latin America. In A. Bebbington and J. Bury (eds) *Subterranean Struggles: New Dynamics of Mining, Oil and Gas in Latin America.* Austin: University of Texas Press, pp. 27–66.

Bebbington, A. J., Hinojosa, L., Humphreys Bebbington, D., Burneo, M. L. and Warnaars, X. (2008) Contention and ambiguity: Mining and the possibilities of development. *Development and Change* 39(6): 965–992.

Bebbington, A. J. and Humphreys Bebbington, D. (2011) An Andean Avatar: Post-neoliberal and neoliberal strategies for promoting extractive industries. *New Political Economy* 16(1): 131–145.

———, Bury, J., Lingan, J., Pablo Muñoz, J. and Scurrah, M. (2008) Mining and social movements: Struggles over livelihood and rural territorial development in the Andes. *World Development* 36(12): 2888–2905.

Bebbington, A. J. and Williams, M. (2008) Water and mining conflicts in Peru. *Mountain Research and Development* 28(3/4): 190–195.

Billo, E. (2012) Competing sovereignties: Oil extraction, corporate social responsibility, and indigenous peoples in Ecuador. Unpublished Ph.D. dissertation, Syracuse University.

———. (2017) Mining, criminalization and the right to protest: Everyday constructions of the post-neoliberal Ecuadorian state. In L. Leonard and S. N. Grovogui (eds) *Governance in the Extractive Industries: Power, Cultural Politics and Regulation*. New York: Routledge, pp. 39–56.

Bridge, G. (2000) The social regulation of resource access and environmental impact: Production, nature and contradiction in the US copper industry. *Geoforum* 31: 237–256.

———. (2004a) Contested terrain: Mining and the environment. *Annual Review of Environmental Resources* 29: 205–259.

———. (2004b) Mapping the bonanza: Geographies of mining investment in an era of neoliberal reform. *Professional Geographer* 56(3): 406–421.

———. (2015) The hole world: Scales and spaces of extraction. *Scenario Journal* Fall. www.scenariojournal.com/article/the-hole-world

Brown, K. W. (2012) *A History of Mining in Latin America: From the Colonial Era to the Present*. Albuquerque: University of New Mexico Press.

Budds, J. (2010) Water rights, mining and indigenous groups in Chile's Atacama. In R. Boelens, D. Getches and A. Guevera-Gil (eds) *Out of the Mainstream: Water Rights, Politics and Identity*. London: Earthscan, pp. 197–211.

——— and Hinojosa, L. (2012) Restructuring and rescaling water governance in mining contexts: The co-production of waterscapes in Peru. *Water Alternatives* 5(1): 119–137.

Bury, J. (2004) Livelihoods in transition: Transnational gold mining operations and local change in Cajamarca, Peru. *The Geographical Journal* 170(1): 78–91.

Castillo, G. (2013) *Spatial Production of the Andes and Mining Historical Development in Peru*. Unpublished manuscript.

Coronil, F. (1997) *The Magical State: Nature, Money, and Modernity in Venezuela*. Chicago: University of Chicago Press.

Davidov, V. (2013) Mining versus oil extraction: Divergent and differentiated environmental subjectivities in 'post-neoliberal' Ecuador. *The Journal of Latin American and Caribbean Anthropology* 18(3): 485–504.

Dougherty, M. (2017) Can mining produce development in Central America? *Georgetown Journal of International Affairs* 16 April. www.georgetownjournalofinternationalaffairs.org/online-edition/can-mining-produce-development-in-central-america?rq=Dougherty (Accessed 21 October 2017).

Frank, A. G. (1967) *Capitalism and Underdevelopment in Latin America: Historical Studies of Chile and Brazil*. New York: Monthly Review Press.

Galeano, E. (1973) *Open Veins of Latin America: Five Centuries of the Pillage of a Continent*. New York: Monthly Review Press.

Gudynas, E. (2011) Alcances y contenidos de las transiciones al post-extractivismo. *Ecuador Debate* 82: 61–79.

Himley, M. (2010) *Frontiers of Capital: Mining, Mobilization, and Resource Governance in Andean Peru*. Unpublished Ph.D. dissertation, Syracuse University.

———. (2013) Regularizing extraction in Andean Peru: Mining and social mobilization in an age of corporate social responsibility. *Antipode* 45(2): 394–416.

———. (2014) Mining history: Mobilizing the past in struggles over mineral extraction in Peru. *Geographical Review* 104(2): 174–191.

Kohl, B. and Farthing, L. (2012) Material constraints to popular imaginaries: The extractive economy and resource nationalism in Bolivia. *Political Geography* 31(4): 225–235.

Le Billon, P. (2013) Resources. In K. Dodds, M. Kuus and J. Sharp (eds) *The Ashgate Research Companion to Critical Geopolitics*. Burlington: Ashgate, pp. 281–304.

Li, F. (2017) Illness, compensation, and claims for justice: Lessons from the Choropampa mercury spill. In L. Leonard and S. N. Grovogui (eds) *Governance in the Extractive Industries: Power, Cultural Politics and Regulation*. New York: Routledge, pp. 176–194.

Li, T. M. (2007) *The Will to Improve: Governmentality, Development and the Practice of Politics*. Durham, NC: Duke University Press.

Marichal, C. (2006) The Spanish-American silver peso: Export commodity and global money of the *Ancien Regime*, 1550–1800. In S. Topik, C. Marichal and Z. Frank (eds) *From Silver to Cocaine: Latin American Commodity Chains and the Building of the World Economy, 1500–2000*. Durham, NC: Duke University Press, pp. 25–52.

Marston, A. and Perreault, T. (2017) Consent, coercion and *cooperativismo*: Mining cooperatives and resource regimes in Bolivia. *Environment and Planning A* 49(2): 252–272.

Montoya, J. C., Amusquívar, J., Guzmán, G., Quispe, D., Blanco, R., and Mollo, N. (2010) *Thuska Uma: Tratamiento de aguas ácidas con fines de riego*. La Paz: PIEB.

Nash, J. (1993) *We Eat the Mines and the Mines Eat Us: Dependency and Exploitation in the Bolivian Tin Mines*. New York: Columbia University Press.

Orihuela, J. C. and Thorp, R. (2012) The political economy of managing extractives in Bolivia, Ecuador and Peru. In A. Bebbington (ed) *Social Conflict, Economic Development and Extractive Industry*. New York: Routledge, pp. 27–45.

Perreault, T. (2008) Custom and contradiction: Rural water governance and the politics of *usos y costumbres* in Bolivia's irrigators' movement. *Annals of the Association of American Geographers* 98(4): 834–854.

———. (2013) Dispossession by accumulation? Mining, water and the nature of enclosure on the Bolivian Altiplano. *Antipode* 45(5): 1050–1069.

———. (2015) Performing participation: Mining, power, and the limits of consultation in Bolivia. *The Journal of Latin American and Caribbean Anthropology* 20(3): 433–451.

———. (2017a) Mining, meaning and memory in the Andes. *The Geographical Journal* online. doi:10.1111/geoj.12239

———. (2017b) Tendencies in tension: Resource governance and social contradictions in contemporary Bolivia. In L. Leonard and S. N. Grovogui (eds) *Governance in the Extractive Industries: Power, Cultural Politics and Regulation*. London: Routledge, pp. 17–38.

Revette, A. C. (2017) This time it's different: Lithium extraction, cultural politics and development in Bolivia. *Third World Quarterly* 38(1): 149–168.

Riofrancos, T. (2017) Scaling democracy: Participation and resource extraction in Latin America. *Perspectives on Politics* 15(3): 678–696.

Robins, N. (2011) *Mercury, Mining and Empire: The Human and Ecological Cost of Colonial Silver Mining in the Andes*. Bloomington: Indiana University Press.

Roca Servat, D. (2014) Injusticias socioambientales en torno al agua y minería a gran escala: Elc caso de la ciudad de Arequipa, Perú. In T. Perreault (ed) *Minería, Agua y Justicia Social en los Andes: Experiencias Comparativas de Perú y Bolivia*. Cusco: CBC; La Paz: PIEB, pp. 125–148.

Ross, M. (1999) The political economy of the resource curse. *World Politics* 51(2): 297–322.

———. (2012) *The Oil Curse: How Petroleum Wealth Shapes the Development of Nations*. Princeton, NJ: Princeton University Press.

Sachs, J. D. and Warner, A. M. (2001) The curse of natural resources. *European Economic Review* 45: 827–838.

Sánchez García, D. P. (2014) El conflict por la productción del territorio en el caso del Proyecto minero La Colosa, Tolima, Colombia. In B. Göbel and A. Ulloa (eds) *Extractivismo minero en Colombia y América Latina*. Bogotá: Universidad Nacional/Ibero-Amerikanisches Institut, pp. 347–388.

Schilling-Vacaflor, A. (2013) Prior consultation in plurinational Bolivia: Democracy, rights, and real life experience. *Latin American and Caribbean Ethnic Studies* 8(2): 202–220.

Sosa Landeo, M. (2014) Justicia ambiental y medidas de mitigación y compensación por impactos en Cajamarca, Perú. In T. Perreault (ed) *Minería, Agua y Justicia Social: Experiencias Comparativas de Perú y Bolivia*. Cusco: Centro Bartolomé de las Casas, pp. 149–168.

Walter, M. and Urkidi, L. (2017) Community mining consultations in Latin America (2002- 2012): The contested emergence of a hybrid institution for participation. *Geoforum* 84: 265–279.

37

TOWERS OF INDIFFERENCE

Water and politics in Latin America[1]

Rutgerd Boelens

Insértese la rama que sea,
pero que sea en tronco fuerte
You can graft any branch you like,
as long as it's on a sturdy trunk

(José Martí, *Nuestra América*, 1891)

Towers and UnGovernance

My first encounter with modern hydraulic dreams in Latin America was with the huge *Canal Nuevo*, cutting straight through the homestead of my host parents, peasant farmers Santiago Quintana and Alberta Pérez. Nearly three decades ago, in the remote Peruvian highlands of Mollepata, they taught me about the many Andean water worlds that are omitted from irrigation textbooks.

When government engineers and international consultants built Canal Nuevo, they entirely ignored these highland water cultures. Santiago explained: "Engineers designed it. . . . They started construction, cutting through farms without anyone's permission. They began digging the canal intrusively, without paying any compensation for land or crops we lost. It made no sense to complain, because it was for development, for people's progress."[2] Our neighbour, Cirilo Hermosa, commented, "We didn't know. No one talked with the people." Nevertheless, the urban, white-*mestizo* elites *did* know. In a powerful alliance with the Lima-based and foreign engineers, they acquired large land properties and re-patterned canals and water flows.

Anthropologist Clifford Geertz (1980: 13) once remarked that irrigation systems are "texts to be read." This text made for sad reading, carving injustice into Mollepata's landscape, combining modern expert knowledge with deep-seated racism. Cirilo says: "The canal would benefit the rich, not us, the poorest. The mayor, the governor, and the judge had grabbed the best land, where the canal would go. Small farmers had no say." Donna Haraway (1991) has argued that scientists use the "god trick" to objectify reality; indigenous farmer Cirilo framed it in his own words: "As if they were gods, they designed the system from the air, with their aerial photographs. And we were forced to turnover our land, so that they could build the canal."

Only three days after it was officially opened for use, the canal dramatically collapsed. It broke down, crumbled, and was abandoned, to this day. Local families nicknamed it *Elefante Blanco* (the White Elephant), a costly failure. As Santiago explains: "The project execution agents vanished. . . . There was no water; it didn't work." In later years, I encountered new expert teams trying to restore the remnants. They failed. They emerged out of the blue, without engaging with the peasant families, leaving them only water scarcity and the White Elephant.

Since these years of living in Latin America, one tale has continually crossed my mind, when listening to such common stories about expert blindness and water injustice:

> When Mrs. Glü peered down from the highest lookout tower, her son appeared in the street, like a tiny little toy. She recognized him by the color of his coat. The next moment a toy truck hit that little toy. But that event of a minute ago was no more than an unreal, brief accident, involving a broken toy. 'I don't want to come down!' she screamed, resisting fiercely as she was being led down the stairway. 'I don't want to go down! I'll go crazy down there!'
>
> *(Anders, 1992 [1932])*

This tale, written by German philosopher Günter Anders in 1932, led an underground existence for many years. Its manuscript was hidden from the Nazis in an old chimney, travelled to Paris, and eventually arrived in the New World. I feel this tale symbolizes the many Towers of Indifference in the modernist water science-policy nexus: it tells how they create distanced views and models, detached from reality, that deny real-life people and nature, transforming them from subjects into objects; objects that experience no suffering.[3]

In Latin America, most international policy models and national water laws are not adapted to the contexts of the local populations, based on the justification that it is these populations who need to adapt, not the plans. These models aim to *create* their own, utopian water world.

To understand this deep, often subconscious neglect of Latin America's diverse water societies, we might disclose a longstanding, dangerous myth. Let me call it *the dark legend of UnGovernance*. In various manifestations, this untold legend claims that local water territories are basically unruled – or at least unruly: disorganized humans, irrational values, unproductive ecologies, inefficient resource use, and continual water conflicts. This "UnGov Legend" disfigures Latin America's water societies by overlooking water users, meanings, values, identities, and rights systems on the ground. It then constructs its own water users, with identities that conveniently fit the models, with needs and rationales matching the imaginations of those in power, shored up in their science, technology, and policy towers. This way, the UnGov Legend justifies dramatic interventions. By simplifying diversity according to water expert notions, Towers of Indifference produce policy models that depoliticize their deeply political choices and dehumanize the people they affect.

Civilizing the unruly New World justified ancient colonization. Or, as John Locke (1970 [1690]: par.49) said: "In the beginning, all the world was America." In this chapter I argue, first, how inventing the *Indios* and labelling their property rules as "*Radically Different*" enabled early political exclusion and distributive injustice. Next, I show how currently water users and societies are, again, artificially invented: now as the "*Potentially Equal.*" Third, I explain how both misinterpretations generate contemporary Indifference. They facilitate modern interventions to combat the "water crisis" by reorganizing "unruly" water cultures and thus justify *Moral Regime Change*. As I argue, the "Dark Legend" has never disappeared. Today, however, it encounters growing resistance from marginalized water user families and networks, fighting for water justice, refusing to be classified and governed as imaginary objects.

Latin America's political ecology of water

Throughout Latin America, growing demand and declining availability bring about escalating water conflicts. Inequality is very deep indeed. From Argentina to Mexico, water is reduced to an economic resource allocated to the "most profitable users and uses" in the win-or-lose market. Water dispossession is now worse than ever. "Water scarcity" is not a natural but a political construct that has been de-politicized, as if it were Nature's fault.

Meanwhile, Latin American governments increasingly invoke the international War against Terrorism to label and imprison villagers who protest water theft and pollution as "environmental terrorists." A Peruvian leader said: "We now have a State that no longer protects people's rights and instead protects investment."[4] In 2014, the Inter-American Human Rights Commission investigated 22 large-scale Canadian mining projects in nine Latin American countries,[5] concluding that they all caused profound environmental impacts, contaminating rivers, displacing people, impoverishing communities, and dispossessing water rights. Protesters have been killed. As the report observes, development cooperation increasingly promotes mining; Canada has advised Latin American governments on how to circumscribe protective laws and curtail civil rights to facilitate mining. China, Australia, Europe, and the United States may follow suit.

Such water dramas are widespread. In many Latin American cities, abundance for the wealthy often means water insecurity for *favelas*. In rural areas, landgrabbing (see Ojeda, this volume) goes hand in hand with water grabbing. Water-scarce subsistence systems export agricultural commodities in massive amounts as "virtual water" to rich, water-abundant countries. In Ecuador, small farmers are 86% of water users but receive only 13% of total flow. The 1% large farmers control 67% of total water flow (Gaybor, 2011). Water extraction impacts water quantity and quality alike. Approximately two-thirds of Ecuador's Amazon region is granted in concessions to oil companies. In Peru, such lease arrangements for oil exploration and exploitation involve nearly three-quarters of the Amazon area (Bebbington, Humphreys-Bebbington, and Bury, 2010).

To examine and understand its underlying layers, relationships, and contradictions, political ecology understands "water governance," beyond humans governing water, as *governing people through water;* as ways of organizing power and decision-making to bring about environmental control and societal order at once (cf. Boelens, 2015b; Bridge and Perreault, 2009; Yacoub, Duarte, and Boelens, 2015). Consequently, to understand marginalized-water-cultures, above all, it is crucial to study the water-cultures-that-marginalize-them, i.e., the different faces of "water governmentality" or the "art of governance through water" (cf. Foucault (1991 [1978], "the conduct of conduct"). The political ecology of water, then, focuses on unequal distribution of benefits and burdens, access to and control over water, winners and losers, and disputed water rights, knowledge, and culture; it can be defined as: "*the politics and power relationships that shape human knowledge of and intervention in the water world, leading to forms of governing nature and people, at once and at different scales, to produce particular hydro-social order.*" Intrinsically joining nature and society, it explains how thoughts and actions concerning water are always political, never neutral. It is also about building alternative water realities.

The empire of inequality: inventing the "Radically Different"

Latin America's empire of inequality builds this large *class-based difference* in access to water and means of production on cultural discrimination and imposed *racial, gender, and ethnic differentiation*. These two connect to *political differentiation*, exclusion from governance decisions. As Fraser

(2007) and Schlosberg (2004) have argued, distributive injustice, cultural injustice, and political injustice are a mutually reinforcing complex.

Inventing radical difference thus justifies inequality and exclusion. This holds true in water governance as well. Local water users are rarely allowed to define their own ways of being different. In Ecuador, for example, my *comadre* Inés Chapi would tell me for hours how she and her peers were purposely constructed by the white elites: as an intermediate race, as *cutos*. This also excluded them from property ownership and political voice. Her neighbour Rosa Guamán explains the locally imposed classification as follows: "We were raised as *cutos*. A *cuto* is a dog whose tail has been cut off. *Cutos* were the *indígenas* brought to serve the whites. They changed our clothes and cut our hair. We were not allowed to identify as indigenous or as white" (cited in Boelens, 2015b).

Throughout Latin American history, "external" and "internal" colonizers use UnGov Legends to domesticate profoundly diverse local water cultures, imposing new identity labels and rules: first, to justify taking over local water resources; next, to enable control of everyday water use by installing the dominant frames of reference and "rational order"; and finally, to govern and control.

Long before Columbus reached the Americas, Europeans had already formulated the imaginaries of the New World. Its inhabitants, property relations, and governance structures would be *Radically Different*. They constructed a convenient mirror to promote their own governance ideas in the West and a utopian garden for experimenting with how to organize people and property: presenting a civilized Self versus a barbarian Other (e.g., Flores Galindo, 1988).[6]

This continued after the Conquest. Indigenous governance and identity were not known but invented to justify invasion and to introduce order to the ungoverned. Aníbal Quijano's (2000) notion "coloniality of power" shows how naturalizing such dualist, racist differentiation in capitalist class relationships became the foundation for Latin America's knowledge and property structures and remains so to this day.

In this respect, a well-known 16th-century debate represented two streams of reasoning that are central here: Juan Ginés de Sepúlveda depicted Indians as radically different dog-like slaves, lacking rationality and property notions (Sepúlveda, 1996). His opponent, Bartolomé de la Casas, argued that they were *potentially equal* and could be civilized to understand property and the rule of law.[7] Both regarded being human and rational as obeying the natural, divine order, i.e., Western religion and property foundations.

Linking dualist difference with UnGovernance was basic to influential European thinkers. For example, Thomas Hobbes (1985 [1651]) invented the indigenous order as the savage "State of Nature." John Locke (1970 [1690]) advocated "possessive individualism," justifying land and water rights dispossession, because Indians would not privately invest in their territories. Occupying and bringing order to the "un-possessed wilderness" became an act of moral progress. Jeremy Bentham (1988 [1781]), founder of utilitarianism, advocated subdividing indigenous territories into private properties to transform chaos into rational order. Liberation hero Simón Bolívar followed his enlightened advisor to liberate the continent from the collective "indigenous burden."

From Thomas More to Voltaire, Rousseau, Smith, Marx, or even Hannah Arendt,[8] the New World led to the invention of dualistic Difference. Actual peoples and their natural resource governance forms were conveniently denied.

Neoliberals like Milton Friedman eagerly extend the legend of UnGovernance, to colonize empty wasteland and put into practice radically free-market utopian ideas (Achterhuis et al., 2010). Using Chile's dictatorship, Friedman preached the link between freedom, private initiative and property rights. Supported by Friedman's (1980) and Hayek's (1944, 1960) groundwork, the Chicago Boys designed the world's most far-reaching water privatization experiment. To realize this "liberal" Water Code, Pinochet controlled all deviant behaviour – imprisoning water users to liberate the water market.

The World Bank praised the dictatorship's 1981 Water Code. In the 1990s, the World Bank forced most Latin American countries to copy the Chilean experiment and to privatize water rights, to "benefit the poor." Ironically, Chile's water policy identified these poor water users as anti-modern nobodies. Following the dark legend, it considered all customary water rights as "non-property."[9]

Regimes of "potential equality": enforcing uniform water rights and identities

Thirty years later, Peru's former president Alan García advocated putting an end to peasant communities' collective rights. He called them anti-modern "lazy, jealous dogs," who refuse to be productive but also refuse to sell their collective land and water to private companies.[10] The future is the Western example: Peruvians are expected to imitate "successful" foreign capitalist companies. García expressed beliefs deeply shared by most regional and international water policy-makers. The UnGov Legend flourishes.

However, governance ideologies gradually shifted, from stressing Radical Difference to depicting Potential Equality. In fact, Bartolomé de Las Casas' historic appeal to see the Other as "miserable but human" already foreshadowed this modern current. Discourse has changed from promoting racial distinction to extending liberal society's values to "include" *campesino, indígena,* and *favela* inhabitants as nation-state citizens.

Modern integrated water policies also favour including all water users as equals, rather than excluding them. Everybody is potentially equal, has the right to be equal, and *should* be equal. In Latin America's water policies, equalizing expansionism is not based on violent conquest but on universal water rationality.[11]

But equal to what? Equal to whom? "Equality" always needs a referential model. Modernist water science and policy projects in Latin America churn out recipes: new water knowledge, rules, and identities for becoming equal, by "certifying good water use." Water cultures are judged by how well they meet these standards.[12] Failure to meet these "self-evident" principles is presented as a lack of capacity for reason and as unwillingness to progress.[13] Indeed, as Mollepata water user Santiago Quintana indicated, questioning the errors of modern water policies and non-correspondence with on-the-ground water societies is portrayed as sticking to superstition, ungovernance, and non-development.

In keeping with international good governance and water citizenship norms, Peru ranks water users as modern or traditional. Modern users – such as capitalist agribusiness companies – are the model to follow. The water crisis is tackled by transferring rights to more "valuable" uses and rewarding "more efficient" users. Farmers who "modernize" get an efficiency certificate,[14] as well as preference for new rights. In practice, however, only a wealthy minority can afford this cutting-edge technology. They use any water saved not to restore aquifers but to expand their area with high-consumptive export crops. Ironically, model farmers' additional water rights and powerful pumps deplete aquifers. The dry Ica Valley's water table is dropping at a pace of nearly one metre per annum. Thousands of small-holders' wells are dry; they cannot compete with these "efficient" farmers' pumps. Even if they modernize, it is not easy to become equal in these circumstances.[15] New, inaccessible standards make them feel permanently backward. As Ivan Illich (1978) once remarked, the equalizing white paradise keeps moving further away, whenever poor people think they may be catching up.

A main source of water conflict in Latin America is the neglect of local rights systems. Although official and local customary water laws are necessarily interlinked and presuppose each other's existence, their relation is often problematic. Official rights are uniform and written in ink. Local water rights are context-based, diverse, and written in blood, sweat, and tears. Local water rights are not "traditional" but dynamically combine rules and principles from diverse normative sources. They combine local, national, and global rules and hybridize indigenous, colonial, and current norms. Water user collectives reconstruct these norms in territorial systems of organized complexity.

This diversity presents serious problems for water bureaucrats, elites, and international companies. Divergent authorities, territorial autonomies, collective property rules forbidding water transfer to third parties, and other local arrangements, complicate state control and make free market operation very difficult. These need uniform playing fields, where "equality" is universal.[16] Water users are therefore expected to abandon their unruly disorder and equalize. Cultural differences would evaporate when people experience rational global water rights systems.

Peru has received US$200 million from the Inter-American Development Bank to battle what the government describes as the country's "limited water culture" and "irrational water use." The aim is to "promote a modern water culture among the people" by neoliberal equalization[17]: not by force but by integrated water management. Highland communities have rejected these formalization projects that in their eyes bring about water security for the rich and insecurity for the poor. Nevertheless, the Program includes education to teach "water culture concepts."

Water neoliberalism in Latin America includes a finely graded cultural program. In Marx' terms, it "equalizes" and "creates a water world in its own image." Differences are allowed, as long as water users behave "as if they were equals" (e.g., Boelens and Gelles, 2005). Following universalistic good governance discourse, governments differentiate "responsible water citizens," who are state and market-compatible, from "irrational water spoilers," who devise their own rights systems.

During one of my visits to the Tulabug communities in Ecuador, I found that Riobamba City had decided to build a large water treatment project. Instead of using fallow *hacienda* lands, however, they confiscated land from many indigenous small-holders who had invested years of hard work to build their irrigation system. The two sides had entirely different views of hydro-social territory, and these communities protested and blocked all the roads. The Mayor of Riobamba expresses his view of the "barren continent": "These folks cling to their ancestors; even if their land is worthless, they defend it to death."[18]

Water leader Héctor Pilataxi explains: "It took 30 years of hard work to get the water here. Many people have only one plot, some died before the water arrived. They need the water." Pacífica Yupa agrees: "We will defend our land and water to the end. I need this little farm for my children, to survive. I will irrigate my land with my own blood, but I will not be put off it!"

Communities express how they are "rooted" in their water territory: the ways they have built it, how it connects humans, nature, gods, and ancestors, how to cherish it for the future. Pacífica: "They don't realize that we feed the city. We use water to produce new crops, livestock, to keep the city alive." Fundamentally, Tulabug families and their forefathers created their collective rights by animating hydro-social territory.

The city's water experts are blind to this water rights-building and rootedness, seeing only non-rights and unruled emptiness. The Waterworks Director says local water rules are for cannibals: "We should not revert to cannibalism and take justice into our own hands; we all can be rational, we have studied. Laws are universal." In his eyes, the problem is that the people to be displaced are too selfish. They lack civilization. "Countries in Europe are developed, not selfish. We all want to be at the same level, obviously."

To persuade indigenous families, officials first talk of integration, participation, and equality. The Waterworks Director says: "We are all Riobamba, we are all equal, the same country." However, some are a bit less equal and will have to suffer: "They should understand that some must make a sacrifice and pitch in, so that the rest can develop. As in any major project, we have to give up something, for the great majority's benefit." Indeed, Jeremy Bentham's universalist, utilitarian ethos is deeply ingrained in Latin America's policies of Integrated Water Management: sacrificing a minority for the "majority's happiness." This often turns local poor *majorities* into outlaws.

But families refused, and 17 fellow communities supported them in solidarity, as did the national indigenous movement. They revealed the invisible links of their hydro-social network. State authorities changed their tune. From Potentially Equals to be involved, communities became Radically Different, to be displaced with military violence. The Waterworks Director explained: "The self-interest of a couple of indigenous peasants without any technical criterion cannot endanger a whole city's welfare."

As we see, this modern "equality for all" commonly does not address the key issue of abolishing inequality in water access and representation; rather, making water users equal entails curtailing their deviation from the formal rules and rights, from Right-ness, whiteness, and assigned identities. All over Latin America we witness this attempt to "*Moral Regime Change*" – the moral mission to replace local water knowledge and rights systems by modern ones, placing them under state and market control; to combat presumed UnGovernance and make them identify with the ruling system.

Indifference and Moral Regime Change: de-humanizing the Latin American water worlds

As a consequence, in their daily practice, water users suffer not only from being categorized as Radically Different but *also* from being labelled as Potentially Equal: both visions deny cultural diversity in water control. As alternating twin forces, racist exclusion complements equalizing inclusion (Figure 37.1). They bring about indifference as to who Latin American water users are, what they want, and how they feel, know, value, and identify.

The water world abounds with such "indifference regimes," produced by pre-modern empires based on inequality, by modernist socialist, liberal, and neoliberal policies forcing abnormal water cultures to become equal and by post-modern schools, a-critically embracing "the right to difference" – which is in turn conducive to indifference and promotes cultural relativist blindness to local water injustices. Next, anti-modernist, indigenist schools or romantic Return-to-Mother-Earth ideologies similarly disregard locally diverse livelihoods and hybrid identities. Enslaved by the same modernist approaches they fight, anti-modernists tend to simply invert Western stereotypes to glorify the indigenous or Latin American water worlds. Schools such as PRATEC blame all evil on the West, ignoring internal injustices. This evades critical power analysis and actively weakens efforts by marginalized water users to challenge injustices.[19] Long ago, Frantz Fanon (1961: 185) warned against such simplifying perspectives: "It is the colonialists who have become the defenders of the native lifestyle."

Today, in particular, the modernist water schools of State-centralization and neoliberalism seek aggressively to *transform* this reality. In other words, essentialization in current Latin American water politics is not just an *academic error* but has an equally important *political purpose* of conveniently reshaping local water rights and identities.[20] Moreover, recognizing local water rights for some groups – commonly in their essentialized expression – often entails illegality and active criminalization of all communities and rules that are *not* recognized (e.g., Boelens and Seemann, 2014). Such disregard subtly affirms and deepens the dark legend.

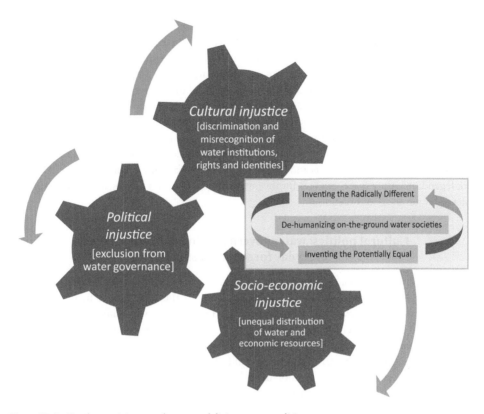

Figure 37.1 De-humanizing on-the-ground, living water realities

A horrifying example is the building of the Chixoy Dam in Guatemala. It required labelling the Achi Maya population living there as people without rights and properties, and the dam site as unruled, empty space. Project documentation ignored the Achi Maya's strong cultural-productive roots in their territory. The project blended participatory jargon – to include the "Potentially Equals" in globalized development – with racist ideas to explain why these "Radically Different" resisted displacement. An Inter-American Development Bank report states: "In the native peoples' worldview, traditional lifestyles and agricultural practices are expected to remain changeless for evermore, which explains why native *campesinos* . . . have proven resistant to change and innovation."[21]

When Achi Maya communities peacefully resisted displacement from their homes, the World Bank, donor governments, and international consultants actively ignored State-sponsored military violence (e.g., Johnston, 2018; Lynch, 2006). To construct this dam, the legend had to be made real, emptying the space. Many years of intimidating, torturing, and raping the local population left 440 men, women, and children dead and displaced thousands of local families. Mrs. Glü refused to descend from the Tower – she would go mad downstairs. The consultants were instructed "not to ask," because hydraulic dams are for everyone's progress. Their water knowledge and epistemology made them oblivious to human suffering.

Latin America's often dramatic water scenery shows how civilizing hydro-political dream schemes separate "all-seeing knowers" from "ignorant users." Modernist rational designs are blinded from *imagining* the human suffering they may provoke, and in this way they make it easy to implement decisions that have far-reaching consequences for other people's lives.

As Günter Anders argued, the ability to *understand* the suffering caused by modernist interventions does not keep pace with the growing ability to *intervene*. We can *do* far more than we are able to *morally understand and justify*. In the face of our intervention technologies, he said, we humans are obsolete, "*antiquiert,*" we are smaller than ourselves (1980 [1956]).

Towers of Indifference perceive and construct "equals" that fit into their models. Perceived from high above, everybody is equal and made equal, far from the diverse worlds of flesh-and-blood water users. Actively constructing ignorance helps explain and thereby intervene in the complex water world. Hovering above context, above the arena of interests, tower builders objectify human and natural subjects. This *de-humanizes* water reality. Rather than revealing the suffering, only governable subjects and objects appear: in perfect alignment with Latin America's dark legend, an empty space emerges for inventing games, rules, and players, for constructing and ordering water users and identities.

Cultural politics: water governance formalization versus everyday forms of resistance

Large-scale water injustices, such as during the Water Wars in Bolivia, the Texaco water pollution in the Amazon, Yanacocha gold mining in Peru, or the Belo Monte hydropower dam in Brazil appear in the newspapers. Here, the Dark Legend supports materially erasing water territories and sometimes entire user populations. Far more widespread, however, is the ongoing, invisible, everyday water dispossession and discrimination. In such cases, the Dark Legend and Indifference Towers erase water rights, norms, and identities.

Often, these water injustices are not intended to be harmful; commonly they are highly moral, rational, and development-oriented. Latin America's water policies, irrigation, drinking water, and hydropower interventions follow a deeply universalist designing rationality, totally diverging from on-the-ground governance realities. Far beyond just hydraulic technology, such designs artificially construct new subjects: individualized water users, arranged in new spatial management units neatly fitting the dominant legal-political framework. New forms of governmentality rearrange people like chess pieces within new "convenient communities" (e.g., Boelens, 2015b; Rodríguez de Francisco and Boelens, 2015; Valladares and Boelens, 2017). Principles embedded in the new hydraulic-legal-political designs fundamentally displace the knowledge and norms of users and erode existing community rationality and rule-making. "Hydro-political dream schemes" forcefully align micro-water control with meso-/macro-governance order.

The universally proclaimed Water Crisis justifies crushing local rights, forcing everyone to speak the same water language: building Babylon's Water Tower. Moral Regime Change thus legitimizes material, social, and discursive violence to make and break water societies. Utopian-inspired projects undermine and seek to erase existing arrangements as a pre-condition for building the new order, reshaping society and nature at once (see Achterhuis, Boelens, and Zwarteveen, 2010). Aside from outright imposition, Latin American history is full of so-called "reciprocity pacts," governance ideologies, and cultural categories, deliberately constructed to fit lines of command. They try to make people identify with formalized water culture: as colonial subjects, nation-state citizens, or water market clients.

Simultaneously, however, deeply challenging this UnGov Legend, Latin America's water cultures are actively (re-)constructed. Water flows, use systems, and hydrological cycles, from micro to macro scales, are mediated by power relations and human intervention. They shape hydrosocial territories (Boelens et al., 2016; cf. Duarte-Abadía et al., 2015; Hommes and Boelens, 2017; Hoogesteger, Boelens, and Baud, 2016). In practice, different parties envision or construct such territories differently, with diverging functions, values, and meanings. Territorial spaces are sites of contested control over socio-natural configuration. Examining these conflicting hydro-social imaginaries offers deep insight into how water benefits and burdens are to be distributed, how humans and non-humans are ordered and assumed to behave, and how this is sustained by political-economic, technological, and symbolic orders in ways that can strengthen or challenge the status quo. Water governmentalities (and counter-conducts) involve the politics of constructing new meanings of and connections among the social, the natural, and the technological.

In user collectives' everyday contestations and counter-conducts, local water cultures appear as rooted in history and schemes of belonging among people, place, and water – very down-to-earth, not as revolutionary abstractions or water utopias. In their struggles, they continually reinvent rules and identities and traditions – in a strange mixture of "returning to tradition" and "building new orders." They see water rights as instruments to arrange their systems and as weapons to defend themselves. Far from egalitarian micro-societies, they are an effort, a process, and a capacity to merge collectivity with diversity and to exercise mutual dependence on nature and on each other.

Therefore, far more important than the open water struggles are the thousands of invisible daily battlefields. In undercurrents, communities build their own rights systems. These question the self-evidence of formal state, science, or market-based governance. They may alternate between overt resistance and disguise.

Beyond an "internal affair," most water user communities try to tie in with national and international policies, markets, and partnerships, embedding local in global and global in local: the transnationalizing scales of extractive industries and policy-making force them to scale up their water struggles as well. From Chile to Nicaragua, from Brazil to Mexico, villagers unite in multi-actor, multi-scalar water defence networks, involving entities as peasant and indigenous federations, the Movement of Dam-affected People, Greenpeace, Amnesty International, and Human Rights Tribunals. Trans-local networks are arising throughout Latin America. In Guatemala, the displaced Achi Maya families have forged alliances with NGOs, scholars, human rights advocates, journalists, and networks such as International Rivers. After three decades of struggle, the US Congress passed a bill forcing the World Bank and other donors to arrange multi-million dollar compensations through Guatemala's government. Although they can never compensate for the suffering, such victories are milestones. These networks also show that state, scientific, and policy-making communities are not monolithic but track records of social conquests. Many state employees, professionals, and scientists struggle "from within" to crumble the Towers of Indifference. They ally with water-user groups to capture cross-scale opportunities, interlace their mutual bodies of water knowledge, and co-design water societies: creating political water societies.

Water justice

In Latin America, struggles over water are battles over resources and legitimacy, to exist as water control communities and self-define the nature of water problems and solutions. Conflicts are therefore often deep and intense. User organizations try to shape their own projects, re-moralizing universalist placeless systems to build rooted hydro-social territories, claiming the freedom to deviate.

Grassroots environmental and water user organizations counter being mislabelled as Radically Different and as Potentially Equals with their *own* paradox. By connecting material with cultural-political struggles, they demand both the right to be equal *and* the right to be different. They combine their struggle against highly unequal resource distribution with their demands for greater autonomy, sharing in water authority and a pluralistic water rights order reflecting context and diversity. Or, as Juan Carlos Ribadeneira (1993: 6) remarked: "The search for equality amidst difference goes hand-in-hand with its opposite: finding difference in the face of the empire of equality."

Here we arrive at the heart of Latin American grassroots battles for water justice, against indifference. Contesting the Dark Legend of UnGovernance, the intimate connection among people, water, space, and identity fuses their struggles for *material* access and control of water-use systems and *ecological defence* of neighbourhoods and territories with their battle over the right to *culturally define* and *politically organize* these socio-natural systems (Figure 37.2).

There is a need to move beyond universalist, descriptive theories, and modernist laws, focusing on what water justice "should be." We need to start by understanding how people on-the-ground *experience and define* water justice, rights, and ways of belonging – not taking them for granted but as starting points. In Latin America's water world, liberal, socialist, or neoliberal models of equality have always tended to reflect the dominant water society's colonial mirror, ignoring *campesino*, *indígena*, and women's interests and views. Beyond abstract de-humanized models *and* beyond localized romanticism, it is urgent to climb down from the Tower and systematically explore the sources of water injustice, local views on fairness, and the impacts of formal laws and justice policies on live human beings. Understanding water justice calls for a contextual, grounded, relational approach.

Appeals for greater water justice demand action against blunt water grabbing *and* highlighting those forms of suffering that are concealed by the modernist water science-policy nexus. This requires combining grassroots, academic, activist, and policy action: engagement across differences. Accordingly, we may define *water justice* as: "*the interactive societal and academic endeavour to critically explore water knowledge production, allocation, and governance and to combine struggles against water-based forms of material dispossession, cultural discrimination, political exclusion, and ecological destruction, as rooted in particular contexts.*"

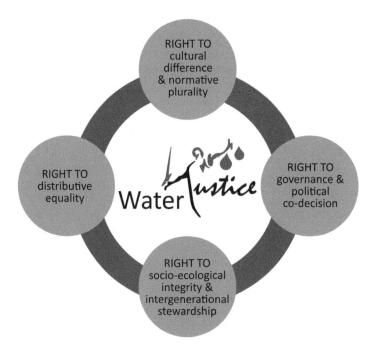

Figure 37.2 Interweaving and balancing the struggles for water justice

Water justice is about breaking open the forms of indifference towards knowing, transforming, and distributing nature. It is about questioning our silence on water science; about research and action engaging diverse water-societal actors, to see multiple water truths and worldviews and to co-create transdisciplinary knowledge. It is about facilitating colonized water cultures to question civilizing water rationality and so-called moral right-ness; to unmask current day expressions of water's "coloniality of power." It is about critical engagement with water movements, dispossessed water societies, and interactive design of alternative hydro-social orders. These alternatives cannot be engineered by scientists or handed out by policy-makers: they result from "unity without uniformity," from interweaving cross-cultural water knowledge and cross-societal pressures from below. It is a dialectical process: people struggle and thereby create the water world in which they live: a conscious and subconscious strategy to create and diversify rooted water cultures. They resist to be able to create, and they create to be able to resist, giving form and substance to Latin America's water societies.

I started this chapter with the White Elephant. It has never been able to walk, and now it belongs to the archaeology of modernist development. By contrast, Santiago, Alberta, and Cirilo are very much alive. They have teamed up with neighbouring families and communities, Peruvian NGOs, and an international support network. As Santiago comments: "We no longer let a project come in just like that. Now we are firmly organized. Users now have rights, a voice to state demands." They have recovered their own canals, rules, and organization. As Cirilo explains: "With the beating we got from this phantom canal, we have learned to stand up and fight!"

A final reflection on the View from the Tower. Yes, the tale is about how modernist water science and policies create distance, objectify reality, and cannot see the suffering they cause in Latin American water societies. It is about the need to understand the real-life other, defeat the Dark Legend, and entwine efforts to achieve water justice. But I think there is more.

When Mrs Glu looks down she sees the objectified other but, at the same time, her son. She sees *herself*. After centuries of (neo-)colonized mirrors, how we see and represent the other relates closely to how we see and represent *ourselves*. Water justice is about decolonizing the mirrors that we have mutually constructed, and that distort what they reflect, questioning our own cultural politics, our modes of categorizing and organizing water and society. Getting closer to on-the-ground water user societies requires getting closer to ourselves. Eduardo Galeano (1998: 111) once remarked: "*Al fin y al cabo, somos lo que hacemos para cambiar lo que somos. Ultimately, we are what we do to change who we are.*"

Notes

1 Chapter adapted from my inaugural lecture at CEDLA/ University of Amsterdam, Chair 'Political Ecology of Water in Latin America' (Boelens, 2015a).
2 Quotes come from my 2015 book *Water, Power and Identity. The cultural politics of water in the Andes.*
3 In Boelens (2015b) I compare this tower-induced "reality-indifference" to Bentham's Panoptical Tower and Sumerian engineers' biblical Tower of Babylon.
4 *The Guardian*, 14–15–2014. www.theguardian.com/environment/andes-to-the-amazon/2014/may/14/canadian-mining-serious-environmental-harm-iachr
5 Ibid; see also Stoltenborg and Boelens (2016).
6 Other thinkers (e.g. Rousseau) constructed the Noble Savage as opposite of the materialistic European. Ancient Greek notions of civilians versus barbarians and Christian mythology (the Golden Age and Earthly Paradise of primitive people in a "natural state") were central in representing Self and Other.
7 Las Casas (1552) owned black slaves himself. Pleas from much of Las Casas' group (and from royals) to "protect" the Indians were based on the need to stabilize the system of plunder.
8 Arendt (1963), misled by the UnGov Legend, portrays the Americas as wilderness-to-be-civilized. She praises the brave settlers, ignoring widespread dispossession of indigenous properties.

9 E.g., Bauer (2004), Prieto (2015)

10 *El Comercio* 28–10–2007.

11 Socialist authorities tried to standardize and abolish water rights of local collectives in the name of "equality"; (neo)liberal nation-States foresaw equalization of water users as players on the water market.

12 Modernist water policies presume equalizing to technocentric models. In Latin American laws "rational water management" abounds with Eurocentric norms about efficiency, social security, effective organization, private ownership, and economic functionality.

13 Western, modernist water knowledge and technologies are strongly promoted by Latin American technical universities: not "imposed" but requested as equalizing measures.

14 Modern farmers need to have installed advanced technologies, e.g. drip and sprinkler irrigation.

15 These norms for becoming equal are not fixed but respond to the urge to belong to the model's mirror community. Modern capillary power (self-) corrects "abnormal" thought and behaviour.

16 Notwithstanding constitutional recognition, L.A. Water Laws continue monistic judicial-political models.

17 Neoliberal water policy sees a clear need to present all water actors as "potentially equal." Markets and transactions require participants who interact "as equals." Existing power differences that favour private companies are discursively neglected or presented as the gap that the potentially equals should close.

18 Quotes: Boelens (2015b).

19 Dichotomizing complex reality, they construct radical We-Other representations that negate cross-cultural political choices. This current also negates today's rapidly changing rural-urban and globalizing contexts, and forgets to accommodate technology- and identity-shopping.

20 Responding racist and modernist regimes, grassroots/indigenous movements *also* (re-)essentialize. Deconstructivist schools often criticize this counter-representation (on "scientific grounds," applying the same objectivist perspective they claim they challenge) but neglect *political* properties in counterdiscourses. Grassroots counter-images require contextualized examination as part of concrete struggles.

21 IDB 1991, annex II-2:1, cited by Lynch (2006: 14).

References

Achterhuis, H., Boelens, R. and Zwarteveen, M. (2010) Water property relations and modern policy regimes: Neoliberal utopia and the disempowerment of collective action. In R. Boelens, D. Getches and A. Guevara-Gil (eds) *Out of the Mainstream. Water Rights, Politics and Identity*. London; Washington, DC: Earthscan, pp. 27–55.

Anders, G. (1980(1956)). *Die Antiquiertheit des Menschen*. Munich: Beck.

———. (1992(1932)). *Die Molussische Katakombe*. Munich: Beck.

Arendt, H. (1990(1963)) *On Revolution*. London: Penguin Books.

Bauer, C. (2004) *Siren Song: Chilean Water Law as a Model for International Reform*. Washington, DC: RFF Press.

Bebbington, A., Humphreys-Bebbington, D. and Bury, J. (2010) Federating and defending: Water, territory and extraction in the Andes. In R. Boelens, D. Getches and A. Guevara-Gil (eds) *Out of the Mainstream. Water Rights, Politics and Identity*. London; Washington, DC: Earthscan, pp. 307–327.

Bentham, J. (1988(1781)) *The Principles of Morals and Legislation*. Amherst, NY: Prometheus.

Boelens, R. (2015a) *Water Justice in Latin America: The Politics of Difference, Equality, and Indifference*. Amsterdam: CEDLA and University of Amsterdam.

———. (2015b) *Water, Power and Identity: The Cultural Politics of Water in the Andes*. London; Washington, DC: Routledge.

——— and Gelles, P. H. (2005) Cultural politics, communal resistance and identity in Andean irrigation development. *Bulletin of Latin American Research* 24(3): 311–327.

——— and Seemann, M. (2014) Forced engagements. Water security and local rights formalization in Yanque, Colca Valley, Peru. *Human Organization* 73(1): 1–12.

———, Hoogesteger, J., Swyngedouw, E., Vos, J. and Wester, P. (2016) Hydrosocial territories: A political ecology perspective. *Water International* 41(1): 1–14.

Boelens, R., Perreault, T. and Vos, J. (2018) *Water Justice*. Cambridge: Cambridge University Press.

Bridge, G. and Perreault, T. (2009) Environmental governance. In N. Castree et al. (eds) *Companion to Environmental Geography*. Oxford: Wiley-Blackwell, pp. 475–497.

Duarte-Abadía, B., Boelens, R. and Roa-Avendaño, T. (2015) Hydropower, encroachment and the repatterning of hydrosocial territory: The case of Hidrosogamoso in Colombia. *Human Organization* 74(3): 243–254.

Fanon, F. (1961) *The Wretched of the Earth*. New York: Grove Press.

Flores Galindo, A. (1988) *Buscando un Inca: Identidad y Utopía en los Andes*. Lima: Horizonte.

Foucault, M. (1991(1978)) Governmentality. In G. Burchell, C. Gordon and P. Miller (eds) *The Foucault Effect: Studies in Governmentality*. Chicago: University of Chicago Press, pp. 87–104.

Fraser, N. (2007) Identity, exclusion, and critique: A response to four critics. *European Journal of Political Theory* 6(3): 305–338.

Friedman, M. (1980) *Free to Choose*. Harcourt: Harvest.

Galeano, E. (1998) *El libro de los abrazos*. Bogotá: Tercer Mundo Editores.

Gaybor, A. (2011) Acumulación en el campo y despojo del agua en el Ecuador. In R. Boelens, L. Cremers and M. Zwarteveen (eds) *Justicia Hídrica*. Lima: IEP, pp. 195–208.

Geertz, C. (1980) *Negara*. Princeton, NJ: Princeton University Press.

Haraway, D. (1991) *Simians, Cyborgs and Women: The Reinvention of Nature*. London: Free Association Books.

Hayek, F. A. (1944) *The Road to Serfdom*. London: Routledge.

———. (1960) *The Constitution of Liberty*. Chicago: University of Chicago Press.

Hobbes, T. (1985(1651)) *Leviathan*. Meppel: Boom.

Hommes, L. and Boelens, R. (2017) Urbanizing rural waters: Rural-urban water transfers and the reconfiguration of hydrosocial territories in Lima. *Political Geography* 57: 71–80.

Hoogesteger, J., Boelens, R. and Baud, M. (2016) Territorial pluralism: Water users' multi-scalar struggles against state ordering in Ecuador's highlands. *Water International* 41(1): 91–106.

Illich, I. (1978) *Toward a History of Needs*. New York: Random House.

Johnston, B. R. (2018) Large-scale dam development and counter movements: Water justice struggles around Guatemala's Chixoy dam. In R. Boelens, T. Perreault and J. Vos (eds) *Water Justice*. Cambridge: Cambridge University Press.

Las Casas, B. de (1999(1552)) *Brevísima relación de la destruición de las Indias*. Madrid: Castalia.

Locke, J. (1970(1690)) *Two Treatises on Government*. Cambridge: Cambridge University Press.

Lynch, B. D. (2006) *The Chixoy Dam and the Achi Maya: Violence, Ignorance, and the Politics of Blame*. Ithaca, NY: Cornell University Press.

Prieto, M. (2015) Privatizing water in the Chilean Andes. *Mountain Research and Development* 35(3): 220–229.

Quijano, A. (2000) Colonialidad del poder, eurocentrismo y América Latina. In E. Lander (ed) *La colonialidad del saber*. Buenos Aires: CLACSO, pp. 201–246.

Ribadeneira, J. C. (1993) *Derecho, pueblos indígenas y reforma del Estado*. Quito: Abya-Yala.

Rodriguez-de-Francisco, J. C. and Boelens, R. (2015) Payment for environmental services: Mobilising an epistemic community to construct dominant policy. *Environmental Politics* 24(3): 481–500.

Schlosberg, D. (2004) Reconceiving environmental justice: Global movements and political theories. *Environmental Politics* 13(3): 517–540.

Sepúlveda, J. G. de (1996(1550)) *Tratado sobre las justas causas de la guerra contra los indios*. Mexico, DF: Fondo de Cultura Económica.

Stoltenborg, D. and Boelens, R. (2016) Disputes over land and water rights in gold mining: Cerro de San Pedro, Mexico. *Water International* 41(3): 447–467.

Valladares, C. and Boelens, R. (2017) Extractivism and the rights of nature: Governmentality, 'convenient communities', and epistemic pacts in ecuador. *Environmental Politics* 26(6): 1015–1034.

Yacoub, C., Duarte, B. and Boelens, R. (2015) *Agua y Ecología Política. El extractivismo en la agro-exportación, la minería y las hidroeléctricas en Latino América*. Quito: Abya-Yala.

38

ENERGY VIOLENCE AND UNEVEN DEVELOPMENT

Mary Finley-Brook and Osvaldo Jordan Ramos

Introduction: energy violence

Energy supply is a cornerstone of economic development and Latin American electricity needs are projected to grow 90% by 2040 (Balsa, Espinasa, and Serebrisky, 2015). Regional power sources demonstrate energy violence, defined as intentional or disproportionate injury or harm during the energy lifecycle (i.e., production, transport, waste disposal, and so on). Our objective is to identify cross-national types of violence in the energy sector (Table 38.1).

We analyze violence in eight hydro and natural gas case studies in six countries (Figure 38.1). In spite of industry claims that natural gas is clean and safe (i.e., YPF, 2014), research documents public health concerns from chemicals in hydraulic fracturing (Colborn et al., 2011; Hays and Shonkoff, 2016; Finley-Brook et al., 2018). Gas transport contributes to toxic water and air pollution, including toxic emissions surrounding compressor stations (Colborn et al., 2011; Hays and Shonkoff, 2016). Gas is promoted as low-carbon even though production and transportation releases methane, a potent greenhouse gas (GHG) (Hays and Shonkoff, 2016). While large hydropower had begun to decline because of criticisms of forced resettlement and biodiversity loss,[1] international firms and donors re-engaged in dam building as part of low-carbon economic growth strategies (Finley-Brook and Thomas, 2011; Finley-Brook, 2017). Dams may erroneously be labelled as climate neutral, since impounded reservoirs in tropical environments release methane (Hall and Branford, 2012; Fearnside, 2014, 2017). Dam building also releases carbon during forest clearing and requires use of cement, a high GHG-emitting construction material. Meanwhile, climate change and seasonal variation impact reliability and efficiency of hydropower, so many dams run at partial capacity. Nonetheless, showing steady increase, 10,000 MW of new Latin American hydropower was produced in 2017 (IHA, 2017). Growth was concentrated in Brazilian megaprojects with labour conflicts and indigenous opposition (Fearnside, 2017; MAB, 2017; Sullivan, 2017). Construction practices violated international agreements like the Declaration of Human Rights and the Declaration on the Rights of Indigenous Peoples, which requires Free, Prior and Informed Consent (FPIC).

Historically, large-scale energy generation created exploitation, conflict, and violence for Indigenous and Afro-descendent Peoples in Latin America. In a well-known example, Guatemala's Chixoy Dam construction led to a massacre and displacement of thousands of Maya in the 1980s. Still today, murders of anti-dam activists continue, particularly in Brazil, Guatemala, and

Table 38.1 Continuum of energy violence

Type of Violence	Description
Everyday/Slow	Ongoing exposure to toxic pollutants or to deprivation; climate disruption
Gender	Harm tied to gender and gender difference, including rape, sex trafficking, and harassment
Instrumental	Deliberate show of force or use of aggression to advance a position; includes energy-related threats, crime, and resistance
Legal	Legislation causing physical harm, such as dispossessing specific populations, or criminalizing activities like protest; lack of due process
Occupational	Workplace transactions creating harm (i.e., explosions, accidents); absence of regulations and oversight; violence conflict involving workers
Organized	Harm resulting in use of force by military, security, paramilitary, or police
Physical	Bodily harm; force intended to hurt, damage, or kill
Political	Discord from power struggles in formal institutions; damage from systemic concealment or lack of transparency
Structural	Systemic discrimination or malapportionment based on race, gender, income, nationality, age, etc.
Territorial	Loss or restriction of land or property rights

Figure 38.1 Case studies of energy violence

Honduras (International Rivers, 2016). Violence is not limited to hydro-development. Wind farms cause land conflict with indigenous communities in Mexico (Baker, 2011). Biofuel production in Colombia and Guatemala creates repression and displacement (Ahmad, 2017; Mingorría, 2017). These examples show violence can occur in renewable energy projects as well as fossil fuel operations (see also Finley-Brook and Thomas, 2011; Vásquez, 2014; Zaitchik, 2017).

Uneven development

Uneven development refers to how capitalism and neoliberalism shape and reshape nature and society to advance inequality in distribution of gains and losses (Smith, 2010). Benefits are often spatially and socially uneven; for example, Belo Monte dam's high voltage lines transmit electricity long distances to urban and industrial areas in southeast Brazil while thousands of indigenous poor were displaced to inhumane settlements without promised compensation. Poor spatial distribution of intense harm exists in fossil fuels as well (i.e., improper disposal of oil waste in Ecuador and Peru caused public health emergencies in Indigenous Peoples' territories) (Kimerling, 2013; San Sebastián and Hurtig, 2014; Zaitchik, 2017). In San Jorge Gulf, the center of Argentina's oil industry, clean water is scarce and cancer patients are poorly attended (Scandizzo, 2017). Now with the expansion of gas in Argentina, hydraulic fracturing, or fracking, brought deliberate dumping and widespread spills and leaks (Alvarez Mullally, 2017; Zaitchik, 2017).

Energy violence exists within broad social and institutional inequality, meaning lack of access to public health services in low-income frontline communities with O&G pollution in their air and water every day. Biofuels can also be dangerous because toxic pesticides persist in local ecosystems. Latin American countries struggle to respond to increasing rates of cancer and advanced disease (Curado and de Souza, 2014). Access to health care is frequently inadequate for populations exposed to various forms of economic, geographic, and ethnic marginalization.

Uneven development expands with unfair retaliation for opposition when people protest atrocities. The state often sides with O&G firms, bringing military and police to defend private interests and assets (Jaskoski, 2013; Vásquez, 2014). Similarly the state protects corporate producers in toxic hotspots harming low-income rural populations following commercial production of biofuels (Oliveira and Hecht, 2016; Finley-Brook, forthcoming). Environmental defenders are stigmatized and targeted with smear campaigns. They risk physical attacks as well as sexual harassment, kidnap, forced disappearance, arrest, arbitrary detention, travel bans, exile, raids and searches, asset freezing, licence revocation, forced closure, illegal surveillance, cyberattacks, criminalization, and judicial harassment (Hallam, 2017: 42).

Research methods

Our comparative case study approach exposes the multiple types of violence that subsume the energy sector. We suggest that many energy projects in Latin America have at least one of these forms of violence. We focus here on highly violent cases with four or more different types of violence (Table 38.2), to demonstrate current systems allow problematic projects to exist. While there were additional cases deserving attention, we selected these cases because they (1) show a wide range of social and ecological problems and (2) received media attention as important national and international examples. Our information comes from Spanish and English sources including scholarly articles, periodicals, interviews, non-governmental organization (NGO) reports, and government documents.

We find structural and territorial violence is ubiquitous. While research on hydropower in Latin America is extensive, gas has received much less attention and some of the openings in Table 38.2 may be due to incomplete information. Documentation of potential disease clusters tied to O&G is rare in Latin America. Current health systems prioritize the needs of wealthy, urban minorities (Curando and de Souza, 2014). O&G production is often rural and access to medical services in the countryside is seldom adequate.

Table 38.2 Energy violence in case studies

Type of Violence	AES	AP	AZ	BB	BM	C	J	VM
Everyday/Slow	X	X		X	X	X	X	X
Gender				X	X			X
Instrumental		X					X	X
Legal				X	X			X
Physical		X	X	X	X	X	X	X
Political	X		X	X	X	X	X	X
Occupational					X			X
Organized			X	X	X	X	X	X
Structural	X	X	X	X	X	X	X	X
Territorial	X	X	X	X	X	X	X	X

AES = AES Colon
AP = Agua Prieta
AZ = Agua Zarca
BB = Barro Blanco
BM = Belo Monte
C = Camisea
J = Jirau
VM = Vaca Muerta

Energy case studies

Our eight cases studies are large and internationally prominent (Table 38.3). Neoliberal LAC energy development involves neo-extractivism, meaning state facilitation of private expansion of infrastructure and production (Raftopoulos, 2017). State political support and public finance often exists alongside assets from International Finance Institutions (IFIs,) technical or consulting consortiums, and international or domestic firms to form an expansive network of powerful actors who collaborate on expensive energy infrastructure.

Hydro cases

We examine four dam projects (Figure 38.2) from Brazil, Honduras, and Panama.

We identify patterns in violence associated with dams (Table 38.4), with entrenched political and structural inequities.

Natural gas case studies

We examine four gas projects (Figure 38.3) in Argentina, Mexico, Panama, and Peru. These key initiatives are likely to shape regional energy circuits for decades: two are functioning and two remain under construction. LAC countries with gas reserves eagerly extract and allow occupational violence with workers exposed to toxic chemicals. Inadequate monitoring and oversight reflects political and structural violence, since gas operations expel carcinogens, mutagens, and teratogens with potentially devastating health consequences (Colborn et al., 2011; Hays and Shonkoff, 2016; Finley-Brook et al., 2018). Gas operations emit elements dangerous to human health including nitrous oxides, BTEX (benzene, toluene, ethylbenzene, and xylenes) compounds, silica, Volatile Organic Compounds (VOCs), and precursors to ground-level ozone (Colborn et al., 2011; Hays and Shonkoff, 2016).

Table 38.3 Select energy violence cases

Project Name	Location	Project Type	Total cost (US$)	Public Owners and Developers	Private Owners and Developers	IFIs
BM	Brazil	Hydro	18.5 Bn	BNDES Eletrobras	Norte Energia, Iberdrola, various utility and mining companies	–
J	Brazil	Hydro	15.6 Bn	Eletrobras	Engie, Mitsui	–
AZ	Honduras	Hydro	44 Mn	–	Desarrollos Enerégticos, SA (DESA)	FMO, Finnfund, CABEI
BB	Panama	Hydro	86 Mn	–	Generadora del Istmo S.A. (Genisa)	CABEI, DEG, FMO
VM	Argentina	Gas	6–20 Bn/ year	Petrobras, Petronas	Chevron, Shell, Exxon-Mobil, Pluspetrol, Tecpetrol, Repsol-YPF, Pan American Energy Group, Dow Chemical, Wintershall, Total	–
AP	Mexico	Gas	10.8 Mn	–	Sempra Energy, ieNova	–
AES	Panama	Gas	1 Bn	–	AES,[1] Inversiones Bahia, Motta Group	IFC, IDB, CABEI
C	Peru	Gas	6.3 Bn	–	Pluspetrol, Hunt Oil Company, SK Corporation, Repsol YPF, Tecpetrol, and Sonatrach[2]	IDB

DEG = German Investment Corporation
FMO = Netherlands Development Finance Institution
IDB = Inter-American Development Bank
IFC = International Finance Corporation (World Bank)
CABEI = Central American Economic Integration Bank
CAF = Andean Development Corporation
BNDES = Brazil's National Development Bank

1 AES is the complete name of the company, previously shortened from Allied Energy Sources.
2 Ownership has changed to include Grupo de Energia de Bogota, Enagás, Engie, Canada Pension Plan Investment Board, and Graña y Montero.

We demonstrate patterns of violence in gas projects (Table 38.5), with widespread everyday, physical, structural, and territorial violence, particularly for low-income populations and communities of colour.

Who is responsible?

Our cross-national research suggests broad responsibility for violence. Projects demonstrate institutional violence with unclear public-private divides and messy, transnational partnerships. There is systemic failure to stop repeated harm, for which state agencies, funders, firms, consumers, and others share blame. International agencies like the United Nations try to mediate conflicts but lack the power to force states to respect human rights.

States

Since the 1980s, LAC electricity provision was increasingly privatized. Yet state agencies set norms for energy development with national oversight roles. States invested in public-private

> **Belo Monte (Brazil):** Plans for the controversial project commenced in 2010 (Fearnside, 2017). The dam has a maximum capacity of 11,233 MW. It has been functional and working since 2015, even though the mega-project's construction continues until 2019.
>
> **Jirau (Brazil):** The Jirau dam broke a taboo of large-scale projects in Amazonia. The project is part of the initiative for the Integration of Regional Infrastructure of South America (IIRSA) (Bank Track, 2015). Opposition from local fisherman and spokesperson killed.
>
> **Aqua Zarca (Honduras):** Honduran company DESA (Desarrollos Energéticos) tried to build the dam from 2008 to 2017. Indigenous Lenca defending their land and livelihood rights faced multiple types of violence, including various assassinations, leading donors to pull funding (Brun-Guzman, 2017; Loperena, 2017; Hallam, 2017).
>
> **Barro Blanco (Panama):** In 2008, the Honduran company GENISA developed the Barro Blanco dam on the Tabasara River (Anaya, 2013; Finley-Brook, 2016). Dam opponents experienced state violence, including murder (Bill et al., 2012; Finley-Brook, forthcoming).

Figure 38.2 Hydro project descriptions

Table 38.4 Examples of energy violence in dams

Types of Violence	Impacts (per project)	Sources
Everyday/Slow	flooding (J, BM); loss of fish (J); illegal deforestation (J); community water source threatened (AZ)	Fearnside, 2014, 2017; Finley-Brook, forthcoming
Instrumental	arson (J)	Romero, 2012
Gender	sexual violence and prostitution in area of dam (BM); harassment of women and girl protestors (BB)	Greenpeace, 2016; MAB, 2017
Legal	lawsuits citing environmental and human rights damages overturned (BM)	Fearnside, 2017
Occupational	violence among migrant labourers (BM); union leaders fired (BM)	MAB, 2017
Organized	military and police defend infrastructure (AZ, J); police employ tear gas and rubber bullets (BB); state restricts telecommunications to isolate protestors (BB)	Bill et al., 2012; Romero, 2012; Loperena, 2017
Physical	repression and authoritarianism (AZ, BM); torture (J) assassination of anti-dam activist/s (AZ, BB, J)	Hall and Branford, 2012; International Rivers, 2016; Finley-Brook, 2017; Hallam, 2017; Loperena, 2017
Political	inadequate consultation of host communities in planning stages (AZ, BB, BM); incomplete environmental assessment (BM); corruption (BM, J); incentives for pollution (J)	Hall and Branford, 2012; Anaya, 2013; Fearnside, 2014, 2017
Structural	resettlement conditions inadequate (BM); 40,000 displaced without promised compensation (BM); impacts for peoples in voluntary isolation (uncontacted) (J); threaten livelihoods of Indigenous Peoples (J)	Greenpeace, 2017; Sullivan, 2017
Territorial	displacement and land conflict (AZ, BB, BM); no FPIC (BB, BM, J); flooding and loss of land (BB)	Anaya, 2013; Bank Track, 2015; Brun-Guzman, 2017; Fearnside, 2017

Agua Prieta Pipeline (US-Mexico Border): 19 proposed or existing pipelines span the US-Mexico border. Agua Prieta gas pipeline crossed indigenous Yaqui territory without consent (Telesur, 2016a, 2016b). Yaqui from Loma de Bacum won a moratorium (Navarro, 2016). Construction resumed with support of government officials. Armed confrontation between communities for and against the pipeline turned fatal (Escobar 2017).[1]

AES Colon (Panama): With tax incentives from the Panamanian government, beginning in 2018 this LNG-to-power project will produce electricity for domestic and international sales (AES, 2015). Impacted low-income residents have little specific project information (International Accountability Project, 2017; Sampaio, 2017; see IIC, 2015a, 2016b). AES violated human rights and harmed biodiversity in other Panamanian projects (Jordan, 2008; Finley-Brook and Thomas, 2011).

Camisea (Peru): Peru's military signed a security contract to protect Camisea, the country's flagship gas project (Jaskoski, 2013; Vásquez, 2014). After a series of pipeline ruptures and leaks, police extinguished local protest in 2005 (Feather, 2014). Negative environmental and health impacts, particularly for Indigenous Peoples, generated an international campaign against the likely expansion of the project.

Vaca Muerta (Argentina): Argentina's gas fields are some of the largest and most productive and global firms flocked to Vaca Muerta. Partnerships demonstrate the complexity of energy investment; for example, the Pan American Energy Group unites Argentinian, British, and Chinese investment (Markova, 2017). Indigenous Mapuche oppose gas extraction and face repression while demanding land rights (Aguirre, 2017).

Figure 38.3 Illustrated cases of natural gas violence

1 Eyewitnesses report police presence for three hours without intervening to stop violence as citizens were injured. A Yaqui leader suggests the firm instigated involvement of the pro-pipeline community who confronted anti-pipeline opponents (Navarro, 2016).

partnerships for energy infrastructure do not always work to support the health and well-being of their own constituencies (Finley-Brook and Thomas, 2011). State O&G firms tied to our cases include Petrobras (Petróleo Brasileiro), PEMEX (Petróleos Mexicanos), and Argentina's YPF (Yacimientos Petrolíferos Fiscales). Re-nationalization of YPF created incentives for Argentina to extract gas aggressively: government ownership compromises independence of oversight to protect the environment and property rights in Vaca Muerta. Petrobras and PEMEX are riddled with charges of corruption and both have been tied to the crisis of violence pervasive in Brazil and Mexico.

IFIs

IFIs fund expensive infrastructure and promote growth in electricity and fuel sectors (Baker, 2011). IFIs consolidate and integrate regional energy infrastructure through gas terminals and pipelines. To drive competition, IFI investment builds large projects. IFIs hesitate to divest once commited to projects, making projects hard to stop once construction commences. Barro Blanco donors distributed all funds, even as the project was embroiled in controversy and petitions circulated the globe. In the case of Agua Zarca financiers took a year to announce

Table 38.5 Examples of energy violence in natural gas

Types of Violence	Description	Sources
Everyday/Slow	spills pollute marine areas or drinking water (AES, AP, C, VM); air pollution (AES, C); harm to biodiversity (AES, C, VM)	Torres-Slimming, 2010; Bianchi, 2014; IIC, 2015b; Markova, 2017; Sampaio, 2017
Gender	prostitution and human trafficking (VM)	Bianchi, 2014
Legal	lawsuit after did not consult Indigenous Peoples (VM)	Markova, 2017
Instrumental	cars burned in conflict between supporters and opponents (AP); occupation of oil wells by workers (VM)	Navarro, 2016; Reuters, 2017
Occupational	state pushes labour flexibilization (VM); union busting (VM)	Reuters, 2017
Organized	police and military contain protestors (VM, C); military contracted to protect private company (C)	Jaskoski, 2013; Feather, 2014; Vásquez, 2014
Political	ineffective regulation leads to legal and illegal dumping of fracking waste (VM)	Alvarez Mullally, 2017
Physical	pipeline supporters assault opponents (AP); kidnapping of lawyer for pipeline opponents (AP); state use of force against protestors (VM); Indigenous mortality from outside disease and mercury poisoning (C)	Hill, 2016; Telesur, 2016a, 2016b; Markova, 2017
Structural	delegitimization of indigenous governance (VM); fracking pads located among towns and along rivers and water bodies (VM); gas infrastructure close to low-income neighbourhoods (AES); high noise levels (AES); after a decade of extraction, poverty remains widespread (C); infrastructure intersects reserve for Indigenous Peoples in voluntary isolation (C); gas equipment disrupts hunting (C)	Markova, 2017; IIC, 2015a, 2015b; Torres-Slimming, 2010; Feather, 2014
Territorial	displacement of Indigenous Peoples (VM); no FPIC (AP, C, VM); land disputes (AP)	Bianchi, 2014; Hill, 2014, 2016; Navarro, 2016

withdrawal,[2] after murder convictions for the Goldman Environmental Award winner Berta Cáceres involved military personnel and a company employee (Loperena, 2017).

Private sector: firms, shareholders, and consumers

Firms aim to control regional infrastructure in strategic locations. Many receive backing from state military forces (Jaskoski, 2013; Vásquez, 2014). Some firms hire private (i.e., permanent or temporary contracts) or irregular security forces (i.e., paramilitaries, hit men) to protect assets. High profits from energy projects contribute to uneven development. There have been strong economic returns for Camisea shareholders (Hill, 2016), but economic benefits extend more broadly; LNG and LNG-powered exports from South America circulate the globe.

National and international courts

In spite of domestic and international complaint mechanisms, states agencies and courts charged with oversight are ineffective in defending rights (Hofbauer, 2017). Since national affairs can be corrupt, it is necessary to strengthen international oversight (Bank Track, 2015). International courts generally intervene after damage has occurred; without enforcement, censorship is largely symbolic.

Discussion

Dispossession and environmental racism, two difficult to remedy problems, exist in dams in countries like Brazil where rule of law is constantly violated (Greenpeace, 2016; Fearnside, 2017). It should be no surprise when areas with ongoing bloodshed and repression experience energy violence. However, in addition, what our research suggests is that networked finance and ownership complicates assigning responsibility. For example, it is hard to mandate accountability for human rights violations when 65 different corporation and institutions from 12 countries are involved in Belo Monte (Greenpeace, 2016). In our cases, factors influencing accountability included the number of actors responsible, their power, and transparency (i.e., reporting protocols). If individual firms or agencies are unaccountable, they may lack ethics.

State agencies and courts have a mixed record on halting problematic projects or demanding improvements (Finley-Brook and Thomas, 2011; Fearnside, 2014, 2017). Nonetheless, domestic and international sanctions exist: of our hydro cases, only one remains without sanction. One project was halted, one was fined and delayed, and another the International Water Tribunal identified as violating rights of displaced populations.

Dispossession or self-determination?

Human rights rapporteurs record FPIC often does not happen in the planning and implementation of LAC energy projects (i.e., Anaya, 2013). Although self-determination is codified in international agreements granting rights to auto-govern customary lands, indigenous peoples are often blocked from controlling or even accessing their historical territories if these have valuable energetic resources or potential (i.e., along rivers, near gas beds) (Finley-Brook and Thomas, 2011; Finley-Brook, 2016, forthcoming). State agencies frequently grant access to land, water, or resources regardless of Indigenous Peoples' long-standing territorial claims. Multinational projects like Belo Monte and Barro Blanco enter aggressively where decades-earlier projects stalled due to indigenous opposition. Donors should intensively vet investments in areas with reoccurring human rights violations (Baker, 2011; Finley-Brook, forthcoming).

Environmental racism and toxic energy

Latin American states inadequately regulate the use and disposal of toxic chemicals and waste tied to energy (Ahmad, 2017; Alvarez Mullally, 2017; Scandizzo, 2017). Private sector actors are often unwilling to take responsibility and remediate pollution (San Sebastián and Hurtig, 2014; Oliveira and Hecht, 2016). Environmental impacts from large-scale infrastructure can be high. For example, impacts of AES Colon's gas project include (1) dredging and suspension of sediments contaminated with heavy metals, (2) displacement of artisanal fishing, and (3) health and safety risks to neighbouring communities, including possibility of explosion (IIC, 2015b).

The gas industry suggests gas is environmentally positive (Finley-Brook et al., 2018). YPF's (2014) Vaca Muerta website highlights chemical additives in fracking are also found in products like cosmetics and beverages to allege these fluids are safe. In international research, Colborn et al. (2011) examine 353 chemicals used during gas operations: they find 75% could affect the skin, sensory organs, and respiratory and gastrointestinal systems; 40–50% could affect the nervous, immune, and cardiovascular systems; 37% could affect the endocrine system; and 25% could cause cancer and mutations. Public health officials in Argentina believe there are cancer clusters around O&G infrastructure and criticize the lack of state record keeping (Scandizzo, 2017). Without specialized medical personnel or equipment, patients with cancer, leukaemia, and other advanced diseases cannot be treated (Curado and de Souza, 2014).

Have improvements begun?

Local-to-global international campaigns pressure energy companies to respect human rights. With decades of experience, massive international campaigns against dams (i.e., International Rivers, Movement of People Affected by Dams) target harmful projects, including the cases covered here. Non-governmental watchdog organizations and legal support teams tirelessly oppose mega-projects and advocate for the rights of the displaced. Anti-gas movements (i.e., the Mexican Anti-Fracking Alliance, Argentina Free of Fracking) are new and fragmented compared to vast anti-dam networks. While progress has been made, in particular by educating citizens about drawbacks from gas and large hydro, energy firms have greater resources to circulate pro-extraction and pro-industry narratives (Finley-Brook et al., 2018).

Conclusion: transition amidst tension

Similarities between the hydro and gas projects include a polycentric array of powerful actors, lack of accountability, and violence. With gas projects, not enough attention has been given to everyday violence, such as toxic emission plumes, waste discharge, or spills and leaks. Across the energy spectrum, powerless groups like communities of colour become collateral damage. As social movements develop and activists fight for change they risk violent repression (International Rivers, 2016; Finley-Brook, forthcoming). Clearly, environmental and human rights protections must increase during expansion of high-stakes energy projects.

Notes

1 Potential for sustainable micro-hydro is beyond the scope of this chapter focusing on large-scale hydropower.
2 Some IFIs (CABEI, FMO, IDB) were involved in two controversial cases.

References

AES. (2015) *Press Release: AES Awarded Panama's First Natural Gas Generation Plant.* www.businesswire.com/news/home/20150911005301/en/ (Accessed 14 May 2018).

Aguirre, C. S. (2017) Pueblo Mapuche, Estado, Economía, y Tierras. Un Conflicto en Vaca Muerta: Nuequen, 2010–2015. *Cuaderno do Ceas* 240: 71–97.

Ahmad, N. B. (2017) Blood Biofuels. *Duke Environmental Law and Policy Forum* 27: 265–317.

Alvarez Mullally, M. (2017) *Basureros petroleros, qué son y dónde se encuentran.* Observatorio Petroleo Sur. http://www.opsur.org.ar/blog/2017/05/23/basureros-petroleros-que-son-y-donde-se-encuentran/ (Acccessed 23 September 2018).

Anaya, J. (2013) *Declaración del Relator Especial sobre los derechos de los pueblos indígenas al concluir su visita oficial a Panamá*. http://unsr.jamesanaya.org/statements/declaracion-del-relator-especial-sobre-los-derechos-de-los-pueblos-indigenas-al-concluir-su-visita-oficial-a-panama (Accessed 20 September 2018).

Baker, S. H. (2011) Unmasking project finance: Risk mitigation, risk inducement, and an invitation to development disaster? *Texas Journal of Oil, Gas, and Energy Law* 6: 273–334.

Balsa, R. H., Espinasa, R. and Serebrisky, T. (2015) *Lights On! Energy Needs in Latin America and the Caribbean Until 2040*. Washington, DC: Inter-American Development Bank.

Bank Track. (2015) *Rio Madeira Dams*. www.banktrack.org/project/rio_madeira_dam (Accessed 14 May 2018).

Bianchi, A. (2014) El Dorado a 3.000 metros bajo tierra. *Nuevo Sociedad* 253: 210–222.

Bill, D., Arce, M., Solis, F.W., Sandoya, W.L., and de Leon, D. (2012) Informe de gira de observación de derechos humanos luego de las protestas contra la minería e hidroeléctricas en la Comarca Ngäbe-Buglé. http://cdn.otramerica.com/OTRAMERICA_web/48/posts/docs/0448033001330859048.pdf (Accessed 23 September 2018).

Brun-Guzman, E. (2017) Conflictividad socioambiental en: Una década de rearticulación y movilización social y política. *Argumentos* 30(83): 43–46.

Colborn, T., Kaiwatkowski, C., Shultz, K. and Bachran, M. (2011) Natural gas operations from a public health perspective. *Human and Ecological Risk Assessment: An International Journal* 17(5): 1039–1056.

Curado, M. P. and de Souza, D. L. B. (2014) Cancer burden in Latin America and the Caribbean. *Annals of Global Health* 80(5): 370–377.

Escobar, A. (2017) Tribunal ordena a secretarías frenar obras de gasoducto. *El Universal* July 27.

Fearnside, P. M. (2014) Impacts of Brazil's Madeira River Dams: Unlearned lessons for hydroelectric development in Amazonia. *Environmental Science and Policy* 38: 164–172.

———. (2017) Belo Monte: Actors and arguments in the struggle over Brazil's most controversial Amazonian dam. *Journal of the Geographical Society of Berlin* 148(1): 14–26.

Feather, C. (2014) *Violating Rights and Threatening Lives: The Camisea Gas Project and Indigenous Peoples in Voluntary Isolation*. Moreton-in-Marsh, UK: Forest People's Program.

Finley-Brook, M. (2016) Justice and Equity in Carbon Offset Governance: Debates and Dilemmas. In S. Paladino and S. Fiske (eds) *The Carbon Fix: Forest Carbon, Social Justice and Environmental Governance*. London: Routledge. pp. 74–88.

———. (2017) Hydropower's fluid geographies. In B. Soloman and K. Calvert (eds) *Handbook on the Geographies of Energy*. Northampton, MA: Edward Elgar Publishing. pp. 119–133.

———. (forthcoming) Extreme Energy Injustice and the Expansion of Capital. In D. Paley and S. Granovsky-Larsen (eds) *Organized Violence and the Expansion of Capital*. Regina: University of Regina Press.

——— and Thomas, C. (2011) Renewable energy and human rights violations: Illustrative cases from indigenous territories in Panama. *Annals of the Association of American Geographers* 101: 863–872.

———, Williams, T. L., Sheppard, J. A. and Jaromin, M. K. (2018) Critical energy justice in US natural gas infrastructuring. *Energy Research and Social Science* 41: 176–190.

Greenpeace. (2016) *Damning the Amazon: The Risky Business of Hydropower in the Amazon*. Sao Paulo: Greenpeace.

Hall, A. and Branford, S. (2012) Development, dams and Dilma: The saga of Belo Monte. *Critical Sociology* 38(6): 851–862.

Hallam, K. (2017) Environmental defenders: Murdered, missing, and at risk. *Socialist Lawyer* February: 40–43.

Hays, J. and Shonkoff, S. B. C. (2016) Toward an understanding of the environmental and public health impacts of unconventional natural gas development: A categorical assessment of the peer-reviewed scientific literature, 2009–2015. *PLoS One* 11(4): e0154164. doi:10.1371/journal.pone.0154164.

Hill, D. (2014) Two lawsuits to stop Peru's biggest gas project in indigenous reserve. *The Guardian*, 25 February.

———. (2016) Pioneer gas project in Latin America fails Indigenous People. *The Guardian*, 2 June.

Hofbauer, J. A. (2017) Operationalizing extraterritorial obligations in the context of climate project finance – the Barro Blanco case. *Journal of Human Rights and the Environment* 8(1): 98–118.

IHA (International Hydropower Association). (2017) *2017 Hydropower Status Report*. www.hydropower.org/2017-hydropower-status-report (Accessed 14 May 2018).

IIC (Inter-American Investment Corporation). (2015a) *Costa Norte Gas-fired Thermal Plant and LNG Terminal Project*. www.iic.org/en/projects/project-disclosure/pn-l1123/costa-norte-gas-fired-thermal-power-plant-and-lng-terminal (Accessed 14 May 2018).

———. (2015b) *Costa Norte Environmental and Social Review Summary*. http://cdn.iic.org/sites/default/files/disclosures/costa_norte_esrs-_eng.pdf (Accessed 14 May 2018).

International Accountability Project. (2017) *Proyecto de Gas Licuado por AES Panamá*. http://accountability project.org/wp-content/uploads/2017/09/Colon-Infographic-28SEP17.pdf (Accessed 14 May 2018).

International Rivers. (2016) *Murdered for Their Rivers: A Roster of Fallen Dam Fighters*. www.international rivers.org/resources/murdered-for-their-rivers-a-roster-of-fallen-dam-fighters-11499 (Accessed 8 May 2018).

Jaskoski, M. (2013) *Military Politics and Democracy in the Andes*. Baltimore: John Hopkins University Press.

Jordan, O. (2008) "I entered during the day and I came out during the night": Power, environment, and indigenous peoples in globalizing Panama. *Tennessee Journal of Law and Policy* 4(2): 467–505.

Kimerling, J. (2013) Oil, contact, and conservation in the Amazon: Indigenous Huaorani, Chevron, and Yasuni. *Colorado Journal of International Law and Policy* 24(1): 43–115.

Loperena, C. A. (2017) Settler Violence? Race and Emergent Frontiers of Progress in Honduras. *American Quarterly* 69(4): 801–807.

MAB (Movimento Dos Atingidos por Barragens). (2017) *Belo Monte Dam Makes Altamira Most Violent City in Brazil*. www.mabnacional.org.br/en/noticia/belo-monte-dam-makes-altamira-most-violent-city-brazil (Accessed 14 May 2018).

Markova, A. (2017) *BP's Fracking Secrets: Pan-American Energy and Argentina's shale mega-project*. London: Platform.

Mingorría, S. (2017) Violence and visibility in palm oil and sugarcane conflicts: The case of Polochic Valley, Guatemala. *The Journal of Peasant Studies* doi:10.1080/03066150.2017.1293046.

Navarro, C. (2016) Disagreement over pipeline causes violent confrontation among Yaqui factions in Sonora. *SourceMex*, 9 November.

Oliveira, G. and Hecht, S. (2016) Sacred groves, sacrifice zones, and soy production: Globalization, intensification, and neo-nature in South America. *The Journal of Peasant Studies* 43(2): 251–285.

Raftopoulos, M. (2017) Contemporary debates on socio-environmental conflicts, extractivism and human rights in Latin America. *The International Journal of Human Rights* 21(4): 387–404.

Reuters. (2017) *Argentina Cinches Deal to Attract Investment in Vaca Muerta Shale*. www.reuters.com/arti cle/us-argentina-gas/argentina-clinches-deal-to-attract-investment-in-vaca-muerta-shale-idUSKBN 14V03N (Accessed 12 May 2018).

Romero, S. (2012) Amid Brazil's rush to develop, workers resist. *The New York Times*, 6 May.

Sampaio, C. A. (2017) *Will a Natural Gas Project in Panama Actually Benefit Local Communities?* https://medium.com/@accountability/will-a-natural-gas-project-in-panama-actually-benefit-local-commu nities-971b69d8ec7f (Accessed 14 May 2018).

San Sebastián, M. and Hurtig, A. K. (2004) Oil exploitation in the Amazon basin of Ecuador: A public health emergency. *Revista Panamerican de Salud Pública* 15(3): 205–211.

Scandizzo, H. (2017) *Water Contamination and Cancer: The Other Impact of Oil Drilling*. www.opsur.org.ar/blog/2017/10/09/water-contamination-and-cancer-the-other-impact-of-oil-drilling/ (Accessed 14 May 2018).

Smith, N. (2010) *Uneven Development: Nature, Capital and the Production of Space*. Athens: The University of Georgia Press.

Sullivan, Z. (2017) *Belo Monte Dam Installation License Suspended, Housing Inadequacy Cited*. https://news.mongabay.com/2017/09/belo-monte-dam-installation-license-suspended-housing-inadequacy-cited/ (Accessed 13 May 2018).

Telesur. (2016a) *Mexico Clash Over Pipeline Leaves 1 Dead in Yaqui Community*. www.telesurtv.net/english/news/Mexico-Clash-Over-Pipeline-Leaves-1-Dead-in-Yaqui-Community-20161022-0015.html (Accessed 13 May 2018).

———. (2016b) *Lawyer for Yaqui Tribe Fighting Mexico's DAPL Kidnapped*. www.telesurtv.net/english/news/Lawyer-for-Yaqui-Tribe-Fighting-Mexicos-DAPL-Kidnapped-20161215-0002.html (Accessed 13 May 2018).

Torres-Slimming, P. (2010) Globalización, El Proyecto Camisea, y la salud de los Matsiguenkas. *Revista Peruana de Medicina Experimental y Salud Pública* 27(3): 458–465.

Vásquez, P. I. (2014) *Oil Sparks in the Amazon: Local Conflicts, Indigenous Populations, and Natural Resources*. Athens: University of Georgia Press.

YPF (Yacimientos Petrolíferos Fiscales). (2014) *The Energy Challenge*. https://www.ypf.com/vacamuerta challenge/paginas/index.html (Accessed 23 September 2018).

Zaitchik, A. (2017) Water is Life. *The Intercept*. https://theintercept.com/2017/12/27/peru-amazon-oil-pollution-indigenous-protest/ (AFVccessed 14 May 2018).

39

THE OIL COMPLEX IN LATIN AMERICA

Politics, frontiers, and habits of oil rule

Gabriela Valdivia and Angus Lyall

Introduction

Latin America is the second largest oil-producing region in the world, following the Middle East. Major producers, such as Brazil, Mexico, and Venezuela, export to other Latin American countries, in addition to the Caribbean, North America, Europe, and, increasingly, Asia. The US, the world's largest oil consumer, is the largest purchaser of Latin American oil; approximately 26% of US crude oil imports come from Latin America. Conversely, about 56% of US exports of refined fuels are shipped to Latin America. These trade networks and patterns offer some perspective on Latin America's global oil connectivity, but say little about how its flows become entangled with (and often deepen) existing structures of economic dependence, colonialism, and geopolitical power in the region (e.g., Bebbington and Bury, 2013; Watts, 2016).

A closer look at how oil firms, states, financiers, environmentalists, and communities govern the distribution of economic benefits and socio-environmental harms allows us to examine how "oil complexes" (Watts, 2004) of resource and population governance take shape in Latin America. Our analysis is necessarily partial – the region is vast and oil politics continue to unfold under diverse politico-economic and geological conditions of possibility. Notwithstanding local specificities, the 20th and 21st centuries suggest some regional generalities in how oil-flow and oil-money link firms, states, and civil society through development imaginaries and initiatives. We start by discussing the consolidation of 20th-century petro-politics in oil-producing countries, tracing the connections between global oil economics (e.g., market prices, investment cycles, and production trends), resource nationalizations, and neoliberal and "post-neoliberal" oil governance. Then, we examine petro-politics in the spaces of oil logistics: oil frontiers. We end with a look at the enduring habits of oil rule and the near future of oil politics in the region.

Petro-politics

Politics, states Fernando Coronil, is "a battle of desires waged on an uneven terrain" (2011: 264). The modern history of Latin American oil politics is forged amid countless battles of desires and aspirations among disparate actors. At the turn of the 20th century, oil became an increasingly coveted source of energy. During World War I, it became a global strategic resource and an object of geopolitical struggle. Latin America became an attractive region for exploration and investment for the major oil firms of the era, Royal Dutch (later Shell), Standard Oil (or Esso, later Exxon), British Petroleum, Mobil, Texaco, Chevron, and Gulf. This handful of international oil companies (IOCs) often sought vertical integration of production, transport, and refining in the pursuit of quasi-monopoly profits. In Peru, Mexico, Venezuela, Colombia, and Argentina, they achieved extraordinary rates of return. One of the world's largest oil companies, Jersey Standard Oil, generated up to 40% of its global profits from Latin American oil reserves in the years following World War I (Philip, 1982).

IOCs dominated the Latin American oil complex through the 1920s. The oil rents that stayed in the region generally flowed from firms to land owners and provincial authorities, bypassing state coffers. Increasing flows of oil and rents to companies based in the Global North reinforced economic inequalities in the hemisphere, as Latin American countries lacked the capital and expertise to engage in exploration, drilling, and refining activities. The oil complex also reproduced inequalities that deepened dynamics of internal colonization, through contamination and the exploitation of local labour in rural spaces of extraction and processing. The 1930s novels *Barrancabermeja* (1934) and *Mancha de Aceite* [Oil Stain] (1935), which describe environmental and health disasters in Colombian oil fields and worker exploitation in Venezuela, respectively, powerfully illustrate the subordination of local communities in the early years of the Latin American oil complex.

Nationalisms

By the 1930s, and more so after World War II, the existing patterns of trade and rent capturing sowed discontent among oil-producing countries. Their efforts to change these patterns aligned with a regional shift towards economic modernization based on developing national industries and eliminating dependencies on imports – or "import substitution industrialization" (ISI) (e.g., Faletto and Cardoso, 1969). Policies to strengthen control over oil varied by country and over time but often led to the creation of national oil companies (NOCs). Argentina created Yacimientos Pretrolíferos Fiscales (YPF) in 1922 to avoid the sort of wartime fuel shortages that had curtailed industrial production (Solberg, 1973) and to better control oil activities in the unincorporated Patagonia. In the late 1930s, Bolivia and Mexico founded Yacimientos Petrolíferos Fiscales Bolivianos (YPFB) and Petróleos Mexicanos (PEMEX), respectively. In Brazil, President Getúlio Vargas created the country's first oil company, Petrobras, in 1953 with the slogan "*O petróleo é nosso*" ("The Oil is Ours"). Chile formed the Empresa Nacional del Petróleo (ENAP) in 1954 after discovering oil in the Magallanes area. National oil companies proliferated amid nationalist movements of the 1960s and 1970s in Cuba, Colombia, Costa Rica, Peru, Ecuador, Venezuela, and Trinidad and Tobago. Paraguay followed suit in 1981.

Throughout the 20th century, some governments directly expropriated and nationalized foreign assets, often justifying such actions as a response to US and European imperialism. In the late 1930s, Mexican President Lázaro Cárdenas expropriated the assets of the US and Anglo-Dutch oil companies and barred IOC operations. Bolivia nationalized the assets of the Standard Oil of New Jersey in the late 1930s, after the Chaco War, and seized control of Gulf Oil operations in 1969. Peru nationalized the assets of the International Petroleum Company in 1968, amidst a politically divided government, the military's concern with the possible election of a Left-leaning president, and threats of US economic sanctions.[1] While many of these expropriations were widely contested, others were not. Venezuela, for example, used a gradual nationalization process that over decades transferred ownership of oil operations from private to state-owned, and guaranteed that oil firms received income throughout the duration of their concessions. Other governments purchased foreign firm assets and restructured tax and royalties systems to increase their take from foreign operations. Several joint ventures or public-private enterprises were formed, with at least 51% state ownership, such as Ecuador's CEPE-Texaco consortium.

Governments with major production also sought to intervene in markets. In 1960, Venezuela, along with oil-producing Middle Eastern countries, founded the Organization of the Petroleum Exporting Countries (OPEC) to shift control over international oil prices from consumers to producers by putting a ceiling on the production quotas of members. Ecuador joined in 1973. In Venezuela and Ecuador, rising international oil prices and nationalist sentiments in the 1970s shaped the emergence of "petro-states," states whose revenues and spending depended heavily on oil exports. Oil rents transformed governmental power. An increase in oil income allowed states to expand public spending, payrolls, and political clientelism, reinforcing elite control of political power and state resources (Gledhill, 2008). In Venezuela, the dominant parties' control of the electoral machinery and patronage networks, infused by oil money, helped maintain a virtual two-party system for decades (Karl, 1997).

Petro-politics in the Caribbean were starkly different from mainland Latin America, in part because of unfavourable geologic conditions for oil and gas production (except in Barbados and Trinidad and Tobago), but also due to the Caribbean's geographic location. From World War II through the 1980s, the Caribbean functioned as a node for US oil networks. Multinational oil firms set up ports, refineries, and processing and storage facilities for oil coming to and from the US and mainland Latin America. These energy infrastructures created oil dependency in the region – today, nearly 90% of the region's electricity is generated from imported oil (Gustafson, 2017).

Neoliberal adjustments

International banks, flush with cash due to high oil prices between 1973 and 1980, "recycled" this money as loans, which Latin American oil-producing countries used for economic development, hoping that high prices would persist and allow them to pay off their debt. By the 1980s, however, falling oil prices led to sluggish economic growth, increasing debt, and hyperinflation in Latin America. This moment is often referred to as an example of the "paradox of plenty" (Karl, 1997) or "*la maldición de la abundancia*" ("curse of abundance") (Acosta, 2009), referring to a correlation between national resource wealth and the persistence of poverty, debt, and weak institutions. Oil-rich countries with rising debt adopted fiscal austerity and sector liberalization, opening oil sectors to IOCs and private investors. NOCs experienced disinvestment and/or privatization. International banks and lending agencies required this reduction in state control over natural resources and more favourable foreign investment terms to continue granting loans. In the early 1990s, for example, the Ecuadorian government opted not to renew Texaco's contract to sustain the CEPE-Texaco consortium and opened up the oil sector to diverse private firms through bidding rounds or auctions. In Brazil, President Fernando Henrique Cardoso privatized Petrobras in 1995 and ended its monopoly rights. According to former congressperson Roberto Campos ("Bobby Fields"), Petrobras had become a "Petrossauro," a dinosaur company plagued by debt and corruption (OGlobo, 2011). In 1998, Argentina's YPF, one of the world's oldest vertically integrated NOCs, was purchased by a transnational oil corporation (YPF S.A., later YPF-Repsol) (Shever, 2012).

These changes prompted a super-cycle of foreign investment in the region, as states relaxed regulations and created more enticing investment conditions. In 1993, the *New York Times* reported on the liberalization of the Latin American oil sector: "The investors are lining up throughout the region. . . . Some want to diversify their portfolios . . . transportation to the

United States is generally quicker and cheaper . . . and environmental standards are generally less stringent. . . . And, finally, the oil is here" (Brooke, 1993).

"Post" neoliberalisms

Neoliberal adjustments and divestments in public services fed widespread discontent. Broad swathes of civil society led anti-neoliberal uprisings across the region, from the 1989 Caracazo protests in Venezuela to the 2001 Argentine rebellion and the expulsion of a series of presidents in Ecuador in the 1990s and 2000s. Likewise, social mobilizations took place in many contexts to "defend" gas and oil as national resources, as in Ecuador and Bolivia (Kohl and Farthing, 2012; Perreault and Valdivia, 2010). Such political uprisings, coupled with rising international oil prices, opened the possibility for a shift in oil governance – a so-called "post" neoliberal moment. Between 1998 and 2008, the international benchmark oil price WTI (West Texas Intermediate) rose from $17 to $157 per barrel, elevated by Chinese demand, war in the Middle East, and a declining US dollar, in addition to speculation in oil markets. Despite a slip following the 2008 financial crisis, prices spiked again through mid-2014.

Many oil-producing states in this period forced contract renegotiations with IOCs; used their NOCs to assert greater control over oil flows and rents; and, in some cases, nationalized reserves to take advantage of higher prices. Hugo Chávez came to power in Venezuela in 1998, but he did not force contract renegotiations with IOCs until after 2005, when prices had ticked up significantly. Bolivia nationalized its oil and gas sector in 2006; Ecuador nationalized a major oil field, cancelling the contract of US-based Occidental Petroleum on a technicality in 2006, and forced contract renegotiations in 2010–2011; and Argentina nationalized YPF in 2012. Venezuela and Ecuador perceived the greatest impacts of rising resource rents. Oil revenues constituted over half and one-third of their respective state budgets at the height of this boom. Bolivia, Brazil, and Argentina also experienced rising resource rents and, in turn, increased social and economic spending.

Not all oil-producers followed these trends. Colombia, Guatemala, and Peru continued liberalization policies during this period. For example, in 2003, Colombia applied an "energy reform" that modified the structure of its NOC, Ecopetrol, to give it greater autonomy from state regulation and access to foreign investment in oil activities (UPME, 2015). In 2007, Ecopetrol began a public stock offering to both finance its continued growth and increase competitiveness.

Post-neoliberal discourses generally echoed nationalist discourses from the 1970s but, in practice, leaders did not rely on NOCs alone. Remaining oil reserves, increasingly marginal and difficult-to-access, required ever greater amounts of capital and specialized technologies. In 2009, Venezuela set out to secure $100 billion USD in financing from IOCs for joint ventures to exploit extra-heavy oil deposits (Monaldi, 2015). Argentina, Ecuador, and Brazil increased the state's take of oil rents mostly through tax and contract reforms, rather than nationalizations. NOCs also turned to new sources of oil financing. China, in particular, invested heavily into strategic sectors in Latin America and increased overall investment in the region from $3.8 billion to $109.5 billion USD between 2005 and 2014 (Espinasa, Marchan, and Sucre, 2015).

Diverse regional initiatives of energy integration emerged during this phase of growth. Chavez's Bolivarian agenda relied heavily on "petro-diplomacy," using oil as an instrument of foreign policy to back financial agreements of regional cooperation, from barter arrangements to deep discounts. These petro-initiatives included a program of subsidies and discounts for members of the Bolivarian Alliance of the Americas (ALBA); Petrocaribe, which offered oil at discounted prices to Caribbean and Central American countries in exchange for commodity exports that were then distributed to Venezuelans through PDVSA's discounted food program;

a program that provided 33,000 barrels of oil per day to Cuba in exchange for medical health care provision; and PetroAmerica, which offered discounts and loans to Latin American countries for mega-projects of energy integration (Hellinger, 2016). Also, in the early 2000s, the Initiative for Regional Infrastructural Integration in South America (IIRSA) involved all 12 South American nations in plans for massive infrastructural investment in transport, energy, and communications, such as transoceanic roads, pipelines, and transmissions lines linking energy sources to consumers, exporters, and industrial sites. The initiative included proposals for Bolivian gas to supply energy needs in Argentina and Brazil and for gas from the Peruvian Amazon to supply energy to mining sectors in Peru and Chile. Furthermore, in the 2004 South American Summit, all South American presidents agreed to integrate an organism that would attempt to parallel the European Union called the South American Community of Nations (CASA), which, in 2007, would become the Union of South American Nations (UNASUR). Several energy integration projects were executed or planned as part of this effort, including a natural gas pipeline between Venezuela and Colombia and plans for a new agency, Petrosur, to coordinate integration between Brazil, Argentina, and Venezuela, including a proposed gas pipeline between these three countries (Kozloff, 2007).

Plans for energy integration faced political and economic difficulties. Venezuela's regional integration efforts weakened after President Chavez' death in 2013. In early 2014, Ecuadorian President Rafael Correa warned of a "conservative restoration" across the region, after his ruling party lost key municipal and regional offices to the opposition. In turn, a glut in international oil supplies driven by the US shale or "fracking" industry led to a collapse in oil prices and generated dire economic conditions for post-neoliberal regimes. The international benchmark oil price WTI plummeted from $109 in mid-2014 to $50 by early 2015 and wavered between $42 and $55 through 2017. This price collapse sent fiscal accounts into crisis in Venezuela, Ecuador, and Colombia, and set the stage for a broad political shift to the Right, and to another cycle of oil sector liberalization. In 2017, Mexico, Cuba, Trinidad and Tobago, Suriname, Peru, Brazil, Uruguay, Argentina, and Ecuador licenced new rounds for IOCs to bid on oil fields, in the hopes that increased oil production volumes would help breach growing fiscal deficits and debt.

Oil frontiers

The political-economic vicissitudes of the oil complex outlined above leave unanswered questions about the place-specific politics of the oil complex. Oil extraction, transport, and processing take place in physical spaces where firms and states engage with existing cultural and political dynamics. As the industry expands geographically, new spaces are transformed into "edges" of oil activity or "oil frontiers," sites where the conditions for new rounds of extractive accumulations are put in place (Watts, 2016).

Governments often justify policy shifts, towards nationalization or liberalization, in terms of the "public good," but avoid questions about who and what is sacrificed by the oil industry (Valdivia, 2015). In Argentina, for example, YPF's widely celebrated expansion and consolidation in Neuquén, Patagonia, occurred only after the 19th-century incorporation of indigenous land as national territory and the genocide of native peoples under the "Campaign of the Desert" (Riffo, 2017). However, oil frontier expansion is generally contested and/or negotiated by local communities and other place-based interests (Breglia, 2013; Schilling-Vacaflor and Eichler, 2017). This is evident in the Amazon, where oil activities currently affect 15% of the region and approximately two million indigenous people (Luna, 2015). Since the 1990s, indigenous peoples have increasingly built transnational alliances that allow them to contest the geographies and terms of oil activities. Their struggles are fortified by growing public awareness of the environmental and health impacts of hydrocarbons, and backed by international accords that recognize indigenous territorial rights (Sawyer, 2004; Stetson, 2012).

Negotiations and contestations at the oil frontier are entangled with postcolonial power relations. During 20th-century liberalization, governments acted more as brokers between communities and oil firms than as guarantors of citizen rights under democratic rule (Reider and Wasserstrom, 2013). In Ecuador, firms provided limited medical, educational, and other services for local citizens as part of their "operating costs" (Navarro, 1995), turning affected communities into "clients" that traded consent to oil activities in exchange for the partial provision of needs. In Mexico, fishing communities in the Campeche Sound protested a 2003 bill restricting fishing in areas surrounding oil platforms, to which the Mexican state responded by creating a development fund (Zalik, 2009). Oftentimes, members of oil-afflicted communities also consent in exchange for compensation payments and oil-related jobs, although research demonstrates that the oil industry exacerbates existing gender, ethnic, and economic inequalities, and generates or renews conflicts within communities and families (Billo, 2015; Breglia, 2013; Cielo, Coba, and Vallejo, 2016; Scott et al., 2013). Oil activities also generate socio-environmental risks (Lu, Valdivia, and Silva, 2016) and affect subsistence resources through contamination and the extension of road, communication networks, and human settlements (Bozigar, Gray, and Bilsborrow, 2016). Sites of refineries, storage facilities, and pipelines experience similar patron-client relations and exchanges, where communities endure toxicity in exchange for urban services (Auyero and Swistun, 2009; Ofrias, 2017).

Fine-grained attention to everyday life in the oil complex reveals how resistance, negotiation, and consent have shaped oil flows in recent decades. In the Ecuadorian Amazon, Cofán communities strategically engage in negotiations with companies and the state (Cepek, 2018); a handful of Kichwa and Shuar communities have demanded rights to extract oil themselves; the Kichwa community of Sarayacu defends territorial autonomy and plurinationality against oil companies (Fajardo, 2016); and other indigenous organizations, including the Sápara, Shiwiar, Shuar, and Achuar indigenous nationalities, have split, as some members oppose oil extraction and others seek out compensation agreements (Vallejo, 2014). In the case of the Sápara, some leaders proposed to strategically support oil *exploration* to obtain schools and to later oppose oil *extraction* to evade its impacts. The political positions that such groups take are manifold and, in some cases, shifting.

Oil prices also shape frontier politics. Following the price surge at the turn of the 21st century, new bidding rounds made over 70% of the Peruvian Amazon (56.1 million hectares) (Bebbington and Bury, 2013) and over half of the Ecuadorian and Bolivian Amazon available for oil and gas exploration (Finer et al., 2008). Many of these new fields, with higher production costs, were located in sites of high biodiversity and indigenous territories (e.g., Guzmán-Gallegos, 2012). USAid (2014: 4) observed a correlation between expanding oil and gas development and "chronic low-grade socio-environmental conflict punctuated by periodic violence in communities in, or adjacent to, extraction zones."

Biopolitical control, or the state's power to influence a population's political agency by providing access to life-enhancing resources such as education, health, and housing, also intensified with the surge in oil prices. This was evident in the allocation of compensation payments. Venezuela's PdVSA increased social spending in sites of oil logistics, from $249 million in 2003 to $13.26 billion in 2006, channelling funds through the Economic and Social Development Fund (FONDESPA) and the National Development Fund (FONDEN), as well as neighbourhood "missions" for targeted socioeconomic interventions (Franco, 2008). The Ecuadorian government created the public company Strategic Ecuador to build public works projects in energy sites, including more than 1,000 projects between 2012 and 2015 alone. Governments often singled out specific communities willing to negotiate and consent in order to fulfil legal require-ments and generate positive press (Lyall, 2017; Schilling-Vacaflor and Eichler, 2017), despite resistance from neighbouring communities, or they weakened or evaded consultation laws. For example, Peru reclassified indigenous communities as peasant-farmer communities to evade its obligations regarding indigenous rights (Poole, 2016). In general, governments exerted greater regulatory control over critical social movements and environmental NGOs and persecuted activists for "sabotage" and "terrorism" (Valdivia and Benavides, 2018; Zibechi, 2012).

Persistent habits of oil rule

Analysts and media commentators often describe the Latin American oil complex as plagued by corruption and socio-environmental conflict. Industry and state actors pose policy and technical solutions to reform the complex, but ultimately fail to resolve persistent, underlying problems. In this section, we examine how habits of petro-rule continue to persist in new contexts.

The turn to the Right

One of the often-cited problems of Latin American oil complexes is the "bad government" of oil money. On the one hand, petro-states seek legitimacy by making promises and plans to use oil rents for civic ends, i.e., to serve the needs of citizens. On the other hand, oil often under-writes centralized and authoritarian forms of power that lead to clientelism and corruption (Karl, 1997). Oil sector privatizations and nationalizations in the 20th and 21st centuries were both offered as political solutions to this problem. Yet, the co-dependence between government and ruling capitalist class persists. Regardless of regime type, governments invest, spend, and dis-tribute resource rents and revenues in ways that have served the interest of the economic ruling class first (Cypher, 2014; Petras and Veltmeyer, 2016).

In the current shift to the Right, newly elected conservative regimes are cultivating wide-spread concern about oil corruption to justify more oil sector liberalization. Ecuador, Argentina, Brazil, and Peru have spearheaded criminal investigations into the post-neoliberal management of resource sectors, singling out technically or economically questionable contracts and activi-ties, as proof of the "moral" illegitimacy of nationalist policies. In 2017, legal charges against individual public-sector oil managers and ministers in Colombia, Brazil, Venezuela, Mexico, and Ecuador led to high-profile firings and imprisonment. Meanwhile, conservative rule and greater reliance on IOCs are framed as both moral and technical solutions to economic malaise, without addressing the underlying links between oil money, clientelism, and corruption. How sustainable this political strategy is remains to be seen, particularly if oil prices, revenues, and public spending remain reduced.

Expanding frontiers

Debt is a recurrent problem among oil-dependent nations, as states over-borrow during rev-enue booms. In the 21st century, the depletion of proven, conventional oil reserves compounds this problem. Amidst declining production of mature fields, states and companies have turned to the exploitation of unconventional reserves, such as deep water, shale, and tar sands, as the new frontier of petro-capitalism. Approximately 40% of the world's unconventional oil reserves are in Latin America. Accessing such unconventional reserves will involve high inputs of capital and specialized technologies, as well as particular socio-environmental risks. In 2012, Mexico opened PEMEX to private investment for the first time in 75 years to promote shale development and risky, deep-water projects. Brazil became the region's second largest producer of crude oil since its 2007 discovery of deep water, "pre-salt" reserves, at four kilometres below the ocean floor. Production of these oil deposits only began in 2013, but they already constitute 90% of Brazil's total proven reserves. In Argentina, the Vaca Muerta region features the second largest shale oil and gas field in the world and trails only the US shale industry. Argentina plans an $800 million USD shale oil and gas plant, despite protests by unions and indigenous Mapuche communities denouncing the associated environmental hazards (Riffo, 2017). Venezuela, Mexico, Brazil, Colombia, and Chile are also investing in shale. Finally, the US Geological Survey estimates that Venezuela's Orinoco tar sands has between nine and 1.4 trillion barrels, constituting the largest single proven reserve of oil in the world – 18% of total global reserves.

Governments face the problem of securing costly investments in unconventional reserves, despite shrinking budgets for social compensation. In some cases, governments have justified diminished compensation by blaming previous governments or the volatility of international oil prices. Concurrently, militarization of oil frontiers has intensified in many areas, such as the southern Ecuadorian Amazon, northern Peru, and southern Argentina. Thus, a shift towards unconventional reserves appears to be accompanied by diminished compensation, more coercive measures of social-territorial control, and more socio-environmental conflict.

Climate justice

Global climate politics centre on the urgent need to reduce carbon emissions in the name of planetary survival. However, oil-dependent governments in Latin America frequently frame the climate politics of oil as an untenable choice between reducing carbon emissions and reducing poverty. Some governments characterize themselves as insignificant players in this global problem – "too small" as hydrocarbon producers to make a difference by curbing production (Hughes, 2017). Meanwhile, links between petrochemical companies and banks have blocked initiatives to control carbon dioxide emissions and pushed the development of unconventional fossil fuel exploitation among big producers like Brazil (Costa, 2017).

Ecuador offers one of the best-known examples of the entanglements between climate change and oil politics. In 2007, Ecuador announced a novel proposal to potentially resolve tensions between climate change, oil production, and poverty reduction by leaving underground 846 million barrels of oil located in the Yasuní National Park, a biodiversity hotspot, if the international community transferred half of the value of these reserves into a development trust. Global financial commitment to this proposal was low. In 2013, President Correa declared that the international community had failed to show sufficient interest and authorized oil extraction in Yasuní, claiming that reducing poverty was a moral necessity. Carbon emissions, he argued, are an unavoidable cost. He also noted that oil rents would enable state investment in forms of "sustainable development," such as ecotourism, and that investments in modern technologies and techniques, such as horizontal drilling, would minimize environmental impacts (Fiske, 2017). In a turnabout from his 2007 position, Correa also argued that climate change was a problem caused by the Global North, one that the Global South should not be expected to subsidize by reducing its relatively minor impacts. Bolivian Vice-President Álvaro Linera García (2012) added that resource extraction in the Global South is necessary to fund counter-hegemonic struggles and a gradual transition towards a truly transformative eco-socialism. Such moral, economic, technological, and political justifications for ongoing oil production evade accounting for the places and people sacrificed by extraction activities and leave in place the habits of petro-rule that oil rents underwrite.

The future

In this chapter, we traced how certain tendencies and practices of oil rule over time have become habits of the Latin American oil complex. As Arturo Escobar (2015) reminds us, however, what we know as "reality" is connected to the frameworks we use to interpret it. The implications of this understanding for oil politics are substantial and make room for possibilities to transform and undo these habits. In Latin America, some oil rule undoings have taken the form of state-led "revolutions" (e.g., Ecuador, Bolivia, and Venezuela in the 21st century). Other undoings have emerged on specific oil frontiers, for example, through civic participation and contentious collective action. While not offering lasting "solutions," both sets of dynamics have questioned the habits of the oil complex and articulated other possible ways of existing vis-à-vis oil flows. Both also underscore the deadly double-movement at the core of oil complexes – the primacy of "economic growth" as a universal good and its disregard for the worlds extinguished by this growth – and interrupt (however, incompletely) dominant framings of power and truth.

Today, emergent petro-politics push for the decolonization of oil governance. Environmental and indigenous movements, critical academics, and progressive NGOs, in different ways and through their own understanding of reality, see the problems of the oil complex as symptomatic of a deeper civilizational crisis that requires immediate transitions to post-extractivism, post-development, or de-growth (Acosta and Brand, 2018; Escobar, 2015; Gudynas, 2011; Paulson, 2017) – paradigms of sufficiency or *buen vivir* [good living]. For climate-conscious social movements at oil frontiers in the 21st century, "talking about oil without mentioning climate change, is totally out of touch with reality" (Acosta, 2017: 450). While at times these alternative frames are discounted as romantic, they are also hopeful and compelling. They call for a radical rethinking of what we know as reality and demand that we take alternative lifeways seriously, as viable possibilities for building more sustainable and equitable futures. They provincialize the existing habits of oil rule and give hope that post-oil futures are possible.

Note

1 Despite nationalist sentiments, the entanglements of expertise and imperialism were hard to avoid. In Peru, personnel of the newly nationalized PetroPerú trained, ideologically and technically, in North American and European centres of oil-flow calculation.

References

Acosta, A. (2009) *La Maldición de la Abundancia*. Quito: Abya-Yala.
——— and Brand, U. (2018) *Salidas al Laberinto Capitalista: Decrecimiento y Postextractivismo*. Quito: Rosa Luxemburgo.
Auyero, J. and Swistun, D. (2009) *Flammable: Environmental Suffering in an Argentine shantytown*. Oxford: Oxford University Press.
Bebbington, A. and Bury, J. (2013) *Subterranean Struggles: New Dynamics of Mining and Oil in Latin America*. Austin: University of Texas Press.
Billo, E. (2015) Sovereignty and subterranean resources: An institutional ethnography of Repsol's corporate social responsibility programs in Ecuador. *Geoforum* 59: 268–277.
Bozigar, M., Gray, C. and Bilsborrow, R. (2016) Oil extraction and indigenous livelihoods in the Ecuadorian Amazon. *World development* 78: 125–135.
Breglia, L. (2013) *Living with Oil: Promises, Peaks, and Declines on Mexico's Gulf Coast*. Austin: University of Texas Press.
Brooke, J. (1993) Latin America's oil rush: Tapping into foreign investors. *New York Times*. www.nytimes.com/1993/07/11/business/latin-america-s-oil-rush-tapping-into-foreign-investors.html?pagewanted=all (Accessed 10 November 2017).

Cepek, M. (2018) *Life in Oil*. Austin: University of Texas Press.

Cielo, C., Coba, L. and Vallejo, I. (2016) Women, nature, and development in sites of Ecuador's petroleum circuit. *Economic Anthropology* 3(1): 119–132.

Coronil, F. (2011) The future in question: History and utopia in Latin America (1989–2010). In C. Calhoun and G. Derluguian (eds) *Business as Usual: The Roots of the Global Financial Meltdown*. New York: New York University Press, pp. 231–292.

Costa, A. A. (2017) As the earth heats up, Brazil digs deeper. *NACLA Report on the Americas* 49(4): 444–450.

Cypher, J. M. (2014) Energy privatized: The ultimate neoliberal triumph. *NACLA Report on the Americas* 47(1): 27–31.

Escobar, A. (2015) Degrowth, postdevelopment, and transitions: A preliminary conversation. *Sustainability Science* 10(3): 451–462.

Espinasa, R., Marchan, E. and Sucre, C. (2015) *Financing the New Silk Road: Asian Investment in Latin America's Energy and Mineral Sector*. Technical Note 824. Washington, DC: Inter-American Development Bank.

Fajardo, A. (2016) *Sarayaku y las TIC: Una Lucha por la Autodeterminación Territorial*. Master's thesis, FLACSO, Quito.

Faletto, E. and Cardoso, F. H. (1969) Dependencia y desarrollo en América Latina. In *Cincuenta Años del Pensamiento de la CEPAL: Textos Seleccionados*. Chile: CEPAL, pp. 475–499.

Finer, M., Jenkins, C., Pimm, S., Keane, B. and Ross, C. (2008) Oil and gas projects in the western Amazon: Threats to wilderness, biodiversity, and indigenous peoples. *PloS One* 3(8): e2932.

Fiske, A. (2017) Natural resources by numbers: The promise of 'el uno por mil' in Ecuador's Yasuní-ITT oil operations. *Environment and Society* 8(1): 125–143.

Franco, R. (2008) Venezuela: Energy, the tool of choice. In C. Arnson, C. Fuentes and F. Rojas Aravena (eds) *Energy and Development in South America: Conflict and Cooperation*. Chile: FLACSO, pp. 35–41.

Gledhill, J. (2008) "The people's oil": Nationalism, globalization, and the possibility of another country in Brazil, Mexico, and Venezuela. *Focaal* (52): 57–74.

Gudynas, E. (2011) Alcances y contenidos de las transiciones al post-extractivismo. *Ecuador DEBATE* 82: 61–80.

Gustafson, B. (2017) The new energy imperialism in the Caribbean. *NACLA Report on the Americas* 49(4): 421–428.

Guzmán-Gallegos, M. (2012) The governing of extraction: Oil enclaves and indigenous responses in the Ecuadorian Amazon. In H. Haarstad (ed) *New Political Spaces in Latin American Natural Resource Governance*. New York: Palgrave Macmillan, pp. 155–176.

Hellinger, D. (2016) Resource nationalism and the Bolivarian revolution in Venezuela. In *The Political Economy of Natural Resources and Development: From Neoliberalism to Resource Nationalism*. New York: Routledge, pp. 204–218.

Hughes, D. (2017) *Energy Without Conscience: Oil, Climate Change, and Complicity*. Durham, NC: Duke University Press.

Jaramillo Arango, R. (1934) *Barrancabermeja: Novela de Proxenetas, Rufianes, Obreros y Petroleros*. Bogotá: Editorial E.S.B.

Karl, T. (1997) *The Paradox of Plenty: Oil Booms and Petro-States*. Berkeley, CA: University of California Press.

Kohl, B. and Farthing, L. (2012) Material constraints to popular imaginaries: The extractive economy and resource nationalism in Bolivia. *Political Geography* 31(4): 225–235.

Kozloff, N. (2007) *Hugo Chávez: Oil, Politics, and the Challenge to the US*. New York: Palgrave Macmillan.

Linera, A. (2012) *Geopolítica de la Amazonía: Poder Hacendal-Patrimonial y Acumulación Capitalista*. La Paz: Vicepresidencia de Bolivia.

Lu, F., Valdivia, G. and Silva, N. L. (2016) *Oil, Revolution, and Indigenous Citizenship in Ecuadorian Amazonia*. New York: Palgrave Macmillan.

Luna, N. (2015) In the shadows of the extractive industry: A hard road for indigenous women. *ReVista: Harvard Review of Latin America: Energy, Oil, Gas and Beyond* 15(1): 70–77.

Lyall, A. (2017) Voluntary resettlement in land grab contexts: Examining consent on the Ecuadorian oil frontier. *Urban Geography* 38(7): 958–973.

Monaldi, F. (2015) Latin America's oil and gas: After the boom, a new liberalization cycle? *ReVista: Harvard Review of Latin America: Energy, Oil, Gas and Beyond*. https://revista.drclas.harvard.edu/book/first-take-latin-america%E2%80%99s-oil-and-gas (Accessed 20 September 2018).

Navarro, M. (1995) Conflictos en políticas de asignación y uso de los fondos de beneficio social y mitigación de impacto ambiental de las petroleras. In A. Varea and P. Ortíz (eds) *Marea Negra en la Amazonia*. Abya Yala: Quito, pp. 241–264.

Ofrias, L. (2017) Invisible harms, invisible profits: A theory of the incentive to contaminate. *Culture, Theory and Critique* 58(4): 435–456.

OGlobo. (2011) De herege a profeta. *O Globo*, 27 November. http://libproxy.lib.unc.edu/login?url=https:// search-proquest-com.libproxy.lib.unc.edu/docview/906113090?accountid=14244 (Accessed 1 December 2017).

Paulson, S. (2017) Degrowth: Culture, power and change. *Journal of Political Ecology* 24(1): 425–448.

Perreault, T. and Valdivia, G. (2010) Hydrocarbons, popular protest and national imaginaries: Ecuador and Bolivia in comparative context. *Geoforum* 41(5): 689–699.

Petras, J. and Veltmeyer, H. (2016) *What's Left In Latin America? Regime Change in New Times*. New York: Routledge.

Philip, G. (1982) *Oil and Politics in Latin America: Nationalist Movements and State Companies*. Cambridge: Cambridge University Press.

Poole, D. (2016) Mestizaje as ethical disposition: Indigenous rights in the neoliberal state. *Latin American and Caribbean Ethnic Studies* 11(3): 287–304.

Reider, S. and Wasserstrom, R. (2013) Undermining democratic capacity: Myth-making and oil development in Amazonian Ecuador. *Ethics in Science and Environmental Politics* 13(1): 39–47.

Riffo, L. (2017) Fracking and resistance in the land of fire. *NACLA* 49(4): 470–475.

Sawyer, S. (2004) *Crude Chronicles: Indigenous Politics, Multinational Oil, and Neoliberalism in Ecuador*. Durham, NC: Duke University Press.

Schilling-Vacaflor, A. and Eichler, J. (2017) The shady side of consultation and compensation: 'Divide-and-rule' tactics in Bolivia's extraction sector. *Development and Change* 48(6): 1439–1463.

Scott, J., Rose D., Heller, K. and Eftimie, A. (2013) Extracting lessons on gender in the oil and gas sector: A survey and analysis of the gendered impacts of onshore oil and gas production in three developing countries. In *Extractive Industries for Development Series* 28. Washington, DC: World Bank.

Shever, E. (2012) *Resources for Reform: Oil and Neoliberalism in Argentina*. Stanford, CA: Stanford University Press.

Solberg, C. (1973) The tariff and politics in Argentina 1916–1930. *The Hispanic American Historical Review* 53(2): 260–284.

Stetson, G. (2012) Oil politics and indigenous resistance in the Peruvian Amazon: The rhetoric of modernity against the reality of coloniality. *The Journal of Environment & Development* 21(1): 76–97.

UPME. (2015) *Plan Energético Nacional Colombia: Ideario Energético 2050*. Bogotá: Ministry of Mines and Energy.

Uribe Piedrahita, C. (1935) *Mancha de Aceite*. Bogotá: Editorial Renacimiento.

USAid. (2014) *Addressing Biodiversity-Social Conflict in Latin America*. Washington, DC: USAid. http://pdf. usaid.gov/pdf_docs/PA00KRV5.pdf (Accessed 10 November 2017).

Valdivia, G. (2015) The sacrificial zones of "progressive" extraction in Latin America. *Latin American Research Review* 50(3): 245–253.

——— and Benavides, M. (2018) "The end of the good fight": Oil workers and neoliberal restructuring in Ecuador. In C. Atabaki, E. Bini and K. Ehsani (eds) *Working for Oil: Comparative Social Histories of Labor in the Global Oil Industry*. New York: Palgrave Macmillan, pp. 159–185.

Vallejo, I. (2014) Petróleo, desarrollo y naturaleza: Aproximaciones a un escenario de ampliación de las fronteras extractivas hacia la Amazonía suroriente en el Ecuador. *Anthropologica* 32(32): 115–137.

Watts, M. (2004) Antimonies of community: Some thoughts on geography, resources, and empire. *Transactions of the Institute British Geographers* 29: 195–216.

———. (2016) Accumulating insecurity and manufacturing risk along the energy frontier. In S. Soederberg (ed) *Risking Capitalism*. Bingley, UK: Emerald Group Publishing Limited, pp. 197–236.

Zalik, A. (2009) Zones of exclusion: Offshore extraction, the contestation of space and physical displacement in the Nigerian Delta and the Mexican Gulf. *Antipode* 41(3): 557–582.

Zibechi, R. (2012) Latin America: A new cycle of social struggles. *NACLA Report on the Americas* 45(2): 37–40.

40

FOOD SECURITY AND SOVEREIGNTY

Beth Bee

Introduction

In 2003, thousands of campesinos marched in Mexico City's Zócalo, the largest public square in the western hemisphere, declaring, "el campo no aguanta más" – the countryside can take no more. The march and subsequent social movement came to signify a critical moment for rural livelihoods, food security, and the pushback against decades of agricultural restructuring that marginalized rural producers; not only in Mexico, but throughout Latin America. Since that time, the world food crisis in 2007–2008 and the global financial crisis have posed further challenges to maintaining rural livelihoods and food security[1] throughout the region. More recently, issues such as climate change, biotechnology, and land grabs present new challenges to maintaining food security. In the face of such upheavals, the food sovereignty movement has come to the fore as several countries have adopted the language of food sovereignty in their national constitutions, although the results have varied considerably.

This chapter aims to highlight several factors that shape food access, choice, and autonomy in Latin America. Beginning with a brief overview of agrarian restructuring and its effects on the agro-food system, it then explores of the rise of and resistance to biotechnology in agriculture and the emergence of food sovereignty discourse and policy. The chapter then turns to highlight current and emerging research on gender, race, and identity. It concludes with a discussion of unpacking the complex interactions between climate change and food security.

Agrarian restructuring in Latin America

The proliferation of agrarian reforms from the 1980s onward, which varied by place in their intensity and outcomes, significantly reshaped the agrarian-food system throughout Latin America and indeed the world. These reforms stood in stark contrast to the previous half-century of state-led economic policies known collectively as import substitution industrialization (ISI). ISI emphasized development through national self-sufficiency and thus, included a range of public services to support domestic food production and peasant agriculture during this time. However, the debt crisis of the 1980s and the resulting structural adjustment policies mandated by the World Bank and the International Monetary Fund ushered in an era of neoliberalism that had profound consequences for both the rural economy and the production and consumption of food.

To make countries more competitive in an international agro-food market, these policies facilitated a shift toward large-scale, export-oriented industrial agriculture. This process also included the reduction or in some cases, elimination of agricultural subsidies and price guarantees for food crops in favour of cash crops, and increased access to Green Revolution technologies, such as improved seeds and chemical inputs. In the case of Mexico, an ideological commitment to neoliberal restructuring opened communally held land to privatization, which then paved the way for the controversial signing of the North American Free Trade Agreement (NAFTA) in 1993 (Cornelius and Myhre, 1998). Further efforts to liberalize land and labour, the opening of economies to world markets, and the increase of free-trade agreements throughout the region led to a commodity export boom and an increased dependency upon imported grains (Kay, 2015). The result has been a continual shift in land use patterns and economic activities, which move rural communities away from small-scale subsistence agriculture and toward non-farm employment, urban migration, and day or contract labour for large agribusinesses.

By the time the 2007–2008 world food crisis took hold, decades of neoliberal policies had significantly undermined smallholder livelihoods, food security, and ecosystems, leaving millions of people vulnerable to food price spikes. While some argue that the root of the food crisis was the result of the iron grip that transnational corporations (TNCs) maintain on the food system (Rosset, 2009), others suggest that the situation was precipitated by the promotion of agro-exports, biofuels, and the resulting land grabs, led in part by TNCs and in part by sympathetic governments (Altieri and Toledo, 2011). Still others posit that it was the predictable outcome of neoliberal food and agriculture policies that propel the overproduction of feed grains for the livestock industry, which are dependent upon and thus susceptible to fossil fuel price fluctuations (Jarosz, 2009).

Now, the region is paradoxically experiencing high rates of agricultural production while hunger and food insecurity remain a concern for vast numbers of individuals and households. Despite national programs and policies to address hunger and food insecurity, research demonstrates that the outcomes of these policies are often spatially uneven (Cuesta, Edmeades, and Madrigal, 2013; Salazar et al., 2016). In many ways, this paradox has been fueled by the widespread expansion of a large-scale industrial agriculture as the most efficient and viable system of production. As Altieri and Toledo (2011) make clear, "the tragedy of industrial agriculture is that a growing human population depends on the ecological services provided by nature (e.g. climate balance, pollination, biological control, soil fertility) which intensive industrial agriculture increasingly pushes beyond the tipping point" (591, citing Perfecto et al., 2009). Increasingly, biotechnology has become a key feature of industrial agriculture and, as many argue, a continued threat to agro-biodiversity and food security.

The neoliberal food regime and bio-hegemonic resistance

The restructuring of the agrarian political economy has gone hand in hand with the introduction of biotechnology and indeed, the Latin American countryside has played a central role in the adoption of this technology. Referring to what he dubs a "neoliberal food regime," Otero (2012) argues that the principal characteristics of such a regime include the state, agribusiness transnational corporations (ATNCs), and biotechnology in the form of genetically modified (GM) crops. Similarly, Newell (2009) coined the term "bio-hegemony" which emerges from a relationship between the state, agribusiness (both national and transnational), and biotechnological agricultural production that neoliberal restructuring made available. These characteristics are particularly pertinent in the case of Argentina.

Often referred to as the "*República Unida de la Soja*" (Newell, 2009), just over half of cultivated land in Argentina is sown with soybeans and 100% of this is GM soy (Leguizamón, 2016). The Argentine seed and agrochemical industry involves over 50 ATNCs and national firms including Monsanto-Bayer, Syngenta, Cargill, Dow-Agrosciences, and DuPont-Pioneer (Regunaga, 2010). Notably, Argentina is the third largest producer and exporter of soybeans and soy-derived products like meal for animal feed, and oilseed, and shares roughly half of the export market with Brazil (Leguizamón, 2016; USDA, 2017). This, of course, has serious implications for food security. Yet, given that the majority of soy exports are destined for animal feed, Argentina has avoided much of the controversy surrounding the human consumption of GM foods (Newell, 2009). Contestations about biotechnology in Argentina tend to centre around access and ownership of the technology, rather than on its social and environmental effects and desirability (Newell, 2008). Nevertheless, as the costs and benefits of GM soy are not equally distributed within Argentina, small-scale farmers, indigenous people, and non-farming rural communities have mobilized to defend their territorial rights against agribusiness and pesticide drift caused by spraying agrochemicals, particularly like glyphosate (Leguizamón, 2016). Although the majority of rural areas in Argentina remain inactive on the issue, residents of Malvinas Argentinas successfully blocked the construction of a Monsanto-owned corn seed plant (Leguizamón, 2016). Additionally, in one of the poorest provinces, Santiago del Estero, a regional campesino movement is actively resisting bio-hegemony to improve the quality of life of small producers (Wald, 2015).

Throughout Latin America, activists and farmers are mobilizing to resist bio-hegemonic agriculture. In southern Chile, such mobilizing manifests as "insurgent territorial configurations" that reimagine local socio-natural spaces by emphasizing ethical consumption, horizontal partnerships, and solidarity economies (Cid-Aguayo and Latta, 2015: 398). Costa Rican activists, on the other hand, have reimagined local spaces by successfully petitioning several municipalities to institute "transgenic-free territory" ordinances (Pearson, 2012). In Brazil, coalition building between local and transnational groups such as MST and La Via Campesina have worked to successfully block the importation of GM produce into the country (Newell, 2008). Furthermore, Peschard (2012) demonstrates that large soybean and seed producers who were once staunch supporters of agricultural biotech, are now becoming increasingly critical and uneasy with the current system. This discontent particularly centres on royalties, overall profitability, and lack of access to conventional varieties (Peschard, 2012). In Colombia, the Red de Semillas Libres, an activist organization network, defends seed sovereignty, or a farmer's control over seed biodiversity, and the ways that farmers are dispossessed of their criollo seeds through the intellectual property rights of GM seeds (Gutiérrez Escobar and Fitting, 2016). In central Mexico, on the other hand, resistance to transgenic maize is situated within the everyday practice of growing criollo maize in small and medium-scale peasant fields, alongside the knowledge systems and cultural identities related to its production (Mullaney, 2014). Resistance to transgenic maize in southern Mexico is also leading to a re-articulation of Mayan social movements and their alliance with environmental and campesino struggles, which in turn shape national debates over the technology and its role in the neoliberal food regime (Klepek, 2012). Furthermore, although peasant farmers in highland Guatemala are neither employing the language of food sovereignty nor directly countering the proliferation of transgenic seeds, their dedication to biodiverse *milpa* agriculture can be viewed as a form of post-capitalist politics that can exist simultaneously within and against the neoliberal food regime (Isakson, 2009).

Challenges to bio-hegemony and the neoliberal food regime in Latin America also include struggles against the development of biofuels. Brazil, Argentina, and Colombia in particular have already, as is the case in Brazil, or are in the process of developing soy and African palm oil-based biofuel industries (Renfrew, 2011). The energy efficiency benefits of biofuel production are often offset by several unintended consequences including the increase in grain and food prices; the displacement of subsistence domestic crops with export-oriented cash crops; land concentration; and threats to communal environmental and territorial rights, among others (Renfrew, 2011). Thus, opposition to biofuel production has been framed as posing risks to both "food security" as well as "food sovereignty," which engages broader debates about locally determined control over food production and distribution that is culturally appropriate, ecologically sustainable, and nutritious (Renfrew, 2011).

The paradox of food sovereignty

Amongst the backdrop of socio-environmental change and the neoliberalization of the agro-food system, the concept of food sovereignty has gained global attention. While the origins of the term can be traced back to government policy and rural advocacy groups from Mexico and Central America in the 1980s (Edelman, 2014), the widespread use of the term was popularized by *La Vía Campesina*; a transnational social movement and network of grassroots organizations (Martínez-Torrez and Rosset, 2010). The concept has evolved since its original emphasis national food sovereignty, as originally defined by *La Vía Campesina* (1996), to a one that emphasizes ecologically sustainable, healthy, culturally appropriate, and locally determined food production and consumption (Nyéléni Declaration, 2007). The concept of "food sovereignty" defies singular definition and is considered a "big tent" word (Patel, 2009), encompassing many things. Yet the fact that it lacks a consistent set of ideas and includes multiple layers of meaning can cause confusion to the detriment of agrarian social movements (Boyer, 2010).

Nonetheless scholars argue that food sovereignty represents a real alternative and break away from neoliberal values (Trauger, 2014). As a reflection of this approach, the concept has become enshrined in several Latin American state constitutions as part of the "Pink Tide," or a shift to leftist or center-left political regimes. Ecuador, Bolivia, and Venezuela have notably been a part of this state-led effort. This is critical, as scholars contend that food sovereignty entails dynamic state-society interactions and the interplay between sovereignty at multiple scales (Agarwal, 2014; Edelman, 2014; McKay, Nehring, and Walsh-Dilley, 2014). Yet, tensions remain between the constitutional inclusion of food sovereignty nationally and the interests of food sovereignty locally.

McKay, Nehring, and Walsh-Dilley (2014: 1177), for example, argue that in order to support the right of the people to determine their own agro-food systems, "state efforts must include some degree of structural reform to distribute power in ways that facilitate such local autonomy." Thus, a central tension of food sovereignty is that it paradoxically requires both a centralized state and a redistribution of power to facilitate control over food systems that may threaten the state itself (McKay, Nehring, and Walsh-Dilley, 2014). Furthermore, the centrally planned economy of such a state has historically limited consumer choice and diversity in the food system, which is antithetical to the goals of food sovereignty (Edelman, 2014).

The institutionalization of food sovereignty in Ecuador was made possible both by a favourable political climate as well as the inclusion of an indigenous worldview that envisioned food sovereignty as a means to resist neoliberal capitalism (Giunta, 2014; Peña, 2016). Yet, there is a gap between the success of social movements to establish new normative frameworks such as *buen vivir* and their actual implementation at the state level. The result is an ongoing ideological conflict between the concept of *buen vivir* and a new form of developmentalism in which markets retain primacy and transformation is based on exploitation of nature and modernization (Giunta, 2014).

Another point of concern, in Bolivia, is a lack of participatory democratic decision-making spaces necessary to sustain food sovereignty (McKay, Nehring, and Walsh-Dilley, 2014). In both

Bolivia and Ecuador, McKay, Nehring, and Walsh-Dilley (2014) contend that food sovereignty is little more than a legitimizing discourse to galvanize consent and popular support, as state actors co-opt or consolidate food sovereignty in ways that favour their own interests. Instead, they suggest that Venezuela is the most promising example of actually promoting food sovereignty for its citizens as significant structural changes in governance systems has provided greater local autonomy over food systems (McKay, Nehring, and Walsh-Dilley, 2014).

However, Enríquez and Newman (2016) argue that Venezuela has now shifted from a national goal of attaining food sovereignty to a nationalist food security policy and the pursuit of domestic food production by "any means necessary," including the use of conventional, petrochemical-based production. They write that, "the state's continued support for and reliance upon private, large-scale agrarian property structures, along with conventional practices, has restricted its capacity to accomplish its transformative development goals" (Enríquez and Newman, 2016: 620). Moreover, despite the fact that Venezuela was once recognized by the FAO for meeting both the first Millennium Development Goal and the World Food Summit goal by reducing the absolute number of undernourished people by half (FAO, 2013a; FAO, 2013b), Venezuela now finds itself in the depths of a food crisis. As a reflection of the severity of this crisis, referred to as "the Maduro diet," three-quarters of adults recently surveyed said they lost weight at an average of 19 pounds in 2016 (Zuñiga and Miroff, 2017). Although the roots of the crisis are complex, one of the problems identified that has contributed to Venezuela's food crisis is the reliance upon imported foods, which has been financed through massive oil revenues and accounts for roughly 96% of Venezuela's exports and half its income (OPEC, 2017). The dependence upon food imports, along with price distortion and input scarcity, has served to undermined land reforms and other efforts to support state-led food sovereignty (Purcell, 2017).

Institutionalizing food sovereignty within Latin America is clearly a slippery slope, rife with tensions and paradoxical issues. Henderson (2017) also illustrates that the counter-hegemonic discourse within the food sovereignty movement does not necessarily translate into practice. In reality, adhesion to food security discourses is often used as a way to find compromise within the neoliberal state. As a result, practices in Latin America tend to be "defensive counter-movements [that] seek protections and entitlements within this [neoliberal] regime," rather than counter-hegemonic movements that would aim to change it (Henderson, 2017: 4). Yet, this does not mean that efforts to create food sovereignty should be abandoned altogether. On the contrary, Edelman (2014) argues that if food sovereignty advocates are to imagine not only a successful small farm in a food-sovereign society, but a successful food-sovereign society built on a dynamic small farm sector, we need to devote considerably more attention to some of the challenges and paradoxes of food sovereignty. As Agarwal (2014) suggests, this is certainly the case if we are to reconcile the possible tensions that can arise out of promoting family farming if intra-household inequities are not also addressed. Understanding the gender and racialized dimensions of food sovereignty is an important area for further exploration.

Gender and race: necessary and emerging directions in research

Despite a wealth of literature on food sovereignty, the specific gendered dimensions remain understudied. A notable exception analyzes how women have built networks within La Via Campesina, which engages collectives from the local to the transnational scales (Pinheiro Machado Brochner, 2014). These networks strategically link issues of food sovereignty to other issues, such as gender-based violence and justice; particularly providing a critique around distribution and recognition that discursively connect claims for food sovereignty as being inseparable from other gender issues (Pinheiro Machado Brochner, 2014). Examining the relationship between fair-trade cooperatives, food sovereignty, and food security, Bacon (2015) illustrates that although these cooperatives improve gender equity, livelihoods, and farmer autonomy, such outcomes do not necessarily result in improved food security.

On the other hand, exploring broader issues of food security and gender has received wider attention (e.g. Palacios Luna, 2013; Valdivia, 2001; Vera Delgado, 2015). Deere (2005) argues that as neoliberal policies have undermined peasant agriculture, production has become increasingly oriented toward household food security and thus, an extension of women's domestic responsibilities. In Mexico, despite the additional work, women are particularly interested in maintaining maize not only because it is an important source of household food security, but also because it is a means to defend their identity as providers (Preibisch, Rivera Herrejón, and Wiggins, 2002). Additionally, as economic stressors push men to migrate, women become primary decision makers in agriculture and thus their knowledge of maize for domestic uses is an important source of *in situ* maize biodiversity conservation (Chambers and Momsen, 2007).

Urban agriculture, and specifically urban home gardens which are primarily tended by women, also plays a significant role in maintaining food security and livelihoods in Latin American cities (Altieri et al., 1999; Keys, 1999; Leitgeb, Schneider, and Vogl, 2016; Premat, 2010). WinklerPrins and Souza (2005), for example, illustrate how home gardens in urban Brazil help facilitate a flow of food and other consumables between rural and urban households that enable newly urban migrants to adapt to their new environment. Urban home gardens and the women who tend them also play a key role in creating a livable space that likewise contributes to the socio-natural production of urban space (Shillington, 2013). As Shillington (2013) demonstrates in the case of Nicaragua, home gardens and home gardeners are thus intimately connected to the socio-natural processes that constitute citizenship, participation, and the appropriation of space.

In addition to gender, issues such as race as well as the intersections between these and other identities, have become an emerging issue in understanding food security throughout Latin America. For example, while not surprising to many, an analysis of Brazil's 2009 National Household Survey reveals that racial inequities as well as gender significantly influence food security issues. In particular, Afro-Brazilians experience higher rates of food insecurity compared to white Brazilians (Wood and Felker-Kantor, 2013). Among indigenous Mayan in the Yucatan state in Mexico, food insecurity is characterized by several factors including cultural changes that have resulted in a devaluing of Mayan foods and subsistence agriculture and an increased preference for high calorie, processed foods (Calix de Dios et al., 2014). This is further facilitated by the migration of Mayan youth to tourist areas in search of money and modernity (Calix de Dios et al., 2014). On the other hand, household swidden gardens or *chacras* remain an important source of ethnic and gender identity as well as food security for Kichwa households in highland Guatemala (Perreault, 2005). How ethnic diversity, class divisions, and gender inequities can enhance or inhibit food security and sovereignty discourses and practices are important areas for exploration. The intersecting racialized and gendered aspects of land access and control (Mollett, 2010; Radcliffe, 2014) and their relationship to food security and sovereignty is one potential avenue of inquiry. Understanding how gender, race, and other categories of difference shape food security and sovereignty will become more urgent in the context of climate change.

Climate change

Concerns that climate change will provoke widespread food shortages have fueled a great deal of research on both the impacts and the technological solutions to maintain agricultural production in the face of a variable climate. This also includes support for the adoption of transgenic crops, or "transgenic adaptation strategies," as a viable means to maintain food security (see Mercer, Perales, and Wainwright, 2012). Increasingly however, scholars are shedding light on significant and complex challenges that climate change presents for agricultural livelihoods and food security that go beyond debates about food availability and agricultural production.

Several scholars argue for a return to agroecological principals and the revitalization of small family farms, which conserve ecological integrity and makes substantial contributions to domestic food security (Altieri and Toledo, 2011). Yet, the challenge for re-scaling the agroecological revolution is a lack of state support. In the case of Bolivia, for example, the state continues to support cattle and large-scale monocultures (Jacobi, 2016), which reproduce the myth that large-scale agribusiness is the only "viable" form of agriculture (Hausermann and Eakin, 2008). The food sovereignty movement also promotes localized, agroecological principals, which although often promoted for social benefits, also challenges large-scale, capitalist export-oriented agriculture that ruptures the link between citizenship, nature, and agriculture. Thus, new forms of agrarian resistance and efforts to promote food sovereignty can rework the "metabolic rift" between society and nature, and promoting food sovereignty can be more easily adaptable to a changing climate than industrial models (Wittman, 2009).

A common theme in the Latin American food security and climate change literature is the need for policies and institutional arrangements that support sustainable food systems and localized adaptation strategies, particularly in the Caribbean and Central American states (Bizikova et al., 2015; Saint Ville, Hickey, and Phillip, 2017; Trotman et al., 2009). Such policies should include stakeholders across scales, from local to regional, in the planning and implementation process (Trotman et al., 2009; Bizikova et al., 2015). At the same time, current policies and the industrial, corporate model of agricultural production arguably works against efforts to create a more sustainable producer-consumer network that is resilient to climate change (Wilson, 2016).

Although limited, a handful of studies look to unpack the complex links between food security, climate change, and locally relevant and context-specific socio-political and economic issues. For example, Milan and Ruano (2014) investigate the circumstances under which migration becomes a risk-management strategy to mitigate increasing rainfall variability and food insecurity in Guatemala. It is often one strategy of several which include diversifying income to include on-farm and non-farm incomes, as well as remittances from family in the US. Another emerging area of research involves an exploration of the complex relationships between climate change, food security, and gender (Bee, 2014; Beuchelt and Badstue, 2013). Both migration and gender, as well as race and ethnicity, land tenure, and a range of other context-specific issues as they relate to the nexus between food security and climate change nexus, warrant further attention. Moreover, the vast majority of this literature focuses on terrestrial environments, while an area deserving more attention is the relationship between fisheries and food security in the face of climate change and shifting identities.

Conclusion

Decades of neoliberal policies have arguably wreaked havoc on Latin American agro-food systems and livelihoods. However, this turmoil has also fueled powerful counter-movements, including the food sovereignty movement, that are working to restore dignity and autonomy to farmers and consumers. Although several countries have incorporated the concept of food sovereignty into their constitutions, gaps persist between the stated goals of food sovereignty and the practical implementation of these goals. Yet rather than abandoning these struggles, attention to these challenges is necessary to address the shortcomings. Likewise, attention to gender, race, and their intersections are important to consider as they cut across multiple issues related to food security and food sovereignty, in both rural and urban contexts. Furthermore, the challenges posed by a changing climate will require attention, not just to the projected impacts, but also to the complex and context-specific connections between the ecological, social, economic, and political dimensions of food security.

Note

1 Food security is most commonly defined as occurring when "all people, at all times, have physical, social and economic access to sufficient safe and nutritious food that meets their dietary needs and food preferences for an active and healthy life" (FAO 1996). The food sovereignty discourse, on the other hand, although defying singular definition is often considered to be an alternative paradigm which emphasizes food self-sufficiency and agroecology, among other things. Yet some scholars argue that these concepts also overlap and interrelate in many ways (Clapp, 2014; Jarosz, 2014).

References

Agarwal, B. (2014) Food sovereignty, food security and democratic choice: Critical contradictions, difficult conciliations. *Journal of Peasant Studies* 41(6): 1247–1268.
Altieri, M. A., Companioni, N., Cañizares, K., Murphy, C., Rosset, P., Bourque, M. and Nicholls, C. I. (1999) The greening of the 'barrios': Urban agriculture for food security in Cuba. *Agriculture and Human Values* 16(2): 131–140.
———— and Toledo, V. M. (2011) The agroecological revolution in Latin America: Rescuing nature, ensuring food sovereignty and empowering peasants. *Journal of Peasant Studies* 38(3): 587–612.
Bacon, C. M. (2015) Food sovereignty, food security and fair trade: The case of an influential Nicaraguan smallholder cooperative. *Third World Quarterly* 36(3): 469–488.
Bee, B. A. (2014) 'Si no comemos tortilla, no vivimos': Women, climate change, and food security in central Mexico. *Agriculture and Human Values* 31(4): 607–620.
Beuchelt, T. D. and Badstue, L. (2013) Gender, nutrition- and climate-smart food production: Opportunities and trade-offs. *Food Security* 5(5): 709–721.
Bizikova, L., Tyler, S., Moench, M., Keller, M. and Echeverria, D. (2015) Climate resilience and food security in Central America: A practical framework. *Climate and Development* 8(5): 1–16.
Boyer, J. (2010) Food security, food sovereignty, and local challenges for transnational agrarian movements: The Honduras case. *Journal of Peasant Studies* 37(2): 319–351.
Calix de Dios, H., Putnam, H., Alvarado Dzul, S., Godek, W., Kissmann, S., Luckson Pierre, J. and Gliessman, S. (2014) The challenges of measuring food security and sovereignty in the Yucatan Peninsula. *Development in Practice* 24: 199–215.
Chambers, K. J. and Momsen, J. H. (2007) From the kitchen and the field: Gender and maize diversity in the Bajío region of Mexico. *Singapore Journal of Tropical Geography* 28(1): 39–56.
Cid Aguayo, B. and Latta, A. (2015) Agro-ecology and food sovereignty movements in Chile: Sociospatial practices for alternative peasant futures. *Annals of the Association of American Geographers* 105(2): 397–406.
Clapp, J. (2014) Food security and food sovereignty: Getting past the binary. *Dialogues in Human Geography* 4(2): 206–211.
Cornelius, W. and Myhre, D. (1998) *The Transformation of Rural Mexico: Reforming the Ejido Sector*. San Diego, CA: Center for US-Mexico Studies, University of California San Diego.
Cuesta, J., Edmeades, S. and Madrigal, L. (2013) Food security and public agricultural spending in Bolivia: Putting money where your mouth is? *Food Policy* 40: 1–13.
Deere, C. D. (2005) The feminization of agriculture? Economic restructuring in rural Latin America, *Gender Equality: Striving for Justice in an Unequal World*. Geneva: United Nations Research Insitute for Social Development (UNRISD).
Edelman, M. (2014) Food sovereignty: Forgotten genealogies and future regulatory challenges. *The Journal of Peasant Studies* 41(6): 959–978.
Enríquez, L. J. and Newman, S. J. (2016) The conflicted state and Agrarian transformation in Pink tide Venezuela. *Journal of Agrarian Change* 16(4): 594–626.
FAO. (1996) *Rome Declaration on World Food Security and World Food Summit Plan of Action*. World Food Summit 13–17 November 1996. Rome.

———— (2002) *The State of Food Insecurity in the World 2001.* Rome: Food and Agriculture Organization.

————. (2013a) *38 Countries Meet Anti-Hunger Targets for 2015.* www.fao.org/news/story/en/item/177728/icode/%0A (Accessed 29 October 2017).

————. (2013b) *Reconocimiento de la FAO a Venezuela.* www.fao.org/americas/noticias/ver/en/c/230150/ (Accessed 29 October 2017).

Giunta, I. (2014) Food sovereignty in Ecuador: Peasant struggles and the challenge of institutionalization. *The Journal of Peasant Studies* 41(6): 1201–1224.

Gutiérrez Escobar, L. and Fitting, E. (2016) The Red de Semillas libres: Contesting Biohegemony in Colombia. *Journal of Agrarian Change* 16(4): 711–719.

Hausermann, H. and Eakin, H. C. (2008) Producing 'Viable' landscapes and livelihoods in central Veracruz, Mexico: Institutional and producer responses to the Coffee commodity crisis. *Journal of Latin American Geography* 7(1): 109–131.

Henderson, T. P. (2017) State – peasant movement relations and the politics of food sovereignty in Mexico and Ecuador. *The Journal of Peasant Studies* 44(1): 33–55.

Isakson, S. R. (2009) No hay ganancia en la milpa: The agrarian question, food sovereignty, and the on-farm conservation of agrobiodiversity in the Guatemalan highlands. *Journal of Peasant Studies* 36(4): 725–759.

Jacobi, J. (2016) Agroforestry in Bolivia: Opportunities and challenges in the context of food security and food sovereignty. *Environmental Conservation* 43(4): 307–316.

Jarosz, L. (2009) Energy, climate change, meat, and markets: Mapping the coordinates of the current world food crisis. *Geography Compass* 3(6): 2065–2083.

————. (2014) Comparing food security and food sovereignty discourses. *Dialogues in Human Geography* 4(2): 168–181.

Kay, C. (2015) The agrarian question and the neoliberal rural transformation in Latin America. *Revista Europea de Estudios Latinoamericanos y del Caribe* 100: 73.

Keys, E. (1999) Kaqchikel gardens: Women, children, and multiple roles of gardens among the Maya of Highland Guatemala. *Conference of Latin Americanist Geographers Yearbook* 25: 89–100.

Klepek, J. (2012) Against the grain: Knowledge alliances and resistance to agricultural biotechnology in Guatemala. *Canadian Journal of Development Studies* 33(3): 310–325.

Leguizamón, A. (2016) Environmental Injustice in Argentina: Struggles against Genetically Modified Soy. *Journal of Agrarian Change* 16(4): 684–692.

Leitgeb, F., Schneider, S. and Vogl, C. R. (2016) Increasing food sovereignty with urban agriculture in Cuba. *Agriculture and Human Values* 33(2): 415–426.

Martínez-Torres, M. E. and Rosset, P. M. (2010) La Vía Campesina: The birth and evolution of a transnational social movement. *The Journal of Peasant Studies* 37(1): 149–175.

McKay, B., Nehring, R. and Walsh-Dilley, M. (2014) The 'state' of food sovereignty in Latin America: Political projects and alternative pathways in Venezuela, Ecuador and Bolivia. *The Journal of Peasant Studies* 41(6): 1175–1200.

Mercer, K. L., Perales, H. R., & Wainwright, J. D. (2012) Climate change and the transgenic adaptation strategy: Smallholder livelihoods, climate justice, and maize landraces in Mexico. *Global Environmental Change,* 22(2): 495–504.

Milan, A. and Ruano, S. (2014) Rainfall variability, food insecurity and migration in Cabricán, Guatemala. *Climate and Development* 6(1): 61–68.

Mollett, S. (2010) Está listo (Are you ready)? Gender, race and land registration in the Río Plátano Biosphere Reserve. *Gender, Place and Culture* 17(3): 357–375.

Mullaney, E. G. (2014) Geopolitical maize: Peasant seeds, everyday practices, and food security in Mexico. *Geopolitics* 19(2): 406–430.

Newell, P. (2009) Bio-hegemony: The political economy of agricultural biotechnology in Argentina. *Journal of Latin American Studies* 41(1): 27–57.

————. (2008) Trade and biotechnology in Latin America: Democratization, contestation and the politics of mobilization. *Journal of Agrarian Change* 8(2–3): 345–376.

Nyéléni Declaration. (2007) *Declaration of the Forum for Food Sovereignty, Nyéléni.* www.nyeleni.org/%0Dspip.php?article290 (Accessed 29 October 2017).

OPEC. (2017) *Annual Statistics Bulletin.* Vienna, Austria: Organization of Petroleum Exporting Countries.

Otero, G. (2012) The neoliberal food regime in Latin America: State, agribusiness transnational corporations and biotechnology. *Canadian Journal of Development Studies/Revue canadienne d'études du dévelopement* 33(3): 282–294.

Palacios Luna, A. P. (2013) Neoliberalism and women's struggle for food security in Tlaucingo. *Latin American Perspectives* 40(5): 93–104.

Patel, R. (2009) What does food sovereignty look like? *The Journal of Peasant Studies* 36(3): 663–706.

Pearson, T. W. (2012) Transgenic-free territories in Costa Rica: Networks, place, and the politics of life. *American Ethnologist* 39(1): 90–105.

Peña, K. (2016) Social movements, the state, and the making of food sovereignty in ecuador. *Latin American Perspectives* 43(1): 221–237.

Perfecto, I, Vandermeer, J. and Wright, A. (2009). *Nature's matrix: linking agriculture, conservation and food sovereignty.* London: Earthscan.

Perreault, T. (2005) Why chacras (Swidden gardens) persist: Agrobiodiversity, food security, and cultural identity in the Ecuadorian Amazon. *Human Organization* 64(4): 327–339.Peschard, K. (2012). Unexpected discontent: Exploring new developments in Brazil's transgenics controversy. *Canadian Journal of Development Studies* 33(3): 326–337.

Pinheiro Machado Brochner, G. (2014) Peasant women in Latin America: Transnational networking for food sovereignty as an empowerment tool. *Latin American Policy* 5(2): 251–264.

Preibisch, K. L., Rivera Herrejón, G. and Wiggins, S. L. (2002) Defending food security in a free-market economy: The gendered dimensions of restructuring in Rural Mexico. *Human Organization* 61(1): 68–79.

Premat, A. (2010) Moving between the Plan and the Ground: Shifting Perspectives on Urban Agriculture in Havana, Cuba. In L. J. Mougeot (ed) *Agropolis: The Social, Political and Environmental Dimensions of Urban Agriculture.* New York: Routledge, pp. 153–185.

Purcell, T. F. (2017) The political economy of rentier capitalism and the limits to agrarian transformation in Venezuela. *Journal of Agrarian Change* 17(2): 296–312.

Radcliffe, S. A. (2014) Gendered frontiers of land control: Indigenous territory, women and contests over land in Ecuador. *Gender, Place and Culture* 21(7): 854–871.

Regunaga, M. (2010) *Implications of the Organization of the Commodity Production and Processing Industry: The Soybean Chain in Argentina.* LCSSD occasional paper series on food prices. Washington, DC: World Bank. https://openknowledge.worldbank.org/handle/10986/18710 (Accessed 30 March 2018).

Renfrew, D. (2011) The curse of wealth: Political ecologies of Latin American neoliberalism. *Geography Compass* 5(8): 581–594.

Rosset, P. M. (2009) Food sovereignty in Latin America: Confronting the 'New' crisis. *NACLA Report on the Americas* 42(3): 16–21.

Saint Ville, A. S., Hickey, G. M. and Phillip, L. E. (2017) How do stakeholder interactions influence national food security policy in the Caribbean? The case of Saint Lucia. *Food Policy* 68: 53–64.

Salazar, L., Aramburu, J., González-Flores, M. and Winters, P. (2016) Sowing for food security: A case study of smallholder farmers in Bolivia. *Food Policy* 65: 32–52.

Shillington, L. J. (2013) Right to food, right to the city: Household urban agriculture, and socionatural metabolism in Managua, Nicaragua. *Geoforum,* 44: 103–111.

Trauger, A. (2014) Toward a political geography of food sovereignty: Transforming territory, exchange and power in the liberal sovereign state. *The Journal of Peasant Studies* 41(6): 1–22.

Trotman, A., Gordon, R. M., Hutchinson, S. D., Singh, R. and McRae-Smith, D. (2009) Policy responses to GEC impacts on food availability and affordability in the Caribbean community. *Environmental Science and Policy* 12(4): 529–541.

USDA. (2017) *Trade.* www.ers.usda.gov/topics/crops/soybeans-oil-crops/trade/%0A (Accessed 29 October 2017).

Valdivia, C. (2001) Gender, livestock assets, resource management, and food security: Lessons from the SR-CRSP. *Agriculture and Human Values* 18:27–39.

Vera Delgado, J. (2015) The socio-cultural, institutional and gender aspects of the water transfer-agribusiness model for food and water security. Lessons learned from Peru. *Food Security* 7(6): 1187–1197.

Wald, N. (2015) In search of alternatives: Peasant initiatives for a different development in northern Argentina. *Latin American Perspectives* 42(2): 90–106.

Wilson, M. (2016) Food and nutrition security policies in the Caribbean: Challenging the corporate food regime? *Geoforum* 73: 60–69.

WinklerPrins, A. M. G. and Souza, P. S. De. (2005) Surviving the city: Urban home gardens and the economy of affection in the Brazilian Amazon. *Journal of Latin American Geography* 4(1): 107–126.

Wittman, H. (2009) Reworking the metabolic rift: La Vía Campesina, agrarian citizenship, and food sovereignty. *Journal of Peasant Studies* 36(4): 805–826.

Wood, C. H. and Felker-Kantor, E. (2013) The color of hunger: Food insecurity and racial inequality in Brazil. *Latin American and Caribbean Ethnic Studies* 8(3): 304–322.

Zuñiga, M. and Miroff, N. (2017) Venezuela says it will quit Organization of American States. *The Washington Post*. http://wapo.st/2pmKqTF.

41

ADAPTING TO CLIMATE CHANGE IN THE ANDES

Changing landscapes and livelihood strategies in the Altiplano[1]

Corinne Valdivia and Karina Yager

Introduction

Globalization and climate change are impacting on the tropical regions of the world in significant ways, especially in rural communities that depend on farming and pastoralism for their livelihoods (Coppock et al., 2017; Seth et al., 2010; Valdivia et al., 2010; Young and Lipton, 2006; Lobell, 2008; Bebbington, 1999). This is specially the case in mountain regions, like the Andean region in South America. Increases in food insecurity are expected, which will require policies to protect vulnerable populations (Brown and Funk, 2008). Coupled with these global drivers are government policies on economic growth that have adverse effects on rural development in Latin America, where very few rural regions have experienced economic development (Berdegue, Bebbington, and Escobal, 2015).

The lack of development in rural regions pushes people to migrate to places where they perceive that there are more opportunities. Economic growth policies and accessibility to markets shape the context in which people in rural communities pursue their livelihoods strategies and wellbeing (Valdivia, 1992, 2004), which depend mostly on agriculture, where families engage in farming, small-scale ranching, and/or pastoralism. Often households will combine activities on the farm with non-agricultural activities in rural areas, or with migration to other regions returning only at key moments such as high labour needs (Valdivia et al., 2001; Turin and Valdivia, 2011) or for community-based *cargo* responsibilities. These agricultural, agropastoral, rural, and migration livelihood strategies (Bebbington, 1999) shape the economic portfolio of the *campesino* household, and the nature/type of diversification (Ellis, 1998; Valdivia, Dunn and Jetté, 1996).

Investments in infrastructure that facilitate exports, and investments in urban centres, have prevailed over investment in infrastructure in rural regions, especially in the Altiplano in the case of Peru (Maletta, 2017). In Bolivia on the other hand, recent investments have focused on infrastructure such as roads and health care centres, and irrigation projects in the Altiplano (World Bank, 2010). Will the concerns with climate change lead countries to plan, develop, and invest systematically in programs and policies that favour sustainable development in rural communities of Andes?

The Andes of Bolivia and Peru

The Tropical Andes are the largest continental mountain range in the world, extending approximately 7,000 kilometres, and containing 99% of the world's tropical glaciers (Kaser, 1999). The Andean region runs from north to south on the western side of South America, encompasses seven countries (Venezuela, Columbia, Ecuador, Peru, Bolivia, Chile, and Argentina), and is home to almost 44% of the over 85 million people that live in these countries (Devenish and Gianella, 2012; FAO, 2012). The Altiplano (High Plateau) spans Peru and Bolivia, starting from the northwest of Lake Titicaca in Puno (southern Peru) extending to the southeast and southwestern corner of Bolivia, and into parts of Chile and Argentina.

In recent decades this region has experienced significant socio-economic and environmental changes. Climate change and markets have an impact on smallholder livelihood decisions and strategies, expressed in changing land use patterns and landscapes (Valdivia et al., 2010). Other important factors driving change in the Altiplano are related to population dynamics. A process of market integration has taken place in the last 20–30 years, which includes an increase in the variety of products being sold in local, regional, and urban markets (Valdivia Gilles and Turin, 2013). Population dynamics resulting from pull or push migration, have also had an effect on production practices, management of natural resources, landscapes, and the marketing strategies of households in this region (Valdivia et al., 2010; Turin and Valdivia, 2011).

Mountain regions are recognized as important water towers that provide critical ecosystem services, but are also characterized by rugged and extreme landscapes spanning vertically stacked zones of climate, hydrology, and biodiversity (Price, 2015; Körner, 2003). These regions are also highly sensitive to climate change, and this is strikingly evident in the increasingly rapid rate of tropical Andean deglaciation (Rabatel et al., 2013). Agriculture and pastoralism in the Andes are vulnerable to droughts, frost and hail events, and floods because of the orography, including high altitude and variable rainfall patterns (Sperling et al., 2008). Across the Altiplano longer dry seasons and more frequent storm events were projected by climate models (Robledo et al., 2004), and observed by farmers in the region (PNCC, 2005). In recent years, traditional crop rotations are being abandoned and production strategies that have sustained populations for hundreds of years no longer function, but climate is not the only cause for these changes. These traditional agro-pastoral practices are difficult to maintain due to migration and the growth of a market economy (Zimmerer, 1993). As a result, agro-ecosystems are becoming less diverse and more vulnerable to shocks.

Climate variability and change are increasing food insecurity in the Andes (Valdivia et al., 2010). Formal institutions to protect against risks in farming are absent – especially in rainfed production systems because of the degree of shared risks is high. Therefore, farmers in regions like the Andes rely on local knowledge and their own institutions to address uncertainty and risks, and these have proven to work for many centuries (Cruz et al., 2017; Gilles and Valdivia, 2009; Mayer, 2002; Orlove et al., 2000; Valdivia, Dunn, and Jetté, 1996).

Here we provide insights about the livelihood strategies of agro-pastoral households in rural communities (*comunidades campesinas*) in the Altiplano in the last 25 years, in particular changes taking place as a result of climate change and market participation in Bolivia and Peru, in both farming and pastoral landscapes. The cases presented summarize studies of diversity and sustainability of livelihoods in peasant communities (Coppock and Valdivia, 2001; Valdivia, 2001b; Valdivia et al., 2003; Valdivia, 2004; Valdivia, Jimenez, and Romero, 2007; Jimenez et al., 2013; Turin and Valdivia, 2010; Valdivia, Gilles and Turin, 2013; Yager, 2015).

A framework for understanding change and adaptation

Contexts are diverse, driven by climate change, global markets, environmental changes, population growth, and policies. The cases documented are informed by a framework to study the dynamics of change in social and ecological systems, and to learn about the agency of individuals, groups, and communities within and across scales, by engaging with local decision makers, their communities, and those organizations they connect with. As Ostrom (2007: 15183) states "The key is assessing which variables at multiple tiers across the biophysical and social domains affect human behaviour and social-ecological outcomes over time." The scales in the context of the Altiplano include: (1) the household and the decision makers within; (2) rural communities (*comunidades campesinas*, formally recognized in Bolivia and Peru) and their norms; (3) landscapes, shaped by social and ecological drivers; (4) markets at each scale; and (5) the national and global contexts in relation to policies that impact on adaptation to climate change.

Systems and institutions at each scale are studied to learn how these impact on local decision makers and communities, to learn how markets, climate, and their interaction impact on livelihood outcomes, which in turn impact on the natural capital, and ultimately the landscapes (Valdivia, 2007). Changes in capitals and assets built that families in rural communities own, and their practices as stewards of their land and environment (their natural capital), are expressed in livelihood strategies embedded in the socio-ecological system (Valdivia and Barbieri, 2014). Transdisciplinary research engages local decision makers in defining the issues and in the research. This approach, along with translational processes (Valdivia et al., 2014) aims, through participatory research, to build actionable knowledge, meaning salient, trusted, in the context of the decision maker, with networks that support the process of adaptation (Valdivia and Gilles, 2001; Valdivia et al., 2010). It engages with decision makers to frame the issues and learn from each other seeking to be transformative; understanding and linking people and their knowledge through participatory process in order to identify what to do; and identifying pathways for how adaptation can happen (Valdivia et al., 2010). Knowledge generated through groups and various forms of collective action engender the ability to act, not only locally but at other scales, horizontally and vertically (see Box 41.3).

In the sustainable livelihoods framework agency – the ability to act – is a function of capitals (tangible and intangible) and the ability to negotiate in markets and influence policy (Chambers and Conway, 1992; Valdivia et al., 2003). The link between livelihoods and markets is the capability or bargaining power of women and men in order to capture the benefits from market access, so families and communities can make investments in their "capitals" (economic, human, social, natural, physical, and political) in order to build resilient livelihoods (Valdivia, 2004; Valdivia, 2001b; de Haan, 2000; Valdivia and Gilles, 2001; Bebbington, 1999). As Wilson (2012: 1219) states "Similarly important is that the implementation of pathways of resilience can only find its most direct expression at the level of the individual/household and the community, as it is only at the most local level that outcomes of policies and decisions . . . are experienced with tangible effects 'on the ground'." Adapting and building social resilience means the ability of households, groups, or communities to negotiate external stresses and disturbances (Adger, 2000). This also is dependent on the specific context, as customary institutions shape change (North, 1993).

Climate change is a driver of change that mostly challenges the ability to build sustainable livelihoods. Andean communities have dealt for centuries with climate variability, and it is the uncertainty of extreme events, warming, and fast paced environmental changes that challenge even further the capabilities of local decision makers and their communities (Cruz et al., 2017). Recent findings on Andean climate change include rapid deglaciation (Bradley et al., 2006;

Thompson et al., 2003; Vergara et al., 2007a; Vergara et al., 2007b; Vuille et al., 2003; Rabatel et al., 2013; Soruco et al., 2009); temperature warming (Diaz and Graham, 1996; Seth et al., 2010); increasing extremes (Arana Pardo et al., 2007; Thibeault et al., 2010); and water deficit (Bradley et al., 2006); as well as consequent ecological changes including species migration and turnover (Seimon et al., 2007, 2017).

Climate change, agriculture, and livelihood strategies in the Altiplano

Projections of climate change for the Altiplano indicate warming and increased extreme events (Seth et al., 2010). The temperature in this region has increased by 1 degree Celsius in the past 50 years, and projections by 2090 in the Altiplano indicate a drier spring, wetter summer, and an average 4 degree Celsius increase in temperature (Seth et al., 2010; Thibeault, Seth, and Garcia, 2010). Even though the rainy season is projected to be wetter, loss of soil moisture by end of the century is also projected (Thibeault, Seth, and Garcia, 2010). These projections indicate a need to adapt at a faster pace and with a long-term view. In addition, we recognize that greater meteorological instrumentation is needed, especially in mountain regions across the Andes (see Box 41.1).

Results from many years of collaborations have identified practices that contribute to resilient livelihoods and environments. These practices matter because the transformational changes create new conditions that can't be dealt solely with local or scientific knowledge. For example, the warming trends will increase soil evapotranspiration impacting on the ability of the soils to absorb water during extreme events. Studies show the need for soil organic matter as a practice that can reduce evapotranspiration in the context of warming climate (see Box 41.2).

The Altiplano across Bolivia and Peru has a diversity of landscapes supporting rural livelihoods (see Figure 41.1); from grasslands sustaining pastoralists raising alpacas and llamas to agropastoral systems incorporating multiple crops and animals, and more intensive crop and livestock production systems. These landscapes are shaped by altitude and distance from the Lake Titicaca. The diversity of land use patterns depends on the altitude of the Andean mountain range, rainfall patterns, and temperature. Farmers combine crop production, mainly potatoes and other tubers and grains, with production of alpacas, llamas, sheep, and/or cattle (Markowitz and Valdivia, 2001). Agropastoral systems are mainly present in semi-arid regions; crop and livestock production are integrated and present in sub-humid regions; and at higher elevations where agriculture is not possible pastoralist systems prevail.

Box 41.1 The importance of data quality for climate studies in the Altiplano region

Several developing countries, especially in Latin America, have sparse ground-based networks of climatic observations. Typically, data collected from those networks suffer problems of completeness, which drastically reduce the number of time series available for different studies. In addition, a large amount of data, especially from the last 30 or 40 years, comes from manned stations which are prone to have more issues than automatic weather stations due to the introduction of an additional external factor: the observer. In fact, common problems detected in these types of networks, as suggested by Hunziker et al. (2017) for the case of the Bolivian stations on the central Altiplano, can be traced back to errors related to the observers. Several problems found on these time series are hardly detected by conventional quality control (QC) procedures, which requires a much more detailed and time-consuming data analysis for obtaining "clean" time series for any given station. Truncation errors in precipitation, which occur when observers do not register the total amount of rainfall for a 24-hour period, or missing temperatures among some interval values, typically around zero degrees Celsius, are common in the Bolivian time series. However, even if those problems are found in data from a station this does not mean that its time series is not useful. Depending on the purpose of the study (climatologies, climatic indices, etc.), these data could be recovered and/or corrected. Increasing the available climatic records is especially important for a country like Bolivia with a relatively big area (>1 million of km²), large altitudinal gradients, and a limited number of stations along its territory.

How important are all the efforts for having a high quality time series? Clearly bad quality data will lead to wrong conclusions, but even data that have passed standard quality control procedures could still have problems. Hunziker et al. (2018) investigated the effects of undetected quality issues left on the time series using data for the Central Altiplano. In this particular case a considerable part of the observations are inappropriate for calculating monthly means or monthly total precipitation. These data with undetected quality issues largely affects climatological analyses, increase the spread of regional trends due to the noise present on the data and could even change or hide real station trends. Applying different QC and homogenization procedures Hunziker et al. (2018) showed that trends in minimum temperature (TN), maximum temperature (TX) and precipitation (PCP) become more coherent spatially when the quality of the data increases and more and more refined procedures are applied to them. For instance, in the case of TN, the calculated decadal trends changed when data with the standard QC (−0.04°C/decade) and homogenized enhanced QC (+0.22°C/decade) were used. Although this is not the case for other variables (TX and PCP), the associated spread (uncertainty) was reduced as the quality of the data improved. In particular, for the Altiplano region the following trends are reported for the period 1981–2010 (Andrade, 2018) using high quality data: +0.22 ± 0.07°C/decade for TN, +0.40 ± 0.10°C/decade for TX, and −34 mm/decade for PCP, with a strong spatial coherence for temperature (all trends are positive). These results show strong, statistically significant, warming in the Altiplano region. At the same time a regional decrease in rainfall is observed, although the individual station trends vary and in some cases show different signs.

Marcos Andrade

Analysis of temperature trends in maximum and minimum temperatures show that warming has happened over the last 50 years, with differences in the north, central, and southern Altiplano of Bolivia (Seth et al., 2010). Significant warming is identified in the central and Northern Altiplano, and significant cooling trends in minimum temperatures in the southern Altiplano (Valdivia et al., 2010). The increasing maximum temperatures relate to increases in evapotranspiration in the region; and the impacts are larger in the southern and Northern Altiplano, and less significant in the central region (*Ibid*). Cooling trends in minimum temperatures in the southern Altiplano of Bolivia might be due to the reported advance of non-sustainable agricultural activities like quinoa production (Valdivia et al., 2010).

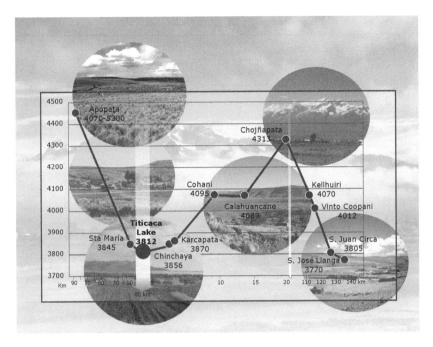

Figure 41.1 Landscape diversity in the Altiplano of Peru and Bolivia

Landscapes and livelihoods in the Bolivian Altiplano

Animals are important in the Altiplano. According to the 2013 Bolivian Ag Census (INE, 2015) there are 245,455 Agricultural Production Units, covering 292,698.8 hectares of land, in the Department of La Paz, with cattle (502,000), and sheep (close to 1.8 million); this is 2.5 times larger in cattle since the 1950s Census, while sheep decreased by 20%. The 2013 census also reported on camelids consisting of llamas and alpacas. Llamas in 21,137 production units manage 448,314 head, and there were 305,462 alpacas in 6,494 production units (INE, 2015).

Aroma in the Central Altiplano and Omasuyos in the North are home to 17.7% of the total in this department. The province of Aroma in Central Altiplano reported 19,246 Ag Production Units, 75% of the ag production units were led by men in 2013. This semi-arid region is important for potato production as well as dairy. Omasuyos (Northern Altiplano), near Lake Titicaca, had in 2013 23,795 Production Units; 70% were led by men, while women were in charge of the remaining 30%. Table 41.1 presents data on the livelihoods and capitals of rural households in the Ancoraimes municipality (North Altiplano) at three elevations in a watershed, going from intensive crop production landscapes, to crop and livestock, and pastoral.

With the rising minimum temperatures frost, a limiting factor in farming in the past, is becoming less of a risk in the North and Central Altiplano. This change in temperature, along with market opportunities in the North near Lake Titicaca in Bolivia, have resulted in shifts in production from the more resilient potato varieties that withstand harsher climates to high value crops like onions and peas, which require access to irrigation. These new important cash crops in the region (Valdivia et al., 2013) are an opportunistic logical response in the near term, but are not sustainable in the long run due to the high demand for irrigation.

The view at the household level is presented in Table 41.1, based on data of 330 households interviewed in North and Central Altiplano of Bolivia. There are three landscapes in the North

Table 41.1 Livelihood strategies and capitals by landscapes and municipalities

Central and North Altiplano of Bolivia

1. Characteristics	Umala – Central Altiplano			Ancoraimes – Northern Altiplano			
	Low	Hillside	Sig	Low	Mid	High	Sig
Seasonal Migration % HH	23.4	40.4	★	13.8	46.0	57.1	★
Number of Households (HHs)	107	52		80	50	28	
2. Economic and Financial Capitals							
Sheep (head)	15.7	28.3	★	14.0	19.9	23.1	★
Cattle (head)	6.1	4.2	★	3.4	2.7	3.0	
Total Cash Income (Bs) Year	15,882	7,323	★	8,633	5,567.0	5,728	
Off-Farm Income (% HH)	21.5	32.7		42.5	48.0	42.9	
Total Income Agriculture (Bs)Year	12,855	4,336	★	4,337.4	1,946.5	1213.1	★
Has food security crop (chuño) %	87.9	84.6		32.5	48.0	85.7	★
3. Human Capital							
Years of Schooling (HH head)	6.7	5.6	★	6.9	6.1	5.6	
Age of Head of Household (years)	52.4	56.9		50.9	51.8	55.0	
Labour Available (Adult equivalents)	3.3	3.8		3.2	4.1	3.9	★
4. Natural Capital							
Hectares (Has.) of Land	12.0	5.9	★	2.9	1.7	2.4	
Alfalfa (Has.)	2.5	0.8	★	0.1	0.1	0.1	
Native pastures (Has.)	1.7	1.0		0.6	0.7	1.3	
Land in fallow (Has.)	4.4	3.4		1.4	0.5	0.7	★
Land in fallow (years)	5.7	5.1		3.2	2.4	3.1	
Number of crops planted	1.8	2.7	★	5.0	3.2	2.7	★
Varieties of potato (number)	3.6	4.5	★	1.8	2.0	2.0	
5. Climate Events							
Did climate events severely impact your production? %HHs = yes	97.2	59.6	★	98.8	100.0	100.0	
Strategies to cope with losses							
Sold their animals % HHs	7.5	5.8		10.0	2.0	14.3	
Used their savings % HHs	35.5	7.7	★	81.3	54.0	42.9	★
Borrowed cash or food % HHs	0.9	0.0		2.5	10.0	10.7	
Had to migrate for work % HHs	0.0	1.9		1.3	6.0	7.1	
6. Use of Financial Services							
Obtained a loan % HHs	11.2	5.8		16.3	14.0	3.6	

★ Differences between groups are significant at p < 0.05; ★★ HH: household
Bs: Boliviano/Exchange rate 6.5Bs = US$ 1.00

Source: Sustainable Agriculture and Natural Resource Management Adapting to Change in the Andes (LRA4) DataBase.

and two in the Central Altiplano. The household survey focused on the livelihood strategies, capitals, and perceptions of risks of farming households (Valdivia, 2007, 2001b).

Income from agriculture was higher in the semi-arid Central Altiplano, at the lower elevation where dairy production is a mayor activity. This region also had a high diversity of potato varieties, sold in the local market of Patacamaya, a growing commercial centre in the region. There was less migration in the central Altiplano, especially at the lower elevation. In the Northern

Altiplano near Lake Titicaca there was another community with high income. A large diversity of crops were planted, and these diversity decreased at higher elevations in this watershed. The share of income from agriculture is low. There was less migration in the Central Altiplano, especially in the lower elevation, while around 40% of households received income from migration.

The households in the Central Altiplano had more *chuño* stored (food security crop), and while being affected by climate events, did not need to use their animals or savings as much as the Northern Altiplano households to cope with events. In Ancoraimes, at the lower elevation onions are an important cash crop. There is less access to land in this region, and therefore more reliance on off-farm income, and more intensive use of land. Less labour also means less ability to invest in organic amendments to improve soil organic matter and less land in fallow (see Box 41.2).

Box 41.2   Soils research in the Altiplano

In the Bolivian Altiplano, management of soil resources has been greatly affected by multiple factors including changes in population density and urban migration, in animal husbandry, in market prices and proximity, in farmer knowledge and cropping practices, and in a high risk climatic environment for agriculture at high elevation which is shifting due to climate change (García et al., 2007; Valdivia et al., 2010). Many of these factors have led to greater but highly spatially variable soil degradation and lower soil quality. Based on surveys of representative agricultural communities in central and northern Bolivia, perceived soil-related problems included lower soil quality and soil fertility, excessive water and wind-induced soil erosion, insufficient soil moisture, and inadequate soil management practices including inappropriate tractor-based tillage practices, overgrazing by sheep, and insufficient soil fertility inputs (Zimmerer, 1993; Aguilera, 2010, 2014).

Cropping practices are diverse and dependent on the specific region and elevation in the Altiplano, but some observed changes that have occurred in communities in the central and north regions include increased cropping intensity for cash crops leading to decreased lengths of fallow periods in the crop rotation, greater use of tractor-based tillage, reduction in the planting of early season crops, such as quinoa (*Chenopodium quinoa* Willd.) and fava bean (*Vicia faba* L.), due to early season rainfall uncertainty, and loss of native shrub species (e.g., thola or *Parastrephia lepidophylla*) that have multiple uses to restore soil fertility during fallow periods (Aguilera et al., 2013a).

Over centuries, Andean farmers have developed successful traditional strategies to adapt to soil degradation and the effects of climate change at high elevations including a well-developed awareness and classification of soil resources, use of terraces and crop rotations, and development of livestock management practices to maximize utilization of manure and other waste materials (LeBaron et al., 1979). Some existing impediments to integrate new sustainable practices with these well-developed traditional systems include lack of uniform scientific-based soil survey information, insufficient affordable soil testing laboratory facilities with research-based soil test interpretations and fertilizer recommendations, inadequate agricultural extension available to all communities through public or non-governmental organizations, and poor infrastructure and resources for fertilizer use. Lack of decision aids for farmers including low-cost in-field crop and soil testing and other decision support information also hampers effective management (Aguilera et al., 2013b).

Among the multiple possible practices to improve soil resiliency and sustainability of agricultural systems in this environment are practices that are acceptable to the communities and increase soil quality and crop and livestock production (Peñaranda et al., 2011; Yucra and Valdivia, 2011). For example, increasing soil organic matter through use of practices, such as improved crop residue and manure management, improved fallow periods with planting of multipurpose native dryland vegetative cover and forages, and reduction in excessive tillage, could potentially improve soil fertility, soil water-holding capacity, soil biological activity, and soil physical properties while reducing soil erosion rates (Aguilera et al., 2012, 2013c; Gomez-Montano et al., 2013; Jintaridth, 2017). Development of simple and rapid low-cost portable soil and plant testing tools, such as the permanganate oxidizable active carbon test, with appropriate research-based interpretation and recommendation information and training could also assist farmers to make well-informed management decisions based on soil and plant status (Aguilera et al., 2013b). The results of such testing would also allow the communities in partnership with government and non-governmental organizations to assess the long-term impact of field management practices on soil quality and environmental stewardship.

Peter P. Motavalli

In the Central Altiplano, which is drier, the later onset of rains is challenging dryland farming systems. The risk-management practice of using early, middle, and late planting is being challenged, and later planting increases the risk of crop failure as these would have not reached maturity before the onset of frosts in February (Valdivia, Dunn, and Jetté, 1996, 2003).

There is need for viable early maturing varieties. Another important challenge related to food security are warming trends that challenge the production of freeze dried potatoes (*chuño*) that can be stored for many years. At the higher elevation the system is agropastoral and based on potato production for food security, and flocks of camelids. Because of the sub-humid climate and increasing minimum temperatures farming new varieties is possible, though challenged by an increase in pests and diseases (Garrett et al., 2011).

In the Peruvian Altiplano these climate trends are also expected to be severe in Puno, with temperature increase projected to reach 6 degrees Celsius (Sanabria and Lhomme, 2012). Using a regional climate model HadRM3P of the Hadley Centre (UK) to 2071–2100, Sanabria and Lhomme (2012) have studied the impact of climate change on Andean potato production and a bitter variety of potato, using A2 high emissions scenario, and find that yield deficits will be higher for the Andean potato, while the bitter varieties have mixed results. Potato is a major crop for food security and markets in this region, and throughout the Altiplano.

At the higher elevations in the Altiplano in Puno, over 4,000 m., pastoralists' livelihoods depend on camelids (alpacas and llamas), and sheep and cattle. Peatlands (*bofedales*) and dry grasslands (*pajonales*) are the main natural forage resource for this livelihood system. Agropastoral systems, combining crops and herding animals, are at lower elevations (starting at 3,800 metres) near Lake Titicaca in Peru and Bolivia, and at 3,700 metres in the Central Altiplano of Bolivia. The increases in minimum temperature are already changing the types of crops and varieties farmed, as is the case in the Northern Altiplano in Bolivia (Valdivia et al., 2013).

The impacts of climate variability on farming systems in the Altiplano are well documented (Sperling et al., 2008; Valdivia, 2004; Coppock and Valdivia, 2001; Valdivia et al., 2003). Climate change is presenting new challenges that reduce crop production, such as increased pressure from pests and plant diseases, and shifts in the onset and intensity of rains (Garrett et al., 2011,

2013). The increased temperature, extreme events such as drought, low temperatures, and snow events are impacting on the peatlands that support the livelihoods of pastoralists, while regulating water and sequestering carbon (Segnini et al., 2011).

Andean pastoralism

For Andean pastoralists living in the *puna*, *bofedales* are the primary and most important pasture zone (Flores-Ochoa, 1977; Palacios-Ríos, 1992; Rocha and Saez, 2003; Yager, 2015). *Bofedales* are vegetation systems characterized by cushion peat–forming plants in the high Andes (Ruthsatz 1993; Squeo et al., 2006; Cooper et al., 2010; Meneses et al., 2014). *Bofedales* are important not only as forage, but also support biodiversity, sequester carbon, and influence mountain hydrology (Cooper et al., 2015; Segnini et al., 2011; Chimner and Karberg, 2008; Earle et al., 2003). *Bofedales*, while integral to pastoral systems, are also dependent on continuous water supply. As trends of diminishing water availability continue, some related to climate change (Urrutia and Vuille, 2009), peatland systems are under threat of loss.

Pastoralism is the main economic activity of communities in the higher elevations in Puno. Families combine this economic activity that defines them as pastoralists, with migration as their other source of income (Turin and Valdivia, 2011). The Huenque-Ilave watershed, between 4,040 and 5,300 metres above sea level, is defined as dry *puna*. Apopata, a peasant community in this watershed, depends on the *bofedales* and dry grasslands to sustain their production and livelihoods. This community, according to the micro-region definition (Maletta, 2017) belongs to the micro-region Puna 5140. It is a region where extreme cold events happen almost every year, freezing the water in the peatlands (*bofedales*) and having a severe impact on the health of people and their animals. Apopata presents a very different reality compared to communities in the lower elevations, near Lake Titicaca, where potato production is important (see Table 41.2).

Santa María, a peasant community near Lake Titicaca in Peru, is in the micro-region *Suni* 4140 (Maletta, 2017). There are a total of 581,072 family farms in these region, where 349,991 family farms live below and at subsistence, accounting for 74% of the farming households. On the other hand, the *Puna-Jalca* region is home to 57,897 family farm units that live at or below subsistence representing 54% of the total in this micro-region. While food insecurity and low levels of income describe the situation of families in this region, Maletta (2017) finds that public investment in this region is lacking, with less infrastructure, potable water, and electricity.

The households in Apopata participate in markets selling their fibre to intermediaries, and their animals and meat at the local market in Mazocruz (Valdivia, Gilles, and Turin, 2013). Seasonal migration is another important source of income for pastoralist families (Turin and Valdivia, 2011). These statistics highlight the fact that along with climate stressors, extreme events, and climate change, poverty is a major constraint in negotiating adaptation. The markets producers sell to are dominated by intermediaries who set the prices for alpaca fibre, as do local buyers in the case of meat (Turin and Valdivia, 2011; Valdivia, 1992). In this context migration becomes an alternative livelihood strategy (see Table 41.2), which often results in women or the elderly remaining in charge of the pastoralist activities and managing and irrigating the *bofedales* (peatlands).

The case of the Sajama National Park in Oruro Bolivia

Around 50% of Bolivia's camelid population is found within the Department of Oruro, in the Central Altiplano, of which approximately 25% (totalling around 50,000 domesticated camelids)

Table 41.2 Characteristics and capitals of communities near the Lake and at high elevation, Huenque Ilave Watershed, Puno Altiplano, Peru (2009)

Household Capitals\Puno Communities	Near the Lake	High Elevation	Sig
1. Characteristics:			
Monthly Household (HH★★) Income (Soles)	163.80	270.60	★
Seasonal Migration (% HH)	51.10	36.8	
Number of households	47	68	
2. Economic and Financial Capitals			
Sheep/ total number (head)	8.9	26.2	★
Cattle/ total number (head)	3.3	0.2	★
Income from sales of crops (% HH)	38.30	2.90	★
Income from livestock products (% HH)	72.30	79.40	
Earns income for work outside the community (% HH)	44.70	33.80	
Prepares *chuño* (% HH)	91.50	–	
3. Human Capital			
Schooling Household Head (years)	8.9	6.8	★
Age of Household Head (years)	52.7	49.2	
Labour Available on the farm (Adult equivalent units)	1.8	2.3	
4. Natural Capital			
Alfalfa planted area (m^2)	0.452	0	
Percent who use fallow fields for pastures	0	79.40	
Has. of land available	17.9	5.4	★
Number of crops	2.3	0	
5. Climate Events			
Households that have been impacted by at least one climate event (frost, flood, freeze, hail) (%)	100.00	89.70	★
Strategies to cope with losses (% HH) of			
Sold their animals	77.30	100.00	★
Used their savings	54.50	45.60	
Received a loan or food	4.50	0	
Had to migrate for work	45.50	13.20	★

★★ HH: household;
★ significant difference between groups
Exchange rate: 3 Soles = $1.00

Prepared with data from the Sustainable Agriculture and Natural Resource Management Collaborative Research Support Program Long Term Research Project LTRA Adapting to Change in the Andes household survey conducted in Peru in 2009 (Household Survey Peru 2009).

are in Sajama National Park (PNS).[2] The Department of Oruro had 62,692 agricultural or farming production units in 2012, on 111,130 hectares of cultivated land (INE, 2015a), 79,950 heads of cattle, 1,187,850 sheep, and 1,069,169 camelids. The three animal species increased, compared to the 1950s census (INEa, 2015). Within PNS, herding animals include llama (approximately 49%), alpaca (~43%), sheep (8%), along with a few equines used for trekking (Yager, 2015). Pastoralism is the primary livelihood practice in Sajama National Park which was created in 1939, and has been for several centuries. Five Aymaran communities (Caripe, Sajama, Lagunas, Manasaya, and Papelpampa) are located in the park. The park covers an area of approximately 1,002 km², with an elevation range from around 4,200 m to 6,542 m, the height of *Nevado* Sajama, the highest mountain in Bolivia. *Nevado* Sajama is an *apu* (local deity), and holds a glacier record of 25,000 years of climate history (e.g. Hardy, Vuille, and Bradley, 1998, 2003; Thompson et al., 1998; Liu et al., 2005, 2007; Rosman et al., 2003; Thompson, 2003; Vuille et al., 2001). The rapid retreat of glacier Sajama and diminished snow cover – a clear indicator to locals of the impact of recent climate change (Agua Sustentable, 2013; Ulloa and Yager, 2007; Yager, 2015).

Although long-term records of meteorological data are scarce for the high Andes (Vuille et al., 2003), short-term datasets indicate average annual temperature to be 4.6° C and mean annual precipitation of 321 mm (maximum 430 mm and minimum 60 mm) (Beck et al., 2010). Over 200 vascular plants and 100 vertebrates are found in the park, including many endemic and endangered species (Beck et al., 2010). Broadly defined, there are six distinct ecological zones in the PNS: alpine/high-andean vegetation, *queñua* forests, *tholares* (shrublands), *pajonales* (tussock grasslands), *bofedales* (peatlands), and *colpares* (salt flats). The other ecological zones provide supplemental forage to varying degrees depending on access and season, but the primary production zones, defined as those areas that are managed by people for their livelihood and economic practices (Mayer, 2002), are the *bofedales* and *pajonales* (Yager et al., 2007).

The changing climate and environments

Andean communities have long managed climate extremes, and have locally adapted to the notably extreme conditions of cold events, arid conditions, and fragile soils characteristic of the altiplano (Brush, 1976; Browman, 1987). Nonetheless, local strategies to manage extreme climate events have become increasingly challenged over recent years. During workshops held in 2007 and in 2016 in Sajama National Park, the communities identified recent climate change and extreme events that include rapid deglaciation, increased temperature extremes, change in vegetation and pastures, expansion of desiccated areas, decrease in water resources, and change in diurnal, seasonal, and inter-annual weather patterns. As a result, the ability of local herders to maintain livestock health and production according to current management practices is becoming increasingly challenged by global climate change and the increased extremity of climate events. The fact that these communities are in the park means that there are rules that prohibit planting exotic species, and wild animals like the vicuña, Andean gato, and puma are protected. It also means there is local development of projects to provide alternative sources of income, like tourism, and selling handicrafts.

Extreme climate events impact on the *bofedales* and animals. Examples include the drought of 2016, and the snow event of June 2018 that resulted in lack of forage for multiple days. Access to markets for forage during this event was very difficult. Most *comunarios* (member of the *comunidad campesina*) sell in their communities rather than participate in other markets because it is too costly. Migration becomes an alternative source of income for the households. Adaptation strategies to extreme events and to their changing ecosystem have to be informed and designed with the communities in the park, as acting collectively is what the ecosystem will require, as the issues go beyond the *sayañas* (homestead) and communities.

Lessons from Altiplano landscapes and livelihoods

There are many and diverse livelihoods and strategies across the Altiplano region. The strategies pursued can be classified as agricultural, rural, and off-farm employment. The cases show that markets and farming that negotiates shocks lead to a greater participation in markets, such as milk, potatoes, and onions, and higher incomes. Distance to cities is a factor in the nature of the labour strategies, working off the farm near the household, or migrating off the farm for long periods of time. The labour market, and employment opportunities outside the family farm, play a major role in the nature of the livelihood strategies of rural households. While migration and labour mobility improve income and reduce the risk level of the economic portfolio in the short term, it may also reduce the likelihood of maintaining or adopting labour-intensive practices that contribute to a resilient ecosystem. There are high costs associated with sustainability, and trade-offs between short-term income, and long-term adaptation. Therefore identifying strategies, practices, and networks of information and collaboration, and engaging local decision makers are needed for sustainable the sustainable management of land and water under uncertain conditions driven by extreme events.

To address the limitations of selling at low prices in local markets, approaches like participatory marketing chains are being developed to tackle existing barriers to reaching high income consumers (Devaux et al., 2009; Figueroa-Armijos and Valdivia, 2017).

The way forward

The design of adaptation policies and programs has to engage local decision makers and their communities in order to be actionable and effective. The transformational element of the framework focuses on how change towards adaptation might take place. Can information about climate and markets improve portfolio selection and household's ability to accumulate assets? Can information about employment opportunities improve the chances of migration that contribute to building assets in Andean communities? And is it possible to address adaptation in the context of these fast paced changes driven by climate change, population dynamics, and ecosystem changes? Institutional, economic, and political constraints on the use of climate forecasts even when these are skillfully conducted (Broad, Pfaff, and Glantz, 2002; Eakin, 2000), highlights the fact that information alone may not address vulnerability, but that institutions and practices need to work together in order to act on this information. The research has shown the need to develop networks of collaboration (see Box 41.3) in order for information on forecasts about climate conditions to be reliable, and that participatory and translational processes are effective for this purpose (Valdivia et al., 2003; Patt Suarez and Gwata, 2005; Gilles and Valdivia, 2013). Information about the changing context combined with information about alternatives practices allow farmers to adjust, as the lessons on soils and climate forecasts show (see the cases in Boxes 41.2 and 41.3). Shortening the fallow cycle has a negative impact on soil organic matter, and diversity (Gomez et al., 2013), while investments in practices such as fallowing and use of organic manure contribute to ecological resilience (Jensen and Valdivia, 2013). Migration, a strategy to diversify the economic portfolio and face the multiple risks and uncertainties about climate, pests and diseases, and changing weather patterns, while contributing to reduced income risk in the short-term are challenging the ability to adapt long term, because it competes with investment in practices that contribute to a resilient ecosystem (Jensen and Valdivia, 2013; Turin and Valdivia, 2011). On the other hand, when off-farm employment is possible near rural towns households have a more diversified portfolio and higher incomes in agricultural and non-agricultural activities, pointing to a rural strategy (Bebbington, 1999) that contributes to both social and ecological resilience, consistent with other studies for Latin America (Berdegue, Bebbington, and Escobal, 2015), and therefore questioning economic growth as the path to development and adaptation to climate change.

Box 41.3   Combining local and scientific forecast knowledge in Bolivian Altiplano

One of the consequences of climate change is the increased frequency of extreme weather events; small farmer adaptation thus requires improved forecasting. In the Bolivian Altiplano the density of existing weather stations is less than 50% of that needed to make agricultural forecasts and a majority of these stations have data quality issues. As a consequence, less than 10% small Altiplano producers listen to meteorological forecasts on the radio or television (Gilles and Valdivia, 2009). Many have much more confidence in local indicators, but the use of this local knowledge is declining (Gilles, Thomas, and Valdivia, 2013; Gilles et al, 2014). Some indigenous farmers in the Altiplano have opportunistically adapted to climate change over the past thirty years by taking advantage of warmer temperatures to plant new crops and respond to market opportunities (Valdvia et al, 2010; Taboada et al., 2017), but their capacity for long-term adaptation is declining due to water shortages. Improved forecasting of extreme events would aid adaptation, informing decisions on production practices. However, because of the limitations of scientific forecasting, combining local knowledge and meteorological forecasting holds the most potential for reducing climate related risks. However, there remains considerable scepticism on the part of researchers, agronomists and many younger indigenous farmers about the utility of local knowledge. To address/evaluate these doubts, the use of traditional indicators for the management of weather related risk was evaluated in eight indigenous communities in the Bolivian Altiplano using meteorological data from small locally placed stations as well as evaluations by individual farmers and traditional forecasting experts. Initially, there was a concern on the part of many younger farmers and some technicians that environmental change was reducing the effectiveness of traditional forecast knowledge. An evaluation of the principal indicator used to predict growing season precipitation and four indicators used to determine optimal planting dates showed they were still valid. The indicator used to predict precipitation (minimum temperature on the 24th of June) explained 64% of annual precipitation and the flowering behaviour of plants was correlated with best planting dates. This appears to be related to relative humidity and changes in the behaviour of jet streams around the time of the winter solstice (Lupo et al., 2017). However, discussions with producers indicated that shorter-term forecasts and traditional forecasts of events like frost were less reliable than desired. In this area, farmers felt they would benefit by receiving both forecasts based on local knowledge and those from national metrological services, as long as they had help in interpreting them and could compare the forecast data to information obtained from weather stations placed in their own communities. Based on this local suggestion, a network was created using cell phones and the Whatsapp application. The network included 8 local experts with smartphones who reported to farmer research groups in their communities as well as agronomists and agro-climatologists from the Universidad Mayor de San Andres in La Paz. Farmers are now sharing and evaluating forecast information and supplementing national weather forecasts with forecasts from WeatherUnderground based on project weather stations.

Jere L. Gilles

The impacts of climate change, expressed in warming, extreme events, and changing weather patterns are a call to action. Public investments targeting rural regions in the Andes can support climate resilient development (Sperling et al., 2008), if these include the local decision makers in the design of what is needed. As this chapter reveals, local knowledge and networks of collaboration matter.

Notes

1 We are thankful to the communities in the Altiplano of Peru and Bolivia who have shared their time and participated in the research that informs this chapter. Shirley Rojas, doctoral student at the University of Missouri, helped with the development of the tables included in this chapter. Patricia Valdivia designed the figure of the landscapes in the Altiplano. Thanks to Mark Andrade, Peter P. Motavalli, and Jere L. Gilles who authored the material presented in Boxes 41.1, 41.2, and 41.3.
2 Parque Nacional Sajama

References

Adger, W. N. (2000) Social and ecological resilience: Are they related? *Progress in Human Geography* 24(3): 347–364.

Agua Sustentable (2013) *Plan de Adaptación al Cambio Climático del Parque Nacional Sajama.* La Paz: Nordic Development Fund, Diakonia and Christian Aid.

Aguilera, J. (2010) *Impacts of Soil Management Practices on Soil Fertility in Potato-Based Cropping Systems in the Bolivian Andean highlands.* Ph.D. dissertation, University of Missouri, Columbia, MO.

Aguilera, J., Motavalli, P., Gonzales, M. A. and Valdivia, C. (2012) Initial and residual effects of organic and inorganic amendments on soil properties in a potato-based cropping system in the Bolivian Andean Highlands. *American Journal of Experimental Agriculture* 2(4): 641–666.

———. (2014) Evaluation of rapid field test method for assessing nitrogen status in potato plant tissue in rural communities in the Bolivian Andean highlands. *Communications in Soil Science and Plant Analysis* 45(3): 347–361.

Aguilera, J., Motavalli, P., Valdivia, C. and Gonzales, M. A (2013a) Impacts of cultivation and fallow length on soil carbon and nitrogen availability in the Bolivian Andean Highland Region. *Mountain Research and Development* 33(4): 391–403.

———. (2013b) Response of a potato-based cropping system to conventional and alternative fertilizers in the Andean Highlands. *International Journal of Plant &Soil Science* 3: 139–162.

Andrade, M. F., Moreno, I., Calle, J. M, Ticona, L., Blacutt, L. Lavado-Casimiro, W., Sabino, E., Huerta, A., Aybar, C., Hunziker, S. and Brönnimann, S. (2018) Atlas – Clima y eventos extremos del Altiplano Central perú-boliviano/Climate and extreme events from the Central Altiplano of Peru and Bolivia 1981–2010. *Geographica Bernensia* 188. doi:10.4480/GB2018.N01

Aranda Pardo, I., García, M., Aparicio, M. and Cabrera, M. (2007) *Mecanismo Nacional de Adaptación al cambio Climático. Programa Nacional de Cambio Climático.* La Paz, Bolivia: Ministerio de Planificación del Desarrollo.

Bebbington, A. (1999) Capitals and capabilities: A framework for analyzing peasant viability, Rural livelihoods and poverty. *World Development* 27(12): 2021–2044.

Beck, S, Domic., A, Garcia, C., Meneses, R. I., Yager, K., Halloy. and S. (2010) *El Parque Nacional Sajama y sus Plantas.* Herbario Nacional de Bolivia-Fundacion PUMA: La Paz.

Berdegue, J., Bebbington, A. and Escobal, J. (2015) Growth, poverty and inequality in sub-national development: Learning from Latin America's Territories. *World Development* 73: 1–138.

Bradley, R. S., Vuille, M., Diaz, H. F. and Vergara, W. (2006) Threats to Water Supplies in the Tropical Andes. *Science*, 312, 1755–1756.

Broad, K., A., Pfaff, S. P. and Glantz, M. H. (2002) Effective and equitable dissemination of seasonal to interannual forecasts: Policy implications from the Peruvian fishery during El Niño 1997–98. *Climatic Change*, 54, 415–438.

Browman, D. (1987) *Arid Land Use Strategies and Risk Management in the Andes: A Regional Anthropological Perspective.* Boulder, CO and London: Westview Press.

Brown, M. E. and Funk, C. C. (2008) Food security under climate change. *Science* 319: 580–581.

Brush, S. B. (1976) Man's use of an Andean ecosystem. *Human Ecology* 4: 147–166.

Chambers, R. and Conway, G. R. (1992) *Sustainable Rural Livelihoods: Practical Concepts for the 21st Century.* Discussion Paper 296. IDS, London. 32pp.

Chimner, R. A., and Karberg, J. M. (2008) Long-term carbon accumulation in two tropical mountain peatlands, Andes Mountains, Ecuador. Mires and Peat 3: Art. 4. http://www.mires-and-peat.net/pages/volumes/map03/map0304.php

Cooper, D. J., Wolf, E. C., Colson, C., Vering, W., Granda, A., and Meyer, M. (2010) Alpine Peatlands of the Andes, Cajamarca, Peru. Arctic, Antarctic, and Alpine Research, 42(1), 19–33. doi:10.1657/1938-4246-42.1.19

Cooper, D, Kaczynski, K, Slayback, D, Yager, K (2015) Growth and organic production in peatlands dominated by Distichia muscoides, Bolivia. Arctic, Antarctic and Alpine Research 47 (3): 505–510. https://doi.org/10.1657/AAAR0014-060

Coppock, D. L. and Valdivia, C. (2001) (eds) *Sustaining Agropastoralism on the Bolivian Altiplano: The Case of San Jose Llanga*. Rangeland Resources Department, Utah State University, Logan, UT. 266 pp. http://digitalcommons.usu.edu/cgi/viewcontent.cgi?article=1051&context=usufaculty_monographs

———, Fernández-Giménez, M., Hiernaux, P. Huber-Sannwald, E., Schloeder, C., Valdivia, C., Arredondo, J. T., Jacobs, M., Turin, C. and Turner, M. (2017) Rangelands in developing nations: Conceptual advances and societal implications. In D. D. Briske (ed) *Rangeland Systems: Processes, Management and Challenges*. Cham: Springer Open, pp. 596–642.

Cruz, P., Winkel, T., Marie-Pierre, L., Bernard, C., Egan, N., Swingedouw, D. and Joffre, R. (2017) Rainfed agriculture thrived despite climate degradation in the pre-Hispanic arid Andes. *Science Advances* 3(12): e1701740 11pp. doi:10.1126/sciadv.1701740

de Haan, L. J. (2000) Globalization, localization and sustainable livelihood. *Sociologia Ruralis* 40(3): 339–365.

Diaz, H.F. and Graham, N.E. (1996) Recent changes in tropical freezing heights and the role of sea surface temperature. Nature 383: 152–155.

Devaux, A., Horton, D., Velasco, C., Thiele, G., López, G., Bernet, T., Reinoso, I. and Ordinola, M. (2009) Collective action for market chain innovation in the Andes. *Food Policy* 34: 31–38.

Devenish, C. and Gianella, C. (2012) (eds) 20 years of Sustainable Mountain Development in the Andes - from Rio 1992 to 2012 and beyond. *Regional Report 2012*. Consorcio para el Desarrollo Sostenible de la Ecorregión Andina -FAO, May. www.fao.org/fileadmin/user_upload/mountain_partnership/docs/ANDES%20FINAL%20Andes_report_eng_final.pdf

Eakin, H. (2000) Smallholder Maize Production and Climatic Risk: A Case Study from Mexico. *Climatic Change* 45: 19–36.

Earle, L., Warner, B., Aravena, R. (2003) Rapid development of an unusual peat-accumulating ecosystem in the Chilean Altiplano. *Quaternary Research*, 59(1), 2–11. doi:10.1016/S0033-5894(02)00011-X

Ellis, F. (1998) Household strategies and rural livelihood diversification. *The Journal of Development Studies* 35(1): 1–38.

FAO (2012) Peru: Nota de Analisis Sectorial. Corporación Andina de Fomento (CAF). Food and Agricultural Organization. Rome. 37pp. http://www.fao.org/3/a-ak169s.pdf

Flores-Ochoa, J. (1977) *Pastores de Puna: Uywamichiq Punarunakuna*. Lima: Instituto de Estudios Peruanos.

Figueroa-Armijos, M. and Valdivia, C. (2017) Sustainable innovation to cope with climate change and market variability in the Bolivian Mountains. *Innovation and Development* 7: 17–35.

García, M., Raes, D., Jacobsen, S. E. and Michel, T. (2007) Agroclimatic constraints for rainfed agriculture in the Bolivian Altiplano. *Journal Arid Environments* 71: 109–121.

Garrett, K. A., Dobson, A., Kroschel, J., Natarajan, B., Orlandini, S., Randolph, S., Tonnang, H. E. Z. and Valdivia, C. (2013) The effects of climate variability and the color of weather tie series on agricultural diseases and pests, and on decisions for their management. *Agricultural and Forest Meteorology* 170: 216–227.

———, Forbes, G. A., Savary, S., Skelsey, P., Sparks, A. H., Valdivia, C., van Bruggen, A. H. C., Willocquet, L., Djurle, A., Duveiller, E., Eckersten, H., Pande, S. Vera Cruz, C. and Yuen, J. (2011) Complexity in climate change impacts: An analytical framework for effects mediated by plant disease. *Plant Pathology* 60: 15–30.

Gilles, J. L., Thomas, J., Valdivia, C. and Yucra Sea, E. (2013) Where are the laggards? Conservers of traditional knowledge in Bolivia. *Rural Sociology* 78(1): 51–74.

Gilles, J. L. and Valdivia, C. (2009) Local forecast communication in the Altiplano. *Bulletin of the American Meteorological Society* 90(1): 85–91.

Gilles, J. L., Yucra, E., Garcia M., Quispe, R. Yana, G. and Fernandez, H. (2014) Factores de pérdida de conocimientos de uso de los indicadores climáticos locales en comunidades del Altiplano Norte y Central. *Revista de Investigación e Innovación Agropecuaria y de Recursos Naturales* 2(1): 7–15.

Gomez- Montano, L., Jumpponen, A., Gonzales, M. A., Cusicanqui, J., Valdivia, C., Motavalli, P., Herman, M. and Garrett, K. (2013) Do bacterial and fungal communities in soils of the Bolivian Altiplano change under shorter fallow periods? *Soil Biology and Biochemistry* 65: 50–59.

Hardy, D. R., Vuille, M., and Bradley, R. (2003). Variability of snow accumulation and isotopic composition on Nevado Sajama, Bolivia. *Journal of Geophysical Research*, 108, 46–93.

Hardy, D. R., Vuille, M., Braun, C., Keimig, F., and Bradley, R. (1998). Annual and daily meterological cycles at high altitude on a tropical mountain. *Bulletin of the American Meteorological Society*, 79, 1899–1913.

Hunziker, S., Brönnimann, S., Calle, J., Moreno, I., Andrade, M., Ticona, L., Lavado, W. and Huerta, A. (2018) Effects of undetected data quality issues on climatological analyses. *Climate of the Past* 14: 1–20.

———, Gubler, S., Calle, J., Moreno, I., Andrade, M., Velarde, F., Ticona, L., Carrasco, G., Castellon, Y., Oria, C., Croci- Maspoli, M., Konzelmann, T., Rohrer, M. and Brönnimann, S. (2017) Identifying, attributing, and overcoming common data quality issues of manned station observations. *International Journal of Climatology* 37: 4131–4145.

INEa. (2015a) *Censo Agropecuario 2013 La Paz*. Instituto Nacional de Estadística, Bolivia. December. 446p.

———. (2015b) *Censo Agropecuario 2013 Oruro*. Instituto Nacional de Estadística, Bolivia. December. 338pp.

Jensen, N. and Valdivia, C. (2013) Exploración de la relación entre las estrategias de vida y la resiliencia social y ecológica en el Altiplano de Bolivia. In E. Jiménez (ed) *Adaptación y Cambio Climático en Bolivia*. Postgrado en Ciencias del Desarrollo CIDES Universidad Mayor San Andrés. La Paz, Bolivia, pp. 219–245.

Jiménez, E., Gilles, J. L. Romero, A. and Valdivia, C. (2013) Cambio climático, diversidad de papa y conocimiento local en el Altiplano noliviano. In G. C. Delgado, M. Espina and H. Sejenovich (eds) *Crisis Socio-Ambiental y Cambio Climático*. Buenos Aires: CLACSO, pp. 174–194.

Jintaridth, B. (2017) *Assessing Soil Carbon and Soil Quality for Sustainable Agricultural Systems in Tropical Hillslope Soils Using Spectroscopic Methods*. Ph.D. dissertation, University of Missouri, Columbia, MO.

Liu, K. B., Reese, C. A., and Thompson, L. G. (2005). Ice-core pollen record of climatic changes in the central Andes during the last 400 yr. *Quaternary Research*, 64(2), 272–278.

Liu, K. B., Reese, C. A., and Thompson, L. G. (2007). A potential pollen proxy for ENSO derived from the Sajama ice core. *Geophysical Research Letters*, 34(9).

Kaser, G. (1999) A review of the modern fluctuations of tropical glaciers. *Global Planetary Change* 22: 93–103.

Korner, C. (2003) *Alpine Plant Life*. 2nd Ed. Berlin: Springer.

LeBaron, A., Bond, L. K., Aitken, P. S. and Michelson, L. (1979) An explanation of the Bolivian highlands grazing: Erosion syndrome. *Journal of Range Management* 32: 201–208.

Lobell, D. B., Burke, M. B., Tebaldi, C., Mastrandrea, M. D., Falcon, W. P. and Naylor, R. L. (2008) Prioritizing climate change adaptation needs for food security in 2030. *Science* 319: 607–610.

Lupo, A. R., Garcia, M., Rojas, K. and Gilles, J. (2017) ENSO related seasonal range prediction over South America. *Proceedings of the Second International Electronic Conference on Atmospheric Sciences*, 16–31 July, online. http://sciforum.net/conference/ecas2017

Maletta, H. (2017) La pequeña agricultura familiar en el Perú. Una tipología microrregionalizada. *IV Censo Nacional Agropecuario 2012: Investigaciones para la toma de decisiones en políticas públicas*. Libro V. Lima, FAO.

Markowitz, L. and Valdivia, C. (2001) Patterns of technology adoption at San José Llanga: Lessons in agricultural change. In L. D. Coppock and C. Valdivia (eds) *Sustaining Agropastoralism on the Bolivian Altiplano: The Case of San Jose Llanga*. Logan, UT: Rangeland Resources Dept., USU, pp. 239–256.

Mayer, E. (2002) *The Articulated Peasant: Household Economies in the Andes*. Oxford: Westview Press.

Meneses, R.I., Loza Herrera S., Lliully A., Palabral, A. and Anthelme, F. (2014). Métodos para cuantificar diversidad y productividad vegetal de los bofedales frente al cambio climático, *Ecología en Bolivia*, 49(3): 42–55.

North, D. (1993) The new institutional economics and development. *Econ Papers*. http://econwpa.repec.org/eps/eh/papers/9309/9309002.pdf (Accessed 8 July 2018).

Orlove, B. S., Chang, J. C. H. and Canelik, M. A. (2000) Forecasting Andean rainfall and crop yield from the influence of Pleiades visibility. *Nature* 403: 68–71.

Ostrom, E. (2007) A diagnostic approach for going beyond panaceas. *PNAS* 104(39): 15181–15187.

Palacios-Ríos, F. (1992) Pastizales de regadío para alpacas en la puna alta (el ejemplo de Chichillapi). In: Morlon, P (ed) *Comprender la agricultura campesina en los Andes Centrales*, Perú y Bolivia. Instituto Francés de Estudios Andinos Centro de Estudios Regionales Andinos, Bartolomé de las Casas, Lima and Cusco, pp. 207–213.

Patt, A., Suarez, P. and Gwata, C. (2005) Effects of seasonal climate forecasts and participatory workshops among subsistence farmers in Zimbabwe. *PNAS* 102(35): 12623–12628. www.pnas.org.cgi. doi/10.1073/pnas.0506125102

Peñaranda, M., Valdivia, C., Cusicanqui, J., Miranda, R., García, M., Navia, F. and Yucra, E. (2011) Prácticas y estrategias en respuesta al riesgo climático en agroecosistemas vulnerables de la región andina. *Proyecto Prácticas y Estrategias de Respuesta a los Cambios Climáticos y del Mercado en Agroecosistemas Vulnerables del Programa SANREM-CRSP Compendio 2006–2009*. Universidad Mayor San Andrés. La Paz, Bolivia, pp. 23–31.

Price, M. (2015) *Mountains: A Very Short Introduction*. Oxford: Oxford University Press.

Rabatel, A., Francou, B., Soruco, A., Gomez, J. Cáceres, B., Ceballos, J. L., Basantes, R., Vuille, M., Sicart, J. E., Huggel, C., Scheel, M., Lejeune, Y., Arnaud, Y., Collet, M., Condom, T., Consoli, G., Favier, V., Jomelli, V., Galarraga, R., Ginot, P., Maisincho, L., Mendoza, J., Ménégoz, M., Ramirez, E., Ribstein, P., Suarez, W., Villacis, M. and Wagnon, P. (2013) Current state of glaciers in the tropical Andes: A multi-century perspective on glacier evolution and climate change. *The Cryosphere* 7(1): 81–102.

Rocha O. and C. Sáez (2003) (eds) *Uso Pastoril en Humedales Altoandinos: Talleres de Capacitación para el Manejo Integrado de los Humedales Altoandinos de Argentina, Bolivia, Chile y Perú*. La Paz: Convención RAMSAR, WCS-Bolivia.

Rosman, K. J. R., Hong, S., Burton, G., Burn, L., Boutron, C. F., Ferrari, C. P., et al. (2003). Pb and Sr isotopes from an ice-core provides evidence for changing atmospheric conditions at the Sajama icecap, South America. *Journal De Physique Iv* 107: 1157–1160.

Ruthsatz, B. (1993) Flora and ecological conditions of high Andean peatlands of Chile between 18°00' (Arica) and 40°30' (Osorno) south latitude. *Phytoenologia* 25: 185–234.

Sanabria, J. and Lhomme, J. P. (2012) Climate change and potato cropping in the Peruvian Altiplano. *Theoretical and Applied Climatology* 112: 3–4.

Segnini, A., Posadas, A., Quiroz, R. Milori, D. M. B. P., Vaz, C. M. P. and Martin-Neto, L. (2011) Soil carbon stocks and stability across an altitudinal gradient in southern Peru. *Journal of Soil and Water Conservation* 66(4): 213–220.

Seimon, T. A., Seimon, A., Daszak, P., Halloy, S., Schloegel, L., Aguilar, C., Sowell, P., Hyatt, A., Konecky, B., and Simmons, J. E. (2007). Upward range extension of Andean anurans and chytridiomycosis to extreme elevations in response to tropical deglaciation. *Global Change Biology* 13: 288–299.

Seimon, T.A., Seimon, A., Yager, K., Reider, K., Delgado, A., Sowell, P., Tupayachi, A., Konecky, B., McAloose, D., and Halloy, S. (2017). Long-term monitoring of tropical alpine habitat change, Andean anurans, and chytrid fungus in the Cordillera Vilcanota, Peru: Results from a decade of study. *Ecology and Evolution* 7: 1527–1540.

Seth, A., Thibeault, J., García, M. and Valdivia, C. (2010) Making sense of 21st century climate change in the Altiplano: Observed trends and CMIP3 projections. *Annals of the Association of American Geographers* 100(4): 835–865.

Soruco, A., Vincent, C., Francou, B., and Gonzalez, J.F. (2009) Glacier decline between 1963 and 2006 in the Cordillera Real, Bolivia. *Geophysical Research Letters* 36(3): 1–6.

Sperling, F., Valdivia, C., Quiroz, R., Valdivia, R., Angulo, L. Seimon, A. and Noble, I. (2008) *Transitioning to Climate Resilient Development – Perspectives from Communities of Peru*. Climate Change Series No. 115. The World Bank Environment Department Papers. Sustainable Development Vice Presidency. Washington, DC. 103pp.

Squeo, F., Warner, B., Aravena, R. and Espinoza, D. (2006) Bofedales: high altitude peatlands of the central Andes. *Revista Chilena de Historia Natural* 79(2): 245–255.

Taboada, C., Gilles, J. L., Garcia, M., Yucra, E., Pozo, O. and Rojas, K. (2017) Can warmer be better? Changing production systems in three Andean eco-systems in the face of environmental change. *Journal of Arid Environments* 147: 144–154.

Thibeault, J. M., Seth, A. and Garcia, M. (2010) Changing climate in the Bolivian Altiplano: CMIP3 projections for temperature and precipitation extremes. *Journal of Geophysical Research* 115(D08103): 1–18.

Thompson, L. G., Mosley-Thompson, E., Davis, M. E., Lin, P. N., Henderson, K., & Mashiotta, T. A. (2003) Tropical glacier and ice core evidence of climate change on annual to millennial time scales. *Climatic Change* 59(1–2): 137–155.

Turin, C. and Valdivia, C. (2011) Off-farm work in the Peruvian Altiplano: Seasonal and geographic considerations for agricultural and development policies. In S. Deveraux, R. Sabates-Wheeler and R. Longhurst (eds) *Seasonality, Rural Livelihoods and Development*. London: Earthscan.

Ulloa, D. and Yager, K. (2007). *Memorias del Taller "Cambio Climático: Percepción Local y Adaptaciones en el Parque Nacional Sajama".* La Paz: Conservation International-Bolivia. Available at: http://www.climate2008.net/?a1=pap&cat=3&e=37.

Urrutia, R. and Vuille, M. (2009) Climate change projections for the tropical Andes using a regional climate model: temperature and precipitation simulations for the end of the 21st century. *Journal of Geophysical Research: Atmospheres,* 114 (D02108), 1–15. https://doi.org/10.1029/2008JD011021

Valdivia, C. (1992) Assessing the impact of policy on peruvian small ruminant production systems. *Development Studies Paper Series.* Winrock International Institute for Agricultural Development. Morrilton, AR. 24pp.

———. (2001a) Gender, livestock assets, resource management, and food security: Lessons from the SR-CRSP. *Agriculture and Human Values* 18(1): 27–39.

———. (2001b) Household socioeconomic diversity and coping response to a drought year at San José Llanga. In D. L. Coppock and C. Valdivia (eds) *Sustaining Agropastoralism on the Bolivian Altiplano: The Case of San Jose Llanga.* Logan, UT: Rangeland Resources Dept., Utah State University, pp. 217–237.

———. (2004) Andean livelihoods and the livestock portfolio. *Culture and Agriculture* 26(1 & 2): 19–29.

———. (2007) LTRA-4: Adapting to change in the Andes: Practices and strategies to address climate and market risks in vulnerable agro-ecosystems. In T. Dillaha and K. Moore (Report Coordinators) *SANREM CRSP Annual Report 2007.* Blacksburg, VA: USAID and Virginia Tech, pp. 93–146.

——— and Barbieri, C. (2014) Experiential agritourism: A sustainable strategy for adapting to climate change in the Andean Altiplano. *Tourism Management Perspectives* 11: 18–25.

Valdivia, C., and Gilles, J. L. (2001) Gender and resource management: Households and groups, strategies and transitions. *Agriculture and Human Values* 18(1): 5–9.

Valdivia, C., Seth, A., Gilles, J. L., García, M., Jiménez, E., Cusicanqui, J., Navia, F. and Yucra, E. (2010) Adapting to climate change in Andean ecosystems: Landscapes, capitals, and perceptions shaping rural livelihood strategies and linking knowledge systems. *Annals of the Association of American Geographers* 100(4): 818–834.

———, Gilles, J. L. and Turin, C. (2013) Andean pastoral women in a changing world: Challenges and opportunities. *Rangelands* 35(6): 75–81.

———, Jetté, C., Markowitz, L., Céspedes, J., de Queiroz, J., Murillo, C. and Dunn, E. (2001) Household economy and community dynamics at San José Llanga. In Coppock, D. L., and C. Valdivia (eds) *Sustaining Agropastoralism on the Bolivian Altiplano: The Case of San Jose Llanga.* Logan, UT: Rangeland Resources Dept., Utah State University, pp. 117–165.

———, Jetté, C., Quiroz, R. and Espejo, R. (2003) Coping and adapting to climate variability: The role of assets, networks, knowledge and institutions. *Insights and Tools for Adaptation: Learning from Climate Variability.* National Oceanic and Atmospheric Administration (NOAA) Office of Global Programs, Climate and Societal Interactions. Washington, DC, pp. 189–199.

Valdivia, C., Danda, M. K. Sheikh, D., James, H. S., Gathaara, V., Mbure, G., Murithi, F., and Folk. W. (2014) Using translational research to enhance farmers' voice: a case study of the potential introduction of GM cassava in Kenya's coast. *Agriculture and Human Values* 31(4): 673–681.

Valdivia, C., Dunn, E. and Jetté, C. (1996) Diversification as a risk management strategy in an Andean agropastoral community. *American Journal of Agricultural Economics* 78: 1329–1334.

Valdivia, C., Thibeault, J., Gilles, J. L., García, M. and Seth, A. (2013) Climate trends and projections for the Andean Altiplano and strategies for adaptation. *Advances in Science and Research* (33): 69–77.

Vergara, W., Deeb, A. M., Valencia, A. M., Bradley, R., Francou, B., Zarzar, A., et al. (2007a) Economic impacts of rapid glacier retreat in the Andes. *EOS, Transactions, American Geophysical Union* 88: 261–268.

Vergara, W., Kondo, H., Pérez Pérez, E., Méndez Pérez, J. M., Magaña Rueda, V., Martínez Arango, M. C., et al. (2007b) *Visualizing Future Climate in Latin America: Results from the applications of the Earth Simulator.* World Bank, Working Paper 30.

Vuille, M., Bradley, R. S., Werner, M., & Keimig, F. (2003). 20th century climate change in the tropical Andes: Observations and model results. *Climate Change* 59: 75–99.

Wilson, G. A. (2012) Community resilience, globalization, and transitional pathways of decision-making. *Geoforum* 43: 1218–1231.

World Bank. (2010) *Adaptation to Climate Change Vulnerability Assessment and Economic Aspects – Plurinational State of Bolivia (English).* http://documents.worldbank.org/curated/en/642891468162260845/Adaptation-to-climate-change-vulnerability-assessment-and-economic-aspects-BR-Plurinational-state-of-Bolivia (Accessed 8 July 2018).

Yager, K. (2015) Satellite imagery and community perceptions of climate change impacts and landscape change. In J. Barnes and M. Dove (eds) *Climate Cultures: Anthropological Perspectives on Climate Change.* New Haven: Yale University Press, pp. 146–168.

Young, K. and Lipton, J. K. (2006) Adaptive governance and climate change in the tropical highlands of Western South America. *Climatic Change* 78: 63–102.

Yucra, E. and Valdivia, C. (2011) La evaluación participativa como herramienta de investigación y desarrollo participativo. *Proyecto Prácticas y Estrategias de Respuesta a los Cambios Climáticos y del Mercado en Agroecosistemas Vulnerables del Programa SANREM-CRSP Compendio 2006–2009.* Universidad Mayor San Andrés. La Paz, Bolivia, pp. 15–22.

Zimmerer, K. S. (1993) Soil erosion and labor shortages in the Andes with special reference to Bolivia 1953–1991: Implications for conservation with development. *World Development* 21(10): 1659–1675.

PART VII

Latin American cities

42

JUST ANOTHER CHAPTER OF LATIN AMERICAN GENTRIFICATION

Ernesto López-Morales
Translated by Anna Holloway

Introduction

The anglicism "gentrification" is a term that the globalized urban societies of Latin America are more and more familiar with, especially those activists fighting for the right to housing (Lees et al., 2016). The phenomenon is linked to a spatial replacement of "recentralised" zones promoted by local and global financial/real estate capital and facilitated by entrepreneurial public policies involving the commodification and commercialization of housing and land.

The economic, social, cultural, and political impact of gentrification – especially on popular and even "middle-class" homes – is well known, for it is a process that leads to displacement and social exclusion (López-Morales et al., 2016). With few exceptions, Latin America is a region where the right of poor urban populations to housing is not being respected, as the upper-class colonization of historically deteriorated neighbourhoods advances in the context of a financialised regional economy (Rolnik, 2015). The Latin American context is even more complex, insofar as it is also the region with the highest global rate of urbanization (80.1%).[1]

The conflictive class shifts in the land and houses of Latin America, as well as their repercussions, are evidence opposing those who observe the course of gentrification in different parts of the world with scepticism and consider it as pertaining only to cities of the North Atlantic cultural region (that is, cities in the USA, the UK, and Canada, traditionally associated to this phenomenon; for this "sceptical" approach see Ghertner, 2015; Maloutas, 2017). However, gentrification has mutated. Phenomena unfolding today in cities of the Global South resemble processes that have taken place in many cities of the Global North, such as public space privatization and securitization, slum removals, state-led gentrification, violent police-led evictions amidst housing privatizations, displacement, and exclusion and so on (Lees, Shin, and López-Morales, 2015).

This chapter offers a brief and comparative approach to four specific cases of gentrification in Latin America; they have been deliberately chosen for their dissimilarities and are the object of analysis in an ongoing comparative research project. The cases are: a) the Rio de Janeiro Port Area, until recently abandoned and now renamed as "Porto Maravilha" by its promoters); b) the industrial neighbourhood of Parque Patricios in central-south Buenos Aires, today transformed into the city's emerging "Technological District"; c) the case of Colonia Juárez in Mexico City, currently very present in the media; and d) the area surrounding the centre of Santiago de Chile, including ten traditionally popular communities[2] which are currently experiencing, to different degrees, a socioeconomic substitution of the population through real estate "verticalization."

The selection criteria for these cases was to avoid established commonplaces relating to gentrification in Latin America, such as their exclusive occurrence in "heritage sites" or old city centres, the "*favela* gentrification" through police pacification taking part in Brazil (Gaffney, 2016; Cummings, 2015), or symbolic gentrification in Mexico and Argentina (Janoschka, Sequera, and Salinas, 2014; even with a certain air of *tango gentrification*); these ideas have been excessively generalized for the rest of the region. There has also been an abuse of the concept of "neo-liberalism" as a general factor that explains all regional cases without going into the details of more concrete mechanisms; this, too, is a limitation that the present chapter wishes to overcome (Janoschka, Sequera, and Salinas, 2014).

In all four cases, gentrification is the result of active entrepreneurial governance from the state as well as a rise in invested capital; and this is a phenomenon occurring more and more in the entire world. All cases, except for that of Buenos Aires, illustrate a market constituted by oversized and politically influential real estate corporations. The four cities are of a different scale (ranging from 7 to 22 million inhabitants), but they are all megacities and – in all cases, except that of Rio de Janeiro – the capital cities and major urban hubs of their respective nations.

In no case is this a representative sample of "Latin American gentrification," for there are other relevant cases (cities and neighbourhoods) in these and other countries in the region. To be more precise, this project of comparative analysis prioritizes the search for differences between the four cases, rather than similarities. This strategy actually allows us to give more importance to the existence of regularities between them (see the interesting approach to this type of research by Smith and Phillips, 2018).

Data have been collected through regular field visits between 2015 and 2017 as well as through consulting diverse sources of secondary material, with the collaboration and guidance of local academics and their teams.[3] Data analysis is focused on three principal themes: a) institutional arrangements, b) social conflict, and c) spatial normalization. These dimensions respond *grosso modo* to political economy approaches: the first one relates to the entrepreneurial state (Harvey, 1989; Lopez-Morales, 2015a), while the second and third are linked to the normative and abstract character of spatial representations and their dialectic relation of opposition with lived and representational space imposed by a private-state apparatus (Lefebvre, 1991).

The scale of the observation does not allow for more specific analysis, although certain variables were detected and developed in some detail: the existence of parastatal governance, the problem of the scale of the corporations involved, and territorial stigmatization as part of public policy. Furthermore, we also detected different formations of subalternity, interclass alliances with the goal of political protest, and different types of evidence of displacement and organized opposition to gentrification. At the level of spatial normalization, the policies implemented in said zones included practices such as city branding, spatial zoning, ad hoc investment in infrastructure to serve gentrification, and financial incentives for corporations. Also, there is a total absence of social housing in the public policies implemented in all four zones, in contrast with the discourses of the state apparatus on development and even privatizations.

Case overview

Colonia Juárez is an area located near the historical centre of Mexico City, in the Cuauhtémoc district. Although it has existed since the end of the 19th century, the elites started to abandon it after the 1910 revolution. By the 1980s, the neighbourhood was suffering from generalized decline and stagnation, a situation that was aggravated by the 1985 earthquake which destroyed around 80 buildings in this area alone (the 2017 earthquakes have been less destructive but have also caused significant damage to more than ten buildings). A significant spatial characteristic of this neighbourhood, which spreads across 160 hectares (below the average of the city's neighbourhoods), is the two-scale intervention that has taken place there. On the one hand is the large-scale real estate intervention at the rims of the neighbourhood (Reforma and Chapultepec Avenues); on the other, the small-scale real estate intervention that modifies the interior of the neighbourhood (Figure 42.1). The gentrification of Colonia Juárez also originates from the expansion of pre-existing dynamics of housing elitization in the famous Roma and Condesa

Figure 42.1　New real estate project in Colonia Juárez

Source: Author.

Figure 42.2   New City Government building in Parque Patricios, Buenos Aires (developed by the Normal Foster architect firm)

Source: Author.

neighbourhoods (Salinas, 2013). At the same time, many cafés, gourmet restaurants, designer stores, art galleries, and other businesses have established themselves in the area.

The Buenos Aires neighbourhood of Parque Patricios has traditionally been a working-class, industrial neighbourhood, spreading across almost 400 hectares bordering the city centre from the south. Historically, Parque Patricios has been a deteriorated neighbourhood with a strong industrial past. In the mid-2000s, the city government tried to breathe new life into the neighbourhood through the creation of the Technological District (as well as of other "thematic" districts, such as the Art District of the Barrio de La Boca neighbourhood, also experiencing processes of gentrification; see Gretel Thomasz, 2016). The aim was for Parque Patricios to become the home for corporations operating in the field of New Technologies of Information and Communication (NTICs) through tax exemptions, significant public investment in infrastructure and transport (new metro line H, bicycle lanes, Bus Rapid Transit corridors), the capitalization of public space, and the relocalization of government buildings (Figure 42.2).

The Port Area of Rio de Janeiro, called "Porto Maravilha" by its promoters, emerges in the context of the mega-events (2014 FIFA World Cup and 2016 Olympics) that took place in Brazil. However, the history of this place can be traced back centuries: the port was the city's entry point for the African slave trade and this has marked its historical identity as a point of fusion, the cradle of samba carioca and the home of African cultural institutions. With the restructuring of port activities during the 1970s, the area was gradually abandoned, and its residents became associated with prostitution, drug trafficking, and "criminal" violence. The large-scale transformation of the area has not changed this condition, particularly with regards to drug trafficking which re-emerged significantly in 2017 (Philips, 2018). The transformation began in 2009, with the updating of infrastructure for environmental sanitation, IT networks and light rail (VLT, Figure 42.3), the demolition of an elevated highway, the complete renovation of street seating, as well as the construction of cultural facilities (such as the Museum of Tomorrow and the Rio de Janeiro Art Museum) as part of the area's symbolic resignification. All of the abovementioned triggered an unprecedented process of valorization and real-estate speculation in the city.

The area surrounding the centre of Santiago was occupied in the mid-20th century by the politicized *Pobladores* movement and the efforts to construct social housing by a state with Keynesian intentions. Today, it is the access point for the so-called emerging middle class and

Figure 42.3　New VLT light rail travelling through a popular area in the Rio de Janeiro Port Area
Source: Author.

Figure 42.4   Gentrifying verticalization in the area surrounding the centre of Santiago de Chile
Source: Author.

home to new families, young professionals, and a certain elite of differentiated cultural consumption, amidst intense activity by private real estate companies that are drastically configuring the landscape (Figure 42.4). At the same time, it also houses tens of thousands of Latin American immigrants arriving to Santiago[4] in search of better opportunities; and it is the place that quickly expels many of them as well. A recent survey showed that 60% of the more than 40,000 homes located in informal settlements in the outskirts of Santiago are there as a direct result of the city's prohibitive high rents (see Techo Chile, 2017); the same situation is affecting the country's low-income houses that are also being expelled.

Gentrification: institutional arrangements

The participation of big transnational capital in consortia spearheading local reactivation through mega-projects is a recent phenomenon, launched at the end of the 1980s in Puerto Madero, Argentina (Garay et al., 2013). However, of the four cases studied here only the Port Area of Rio de Janeiro and Mexico's Colonia Juárez have followed this path and involve some type of parastatal association between the national or local government and global transnational real estate corporations. In the case of Rio we find, on the one hand, the consortium Porto Novo (created by the controversial transnational corporation Odebrech Infraestructura, OEA, and Carioca Christiane Nilsen Engenharia) with a state contribution in infrastructure valued at BRL 8 billion (USD 2.5 billion) and, on the other, the Secretary for Urban Development; this is the largest association of public and private sector to ever take place in Brazil. In the case of Mexico City, renamed as CDMX by the authorities in 2016, parastatal association takes place through the agency PROCDMX, a private holding created by the city Mayor Miguel Mancera whose major shareholder is the local government. The role of the agency is to create attractive and relatively safe conditions for investment, including the privatization of extensive areas of public space.

In the case of the urban interventions performed through public-private partnership in Rio de Janeiro, institutional guidelines dictate these should be financed through the sale of Certificates of Potential Additional Construction (CEPAC) for a value of BRL 3.5 billion (USD 1.1 billion). The CEPAC define the construction potential according to the floor area ratio (Smolka, 2013). However, history unfolded differently: in 2015, the limited purchase of these additional construction permits by private developers reflected the state of national political instability and the low levels of trust towards the public sector (Werneck, 2016); it also turned this operation into a financial failure, at least for the time being, pending a global reactivation of the economy.

The case of Mexico is different. The country's real estate market has been very dynamic during the past decade, attracting large-scale transnational corporations such as BBVA Bancomer, Grupo Dahnos, Abilia, Tresalia Capital, Desarrolladora del Parque, New York Life, and Grupo Sordo Madaleno, who have constructed high rise buildings mostly along Avenida Reforma, which borders with Colonia Juárez. However, there are developers operating within Colonia Juárez with investments of a much smaller scale: ReUrbano, Punto Destino, and Aransa Grupo Inmobiliario, who receive bank financing and more local and smaller financial contributions (Romero, 2016). However, these small real estate agents are very aggressive: they have renovated approximately one hundred buildings, supposedly rescuing architectural heritage; while in practice, they have resold central urban space at two, three, or four times the price it had just five years ago (Romero, 2016; Miranda, 2017). In fieldwork conducted in only 11 residential buildings, Romero (2016) identified 60 homes that had suffered direct displacement or were under pressure of displacement. On many occasions, they had been living in dangerous conditions as a result of a renewed interest of the landlords (often fuelled by real estate agencies), some of whom had been absent for a long time or were not even collecting rent. Pressure of displacement occurs because today, under the prevailing norms, the potential rent of a property can be much higher than the rent that is actually being charged. It costs approximately MXN 8,000 (USD 425) to rent a 60m² apartment; however, when renovated, real estate agencies can rent the same property at MXN 16,000 (USD 850) or even MXN 20,000 (USD 1,062). This brings to the neighbourhood new residents who are young, have no children, are wealthier and sometimes foreign, and can afford to pay higher rents than the original residents.

In Argentina there is less social trust in the banking sector (a result of the 2001 crisis), therefore the real estate industry has less financial leverage and requires a higher degree of state support through strong tax incentives and infrastructures, pinpointed by the government through the logic of the neighbourhood Economic Districts. Chile, on the other hand, has "mega-corporations" that greatly condition state actions (Garreton, 2017) following monopolistic practices and administrate the supply of middle-class housing in order to create a relative scarcity which will, in turn, lead to a rise in housing prices. As prices go up, the state is expected to broaden its subsidy mechanisms and ease restrictions on land use and construction. This is somewhat similar to the situation in Brazil, where the lack of corporate interest for CEPAC bonds forced the state to make regulations more flexible.[5] The economy of gentrification evolves in many different ways and involves a diverse group of agencies and stakeholders. However, from a structural logic, the goal is always the same: the accumulation and preservation of power by the dominant classes.

Unsurprisingly, in two out of the four cases studied, territorial stigmatization is used as a powerful mechanism to cause devaluation and create a rent gap (Slater, 2017). The Port Area of Rio de Janeiro has for decades been accused of being a nest for conflict and crime (although

Morro da Conceição, located at the heart of the area, is officially registered as "middle-class"). In Mexico City's Colonia Juárez, the stigma is expressed through the intentional concealment of the neighbourhood's popular character. This is reflected in the discourse of public authorities and private developers, who refer to the affluent white and middle classes as the only saviours of an area plunging into chaos and "going to waste." This attitude displays the disregard for the neighbourhood's current inhabitants, many of whom live in overcrowded and dangerous conditions which have been further aggravated by the recent 2017 and 2018 earthquakes.

In Santiago de Chile too we can observe an interesting discourse circulated by the state and the corporate elite: the private market is presented by its champions (Poduje et al., 2015) not only as more efficient but also as the guarantor of the right of the middle classes to the city. This narrative ignores the popular classes already living in the area, or those who could potentially be eligible for accommodation in the same space if different forms of access to housing were in place. At the same time, the affluent classes (including real estate entrepreneurs) surprisingly appear to be appealing to the same right to the city. In fact, this is a "right" to invest and speculate on housing and land, in a context where both actions are only moderately regulated.

The discourse of state entrepreneurialism in Buenos Aires is somewhat different. Part of the responsibility of a new state-led effort towards national productive and economic reactivation (in the form of the "Technological District") is placed on the location and its residents. And, in this, it finds a direct neighbourhood correlate, a dream whereby everyone (including the neighbourhood's middle-class residents) can one day become entrepreneurs through the commercial exploitation of their homes.

Social conflict

On many occasions, urban conflict in these cities is expressed through the stigmatization of the indigenous or black subject, who is seen as backward, conflictive, dirty, or primitive, compared to the white (to be more precise, *mestizo*) resident who is proper, well-educated, and a "citizen of the world." With the exception of Buenos Aires, all case studies have a high percentage of pre-Hispanic indigenous and/or Afro-descendant population. Therefore, it is in the "new urban forms" acquired by Mexico's Colonia Juárez and in the outskirts of the *favelas* confined by Brazil's Porto Maravilha that we find the clearest expression of what Mexican anthropologist Pablo Gaytan (2016) calls "whitening": the normalization of a space oriented towards the white consumer, eliminating deeply rooted cultural activities such as street food (truly an all-pervasive institution in Mexican public space). New urban space standards are created for new users who move around in an efficient, aseptic, high-tech (and more expensive) public transport system.

The concept of whitening also refers to the polysemy implied by the laundering of money,[6] of facades, of faces appearing in the public and residential space (Lopez-Morales, 2015a). It is a whitening by dispossession, for it deprives the disadvantaged of the right to the city.

Meanwhile, the average resident of Santiago de Chile does not recognize their *mestizo* indigenous/Hispanic heritage, expressed in the fairer complexion of the country's dominant classes. As a result, this difference is not considered an issue at the level of political conflict as occurs in the other Andean countries. This is, undoubtedly, a process of racial denial, the same phenomenon that Swanson (2007) observed in the Ecuadorian elite. Subalternity is a social configuration that differs greatly from one place to another: the four cities have very different historical backgrounds and have experienced dissimilar levels of industrialization and indigenous or black migration to the city.

At this point, a few lines must be dedicated especially to the case of Parque Patricios in Buenos Aires. It is hard to talk of displacement – be it direct or exclusionary (on this concept, see Slater, 2017) – when there is such a high availability of industrial land and a comparatively low presence of working-class housing. This explains the inactivity of local social movements in Parque Patricios in the face of the aggressive spatial restructuring that is taking place. In fact, specific, documented evidence of the displacement of low-income homes exists only in the cases of Mexico City (Romero, 2016), Santiago (López-Morales, 2015b), and Rio de Janeiro (Observatório das Metropoles, 2016). The first two are dominated by the intensive construction of private houses that exclude a big part, if not all, of the subaltern strata. In Rio de Janeiro, while the construction of high-end housing units is still incipient, we were able to detect 64 evictions of small properties which where until recently deliberately understudied by the city's government.

Special mention must be made of the strategic effectiveness of neighbourhood organization in the case of Colonia Juárez. This case reflects the mixed composition of Mexican society, in a country where the system of universal public education and the founding and revolutionary myths of the Mexican state value *mestizo* identity and, in a way, mitigate the rejection of the cohabitation of the numerous well-educated middle class and the indigenous and *mestizo* population in the same neighbourhood, a rejection very much present in other places. Plataforma 06600 is a characteristic exponent of the leadership role assumed by white middle-class tenants motivated by discourses of self-management and visions of "civic responsibility": they are in favour of transparency, against state corruption and the violent practices of real estate agencies, they act for the defence of patrimony and neighbourhood features, and defend the right of populations – particularly the poorer ones – to call a place their home. The platform has created initiatives that range from providing legal assistance to neighbours under pressure of displacement to proposing the inclusion of tenant rights in local regulations, such as the *derecho al tanto* (first choice to buy for tenants).

Damaris Rose (1984) was the first to speak of "marginal gentrification" as a possible explanation for this type of socio-political milieu rendered an externality of gentrification. In Colonia Juárez, residential succession has been dynamic and multifarious: today there is a coexistence and blending (and also a struggle to maintain this character) of identities of artists, indigenous inhabitants, homosexuals, electricians, pensioners, single mothers, plumbers, small merchants, young architects and/or designers and, generally speaking, individuals of a moderate level of income. We are talking of old and not so old residents in a city with a residential sector that is more dynamic than that of other regions. This platform also includes researchers associated to universities and research centres of all types, who articulate and bring to the table mobilizing theoretical viewpoints: the right to the city, the racial component that is inherent in "whitening" capitalism, displacement, normalizing imaginaries and the practices and policies of privatization, the aggressive operation of the real estate market, corruption within the state and private corporations, class conflict, gender-based violence in the city, and gentrification. Certain tactics of dissemination and efforts towards urbanism and collaborative or co-administered political participation could indicate increasingly interesting links to Zapatismo or other movements steering towards autonomy and self-government,[7] mostly through the use of participatory techniques and the reclaiming of urban space as a common good. There are also efforts to create an urban equivalent to the old Law of Customs and Traditions that is still in place in the country's rural indigenous areas, where each community elects its own authorities with the capacity of removal from office by general assembly mandate, a modality of direct and radical democracy that Plataforma 06600 is seeking to implement in the medium and long-term in the city.

The "organic" connection to researchers has been fruitful. An example is the activism associated with the invention of Santa Mari La Juarica, an icon that combines Mexico's cultural religious past which appeals to indigenous syncretism with a hipster visual element. It is on the one hand an emblem of the struggle against gentrification and real estate speculation (Delgadillo, 2017) and, on the other (paradoxically), a product of the artistic and marketing talent of certain leading members of this platform who possess high cultural capital. In fact, we are before a reification of the cynicism that characterizes the situation, for one could say that the "pioneering" gentrificators are defending the neighbourhood from a subsequent phase of corporate gentrification. The icon of Santa Mari La Juarica was then promoted through performances such as public processions, drawing the attention of international media to the gentrification taking place in the neighbourhood (Camhaji, 2017).

Spatial normalization

It was Henri Lefebvre (1991) who years ago – in his thesis on the spatial representations of bureaucracy, real abstractions of everyday life lived at street level – argued that urban space is permanently torn between preserving its social attributes and serving the desires of capital and the ruling power that shape it. The city branding applied to the three case studies has been an interesting strategy that in all cases has secured a certain degree of social and spatial control.

The designation of Parque Patricios in Buenos Aires as a "Technological District" effectively urges residents to assume individual responsibility and eventually turn into small entrepreneurs, using their home to support the neighbourhood's revival (two years ago, Parque Patricios appeared for the first time amongst Airbnb listings). For its part, the term Porto Maravilha[8] sounds more like the over-the-top imagination of major publicists in charge of signifying and staging this gigantic real estate operation for the media. This operation is also linked to the investment of mega-events organized in Rio de Janeiro, and links to similar mega-operations taking place in other global capitals with a longer history constantly emerge. Dockland Ocean View (http://riodockland.com.br) is the stellar example of this type of imaginary of toponymic association. For its part, the recently inaugurated (2016) CDMX acronym (an abbreviation of CiudaD de MeXico – Mexico City) is an even more far-reaching operation that affects the entire city space. The city leaves behind its former designation as Federal District (originating from the 19th century) and dons a striking name as the new city of business and surveilled security that brings to mind the I♥NY imaginary, that was revolutionary in the 1980s. This process is also associated to the custom-made political strategies of Mexico City Mayor Miguel Angel Mancera, who wishes to portray himself as the "modernizing" candidate for the nation's presidency.

The case of Santiago involves incentives for real estate development that do not require practices such as city branding, special zoning, tailor-made infrastructure investments, spatial securitization, and others present in the rest of the cases. The city's Metro network, constantly expanding since its inauguration in 1976, offers ease of mobility towards new areas. Nevertheless, the expansion of this network responds to an increasing demand for public urban mobility and is not necessarily a strategy designed specifically for the benefit of private housing projects. In fact, developers "pick" their intervention of choice amongst a great variety of possible locations close to Metro stations. However, they do so in packs, consuming the rent gaps of different neighbourhoods within five or ten years and saturating them through the elimination of desired environmental attributes, given the intense verticalization and scale of the construction projects (Lopez-Morales, 2015b). This is clear proof that, in Chile, the conditions benefiting the operation of real estate neoliberalism in the areas surrounding the city centre are not something new; they were defined decades ago, in 1990, when democracy was restored after the Pinochet era (1973–1990). Since then, the rules of the urban game have kept operating in favour of real estate capital, allowing for a system of regulation with significant environmental effects given its intense verticalization. "Frog leaps" are also very common in these residential agglomerations, as corporations easily abandon municipal districts when stricter regulation is implemented and establish themselves in other areas offering equal or higher flexibility.

There is also a financial market that is totally calibrated to avoid subprime mortgages; however, it displays a perilous tendency towards price inflation due to the high participation of the more affluent middle classes (supported by the banking sector) as "ant investors,"[9] leading to the exclusion of the two lower-income quintiles from any new residential unit on offer (Lopez-Morales, 2015b).[10] With the exception of deregulated supply through slum tenements or the so-called vertical ghettos (tall and crowded buildings with tiny residential units and a relatively low price per m^2), the process of widespread gentrification led by Chilean developers could be the main cause for the skyrocketing of informal settlements in the outskirts of the city (Techo Chile, 2017), currently accounting for 10% of the national housing deficit.

The case of Santiago, where the dynamic of residential verticalization in the areas surrounding the city centre walks hand-in-hand with a tendency to ascend the social ladder, clearly and physically shows what can happen when "classist" real estate capital prevails in a city and acts as a monopoly: there is an overwhelming change in scale and speed and a greater impact in terms of displaced homes, all with a less direct intervention of the state given that the conditions for the operation of this model were determined at least 25 years ago.

Conclusions

Gentrification in Latin America is not limited to the *favelas*, nor is it mostly "symbolic," "typical," or "cultural." It is, rather, the result of aggressive technocracies of spatial restructuring that aim at expanding real estate financial capital and, to a lesser extent, the service industry capital. While social "cleansing" is not necessarily articulated as an explicit or implicit goal within such technocracies, it is almost always an outcome. Following the tendency that began in the 1970s, the main objective of multiscalar framework of the neoliberal state is still the protection of private corporations. Furthermore, the media still reproduce ideological atavisms such as the "inefficiency" or "extreme corruption" of the state in order to justify the free flow and sacrosanct locational decisions made by private capital, very often without any kind of public scrutiny.

Rather than aiming at generalizations, this essay has discussed four dissimilar cases of gentrification in order to provide an analysis narrative that goes deeper into the particularities of each case. Today we see that partnerships between public authorities and private developers (PROCDMX, Porto Novo) are a rare combination, not present in all cases. However, what these structures have in common is the lack of transparency regarding the origin of the capital employed, the extent of private influence on public decision-making, and the privatization of land as the state's main asset put at the disposal of these corporate partnerships.

In at least three of the cases (with the exception of Santiago), the imagination of publicists runs wild as they officially rebaptize space through a process of city branding and, in doing so, downgrade the power of traditional street cultures. The slave trade and the old working-class, revolutionary, and *mestizo* traditions are seen as a thing of the past. New times call for all things technological, spectacular, marvellous; an iconic, post-modern architecture of overspending and an urban, postpolitical way of governing. Joint public-private administration displays limited transparency and accountability, while the discourse on the Right to the City is used as an instrument for entrepreneurs, new residents, investors, and developers big and small, but not for those who have a limited spending capacity or do not intend to obey the rules of the global market imposed by the urban state.

The differences between the case studies speak of the complex structuring of the imaginaries related to urban development in the subcontinent today. In no case has this intended to be a representative sample of "Latin American gentrification," although certain coincidences are indeed surprising. Once the make up of the pseudo-efficient city with the aseptic sensuality is removed, the face of a neoliberal, repressive, and authoritarian state emerges. The "whitening" of the urban space by dispossession, the conversion of any popular urban space into a replica of an international airport or a mall is a threat that everyone in these areas at least perceives. Real estate money is male, white, hyperbolic, and circulates quickly, and the urban space is being reconfigured in order to adjust to these conditions. The generalized lack of interest or of action on behalf of the state in the direction of producing space for the subaltern populations, be it social or popular housing, is also surprising. Architecture and urban design are two disciplines that are subjugated to satisfying fewer users: specific, global, and powerful ones and often outsiders. And the direct role of the entrepreneurial state through a parastatal consortium or an operation that is especially orchestrated to attain the imposition of gentrification as a public policy seems inevitable. Almost the only exception to this is the area surrounding the centre of Santiago, where – following decades of social disciplining in the market economy – institutional and social restrictions on private real estate are still very limited and the direct role of the state is much more ambiguous than in the rest of the cases.

Notes

1 World Bank (2018) World Urbanization Prospects: The 2018 Revision.
2 The municipalities of: Cerrillos, Estación Central, Independencia, Macul, Ñuñoa, Pedro Aguirre Cerda, Quinta Normal, Recoleta, San Joaquin, and San Miguel.
3 I would like to thank Jorge Blanco of the Universidad de Buenos Aires, Orlando Santos of the Universidad de Rio de Janeiro, and Luis Salinas of the Universidad Autónoma de México, as well as their respective teams. Any errors in the final text are my exclusive responsibility.
4 According to the 2015 survey on the country's socioeconomic situation (CASEN) 88.8% of the country's immigrants are Latin American (2.7% of the country's total population, compared to 1.7% in 2006) of which 30% were born in Peru, 13.6% in Colombia, and 11% in Argentina. 321,561 live in the Santiago Metropolitan Region. At the national level, 56.9% of the country's immigrants belong to the two lower-income quintiles (Ministerio de Desarrollo Social, 2015).
5 The CEPAC were finally purchased by the city government with the savings funds of civil servants, in what is one of the greatest recent scandals linked to this mega-operation.
6 The Spanish word *blanqueamiento* literally means "whitening" but is also used to refer to money laundering.
7 The Ruptura Colectiva blog and the work of reporter Demian Revart have been very revealing in this sense. See http://rupturacolectiva.com.
8 Literally Port Wonder.
9 And this in a country where pension prospects depend on personal savings (including the accumulation of properties).
10 *Inversionistas hormiga* ("Ant investors") have appeared in Chile in the past five years and have grown to the point of becoming an important category of homebuyer in the country's real estate market. They are individuals who purchase several properties to subsequently rent them out in order to receive high investment returns or even save for a pension plan.

References

Camhaji, E. (2017) *La santa que ahuyenta a los 'hipsters' del corazón de la Ciudad de*. México: Devotos de Santa Mari La Juaricua se movilizan contra el aburguesamiento y el blanqueamiento de Santa

María La Ribera y la Colonia Juárez. *El Pais*, 7 June. https://elpais.com/internacional/2017/06/02/mexico/1496433731_922169.html (Accessed 23 October 2017).

Cummings, J. (2015) Confronting favela chic: The gentrification of informal settlements in Rio de Janeiro, Brazil. In L. Lees, H. B. Shin and E. L. d ments in (eds) *Global Gentrifications: Uneven Development and Displacement*. Bristol: Policy Press, pp. 81–99.

Delgadillo, V. (2017) Una santa anti-gentrificadora en la Ciudad de México: Artistas y activistas sociales recurren a Santa Mari La Juarica para implorar que los cure del "blanqueamiento social". *El Pais*, 23 February. https://elpais.com/elpais/2017/02/23/seres_urbanos/1487839044_932988.html (Accessed 23 October 2017).

Gaffney, C. (2016) Gentrifications in pre-Olympic Rio de Janeiro. *Urban Geography* 37(8): 1132–1153.

Garay, A., Wainer, L., Henderson, H. and Rotbart D. (2013) *Puerto Madero: Análisis de un Proyecto, Land Lines*, July. www.lincolninst.edu/pubs/2289_Puerto-Madero − An%C3%A1lisis-de-un-proyecto (Accessed 2 April 2017).

Garreton, M. (2017) City profile: Actually existing neoliberalism in Greater Santiago. *Cities* 65: 32–50.

Gaytán, P. (2016) Espacio público: entre el yosmart y la invención urbanita, *Metapolítica* 20(95): 49–55.

Ghertner, A. (2015) Why gentrification theory fails in 'much of the world. *City* 19(4): 552–563.

Gretel Thomasz, A. (2016) Los nuevos distritos creativos de la Ciudad de Buenos Aires: la conversión del barrio de La Boca en el Distrito de las Artes. *EURE* 42(126): 245–267.

Harvey, D. (1989) From managerialism to entrepreneurialism: The transformation of urban governance in late capitalism. *Geografiska Annaler* 71B, 3–41.

Janoschka, M., Sequera, J. and Salinas, L. (2014) Gentrification in Spain and Latin America − a Critical Dialogue. *International Journal of Urban and Regional Research* 38(4): 1234–1265.

Lees, L., Shin, H. B. and López-Morales, E. (2015) *Global Gentrifications*. Bristol: Polity Press.

———. (2016) *Planetary Gentrification*. Cambridge: Polity Press.

Lefebvre, H. (1991) *The Production of the Space*. Oxford: Wiley-Blackwell.

López-Morales, E. (2015a) Gentrification in the global South. *City* 19(4): 557–566.

———. (2015b) Assessing exclusionary displacement through rent gap analysis in the urban redevelopment of inner Santiago, Chile. *Housing Studies* 31(5): 540–559.

———, Shin, H. B. and Lees, L. (2016) Introduction: Latin American gentrifications. *Urban Geography* 37(8): 1091–1108.

Maloutas, T. (2017) Travelling concepts and universal particularisms: A reappraisal of gentrification's global reach. *European Urban and Regional Studies* 25(3): 250–265.

Ministerio de Desarrollo Social. (2015) *Resultados Encuesta Casen*. http://observatorio.ministeriodesarrollosocial.gob.cl/casen-multidimensional/casen/casen_2015.php (Accessed 1 November 2017).

Miranda, F. (2017) La Colonia Juárez se "acondesa". *El Universal*, 30 May. www.eluniversal.com.mx/articulo/periodismo-de-investigacion/2017/05/30/la-colonia-juarez-se-acondesa (Accessed 1 November 2017).

Observatório das Metropoles. (2016) *Projeto Prata Preta: mapa inédito dos cortiços na zona portuária do Rio*. http://observatoriodasmetropoles.net/index.php?option=com_k2&view=item&id=1665:projeto-prata-preta-mapa-in%C3%A9dito-dos-corti%C3%A7os-na-zona-portu%C3%A1ria-do-rio&Itemid=164# (Accessed 1 November 2017).

Philips, D. (2018) Brazil military's growing role in crime crackdown fuels fears among poor. *The Guardian*, 27 February. www.theguardian.com/world/2018/feb/27/brazil-military-police-crime-rio-de-janeiro-favelas (Accessed 3 March 2018).

Poduje, I., Martínez, J. P., Santa Cruz, D. and Jobet, N. (2015) *Infilling: Cómo cambió SANTIAGO y nuestra forma de vivir la ciudad*. Santiago: Almagro & Atisba.

Rolnik, R. (2015) *A guerra dos lugares: a colonização da terra e da moradia na era das finanças*. Sao Paulo: Boitempo.

Romero, E. (2016) *Gentrificación en la Ciudad de México: El caso de la colonia Juárez*. Undergraduate thesis, School of Geography. National Autonomous University of Mexico (UNAM).

Rose, D. (1984) Rethinking gentrification: Beyond the uneven development of Marxist urban theory. *Environment and Planning D: Society and Space* 2: 47–74.

Salinas, L. (2013) La gentrificación de la Colonia Condesa, Ciudad de México. Aporte para una discusión desde Latinoamérica. *Revista Geográfica de América Central* 2(51): 116–138.

Slater, T. (2017) Planetary rent gaps. *Antipode* 49(1): 114–137.

Smith, D. P. and Phillips, M. (2018) Comparative approaches to gentrification: Lessons from the rural. *Dialogues in Human Geography* (online first).

Smolka, M. (2013) *Implementing Value Capture in Latin America: Policies and Tools for Urban Development.* Policy Focus Report Series. Cambridge, MA: Lincoln Institute of Land Policy.

Swanson, K. (2007) Revanchist urbanism heads South: The regulation of indigenous beggars and street vendors in Ecuador. *Antipode* 39(4): 708–728.

Techo Chile. (2017) *A un paso del campamento (#ni pal arriendo): Encuesta Techo 2017, Resumen Ejecutivo.* Santiago: Techo. https://issuu.com/techochile/docs/folleto_a_un_paso_del_campamento (Accessed 1 November 2017).

Werneck, M. (2016) Os Infames Termos Aditivos e o Mico do Porto Maravilha. *Observatório das Metrópoles.* http://observatoriodasmetropoles.net/index.php?option=com_k2&view=item&id=1956%3Aa-fal%C3%A1cia-do-porto-maravilha-ppps-cepacs-eo-%C3%B4nus-para-op-Power-%C3%BAblico itemid = & # 180 (Accessed 1 November 2017).

43

GANG VIOLENCE IN LATIN AMERICA

Dennis Rodgers

Introduction

Although gangs are not a new feature of the violence panorama in Latin America, they have come to the fore in an unprecedented manner in the post-Cold War period, to the extent that they are now widely considered to epitomize the dynamics of contemporary regional brutality. This is largely because in addition to their exponential growth, as fundamentally urban, socially sovereign, and criminal phenomena (see Hazen and Rodgers, 2014), gangs can be said to paradigmatically reflect four critical transformations that have affected the dynamics of Latin American violence in the post-Cold War period, including: (1) the fact that regional levels of violence have (paradoxically) increased (Pearce, 1998); (2) that violence has shifted from being predominantly rural to becoming overwhelmingly urban (Rodgers, 2009); (3) that it has "democratized," "ceas[ing] to be the resource of only the traditionally powerful or the grim uniformed guardians of the nation . . . [and] increasingly appear[ing] as an option for a multitude of actors in pursuit of all kinds of goals" (Koonings and Kruijt, 1999: 11); and finally (4) that the political economy of violence has changed, with the phenomenon associable both with the rise of "governance voids" consequent to widespread processes of state erosion (Kruijt and Koonings, 2007: 138), as well as the institutionalization of "violent democracies" in the region, based on collusion and the "out-sourcing" of authority to non-state actors (Arias and Goldstein, 2010; see also Rodgers, 2006b).

The multifaceted and epiphenomenal nature of gangs is the principle reason why they have become a major focus of much contemporary research on violence in Latin America, whether directly or indirectly (see Rodgers, 2016a). At the same time, however, as Schneider and Schneider (2008: 352) have pointed out in relation to crime more generally, it is important to distinguish between gangs as a specific social phenomenon and the processes of "criminalization" that frame their representation, as the latter can have important distorting effects. Certainly, gangs have been the focus of a steady stream of highly sensationalist publications over the past few years, which have for example denounced "how street gangs took over Central America" (Arana, 2005), attempted to link gangs with global terrorism (Sullivan, 2006), and even represented them as the protagonists of "another kind of war . . . being waged . . . within the context of a 'clash of civilizations'" (Manwaring, 2008: 1). More generally, they have been identified as fundamentally "anti-developmental" in nature, both intrinsically, as criminal organizations, as well as more indirectly, by undermining the rule of law and therefore the basis for coherent capital accumulation and economic growth (see World Bank, 2011).

As Hagedorn (2008: *xxx*) has remarked, "branding gangs a 'national security threat' or 'new urban insurgency' [or 'anti-developmental'] is consistent with a[n] ... attitude that divides the world into good and evil," and establishes the basis for highly distorted interpretations. The best means through which to guard against such biases is by drawing on rigorous primary research, but there is a paucity of the latter on Latin American gangs, partly for obvious methodological reasons (see Rodgers, 2007a), but also precisely because the politics of representation surrounding gangs constitute them as convenient scapegoats to justify a whole range of generally repressive policies by those in power (Rodgers, 2016a). Moreover, even if there exists primary research on gangs in almost every country in the region (see Rodgers and Baird, 2015), the overwhelming majority of studies have been carried out in isolation, on a one-off basis. There only really exist two well-developed traditions of primary research on gangs in Latin America, the first focused on Central America, and the second on Brazil. The fact that studies within these two corpuses often explicitly draw on each other make them particularly insightful bodies of work to consider in order to get to grips with the underlying nature of contemporary violence in Latin America.

Gangs in Central America

Although gangs can be traced back to the 1940s and 1950s in Central America – and the region's industrialization and concomitant urbanization – the first primary investigation of the phenomenon was not carried out until the late 1980s. This was Levenson's (1988) pioneering research on Guatemalan street gangs, which subsequently directly informed Núñez's (1996) comparative study of street gangs in Guatemala, El Salvador, and Nicaragua, as well as Cruz and Portillo Peña's (1998) study in El Salvador, Salomón et al.'s (1999) study in Honduras, and DeCesare's (1998) study in Guatemala (see also DeCesare, 2013). Similarly, the first ethnographic study of a Central American youth gang, carried out in Nicaragua in 1996–1997 by Rodgers (1997, 2000, 2006a, 2007a), was built upon by Rocha (2000, 2003, 2005) during his research with a different Managua youth gang in 1999–2000, with his results then taken up by Rodgers (2006b, 2007b) in order to calibrate new field research in 2002–2003, and further exchanges occurring for Rocha's continuing research in 2005–2006, 2009, and 2012 (Rocha, 2007a, 2007b, 2008, 2010, 2013), and Rodgers' in 2007, 2009, 2014, 2012, and 2016 (Rodgers, 2015, 2016a, 2016b, 2017a, 2017b), as well as joint research carried out in 2012 (Rodgers and Rocha, 2013).[1] Since this first wave of scholarly investigations, there has been a proliferation of studies which have almost always drawn on this initial body of work in order to inform their studies, including Savenije and Andrade-Eekhoff (2003), Cruz (2005, 2010, 2014), Hume (2007a, 2007b), Savenije (2009), Zilberg (2011), Wolf (2012, 2017), and Martinez D'aubuisson (2015) in El Salvador, Castro and Carranza (2001), Gutiérrez Rivera (2013), Wolseth (2011), Brenneman (2012), and Ayuso (2017) in Honduras, Weegels (2018) in Nicaragua, Merino (2001), Winton (2004), O'Neill (2010, 2011), and Levenson (2013) in Guatemala.

From this body of work, a number of key insights regarding Central American gangs and their violence have emerged, including the fact that gangs are contextually but not necessarily causally associated with poverty, that spatial inequality is often directly correlated with gang formation, that structural links can be made between the post-Cold War emergence of gangs and the long history of insurrection and resistance to oppression in the region, and that no single factor explains gang membership or the life course of individuals after they have left the gang. Stereotypical "determinants" such as family fragmentation, domestic abuse, or a particular psychological make up have been shown not to be consistently significant, and the only factor that has been reported as systematically affecting gang membership is religious, insofar as evangelical Protestant youths tend not to join gangs, and conversion is a common means of leaving the gang, whether in terms of being one of the few acceptable exit strategies in El Salvador, Guatemala, and Honduras, or as a means of adopting a new persona and a definite set of non-gang behaviour patterns in Nicaragua (see Brenneman, 2012).

The most important insight, however, is that there exists significant diversity between both gangs and countries within the region. A critical distinction has to for example be made between "*pandillas*" on the one hand, and "*maras*" on the other. *Maras* are a phenomenon with transnational roots, while *pandillas* are more localized, home-grown gangs that are the direct inheritors of the youth gangs that have long been a historic feature of Central American societies, albeit turbo-charged by the legacy of war and insurrection in the region. *Pandillas* were initially present throughout the region in the post-Cold War period but are now only significantly visible in Nicaragua – and to a lesser extent in Costa Rica – having been almost completely supplanted by *maras* in El Salvador, Honduras, and Guatemala (although it should be noted that *maras* in the latter also involve gangs that elsewhere in the region might be more associated with *pandillas* – see Grassi, 2011).

The origins of the *maras* lie in the 18th Street or *Dieciocho* gang in Los Angeles, a gang founded by Mexican immigrants in the Rampart section of the city in the 1960s, although it rapidly began to accept Hispanics indiscriminately, and grew significantly during the late 1970s and early 1980s as a result of the influx of – mainly Salvadoran and Guatemalan – Central American refugees, who sought to join the gang in order to feel included as outsiders in the United States. In the latter half of the 1980s, a rival – possibly splinter – group founded by a second wave of Salvadoran refugees emerged and became known as the *Mara Salvatrucha* (a combination of *salvadoreño* and *trucha*, meaning "quick-thinking" or "shrewd" in Salvadoran slang). The *Dieciocho* and the *Salvatrucha* rapidly became bitter rivals and frequently fought each other on the streets of Los Angeles. After the US Congress passed the Illegal Immigration Reform and Immigrant Responsibility Act in 1996, whereby non-US citizens sentenced to one year or more in prison were to be repatriated to their countries of origin, almost 50,000 convicts were deported to Central America – in addition to 160,000 illegal immigrants caught without the requisite permit – over the course of the next decade (UNODC, 2007: 40–42).

Partly because Nicaraguan migrants generally emigrated to Miami or Houston rather than Los Angeles (Rocha, 2008), the overwhelming majority of deportees were sent to the three Northern Central American countries (El Salvador, Guatemala, Honduras). Arriving in countries of origin that they barely knew, they rapidly reproduced the structures and behaviour patterns that had provided them with support and security in the United States, including in particular founding local *clikas*, or chapters, of the *Dieciocho* and *Salvatrucha* gangs. These *clikas* rapidly attracted local youths and either supplanted or absorbed local *pandillas*. While each *clika* was explicitly affiliated with either the *Dieciocho* or the *Salvatrucha*, neither gang could ever be characterized a federal structure, and much less by a transnational one. Neither answered to a single chain of command, and their "federated" nature was in many ways more symbolic of a particular historical origin than demonstrative of any real unity, be it of leadership or action.

At the same time, however, the transnational transposition of US gang culture to Central America had brutal effects due to the fact that it was less embedded within a local institutional context than traditional Central American *pandilla* culture, and therefore less rule-bound and constrained. Having said this, the *mara* phenomenon is not simply a foreign problem imported by deportees, but rather one that has evolved and grown in response to domestic factors and conditions, with the majority of *mara* gang members no longer deportees (Demoscopía, 2007).

It is also clear that contrarily to the numerous sensationalist accounts linking Central American gangs to migrant trafficking, kidnapping, and international organized crime, these were initially mainly involved in small-scale, localized crime and delinquency such as petty theft and muggings (although these often resulted in murder). These were most often carried out on an individual basis, although the *maras* in El Salvador, Guatemala, and Honduras also rapidly became involved in the extortion of protection money from local businesses and the racketeering of buses and taxis as they passed through the territories they controlled. In Nicaragua, gang violence in the 1990s had significant structuring effects, and was a reaction the state erosion consequent to the demise of the *Sandinista* revolution (Rodgers, 2006a).

There is however clear evidence that both *pandillas* and *maras* have become more and more involved in drug trafficking and dealing over the past two decades, and that this significantly changed their dynamics. This is perhaps not surprising considering that Central America has become a transit point for over 80% of the total cocaine traffic between the Andean countries and North America. Drug trafficking in Central America was however until recently relatively decentralized, with shipments passing from one small, local cartel to another, with each taking a cut in kind in order to make a profit as the drugs were passed from the much more organized Colombian cartels to the Mexican cartels. The role that *maras* and *pandillas* played in this process was mainly as the local security apparatus of these small cartels, or as small-time street vendors informally connected to them. Certain studies have however suggested that the leaders of these small, local cartels were often ex-gang members who had "graduated" (Rodgers, 2016b). Over the past decade or so, there has been a significant shift in the nature of the drugs trade in Central America, which has increasingly professionalized. In most Central American countries this has led to a decline in gang involvement, but in Honduras, the *Salvatrucha* effectively converted into a drug trafficking organization (Ayuso, 2017).

Another factor that significantly pushed Central American gangs – and in particular the *maras* in the Northern Triangle countries – towards a greater degree of organization was the introduction in the early 2000s of a series of extremely repressive anti-gang policies, often referred to generically as "*Mano Dura*," or "Hard Hand." These led to a series of extremely brutal "tit-for-tat" confrontations between state authorities and gangs in El Salvador, Honduras, and Guatemala that contributed to gangs become more professionalized and less visible than previously, aiding their involvement with the drugs trade. In Nicaragua, the dynamics are very different. Following an initial flirtation with a "soft" *Mano Dura* (see Weegels, 2017), *pandillas* have been effectively isolated through a series of urban infrastructural transformations that have socio-spatially excluded the poor neighbourhoods within which they arise, in order to prevent violence from spilling out into more affluent areas (Rodgers, 2012). This has provoked something of an "involution" of gangs, which have become increasingly predatory of their local neighbourhood communities. Whether this policy is sustainable in the long term is highly doubtful, however, especially as this infrastructural isolation also impacts on other spheres of life including rendering participation in the labour market by youth generally more difficult.

Much has been made of the putative "professionalization" of Central American gangs, as a result of both the rising drugs trade as well as in reaction to state repression, with some commentators going so far as to argue that gangs have begun to morph into more organized forms of transnational criminality (see Sullivan, 2006). It is however increasingly becoming clear that the reality is much more complex. While there is no doubt that *Mano Dura* policies have pushed gangs towards becoming more organized, the spread of drugs has led to their encountering competition in the form of drug cartels and other less territorially based criminal groups. These more powerful organizations have sometimes absorbed gangs – or more accurately, some gang members – but more often than not, have either ignored or repressed local gangs (in this regard, the *Salvatrucha* in Honduras constitutes an exception). As a result, while gangs continue to exist throughout Central America, recent research has suggested that they are increasingly being "squeezed" as significant actors within the regional political economy of violence (see Rodgers and Rocha, 2013). This is a consequence of state action in El Salvador and Nicaragua, and the rise of drug cartels in Guatemala and Honduras. In other words, while Central American gangs emerged in the "governance voids" precipitated by a decade of civil war and post-Cold War regime change, over the past two decades there has been a return of the state, and other criminal actors in the form of drug cartels have also emerged, leading to a transformation – and in many cases a decline – of gangs. Having said this, the evolutionary trajectory of gangs in Central America over the past three decades suggests that this trend could change very rapidly.

Gangs in Brazil

Brazil, perhaps more so than any other country in Latin America, has long been a hotbed of research on violence and exclusion. In recent years, gangs have become a major focus of interest, although it should be noted that work on the topic is by no means recent, with Zaluar's (1983, 1994) foundational research, in particular, the major initial reference point for almost all subsequent primary research on the phenomenon in Brazil (e.g. Abramovay et al., 2002; Arias, 2006; Dowdney, 2003; Gay, 2005, 2009, 2015; Leeds, 1996; Penglase, 2008, 2014; Perlman, 2010; Vianna, 1997). The majority of the literature tends to focus on gangs in Rio de Janeiro, but there is increasingly an emerging literature exploring violence in other cities such as Sao Paulo (e.g. Denyer Willis, 2009, 2015; Holston, 2008), Fortaleza (e.g. Garmany, 2011), or Timbauba, Pernambuco (e.g. Scheper-Hughes, 2015), although most of this literature does not focus specifically on the gang phenomenon.

Although gangs have long history in Brazil, going back to the 19th-century *maltas* that congregated in Rio de Janeiro's port (Soares, 2001), as well as the music-centred *galeras cariocas* and more violent *quadrilhas* of the 1970s and 1980s (Vianna, 1997), since the beginning of the 1990s, the Brazilian gang scene has been dominated by more violent drug gangs often known as *comandos*. Their origins in Rio de Janeiro go back to the Cándido Mendes prison on Ilha Grande in Rio state in the 1970s, where common criminals mixed with left-wing political prisoners arrested by the military dictatorship. Non-political inmates subsequently established the *Falange Vermelho*, later dubbed the *Comando Vermelho*, when they returned to their homes in Rio's 800 or so *favelas*. In its formative years, the *Comando Vermehlo* provided "law and order," "protection" and "social welfare services" for *favela* residents, such as financing child day-care facilities and giving handouts for cooking gas (Penglase, 2008, 2014; Gay, 2009). This was established through a set of symbols, discourses, and tactics, linked to the group's left-wing organizational origins, and included normative projects such as the promotion *boa vizinhança*, or neighbourliness, reflecting the *o colective* prison (Penglase, 2008: 126, 129). In this capacity, the *Comando Vermehlo* took advantage of the widespread distrust of the police and the abandonment of the *favelas* by state welfare services, hence earning their keep and embedding themselves in the communities.

In 1982, partly in response to the increasing drugs trade in Colombia, the *Comando Vermehlo* took the decision to become involved and fund their activities by means of this business (Gay, 2009). Rio became a significant cocaine transhipment point for the Andean product *en route* to Europe, although as much as 20% of shipments are sold nationally, thereby creating large local markets in a way that was not the case in Central America. Much of this retail trade circulates within the *favelas*, sold at *bocas de fumo*, drug-corners run by gangs. Drug penetration in Rio's *favelas* has had a profound impact (Arias, 2006; Leeds, 1996). Perlman (2010) observes that the greatest change in the years following the 1970s has been the domination of many communities by such drug gangs. Dowdney (2007) describes how the rise of gangs in Rio de Janeiro can be divided into three periods: (1) the pre-cocaine, pre-*Comando Vermehlo* era, when drugs trickled into the city in an *ad hoc* manner; (2) the 1980s cocaine and *Comando Vermehlo* phase of territorial definition and "narco-culture," when the group held a monopoly and dominated the business; and (3) the 1990s onwards, which saw the emergence of bloody territorial disputes as a result of conflicts between rival *Comando Vermehlo* leaders – or *donos* – jostling for domination. This led to the splintering of the *Comando Vermehlo* in the late 1990s, which moved away from its "old-style" organizational roots where "respectful" gangsters provided social services (Penglase, 2008), and led to the emergence of a range of new gangs including the *Terceiro Comando, Comando Vermelho Jovem*, and *Amigos dos Amigos*, as well as the escalation of territorial disputes, or turf-wars, over retail drugs sales.

This latter development led to spiralling violence in Rio de Janeiro. Not surprisingly, perhaps, these "gang wars" led to a brutal state-driven "war on gangs," which further increased the bloodshed. Most notoriously, in 2007 military police invaded a group of *favelas* collectively known as *Complexo de Alemão* to expel the *Comando Vermehlo*, leaving scores dead (Gay, 2009). The response by the *Comando Vermehlo* led to the militarization of the *favela*, as well as the increasing use of children and youths in armed combat (Dowdney, 2003). This also changed the already complex and ambiguous relationships that the gangs had with their host communities. In the early years the gangs were interwoven with the social fabric of local communities; however, as they became drugs based, militarized, and ruthless, they also became predatory upon local communities (Leeds, 1996; Perlman, 2010). Drug gangs began to control neighbourhood associations and those not acting in their interests were forcibly removed.

This escalating violence led to the implementation of one of the best-known contemporary examples of urban violence prevention, the Urban Pacification Programme (UPP). Begun in 2008 with the deployment of specially trained Police Pacification Units in gang-affected *favelas*, the UPP aimed to be different from the short-term operations of the military police units. The programme has used army units and special military police to occupy a *favela* (intervention phase), before introducing UPP officers (stabilization), and finally establishing state and private infrastructure and social projects (consolidation phase). The UPP claims legitimacy as a break with the past, both bringing security and development initiatives. It is argued that unlike previous attempts to "integrate Rio" or "normalise" the *favela* through upgrading projects the UPP makes *favela* residents into full citizens, and "reclaims" state sovereignty over the "ungoverned' spaces of the city. In so doing, however, one might argue that pacification renders the *favela* a more "legible" space, one in which the state and private sector can therefore act, and there have also been significant critiques of the UPP programme and its long-term sustainability (see Jones and Rodgers, 2015).

Gang dynamics in Sao Paulo have to a large extent been comparable to those in Rio. The city's dominant gang is the *Primeiro Comando da Capital* (PCC), which in a similar fashion to the *Comando Vermelho* but many years later in 1993, originated as a small-scale gang in *Carandiru* and *Taubaté* prisons, and grew to become a ubiquitous fixture of the city's poor neighbourhoods. The PCC's beginnings were not linked to radical politics, however, but rather were a collective response to prison conditions and violence. One notorious incident in 1991 saw the massacre of 111 inmates in *Carandiru* detention centre, which led to the creation of the PCC in the nearby *Taubaté* prison two years later (Denyer Willis, 2009). The PCC made a variety of demands framed in human rights terms, as well as establishing a command and control structure over Sao Paulo prisons that allowed it to bring a relative calm to prison life. At the same time, however, Holston (2008: 273) has argues that the PCC is nothing more than a "prison-based gang" that has adopted the language of human rights to state repression and prison conditions as a smokescreen to the organization's criminal nature (see also Macaulay, 2007: 638). Certainly, after expanding to a network of thousands of members within the São Paulo prison system, the PCC quickly gained a foothold in the city's poor urban neighbourhoods as most prisoners came from such neighbourhoods and returned home when released. This provided the operational means for the PCC to gain in influence and control the drugs trade, to the extent that some scholars have described the PCC as a "parallel power" to the state (Macaulay, 2007).

Other scholars have however highlighted that the PCC grew very much in symbiosis with corrupt and clientelist elements of the state. Indeed, Denyer Willis (2015) goes so far as to talk of there being a "consensus" between the authorities and the PCC over the exercise of violence in Sao Paulo, in particular in relation to killings, which are governed by particular spatial, cultural, and behavioural codes. In this regard, Denyer Willis' analysis is reminiscent of Arias' (2006) exploration of gang dynamics in Rio de Janeiro, where he argues that there exists widespread collusion between gangs and the authorities. Rather than there simply being a "consensus," however, he contends that the former act to mobilize voters in favour of incumbent politicians, as well as help control the populations of *favelas*, in exchange for the latter letting them conduct their drug trafficking in relative impunity. In this sense, there has been an institutionalization of a particular form of "criminal governance" in Rio de Janeiro.

Conclusion

The evolution of gangs in both Central America and Brazil over the past three decades clearly highlight the constantly changing dynamics of contemporary Latin American violence. The nature of the latter is influenced by changes in both local political economies, with the evolving actions of both the state and other violent actors in particular playing key roles in transforming patterns of gang violence. Another factor that clearly impacts on the nature of gang violence is drugs, and more specifically the involvement of gangs in the narcotics trade. Ultimately, however, the potential wealth that this business can bring either changes gangs, pushing them towards transforming into drug trafficking organizations, or else brings new actors into play, including the state. At the same time, however, the gang phenomenon also highlights how there exist critical continuities in the underlying dynamics of contemporary violence in Latin America. One element that has not changed over the past few decades is its fundamentally gendered nature, something that is reflected in the fact that gang violence remains overwhelmingly male, both in terms of its participants and its victims (Baird, 2015; Hume, 2004, 2009; Hume and Wilding, 2015).

Another enduring aspect of gangs is the importance of processes of scapegoating surrounding them, including in particular the regular representation of gangs as an inherent "other" (Moodie, 2010; Rodgers, 2016a). Where once the "enemy" was communist – and in the colonial period, indigenous – now it is a gang member . . . When considered from a developmental perspective, this puts gangs in a very different light. In many ways, what this persistent scapegoating points to is that many of the fundamental underlying drivers of violence in Latin America have not necessarily changed as much as has widely been reported and can be linked to the enduring unequal development in the region. Certainly, much of the existing research on gangs in Central America and Brazil highlights the persistent connections between their violence and critical questions of socio-spatial inequality and exclusion (Rodgers, 2009). Seen from this perspective, then, what gangs arguably teach us most surely about the nature of violence in Latin America today is that rather than being fundamentally anti-developmental, they are a phenomenon that is inextricably linked to long-standing political economy issues that have still not been resolved in the current epoch.

Note

1 Nicaragua is the Central American country that has probably been studied in greatest empirical depth as a result of this rather unique process of longitudinal ethnographic conversation between Rocha and Rodgers (see Rocha and Rodgers, 2008).

References

Abramovay, M., Waiselfisz, J. J., Andrade, C. C. and das Graças Rua, M. (2002) *Guangues, galeras, chegados e rappers: Juventude, violência e cidadania nas cidades da periferia de Brasília*. Rio de Janeiro: Garamond.

Arana, A. (2005) How the street gangs took Central America. *Foreign Affairs* 84(3): 98–110.

Arias, E. D. (2006) *Drugs and Democracy in Rio de Janeiro: Trafficking, Social Networks and Public Security*. Chapel Hill: University of North Carolina Press.

———— and Goldstein, D. M. (2010) (eds) *Violent Democracies in Latin America*. Durham, NC: Duke University Press.

Ayuso, T. (2017) *Why Did the Game Change in Honduras?* Paper presented to the ERC Dynamics of Civil War MXAC project workshop, Paris, 14–15 December.

Baird, A. (2015) Duros and gangland girlfriends: Male identity and gang socialisation in Medellín. In J. Auyero, P. Bourgois and N. Scheper-Hughes (eds) *Violence at the Urban Margins*. Oxford: Oxford University Press, pp. 112–133.

Brenneman, R. (2012) *Homies and Hermanos: God and Gangs in Central America*. Oxford: Oxford University Press.

Castro, M. and Carranza, M. (2001) Las Maras en Honduras. In ERIC, IDESO, IDIES and IUDOP (eds) *Maras y Pandillas en Centroamérica*, vol. 1. Managua: UCA Publicaciones.

Cruz, J. M. (2005) Los factores asociados a las pandillas juveniles en Centroamérica. *Estudios Centroamericanos* 685–686: 1155–1182.

————. (2010) Central American Maras: From street youth gangs to transnational protection rackets. *Global Crime* 11(4): 279–298.

————. (2014) Maras and the politics of violence in El Salvador. In J. M. Hazen and D. Rodgers (eds) *Global Gangs: Street Violence Across the World*. Minneapolis: University of Minnesota Press, pp. 123–144.

———— and Portillo Peña, N. (1998) *Solidaridad y violencia en las pandillas del gran San Salvador: Más allá de la vida loca*. San Salvador: UCA Editores.

DeCesare, D. (1998) The children of war: Street gangs in El Salvador. *NACLA: Report on the Americas* 32(1): 21–29.

————. (2013) *Unsettled/Desasosiego: Los niños en un mundo de las pandillas*. Austin: University of Texas Press.

Demoscopía, S. A. (2007) *Maras y pandillas, comunidad y policía en Centroamérica*. San José: Demoscopía.

Denyer Willis, G. (2009) Deadly symbiosis? The PCC, the state and the institutionalization of violence in São Paulo. In G. A. Jones and D. Rodgers (eds) *Youth Violence in Latin America: Gangs and Juvenile Justice in Perspective*. New York: Palgrave Macmillan, pp. 167–182.

————. (2015) *The Killing Consensus: Police, Organized Crime and the Regulation of Life and Death in Urban Brazil*. Berkeley, CA: University of California Press.

Dowdney, L. (2003) *Children of the Drug Trade: A Case Study of Children in Organised Armed Violence in Rio de Janeiro*. Rio de Janeiro: 7 Letras.

Dowdney, L. (2007) *Neither War nor Peace: International Comparisons of Children and Youth in Organised Armed Violence*. Rio de Janeiro: Viva Rio/COAV/IANSA.

Garmany, J. (2011) Drugs, violence, fear, and death: The necro and narco-geographies of contemporary urban space. *Urban Geography* 32(8): 1148–1166.

Gay, R. (2005) *Lucia: Testimonies of a Brazilian drug dealer's woman*. Philadelphia, PA: Temple University Press.

————. (2009) From popular movement to drug gangs to militias: An anatomy of violence in Rio de Janeiro. In K. Koonings and D. Kruijt (eds) *Mega-Cities: The Politics of Urban Exclusion and Violence in the Global South*. London: Zed Books.

————. (2015) *Bruno: Conversations with a Brazilian Drug Dealer*. Durham, NC: Duke University Press.

Grassi, P. (2011) La Zona Roja: Potere ed utilizzo del territorio urbano a Città del Guatemala. *Confluenze* 3(2): 181–196.

Gutiérrez Rivera, L. (2013) *Territories of Violence: State, Marginal Youth, and Public Security in Honduras*. New York: Palgrave Macmillan.

Hagedorn, J. M. (2008) *A World of Gangs: Armed Young Men and Gangsta Culture.* Minneapolis: University of Minnesota Press.

Hazen, J. M. and Rodgers, D. (2014) (eds) *Global Gangs: Street Violence Across the World.* Minneapolis: University of Minnesota Press.

Holston, J. (2008) *Insurgent Citizenship: Disjunctions of Democracy and Modernity in Brazil.* Princeton, NJ: Princeton University Press. Hume, M. (2004). "It's as if you don't know, because you don't do anything about it": *Gender and violence in El Salvador. Environment and Urbanization,* 16(2): 63–72.

———. (2007a) (Young) men with big guns: Reflexive encounters with violence and youth in El Salvador. *Bulletin of Latin American Research* 26(4): 480–496.

———. (2007b) Mano Dura: El Salvador responds to gangs. *Development in Practice* 17(6): 739–751.

———. (2009) *The Politics of Violence: Gender, Conflict and Community in El Salvador.* Oxford: Wiley-Blackwell.

——— and Wilding, P. (2015) Es que para ellos el deporte es matar: Rethinking the scripts of violent men in El Salvador and Brazil. In J. Auyero, P. Bourgois and N. Schepper-Hughes (eds) *Violence at the Urban Margins.* Oxford: Oxford University Press, pp. 93–111.

Jones, G. A. and Rodgers, D. (2015) Gangs, guns, and the city: Urban policy in dangerous places. In C. Lemanski and C. Marx (eds) *The City in Poverty.* London: Palgrave Macmillan.

Koonings, K. and Kruijt, D. (1999) Introduction: Violence and fear in Latin America. In D. Kruijt and K. Koonings (eds) *Societies of Fear: The Legacy of Civil War, Violence and Terror in Latin America.* London: Zed Books, pp. 1–30.

Kruijt, D. and Koonings, K. (2007) Epilogue: Latin America's urban duality revisited. In K. Koonings and D. Kruijt (eds) *Fractured Cities: Social Exclusion, Urban Violence, and Contested Spaces in Latin America.* London: Zed Books, pp. 7–22.

Leeds, E. (1996) Cocaine and the parallel polities in the Brazilian urban periphery: Constraints on local-level democratization. *Latin American Research Review* 31(3): 47–83.

Levenson, D. (1988) *Por sí mismos: Un estudio preliminar de las "maras" en la Ciudad de Guatemala.* Guatemala: Asociación para el Avance de las Ciencias Sociales en Guatemala (AVANCSO).

———. (2013) *Adiós Niño: The Gangs of Guatemala City and the Politics of Death.* Durham, NC: Duke University Press.

Macaulay, F. (2007) Knowledge production, framing and criminal justice reform in Latin America. *Journal of Latin American Studies* 39(3): 627–651.

Manwaring, M. G. (2008) *A Contemporary Challenge to State Sovereignty: Gangs and Other Illicit Transnational Criminal Organizations (TCOs) in Central America, El Salvador, Mexico, Jamaica, and Brazil.* Carlisle: Strategic Studies Institute, US Army War College.

Martínez D'aubuisson, J. J. (2015) *Ver, oír y callar: Un año con la Mara Salvatrucha 13.* Logroño: Pepitas de Calabaza.

Merino, J. (2001) Las maras en Guatemala. In ERIC, IDESO, IDIES, y IUDOP (eds) *Maras y pandillas en Centroamérica,* vol. 1. Managua: UCA Publicaciones.

Moodie, E. (2010) *El Salvador in the Aftermath of Peace: Crime, Uncertainty, and the Transition to Democracy.* Philadelphia, PA: University of Pennsylvania Press.

Núñez, J. C. (1996) *De la ciudad al barrio: Redes y tejidos urbanos en Guatemala, El Salvador y Nicaragua.* Ciudad de Guatemala: Universidad Rafael Landívar/PROFASR.

O'Neill, K. L. (2010) The reckless will: Prison chaplaincy and the problem of Mara Salvatrucha. *Public Culture* 22(1): 67–88.

———. (2011) Delinquent realities: Christianity, formality, and security in the Americas. *American Quarterly* 63(2): 337–365.

Pearce, J. (1998) From civil war to "civil society": Has the end of the Cold War brought peace to Central America? *International Affairs* 74(3): 587–615.

Penglase, R. B. (2008) The bastard child of the dictatorship: The Comando Vermelho and the birth of "narco-culture" in Rio de Janeiro. *The Luso-Brazilian Review* 45(1): 118–145.

———. (2014) *Living with Insecurity in a Brazilian Favela: Urban Violence and Daily Life.* New Brunswick, NJ: Rutgers University Press.

Perlman, J. (2010) *Favela: Our Decades of Living on the Edge in Rio de Janeiro.* Oxford: Oxford University Press.

Rocha, J-L. (2000) Pandillas: Una cárcel cultural. *Envío* 219: 13–22.

———. (2003) Tatuajes de pandilleros: Estigma, identidad y arte. *Envío* 258: 42–50.

———. (2005) El traido: Clave de la continuidad de las pandillas. *Envío* 280: 35–41.

———. (2007a) *Lanzando piedras, fumando "piedras": Evolución de las pandillas en Nicaragua 1997–2006.* Managua: UCA Publicaciones.

———. (2007b) Del telescopio al microscopio: Hablan tres pandilleros. *Envío* 303: 23–30.

———. (2008) La Mara 19 tras las huellas de las pandillas políticas. *Envío* 321: 26–31.

———. (2010) Un debate con muchas voces: Pandillas y estado en Nicaragua. *Temas* 64: 9–37.

———. (2013) *Violencia juvenil y orden social en el Reparto Schick: Juventud marginada y relación con el estado.* Washington, DC: Inter-American Development Bank.

———, y Rodgers, D. (2008) *Bróderes descobijados y vagos alucinados: Una década con las pandillas nicaragüenses, 1997–2007.* Managua: Envío.

Rodgers, D. (1997) Un antropólogo-pandillero en un barrio de Managua. *Envío* 184: 10–16.

———. (2000) *Living in the Shadow of Death: Violence, Pandillas, and Social Disintegration in Contemporary Urban Nicaragua.* Unpublished Ph.D thesis, Department of Social Anthropology, University of Cambridge.

———. (2006a) Living in the shadow of death: Gangs, violence, and social order in urban Nicaragua, 1996–2002. *Journal of Latin American Studies* 38(2): 267–292.

———. (2006b) The state as a gang: Conceptualising the governmentality of violence in contemporary Nicaragua. *Critique of Anthropology* 26(3): 315–330.

———. (2007a) Joining the gang and becoming a broder: The violence of ethnography in contemporary Nicaragua. *Bulletin of Latin American Research* 26(4): 444–461.

———. (2007b) When vigilantes turn bad: Gangs, violence, and social change in urban Nicaragua. In D. Pratten and A. Sen (eds) *Global Vigilantes.* London: Hurst, pp. 349–370.

———. (2009) Slum wars of the 21st century: Gangs, Mano Dura, and the new urban geography of conflict in Central America. *Development and Change* 40(5): 949–976.

———. (2012) Haussmannization in the Tropics: Abject urbanism and infrastructural violence in Nicaragua. *Ethnography* 13(4): 411–436.

———. (2015) The moral economy of murder: Violence, death, and social order in Nicaragua. In J. Auyero, P. Bourgois and N. Scheper-Hughes (eds) *Violence at the Urban Margins.* Oxford: Oxford University Press, pp. 21–40.

———. (2016a) La Mystique mara: Bandes, barbarisme, et 'antipolitique' en Amérique Centrale. In C. Duterme, A. Mira and M. Giraldou (eds) *Mauvais Sujets dans les Amériques.* Toulouse: Presses universitaires du Midi, pp. 25–33.

———. (2016b) Critique of urban violence: Bismarckian transformations in contemporary Nicaragua. *Theory, Culture, and Society* 33(7–8): 85–109.

———. (2017a) Of pandillas, pirucas, and Pablo Escobar in the barrio: Change and continuity in Nicaraguan gang violence. In S. Huhn and H. Warnecke (eds) *Politics and History of Violence and Crime in Central America.* New York: Palgrave Macmillan, pp. 65–84.

———. (2017b) *Bróderes* in arms: Gangs and the socialization of violence in Nicaragua. *Journal of Peace Research* 54(5): 648–660.

——— and Baird, A. (2015) Understanding gangs in contemporary Latin America. In S. Decker and D. Pyrooz (eds) *Handbook of Gangs and Gang Responses.* New York: Wiley-Blackwell, pp. 478–502.

Rodgers, D. and Rocha, J-L. (2013) The evolution of gang violence in post-revolutionary Nicaragua. In *Small Arms Survey 2013: Everyday Dangers.* Cambridge: Cambridge University Press, pp. 46–73.

Salomón, L., Castellanos, J. and Flores, M. (1999) *La delincuencia juvenil: Los menores infractores en Honduras.* Tegucigalpa: CEDOH.

Savenije, W. (2009) *Maras y barras: Pandillas y violencia juvenil en los barrios marginales de Centroamérica.* San Salvador: FLACSO.

——— and Andrade-Eekhoff, K. (2003) (eds) *Conviviendo en la orilla: Violencia y exclusión social en el Area Metropolitana de San Salvador.* San Salvador: FLACSO.

Scheper-Hughes, N. (2015) Death squads and vigilante politics in democratic Northeast Brazil. In J. Auyero, P. Bourgois and N. Scheper-Hughes (eds) *Violence at the Urban Margins.* Oxford: Oxford University Press, pp. 266–304.

Schneider, J. and Schneider, P. (2008) The anthropology of crime and criminalization. *Annual Review of Anthropology* 37: 351–373.

Soares, C. E. L. (2001) *A capoeira escrava e outras tradições rebeldes no Rio de Janeiro (1808–1850).* Campinas, SP: UNICAMP.

Sullivan, J. P. (2006) Maras morphing: Revisiting third generation gangs. *Global Crime* 7(3–4): 489–492.

UNODC. (2007) *Crime and Development in Central America: Caught in the Crossfire.* Vienna: United Nations Office on Drugs and Crime.

Vianna, H. (1997) (ed) *Galeras cariocas: Territórios de conflitos e encontros culturais.* Rio de Janeiro: Editora UFRJ.

Weegels, J. (2018) Implementing social policy through the criminal justice system: Youth, prisons, and community-oriented policing in Nicaragua. *Oxford Development Studies* 46(1): 57–70.

Winton, A. (2004) Young people's view on how to tackle gang-violence in "post-conflict" Guatemala. *Environment and Urbanization* 16(2): 83–99.

Wolf, S. (2012) El Salvador's pandilleros calmados: The challenges of contesting Mano Dura through peer rehabilitation and empowerment. *Bulletin of Latin American Research* 31(2): 190–205.

———. (2017) *Mano Dura: The Politics of Gang Control in El Salvador.* Austin: University of Texas Press.

Wolseth, J. (2011) *Jesus and the Gang: Youth Violence and Christianity in Urban Honduras.* Tucson: University of Arizona Press.

World Bank. (2011) *Word Development Report 2011: Conflict, Security, and Development.* Washington, DC: World Bank.

Zaluar, A. (1983) Condomínio do diabo: As classes populares urbanas e a lógica do "ferro e fumo. In P. S. Pinheiro (ed) *Crime, violência e poder.* São Paulo: Brasiliense, pp. 251–277.

———. (1994) *Condomínio do diabo.* Rio de Janeiro: Editora Revan, UFRJ.

Zilberg, E. (2011) *Spaces of Detention: The Making of a Transnational Gang Crisis Between Los Angeles and San Salvador.* Durham, NC: Duke University Press.

44

INFORMAL SETTLEMENTS

Melanie Lombard

Introduction

A recent report by the Latin American Development Bank (CAF, 2017) estimated that around 20–30% of the population in Latin America live in informal settlements, compared to a global average of around 10%. Thus, according to the authors of the report, urban growth in the context of informality represents a challenge as much as an opportunity for Latin America. Such concerns are not new. There is a long tradition of research on informal settlements in Latin America, since the phenomenon of rapid urbanization was first observed in the region in the 1950s (see e.g. Lewis, 1952; Mangin, 1967; Portes, 1972; Turner, 1972; Montaño, 1976; Lomnitz, 1977; Durand, 1983; Peattie, 1990; Azuela and Tomas, 1996; Hernández, Kellett, and Allen, 2010). From early responses of eviction and displacement, based on notions of a "culture of poverty" which blighted these apparently transient communities, to the ideas of "self-help" and the "myth of marginality," ideas about such places have evolved in tandem with the physical and material evolution of cities.

Because of this long tradition, it has been noted that research on informal settlements in Latin America has significantly influenced understandings of urban informality globally (AlSayyad, 2004). Despite this, there is still much we do not know about informal settlements, and this is equally as true in the Americas as in other regions. The diversity of experiences among different Latin American countries – ranging from those with the largest economies and populations in the region such as Mexico, Brazil, and Colombia, to much smaller and/or poorer nations such as Honduras, Guatemala, and Bolivia – underpins a lack of consistency in knowledge at regional level. Informal settlement characteristics may also change and develop depending on wider social, economic, and political shifts. While declining income inequality has been observed in some of the larger Latin American economies (Lustig, Lopez-Calva, and Ortiz-Juarez, 2012), the neoliberalization which has often accompanied democratization and decentralization in the region has played a significant role in shaping informal urban development.[1] According to some observers (e.g. Duhau, 2008; Cordera, Ramírez, and Ziccardi, 2008; Bayón and Saraví, 2013), multiple vulnerabilities and urban fragmentation have contributed to a situation in which informal settlements are no longer seen as a step towards integration in the formal city, but as representing an isolated environment, disconnected from the rest of the city, where social mobility has stagnated.

Certainly, in many Latin American cities – for example, Mexico City, Buenos Aires, Bogotá, and San Salvador, among others – deindustrialization in the context of globalization has led to the exclusion of large sectors of the urban population, and a concomitant expansion of informal activities (Cordera, Ramírez, and Ziccardi, 2008). Meanwhile, older patterns of informal urbanization have been accompanied by massive developments of state-built social housing on the urban periphery, exacerbating existing inequalities through the location of urban poor communities on cheap land which is increasingly distant from the goods and services of the city centre, often side-by-side with privately constructed and -serviced elite enclaves (*ibid*). In particular, the rise of mass housing on the outskirts of cities raises important questions about the relationship between formal and informal modes of urban development, as such housing, intended to supplant informal shelter production processes for low and middle-income households, has been subject to poor quality construction, urban insecurity, and high levels of vacancy (see e.g. Boils Morales, 2008). At the same time, there is a renewed imperative to understand informal settlements in their own right, in terms of how and why these places develop, as well as how they are changing in response to external contingencies. This is a particularly opportune moment given the current interest in urban issues from international development circles (Van Lindert, 2016), and renewed interest more generally in Latin America, seen for decades as a "forgotten continent" in terms of Western foreign policy and international development (Reid, 2007).

This chapter presents a brief overview of the Latin American experience of informal settlements. It focuses initially on the historic development of informal settlements and the evolution of policy responses, followed by a consideration of some of the main theoretical approaches that have informed and interpreted these developments. The penultimate section presents a snapshot of current issues in this field, and the conclusion reflects briefly on further avenues for exploration. While generalizing at a regional level risks a high degree of abstraction from settlements that may differ across and even within cities – depending on factors including government attitudes, land ownership structures, and legal frameworks – it is possible to detect "some common threads by which cities develop and how this affects their poorer citizens" (Hardoy and Satterthwaite, 1989: 18). Regional commonalities thus play at least a partial role in determining the nature of such places, as ultimately, "informality cannot be disentangled from geography" (AlSayyad, 2004: 27).

Informal settlements in Latin America: policy and practice

UN-Habitat, or the United Nations Human Settlement Programme, has formulated a definition of informal settlements which offers a useful starting point, although it has also been subject to debate. This definition sees informal settlements as characterized by inadequate conditions of housing and/or basic services, specifically relating to five characteristics, namely: inadequate access to drinking water; inadequate access to sanitation/other infrastructure; low structural quality of houses; excessive density; and insecure residential status (UN-Habitat, 2003). Informal settlements may have varying combinations and degrees of these factors, although critics have noted that such globally applicable definitions of informal settlements are not nuanced enough, and do not allow for context-specific factors, or the interaction of different factors (Gulyani and Talukdar, 2008).

Similarly, there is little consensus on the causal factors involved in the generation of informal settlements. While they are often associated with rapid urbanization, the conditions in which this occurs are paramount. These may include: high overall levels of poverty; prevailing urban policies which reflect "status quo" interests, in other words those of the ruling national and metropolitan political elite (Fox, 2014), and thus lack provision for low-income housing;

a diminished role for the state (Davis, 2006), particularly relating to the capacity of planning (Fernandes, 2011) and terms of housing production; and other factors in a given context, such as the availability of land, levels of official tolerance, opportunities for political clientelism, and traditions of self-help. Hardoy and Satterthwaite (1989: 35) argue that if informality consists of illegality and a lack of adherence to standards, then it may be the law itself that is the problem, as "laws are unjust when the poverty of the majority of people makes it impossible for them to comply with them."

In Latin America, informal settlements have historically tended to develop in peripheral locations, on cheap, unserviced agricultural land which has been squatted or acquired informally (usually through illegal subdivision),[2] often by the same communities that live there. Mangin (1967: 69) describes the process of organized invasion in Lima, where "after months of planning, thousands of people moved during one night to a site that had been secretly surveyed and laid out," in the face of police opposition. It has been noted that the distinction between central urban areas colonized by the elite and the self-built nature of the rest of the city, occupied by low-income communities, has existed since pre-Columbian times (Hardoy, 1982). However, in the context of rapid urbanization that often accompanied the industrial development engendered by economic policies of import substitution industrialization from the 1950s onwards, informal settlements came to characterize many cities, housing manual and domestic workers, as well as those in the informal economy, who were unable to access formal shelter options via the market or public housing schemes due to their high costs or entry thresholds.

Such places were often characterized by the use of semi-permanent materials such as breezeblock, along with wood and tin, with strong prospects for formalization and/or regularization, and hence consolidation and integration with the formal city in the long-term (Gough and Kellett, 2001). In the Latin American city, informal settlements are often associated with the image of favelas, whose distinctive topography has come to symbolize urban informality; and their formation has been associated with wider politics of populist mobilization and state power. However, Varley (2013) cautions against simplistic assumptions, regarding both the distinction between *morro* (informal settlements) and *asfalto* (formal city), and the association of informality with resistance. These points are discussed further below.

The evolving discourse on urban development in Latin America has reflected a growing interest in urban poverty from within international development, in terms of both its conceptual framing and related policy formulation (Van Lindert, 2016); however, the phenomenon of urban informality was first observed and documented by urban practitioners in the 1950s and 1960s. Confronting official responses of eviction and displacement (Mangin, 1967) and stereotypical framings of slums full of social problems and a "culture of poverty" (Lewis, 1967), these researchers showed how they were in fact sites of social mobility with communities who were motivated by educational aspirations for their children. The notion of "self-help," formulated by the architect John Turner (1972) based on his experiences in the *barriadas* of Lima, suggested that low-income residents were best-placed to provide their own housing solutions through self-build and self-financing practices, facilitated by local government. Building on this, in the 1970s, Janice Perlman's (1976) research in the *favelas* of Rio de Janeiro formulated the idea of the "myth of marginality" to suggest that while these communities were *marginalized* by the rest of the city, they were not *marginal* but integral to urban life, for example as a source of labour. At around the same time, important sociological studies were being undertaken by Latin American researchers. Alejandro Portes (1972) demonstrated how rationality was a defining characteristic in informal settlements residents' pursuit of improving their circumstances in Latin American cities, in the context of adverse structural conditions. Meanwhile, Jorge Montaño (1976) showed how in Mexico City, the "reciprocal obligation" of clientelistic relations between the urban poor and the Mexican authorities worked to respond to the demands of these communities while maintaining stability and avoiding conflict.

Such works coming out of Latin America responded to some degree to hopes or fears (depending on one's ideological perspective) of the radicalization of informal settlement dwellers, linked to the frequent presence or coalescence of activist organizations in such contexts. Certainly, activist organizations have played an important role in the development of urban informal settlements in Latin American cities, particularly helping to obtain land, housing, and services for the residents of specific neighbourhoods. Well-known examples include Villa El Salvador in Peru (Peattie, 1990); Nezahualcóyotl in Mexico City (Bredenoord and Verkoren, 2010); San Martin in Buenos Aires; and Brasilia Teimosa in Recife (Hardoy and Satterthwaite, 1989). In such cases, residents organized, or were supported to organize, and demand or negotiate for land for self-building, and basic services such as water, electricity, and rubbish collection. As one resident of San Martin settlement commented in October 1982: "Twenty thousand people, driven by hunger, driven by high rental costs and unemployment, searched for abandoned lands . . . and that is why, today, thank God, we are here, we're organized, we're united" (*ibid:* 13). One of the most celebrated cases was the Tierra y Libertad movement in Monterrey, Mexico, which succeeded in establishing several autonomous settlements over a period of years. At national and regional scale, the role of Latin American urban social movements such as the Movimiento Urbano Popular in Mexico, which often drew their support base from local level organizations within informal settlements, was seen as holding the potential to radically restructure society in the 1960s and 1970s (Castells, 1983).

However, observers have cautioned against exaggerating the significance of what were ultimately exceptional and short-lived cases (Lombard, 2016; Azuela and Tomas, 1996). In the case of Tierra y Libertad, community organization declined drastically when residents were offered formal land titles by the state government, leading to fragmentation among the groups of residents and their leaders (Vellinga, 1989). More generally, obstacles to sustained mobilization include a lack of class consciousness among the urban poor (Montaño, 1976); the evolving nature of settlements, which requires greater levels of resident participation at certain key stages such as neighbourhood formation and when faced with external threats (Gilbert, 1994); the social aspirations and ultimately conservative nature of many low-income communities (e.g. Mangin, 1967; Perlman, 1976); and the strategic nature of state interventions, which have for example successfully coopted community mobilization in exchange for benefits for the neighbourhood, often backed up with the threat of repression (Gilbert, 1994). More recently, the role of activism has experienced a resurgence, explored further below.

Meanwhile, the radical potential of self-help housing seemed to be significantly diminished as it came to be embraced by international institutions such as the World Bank and the United Nations in the 1970s (Davis, 2006). As a consequence, the World Bank encouraged national governments to pursue programmes of upgrading and sites-and-service provision in the 1970s and 80s, marking a shift from previous policies of eviction and displacement (Imparato and Rusler, 2003). However, the limitations of this approach related to the difficulty of scaling up, as well as that of addressing poverty more broadly (Van Lindert, 2016). This was superseded by the enabling approach which saw a diminished role for the state in urban poverty reduction and especially housing provision, as one among a plurality of actors which also included civil society and the private sector. Alongside the promotion of Urban Management Programmes (Jones and Ward, 1994) as part of wider moves towards decentralization in the context of democratization, this also suggested that the appropriate level of intervention in urban poverty reduction was local, presenting a new role for local governments. That these shifts occurred in the context of neoliberalization in many Latin American countries has led to a "perverse confluence" of ideas

(Dagnino, 2007), relating particularly to the increasing importance of citizen participation in (urban) governance, even while the role of the state shrinks or is transformed (Lombard, 2013).

Emblematic of this paradigm shift in international and national development discourse, the policy of tenure legalization has come to the fore, relating to the provision of legal titles for land, housing, or other property held informally. The widespread take-up of this approach is often associated with Peruvian economist Hernando De Soto's (2000) suggestion that titling offers a solution to poverty and hence informality, as the security that titles offer allows poor people to invest in their homes and businesses, and hence "brings to life" otherwise "dead" capital. While policies of land titling predate De Soto's work in some countries in the region, such as Peru in the 1960s and Mexico in the 1970s, their prevalence in Latin America symbolizes the increasingly tolerant attitude on the part of the state towards informal settlements. However, these programmes have failed to "solve" informality, leading critics of the approach to suggest that it oversimplifies urban poverty and overlooks its political causes (Miranda, 2002). In fact, titling may lead to greater insecurity for informal settlement residents due to property value increases, increased speculation, and potential market evictions, although this contention is subject to debate (Varley, 2016). Moreover, evidence suggests that eviction has been recurring in rural and urban areas in the region in recent decades, with nearly 150,000 people evicted in 15 Latin American countries between 2004 and 2006 (Fernandes, 2011: 7).

Key theoretical approaches to informal settlements in Latin America

As suggested above, urban theorists writing both on and from the region have taken a critical stance towards policy and practice, in an attempt to interpret but also sometimes to influence these. An early attempt to conceptualize responses to informal settlements can be seen in Rakowski's (1994) structuralist/legalist framework, drawing on research from diverse Latin American contexts. Rakowski argues that these two distinct analytical positions tended to inform policy responses on urban informality (both in economic and shelter terms) in the 1980s and 1990s. The structuralist school sees urban informality as an inherent outcome of uneven capitalist development, which depends on state responses to equalize differences, while the legalist school argues that informality is a product of state bureaucratic regulation, thus requiring lower levels of regulation and state intervention as a response. This framework has had relevance for wider debates on urban informality, beyond Latin American contexts (see e.g. AlSayyad, 2004; Roy, 2005; Kudva, 2009).

More recently, theoretical debates on urban informality have taken a "postcolonial" turn, based on the critique that dominant urban theories tend to rest on the experiences of a few Western cities. In calling for the "decolonization" of urban studies, Robinson (2002, 2006) highlights two current and prevalent approaches to contemporary urbanization – global cities, and the "third-world city" – and suggests that these impose limitations on future planning for cities around the world. At the same time, the slum "has become metaphor for the Third World city" (Roy, 2011: 225). In response, Robinson (2006) suggests constructing alternative urban theories reflecting the experiences of a wider range of cities, which are often "off the map," in order to creatively address the situation of low-income urban populations in cities around the world. Taking a postcolonial urban perspective therefore means "to pay attention to forms of urban political life, everyday survival, insurgency, and creative practice that animate the city" (Till, 2012: 6); and this has been applied in studies of urban informality in diverse contexts (see e.g. Legg and McFarlane, 2008; McFarlane, 2008).

However, as Varley (2013) points out, the interaction between postcolonial interpretations of urban informality and debates from Latin America is relatively limited, despite the potential fruitfulness of this interface. This is based in part on Latin American scholars' critique of post-colonialism's "universal historicism" which reflects primarily South Asianist and Africanist per-spectives (*ibid*). However, in keeping with postcolonial urban geographies from other contexts, postcolonial approaches to urban informality in Latin America often encompass both a histori-cal dimension – based on the parallels that exist between today's urban informality and preco-lonial urban forms – as well as a conceptual one, inverting entrenched categories and narrow understandings equating informality with poverty and marginalization, to celebrate the urban informal (Hernández, Kellett, and Allen, 2010). Varley's (2013) critical analysis of this concep-tual move draws on Brillembourg's (2004 in Varley, 2013)[3] concept of "new slum urbanism," which suggests that the apparent disorder found in informal settlements can be read as resistance, and the alchemic use of waste materials for building as a response to capitalism's vicissitudes. In fact, the trope of "informality as resistance" risks generalizing about the form and causes of informality, which may include the creation and use of informality by urban elites as well as by low-income groups (see also Roy's, 2011 critique of "subaltern urbanism"). Varley's discus-sion reflects wider debates about the problematic binary conception of formal and informal (e.g. Angotti, 2013), and shows how attempts to overcome this dualism nevertheless risk repro-ducing it through characterizing informality as "other." Even accounts which emphasize the "heroic" nature of informality risk romanticizing poverty, with problematic implications for the provision of services and support.

In order to move beyond simplifying theoretical interpretations of informality, there is a need to acknowledge its heterogeneity and hybridity (Varley, 2013). Some newer works on urban informality in Latin America attempt to do this, for example by highlighting "the borderlands between the formal and the informal" (McCann, Fischer, and Auyero, 2014: 5). Indeed, it may be that informality in Latin American cities is "constitutive of the urban condition itself" (Hernán-dez, Kellett, and Allen, 2010: 184). Rodgers et al. (2012) respond to prevailing conceptions of the "fractured" Latin American city by arguing for a more systemic, multidisciplinary engage-ment with cities in the region. Meanwhile, Cordera, Ramírez, and Ziccardi (2008) explore new forms of social exclusion, particularly relating to access to employment, education, and other urban goods and services, highlighting the critical and still under-researched link between spatial and other forms of inequality. This work draws on and extends debates on the "new marginal-ity" (e.g. LARR, 2004), which suggest that while marginality implied the existence of places and people beyond the reach of formal institutions, a more recent focus on social exclusion refers to disadvantage as "produced by the institutions of the state" (Roberts, 2004: 196). Exemplify-ing this approach, Salazar (2012) brings together a collection of papers on formal and informal access to land and housing in Latin American cities, in the context of the neoliberalization of government policy. She concludes that the persistent use of informality as a strategy is based on urban inequality and the fundamental asymmetry between those who access formal and infor-mal land markets.

Contemporary issues

It has been suggested that new global, national, and urban forces are shaping the manifestations of poverty, inequality, and social exclusion in Latin American cities today (Cordera, Ramírez, and Ziccardi, 2008). New forms of residential division are replacing and negatively impacting on old models of informal urbanization as a platform for social mobility; thus, as urban devel-opment becomes synonymous with inequality, the question is raised as to what level of social fragmentation urban dwellers will tolerate (Duhau, 2008). Previous interpretations of urban informality as supporting social mobility, with a focus on the individual agency of informal settlement residents, have therefore given way to a resurgent interest in the structural constraints within which these residents are operating under neoliberal globalization.

Such moves towards a more neoliberal urban environment can be seen, for example, in housing policy shifts away from the regularization of informal neighbourhoods towards facilitating the mass provision of formal housing for low-income urban populations, especially in the larger economies of the region such as Mexico, Brazil, and Colombia. This has been underpinned by the assumption that mortgage finance systems are better suited to efficiently meeting demand for low-income housing than incremental and often informal housing development processes. Mexico's housing policy trajectory illustrates this shift. Since the 1990s, the dominant form of urban shelter production has shifted from incremental, usually informal, self-built housing, to one where "housing is built on speculation by private-sector homebuilders and purchased with mortgages" (Monkkonen, 2011: 2). The state's role has thus changed from one of constructing or facilitating low-income housing, to acting as a "simple individual mortgage financier," with the construction process dominated by private sector developers (such as Casas Homex and Casas Geo) supported by international finance (Puebla, 2002 in Bayón and Saraví, 2013: 6). These private developers provide low-cost, low-quality housing at scale which is purchased by low- to middle-income households using mortgage credits accessed through workers' provident funds.[4]

However, this shift has not necessarily had the desired effect of preventing further informal housing production through the provision of low-cost formal housing. While some low-income households who may previously have resorted to informal housing have been assisted to purchase a home with housing finance, such mortgages are usually inaccessible to those working in the informal sector, or earning less than four minimum salaries (Boils Morales, 2008), meaning the poorest sectors of society continue to access land for housing through informal mechanisms (Lombard, 2016). Additionally, such developments have led to urban sprawl due to their construction on cheap peripheral land, thus inflating land values in areas formerly accessible to low-income communities. Skewes' (2005) analysis of the shift from informal to formal housing in Chile suggests that the new design of formal neighbourhoods erases the rich texture of informal settlements: in contrast to previously close-knit social relations, people are strangers to their neighbours, and the sense of social regulation through community is lost as people shut themselves up in their new homes.

Such development patterns have arguably also reinforced the pervasive tendency towards urban fragmentation in Mexican cities, as elsewhere in Latin America. Bayón and Saraví's (2013: 2) study of Mexico City suggests that "[t]he process of social commodification that inspired the neoliberal experiment . . . has left its imprint on urban space," through new patterns of urbanization in the context of inequality. This is particularly seen in the gentrification of central areas and the emergence of exclusive residential areas, accompanied by the continuing growth of poor settlements at the urban periphery through informal expansion, and over the last decade, mass housing complexes. Thus the consolidation of urban informal settlements is no longer "the dominant form of urban integration for the working classes" (Duhau, 2008); the stagnation of socioeconomic indicators in these areas suggests the concentration and expansion of urban poverty, and diminishing prospects of social mobility for their residents (Bayón and Saraví, 2013).

Such suggestions resonate with a more pessimistic strand of debates examining the effects of neoliberalization in Latin America on informal settlements. Auyero (1999) explores how informal neighbourhoods in Buenos Aires, which have experienced material improvements, continue to attract stigmatization from both outside and within, as residents discriminate internally against specific groups such as youths and immigrants, leading one to reflect that "this is a lot like the Bronx." This suggests a weakening of links between informal settlements and the rest of the city, as they become "spaces of survival for those 'excluded' . . . a desolate space of despair, of social immobility and of pervasive physical and social insecurity" (Auyero, 1999: 47). Such views can also be detected in two important longitudinal ethnographic studies of informal settlements. Moser's (2009) 30-year study in Guayaquil, Ecuador, explores the dynamic nature of urban poverty, relating to changes in urban dwellers' perceptions, aspirations, and expectations, to suggest that escaping poverty is increasingly subject to structural factors. Meanwhile, Perlman (2010: 165) suggests that "the most dramatic and devastating change for Rio's poor over the last three decades has been the growth of lethal violence," as favelas have gone from a relatively peaceful and socially cohesive environment to one suffering from an "epidemic of violence." Importantly, these works also start to redress the lack of informal settlement residents' voices in accounts of these places which characterizes much of the research of earlier decades.

Conclusion

This somewhat pessimistic outlook is tempered to some extent, by researchers suggesting alternative narratives, both for ongoing investigation into this area, and for generating improved prospects for residents in the "post-noeliberal" context. This term, which is often associated with leftist regimes such as that of Hugo Chávez in Venezuela, Evo Morales in Bolivia, Rafael Correa in Ecuador, and Luiz Inácio 'Lula' da Silva in Brazil, refers to their diverse responses to neoliberalism's perceived excesses; it therefore denotes "both a utopian project, and a set of emancipatory political projects aimed at overcoming the ideological and institutional heritage of neoliberalism" (Yates and Bakker, 2014: 64). The policies and programmes that have tended to characterize these regimes, such as anti-poverty (and often cash transfer) programmes, a strengthened state, and increasing investment in public goods including housing, suggest positive changes for the residents of urban informal settlements. However, more recent electoral shifts and leadership changes are likely to bring about further re-evaluation of the long-term effects of such regimes and their policies, including in urban contexts.

The post-neoliberal era has also seen a revived interest in activism both within and around informal settlements in Latin America, in the context of a resurgent social mobilization and the "new politics of participation" (Yates and Bakker, 2014: 64). In particular, the ongoing engagement in low-income housing of activist organizations, most notably the Habitat Internacional Coalición Latin America, has provided a focal point for researchers in the region (e.g. Guerrero, 2010) but also a significant channel for input into regional and global policy formulation, particularly via the New Urban Agenda formalized at the Habitat III conference in Quito in 2016, and the new "urban" SDG. These movements concur with a renewed emphasis on the enduring vitality and solidarity that can be encountered in urban informal settlements, found in some of the newer works in the field (e.g. McCann, Fischer, and Auyero, 2014). With this in mind, activism seems to represent an enduringly important area of study for those interested in improved prospects for the region's millions of informal settlement residents.

Finally, alternative narratives have been suggested from both a theoretical and a longitudinal perspective. Taking up the imperative to "decolonize" urban studies found in postcolonial urban geography, recent research has employed a comparative urbanism lens to bring to bear long-standing conceptualizations of urban informality grounded in the Latin American experience on burgeoning informal urban practices in the context of neoliberal economic policies in the Global North (e.g. Durst and Wegmann, 2017; see also Auyero, 2011; Porter, 2011). Meanwhile, building on earlier longitudinal studies, Ward et al. (2011) have explored the prospects for inter-generational wealth transfer in the context of neighbourhoods where intestacy and disputes over property are high, but where asset transfer can represent a real prospect for social change.

Such efforts suggest that the field of research and policy relating to informal settlements in Latin America will retain significance well into the 21st century.

Notes

1 Neoliberalism has been defined as an economic and political project based on trade liberalization, privatization, and market-orientated management of the public sector (Perreault and Martin, 2005: 192). While often understood as a global phenomenon with generalized ideologies and practices, its local and regional variations may be contextually driven and diverse (ibid).

2 Subdivision refers to the division of land into plots and its subsequent sale. Occupants normally buy the land from an informal developer, the landowner, or an intermediary, and although the sale may be legal or semi-legal, construction often contravenes local regulations, due to the land being considered unsuitable for development, or lack of compliance with planning laws or building standards (Durand-Lasserve and Royston, 2002: 5).

3 See also Brillembourg, Feireiss, and Klumpner (2005).

4 Workers provident funds are defined by Monkkonen (2011) as "specialized financial institutions that both are lenders for housing and pension funds" based on mandatory contributions. Institutions such as INFONAVIT and FOVISSSTE are effectively government-sponsored public housing funds for public sector workers, based on contributions.

References

AlSayyad, N. (2004) Urban informality as a "new" way of life. In A. Roy and N. Alsayyad (eds) *Urban Informality: Transnational Perspectives from the Middle East, Latin America and South Asia*. Oxford: Lexington, pp. 7–30.

Angotti, T. (2013) Urban Latin America: Violence, enclaves and struggles for land. *Latin American Perspectives* 40(2): 5–20.

Auyero, J. (1999) 'This is a lot like the Bronx, isn't it?' Lived experiences of marginality in an Argentine slum. *International Journal of Urban and Regional Research* 23: 45–69.

———. (2011) Researching the urban margins: What Can the United States Learn from Latin America and Vice Versa? *City and Community* 10: 431–436.

Azuela, A. and Tomas, F. (1996) (eds) *El acceso de los pobres al suelo urbano*. New Ed, online. Mexico City: Centro de Estudios Mexicanos y Centramericanos.

Bayón, M. and Saraví, G. (2013) The cultural dimensions of urban fragmentation: Segregation, sociability, and inequality in Mexico City. *Latin American Perspectives* 40(2): 35–52.

Boils Morales, G. (2008) Segregación y modelo habitacional en grandes conjuntos de vivienda en México. In R. Cordera, P. Ramírez and A. Ziccardi (eds) *Pobreza, Desigualdad y Exclusión Social en la Ciudad del Siglo XXI*. Mexico City: Siglo XX/UNAM, pp. 273–287.

Bredenoord, J. and Verkoren, O. (2010) Between self-help – and institutional housing: A bird's eye view of Mexico's housing production for low and (lower) middle-income groups. *Habitat International* 34: 359–365.

Brillembourg, A., Feireiss, K. and Klumpner, H. (2005) (eds) *Informal City: Caracas Case*. Munich: Prestel.

Castells, M. (1983) *The City and the Grassroots: A Cross-Cultural Theory of Urban Social Movements*. London: Edward Arnold.

Cordera, R., Ramírez, P. and Ziccardi, A. (2008) (eds) *Pobreza, Desigualdad y Exclusión Social en la Ciudad del Siglo XXI*. Mexico City: Siglo XXI.

Corporación Andina de Fomento/Banco de Desarrollo de América Latina. (2017) *Crecimiento urbano y acceso a oportunidades: Un desafío para América Latina*. Bogota: CAF.

Dagnino, E. (2007) Citizenship: A perverse confluence. *Development in Practice* 17(4–5): 549–556.

Davis, M. (2006) *Planet of Slums*. London: Verso.

De Soto, H. (2000) *The Mystery of Capital: Why Capitalism Triumphs in the West and Fails Everywhere Else*. New York: Basic Books.

Duhau, E. (2008) División social del espacio y exclusión social. In R. Cordera et al. (eds) *Pobreza, Desigualdad y Exclusión Social en la Ciudad del Siglo XXI*. Mexico City: Siglo XXI.

Durand-Lasserve, A. and Royston, L. (2002) *Holding their Ground: Secure Land Tenure for the Urban Poor in Developing Countries*. London: Earthscan.

Durand, J. (1983) *La ciudad invade al ejido. Proletarización, urbanización y lucha política en el Cerro del Judío, D.F.* Mexico City: Ediciones de la Casa Chata.

Durst, N. and Wegmann, J. (2017) Informal housing in the United States. *International Journal of Urban and Regional Research* 41(2): 282–297.

Fernandes, E. (2011) *Regularization of Informal Settlements in Latin America.* Policy Focus Report. Cambridge, MA: Lincoln Institute of Land Policy.

Fox, S. (2014) The political economy of slums: Theory and evidence from Sub-Saharan Africa. *World Development* 54: 191–203.

Gilbert, A. (1994) *The Latin American City.* London: Latin American Bureau.

Gough, K. and Kellett, P. (2001) Housing consolidation and home-based income generation: Evidence from self-help settlements in two Colombian cities. *Cities* 18(4): 235–247.

Guerrero, R. (2010) El problema de la vivienda y hábitat popular en América Latina: Análisis de contribuciones conceptuales y metodológicas de la red HIC-AL. *Revista INVI* 68(25): 185–208.

Gulyani, S. and Talukdar, D. (2008) Slum real estate: The low-quality high-price puzzle in Nairobi's slum rental market and its implications for theory and practice. *World Development* 36(10): 1916–1937.

Hardoy, J. (1982) The building of Latin American cities. In A. Gilbert, J. Hardoy and R. Ramírez (eds) *Urbanization in Contemporary Latin America, Critical Approaches to the Analysis of Urban Issues.* Chicester: John Wiley & Sons, pp. 19–33.

——— and Satterthwaite, D. (1989) *Squatter Citizen.* London: Earthscan.

Hernández, F., Kellett, P. and Allen, L. (2010) *Rethinking the Informal City: Critical perspectives from Latin America.* Oxford: Berghahn.

Imparato, I. and Rusler, J. (2003) *Slum Upgrading and Participation: Lessons from Latin America.* Washington, DC: World Bank.

Jones, G. and Ward, P. (1994) *Methodology for land & housing market analysis.* London: UCL Press.

Kudva, N. (2009) The everyday and the episodic: The spatial and political impacts of urban informality. *Environment and Planning A* 41(7): 1614–1628.

LARR. (2004) From the marginality of the 1960s to the "new poverty" of today: A LARR Research Forum. *Latin American Research Review* 39(1): 183–197.

Legg, S. and McFarlane, C. (2008) Ordinary urban spaces: Between postcolonialism and development. *Environment and Planning A* 40: 6–14.

Lewis, O. (1952) Urbanization without breakdown: A case study. *The Scientific Monthly* 75(1): 31–41.

———. (1967) *La Vida: A Puerto Rican Family in the Culture of Poverty – San Juan and New York.* London: Secker and Warburg.

Lombard, M. (2013) Citizen participation in urban governance in the context of democratization: Evidence from low-income neighbourhoods in Mexico. *International Journal of Urban and Regional Research* 37(1): 135–150.

———. (2016) *Land tenure and conflict in urban Mexico, GURC Working Paper 11.* Manchester: Global Urban Research Centre.

Lomnitz, L. (1977) *Network and Marginality: Life in a Mexican Shanty Town.* New York: Academic Press.

Lustig, N., Lopez-Calva, L. and Ortiz-Juarez, E. (2012) Declining inequality in Latin America in the 2000s: The cases of Argentina, Brazil, and Mexico. *World Development* 44: 129–141.

Mangin, W. (1967) Latin American squatter settlements: A problem and a solution. *Latin American Studies* 2: 65–98.

McCann, B., Fischer, B. and Auyero, A. (2014) *Cities from Scratch, Poverty and Informality in Urban Latin America.* London: Duke University Press.

McFarlane, C. (2008) Urban shadows: Materiality, the "Southern City" and urban theory. *Geography Compass* 2(2): 340–358.

Miranda, L. (2002) A new mystery from de Soto? A review of *The Mystery of Capital*, Hernando de Soto, 2001. *Environment and Urbanization* 14(1): 263–264.

Monkkonen, P. (2011) The housing transition in Mexico: Expanding access to housing finance. *Urban Affairs Review*, 8 March.

Montaño J. (1976) *Los Pobres de la Ciudad en los Asentamientos Espontáneos.* Mexico City: Siglo Veintiuno.

Moser, C. (2009) *Ordinary Families, Extraordinary Lives: Assets and Poverty Reduction in Guayaquil, 1978–2004.* Washington, DC: Brookings Institution Press.

Peattie, L. (1990) Participation: A case study of how invaders organize, negotiate and interact with government in Lima, Peru. *Environment and Urbanization* 2(1): 19–30.

Perlman, J. (1976) *The Myth of Marginality: Urban Poverty and Politics in Rio de Janeiro*. Berkeley, CA: University of California Press.

———. (2010) *Favela: Four Decades of Living on the Edge in Rio de Janeiro*. Oxford: Oxford University Press.

Perreault, T. and Martin, P. (2005) Geographies of neoliberalism in Latin America. *Environment and Planning A* 37(2): 191–201.

Porter, L. (2011) Informality, the commons and the paradoxes for planning: Concepts and debates for informality and planning. *Planning Theory and Practice* 12(1): 115–153.

Portes, A. (1972) Rationality in the slum: An essay on interpretive sociology. *Comparative Studies in Society and History* 14(3): 268–286.

Rakowski, C. (1994) The informal sector debate, part 2: 1984–1993. In C. Rakowski (ed) *Contrapunto: The Informal Sector Debate in Latin America*. Albany, NY: State University of New York Press, pp. 31–50.

Reid, M. (2007) *Forgotten Continent: The Battle for Latin America's Soul*. London: Yale University Press.

Roberts, B. (2004) From marginality to social exclusion: From laissez faire to pervasive engagement. *Latin American Research Review* 39(1): 195–197.

Robinson, J. (2002) Global and world cities: A view from off the map. *International Journal of Urban and Regional Research* 26(3): 531–554.

———. (2006) *Ordinary Cities: Between Modernity and Development*. Abingdon: Routledge.

Rodgers, D., Beall, J. and Kanbur, R. (2012) (eds) *Latin American Urban Development into the 21st Century: Towards a Renewed Perspective on the City*. Basingstoke: Palgrave Macmillan.

Roy, A. (2005) Urban informality: Toward an epistemology of planning. *Journal of the American Planning Association* 71(2): 147–158.

———. (2011) Slumdog cities: Rethinking subaltern urbanism. *International Journal of Urban and Regional Research* 35(2): 223–238.

Salazar, C. (2012) (ed) *Irregular. Suelo y mercado en América Latina*. Mexico City: Colegio de Mexico.

Skewes, J. (2005) De invasor a deudor: el éxodo desde los campamentos a las viviendas sociales en Chile. In A. Rodríguez and A. Sugranyes (eds) *Los con techo: un desafío para la política de vivienda social*. Santiago: Ediciones Sur, pp. 101–122.

Till, K. (2012) Wounded cities: Memory-work and a place-based ethics of care. *Political Geography* 31: 3–14.

Turner, J. (1972) Housing as a Verb. In J. Turner and R. Fichter (eds) *Freedom to Build: Dweller Control of the Housing Process*. New York: Collier-Macmillan, pp. 148–175.

UN-Habitat. (2003) *The Challenge of the Slums: Global Report on Human Settlements*. Nairobi: United Nations Human Settlements Programme.

Van Lindert, P. (2016) Rethinking urban development in Latin America: A review of changing paradigms and policies. *Habitat International* 54: 253–264.

Varley, A. (2013) Postcolonialising informality? *Environment and Planning D* 31(1): 4–22.

———. (2016) Property titles and the urban poor: From informality to displacement? *Planning Theory and Practice* 18(3): 385–404.

Vellinga, M. (1989) Power and independence: The struggle for identity and integrity in urban social movements. In F. Schuurman and T. van Naerssen (eds) *Urban Social Movements in the Third World*. London: Routledge, pp. 151–176.

Ward, P., Jiménez, E., Grajeda, E. and Ubaldo, C. (2011) Self-help housing policies for second generation inheritance and succession of "The House that Mum & Dad Built". *Habitat International* 35: 467–485.

Yates, J. and Bakker, K. (2014) Debating the 'post-neoliberal turn' in Latin America. *Progress in Human Geography* 38(1): 62–90.

45

URBAN MOBILITY IN LATIN AMERICA

Fábio Duarte

Introduction

Since 2000, more than 115 million people came to live in urban areas in Latin America. Migration from rural to urban areas is still a major cause of population growth in cities in the region – by 2050, 90% of Latin Americans will be living in cities (UN-Habitat, 2016). This phenomenon is far from new. Throughout the 20th century, in particular after the urbanization boom starting in the 1950s, this trend caused infrastructural, economic, environmental, and social disruptions in cities throughout the continent. Uncontrolled urban expansion destroyed large swaths of natural areas, including forests, animal habitats, and rivers – often jeopardizing the lives of those who moved to cities in the first place.

People move to cities in search for a better quality of life, which in contemporary society implies access to jobs, education, health services, and social welfare. Urbanization is characterized by a diversity of interdependent social, economic, and cultural activities. Therefore, each of the abovementioned aspects are interlocked: for instance, better jobs come with better education, and better jobs and better salaries create an economic positive cycle that brings more services to specific areas.

Life in cities is heavily dependent on two spatial characteristics: density and mobility. Density promotes more encounters among people, creates the need for more services (and also more job opportunities), and at the same time density optimizes the provision of services and infrastructures. Noulas et al. (2012) have shown, based on the movements of 925,030 users of the location-based social network Foursquare in 11 countries in four continents, that "intervening opportunities," which is tied to density, is the key predictor of movements and interactions. Density and physical proximity, however, have limits, sometimes dictated by the lack of space or resources, sometimes dictated by insalubrity, or by the exhaustion of infrastructure. Thus, human settlements sprawl over larger areas. However, in order to keep the advantages of urban life and social interactions that rely upon "intervening opportunities" such as access to jobs and housing, social activities and health services, all these elements must be connected. Therefore, when density and proximity do not guarantee a good urban life, mobility becomes a key factor to give people access to such activities.

Not surprisingly, mobility has become a paramount measure of urban quality of life. Access to good mobility services and infrastructures determines who has access to what in the city. Consequently, mobility has become a highly contentious social topic.

On the one hand, Latin America is an already highly urbanized region whose urban population is still growing, putting constant pressure on urban areas; on the other hand, most countries in the region do not have enough financial resources to invest in urban services and infrastructures in general, and in urban mobility in particular.[1] These two characteristics mean that most Latin Americans experience insufficient urban mobility. It is now clear that inadequate urban mobility is not only a matter of low-quality roads, vehicles, infrastructures, and services, but a matter of depriving a large number of people of life opportunities, some of which are central for survival.

Fast-growing cities and available resources are critical to the success of urban mobility provision. Take China, as a contrasting example to Latin America. Until recently a markedly rural country, China has recently experienced unprecedented rates of urbanization. In 1960, 16% of the population lived in cities; in 2015 this rate had surpassed 55% (World Bank, 2017). At the same time, the subway network in China expanded dramatically: in 1990, only three Chinese cities had subways. By 2020, this number will have reached 41. Beijing alone expanded its subway network from two lines and around 50 kilometres in 2000, to 19 lines and more than 570 kilometres in 2016 – with expansion projected to reach 1,000 kilometres in 2020 (Riedel, 2014).

São Paulo, economically the most important city in South America (Hoornweg et al., 2010), has implemented only 72 kilometres of subway lines since it opened the first line in 1974. Mexico City, which inaugurated its first line in 1969, has the most extensive subway network in Latin America, and reached 227 kilometres in 2015.[2] Considering the overall economic and political scenario in Latin America over the past decades, such an aggressive investment in infrastructure is not on the horizon in any country.

Solving the mobility problem is far beyond the scope of a single chapter and it is a theme that is vexing some of Latin America's largest institutions. It is also not the goal of this chapter to point out all the malaises people endure due to bad transportation – others have done so in a very compelling way (see for example Vasconcellos, 2001). The goal of this chapter is twofold: on the one hand, to highlight that, despite all the major and undeniable problems, Latin America has also implemented quite innovative urban mobility solutions; and on the other hand, some challenges Latin American cities are facing help us to understand global mobility challenges in critical and novel ways.

Some of the solutions first proposed in Latin America have even been influential worldwide. This is the case of the Bus Rapid Transit (BRT), a massive public bus transportation system. BRT's goal is to achieve performance levels similar to subways or other rail systems but using a much cheaper technology: buses. In this chapter, I highlight a few cities, such as Curitiba, Bogotá, Santiago, and Mexico; some of them for the innovations they have proposed, others because there are examples of key challenges BRT might face when integrated with other transportation systems. In other cases, the examples highlighted here are less related to the novelty of the systems, and more because their implementation bring to the forefront sociotechnical aspects of public transportation worth discussing. This is the case of a recent boom in investments in cycling – from the creation of large networks of bicycle lanes to the implementation of bike-sharing schemes.

It might be tempting to assess Latin American transportation initiatives simply as attempts to keep up the pace with cities around the world. Such initiatives to foster the use of bicycles, either led by national governments or grassroots-based cycling events, do however result in material outcomes specific to Latin America that are worthy of analysis. What seems to be interesting, however, is how cycling mobility has become part of a broad strategy to emphasize political agendas marked by environmental sustainability; and how, in turn, bicycle activists used this moment to foster more socially oriented mobility policies in different cities.

Another recent transformation in urban mobility that has been happening at a global scale is the emergence of ride-sharing apps. Here I argue that such apps are less the result of a search for transportation solutions, and more the outcome of a sociotechnical combination of available and reliable communication infrastructures, and the widespread use of smartphones and social media. For this reason, ride-sharing apps have disrupted the taxi industry in many cities in Latin America – similarly to what has happened in many other countries. What is different here is that the informality of the transactions between drivers and passengers that characterizes ride-sharing apps has characterized point-to-point transportation solutions in Latin America for decades. Therefore, looking at this phenomenon in Latin America might shed new light to the ride-sharing industry and how it might be integrated to comprehensive mobility schemes.

Bus Rapid Transit: when Latin America is innovative

Bus Rapid Transit (BRT) proposes to offer the same level of service of subway or trams, but using buses. The underlying argument is that BRT is cheaper to build and operate, faster to cope with the expansion of the cities in developing countries, and is flexible to adapt to further changes in the cities due to fast urbanization rates. The emergence of BRT as a successful transportation system is incontrovertibly linked to Curitiba in Brazil. What makes the case of Curitiba even more special is that the city's recent history and socioeconomic features is arguably dependent on the BRT. Without the BRT, Curitiba would be another ordinary Brazilian state capital city.

Brazil experienced rapid urbanization between the 1940s and 1970s, mostly due to the migration of the population from rural areas to the cities. In Curitiba, the population grew from 140,000 to 600,000 people between 1940 and 1970 (IBGE, 2010). The city was growing radiocentrically, roughly following a plan proposed in 1943. Consequently, in the 1960s the road system in the city centre was constantly clogged. To aggravate the situation, public transportation was in the hands of many private bus companies, which decided freely where and when to operate. In the early 1970s, Curitiba implemented the first bus corridor of what would become the first known BRT. The plan encompassed five bus corridors with dedicated lanes, sided by slow-traffic lanes – fast-traffic lanes were laid down one block far from the bus corridor, in both directions. In part, decoupling the BRT system in three parts (comprised by the bus corridor with the slow-traffic lanes, plus the two fast-traffic lanes) had been done to avoid creating wide avenues inside the city, which often become urban barriers; in part, it was the consequence of a constraint: the city did not have money to buy land along the corridors and had to adapt the system to the existing road network (Duarte and Rojas, 2012).

In addition to the main bus lines running in segregated bus corridors, the system has continuously been complemented by a series of feeder lines, each with a specific function: for instance, the inter-neighbourhoods line, as its name says, links the neighbourhoods with BRT terminals without crossing the city centre; and the express line has stops only every few kilometres – with fewer stops, it makes the trip faster but only serves key points in the city.

In order to increase boarding speed, fares were supposed to be paid prior to boarding, and bus stops in the main corridors were elevated at the same level of the bus floor. The technology to lower the bus floor was not available when the system was implemented, and, in any case, it would take precious seconds of the boarding process, which is critical in massive transportation systems. Indeed, Curitiba's BRT bus stop created something arguably rare in the history of transportation: a bus stop, known as the tube station, became iconic of Curitiba – as the yellow cabs of New York, or the double-deck buses of London.

There was, however, a more complex aspect of Curitiba's BRT: rather than responding to demand, the BRT should prompt urban development. Newcomers and residents should be attracted to live along the BRT corridors. A good transportation system would help, but it would not be enough. Concomitant with the transportation system, Curitiba implemented a new land-use zoning that restricted high-density and vertical areas to the blocks adjacent to the BRT corridors. This land-use policy induced densification, which, in turn, would provide enough users to make the system successful.

Curitiba, now a 1.8-million-inhabitant city, has been operating its BRT for more than 40 years, although only a small fraction of its residents refer to the transport system in this way. In fact, it was not until the early 2000s that public officials knew that Curitiba had what was called elsewhere a BRT. For the city's officials, what existed was simply a bus transportation system: red bi-articulated buses in the structural lines, silver regular buses in the express lines, and yellow micro-buses serving specific areas of the city. With regards to having denser areas along the bus corridors, it simply made sense. Arguably, it was Bogotá that reminded Curitiba that it had implemented an innovative transportation system long ago. Bogotá paid homage to Curitiba by also making its bi-articulated buses red. While similarities abound, Bogotá faced a different set of challenges.

Bogotá implemented its BRT in 2000, when more than six million people already lived in the urban agglomeration (United Nations, 2001). Although the city's population was still growing around 2% annually, Transmilenio, as Bogotá's BRT is called, was implemented in consolidated central urban areas. And the implementation had to be done fast: with mayors serving only one four-year term, either the system was robust by the end of the term, or it would be easily abandoned by the next mayor who might have other priorities. Because of the large population and extensive urban area, the subway was the preferred solution by many. Smaller cities in the continent had subways, including Santiago de Chile. Even Medellin had a subway, which created a struggle between the cities over national pride.

BRT systems as cheaper transportation solutions have also been appealing in the context of economic constraints, and since 2000 Bogotá has implemented more than 80 kilometres of BRT corridors in addition to 600 kilometres of feeder routes, carrying 1.7 million passengers daily (Hidalgo et al., 2013). Also echoing a program adopted in Curitiba, in Bogotá some of the key BRT stations have important cultural facilities either as part of the same structure, or easily accessed within a walking distance. The idea is to transform BRT stations into regional nodes, by integrating different municipal services.

Despite having the largest subway network in Latin America, Mexico City also saw in the BRT a solution to complement its mobility infrastructure and to serve the increasing mobility demands of its 19 million inhabitants in the metropolitan area (ITDP, 2017). Mexico City's BRT, initially implemented in 2005, comprises six corridors, extending 125 kilometres, and carrying 1.1 million passengers per day. When first implemented, the system had to face a challenge common in developing countries: namely it had to deal with a myriad of poorly regulated private operators, who usually put up a determined show of resistance and fuel strong public opposition, by claiming that the new system will increase fares and offer fewer route options. Mexico City decided therefore to incorporate rather than replace existing operators – which involved the replacement of 900 minibuses with 230 articulated buses.

A public agency, Metrobús, was created to oversee the system's overall governance, while the operation remained with private operators, with a third-party collecting fares and distributing the revenues. With the expansion of the system, the complexity of creating more routes, with smaller passenger demand to be shared among operators, negotiations between incumbents and the government became ever more difficult, eventually leading to a more "forceful" strategy.

Still, Mexico City is an important case study about how to incorporate existing players when implementing a new transportation system (Flores-Dewey and Zegras, 2012).

Santiago de Chile also had a subway network when it implemented its BRT system in 2007 (ITDP, 2017). Currently, Santiago's BRT, called Transantiago, carries more than 350,000 passengers daily and is integrated with the subway – which extends for over 85 kilometres and is still expanding. Looking now, one could argue that overall the system is well fitted into the general mobility scheme of the city. However, what made Santiago's BRT first known among the general public and public transportation experts and scholars, were the multiple failures the system presented during the launch period. Bus schedules, routes structures, payment methods, and contractual relations with operators changed abruptly, without a clear plan – or, at least, a clear communication plan. The immediate result was "catastrophic" (Gómez-Lobo, 2012), and in the first weeks of operation buses received hideous graffiti or were burned down. I visited the city during the launch period, it was hard to understand how the system worked, and even harder to make people understand why I had come to see it.

In the following years, the system has proven to be beneficial, and has resulted in a reduction in overall travel time and accidents involving buses and cars (Gómez-Lobo, 2012). However, the system was tainted to the point that the Chilean president came on television to promise the system would be reformed – in fact, for its critics, Transantiago represented the failure of a technocratic policy-making approach adopted by the national government (Ureta, 2014).

While Bogotá created a strong marketing strategy in support of its BRT, Santiago did not pay enough attention to how the public would react to the system. It did not invest in an education campaign on the values and functions of Transantiago, and it failed to address its multiple failures quickly enough. What both cases show is that beyond the functional aspects of the system, public transportation systems must rely on strong public communication strategies to attract passengers and make them active participants in the success of the system.

Whether facing intermodal competition with different politicians supporting alternately the BRT or the subway, as in Lima (Chayacani Mallqui and Pojani, 2017), or as a long-awaited solution to a chronically under-invested transportation infrastructure, such as in Havana, where lack of maintenance puts 50% of the fleet of articulated buses out of operation in any given day (Warren and Ortegon-Sanchez, 2015), BRT is undoubtedly an important transportation technology in many Latin American major cities. However, what the examples of BRT systems discussed here show is how multiple aspects beyond the technical and narrowly focused transportation solutions play a decisive role in the success of a system. For example, Curitiba's BRT would probably not be as successful if the land-use zoning was not tied to the BRT infrastructure, forcing denser areas to be created exclusively along bus corridors. Bogotá's Transmilenio reinforced the idea that a comprehensive plan is necessary to have the system implemented fast, as well as the development of a communication campaign to gain passengers' support. This support is required not only to use the system, but also for sometimes tough technical and political decisions, such as when the implementation of Transmilenio disrupted some of the traditional minibuses' companies. Understanding that the success of a transportation system is tied to other aspects of urban life, from land use to economic development and community building, is one of the key lessons we can learn from BRT projects developed in Latin America.

Making cycling visible in Latin America

Latin Americans are frequent cyclists. The problem is that most Latin Americans do not cycle out of choice, but rather as a result of a lack of alternative transport options. Those who cycle often live in areas not served by public transportation, or they cannot afford transit fares.

The recent history of the use of bicycles in Latin American cities can be seen as going through four phases, which sometimes overlap: bicycles as a toy, as a step towards motorcycles, as an environmental-friendly mode, and more recently, following the global trend of bike-sharing, as symbols of civic engagement.

The bicycle is ingrained in the Latin American imagination either as a working-class mode of transport, or as a children's toy (Medeiros and Duarte, 2013). In the 1970s, TV and magazines ran a few advertisements targeted at workers, in which they depicted the bicycle as a step toward motorcycles and cars. Omnipresent, though, was the bicycle as a children's toy, with some adverts focusing on sporting bicycles. Seldom, if ever, were bicycles shown as a daily mode of transport.

Things did not change much until the early 1990s when Curitiba implemented almost 100 kilometres of bike paths, most of them linking recently created parks. Without diminishing the positive aspects of both initiatives – the creation of parks and bicycle paths linking them – it is worth noting that this happened around 1992, when Brazil hosted the United Nations Conference on Environment and Development in Rio Janeiro.

In the early 2000s, bicycles began to regain some importance in major Latin American cities. Some of the most important cities created initiatives to promote the use of bicycles, several of them within the *Ciclovías* program, in which cities close streets to motorized traffic and open them to non-motorized traffic, usually during weekends. Bogotá became a flagship city of ciclovías, with 113 kilometres of roads open exclusively to active modes more than 60 days per year, reaching up to 1.5 million users per event, with city officials travelling the world to promote the program (Sarmiento et al., 2017).

Amid these initiatives, led mostly by municipal authorities, another phenomenon emerged: bike-sharing systems. They frequently work as a set of bicycles available to the public at dock stations spread across an urban area, and bicycles that can be used after getting access to them via a pay-as-you-use system, which works on top of an annual membership that includes any number of trips below 30 minutes. The number of bicycles available in bike-sharing systems around the world went from 13 in 2004 to 855 in 2014 – which arguably makes it the fastest-growing transportation system ever.

Dozens of cities in Latin America have also implemented bike-sharing systems, with Mexico City (6,500 bicycles), Buenos Aires (intends to have 3,000 bicycles), and Rio de Janeiro (2,600 bicycles) as important examples. Similar to what has been happening in other cities that have implemented bike-sharing systems, in Latin America it can be difficult to separate what is transportation planning and marketing strategy.

Let us take Rio de Janeiro as an example. In 2011, the city began the implementation of its bike-sharing system. Virtually all the stations were located along the shore in the most touristic areas. The bike-sharing system has been expanded since. Still, the distribution of the docks reveals two socioeconomic aspects. The first is that in general the docks are located in high- and medium-high income areas in the most central areas of the city, not in the poor and more peripheral areas, which are deprived of good public transportation and whose inhabitants either use bicycles more often for commuting or daily errands, or would benefit from a cheaper transportation solution. The second is that the city has several restrictions regarding outdoor media and advertising, especially in areas along the shore visited by tourists and higher-income residents. Not surprisingly, the docks in the initial phases of the bike-sharing systems were located there. Looking at some of the big sponsors of bike-sharing systems in the world we have private banks, which use the bike-sharing systems as an advertising outlet. Indeed, some of the biggest bike-sharing operators worldwide are media companies, such as Clear Channel and JCDecaux – which is also the case in Latin America; while banks are the key sponsors of the bike-sharing system in São Paulo, Rio de Janeiro, and Santiago.

Bicycles have also been playing a political role in Latin America. Echoing what has been happening in other parts of the world, politicians have been embracing bicycles to demonstrate their commitment to sustainable policies. The mayors of São Paulo (Fernando Haddad), Mexico City (Marcelo Ebrard), and Buenos Aires (Mauricio Macri) have been depicted riding bicycles – symbolically, this scene usually happens only during the political campaign and the inauguration day, not during regular working days and certainly not on a daily basis.

Cycling activists are taking advantage of the heightened interest in cycling to make the point that more serious policies towards bicycles and active modes of transport are necessary. This is what happened in Curitiba in 2013. The year was marked by nationwide municipal elections, and by a series of public manifestations that occurred throughout Brazil, triggered by the increase in the bus fare in many cities. The hundreds of thousands of people took to the streets to demand better and more comprehensive public transportation policies. As a response, the then mayor of Curitiba implemented a 4-kilometer bicycle lane in the city centre, which had two particular characteristics: despite the 100-kilometer bicycle network in the city, the newly implemented one was not linked to the network; and it opened only on Sundays. On the opening day, hundreds of cyclists cycled outside the bike lane, demanding a serious commitment from the next mayor. And as it happened the newly elected mayor cycled to his inauguration and created a special unit dedicated to bicycle planning (Duarte and Rojas, 2012).

While the future of bicycle transportation in Latin America is still unclear, so far, it has been experiencing phases of strong political support, followed by complete neglect. Still, signs throughout the continent, backed by global trends, seem to show that bicycles are slowly becoming part of mobility plans and initiatives.

New technologies: disruptive or adaptive?

As in many countries, ride-sharing apps have changed the taxi industry in Latin America. Taxi companies lost passengers when people began to rely on ride-sharing apps to request point-to-point and frequently single-occupancy rides. The emergence of ride-sharing apps was unexpected, mainly because it is less a transportation solution and more the outcome of the technological combination of faster, omnipresent, and more reliable cell-phone connections; the widespread use of smartphones; years of online shopping that had created secured payment systems; and social media, which created a general feeling of trust in sharing personal data. Rather than being specific to each market, the underlying logic of coach-sharing, apartment-sharing, or ride-sharing is the same.

For a long time, taxis have operated as the most common (if not the single) regulated individual point-to-point transportation in cities. If public transportation or private-owned car were not options, taxis would transport you. In the 2010s, private companies such as Uber launched smartphone apps that connect people willing to travel with other people willing to use their own car to drive these passengers, for a fee. Since people driving are not professional drivers, and these companies do not own or manage any fleet, what these companies provide is a social media platform to connect passengers and people driving these passengers. Usually these services are cheaper for users, and provide a seamless experience of calling a ride – without the inconveniences of hailing cabs in the streets or calling radio-taxi companies.

In recent years, there have been several clashes between taxi companies and drivers using ride-sharing apps, and many cities have blocked the use of these apps by either making them illegal, or approving them temporarily or restricting their use within delimited areas of the city. Soon taxi companies, struggling with the apps, created their own apps – but usually to little avail: in general, taxi companies did not understand the fact that the apps are a social media as well as a transportation solution. A social media where passengers and drivers know their names beforehand, rate each other, negotiate prices in advance (automatically, through surges) and after the trip (through app-based tips), and have a dual role of potentially being, at different times, driver and passenger.

This description could describe virtually any city where ride-sharing apps are used. What makes Latin America special is the presence of informal taxis in several of its cities. It could also fall into the definition of paratransit, a transportation option that fits between private cars and buses, that is flexible and ubiquitous, and, as advanced by Robert Cervero (1997: 235), could be improved by high-technology enhancements creating "more door-to-door, short-wait, effortless-transfer service," using automobiles to perform collective-ride modes. In some Latin American cities, paratransit responds for way over 50% of the public road-based modal splits, such as Caracas (74%) and Quito (90%) (Salazar Ferro and Behrens, 2015).

Similar to ride-sharing apps, informal taxis are also based in the momentary relation established between someone driving and another person riding the car, of a price negotiation between driver and passenger that varies constantly even along the same route. Indeed, even recent innovations in the ride-sharing industry, such as pooling systems, in which different passengers, unknown to each other, share the same ride, have a long history in Latin American metropolises.

The potential synergies between mobility schemes based on ride-sharing apps and public transportation is yet to be assessed. Similar to what happens with taxis, ride-sharing apps can be an option to avoid the use of private cars – either based on an economic decision or to bend car-use restrictions (Davis, 2017). Also, similar to what has been tried with paratransit with diverse levels of success (Salazar Ferro and Behrens, 2015), ride-sharing apps could help to improve the integration with massive transportation systems, such as regional trains, subway, or BRT, being an alternative solution to the last-mile problem, in which creating a bus feeder system is costly, and the absence of any option to cover the last mile between the station and passengers' homes discourages transit use.

In any case, the ride-sharing apps have been challenging public transportation authorities all over the world, and at the core of the issue is that this emergence of mobility networks that are flexible both in time and space, connecting users directly who perform alternately as providers and users of transportation, challenged the institutional aspects of current public transportation planning and operating authorities. Perhaps the large experience of Latin America with paratransit, which also performs in-between formal institutional structures, might propitiate a fertile environment for mobility innovations, in flexible and adaptive ways.

Conclusions

Latin America is a special case when it comes to public transportation. On the one hand, its cities resemble thousands of other cities across the world, with an important share of informal transportation modes sharing the roads with private cars, taxis, buses, pedestrians, and bicycles. On the other hand, the continent has been fostering some innovative transportation solutions. This is the case of the Medellín's aerial cable car, which has improved access to deprived communities settled in hilly areas of the city. Though innovative, this chapter has left this and other innovative projects aside, and has focused on a transportation solution that has been advocated for decades and that has seen important global expansion in recent years: the Bus Rapid Transit. The difference of projects such as Medellín's Metrocable and the BRT is that the former has been implemented in very particular urban contexts, whereas the latter has been praised and adopted in dozens of cities around the world.

All of this has been happening in densely urbanized areas. As noted above, in 30 years, 90% of Latin American population will be living in cities. In such a context, mobility is a crucial element to promote urban quality of life. Most Latin American cities are still very far from providing a fair public transportation system for their population. Moving in cities in this part of the world is a daily struggle. Although I have focused on major Latin American cities, medium-size cities have also adopted BRT. Likewise, programs to promote the use of bicycles have been expanded throughout the continent, and ride-sharing apps have been at the same time disrupting traditional modes (such as taxis) and creating novel forms of mobility in Latin America. Overall, this chapter argues that Latin America is an excellent example of how technocratic and technical transportation solutions (such as the BRT) combined with grass-root movements and app-based solutions are challenging the status quo of car-centric policy-makers. Whether these changes will come quickly enough for most people, is an open question. In the meantime, Latin Americans keep moving.

Notes

1 For detailed investments in infrastructure in Latin American countries, consult http://en.infralatam. info/dataviews/227373/transport/
2 For an overview of subway systems worldwide, consult http://mic-ro.com/metro/table.html. Official numbers for Mexico City, check www.metro.cdmx.gob.mx/operacion/cifras-de-operacion, and official numbers for São Paulo, check www.metro.sp.gov.br/metro/institucional/quem-somos/index.aspx

References

Cervero, R. (1997) *Paratransit in America: Redefining Mass Transportation.* Westport, CT: Praeger.

Chayacani Mallqui, Y. and Pojani, D. (2017) Barriers to successful Bus Rapid Transit expansion: Developed cities versus developing megacities. *Case Studies on Transport Policy* 5(2): 254–266.

Davis, L. (2017) Saturday driving restrictions fail to improve air quality in Mexico City. *Scientific Reports* 7: 1–8.

Duarte, F. and Rojas, F. (2012) Intermodal connectivity to BRT: A comparative analysis of Bogotá and Curitiba. *Journal of Public Transportation* 15(2): 1–18.

Flores-Dewey, O. and Zegras, C. (2012) The costs of inclusion: Incorporating existing bus operators into Mexico City's emerging bus rapid transit system. *Unpublished – Department of Urban Studies and Planning, Massachusetts Institute of Technology*, 26. Cambridge, MA: Department of Urban Studies and Planning, Massachusetts Institute of Technology.

Gómez-Lobo, A. (2012) *The Ups and Downs of a Public Transport Reform: The Case of Transantiago.* Facultad Economía y Negocios. Santiago de Chile: Universidade de Chile.

Hidalgo, D., Pereira, L., Estupiñan, N. and Jimenez, P. (2013) TransMilenio BRT system in Bogota, high performance and positive impact: Main results of an ex-post evaluation. *Research in Transportation Economics* 38(1): 133–138.

Hoornweg, D., Bhada, P., Freire, M., Trejos Gomez, C. and Dave, R. (2010) *Cities and Climate Change: An Urgent Agenda.* Washington, DC: World Bank.

IBGE. (2010) *Sinopse do Censo Demográfico 2010.* https://censo2010.ibge.gov.br/sinopse/index. php?dados=6 (Accessed 10 October 2017).

ITDP. (2017). *The BRT Planning Guide.* Washington, D.C.: Institute for Transportation and Development Policy.

Medeiros, R. and Duarte, F. (2013) Policy to promote bicycle use or bicycle to promote politicians? Bicycles in the imagery of urban mobility in Brazil. *Urban, Planning and Transport Research* 1(1): 28–39.

Noulas, A, Scellato, S, Lambiotte, R, Pontil, M, ans Mascolo, C (2012) Correction: A tale of many cities: Universal patterns in human urban mobility. *PLOS ONE* 7(9): e37027.

Riedel, H. U. (2014) Chinese metro boom shows no sign of abating. *International Railway Journal* 19 November. www.railjournal.com/index.php/metros/chinese-metro-boom-shows-no-sign-of-abating. html (Accessed 11 January 2018).

Salazar Ferro, P. and Behrens, R. (2015) From direct to trunk-and-feeder public transport services in the Urban South: Territorial implications. *Journal of Transport and Land Use* 8(1): 123–136.

Sarmiento, O., Diaz del Castillo, A., Triana, C., Acevedo, M., Gonzalez, S. and Pratt, M. (2017) Reclaiming the streets for people: Insights from Ciclovías Recreativas in Latin America. *Preventive Medicine* 103: 34–40.

UN-Habitat. (2016) *World Cities Report: Urbanization and Development: Emerging Futures.* Nairobi: United Nations Human Settlements Programme.

United Nations. (2001) *World Urbanization Prospects: The 2001 Revision.* Washington, DC: United Nations Population Division.

Ureta, S. (2014) Normalizing Transantiago: On the challenges (and limits) of repairing infrastructures. *Social Studies of Science* 44(3): 368–392.

Vasconcellos, E. (2001) Urban *Transport Environment and Equity: The Case for Developing Countries.* Abingdon, UK: Earthscan.

Warren, J. and Ortegon-Sanchez, A. (2015) Designing and modelling Havana's future bus rapid transit. *Proceedings of the Insitution of Civil Engineers – Urban Design and Planning* 169(2): 104–119.

World Bank. (2017) *The World Bank Data.* https://data.worldbank.org/indicator/SP.URB.TOTL.IN.ZS?end=2016&locations=ZJ&start=1961 (Accessed 8 October 2017).

46

OPPRESSED, SEGREGATED, VULNERABLE

Environmental injustice and conflicts in Latin American cities

Marcelo Lopes de Souza

Introduction: cities as symbols of 'development'? Environmental injustice in the urban context

Urbanization and (large) cities have been long seen as symbols of 'progress' or, as it has become more usual to say from the 1950s onwards, of 'development.' Growth and modernization theorists, for instance, usually overestimated the positive character of capitalist urbanization in the 'developing countries' (or 'underdeveloped countries,' as people used to say in the less euphemistic parlance of the 1960s), arguing that "continued concentration of economic growth in large cities is necessary to achieve economies of scale and increase externalities in the form of indirect costs and social and economic infrastructure because these are, in turn, the prerequisites for the subsequent growth needed to provide the resources required to overcome social deficiencies" (Berry, 1978: 51).

Even a modernization theorist as sophisticated as sociologist Bert Hoselitz emphasized that urbanization is a pre-condition for development, though he conceded that cities can be problematic under some circumstances (1960b: 185). He was among the first to acknowledge, for instance, that it would be too simplistic (and, in the end, wrong) to imagine that there is always a positive and strong correlation between industrialization and urbanization. Hoselitz suggested considering certain cities as 'parasitic,' while some others should be seen as 'generative':

> A city will be designated as generative if its impact on economic growth is favorable, i.e., if its formation and continued existence and growth is one of the factors accountable for the economic development of the region or country in which it is located. A city will be considered as parasitic if it exerts an opposite impact.
>
> *(Hoselitz, 1960b: 186–187)*

This attempt to classify cities, however, is itself too simplistic. Hoselitz was, admittedly, aware of the circumstance that the environment outside the city plays an important role and that cities take a particularly relevant position in this context: namely, the 'chief centers of cultural contact' between the so-called 'developed countries' and the 'underdeveloped' ones (Hoselitz, 1960a). What was of interest to him, however, was to know how cities are capable of assimilating cultural values conducive to development (in the Western, capitalist and, moreover, economistic sense) coming from abroad and through their influence contribute to overcoming values regarded as traditional (those of the 'folk-like society'). Hoselitz views the problems of 'underdeveloped countries' as linked to traditions that hinder capitalist economic 'development' and which will be overcome in the wake of the process of Westernization and modernization, rather than as problems whose factors are to be sought in the historical formation and reproduction of the capitalist world system. The process of urban development in 'underdeveloped countries' would ultimately repeat the trajectory of the 'developed' ones (Hoselitz, 1960a: 172). He was not able therefore to see or conceive how the transfer of 'modern' capitalist values does not necessarily contribute to an increase in human welfare, much less in a linear, contradiction-free way. Nor was he fully aware of the extent to which the centres of the capitalist world system and the capitalist elites of the 'underdeveloped' countries profit from certain 'parasitic' processes and from socio-spatial inequality at several levels – which do not always prevent technological modernization and economic growth, as historically shown by several Latin American countries in a very eloquent way.

In general, it can be said that the proponents of theories of modernization believed that urbanization is ultimately essentially positive, like the capitalist status quo itself (an interpretation to which W. W. Rostow's *The Stages of Economic Growth: A Non-Communist Manifesto* [Rostow, 1960] eloquently testifies). Often, however, the question arises of directing and disciplining the process of urbanization and 'development' (understood in its economic/capitalist sense), in order to minimize certain problems. The problems – such as poverty, housing shortage or lack of adequate housing, unemployment, violent crime, environmental pollution – were and are not seen as what they in fact are in the context of the capitalist mode of production, that is, as something inevitable and essential, but rather as mere 'imperfections,' which would be, for neoclassical and other conservative economists, reportable to 'market imperfections,' therefore as problems almost natural, rather than historically conditioned.

Be that as it may, growth and modernization theorists were only a subset of the legion of academics and researchers that have regarded urbanization and urban growth from an overoptimistic and insufficiently critical perspective in the context of an intellectually more or less conventional and politically more or less conservative understanding of the processes covered by the label 'development.' Now, if we take the word *development* simply as a convenient replacement for 'a process of socio-spatial change for the better,' and if we accept the economic process as something necessarily and logically subordinated (for ethical-political reasons) to goals such as *an improvement in quality of life of vulnerable and disadvantaged people* and above all *social justice*, then Latin American urbanization has been anything but an expression of true development. However, as we know, the word 'development' has been widely used as a synonym for capitalist 'economic development' – and from the perspective of capitalist 'economic development' ideology, growing and 'modernizing' cities have always been among the best symbols of a country's or region's prosperity, regardless of the acute contradictions this presupposition embodies.

This chapter aims to show the extent to which the dimension of *environmental injustice* corresponds to a key parameter of analysis of the grotesque contradictions of Latin American urbanization – and at the same time a parameter of reflection about the incongruences and weaknesses of the 'economic development' discourse as such. Conceptually, environmental injustice refers to the social and spatial inequality of the distribution of the burden represented by the generation of contaminants as by-products of industrial processes. More broadly, it concerns any process in which the potential harms arising from the exploitation and use of resources and the generation of undesirable residues are socio-spatially distributed asymmetrically, due to class cleavages and other social hierarchies.

The theme of environmental injustice emerged and consolidated in the United States in the 1980s. Its roots, however, are older: they refer, according to several authors (see for example Bullard, 2000: 14, 29), to the struggles for civil rights of African-Americans that became famous from the 1950s onwards. In Latin America, it was at the beginning of the 21st century that debates and activism explicitly around environmental justice (*justiça ambiental, justicia ambiental*) began to be articulated in several countries (see for example Acselrad [2010] specifically about Brazil).

In the remainder of this text, I concentrate on three main types of urban environmental problem or conflict: a) waste dumping and its links with residential segregation; b) exclusionary forms of 'environmental protection'; c) unequal access to water. Although there are other kinds of conflict that could be understood as 'environmental' in Latin America's cities, these three types are representative of the very problematic landscape people (especially poor people) have faced for generations in this continent.

1. Waste dumping and its links with residential segregation

We can define waste dumping as the practice of transferring waste (industrial, household, bio-medical, or nuclear waste) from one place to another, in order to benefit from less tough laws, weak law enforcement, or a supposed weaker resistance potential of the residents of a specific place. The phenomenon of waste dumping, which has been extensively studied at an international level (see for example Pellow, 2007), can also be observed and has great importance in relation to other scale levels, including the local one. In fact, much of the public visibility of the problem has been linked from the outset to the intra-urban scale: the environmental justice movement, which began in the United States in the 1980s (Bullard, 2005) and later spread throughout the world, had in waste dumping its very fulcrum.

On the local scale, the relationship between residential segregation and waste dumping is very evident. But there is a further element: as David Pellow (2007) summarized, *racism* is also a considerable part of the problem. Reverberating Charles Mills' work on the connections between the images of people of colour and ideas such as 'barbarism,' 'filth,' 'dirt,' and 'pollution,' Pellow draws our attention to the fact that "African peoples themselves are viewed by many whites as a form of pollution, hence making it that much easier to contain industrial waste and factory pollution in their nations and segregated neighborhoods"; and as he almost immediately adds, "[i]mmigrants, indigenous populations, and peoples of colour are viewed by many policy makers, politicians, and ecologists as a source of environmental contamination, so why not place noxious facilities and toxic waste in the spaces these population occupy or relegate these groups to spaces where environmental quality is low and undesirable?" (Pellow, 2007: 97–98).

In Latin American cities, waste dumping and its links with residential segregation are no less perceptible than in the United States. In reality, showing patterns of social inequality and spatial disparities that are even more brutal than those in the United States, this relationship is in Latin America often much sharper and more dramatic. The question of racism has a complexity that varies from country to country and from region to region in that very heterogeneous universe imprecisely and Eurocentrically called 'Latin America'; but it is obviously also present there. Be it the fate of that part of the urban poor in Brazilian cities (largely phenotypically Afro-descendants, especially in certain regions) who, in the largest urban centres, make a living collecting garbage in *lixões* (literally 'garbage dumps,' i.e. sanitary landfills); be it the people derogatorily called *cabecitas negras*[1] who, in the outskirts and *villas* (shantytowns) of Buenos Aires, particularly in the Matanza-Riachuelo basin, bear an enormous *sufrimiento ambiental* (environmental suffering) because of the contamination of air, water, and soil with heavy metals. In these and in many other cases, waste dumping lies at the heart of a framework of environmental injustice that is very often tragic in Latin American cities.

While a considerable portion of the Latin American literature about topics such as environmental contamination in cities ignore or avoid the links between urban environmental problems and residential segregation, the authors who are clearly committed to an environmental justice perspective obviously put their concerns regarding contamination of the air, water, and soil into a broader and socially more critical frame. The several studies carried out by Argentine researchers on the emblematic case of Villa Inflamable (municipality of Dock Sud, Great Buenos Aires) and similar situations can be mentioned here as illustrations of such a socially critical and more concrete treatment of urban environmental problems (see for instance Merlinsky, 2013a, 2013b; Scharager, 2016). Sometimes, even official documents do some justice to the proverbially bad environmental conditions that affect the health of hundreds of thousands of poor people in the Great Buenos Aires area. For instance, a report published by the Auditor General of the City of Buenos Aires shows in some detail that many *villas* located in the Matanza-Riachuelo basin are so polluted that diseases caused by contamination by lead and other heavy metals are usual among the residents of that area, a problem particularly severe in the case of children (Gaiso, 2014).

The social mobilization around the problems of the Matanza-Riachuelo basin had a historic meaning for the environmental struggles in Argentina, as several authors have shown. After decades of environmental contamination and years of socio-political pressure, a group of residents finally filed a suit before the Supreme Court of Argentina in 2004 against both the City and the Province of Buenos Aires, the national government and more than 40 private enterprises, in which they demanded compensation for damages resulting from pollution of that basin as well as the ceasing of polluting activities and a solution for environmental contamination. Several civil society groups – both activists' organizations and NGOs – were responsible for putting the state apparatus and the companies under such pressure, among them the Asociación Ciudadana por los Derechos Humanos, the Asociación de Vecinos de la Boca, the Centro de Estudios Legales y Sociales (CELS), and the Fundación Ambiente y Recursos Naturales. Four years later, in July 2008, Argentina's Supreme Court issued a decision in which it required the City and Province of Buenos Aires and the national government to develop and implement measures to remedy the environmental contamination and prevent future damage. The problem is still far from solved at the time of writing, but the Court's decision was nonetheless an institutional watershed.

Although most of the literature on environmental injustice in Latin American cities is relatively recent, some case studies and reflections appeared in the 1980s. Two of the most important sources of inspiration for these early discussions on environmental problems and struggles in Latin America's cities were the *favelas* Vila Socó and Vila Parisi, both in the city of Cubatão (state of São Paulo, Brazil). Vila Socó, built on stilts over a mangrove and dangerously located very close to a pipeline that carried oil from the nearby port of Santos to a refinery owned by Petrobras (Brazil's state oil company), experienced a huge industrial oil spill fire in 1984. The *favela* was engulfed by a fireball; between 100 (according to official data) and more than 500 (according to other sources) people lost their lives, and in the wake of the tragedy 2,500 people were left homeless, as their shacks were devoured by the flames (see a short essay on the tragedy in Porto-Gonçalves, 1984).

Vila Parisi also experienced a tragedy when an enormous quantity of ammonia gas was released after an accident with a pipeline in 1985, and ca. 8,000 dwellers had to leave their homes (Gutberlet, 1996: 89). However, Vila Parisi's problems did not restrict themselves to that historic accident. As several authors have shown, Cubatão and especially Vila Parisi were for many years in Brazil and actually across the world symbols of an intolerable air pollution – and of the health problems caused by it, from lung diseases to anencephaly (Spektor et al., 1991; Gutberlet, 1996).

The link between pollution and residential segregation is actually a multifaceted one. Although all people who live in segregated (usually peripherally located) areas suffer from discomfort or diseases caused by contamination of the water, soil, and/or air, some groups suffer in an especially intense way. Children, stay-at-home mothers (and other women who do not work outside their homes), elderly, and disabled people spend almost all their time in their neighbourhoods, in the context of restricted circuits that include the home, the school, and local commerce, meaning that they are much longer exposed to local contamination than adults who work far away from home. Working far away from home is however certainly no blessing, as workers who live at the periphery of large cities and metropolises usually spend several hours in (commonly poor quality) public transportation vehicles.

2. Exclusionary forms of 'environmental protection'

From a socially critical viewpoint, 'environmental protection' is a dangerously vague expression, as long as the question regarding *which* environment should be protected, *how* and *for the benefit of whom* is not adequately clarified. In the past decade in Rio de Janeiro, a pro-environment and at the same time clearly anti-poor alliance was formed. Located right in the heart of the city, the slopes of the Tijuca massif mark the landscape of many neighbourhoods of Rio de Janeiro – ranging from the privileged areas of the South Zone to many *favelas*. Of enormous relevance is the fact that the Tijuca massif comprises a national park, the Tijuca National Park (established in 1961). With an area of 39.5 square kilometres, it is the largest replanted urban forest in the world, and it is the most visited national park in the country. The land strip that corresponds to the most densely populated portion of the buffer zone of the park is so to speak a perfect 'laboratory' for watching the (geo)political instrumentalization of the ecological discourse by agents directly or indirectly interested in some sort of 'non-murderous social cleansing.'

The media has played a decisive role with regard to tacitly promoting an asymmetrical treatment of social classes by the state apparatus in Rio de Janeiro. Although the public prosecutor's office (*Ministério Público*) for environmental and cultural heritage issues of the state of Rio de Janeiro has been the main institutional agent of the current attempt to promote the total or partial removal of the *favelas* located in the Tijuca massif, it can be said that its role has not only been made visible and highlighted but perhaps also stimulated by the corporate media. A veritable crusade has been carried out by the public prosecutor's office and the media against the permanence of *favelas* in the aforementioned buffer zone. According to that office, the *favelas* located there are expanding rapidly and tend to form "a single spot comparable to Rocinha" (one of the largest *favelas* in Brazil and the largest one in Rio de Janeiro, whose population has been estimated at 200,000 inhabitants). However, a comparison of such statements with census data and even the data offered by the Pereira Passos Institute of Rio de Janeiro's Municipality itself makes clear that the spatial growth of almost all *favelas* ranged from nothing to very little over a period of a decade and a half (1999–2013), as demonstrated by satellite monitoring carried out by Pereira Passos Institute (see Souza, 2016: 791–792). The available data therefore suggest that the idea according to which the *favelas* of the Tijuca National Park's buffer zone are expanding rapidly amounts to a rather distorted assessment of reality. More than ten years after it was proclaimed, in 2006, this forecast can finally be declared wrong. Likewise, the contention that those mostly small and very small favelas represent a 'threat to biodiversity' seems to be rather an excuse than an unquestionable fact. Moreover, interestingly, while the *Ministério Público* and the corporate media continue their anti-*favela* crusade, residential encroachment of the buffer zone by the middle class is left undisturbed, even in those situations where it occurs close to a *favela* targeted for removal.

In Brazil, in contrast to mayors and governors, public prosecutors are not chosen through elections, but are instead civil servants who must be chosen through a public tendering procedure. This means that prosecutors can be basically accountable only to their own conscience (although not few of them have also tried to attract media and public attention for several reasons), while a mayor or governor must take into account the feelings, needs, and desires of potential voters directly and strongly – and *favela* dwellers are voters. In light of this, it is not difficult to see why in the Tijuca massif's case the *Ministério Público* has recently been a more important protagonist than the *Prefeitura* (City Hall) as far as pressures on *favelados* are concerned, even pushing and prosecuting a former mayor himself for (according to the prosecutor's office) not doing enough to protect the environment from the threat represented by poor, illegal settlements. Nonetheless, the *Prefeitura* had tried to 'contain' the expansion of favelas through highly controversial so-called '*ecolimites*' (fences or more usually walls surrounding *favelas*) since the beginning of last decade, allegedly for the purpose of protecting the remnants of the Atlantic Forest. The state government of Rio de Janeiro followed the same steps and its attempt to build a wall around Rocinha, Rio de Janeiro's biggest *favela*, ended in what could be regarded as a media and public relations disaster for the then governor Sergio Cabral in 2009: strong criticism came not only from Brazilian society but also from abroad, for instance the United Nations.

The use of environmental protection arguments or the risk discourse as an excuse to implement measures leading to further social discrimination and residential segregation has been by no means a privilege of Rio de Janeiro, but Rio can be seen as possessing a 'paradigmatic' character (Souza, 2016), in the sense that it corresponds to a typical example of instrumentalization of the environmental protection discourse as an excuse for *favela* removal. Indeed, in Rio de Janeiro we can find the most usual elements of similar situations (such as the anti-*favela* perspective espoused by local and state government and corporate media) as well as some characteristics that are not very usual (such as the subsequent protagonism on the part of the *Ministério Público*).

In a book chapter on 'green evictions' in New Delhi, Asher Ghertner (2011: 146–147) points out the "metonymic association between slums and pollution," what seems to justify for Indian courts "slum removal as a process of environmental improvement." If we expand the first remark a little bit – by means of including things such as 'environmental degradation' and related ideas as part of the second term of that metonymic association – we easily arrive at a description of Rio de Janeiro and many other, similar cases. Aspects such as the immediate excuse, the role of specific organs of the state apparatus and other agents (media, middle-class residents, and so on), and the way how affected people react to threats of removal and 'contention' will surely vary from case to case, but one thing is certain: 'green evictions' will increasingly plague segregated spaces and poor communities, in Latin America's cities and elsewhere.

An interesting type of 'green eviction' is the one related to the possible effects of global change. Similar to what occurs in relation to urban environmental problems in general, the connections with residential segregation and poverty have remained largely unexplored by conventional and more or less conservative-minded environmental scientists. Even Mexico City, famous for some of the worst air contamination in the world, has been sometimes analyzed without reference to the link between unregulated factories, poor neighbourhoods, and air pollution-related diseases at the periphery of the metropolis, while car traffic – surely a major problem that also deserves critical considerations from a perspective broader than usual – seems a little over-blamed. Such conventional approaches to urban environmental problems lack an adequate concept of social space and its production – and this necessarily includes the (re)production of socio-spatial inequality. As one could surely expect, also in relation to global change scenarios geographical space has not seldom been regarded from a socially rather abstract view-point (for a Mexican example, see Rodríguez, 2010).

Regardless of the impression given by depoliticized and technocratic 'global change science,' there has always been a connection between social problems such as poverty and residential segregation (or in more deep terms, exploitation and oppression due to social class or ethnicity), on the one hand, and vulnerability to 'natural' disasters such as floods and landslides as well as to environmental problems like pollution or formation of heat islands, on the other. The lower the social status, the higher the social vulnerability to environmental hazards. Not to mention the fact that natural disasters have been increasingly *socially produced* through inadequate land use, overexploitation of resources, lack of ecological prudence, and so on. However, within the framework of global change 'extreme climate events' have become more frequent and severe, so that the vulnerability particularly of the poor has become evident as never before. This kind of link between social problems and human-induced global change is *indirect*, however – and actually we can never say when and to what extent a specific phenomenon like a tornado, flood, or heat wave is related to global warming. A *direct* connection between global change and an increased social vulnerability or environmental suffering of specific groups is exemplified by the rise of sea level and its potential consequences. In Xochimilco (one of the 16 boroughs of Mexico City), for instance, plans have been made to relocate residents of shantytowns whose dwellings would supposedly be affected by floods as a consequence of climate change – but interestingly, in spite of the 'vulnerability to global change' discourse only so-called irregular settlements would be relocated (Pashley, 2015). The implementation of such relocation measures would constitute an undeniable violation of human rights and environmental justice principles.

3. Unequal access to water

Although we can list a whole set of resources easily defined as 'vital,' there is a difference between those resources that are vital primarily from the viewpoint of governments and big business and whose geopolitical importance has been stressed for a long time – such as *oil* – and those resources that are vital from the viewpoint of the population in a much more direct way – such as *water*. Scarcity of oil obviously could affect the life of millions or dozens of millions of people of all social classes if transportation and distribution systems collapse and the energy supply is cut off in a specific country, but the way in which the burden of such a shortage is distributed according to the class reveals a clear inequality. Oil has been directly related to the interests of big business and of those agents involved with 'national security' (governments in general and military in particular), so that an acute oil shortage cannot simply be tolerated; not to mention the fact that in the case of such an acute shortage of oil, rich people would be probably less affected as they can pay for short-term alternatives in a way that poor workers who depend on public transportation cannot. In contrast to that, scarcity or lack of freshwater is a problem that has plagued many poor neighbourhoods in the cities of (semi)peripheral countries, a situation that has been met with indifference on the part of the state apparatus and the urban elites – at least until a serious conflict occurs, such as the Water Wars in Cochabamba (1999–2000) and El Alto (2005) in Bolivia. While scarcity of oil is commonly just hypothetical or episodic, scarcity of water has been a chronic problem for the poor.

Neo-anarchist Murray Bookchin defended the thesis of a *post-scarcity anarchism*. For him, the "prospect of material abundance for all to enjoy life without the need for grinding, day to day toil" (Bookchin, 2004: 12) is a *historical* possibility, as much as scarcity is a *historical* product: none of them is a 'natural' fact. According to his perspective, modern technologies could liberate humanity from toil and suffering – provided they are restructured and re-contextualized in a way that converts them from technologies developed by and for capitalism into truly *liberatory technologies*, capable of allowing space for humane social relations and balanced society-nature relations. However, in the context of heteronomy and especially in a world where human-made 'extreme weather events' are likely to occur more and more frequently and with increasingly devastating effects, the spectre of scarcity still haunts us.

Neoliberalism opened a new chapter for the very old and long story of historically produced scarcity. The concrete access to vital resources such as water has been increasingly mediated by the 'world of commodities.' The Water War in Cochabamba took place when the city's municipal water supply company was privatized, a move that led to higher water rates for the citizens; this 'war' lasted for several months (see Linsalata, 2015). In reality, many similar protests have occurred across the continent as a result of 'structural adjustment programmes' imposed by the International Monetary Fund and the World Bank on (semi)peripheral countries and of the hegemony of the neoliberal agenda more generally. With or without the context of 'structural adjustment,' privatization of state companies is one of the pillars of neoliberal 'reform.' And while 'more efficiency' (central argument of neoliberal ideology) is a very uncertain outcome of such a process, increased costs for the poor are almost inevitable.

In Latin America, the 'age of privatization' began in the late 1980s and early 1990s. An early and very representative case is the privatization of water supply in Buenos Aires in 1993. After *Aguas Argentinas* (subsidiary of *Suez*, one of the global giants of the 'water industry') became the owner of the former state company, there was a huge impact on the water price that increased by 88.2% between 1993 and 2002. Tobías (2016) provided an interesting analysis of the problems related to access to water as exemplified by Buenos Aires, showing how the frustration about privatization eventually led to the renationalization of the water supply. Her discussion of that case has a much wider applicability, as she notes that renationalization provides no guarantee that the human right to clean water can be adequately protected.

We cannot deny that *corruption* and *inefficiency* have often – though by no means always – been characteristics of state-led companies, particularly in (semi)peripheral countries. Nevertheless, the *even worse alternative* (at least for the poor) represented by privatization is *not* the only alternative, and the ideological hegemony of a statist Left in this kind of debate has made us blind to the fact that there are other possibilities for the management of the commons beyond the opposition state ownership/management *versus* private ownership/management. Besides older, more traditional forms of *communal* ownership and management (still found in the countryside in many countries and regions), an interesting alternative both to privatization and to nationalization is what Bookchin (1995) called '*municipalization*,' understood as non-authoritarian, radically democratic form of collectivization and popular control over infrastructure and services supply. Although bottom-up collectivization has been part of the programme of the non-authoritarian Left for generations (see for instance the collectivizations successfully implemented by the anarchists in Spain after the 1936 revolution), 'municipalization' seems to be a suitable strategy considering the reality of contemporary big cities and metropolises.

Conclusion: *buen vivir/bem viver* versus capitalist urbanization

In Latin America (as well as Africa and Asia) the struggle for 'social re-appropriation of nature' and for territories (and not only for *land*) cannot be reduced to a simplistic quest for 'economic development' and 'modernisation' in a capitalist way; in fact, these have been more often than not a problem rather than a solution (Porto-Gonçalves, 2012). Furthermore, the struggles that have been carried out in that continent – not only the movements and protests whose protagonists are peasants, indigenous people in rural areas etc., but to some extent even the movements and protests of the urban poor – go far beyond a simple 'right to the city,' subordinated as this motto usually is to an interpretive key controlled by the Western cultural matrix (albeit in a specifically Marxist-Lefebvrian version).

Even in the cities it is urgently necessary to foster specific variants of what many Latin Americans have termed *buen vivir* (from the Quechua *Sumak Kawsay*, or *Suma Qamaña* in the Aymara language) and less frequently *bem viver* (Portuguese): the 'good living,' in a broad sense that encompasses from communitarian values to environmental protection, all subordinated to culturally rooted social justice parameters. There are, however, several ways of interpreting *buen vivir/bem viver*, and there is a lively debate underway about the most appropriate and authentic way of doing it. Atawallpa Oviedo (2013) has been one of the outstanding intellectuals committed to the effort of presenting the cultural heritage of the Andean peoples as a solution to the current 'civilisatory,' ethical and political impasses in Latin America; and he is certainly right in raising objections against what he sees as forms of opportunism on the part of sectors of contemporary Latin American Left, which seek legitimacy in some kind of dialogue with traditions but, in essence, do not intend to give up a strategy embedded in the ideologies of Eurocentrism and 'modernization.' On the other hand, from a humanist and emancipatory perspective, it makes little sense to cultivate a rigid and absolute view of *buen vivir/bem viver*. As a source of direct and often perhaps only indirect inspiration, the conceptions of those whom Oviedo calls 'Peoples of Tradition' (*Pueblos de Tradición*) can be extremely useful in the struggle against capitalism, consumerism, and cultural alienation. But only on the basis of a certain autonomy will it be possible for particular societies, in their specific spatial contexts, to define the concrete contents of *buen vivir/bem viver*.

Having been developed above all among indigenous people in (semi-)rural areas, *Sumak Kawsay* makes reference primarily to customs and a way of life that are not primarily related to an urban *Lebenswelt*. Indeed, life in contemporary large Latin American cities is to a large extent the very reversal of 'good living.' Nevertheless, 'translating' this 'good living' as far as possible not only into the 'language' of large cities – not to uncritically 'adjust to reality,' but to develop and implement realistic *alternatives* to capitalist, real-existing urban (hyper)precariousness and alienation in the urban context itself – but also into languages and cultures other than the Andean ones is a crucial task. 'Good living' understandings in El Alto and Buenos Aires can and should share some basic assumptions, but concrete spatialities, habits, cultures, and values are very diverse. Be that as it may, capitalist urbanization is a common threat and challenge, and within this framework environmental injustice is a common feature that can be eliminated probably only if inspired by processes of socio-spatial development that radically defy heteronomy and therefore contest the premises of capitalist 'development' (capital accumulation, labour exploitation, 'cheap nature,' technolatry, and so on).

The three sets of problems explored in the previous sections of this chapter – waste dumping and its links with residential segregation, exclusionary forms of 'environmental protection,' and unequal access to water – require, at the same time, very broad scales of analysis and action (the country and, ultimately, the world), but also very specific, culturally situated scales (local and regional). After all, it is imperative to take into account the differences in terms of geographic-cultural context – the *indígena* and *campesino* universe in contrast to that of large cities, the different degrees of Westernization, and so on –, so that environmental justice, which is of course an essential part of a *buen vivir/bem viver*, can be effectively achieved in a place-sensitive way.

As long as environmental injustice, resource depletion detrimental to place and community-based ways of living, and aggressions against ecosystems and the commons occur, environmental conflicts will be inevitable. Open environmental struggles in Latin American cities can be delayed by means of repression and/or ideology, but they will take place sooner or later. The only stable and ethically justifiable way to avoid them is to eradicate their *structural causes*. As Bookchin (1995) argued some decades ago, even in the 'Global North' capitalist *urbanization* has been the antithesis of true *cities*, provided one understands the latter as spaces that are symbols of dense and free political life and cultural diversity, and which therefore obviously cannot be confounded with present-day metropolises and megalopolises – huge, anti-ecological and alienation-producing demographic agglomerations for the sake of capitalist production and mass-consumption. The interesting thing is that in continents such as Latin America, due to the existence of a diversity of cultures and multiple forms of cultural resistance, the room for manoeuvre to develop alternatives to capitalist urbanization is probably bigger than in Europe or the United States, in spite of the fact that the material problems and challenges are also much bigger and more acute. This circumstance should not be underestimated.

Note

1 Literally 'little black heads,' that is people who have black hair and slightly dark skin, usually working-class, poor people with indigenous ancestry.

References

Acselrad, H. (2010) Ambientalização das lutas sociais – O caso do movimento por justiça ambiental. *Estudos Avançados* 24(68): 103–119.

Berry, B. (1978) Tamanho das cidades e desenvolvimento econômico: síntese conceitual e problemas de política, com especial referência ao Sul e Sudeste da Ásia. In S. Faissol (ed) *Urbanização e regionalização: Relações com o desenvolvimento econômico*. Rio de Janeiro: IBGE.

Bookchin, M. (1995) *From Urbanization to Cities: Toward a New Politics of Citizenship*. London: Cassel.

———. (2004 [1971, based on essays written between 1965 and 1970]) *Post-Scarcity Anarchism*. 3rd Ed. Edinburgh; Oakland, CA: AK Press.

Bullard, R. (2000 [1990]) *Dumping in Dixie: Race, Class, and Environmental Quality*. 3rd Ed. Boulder, CO: Westview Press.

———. (2005) (ed) *The Quest for Environmental Justice: Human Rights and the Politics of Pollution*. San Francisco: Sierra Club Books.

Gaiso, F. del. (2014) *Contaminación por plomo en niños de las villas de la Ciudad Autónoma de Buenos Aires*. Buenos Aires: Auditoría General de la Ciudad de Buenos Aires.

Ghertner, D. A. (2011) Green evictions: Environmental discourses of a slum-free Delhi. In R. Peet et al. (eds) *Global Political Ecology*. London; New York: Routledge.

Gutberlet, J. (1996 [1991]) *Cubatão: Desenvolvimento, exclusão social, degradação ambiental*. São Paulo: EDUSP.

Hoselitz, B. (1960a) The role of cities in the economic growth in underdeveloped countries. In B. Hoselitz (ed) *Sociological Aspects of Economic Growth*. Glencoe, IL: Free Press.

———. (1960b) Generative and parasitic cities. In B. Hoselitz (ed) *Sociological Aspects of Economic Growth*. Glencoe, IL: Free Press.

Linsalata, L. (2015) *Cuando manda la asamblea: Lo comunitario-popular en Bolívia: Una mirada desde los sistemas comunitarios de agua de Cochabamba*. La Paz: SOCEE, Autodeterminación and Fundación Abril.

Merlinsky, G. (2013a) *Política, derechos y justicia ambiental: El conflicto del Riachuelo*. Buenos Aires: Fondo de Cultura Económica.

———. (2013b) La espiral del conflicto: Una propuesta metodológica para realizar estudios de caso en el análisis de conflictos ambientales. In G. Merlinsky (ed) *Cartografías del conflicto ambiental en Argentina*. Buenos Aires: Ciccus.

Oviedo, A. (2013) *Buen vivir versus Sumak Kawsay: Reforma capitalista y revolución alter-nativa. Una propuesta desde los Andes para salir de la crisis global*. 3rd Ed. Buenos Aires: CICCUS.

Pashley, A. (2015) Mexico uses climate threat to justify 'slum clearance'. *Climate Home News*, 4 May. www.climatechangenews.com/2015/05/04/mexico-uses-climate-threat-to-justify-slum-clearance/ (Accessed 15 January 2018).

Pellow, D. N. (2007) *Resisting Global Toxics: Transnational Movements for Environmental Justice*. Cambridge, MA and London: MIT Press.

Porto-Gonçalves, C. W. (1984) Acidente ecológico: Os casos Tucuruí, Rio São Francisco e Vila Socó ou o dilema entre ecologia e política. In *Paixão da Terra: Ensaios críticos de ecologia e Geografia.* Rio de Janeiro: Rocco and Socii.

———. (2012) A ecologia política na América Latina: Reapropriação social da natureza e reinvenção dos territórios. *INTERthesis* 9(1): 16–50, online, June 16. https://periodicos.ufsc.br/index.php/interthesis/article/view/1807-1384.2012v9n1p16/23002

Rodríguez, R. S. (2010) El cambio climático y la Ciudad de México: Retos y oportunidades. In J. L. Lezama and B. Graizbord (eds) *Los grandes problemas de México* (= IV: Medio Ambiente). Mexico, DF: El Colégio de México.

Rostow, W. W. (1960) *The Stages of Economic Growth: A Non-Communist Manifesto.* Cambridge: Cambridge University Press.

Scharager, A. (2016) La "eliminación de obstáculos" en la causa Riachuelo: Controversias en torno a la relocalización de la Villa 21–24. In G. Merlinsky (ed) *Cartografías del conflicto ambiental en Argentina 2.* Buenos Aires: Ciccus and CLACSO.

Souza, M. L. de (2016) 'Urban eco-geopolitics: Rio de Janeiro's paradigmatic case and its global context'. *City* 20(6): 765–785.

Spektor, D. et al. (1991) Effects of heavy industrial pollution on respiratory function in the children of Cubatão, Brazil: A preliminary report. *Environmental Health Perspectives* 94: 51–54.

Tobías, M. (2016) El accesso al agua en Buenos Aires durante la era posneoliberal: ¿Derecho humano o *commodity*? In G. Merlinsky (ed) *Cartografías del conflicto ambiental en Argentina 2.* Buenos Aires: Ciccus and CLACSO.

47

RETHINKING THE URBAN ECONOMY

Women, protest, and the new commons

Natalia Quiroga Díaz
Translated by Anna Holloway

Introduction

In this chapter I reflect on how the massive mobilization of women around the strikes that took place in Latin America in 2017 and 2018 speaks of a new social articulation that imposes limits on extractivism and on the financialization of the economy in all spheres of value production, and fights against a system that shows contempt for the female body. It also points towards alternatives for the production of new ways of understanding the urban economy that are fuelled by struggles around reproduction.

Adopting the dynamics of the cities in the region and the processes of change they are undergoing as my standpoint, I argue that popular struggles against the advancement of the Right – which promotes and intensifies the logic of commodification in the field of care – have a memory that feeds off the so-called *feminization* of politics, namely processes that focus on the reproduction of life and break with the domesticity imposed by neoliberalism. The women's movement produces mobilizations that socialize the reproductive and, in doing so, prevent the female body from being configured as a sacrifice zone for the implementation of structural adjustment policies.

I suggest an approach to the 8M[1] from the feminization of the urban economy, which refers to the spreading of the struggles around reproduction across the continent. These mobilizations challenge the economic violence against women taking place within the framework of structural adjustment, particularly through ruling institutions that cultivate a femininity of silenced and obedient bodies suffering the cost of the contraction at home. We reflect on the effects of the feminist wave in the city and analyze the creation of the 8M as a new urban commons.

In this work, I draw from the theoretical contributions and experiences of Latin America, which provide a particular lens through which to understand the urban economy. I seek to link the feminization of politics to the continent's long-standing struggles against neoliberal dispossession which have, at different moments in recent history, imposed limits on the voracity of capital and produced new spaces for the commons.

The community popular (*comunitario popular*) in the construction of the city

Academic research at the beginning of the 1960s raised the question of what kind of city is created by the populations that migrate from rural areas to urban centres and inhabit their margins.

This led to studies on the culture of poverty and to emblematic works such as that of Oscar Lewis (1961), who describes the poor and their "particular worldview." From this perspective emerged a critique that, in turn, opened up a space of reflection on the place of women in the new cities. An example of this are the studies on *survival strategies* conducted in the 1970s, which show family strategies in the struggle for the urban habitat, a relationship in which "survival and the study of gender is established" (Anderson, 1991). The women/survival strategies/poverty trinity establishes a perspective on women that considers the family as key to understanding the capacity of the popular sectors to produce and obtain the necessary resources for covering their needs, and for making urban life at the margins and interstices of capital possible.

To consider that reproduction only concerns the poor is to ignore the contributions of the triple working day, of production/reproduction, and of community management to the economy. It is also to define the issue of reproduction in the city as a matter of class, for only there does the leading role of women become visible. The perspective of poverty outlines a city that equates the poor, women, and reproductive and community work, and treats the latter as second-class activities, failing to acknowledge the fact that the reproductive is a fundamental dimension of the city.

This perspective was consolidated in the 1990s by studies promoted by the World Bank through the work of Caroline Moser, who deployed the theoretical arguments for reading women's resources through the logic of (human) capital. Moser (1996) developed the concept of "assets" to refer to the resources or "means of resistance" that individuals, households, or communities develop in order to deal with the privations imposed by the context. As Moser (1996) writes "the more assets one has, the lower the degree of vulnerability, and the greater the erosion of assets, the greater the degree of insecurity." The author adds that the ability to avoid or reduce vulnerability does not only depend on the initial assets, but also on the capacity to administrate them (transform them into income, food, or other means for the satisfaction of basic necessities).

These studies take an interest in popular economies in order to empirically observe the responses developed by women at home and in their communities in the face of the economic crisis; hence the emphasis on "empowerment" which has been instrumental in the sustainability of structural adjustment policies in the 1980s and 1990s (Quiroga, 2011). Thus, the power of female labour and its capacity to organize the community is considered a buffer for the social tensions caused by the implementation of the policies of the Washington Consensus, policies that led to a remarkable rise of poverty in the region and to the withdrawal of the state in the provision of essential living rights.

It is important to stress that cities express a diversity of economic processes where women play a central role, given their capacity to multitask and make urban life a reality but also create new forms of understanding the political. This is particularly so in Latin America, where the perspective of popular economy and of social and solidarity economy is developed in the 1980s. To approach the economy through solidarity and the popular in the city is to challenge the generalized use of the concept of informality by academia and government institutions, a concept that views unpaid work as a manifestation of underdevelopment, and also as a theoretical response to the profound social inequality and persistent erosion of the living conditions of the urban population caused by dictatorships and neoliberalism. It is in this context that the polysemic concept of popular economy is coined, valorizing the different expressions of labour in the household, allowing for an account of the self-production and the reproductive labour conducted predominantly by women, and going beyond the individualist approach entailed by the concept of "labour market."

The vitality of these economic forms oriented by a sense of reproduction combines the production of exchange values with a variety of economic practices and institutions that are not limited to the market, and shows that the city is sustained by diverse rationalities and multiple ties that do not aim exclusively at profit, nor at mere survival. The popular economy stands against the commodified and capitalist perception of the economic and, therefore, opens up new ways of thinking the city.

Within this perspective, the work of Razeto (1983) is fundamental. He highlights the contribution of the impoverished sectors in finding solutions to problems of subsistence through a popular and solidarity economy whose main objective is to overcome poverty. Later, Coraggio (1987) writes on the diversity of forms of labour and on the fragmentation imposed by capital, stressing the need to overcome individuality and fight for a reproductive rationality of life as the main goal of an economy that creates alternatives of transition against capital. Quijano (1989) proposes the concept of "marginal pole" which refers to the heterogeneity of economic activities, forms of organization, uses and levels of resources, technology, and productivity. The author points out that demands focus on the conditions for reproductive autonomy (land, services, etc.) and not on work or salary-related conditions. These writings emphasize the diversity of the forms of organization of the economy, forms that go beyond the insertion in the labour market and speak of the existence of community, indigenous, and peasant institutions that organize their everyday life creatively to satisfy their needs.

In the urban dimension, this creative capacity creates circuits which flexibly spread from production to reproduction and often suffer expropriation, but also politicise the conditions of poverty – even at moments of acute social repression – in order to assemble the state and community resources that guarantee life. At the same time, these circuits pose limits on the conditions of exploitation imposed by a commodifying logic of life. The political force that springs from popular and social economies challenges the individualist vision of income-oriented, formal and informal types of labour, as well as the instrumental approach to families and communities that aims at their obedient accession to the conditions of the market imposed by multilateral and government institutions.

What these founding works lack is a reflection on the economic implications of the bodies of those who create this economy: women. They appear as protagonists of community kitchens and the popular habitat, but their initiatives do not lead to a theoretical interrogation that permits a richer understanding of the contributions of feminism to the social and popular economy. Therefore, one of the main questions I try to answer is the following: how do women's protests in recent years, and particularly the 8M, challenge the rhythm of urban accumulation? What forms of rethinking the city do they propose?

"Feminism" in multilateral institutions: control and security

UN-Habitat (ONU-Habitat, 2010, 2013) highlights that there has been a process of sustained urbanization in Latin America in recent decades, one that places life in the city at the centre of the dynamic of the economy. It points out that a main obstacle for the increase of wealth is the inequality suffered by women in the labour market, where gender gaps still persist despite the increasing feminization of the urban population and particularly of the lower-income segment.

The report asserts that the participation of women in paid employment has increased considerably, but has not been combined with an equal increase in the number of hours that men spend in care and household work. Therefore, women continue to bear the burden of a "reproductive tax" that is combined with other discriminatory processes at home and in the labour market.

According to these analyses, multilateral institutions try to tackle the inequality affecting women by creating a set of policies that revolve around two axes: the first one stresses the contribution of female labour to the economy of the low-income sectors; the second speaks of insecurity as the main problem for women in the city (Moser, 1996; Moser and Felton, 2010; Hábitat, 2010, 2013).

These policies attend to the demographic change of the increasing urban population, one that has not been combined with integration into the labour market to guarantee adequate living conditions or consumption capacity for all sectors. Therefore, multilateral institutions view the sustained feminization of the low-income sectors as something positive, given the capacity of women to simultaneously generate an income and provide care. The idea of the city that emerges from the vision of multilateral organizations is focused on the concept of prosperity, where women play an instrumental role in the attainment of growth and development. According to these perspectives, the increase of wealth in the cities supposedly amounts *per se* to wellbeing, in something like a progressive overflowing of growth; therefore, the dynamics of the market that produce poverty and inequality are not questioned.

The proposal of safe cities

The approach that encompasses the idea of Safer Cities and of policies to tackle violence against women was launched in 1996 in Africa, as a response to increasing criminality and urban violence. In Latin America, the programme was adopted in 2004 under the name of "Ciudades sin violencia hacia las mujeres" (Cities without violence against women) with the participation of Rosario in Argentina, Santiago in Chile, Bogotá in Colombia, Recife in Brazil, Guatemala and El Salvador (see UNIFEM, 2007).

Multilateral agencies view the relationship between gender and urbanism in the context of security, where the link to public space is marked by a greater fear of violence and physical aggresion and is posited as a fundamental obstacle for women's access to the labour market and their free circulation in the city. Initiatives include a set of policies that promote dedicated transport networks for women and children; the reduction of the distance they need to cover in order to access goods and services; the improvement of public transport with train carriages destined exclusively for women; and adequate lighting to facilitate their moving around in insecure areas, amongst others.

As a result, local governments have limited their initiatives to developing specific actions to minimize harassment in public transport, improve lighting, develop training sessions, and provide materials to create awareness amongst public servants. They have encouraged social organizations and NGOs to adopt these approaches and to ensure that they too can contribute to these policies limited to neighbourhood improvement.

I believe however that this connection between women and the city from the viewpoint of "security" contributes to the victimization of women and, thus, defines them as objects of guardianship by the state or the police, by institutions ready to tell women where to travel, where to walk, how to produce wealth, how to have fun, where to protest, and so on. This approach contrasts with the feminist perspective that is deployed in everyday life and places the organization of the city at the centre of the problematic, beyond difficulties observed in specific areas where insecurity is the result of the position of reproduction within the capitalist economy and of the consideration of the female body as a sacrifice zone, as we see further on.

I believe the prosperity-security binomial debilitates the discussion on how space is produced and obstructs the politicization of the organization of cities against wealth in its commodified-dispossessing form. Consequently, it reduces the possibility of organizing spaces and their economies on the basis of care.

The city in feminine: proposals of the women's strike for rethinking the urban economy

In many countries, we women took to the streets united under the slogan "If women stop, the world stops." This slogan marks a shift from classic demands for more rights, moving instead to a call to construct another society and, with it, another economy.

In Latin America, the "feminist wave" ripples through classes, races, and generations; it is not about short-term processes, but rather about the consolidation of a subjectivity that speaks of an accumulation of struggles whose history is lost in popular contributions and protests: for the right to live in the city, for land, housing, and public services up to the 1980s and, as of the 1990s, against the neoliberal state and its adjustment policies. More specifically, the intensification of processes of economic liberalization promoted by the Washington Consensus consolidated dynamics of social inequality, with the resulting territorialization. This caused profound social crisis, met with a broad cycle of popular protests that peaked during the mobilizations in Bolivia (2000–2005) and Argentina (2001), with the Zapatista uprising in Mexico as a fundamental antecedent (1994).

More recently, the mobilizations of students from the secondary schools of Chile demanding tuition-free education (2006), demonstrations in Brazil against the increase in public transport prices (2013), the agrarian strike in Colombia (2014), protests in Argentina against the pension reform and the increase in public service fees (2017–2018), popular insurrection against the social security reform in Nicaragua (2018), to name but a few are all protests that to varying degrees have defended the conditions for reproduction. These struggles constitute an ongoing process of contestation for the spheres of reproduction against the advance of the market; in this context, social movements demand the de-privatization and de-familiarization of the conditions of existence.

At the same time, a new way of thinking of the public sphere is invented. In fact, from the moment the so-called domestic bursts into the sphere of social protest, different authors begin to perceive these changes as an expression of the "feminisation of politics." In Argentina in 2001, pickets (*piquetes*) are organized on the main access roads to the cities, and people occupy the streets and prepare food for the protesters in what became known as the "popular cooking pots" (*ollas populares*). The intention was to show that reproductive tasks are direct producers of social value in the urban sphere, capable of producing different logics for the organization of the city. Therefore, the contemporary vibrancy of feminism must be understood in the context of long-standing processes of struggle and the participation of different generations, without, however, historical experience imposing limits on the novelty and contingency of what is being produced today.

At home, in bed, in the square

One of the main new features of women's mobilizations has been to place femicide and violence at the centre of the protest. Demonstrations have shown that these acts cannot be understood as a problem pertaining to the domestic sphere, even if that is where they take place. In the urban context, the feminist economy shows that to give centre stage to the market and its logic of profit is to promote the encapsulation of the conditions for reproduction within the family; this reinforces the power given to men by patriarchy by making the duty of care a women's responsibility. This approach is a vehement critique of the separation of the so-called public and private spheres which continues to be hegemonic in urban organization and results in the exclusion of all dimensions of care, with the vulnerability that this entails.

For all the aforementioned reasons, practices that bring the reproductive out into the public domain challenge the confinement of the so-called domestic sphere and unveil the capitalist exploitation of the labour of care. Such is the example of the many collective acts that articulated with the 8M and women's strikes: the *tetazo*, a call in different cities of Argentina to women to breastfeed their children and show their breasts in public, in repudiation of attempts to imprison women for having done precisely that. Another example is the *besatón* (kissing marathon) organized by collectivities for sexual diversity in protest against police violence and exclusion from commercial spaces suffered by same-sex couples in Paraguay, Colombia, Chile, and Mexico.

The kissing marathon revealed the heteronormative mandate that rules the public sphere, even for those who act as "consumers." Both acts challenged the gender-based order promoted by capitalism in the urban space and the violent regulation of what is and is not allowed on the streets. Furthermore, they also questioned the alleged freedom of the market, existing only for those who embody the mandate of the "homoeconomicus" individual: white, high-income, male, young, heterosexual, and self-sufficient.

Therefore, the 8M is a locus for the expression and reinforcement of apparently disperse processes that contribute to a new way of inhabiting the city. The slogan "If they touch one of us, they touch all of us" breaks with the domesticity imposed on different forms of violence suffered by women, including harassment and abuse in the working space, in public transport, and at home.

The great participation in the movement and the challenges it deploys are efficiently challenging patriarchal mandates, and their consideration of violence against women as common-sense, through collective actions that erode the privileges of macho masculinity. The resulting social transformations show that the path to the construction of "cities that are safe for women" is far from state guardianship and the reclusion fostered by pink carriages in public transport. It is rather about breaking the foundations of an urban economy that shows contempt for the reproductive and raises the homo economicus to the main city actor, the expression of a masculinity that is all about individualism, competition, and consumption. These mobilizations constitute a female "we" (*nosotras*) that occupies the city and gives us the voice and strength to find, together with other women, alternatives that de-institutionalize patriarchal oppression and, in their politicization of the reproductive, challenge neoliberal hegemony.

Machismo kills and so does capitalism!

The effects of the Women's Strike (*Paro de Mujeres*) around the world show that, just as we impose limits on the different forms of violence suffered by the female body, so can we impose limits on the policies that devastate the conditions for reproduction. There are intense struggles in Latin America against the ongoing process of primitive accumulation maintained by capitalism for the commodification of the commons and the destruction of forms of life that disrupt the conditions imposed by capitalism. These struggles against extractivism unveil a strategy of economic growth that is founded on the unlimited exploitation of nature and of the spaces of the reproduction of life sustained by the communities. They denounce the unfinished process of primitive accumulation, hence the timely character of the work of Silvia Federici (2010) that illustrates how the witch hunt in Europe allowed for the establishment of the capitalist system. The author argues that the first primitive accumulation ocurred on the bodies of women and reveals the connection between the processes of enclosure and the persecution of "witches."

The regulation of the female body, with the establishment of capitalism, meant that the reproductive and labour capacities were expropriated with the control of the State and the market in the form of economic resources and, ultimately, of men in their immediate environment . . .

In economies where subsistence was guaranteed by access to land and knowledge, it was necessary to break with the force coming from kinship and neighbourhood relations. A functional mechanism for this purpose was the degradation of women and the vagueness of witchcraft accusations. This produced a climate of generalized terror and contributed to break the ties of community solidarity. The denigration of feminine knowledge weakened the resistance capacity of women themselves and their communities (Quiroga and Gago, 2014: 7).

The persecution of women or – even better, in words of Silvia Federici – the witch hunt still walks hand in hand with the strategies of dispossession. In many indigenous, peasant, black, and popular communities, women have been able to articulate an effective defence of their territory against extractivist projects. Female leadership in these collectivities entails a political stance that posits a radical alternative to the order promoted by capital, according to which nature is but a mere resource for production. In these communities, reproduction interacts with nature and speaks of other possibilities of living in a world that effectively goes against the logic of capitalist accumulation.

The corporate and state-led femicides of Marielle Franco in Brazil, Berta Cáceres in Honduras, Macarena Valés Muñoz in Chile, the assassination of 28 women defending their communities in the last two years in Colombia, the imprisonment and permanent harassment suffered by Machi Francisca Linconao in Chile, Milagro Sala in Argentina, and Máxima Acuña in Peru, to name but the most famous cases, show that the persecution of women remains an effective way to terrorize populations and erode their capacity to defend themselves.

Although the 8M is primarily urban, networks were created during the preparation events and women's meetings that transcended frontiers and unravelled the strategies of capital that seek to destroy the power and autonomy of women in order to undermine resistance in their territories. Women's meetings acknowledge other forms of authority and political negotiation that contest the channels offered by democracy in capitalism. They deploy more collective forms of participation, where each woman participates in the debate according to her own vital moment and territory; however, they remain connected to the broader struggles taking place in the region in defence of life projects that obstruct the subjective and material reproduction of capitalism.

Women's strikes articulate forms of protection and protest against the attacks of extractivist and financial capital on community leaders; they rewrite a history that rejects the narrative of development and shows the ways in which the female body and female power are sacrificed. When we take to the streets and shout "We want us alive" (*Vivas nos queremos*), we reconstruct a collective body that is capable of standing up against the death caused by the pursuit of more profit, challenging the separation between the so-called rural and urban, acknowledging that the struggle against the predation of nature disrupts an economy that outsources the cost of the planet's destruction to the communities.

The female body as a sacrifice zone for structural adjustment

Struggles for environmental justice have collectively produced the concept of "sacrifice zone" (see Bullard, 1990; Di Chiro, 1999) in reference to the unequal geographical distribution of polluting companies. Sacrifice zones are places left to the mercy of the predatory dynamic of industries that cause disease, poison the earth, the water, and the air. States hand over these territories and the bodies that inhabit them to the rapacious appetite of capital so as to safeguard the logic of increasing profit.

Drawing from this concept, I propose that we should understand the female body is a sacrifice zone for the policies of structural adjustment. Once again, I turn to Federici (2013) to illustrate the strategies of expropriation of the female body that preceded the enclosure of the commons, including the loosening of laws against individual and collective violence in the cities, the establishment of public brothels, and, finally, the inquisition that destroyed their sovereignty over their own bodies and criminalized community practices that acknowledged their power, turning them into a common good offered to men in compensation for the loss of land.

Latin America is witnessing an intensification of neoliberal policies even through military coups, such as the destitution of Brazil's democratically elected president, Dilma Rousseff. The wave of conservative governments is cutting back on health care and education, increasing the price of water, energy, transport, and raising the taxes on the consumption of essential goods. At the same time, laws of labour flexibilization are passed, the age of retirement is extended in order to bring it as close as possible to life expectancy, and pensions are reduced. These contractive reforms are combined with tax reductions for big corporations and land owners.

In the context of the implementation of adjustment measures, the extractivist strategy is radicalized, appropriating spaces from the commons. In the urban sphere, where an economy at the service of the accumulation of profit has been put in place, the financialization of non-commodified spaces acquires strength in the field of social rights, public goods, the popular, and the solidary. The intention is not only to subordinate the conditions for reproduction to the logic of profit, but also to make sure that even the most basic needs of everyday life are up for financial extraction.

In this context, women are under pressure to maintain the quality of everyday life; when incomes drop, indebtedness is the only way to satisfy needs. The bodies of adjustment bear the destruction of labour, impoverishment, and debt, factors that discipline families and reinforce isolation, shifting the problem from the sphere of reproduction to that of consumption. As some authors have pointed out (Hinkelammert, 2013; Lazzarato, 2013), debt cultivates a subjectivity of guilt and shame that depoliticizes and promotes the confinement of responses within the family.

The call to strike on the 8th of March is a call against women maintaining with their time and body a life that is constantly besieged by the logic of business. It also breaks with the social fragmentation imposed by measures that – to varying degrees and in different sectors – attack the world of labour and social rights so as to reveal the extractivist logic that permeates them. Thus debt stops being an individual problem; it is reestablished as a continuation of the macro-economic conditions imposed by institutions such as the IMF and the World Bank, whose programmes indebt and destroy the conditions for the reproduction of life and require the versatility of female labour for their implementation. The discourse of "prosperity" they deploy as the key to a successful urban economy demands obedient bodies, willing to be sacrificed in order to sustain an economy at the service of the market. The policies of debt and adjustment are imposed within a context of severe state repression of all expressions of protest and social discontent. Therefore, struggles in defence of the conditions for reproduction acquire an emancipating character that delegitimizes the institutional violence which supports the belief that capitalist profit must always be on the rise.

The diversity of subjects that converge in the 8M and the gender of those who make the call turn the protest into a devastating political challenge for neoliberalism, a neoliberalism that bases its legitimacy on the understanding of the reproductive as something outside the economy, something female and private. These political struggles make it more difficult for adjustment to be perceived as an inevitable condition for the profitability of an extractivist and financial economy that is presented as natural and indisputable.

At this moment of crisis, in which political parties fail to present an alternative to savage capital-ism, the feminist movement is managing to articulate the demands of different social sectors. Social legitimacy is being woven from below and a new way of looking at the world is in the making, one that denaturalizes the role of the feminine as a catalyst of social frustrations and proposes a new form of authority that involves the acknowledgement of different voices and of the need to give space to a multiplicity of demands. As a result, the economy becomes imbued with this diversity and connects to the everyday realities that confront women. It stops being a space of general laws, where matters are discussed in specific places on the basis of needs, and thus weakens the neoliberal perception of the economy as a sphere for technical experts that is separate from politics.

The 8M and its massive mobilizations are constructing spaces of coming together and of valorizing specific economic practices and experiences. They show that there are vital economies in the cities which exist by virtue of the multiplicity and complexity of interactions and processes conducted by women in the sphere of the social and popular economy. It becomes clear that these circuits provide the foundations for everyday life, and this opens the possibility of organizing the city as an economy that is focused on reproduction. The centrality of the popular and feminine solidarity disrupts the false centrality of capital in the organization of the economy and grants legitimacy to the demand that the market be subordinated to social needs. It also contributes to politicizing the violence suffered by women at all levels of existence as a condition imposed by capital upon societies. The 8M denounces the unlimited expansion of the market, which necessarily becomes the locus for the resolution of needs given the permanent dispossession of community goods and of spaces of collective life. The mobilization of women is managing to weave collective capacities that run through society in its immense diversity, with multiple and transitory leaderships that explain the large numbers and horizontality in the management of the protest.

Conclusion

This chapter has considered the 8M and women's protests as a *new urban commons* that I define as the creation of a disperse, yet durable and escalating, space of assembly which convokes a multitude and creates a new way of understanding the economic on the basis of the reconstruction of a new urban social fabric. This is a fabric that is capable of linking the power of the female body and female subjectivity to everyday life with the construction of alternatives, so as to erode the legitimacy of an economic program that focuses on indebtedness, extractivism, and financialization, presented by governments as "the only possible way."

A space is created in which the rationality of care is not only valid in the domestic sphere, but also constitutes an alternative way of inhabiting the city and occupying the streets. It represents a form of understanding the economic on the basis of reproduction, which stands up against the narrative of multilateral organizations that reinforce the separation of the public from what is considered private. The city is occupied by women who, in their diversity, forcefulness, and joy show us there are other ways of inhabiting it. The mobilizations and the different forms of violence they expose destabilize an order based on gender, age, and race that is implicit in state policies and in the organization of the neoliberal economy. The political might of this space goes against the devaluation of the feminine, a devaluation that keeps supporting the capitalist market and its destruction of the foundations of human life and life on the planet.

In Latin America, women have led a process that is already decades long and is known as the *feminisation of politics*. The long history of accumulated and diverse struggles around reproduction is consolidated in a set of feminist mobilizations that run through society and allow for a multiplicity of voices to express themselves. In this process, a new way of understanding the place of women in the economic, political, and social sphere is materializing, imposing limits on neoliberal devastation and preventing the female body from, once again, being the sacrifice zone for the violence of structural adjustment and other capitalist practices.

To think of the 8M and women's strikes from the viewpoint of the commons allows us to see that there is an urgent need for a counter-power that can articulate the struggles against the violence of the adjustment programmes, extractivism, financialization, and debt, and prevent the dispossession of the female body through strategies that are at times subtle, but more often are violent and bloody. The politicization of the reproductive reveals these strategies in the eyes of society. At the same time, it disseminates a multiplicity of urban popular and solidarity economies that revolve around reproduction and reveal the alternatives currently available. The women's movement remakes a community that is creating a new commons, proposing multiple paths towards the construction of an urban economy in feminine.

Note

1 8 March – International Women's Day.

References

Anderson, J. (1991) Estrategias de sobrevivencia revisitadas. en M. Feijoo y H. Herzer (eds) *Las mujeres y la vida de las ciudades*. Instituto Internacional del Medio Ambiente IIED-América Latina; Grupo Editor Latinoamericano, Buenos Aires. pp. 37–60.

Bullard, R. (1990) *Dumping in Dixie: Race, Class, and Environmental Quality*. London: Routledge.

Coraggio, J. (1987) *Política económica, comunicación y economía popular*. Ecuador Debate, CAAP, 17, Quito.

Di Chiro, G. (1999) La justicia social y la justicia ambiental en los Estados Unidos la naturaleza como comunidad. *Ecología política* 17: 105–118.

Federici, S. (2010) *Calibán y la bruja: Mujeres, cuerpo y acumulación originaria*. Buenos Aires: Tinta Limón.

———. (2013) *Revolution at Point Zero: Housework, Reproduction and Feminist Struggle*. Oakland, CA: Common Notions, PM Press.

Harvey, D. (2004) *El nuevo imperialismo: Acumulación por desposesión*. Buenos Aires: Socialist Register and CLACSO.

Hinkelammert, F. (2013) La rebelión de los límites, la crisis de la deuda, el vaciamiento de la democracia y el genocidio económico-social. In J. L. Coraggio and J. L. Laville (eds) *Reinventar la izquierda en el siglo XXI. Hacia un dialogo Norte-Sur*. Buenos Aires: Universidad Nacional de General Sarmiento, Clacso and IAEN, pp. 223–238.

Lazzarato, M. (2013) *La fábrica del hombre endeudado: Ensayo sobre la condición neoliberal*. Madrid: Amorrortu Editores.

Lewis, O. (1961) *Antropología de la pobreza*: México: Cinco familias and FCE.

Moser, C. (1996) *Situaciones Críticas. Reacción de las familias de cuatro comunidades Urbanas Pobres ante la Vulnerabilidad y la Pobreza*. Washington, DC: World Bank.

——— and Felton, A (2010) The gender nature of asset accumulation in urban contexts: Longitudinal results from Guayaquil, Ecuador. In J. Beall, B. Guha-Khasnobis and R. Kanbur (eds) *Urbanisation and Development: Multidisciplinary Perspectives*. Oxford: Oxford University Press, pp. 183–201.

ONU-Habitat. (2010) *Igualdad de género para ciudades más inteligentes desafíos y avances*. Nairobi: United Nations Human Settlements Programme.

———. (2013) *State of Women.in Cities 2012–2013 Gender and the Prosperity of Cities*. Nairobi: United Nations Human Settlements Programme.

Quijano, A. (1989) *La economía popular y sus caminos en América Latina*. Lima: Mosca Azul Editores.

Quiroga, N. (2011) ¿De qué crisis estamos hablando? Cuestionamientos y propuestas de la política de activos desde la economía feminista y la economía social. In J. L. Coraggio and V. Costanzo (eds) *Mentiras y verdades del "Capital de los Pobres". Perspectivas de la economía social y solidaria*. Buenos Aires: Universidad Nacional de General Sarmiento and Imago Mundi.

——— and Gago, V. (2014) Los comunes en femenino: Cuerpo y poder ante la expropiación de las economías para la vida. *Revista Economía y Sociedad* 19(45): 1–18.

Razeto, L. (1983) *Las Organizaciones Económicas Populares*. Santiago: Ediciones PET.

UNIFEM. (2007) *Ciudades para convivir: Sin violencias hacia las mujeres*. Santiago de Chile: Ediciones Sur.

INDEX